Student Solutions Manual

AF709714

Intermediate Algebra
A Guided Approach

TENTH EDITION

Rosemary M. Karr
Collin College

Marilyn B. Massey
Collin College

R. David Gustafson
Rock Valley College

Prepared by

Michael G. Welden
Mt. San Jacinto College

CENGAGE Learning

Australia • Brazil • Mexico • Singapore • United Kingdom • United States

CENGAGE Learning

© 2015 Cengage Learning

ALL RIGHTS RESERVED. No part of this work covered by the copyright herein may be reproduced, transmitted, stored, or used in any form or by any means graphic, electronic, or mechanical, including but not limited to photocopying, recording, scanning, digitizing, taping, Web distribution, information networks, or information storage and retrieval systems, except as permitted under Section 107 or 108 of the 1976 United States Copyright Act, without the prior written permission of the publisher.

For product information and technology assistance, contact us at **Cengage Learning Customer & Sales Support, 1-800-354-9706**.

For permission to use material from this text or product, submit all requests online at **www.cengage.com/permissions**
Further permissions questions can be emailed to **permissionrequest@cengage.com**.

ISBN-13: 978-1-285-84624-8
ISBN-10: 1-285-84624-9

Cengage Learning
200 First Stamford Place, 4th Floor
Stamford, CT 06902
USA

Cengage Learning is a leading provider of customized learning solutions with office locations around the globe, including Singapore, the United Kingdom, Australia, Mexico, Brazil, and Japan. Locate your local office at: **www.cengage.com/global**.

Cengage Learning products are represented in Canada by Nelson Education, Ltd.

To learn more about Cengage Learning Solutions, visit **www.cengage.com**.

Purchase any of our products at your local college store or at our preferred online store **www.cengagebrain.com**.

Printed in the United States of America
1 2 3 4 5 6 7 17 16 15 14 13

Contents

© 2015 Cengage Learning. All Rights Reserved. May not be scanned, copied or duplicated, or posted to a publicly accessible website, in whole or in part.

Preface

This manual contains detailed solutions to the odd exercises of the text *Intermediate Algebra: A Guided Approach*, tenth edition, by Rosemary Karr, Marilyn Massey, and R. David Gustafson. It also contains solutions to all of the exercises in the Chapter Review and Chapter Test sections of the text.

Many of the exercises in the text may be solved using more than one method, but it is not feasible to list all possible solutions in this manual. Also, some of the exercises may have been solved in this manual using a method that differs slightly from that presented in the text. There are a few exercises in the text whose solutions may vary from person to person. Some of these solutions may not have been included in this manual. For the solution to an exercise like this, the notation "answers may vary" has been included.

Please remember that only reading a solution does not teach you how to solve a problem. To repeat a commonly used phrase, mathematics is not a spectator sport. You MUST make an honest attempt to solve each exercise in the text without using this manual first. This manual should be viewed more or less as a last resort. Above all, DO NOT simply copy the solution from this manual onto your own paper. Doing so will not help you learn how to do the exercise, nor will it help you to do better on quizzes or tests.

I would like to thank Cynthia Ashton of Brooks/Cole Publishing for her help and guidance.

This book is dedicated to my parents, Ed and Carol Welden, for their constant encouragement and support throughout my life.

May your study of this material be successful and rewarding.

Michael G. Welden

© 2015 Cengage Learning. All Rights Reserved. May not be scanned, copied or duplicated, or posted to a publicly accessible website, in whole or in part.

Exercises 1.1 (page 13)

1. 5, −5

3. 7 is divisible only 1 and itself. Yes.

5. $\frac{6}{8} = \frac{\cancel{2} \cdot 3}{\cancel{2} \cdot 4} = \frac{3}{4}$

7. $\frac{32}{40} = \frac{\cancel{8} \cdot 4}{\cancel{8} \cdot 5} = \frac{4}{5}$

9. $\frac{1}{4} \cdot \frac{3}{5} = \frac{1 \cdot 3}{4 \cdot 5} = \frac{3}{20}$

11. $\frac{2}{7} \div \frac{7}{4} = \frac{2}{7} \cdot \frac{4}{7} = \frac{2 \cdot 4}{7 \cdot 7} = \frac{8}{49}$

13. $\frac{5}{9} + \frac{4}{9} = \frac{5+4}{9} = \frac{9}{9} = 1$

15. $\frac{4}{5} + \frac{3}{4} = \frac{4 \cdot 4}{5 \cdot 4} + \frac{3 \cdot 5}{4 \cdot 5} = \frac{16}{20} + \frac{15}{20} = \frac{31}{20}$

17. variable

19. natural

21. integers

23. even

25. natural; 1; itself

27. 0

29. rational

31. $<$

33. $\approx$

35. inequality; compound inequality

37. union; intersection

39. $\frac{7}{8} = 7 \div 8 = 0.875$; terminating

41. $-\frac{7}{15} = -7 \div 15 = -0.4\overline{6}$; repeating

43. natural: $1, 2, 9$

45. integer: $-3, 0, 1, 2, 9$

47. irrational number: $\sqrt{3}$

49. even natural number: 2

51. prime number: 2

53. odd composite number: 9

55. $4 < 7$

57. $-5 > -10$

59. $-7 < 7$

61. $6 > -6$

63. $19 > 12 \Rightarrow 12 < 19$

65. $-6 \leq -5 \Rightarrow -5 \geq -6$

67. $5 \geq -3 \Rightarrow -3 \leq 5$

69. $-3 < 0 \Rightarrow 0 > -3$

71.

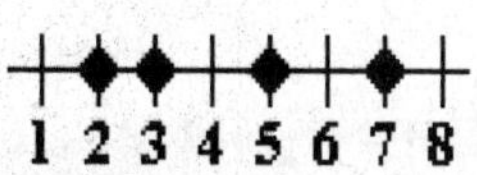

73.

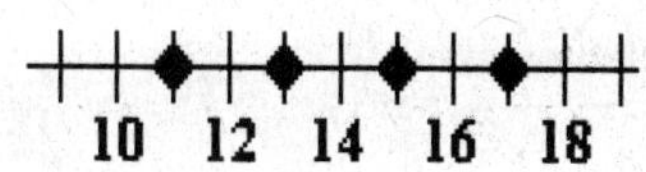

75. $\{x|x > 3\}$

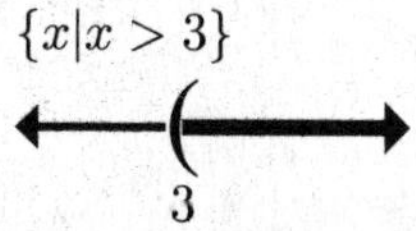

77. $[-5, \infty)$

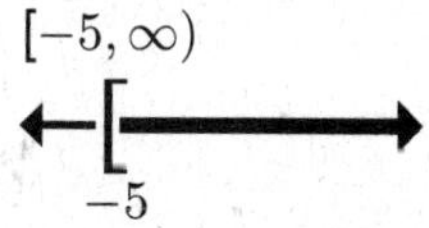

79. $\{x|x < -3 \text{ or } x > 3\}$

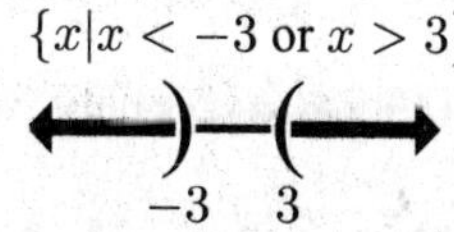

81. $(-\infty, -6] \cup [5, \infty)$

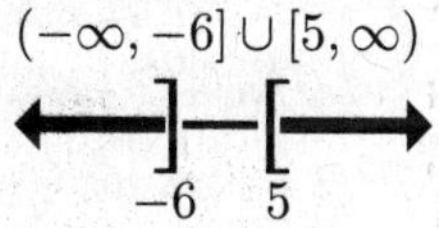

83. $\{x|6 < x \leq 10\}$

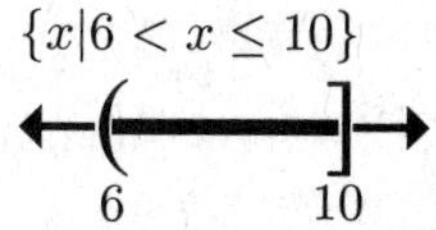

85. $\{x|-1 < x \leq 3\}$

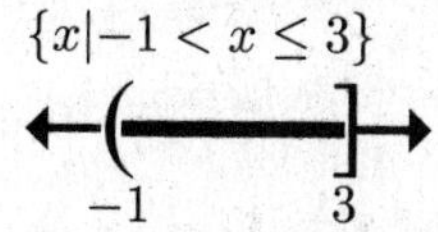

87. $|25| = 25$

89. $-|-5| = -(+5) = -5$

91. $|-5| + |-2| = 5 + 2 = 7$

93. $|-5| \cdot |4| = 5 \cdot 4 = 20$

95. $|x| = 10 \Rightarrow x = 10$ or $x = -10$

© 2015 Cengage Learning. All Rights Reserved. May not be scanned, copied or duplicated, or posted to a publicly accessible website, in whole or in part.

97. If $x = |x|$, then $x \geq 0$.

99. $\{x|x \leq 7\} \Rightarrow$ 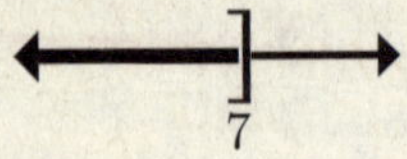

101. $[0, 5) \Rightarrow$ 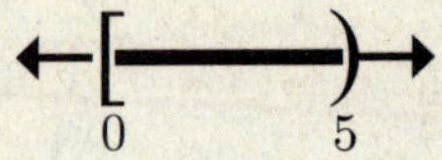

103. $(6, \infty) \cup (9, \infty)$

$(6, \infty)$ 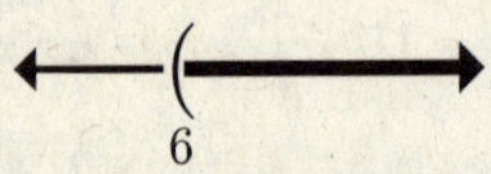

$(9, \infty)$ 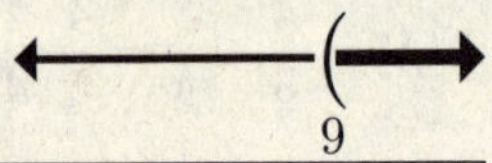

$(6, \infty) \cup (9, \infty)$ 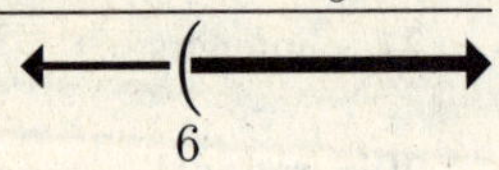

105. $\{x|x > 7 \text{ and } x \geq 11\}$

$x > 7$ 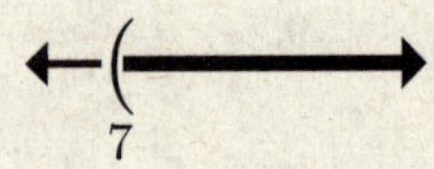

$x \geq 11$ 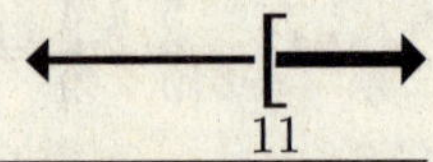

$\{x|x > 7 \text{ and } x \geq 11\}$ 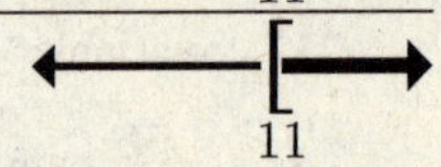

107.

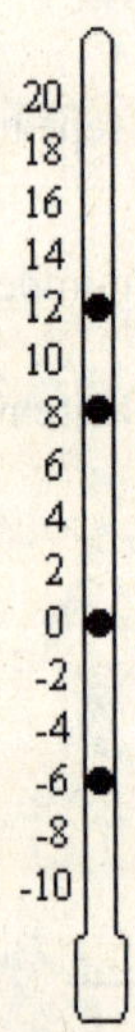

109. Answers may vary.

111. Answers may vary.

113. If $|x| < 50$, then $x \in \{-49, -48, -47, ..., -1, 0, 1, ..., 47, 48, 49\}$. There are 99 integers.

115. Answers may vary.

Exercises 1.2 (page 25)

1. $-5 + (-3)$: addition

3. $-7 - 4$: subtraction

5. $\frac{-10}{2}$: division

7. $\{x|x > 4\} \Rightarrow$ 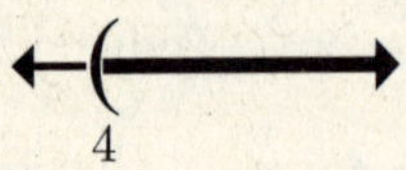

9. $(2, 10] \Rightarrow$ 

© 2015 Cengage Learning. All Rights Reserved. May not be scanned, copied or duplicated, or posted to a publicly accessible website, in whole or in part.

SECTION 1.2

11. Cost of gasoline $= 32(2.29) = 73.28$; Cost of oil $= 3(1.35) = 4.05$
Tax on oil $= 0.05(4.05) = 0.2025 \approx 0.20$
Total cost $= 73.28 + 4.05 + 0.20 = 77.53 \Rightarrow$ The total cost is \$77.53.

13. absolute; common

15. change; add

17. negative

19. product; quotient

21. 0

23. $C = \pi D$; $C = 2\pi r$

25. $a + b = b + a$

27. reciprocals

29. 5; $\frac{1}{7}$

31. $-6 + (-3) = -(6 + 3) = -9$

33. $-7 + 2 = -(7 - 2) = -5$

35. $\frac{1}{2} + \left(-\frac{1}{3}\right) = \frac{3}{6} + \left(-\frac{2}{6}\right) = \frac{1}{6}$

37. $-\frac{5}{9} + \left(\frac{3}{4}\right) = -\frac{20}{36} + \left(\frac{27}{36}\right) = \frac{7}{36}$

39. $-5 - 7 = -5 + (-7) = -12$

41. $-33 - (-33) = -33 + (+33) = 0$

43. $\frac{1}{2} - \left(-\frac{3}{5}\right) = \frac{5}{10} + \frac{6}{10} = \frac{11}{10}$

45. $\frac{1}{3} - \frac{1}{2} = \frac{2}{6} - \frac{3}{6} = -\frac{1}{6}$

47. $-2(6) = -12$

49. $-9(-3) = 27$

51. $\left(-\frac{2}{9}\right)\left(\frac{18}{7}\right) = -\frac{36}{63} = -\frac{4}{7}$

53. $\left(\frac{9}{8}\right)\left(-\frac{6}{7}\right) = -\frac{54}{56} = -\frac{27}{28}$

55. $\frac{-8}{4} = -2$

57. $\frac{-49}{-7} = 7$

59. $\frac{3}{4} \div \left(-\frac{3}{8}\right) = \frac{3}{4}\left(-\frac{8}{3}\right) = -\frac{24}{12} = -2$

61. $-\frac{16}{5} \div \left(-\frac{10}{3}\right) = -\frac{16}{5}\left(-\frac{3}{10}\right)$
$= \frac{48}{50} = \frac{24}{25}$

63. $3 + 4 \cdot 5 = 3 + 20 = 23$

65. $2 - 3 \cdot 5 = 2 - 15 = -13$

67. $8 - 4 - 3 = 4 - 3 = 1$

69. $8 - (4 - 3) = 8 - 1 = 7$

71. $18 \div 6 \div 3 = 3 \div 3 = 1$

73. $18 \div (6 \div 3) = 18 \div 2 = 9$

75. mean $= \frac{7 + 5 + 9 + 10 + 8 + 6 + 6 + 7 + 9 + 12 + 9}{11} = \frac{88}{11} = 8$

77. In order: $5, 6, 6, 7, 7, 8, 9, 9, 9, 10, 12$. The mode is the most frequent value, 9.

79. In order: $8, 8, 10, 12, 12, 12, 14, 16, 16, 23, 23, 26$. The mode is the most frequent value, 12.

© 2015 Cengage Learning. All Rights Reserved. May not be scanned, copied or duplicated, or posted to a publicly accessible website, in whole or in part.

SECTION 1.2

81. $ab + cd = (3)(-2) + (-1)(2)$
$= -6 + (-2) = -8$

83. $\dfrac{ad + c}{cd + b} = \dfrac{3(2) + (-1)}{-1(2) + (-2)} = \dfrac{6 + (-1)}{-2 + (-2)}$
$= \dfrac{5}{-4} = -\dfrac{5}{4}$

85. $C = \dfrac{N(N-1)}{2} = \dfrac{200(200-1)}{2} = \dfrac{200(199)}{2} = \dfrac{\not{2} \cdot 100(199)}{\not{2}} = 19{,}900$ comparisons

87. $P = a + b + c = (23.5 + 37.2 + 39.7)$ ft $= 100.4$ ft

89. commutative property of addition

91. distributive property

93. additive identity property

95. multiplicative inverse property

97. associative property of addition

99. commutative property of multiplication

Problems 101-103 are to be solved using a calculator. The keystrokes needed to solve each problem using a TI-83/84 graphing calculator appear in each solution. There may be other solutions. Keystrokes for other calculators may be slightly different.

101. [(] [3] [7] [.] [9] [+] [2] [5] [.] [2] [)] [+] [1] [4] [.] [3] [ENTER] {77.4}
[3] [7] [.] [9] [+] [(] [2] [5] [.] [2] [+] [1] [4] [.] [3] [)] [ENTER] {77.4}
associative property of addition

103. [2] [.] [7] [3] [×] [(] [4] [.] [5] [3] [4] [+] [5] [7] [.] [1] [2] [)] [ENTER]
{168.31542}
[2] [.] [7] [3] [×] [4] [.] [5] [3] [4] [+] [2] [.] [7] [3] [×] [5] [7] [.] [1] [2]
[ENTER] {168.31542} distributive property

105. $98 - 14 - 14 = 84 - 14 = 70$

107. $24 - (-3) - 5 = 27 - 5 = 22$

109. $-2(21)(-4) = (-42)(-4) = 168$

111. $56 \div 7 \div 2 = 8 \div 2 = 4$

113. $2 + 6 \div 3 - 5 = 2 + 2 - 5 = 4 - 5 = -1$

115. $(3 + 9) \div (4 - 8) = 12 \div (-4) = -3$

117. $\dfrac{3(8+4)}{2 \cdot 3 - 9} = \dfrac{3(12)}{6 - 9} = \dfrac{36}{-3} = -12$

119. $\dfrac{100(2-4)}{1{,}000 \div 10 \div 10} = \dfrac{100(-2)}{100 \div 10} = \dfrac{-200}{10}$
$= -20$

121. $2b - 5ac = 2(-2) - 5(3)(-1) = -4 - (-15) = -4 + 15 = 11$

123. $c - d(2 - a) = -1 - 2(2 - 3) = -1 - 2(-1) = -1 + 2 = 1$

125. $a(b + c) = 3[-2 + (-1)] = 3(-3) = -9$

127. $\dfrac{ad + bc}{cd - ad} = \dfrac{3(2) + (-2)(-1)}{(-1)(2) - 3(2)} = \dfrac{6 + 2}{-2 - 6} = \dfrac{8}{-8} = -1$

© 2015 Cengage Learning. All Rights Reserved. May not be scanned, copied or duplicated, or posted to a publicly accessible website, in whole or in part.

SECTION 1.2

129. $(+22.25) + (+39.75) = +62$
He earned \$62.

131. $(+17) + (-13) = +4$
The temperature rose 4°.

133. $-3(-4) = +12$
It was 12° warmer 3 hours ago.

135. $-(5 \cdot 60)(+23) = -6900$
There were 6900 fewer gallons of water.

The keystrokes needed to solve problems 137-145 using a TI-83/84 graphing calculator appear in each solution. There may be other solutions. Keystrokes for other calculators may be slightly different.

137. [(−)] [2] [3] [0] [0] [+] [1] [7] [5] [0] [+] [1] [8] [7] [5] [ENTER] {1325.}
The army gained +1325 meters.

139. [4] [3] [7] [.] [3] [7] [+] [1] [2] [5] [.] [1] [8] [+] [1] [3] [7] [.] [2] [6] [+] [1] [4] [5] [.] [5] [6] [−] [1] [1] [7] [.] [1] [1] [−] [1] [8] [3] [.] [4] [9] [−] [1] [2] [2] [.] [8] [9] [ENTER] {421.88} Her ending balance is \$421.88.

141. [(] [1] [5] [2] [5] [+] [7] [8] [5] [+] [1] [6] [2] [8] [+] [1] [2] [1] [4] [+] [9] [1] [7] [+] [1] [1] [9] [7] [)] [÷] [6] [ENTER] {1211.} The mean of daily sales is \$1211.

143. [(] [7] [7] [+] [7] [9] [+] [8] [9] [+] [8] [4] [+] [8] [6] [)] [÷] [5] [ENTER] {83.}
His mean score is 83.

145. [(] [1] [0] [0] [0] [0] [0] [+] [4] [×] [1] [0] [0] [0] [0] [)] [÷] [5] [ENTER]
{28000.} The ad is misleading. Although the mean wage of all workers including the businessman is \$28,000 as stated, a better measure of the average would be the median wage, which is \$10,000.

147. $P = 4s = 4(7.5)\text{ cm} = 30\text{ cm}$

149. **Answers may vary.**

151. The mean increases by 7 as well. This property is always true.

153. The median is often the most appropriate average to use when one or two extremely high or low values occur. The following are situations when computing the median might be appropriate:
- finding an average salary of all employees when there are a few highly paid employees
- finding an average household income for a city
- finding an average score of all students on a test when there is one extremely high score

Exercises 1.3 (page 37)

1. $(5)(5)(5) = 5^3$

3. $(-y)(-y) = (-y)^2$

5. $(a)(a)(b)(b)(b) = a^2b^3$

7. $(x^2)(x^2)(x^2) = (x^2)^3$

© 2015 Cengage Learning. All Rights Reserved. May not be scanned, copied or duplicated, or posted to a publicly accessible website, in whole or in part.

SECTION 1.3

9. $\left(\frac{2}{3}\right)\left(\frac{2}{3}\right)\left(\frac{2}{3}\right)\left(\frac{2}{3}\right)=\left(\frac{2}{3}\right)^4$

11. $-(ab)(ab)(ab)=-(ab)^3$

13. $a+b+c=4+(-2)+5=2+5=7$

15. $\frac{ab+2c}{a+b}=\frac{4(-2)+2(5)}{4+(-2)}=\frac{-8+10}{2}$
$=\frac{2}{2}=1$

17. base; exponent; power

19. x^{m+n}

21. x^ny^n

23. 1

25. x^{m-n}

27. $A=s^2$

29. $A=\frac{1}{2}bh$

31. $A=\pi r^2$

33. $V=lwh$

35. $V=Bh$

37. $V=\frac{1}{3}Bh$

39. base $=5$, exponent $=3$

41. base $=x$, exponent $=5$

43. base $=b$, exponent $=6$

45. base $=-mn^2$, exponent $=3$

47. $3^2=3\cdot3=9$

49. $-3^2=-1\cdot3^2=-1\cdot3\cdot3=-91$

51. $(-3)^2=(-3)(-3)=9$

53. $\left(-\frac{2}{5}x\right)^4=\left(-\frac{2}{5}x\right)\left(-\frac{2}{5}x\right)\left(-\frac{2}{5}x\right)\left(-\frac{2}{5}x\right)=\frac{16}{625}x^4$

55. $x^3x^5=x^{3+5}=x^8$

57. $x^4y^2x^5y^6=x^{4+5}y^{2+6}=x^9y^8$

59. $-12x^6\left(\frac{1}{3}x^3y^5\right)=-12\left(\frac{1}{3}\right)x^6x^3y^5$
$=-4x^{6+3}y^5=-4x^9y^5$

61. $p^9pp^0=p^9p^1\cdot1=p^{9+1}=p^{10}$

63. $(x^6)^3=x^{6\cdot3}=x^{18}$

65. $(-x)^2y^4x^4=(-1)^2x^2y^4x^4=x^{2+4}y^4=x^6y^4$

67. $(a^2a^3)^4=(a^{2+3})^4=(a^5)^4=a^{5\cdot4}=a^{20}$

69. $(x^3)^4(x^5)^2=x^{3\cdot4}x^{5\cdot2}=x^{12}x^{10}=x^{22}$

71. $(-2x)^6=(-2)^6x^6=64x^6$

73. $(x^3y^2)^4=(x^3)^4(y^2)^4=x^{3\cdot4}y^{2\cdot4}=x^{12}y^8$

75. $\left(\frac{a^4}{b^5}\right)^7=\frac{(a^4)^7}{(b^5)^7}=\frac{a^{28}}{b^{35}}$

77. $\left(\frac{1}{2}a^2b^5\right)^4=\left(\frac{1}{2}\right)^4(a^2)^4(b^5)^4=\frac{1}{16}a^{2\cdot4}b^{5\cdot4}$
$=\frac{1}{16}a^8b^{20}$

79. $8^0=1$

81. $-(5x^3y^6)^0=-1\cdot1=-1$

© 2015 Cengage Learning. All Rights Reserved. May not be scanned, copied or duplicated, or posted to a publicly accessible website, in whole or in part.

SECTION 1.3

83. $7x^0 = 7 \cdot 1 = 7$

85. $7^0 + 7^2 = 1 + 49 = 50$

87. $5^{-2} = \frac{1}{5^2} = \frac{1}{25}$

89. $(3x)^{-4} = \frac{1}{(3x)^4} = \frac{1}{3^4x^4} = \frac{1}{81x^4}$

91. $x^{-7}x^3 = x^{-7+3} = x^{-4} = \frac{1}{x^4}$

93. $\frac{1}{x^{-5}} = x^5$

95. $\frac{a^8}{a^3} = a^{8-3} = a^5$

97. $\frac{x^{-6}x^5}{x^{10}} = \frac{x^{-1}}{x^{10}} = x^{-1-10} = x^{-11} = \frac{1}{x^{11}}$

99. $\frac{5a^7b^{-4}}{a^{-2}b^{-6}} = 5a^{7-(-2)}b^{-4-(-6)} = 5a^9b^2$

101. $\frac{(x^{-2})^9}{(x^4)^{-2}} = \frac{x^{-18}}{x^{-8}} = x^{-18-(-8)} = x^{-10} = \frac{1}{x^{10}}$

103. $\left(\frac{a^{-3}}{b^{-2}}\right)^{-2} = \frac{(a^{-3})^{-2}}{(b^{-2})^{-2}} = \frac{a^6}{b^4}$

105. $\left(\frac{3x^5y^2}{6x^5y^{-2}}\right)^{-4} = \left(\frac{6x^5y^{-2}}{3x^5y^2}\right)^4 = \left(\frac{2}{y^4}\right)^4 = \frac{2^4}{(y^4)^4} = \frac{16}{y^{16}}$

107. $\frac{a^na^3}{a^4} = \frac{a^{n+3}}{a^4} = a^{n+3-4} = a^{n-1}$

109. $\left(\frac{b^n}{b^3}\right)^3 = \frac{(b^n)^3}{(b^3)^3} = \frac{b^{3n}}{b^9} = b^{3n-9}$

111. $x^2y^3 = (-2)^2(3)^3 = 4(27) = 108$

113. $\frac{x^{-3}}{y^3} = \frac{1}{x^3y^3} = \frac{1}{(-2)^3(3)^3} = \frac{1}{-8(27)}$
$= -\frac{1}{216}$

115. $(xy^2)^{-2} = x^{-2}y^{-4} = \frac{1}{x^2y^4} = \frac{1}{(-2)^2(3)^4} = \frac{1}{4(81)} = \frac{1}{324}$

117. $(-yx^{-1})^3 = (-1)^3y^3(x^{-1})^3 = -y^3x^{-3} = -\frac{y^3}{x^3} = -\frac{3^3}{(-2)^3} = -\frac{27}{-8} = \frac{27}{8}$

119. $A = lw = (3\,\text{m})(5\,\text{m}) = 15\,\text{m}^2$

121. $A = \pi r^2 = \pi(6\,\text{cm})^2 = \pi(36\,\text{cm}^2)$
$\approx 113\,\text{cm}^2$

123. $V = s^3 = (7\,\text{m})^3 = 343\,\text{m}^3$

125. $V = lwh = (6\,\text{ft})(6\,\text{ft})(10\,\text{ft}) = 360\,\text{ft}^3$

127. $aba^3b^4 = a^{1+3}b^{1+4} = a^4b^5$

129. $(-3p^2q^3)^5 = (-3)^5(p^2)^5(q^3)^5 = -243p^{2\cdot5}q^{3\cdot5} = -243p^{10}q^{15}$

131. $(x^2)^3(x^3)^2 = x^{2\cdot3}x^{3\cdot2} = x^6x^6 = x^{6+6} = x^{12}$

133. $(r^{-3}s)^3 = (r^{-3})^3s^3 = r^{-3\cdot3}s^3 = r^{-9}s^3 = \frac{s^3}{r^9}$

© 2015 Cengage Learning. All Rights Reserved. May not be scanned, copied or duplicated, or posted to a publicly accessible website, in whole or in part.

SECTION 1.3

135. $(b^{-8})^9 = b^{-8\cdot 9} = b^{-72} = \frac{1}{b^{72}}$

137. $(-d^2)^3(d^{-3})^3 = (-1)^3(d^2)^3d^{-3\cdot 3} = -1\cdot d^{2\cdot 3}d^{-9} = -1\cdot d^6d^{-9} = -1\cdot d^{6+(-9)} = -d^{-3} = -\frac{1}{d^3}$

139. $\left(\frac{3a^{-2}b^2}{17a^2b^3}\right)^0 = 1$

141. $\frac{(3x^2)^{-2}}{x^3x^{-4}x^0} = \frac{3^{-2}(x^2)^{-2}}{x^{-1}} = \frac{3^{-2}x^{-4}}{x^{-1}} = 3^{-2}x^{-4-(-1)} = 3^{-2}x^{-3} = \frac{1}{3^2x^3} = \frac{1}{9x^3}$

143. $\frac{a^{-n}a^2}{a^3} = \frac{a^{-n+2}}{a^3} = a^{-n+2-3} = a^{-n-1}$ $\left(\text{an equivalent form is } \frac{1}{a^{n+1}}\right)$

145. $\frac{a^{-n}a^{-2}}{a^{-4}} = \frac{a^{-n-2}}{a^{-4}} = a^{-n-2-(-4)} = a^{2-n}$ $\left(\text{an equivalent form is } \frac{1}{a^{n-2}}\right)$

147. $\left(\frac{4a^{-2}b}{3ab^{-3}}\right)^3 = \left(\frac{4bb^3}{3aa^2}\right)^3 = \left(\frac{4b^4}{3a^3}\right)^3 = \frac{4^3(b^4)^3}{3^3(a^3)^3} = \frac{64b^{12}}{27a^9}$

149. $\left(\frac{-2a^4b}{a^{-3}b^2}\right)^{-3} = \left(\frac{a^{-3}b^2}{-2a^4b}\right)^3 = \frac{(a^{-3})^3(b^2)^3}{(-2)^3(a^4)^3b^3} = \frac{a^{-9}b^6}{-8a^{12}b^3} = \frac{b^3}{-8a^{21}} = -\frac{b^3}{8a^{21}}$

151. $A = \frac{1}{2}h(b_1+b_2) = \frac{1}{2}(5\text{ cm})(6\text{ cm}+12\text{ cm}) = \frac{1}{2}(5\text{ cm})(18\text{ cm}) = \frac{1}{2}(90\text{ cm}^2) = 45\text{ cm}^2$

153. $A = s^2 + \frac{1}{2}bh = (15\text{ cm})^2 + \frac{1}{2}(15\text{ cm})(10\text{ cm}) = (225+75)\text{ cm}^2 = 300\text{ cm}^2$

155. $V = \frac{1}{3}Bh = \frac{1}{3}\pi r^2h = \frac{1}{3}\pi(4\text{ ft})^2(10\text{ ft}) = \frac{1}{3}(160\text{ ft}^3)\pi \approx 168\text{ ft}^3$

157. $V = Bh + \frac{1}{2}\cdot\frac{4}{3}\pi r^3 = \pi r^2h + \frac{2}{3}\pi r^3 = \pi(6\text{ m})^2(20\text{ m}) + \frac{2}{3}\pi(6\text{ m})^3 \approx 2714\text{ m}^3$

Problems 159-171 are to be solved using a calculator. The keystrokes needed to solve each problem using a TI-83/84 graphing calculator appear in each solution. There may be other solutions. Keystrokes for other calculators may be slightly different.

159. [1] [.] [2] [3] [^] [6] [ENTER]
{3.462825992}

161. [(–)] [6] [.] [2] [5] [^] [3] [ENTER]
{−244.140625}

163. [3] [.] [6] [8] [^] [0] [ENTER] {1.}

165. [7] [.] [2] [x^2] [×] [2] [.] [7] [x^2] [ENTER] {377.9136}
[(] [7] [.] [2] [×] [2] [.] [7] [)] [x^2] [ENTER] {377.9136}

167. [3] [.] [2] [x^2] [×] [3] [.] [2] [^] [(–)] [2] [ENTER] {1}

© 2015 Cengage Learning. All Rights Reserved. May not be scanned, copied or duplicated, or posted to a publicly accessible website, in whole or in part.

SECTION 1.3

169. [7] [.] [2] [3] [^] [(-)] [3] [ENTER] {0.002645971171398}
[1] [÷] [(] [7] [.] [2] [3] [^] [3] [)] [ENTER] {0.002645971171398}

171. [5] [0] [0] [0] [(] [1] [+] [.] [1] [1] [)] [^] [5] [0] [ENTER]
{922824.1337} The balance will be $922,824.13.

173-175. Answers may vary.

177. $2^{-1}+3^{-1}-4^{-1}=\frac{1}{2}+\frac{1}{3}-\frac{1}{4}=\frac{6}{12}+\frac{4}{12}-\frac{3}{12}=\frac{7}{12}$

179. Let $x=2$, $m=3$ and $n=4$: $x^m+x^n=2^3+2^4=24$ while $x^{m+n}=2^{3+4}=2^7=128$.

Exercises 1.4 (page 46)

1. $2.3(10{,}000)=23{,}000$

3. $-5.7(1000)=-5700$

5. $\frac{3.2}{100}=0.032$

7. $\frac{-5.1}{1000}=-0.0051$

9. $\frac{3}{4}=3\div 4=0.75$

11. $\frac{13}{9}=13\div 9=1.\overline{4}$

13. $3^2+4^3+2^4=9+64+16=89$

15. scientific; 10^n

17. left

19. $5200=5.2\times 10^3$

21. $17{,}600{,}000=1.76\times 10^7$

23. $0.0059=5.9\times 10^{-3}$

25. $0.0000096=9.6\times 10^{-6}$

27. $-45{,}000=-4.5\times 10^4$

29. $-0.000037=-3.7\times 10^{-5}$

31. $4.6\times 10^2=460$

33. $7.96\times 10^5=796{,}000$

35. $6000\times 10^{-7}=6.0\times 10^3\times 10^{-7}$
$=6.0\times 10^{-4}$

37. $0.0527\times 10^5=5.27\times 10^{-2}\times 10^5$
$=5.27\times 10^3$

39. $\frac{(4000)(30{,}000)}{0.0006}=\frac{(4\times 10^3)(3\times 10^4)}{6\times 10^{-4}}=\frac{(4)(3)}{6}\cdot\frac{10^3 10^4}{10^{-4}}=2\times 10^{11}$

41. $\frac{(640{,}000)(2{,}700{,}000)}{120{,}000}=\frac{(6.4\times 10^5)(2.7\times 10^6)}{1.2\times 10^5}=\frac{(6.4)(2.7)}{1.2}\cdot\frac{10^5 10^6}{10^5}=14.4\times 10^6$
$=1.44\times 10^7$

43. $5.26\times 10^{-3}=0.00526$

45. $3.7\times 10^{-4}=0.00037$

47. $23.65\times 10^6=23{,}650{,}000$

49. $323\times 10^5=32{,}300{,}000$

51. $0.0317\times 10^{-2}=3.17\times 10^{-2}\times 10^{-2}$
$=3.17\times 10^{-4}$

53. $52.3\times 10^0=52.3=5.23\times 10^1$

© 2015 Cengage Learning. All Rights Reserved. May not be scanned, copied or duplicated, or posted to a publicly accessible website, in whole or in part.

SECTION 1.4

55. $\dfrac{(0.006)(0.008)}{0.0012} = \dfrac{(6 \times 10^{-3})(8 \times 10^{-3})}{1.2 \times 10^{-3}} = \dfrac{(6)(8)}{1.2} \cdot \dfrac{10^{-3}10^{-3}}{10^{-3}} = 40 \times 10^{-3} = 4 \times 10^{-2}$

57. $\dfrac{(220{,}000)(0.000009)}{0.00033} = \dfrac{(2.2 \times 10^{5})(9 \times 10^{-6})}{3.3 \times 10^{-4}} = \dfrac{(2.2)(9)}{3.3} \cdot \dfrac{10^{5}10^{-6}}{10^{-4}} = 6 \times 10^{3}$

59. $23{,}437^{3} \approx 1.2874 \times 10^{13}$

61. $(63{,}480)(893{,}322) \approx 5.671 \times 10^{10}$

63.
$$\begin{aligned}\frac{(320{,}000)^{2}(0.0009)}{12{,}000^{2}} = \frac{(3.2 \times 10^{5})^{2}(9 \times 10^{-4})}{(1.2 \times 10^{4})^{2}} &= \frac{[(3.2)^{2} \times 10^{10}](9 \times 10^{-4})}{(1.2)^{2} \times 10^{8}} \\ &= \frac{(3.2)^{2}(9)}{(1.2)^{2}} \cdot \frac{10^{10}10^{-4}}{10^{8}} \\ &= 64 \times 10^{-2} = 0.64\end{aligned}$$

65. $\dfrac{(69.4)^{8}(73.1)^{2}}{(0.0043)^{3}} \approx 3.6 \times 10^{25}$

67. gamma ray, x-ray, visible light, infrared, radio wave

69.
$$\begin{aligned}3.31 \times 10^{4}\text{ cm/sec} = \frac{3.31 \times 10^{4}\text{ cm}}{1\text{ sec}} \cdot \frac{60\text{ sec}}{1\text{ min}} \cdot \frac{60\text{ min}}{1\text{ hr}} &= \frac{(3.31 \times 10^{4})(6 \times 10^{1})(6 \times 10^{1})\text{ cm}}{1\text{ hr}} \\ &= (3.31)(6)(6) \cdot (10^{4}10^{1}10^{1})\text{ cm/hour} \\ &= 119.16 \times 10^{6}\text{ cm/hour} \\ &= 1.1916 \times 10^{8}\text{ cm/hour} \\ &\approx 1.19 \times 10^{8}\text{ cm/hour}\end{aligned}$$

71.
$$\begin{aligned}\text{Mass of one million protons} = \text{Mass of one proton} \cdot 1{,}000{,}000 &= 1.67248 \times 10^{-24} \cdot 1 \times 10^{6}\text{ g} \\ &= 1.67248 \times 10^{-18}\text{ g}\end{aligned}$$

73.
$$\begin{aligned}239{,}000\text{ miles} = \frac{2.39 \times 10^{5}\text{ mile}}{1} \cdot \frac{5280\text{ ft}}{1\text{ mile}} \cdot \frac{12\text{ in}}{1\text{ ft}} &= \frac{(2.39 \times 10^{5})(5.28 \times 10^{3})(1.2 \times 10^{1})\text{ in}}{1} \\ &= (2.39)(5.28)(1.2) \cdot 10^{5}10^{3}10^{1}\text{ in.} \\ &= 15.14304 \times 10^{9}\text{ in.} \\ &= 1.514304 \times 10^{10}\text{ in.} \approx 1.51 \times 10^{10}\text{ in.}\end{aligned}$$

75.
$$\begin{aligned}95{,}000\text{ km} = \frac{9.5 \times 10^{4}\text{ km}}{1} \cdot \frac{0.6214\text{ mile}}{1\text{ km}} &= (9.5 \times 10^{4})(6.214 \times 10^{-1})\text{ miles} \\ &= (9.5)(6.214) \cdot 10^{4}10^{-1}\text{ miles} \\ &= 59.033 \times 10^{3}\text{ miles} \approx 5.9 \times 10^{4}\text{ miles}\end{aligned}$$

77. $1\text{ in.} = \dfrac{1\text{ in.}}{1} \cdot \dfrac{25.4\text{ mm}}{1\text{ in.}} \cdot \dfrac{1\text{ angstrom}}{0.0000001\text{ mm}} = \dfrac{2.54 \times 10^{1}}{1 \times 10^{-7}} = 2.54 \times 10^{8}\text{ angstroms}$

© 2015 Cengage Learning. All Rights Reserved. May not be scanned, copied or duplicated, or posted to a publicly accessible website, in whole or in part.

79. $\dfrac{3{,}574{,}000{,}000}{18{,}000}\text{ hr} = \dfrac{3.574 \times 10^9}{1.8 \times 10^4}\text{ hr} = \dfrac{3.574}{1.8} \cdot \dfrac{10^9}{10^4}\text{ hr} = 1.98555555556 \times 10^5\text{ hr}$

$$\begin{aligned} 1.98555555556 \times 10^5\text{ hr} &= \frac{1.98555555556 \times 10^5\text{ hr}}{1} \cdot \frac{1\text{ day}}{24\text{ hr}} \cdot \frac{1\text{ yr}}{365\text{ day}} \\ &= \frac{(1.98555555556 \times 10^5)}{(2.4 \times 10^1)(3.65 \times 10^2)}\text{ years} \\ &= \frac{1.98555555556}{(2.4)(3.65)} \cdot \frac{10^5}{10^1 10^2}\text{ years} \\ &= 0.2266615930999 \times 10^2\text{ years} \approx 23\text{ years} \end{aligned}$$

81.
$$\begin{aligned} 186{,}000\text{ mi/sec} &= \frac{1.86 \times 10^5\text{ mi}}{1\text{ sec}} \cdot \frac{60\text{ sec}}{1\text{ min}} \cdot \frac{60\text{ min}}{1\text{ hr}} \cdot \frac{24\text{ hr}}{1\text{ day}} \cdot \frac{365\text{ day}}{1\text{ yr}} \\ &= (1.86 \times 10^5)(6 \times 10^1)(6 \times 10^1)(2.4 \times 10^1)(3.65 \times 10^2)\text{ mi/yr} \\ &= (1.86)(6)(6)(2.4)(3.65) \cdot 10^5 10^1 10^1 10^1 10^2\text{ mi/yr} = 586.5696 \times 10^{10}\text{ mi/yr} \end{aligned}$$

$$1.3\text{ p} = \frac{1.3 \times 10^0\text{ p}}{1} \cdot \frac{3.26\text{ ly}}{1\text{ p}} = 1.3(3.26 \cdot 586.5696 \times 10^{10})\text{ mi} \approx 2.5 \times 10^{13}\text{ miles}$$

83. **Answers may vary.**

85. Answer will depend on calculator.

Exercises 1.5 (page 57)

1. $3x + 1 = 3(-2) + 1 = -6 + 1 = -5$

3. $-x + 5 = -(-2) + 5 = 2 + 5 = 7$

5. $3 - x = 3 - (-2) = 3 + 2 = 5$

7. $4 \cdot \dfrac{1}{4}x = x$

9. $4 \cdot \dfrac{x+1}{4} = x + 1$

11. $4 \cdot \left(\dfrac{x}{4} - \dfrac{3}{2}\right) = 4 \cdot \dfrac{x}{4} - 4 \cdot \dfrac{3}{2} = x - 6$

13. $\left(\dfrac{x+y}{x-y}\right)^0 = 1$

15. $\left(\dfrac{x^2x^5}{x^3}\right)^2 = \left(\dfrac{x^7}{x^3}\right)^2 = \left(x^4\right)^2 = x^8$

17. $(-4x)^{-3} = \dfrac{1}{(-4x)^3} = \dfrac{1}{-64x^3} = -\dfrac{1}{64x^3}$

19. equation

21. equivalent

23. c; c

25. term

27. numerical coefficients; variables

29.
$$\begin{aligned} 3x + 2 &= 17 \\ 3(5) + 2 &\overset{?}{=} 17 \\ 15 + 2 &\overset{?}{=} 17 \\ 17 &= 17 \end{aligned}$$
5 is a solution.

31.
$$\begin{aligned} \frac{3}{5}x - 5 &= -2 \\ \frac{3}{5}(5) - 5 &\overset{?}{=} -2 \\ 3 - 5 &\overset{?}{=} -2 \\ -2 &= -2 \end{aligned}$$
5 is a solution.

33.
$$\begin{aligned} x + 6 &= 8 \\ x + 6 - 6 &= 8 - 6 \\ x &= 2 \end{aligned}$$

35.
$$\begin{aligned} -5y &= 15 \\ \frac{-5y}{-5} &= \frac{15}{-5} \\ y &= -3 \end{aligned}$$

© 2015 Cengage Learning. All Rights Reserved. May not be scanned, copied or duplicated, or posted to a publicly accessible website, in whole or in part.

SECTION 1.5

37.
$$\begin{aligned} 3a - 27 &= 0 \\ 3a - 27 + 27 &= 0 + 27 \\ 3a &= 27 \\ \frac{3a}{3} &= \frac{27}{3} \\ a &= 9 \end{aligned}$$

39.
$$\begin{aligned} \frac{x}{4} &= 7 \\ 4 \cdot \frac{x}{4} &= 4 \cdot 7 \\ x &= 28 \end{aligned}$$

41.
$$\begin{aligned} 3x + 1 &= 3 \\ 3x + 1 - 1 &= 3 - 1 \\ 3x &= 2 \\ \frac{3x}{3} &= \frac{2}{3} \\ x &= \tfrac{2}{3} \end{aligned}$$

43.
$$\begin{aligned} 6y - 10 &= 14 \\ 6y - 10 + 10 &= 14 + 10 \\ 6y &= 24 \\ \frac{6y}{6} &= \frac{24}{6} \\ y &= 4 \end{aligned}$$

45.
$$\begin{aligned} 3(x - 4) &= -36 \\ 3x - 12 &= -36 \\ 3x - 12 + 12 &= -36 + 12 \\ 3x &= -24 \\ \frac{3x}{3} &= \frac{-24}{3} \\ x &= -8 \end{aligned}$$

47.
$$\begin{aligned} -3(r + 6) &= -2 \\ -3r - 18 &= -2 \\ -3r - 18 + 18 &= -2 + 18 \\ -3r &= 16 \\ \frac{-3r}{-3} &= \frac{16}{-3} \\ r &= -\tfrac{16}{3} \end{aligned}$$

49.
$$\begin{aligned} 6(2x - 5) + 3(x + 1) &= 12x - 30 + 3x + 3 \\ &= 15x - 27 \end{aligned}$$

51.
$$\begin{aligned} 3(x + 4) + 2(x - 1) &= 3x + 12 + 2x - 2 \\ &= 5x + 10 \end{aligned}$$

53. $8 + 2(x - 4) - (6x - 9) = 8 + 2x - 8 - 6x + 9 = -4x + 9$

55. $6x - 3(9 - 5x) + 8 = 6x - 27 + 15x + 8 = 21x - 19$

57.
$$\begin{aligned} 2(a - 5) - (3a + 1) &= 0 \\ 2a - 10 - 3a - 1 &= 0 \\ -a - 11 &= 0 \\ -a - 11 + 11 &= 0 + 11 \\ -a &= 11 \\ a &= -11 \end{aligned}$$

59.
$$\begin{aligned} 3(y - 5) + 10 &= 2(y + 4) \\ 3y - 15 + 10 &= 2y + 8 \\ 3y - 5 &= 2y + 8 \\ 3y - 2y - 5 &= 2y - 2y + 8 \\ y - 5 &= 8 \\ y - 5 + 5 &= 8 + 5 \\ y &= 13 \end{aligned}$$

61.
$$\begin{aligned} 9(x + 2) &= -6(4 - x) + 18 \\ 9x + 18 &= -24 + 6x + 18 \\ 9x + 18 &= 6x - 6 \\ 9x - 6x + 18 &= 6x - 6x - 6 \\ 3x + 18 &= -6 \\ 3x + 18 - 18 &= -6 - 18 \\ 3x &= -24 \\ \frac{3x}{3} &= \frac{-24}{3} \\ x &= -8 \end{aligned}$$

63.
$$\begin{aligned} -4p - 2(3p + 5) &= -6p + 2(p + 2) \\ -4p - 6p - 10 &= -6p + 2p + 4 \\ -10p - 10 &= -4p + 4 \\ -10p + 4p - 10 &= -4p + 4p + 4 \\ -6p - 10 &= 4 \\ -6p - 10 + 10 &= 4 + 10 \\ -6p &= 14 \\ \frac{-6p}{-6} &= \frac{14}{-6} \\ p &= -\tfrac{7}{3} \end{aligned}$$

© 2015 Cengage Learning. All Rights Reserved. May not be scanned, copied or duplicated, or posted to a publicly accessible website, in whole or in part.

SECTION 1.5

65.

$$\begin{aligned}
\frac{x}{6}+1&=\frac{x}{3}\\
6\left(\frac{x}{6}+1\right)&=6\left(\frac{x}{3}\right)\\
6\left(\frac{x}{6}\right)+6(1)&=2x\\
x+6&=2x\\
x-2x+6&=2x-2x\\
-x+6&=0\\
-x+6-6&=0-6\\
-x&=-6\\
x&=6
\end{aligned}$$

67.

$$\begin{aligned}
5-\frac{x+2}{3}&=7-x\\
3\left(5-\frac{x+2}{3}\right)&=3(7-x)\\
3(5)-3\left(\frac{x+2}{3}\right)&=3(7)+3(-x)\\
15-(x+2)&=21-3x\\
15-x-2&=21-3x\\
13-x&=21-3x\\
13-x+3x&=21-3x+3x\\
13+2x&=21\\
13-13+2x&=21-13\\
2x&=8\\
\frac{2x}{2}&=\frac{8}{2}\\
x&=4
\end{aligned}$$

69.

$$\begin{aligned}
\frac{4x-2}{2}&=\frac{3x+6}{3}\\
6\left(\frac{4x-2}{2}\right)&=6\left(\frac{3x+6}{3}\right)\\
3(4x-2)&=2(3x+6)\\
12x-6&=6x+12\\
12x-6x-6&=6x-6x+12\\
6x-6&=12\\
6x-6+6&=12+6\\
6x&=18\\
\frac{6x}{6}&=\frac{18}{6}\\
x&=3
\end{aligned}$$

71.

$$\begin{aligned}
\frac{a+1}{3}+\frac{a-1}{5}&=\frac{2}{15}\\
15\left(\frac{a+1}{3}+\frac{a-1}{5}\right)&=15\left(\frac{2}{15}\right)\\
15\left(\frac{a+1}{3}\right)+15\left(\frac{a-1}{5}\right)&=2\\
5(a+1)+3(a-1)&=2\\
5a+5+3a-3&=2\\
8a+2&=2\\
8a+2-2&=2-2\\
8a&=0\\
\frac{8a}{8}&=\frac{0}{8}\\
a&=0
\end{aligned}$$

73.

$$\begin{aligned}
\frac{1}{2}x-4&=-1+2x\\
2\left(\frac{1}{2}x-4\right)&=2(-1+2x)\\
2\left(\frac{1}{2}x\right)+2(-4)&=2(-1)+2(2x)\\
x-8&=-2+4x\\
x-4x-8&=-2+4x-4x\\
-3x-8&=-2\\
-3x-8+8&=-2+8\\
-3x&=6\\
\frac{-3x}{-3}&=\frac{6}{-3}\\
x&=-2
\end{aligned}$$

© 2015 Cengage Learning. All Rights Reserved. May not be scanned, copied or duplicated, or posted to a publicly accessible website, in whole or in part.

SECTION 1.5

75.
$$\begin{aligned} \frac{x}{2}-\frac{x}{3}&=4\\ 6\left(\frac{x}{2}-\frac{x}{3}\right)&=6(4)\\ 6\left(\frac{x}{2}\right)-6\left(\frac{x}{3}\right)&=24\\ 3x-2x&=24\\ x&=24 \end{aligned}$$

77.
$$\begin{aligned} \frac{5a}{2}-12&=\frac{a}{3}+1\\ 6\left(\frac{5a}{2}-12\right)&=6\left(\frac{a}{3}+1\right)\\ 6\left(\frac{5a}{2}\right)-6(12)&=6\left(\frac{a}{3}\right)+6(1)\\ 3(5a)-72&=2a+6\\ 15a-72&=2a+6\\ 15a-2a-72&=2a-2a+6\\ 13a-72&=6\\ 13a-72+72&=6+72\\ 13a&=78\\ \frac{13a}{13}&=\frac{78}{13}\\ a&=6 \end{aligned}$$

79.
$$\begin{aligned} \frac{a+1}{4}+\frac{2a-3}{4}&=\frac{a}{2}-2\\ \frac{3a-2}{4}&=\frac{a}{2}-2\\ 4\left(\frac{3a-2}{4}\right)&=4\left(\frac{a}{2}-2\right)\\ 3a-2&=4\left(\frac{a}{2}\right)-4(2)\\ 3a-2&=2a-8\\ 3a-2a-2&=2a-2a-8\\ a-2&=-8\\ a-2+2&=-8+2\\ a&=-6 \end{aligned}$$

81.
$$\begin{aligned} 0.45&=16.95-0.25(75-3a)\\ 0.45&=16.95-18.75+0.75a\\ 0.45&=-1.8+0.75a\\ 0.45+1.8&=-1.8+1.8+0.75a\\ 2.25&=0.75a\\ \frac{2.25}{0.75}&=\frac{0.75a}{0.75}\\ 3&=a \end{aligned}$$

83.
$$\begin{aligned} 0.09x+0.14(10000-x)&=1275\\ 0.09x+1400-0.14x&=1275\\ -0.05x+1400&=1275\\ -0.05x+1400-1400&=1275-1400\\ -0.05x&=-125\\ \frac{-0.05x}{-0.05}&=\frac{-125}{-0.05}\\ x&=2500 \end{aligned}$$

85.
$$\begin{aligned} 4(2-3t)+6t&=-6t+8\\ 8-12t+6t&=-6t+8\\ 8-6t&=-6t+8\\ -6t+8&=-6t+8 \end{aligned}$$
Identity, $\mathbb{R}$

87.
$$\begin{aligned} 2x-6&=-2x+4(x-2)\\ 2x-6&=-2x+4x-8\\ 2x-6&=2x-8\\ 2x-2x-6&=2x-2x-8\\ -6&=-8 \quad \text{Contradiction, } \emptyset \end{aligned}$$

89.
$$\begin{aligned} A&=lw\\ \frac{A}{l}&=\frac{lw}{l}\\ \frac{A}{l}&=w, \text{ or } w=\frac{A}{l} \end{aligned}$$

© 2015 Cengage Learning. All Rights Reserved. May not be scanned, copied or duplicated, or posted to a publicly accessible website, in whole or in part.

SECTION 1.5

91.
$$V = \frac{1}{3}Bh$$
$$3V = 3 \cdot \frac{1}{3}Bh$$
$$3V = Bh$$
$$\frac{3V}{h} = \frac{Bh}{h}$$
$$\frac{3V}{h} = B, \text{ or } B = \frac{3V}{h}$$

93.
$$I = prt$$
$$\frac{I}{pr} = \frac{prt}{pr}$$
$$\frac{I}{pr} = t, \text{ or } t = \frac{I}{pr}$$

95.
$$p = 2l + 2w$$
$$p - 2l = 2l - 2l + 2w$$
$$p - 2l = 2w$$
$$\frac{p-2l}{2} = \frac{2w}{2}$$
$$\frac{p-2l}{2} = w, \text{ or } w = \frac{p-2l}{2}$$

97.
$$A = \frac{1}{2}h(B+b)$$
$$2A = 2 \cdot \frac{1}{2}h(B+b)$$
$$2A = h(B+b)$$
$$\frac{2A}{h} = \frac{h(B+b)}{h}$$
$$\frac{2A}{h} = B+b$$
$$\frac{2A}{h} - b = B + b - b$$
$$\frac{2A}{h} - b = B, \text{ or } B = \frac{2A}{h} - b$$

99.
$$y = mx + b$$
$$y - b = mx + b - b$$
$$y - b = mx$$
$$\frac{y-b}{m} = \frac{mx}{m}$$
$$\frac{y-b}{m} = x, \text{ or } x = \frac{y-b}{m}$$

101.
$$S = \frac{a - lr}{1-r}$$
$$(1-r)S = (1-r) \cdot \frac{a-lr}{1-r}$$
$$S - rS = a - lr$$
$$S - Sr - a = a - a - lr$$
$$S - Sr - a = -lr$$
$$\frac{S - Sr - a}{-r} = \frac{-lr}{-r}$$
$$\frac{-S + Sr + a}{r} = l, \text{ or } l = \frac{a - S + Sr}{r}$$

103.
$$S = \frac{n(a+l)}{2}$$
$$2S = 2 \cdot \frac{n(a+l)}{2}$$
$$2S = n(a+l)$$
$$2S = na + nl$$
$$2S - na = na - na + nl$$
$$2S - na = nl$$
$$\frac{2S - na}{n} = \frac{nl}{n}$$
$$\frac{2S - na}{n} = l, \text{ or } l = \frac{2S - na}{n}$$

105. $4(x+1) + (x+6) = 4x + 4 + x + 6 = 5x + 10$

107. $-2(x+1) - (x+3) = -2x - 2 - x - 3 = -3x - 5$

© 2015 Cengage Learning. All Rights Reserved. May not be scanned, copied or duplicated, or posted to a publicly accessible website, in whole or in part.

SECTION 1.5

109.
$$\begin{aligned} 5(2x-3) &= 3x+6 \\ 10x-15 &= 3x+6 \\ 10x-3x-15 &= 3x-3x+6 \\ 7x-15 &= 6 \\ 7x-15+15 &= 6+15 \\ 7x &= 21 \\ \frac{7x}{7} &= \frac{21}{7} \\ x &= 3 \end{aligned}$$

111.
$$\begin{aligned} -2(4x-5) &= 4x-3 \\ -8x+10 &= 4x-3 \\ -8x-4x+10 &= 4x-4x-3 \\ -12x+10 &= -3 \\ -12x+10-10 &= -3-10 \\ -12x &= -13 \\ \frac{-12x}{-12} &= \frac{-13}{-12} \\ x &= \tfrac{13}{12} \end{aligned}$$

113.
$$\begin{aligned} 3a-22 &= -2a-7 \\ 3a+2a-22 &= -2a+2a-7 \\ 5a-22 &= -7 \\ 5a-22+22 &= -7+22 \\ 5a &= 15 \\ \frac{5a}{5} &= \frac{15}{5} \\ a &= 3 \end{aligned}$$

115.
$$\begin{aligned} 3(y-4)-6 &= y \\ 3y-12-6 &= y \\ 3y-18 &= y \\ 3y-y-18 &= y-y \\ 2y-18 &= 0 \\ 2y-18+18 &= 0+18 \\ 2y &= 18 \\ \frac{2y}{2} &= \frac{18}{2} \\ y &= 9 \end{aligned}$$

117.
$$\begin{aligned} 5(5-a) &= 37-2a \\ 25-5a &= 37-2a \\ 25-5a+2a &= 37-2a+2a \\ 25-3a &= 37 \\ 25-25-3a &= 37-25 \\ -3a &= 12 \\ \tfrac{-3a}{-3} &= \tfrac{12}{-3} \\ a &= -4 \end{aligned}$$

119.
$$\begin{aligned} 4(y+1) &= -2(4-y) \\ 4y+4 &= -8+2y \\ 4y-2y+4 &= -8+2y-2y \\ 2y+4 &= -8 \\ 2y+4-4 &= -8-4 \\ 2y &= -12 \\ \tfrac{2y}{2} &= \tfrac{-12}{2} \\ y &= -6 \end{aligned}$$

121.
$$\begin{aligned} 2(2x+1) &= 15+3x \\ 4x+2 &= 15+3x \\ 4x-3x+2 &= 15+3x-3x \\ x+2 &= 15 \\ x+2-2 &= 15-2 \\ x &= 13 \end{aligned}$$

123.
$$\begin{aligned} 9.8-15z &= -15.7 \\ 9.8-9.8-15z &= -15.7-9.8 \\ -15z &= -25.5 \\ \frac{-15z}{-15} &= \frac{-25.5}{-15} \\ z &= 1.7 \end{aligned}$$

© 2015 Cengage Learning. All Rights Reserved. May not be scanned, copied or duplicated, or posted to a publicly accessible website, in whole or in part.

SECTION 1.5

125.
$$\begin{aligned}-3x &= -2x+1-(5+x)\\ -3x &= -2x+1-5-x\\ -3x &= -3x-4\\ -3x+3x &= -3x+3x-4\\ 0 &= -4 \quad \text{Contradiction, } \emptyset\end{aligned}$$

127.
$$\begin{aligned}4x-2(3x+2) &= 2(x+3)\\ 4x-6x-4 &= 2x+6\\ -2x-4 &= 2x+6\\ -2x-2x-4 &= 2x-2x+6\\ -4x-4 &= 6\\ -4x-4+4 &= 6+4\\ -4x &= 10\\ \frac{-4x}{-4} &= \frac{10}{-4}\\ x &= -\frac{5}{2}\end{aligned}$$

129.
$$\begin{aligned}l &= a+(n-1)d\\ l &= a+nd-d\\ l-a+d &= a-a+nd-d+d\\ l-a+d &= nd\\ \frac{l-a+d}{d} &= \frac{nd}{d}\\ \frac{l-a+d}{d} &= n, \text{ or } n = \frac{l-a+d}{d}\end{aligned}$$

131.
$$\begin{aligned}F &= \frac{GmM}{d^2}\\ d^2F &= d^2 \cdot \frac{GmM}{d^2}\\ Fd^2 &= GmM\\ \frac{Fd^2}{GM} &= \frac{GmM}{GM}\\ \frac{Fd^2}{GM} &= m, \text{ or } m = \frac{Fd^2}{GM}\end{aligned}$$

133.
$$\begin{aligned}F &= \frac{9}{5}C+32\\ F-32 &= \frac{9}{5}C+32-32\\ F-32 &= \frac{9}{5}C\\ \frac{5}{9}(F-32) &= \frac{5}{9}\cdot\frac{9}{5}C\\ \frac{5}{9}(F-32) &= C, \text{ or } C = \frac{5}{9}(F-32)\end{aligned}$$

$F = 32°$: $C = \frac{5}{9}(32-32) = \frac{5}{9}(0) = 0°$

$F = 70°$: $C = \frac{5}{9}(70-32) = \frac{5}{9}(38) \approx 21.1°$

$F = 212°$: $C = \frac{5}{9}(212-32) = \frac{5}{9}(180) = 100°$

135.
$$\begin{aligned}C &= 0.07n+6.50\\ C-6.50 &= 0.07n+6.50-6.50\\ C-6.50 &= 0.07n\\ \frac{C-6.50}{0.07} &= \frac{0.07n}{0.07}\\ \frac{C-6.50}{0.07} &= n, \text{ or } n = \frac{C-6.50}{0.07}\end{aligned}$$

To avoid decimals in the denominator:

$$n = \frac{100(C-6.50)}{100(0.07)} = \frac{100(C-6.50)}{7}$$

$C = 49.97$: $n = \frac{49.97-6.50}{0.07} = \frac{43.47}{0.07} = 621$ kwh

$C = 76.50$: $n = \frac{76.50-6.50}{0.07} = \frac{70}{0.07} = 1000$ kwh

$C = 125$: $n = \frac{125-6.50}{0.07} = \frac{118.50}{0.07} \approx 1692.9$ kwh

© 2015 Cengage Learning. All Rights Reserved. May not be scanned, copied or duplicated, or posted to a publicly accessible website, in whole or in part.

SECTION 1.5

137. $E = IR$

$$\frac{E}{I} = \frac{IR}{I}$$

$$\frac{E}{I} = R, \text{ or } R = \frac{E}{I}$$

$$R = \frac{56}{7} = 8 \text{ ohms}$$

139.

$$a = 180\left(1 - \frac{2}{n}\right)$$

$$a = 180 - \frac{360}{n}$$

$$a - 180 = 180 - 180 - \frac{360}{n}$$

$$a - 180 = -\frac{360}{n}$$

$$n(a - 180) = n\left(-\frac{360}{n}\right)$$

$$n(a - 180) = -360$$

$$\frac{n(a - 180)}{a - 180} = \frac{-360}{a - 180}$$

$$n = \frac{360}{180 - a}$$

$$n = \frac{360}{180 - a} = \frac{360}{180 - 135} = 8 \text{ sides}$$

141. Answers may vary.

143. The distributive property was not used correctly.

Exercises 1.6 (page 66)

1. $0.30(650) = 195$

3. $14(\$54) = \$756$

5. $A = lw = (8 \text{ m})(5 \text{ m}) = 40 \text{ m}^2$

7.

$$\left(\frac{3x^{-3}}{4x^2}\right)^{-4} = \left(\frac{4x^2}{3x^{-3}}\right)^4 = \left(\frac{4x^5}{3}\right)^4 = \frac{256x^{20}}{81}$$

9.

$$\frac{a^n a^5}{a^6} = \frac{a^{n+5}}{a^6} = a^{n+5-6} = a^{n-1}$$

11. $5x + 4$

13. $89x$

15. acute

17. $180°$

19. isosceles

21. equilateral

23. Let $x =$ length of 1st piece.
Then $2x$, $4x$ and $8x =$ the other lengths.

[Sum of lengths] = [total length]

$$x + 2x + 4x + 8x = 90$$

$$15x = 90$$

$$x = 6$$

The longest piece has a length of $8x$, or $8(6) = 48$ feet.

25. Let $x =$ the length of the shorter piece.
Then $2x + 1 =$ the length of the other piece.

[shorter length] + [other length] = [total length]

$$x + 2x + 1 = 22$$

$$3x + 1 = 22$$

$$3x = 21$$

$$x = 7$$

The pieces have lengths of 7 feet and 15 feet.

© 2015 Cengage Learning. All Rights Reserved. May not be scanned, copied or duplicated, or posted to a publicly accessible website, in whole or in part.

SECTION 1.6

27. Percent markdown $= \dfrac{\text{amount of markdown}}{\text{regular price}} = \dfrac{726 - 580.80}{726} = \dfrac{145.20}{726} = 0.20 = 20\%.$

29. Percent markup $= \dfrac{\text{amount of markup}}{\text{regular price}} = \dfrac{78 - 18}{18} = \dfrac{60}{18} = \dfrac{10}{3} = 3\dfrac{1}{3} = 333\dfrac{1}{3}\%.$

31. Let $B = \#$ of shares in Big Bank. Then $500 - B = \#$ of shares in Safe Savings.

$$\boxed{\text{Big Bank shares}} + \boxed{\text{Safe Savings shares}} = \boxed{\text{Total value}}$$
$$115B + 97(500 - B) = 53900$$
$$115B + 48500 - 97B = 53900$$
$$18B = 5400$$
$$B = 300$$

The student owns 300 shares of Big Bank and 200 shares of Safe Savings.

33. Let $B = \#$ of scientific models sold. Then $85 - B = \#$ of graphing models sold.

$$\boxed{\text{Value of scientific}} + \boxed{\text{Value of graphing}} = \boxed{\text{Total value}}$$
$$15B + 67(85 - B) = 3875$$
$$15B + 5695 - 67B = 3875$$
$$-52B = -1820$$
$$B = 35$$

The store sold 35 scientific calculators and 50 graphing calculators.

35. Let $x =$ the measure of each angle.

$$\boxed{\text{Sum of measures of all angles}} = 180°$$
$$3x = 180°$$
$$x = 60°$$

Each angle has a measure of 60°.

37. $$\boxed{\text{Measure of } \angle 1} = \boxed{\text{Measure of } \angle 3}$$
$$3x + 10 = 5x - 10$$
$$-2x = -20$$
$$x = 10$$

39. Let $x =$ the measure of the smaller angle. Then $x + 35 =$ the other measure.

$$\boxed{\text{The sum of the angles}} = 180°$$
$$x + x + 35 = 180$$
$$2x + 35 = 180$$
$$2x = 145$$
$$x = 72.5$$

The smaller angle has a measure of 72.5°.

41. Let $x =$ the measure of the smaller angle. Then $x + 22 =$ the other measure.

$$\boxed{\text{sum of angles}} = 90°$$
$$x + x + 22 = 90$$
$$2x + 22 = 90$$
$$2x = 68$$
$$x = 34$$

The larger angle has a measure of 56°.

43. Let $w =$ the width. Then $2w =$ the length.

$$\boxed{\text{Perimeter of garden}} = 72$$
$$w + 2w + w + 2w = 72$$
$$6w = 72$$
$$w = 12$$

The dimension are 12 m by 24 m.

45. Let $w =$ the width. Then $2w =$ the length.

$$\boxed{\text{Total amount of fence}} = 624$$
$$w + 2w + w = 624$$
$$4w = 624$$
$$w = 156$$

The dimension are 156 feet by 312 feet.

© 2015 Cengage Learning. All Rights Reserved. May not be scanned, copied or duplicated, or posted to a publicly accessible website, in whole or in part.

SECTION 1.6

47.

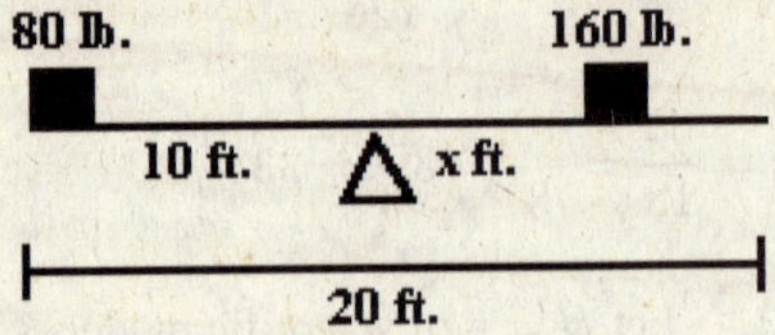

$$\boxed{\text{Boy's force times distance}} = \boxed{\text{Father's force times distance}}$$

$$80(10) = 160x$$
$$800 = 160x$$
$$5 = x$$

The father should sit 5 feet from the fulcrum.

49.

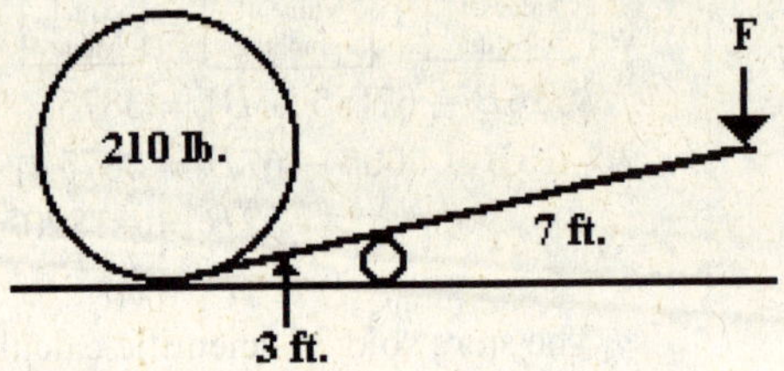

$$\boxed{\text{Stone's force times distance}} = \boxed{\text{Woman's force times distance}}$$

$$210(3) = F(7)$$
$$630 = 7F$$
$$90 = F$$

The woman needs to exert 90 pounds of force.

51. Let $r =$ the number of roses he buys.

$$\boxed{\text{Cost of roses}} + \boxed{\text{Delivery charge}} = \boxed{\text{Total cost}}$$

$$1.50r + 9.50 = 27.50$$
$$1.50r = 18$$
$$r = 12$$

The man can buy 12 roses.

53. Let $m =$ the number of miles he drives.

$$\boxed{\text{Mileage cost}} + \boxed{\text{Daily cost}} = \boxed{\text{Total cost}}$$

$$0.30m + 2 \cdot 12 = 42$$
$$0.30m = 18$$
$$m = 60$$

$$\boxed{\text{Mileage cost}} + \boxed{\text{Daily cost}} = \boxed{\text{Total cost}}$$

$$0.30m + 2 \cdot 12 = 60$$
$$0.30m = 36$$
$$m = 120$$

He can drive 60 miles for \$42 and 120 miles for \$60.

55. Let $x =$ the cost of the player.
Then $x + 501 =$ the cost of the TV.

$$\boxed{\text{Player cost}} + \boxed{\text{TV cost}} = \boxed{\text{total cost}}$$

$$x + x + 501 = 655$$
$$2x + 501 = 655$$
$$2x = 154$$

$x = 77$; The TV costs $x + 501 = 77 + 501 = \$578$.

57. Let $x =$ the original height. Then $3x =$ the new height.

$$\boxed{\text{Old Area}} + 84 = \boxed{\text{New Area}}$$

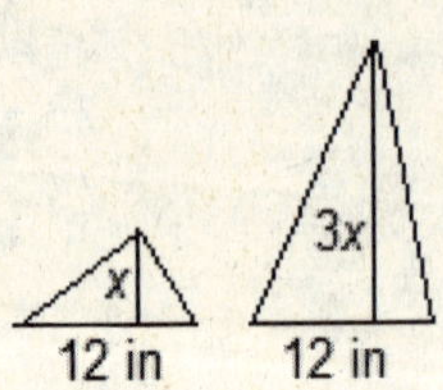

$$\tfrac{1}{2}(12)(x) + 84 = \tfrac{1}{2}(12)(3x)$$
$$6x + 84 = 18x$$
$$84 = 12x$$
$$7 = x$$

The original height is 7 inches.

© 2015 Cengage Learning. All Rights Reserved. May not be scanned, copied or duplicated, or posted to a publicly accessible website, in whole or in part.

59. $\boxed{\text{The sum of the measures of } \angle 1 \text{ and } \angle 2} = 180°$

$$\begin{aligned} x + 50 + 2x - 20 &= 180° \\ 3x + 30 &= 180° \\ 3x &= 150° \\ x &= 50° \end{aligned}$$

61.

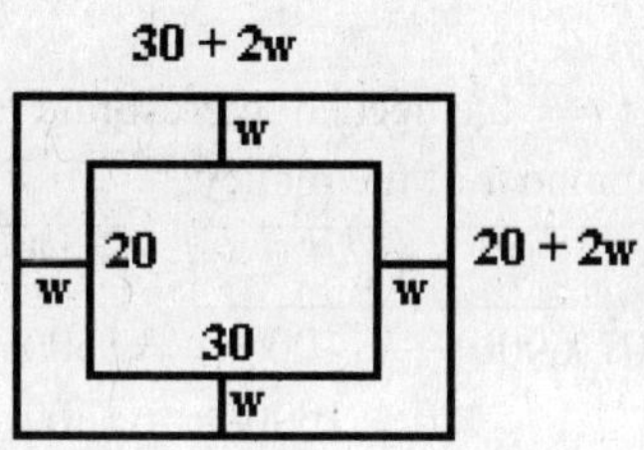

$\boxed{\text{Total fencing}} = 180$

$$\begin{aligned} 2(30 + 2w) + 2(20 + 2w) &= 180 \\ 60 + 4w + 40 + 4w &= 180 \\ 8w + 100 &= 180 \\ 8w &= 80 \\ w &= 10 \end{aligned}$$

The walkway will be 10 feet wide.

63. $\boxed{\text{Sum of forces times distances on left}} = \boxed{\text{Sum of forces times distances on right}}$

$$\begin{aligned} 100(12) + 70(8) &= 40(x) + 200(8) \\ 1200 + 560 &= 40x + 1600 \\ 160 &= 40x \\ 4 &= x \end{aligned}$$

The distance should be 4 feet.

65.

$$\begin{aligned} F &= \frac{5}{9}(F - 32) \\ 9F &= 5(F - 32) \\ 9F &= 5F - 160 \\ 4F &= -160 \\ F &= -40 \end{aligned}$$

The temperatures are the same at $-40°$.

67. **Answers may vary.**

69.

$$\begin{aligned} 40(6) + 10x &= 50(4) \\ 240 + 10x &= 200 \\ 10x &= -40 \\ x &= -4 \end{aligned}$$

Exercises 1.7 (page 74)

1. $i = prt = 1000(0.03)(1) = \$30$

3. $i = prt = (12000 - x)(0.04)(1) = \$(480 - 0.04x)$

5. $i = prt = 2500(0.03)(1) = \$75$

7.

$$\begin{aligned} 10x - 6 &= 4x \\ 10x - 10x - 6 &= 4x - 10x \\ -6 &= -6x \\ 1 &= x \end{aligned}$$

© 2015 Cengage Learning. All Rights Reserved. May not be scanned, copied or duplicated, or posted to a publicly accessible website, in whole or in part.

9.
$$\begin{aligned} \frac{8(y-5)}{3} &= 2(y-4) \\ 3 \cdot \frac{8(y-5)}{3} &= 3 \cdot 2(y-4) \\ 8(y-5) &= 6(y-4) \\ 8y-40 &= 6y-24 \\ 2y &= 16 \\ y &= 8 \end{aligned}$$

11. principal; rate; time

13. value; price; number

15. Let $x =$ amount invested at 4%. Then $12{,}000 - x =$ amount invested at 6%.

$\boxed{\text{Interest at 4\%}} + \boxed{\text{Interest at 6\%}} = \boxed{\text{Total interest}}$

$$\begin{aligned} 0.04x + 0.06(12000 - x) &= 680 \\ 0.04x + 720 - 0.06x &= 680 \\ -0.02x &= -40 \\ x &= \frac{-40}{-0.02} \\ x &= 2000 \end{aligned}$$

She invested \$2000 in the money market account (4%) and \$10,000 in the CD (6%).

17. Let $r =$ the needed interest rate for the remainder of the money.

$\boxed{\text{Interest at 7\%}} + \boxed{\text{Interest at } r\%} = \boxed{\text{Total interest}}$

$$\begin{aligned} 0.07(6000) + r(10000) &= 1500 \\ 420 + 10000r &= 1500 \\ 10000r &= 1080 \\ r &= \frac{1080}{10000} \\ r &= 0.1080 \end{aligned}$$

She needs to earn 10.8% interest on the remainder of the money.

19. Let $x =$ the amount she has to invest. She needs $x + 3000$ to earn 11% interest.

$\boxed{\text{Interest at 11\%}} = 2 \cdot \boxed{\text{Interest at 7.5\%}}$

$$\begin{aligned} 0.11(x + 3000) &= 2(0.075)(x) \\ 0.11x + 330 &= 0.15x \\ -0.04x &= -330 \\ x &= \frac{-330}{-0.04} \\ x &= 8250 \end{aligned}$$

She has \$8,250 on hand to invest.

21. Let $t =$ time for the cars to meet.

	Rate	Time	Dist.
First car	50	t	$50t$
Second car	48	t	$48t$

$\boxed{\text{1st dist.}} + \boxed{\text{2nd dist.}} = \boxed{\text{Total dist.}}$

$$\begin{aligned} 50t + 48t &= 343 \\ 98t &= 343 \\ t &= \tfrac{343}{98} \\ t &= \tfrac{7}{2} = 3\tfrac{1}{2} \end{aligned}$$

It will take the cars $3\frac{1}{2}$ hours to meet.

© 2015 Cengage Learning. All Rights Reserved. May not be scanned, copied or duplicated, or posted to a publicly accessible website, in whole or in part.

23. Let t = time car travels.
Then $t+1$ = time cyclist travels.

	Rate	Time	Dist.
Cyclist	18	$t+1$	$18(t+1)$
Car	45	t	$45t$

Cyclist distance = Car distance

$$18(t+1) = 45t$$
$$18t + 18 = 45t$$
$$-27t = -18$$
$$t = \frac{-18}{-27}$$
$$t = \frac{2}{3}$$

It will take $\frac{2}{3}$ of an hour to pass the cyclist.

25. Let t = time for them to run.

	Rate	Time	Dist.
1st Runner	12	t	$12t$
2nd Runner	10	t	$10t$

1st dist. − 2nd dist. = Dist. between them

$$12t - 10t = 0.25$$
$$2t = \frac{1}{4}$$
$$t = \frac{1}{8}$$

It will take $\frac{1}{8}$ hour to be $\frac{1}{4}$ mile apart.

27. Let x = pounds of cheaper candy. Then $30 - x$ = pounds of other candy.

	Price per pound	# pounds	Value
Cheaper	2.95	x	$2.95x$
Other	3.10	$30 - x$	$3.10(30 - x)$
Mixture	3.00	30	$3.00(30)$

Value of cheaper + Value of other = Value of mixture

$$2.95x + 3.10(30 - x) = 3.00(30)$$
$$2.95x + 93 - 3.1x = 90$$
$$-0.15x = -3$$
$$x = \frac{-3}{-0.15} = 20 \Rightarrow 30 - x = 30 - 20 = 10$$

The owner should use 20 pounds of the cheaper candy and 10 pounds of the other candy.

29. Let x = the number of additional problems the student gets correct.

Original # correct + Additional # correct = Total # correct

$$0.70(30) + x = 0.80(45)$$
$$21 + x = 36$$
$$x = 15$$

The student must get all 15 of the additional problems correct to receive a grade of 80%.

31. Let x = gallons of cream used.
Then $42 - x$ = gallons of 1% milk used.

Butterfat in cream + Butterfat in 1% milk = Butterfat in 2% milk

$$0.22x + 0.01(42 - x) = 0.02(42)$$
$$0.22x + 0.42 - 0.01x = 0.84$$
$$0.21x = 0.42$$
$$x = \frac{0.42}{0.21} = 2$$

2 gallons of cream should be used.

© 2015 Cengage Learning. All Rights Reserved. May not be scanned, copied or duplicated, or posted to a publicly accessible website, in whole or in part.

33. Let x = number of points earned on the final.

[Original points] + [Points on final] = [Points at end]

$$375 + x = 0.90(450)$$
$$375 + x = 405$$
$$x = 30$$

She must earn 30 points on the final to receive a grade of 90% (A).

35. Let w = the wholesale price of the textbook.

[Wholesale price] + [Markup] = [Retail price]

$$w + 0.35w = 108$$
$$1.35w = 108$$
$$w = \frac{108}{1.35} = 80$$

The bookstore pays \$80 for the book.

37. Let t = time to return.

	Rate	Time	Dist.
Upstream	8	3	24
Downstream	16	t	$16t$

[Downstream distance] = [Upstream distance]

$$16t = 24$$
$$t = \tfrac{24}{16} = \tfrac{3}{2} = 1\tfrac{1}{2}$$

It will take the rider $1\frac{1}{2}$ hours to return.

39. Let s = the # of student tickets sold. Then $225 - s$ = the # of adult tickets sold.

[Student \$] + [Adult \$] = [Total \$]

$$4s + 8(225 - s) = 1300$$
$$4s + 1800 - 8s = 1300$$
$$-4s + 1800 = 1300$$
$$-4s = -500$$
$$s = \tfrac{-500}{-4}$$
$$s = 125$$

125 student tickets were sold.

41. Let t = her time walking north.

	Rate	Time	Dist.
North	3	t	$3t$
Return	4	$3.5 - t$	$4(3.5 - t)$

[North distance] = [Return distance]

$$3t = 4(3.5 - t)$$
$$3t = 14 - 4t$$
$$7t = 14$$
$$t = 2$$

[Total distance] $= 3t + 4(3.5 - t)$
$= 3(2) + 4(3.5 - 2) = 12$

She walked a total of 12 miles.

43. Let x = ounces of water added (0% alcohol).

[Alcohol at start] + [Alcohol added] = [Alcohol at end]

$$0.15(20) + 0(x) = 0.10(20 + x)$$
$$3 + 0 = 2 + 0.10x$$
$$1 = 0.10x$$
$$x = \frac{1}{0.10} = 10$$

10 ounces of water should be added.

45. Let x = space in front of first partition. Then $x + 3$ and $x + 6$ are the other spaces.

[Lengths of spaces] + [Lengths of partitions] = [Total length]

$$x + x + 3 + x + 6 + 2(0.5) = 28$$
$$3x + 10 = 28$$
$$3x = 18$$
$$x = 6$$ The first space should be 6 inches.

47. **Answers may vary.**

49. We need to know how far or how long he drove while at the other rate in order to solve the problem.

© 2015 Cengage Learning. All Rights Reserved. May not be scanned, copied or duplicated, or posted to a publicly accessible website, in whole or in part.

Chapter 1 Review (page 80)

1. whole numbers: $0, 1, 2, 4$

2. natural numbers: $1, 2, 4$

3. rational numbers: $-4, -\frac{2}{3}, 0, 1, 2, 4$

4. integers: $-4, 0, 1, 2, 4$

5. irrational numbers: π

6. real numbers: $-4, -\frac{2}{3}, 0, 1, 2, \pi, 4$

7. negative numbers: $-4, -\frac{2}{3}$

8. positive numbers: $1, 2, \pi, 4$

9. prime numbers: 2

10. composite numbers: 4

11. even integers: $-4, 0, 2, 4$

12. odd integers: 1

13.

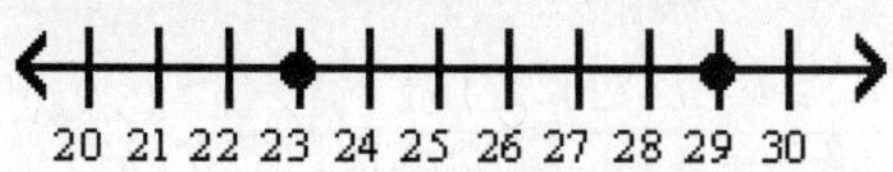

14.

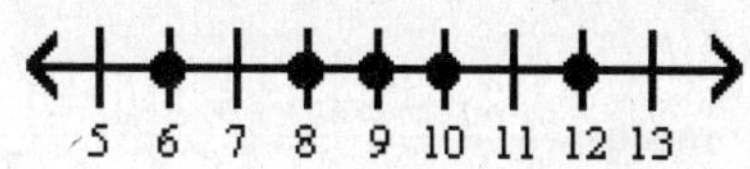

15. $\{x|x \geq -2\} \Rightarrow$ 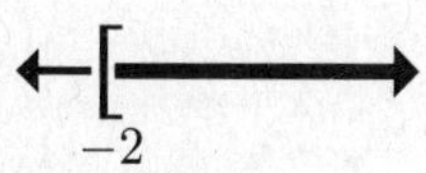

16. $\{x|-2 < x \leq 6\} \Rightarrow$ 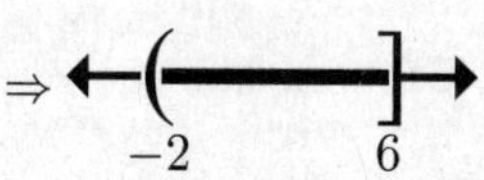

17. $[-4, 1) \Rightarrow$ 

18. $(-3, 5] \Rightarrow$ 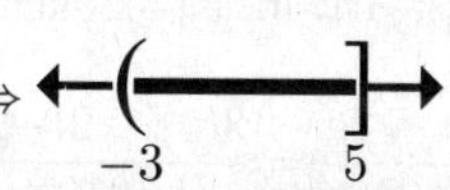

19. $\{x|x > 2\} \Rightarrow$ 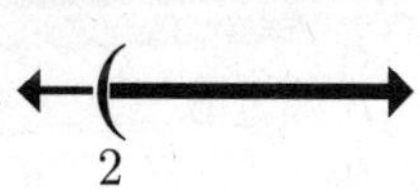

20. $(-\infty, -1) \Rightarrow$ 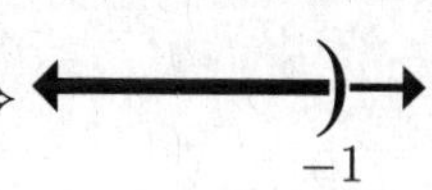

21. $(-\infty, 0] \cup (2, \infty) \Rightarrow$ 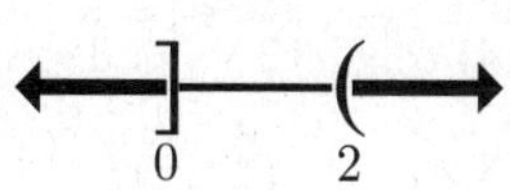

22. $|0| = 0$

23. $|-1| = 1$

24. $-|-5| = -5$

25. $-|6| = -6$

26. $3 + (+5) = +8$

27. $17 + (+5) = +(17 + 5) = 22$

28. $-15 + (-13) = -(15 + 13) = -28$

29. $-9 + (-18) = -(9 + 18) = -27$

30. $-2 + 5 = +(5 - 2) = 3$

31. $3 + (-12) = -(12 - 3) = -9$

32. $8 + (-3) = +(8 - 3) = +5$

33. $11 + (-15) = -(15 - 11) = -4$

34. $-25 + 12 = -(25 - 12) = -13$

35. $-30 + 35 = +(35 - 30) = +5$

36. $-3 - 10 = -3 + (-10) = -13$

37. $-7 - 8 = -7 + (-8) = -15$

38. $27 - (-12) = 27 + 12 = 39$

39. $38 - (-15) = 38 + 15 = 53$

© 2015 Cengage Learning. All Rights Reserved. May not be scanned, copied or duplicated, or posted to a publicly accessible website, in whole or in part.

40. $(-3)(-7) = 21$ **41.** $(-6)(-7) = 42$ **42.** $\frac{-16}{-4} = 4$ **43.** $\frac{-25}{-5} = 5$

44. $4(-3) = -12$ **45.** $-3(8) = -24$ **46.** $\frac{-8}{2} = -4$ **47.** $\frac{12}{-3} = -4$

48. $-4(3-6) = -4(-3) = 12$ **49.** $5[9-(-2)] = 5(11) = 55$

50. $-[4-2(6-4)] = -[4-2(2)] = -[4-4] = -[0] = 0$

51. $3[-5+3(2-7)] = 3[-5+3(-5)] = 3[-5+(-15)] = 3[-20] = -60$

52. $\frac{3-8}{10-5} = \frac{-5}{5} = -1$ **53.** $\frac{-32-8}{6-16} = \frac{-40}{-10} = 4$

54. $\text{mean} = \frac{12+13+14+14+15+15+15+17+19+20}{10} = \frac{154}{10} = 15.4$

55. $\text{median} = \frac{15+15}{2} = \frac{30}{2} = 15$ **56.** $\text{mode} = 15$

57. Yes. The mean, median and mode **could** be the same value for a group of numbers.

58. $\frac{3a-2b}{cd} = \frac{3(5)-2(-2)}{(-3)(2)} = \frac{15-(-4)}{-6}$
$= \frac{19}{-6} = -\frac{19}{6}$

59. $\frac{3b+2d}{ac} = \frac{3(-2)+2(2)}{5(-3)} = \frac{-6+4}{-15}$
$= \frac{-2}{-15} = \frac{2}{15}$

60. $\frac{ab+cd}{c(b-d)} = \frac{5(-2)+(-3)(2)}{-3(-2-2)} = \frac{-10+(-6)}{-3(-4)} = \frac{-16}{12} = -\frac{4}{3}$

61. $\frac{ac-bd}{a(d+c)} = \frac{5(-3)-(-2)(2)}{5[2+(-3)]} = \frac{-15-(-4)}{5(-1)} = \frac{-11}{-5} = \frac{11}{5}$

62. distributive property **63.** commutative property of addition

64. associative property of addition **65.** additive identity property

66. additive inverse property **67.** commutative property of multiplication

68. associative property of multiplication **69.** multiplicative identity property

70. multiplicative inverse property **71.** double negative rule

72. $5^4 = 5 \cdot 5 \cdot 5 \cdot 5 = 625$ **73.** $-3^5 = -1 \cdot 3^5 = -1 \cdot 3 \cdot 3 \cdot 3 \cdot 3 \cdot 3$
$= -1 \cdot 243 = -243$

© 2015 Cengage Learning. All Rights Reserved. May not be scanned, copied or duplicated, or posted to a publicly accessible website, in whole or in part.

74. $(-4)^3 = (-4)(-4)(-4) = -64$

75. $-5^{-4} = -1 \cdot 5^{-4} = -1 \cdot \dfrac{1}{5^4} = -\dfrac{1}{625}$

76. $(3x^4)(-2x^2) = 3(-2)x^4x^2 = -6x^{4+2} = -6x^6$

77. $(-x^5)(3x^3) = -1(3)x^5x^3 = -3x^{5+3} = -3x^8$

78. $x^7x^{-10} = x^{7+(-10)} = x^{-3} = \dfrac{1}{x^3}$

79. $y^{-12}y^7 = y^{-12+7} = y^{-5} = \dfrac{1}{y^5}$

80. $(3x^2)^3 = 3^3(x^2)^3 = 27x^6$

81. $(4x^4)^4 = 4^4(x^4)^4 = 256x^{16}$

82. $(-2x^2)^5 = (-2)^5(x^2)^5 = -32x^{10}$

83. $-(-3x^3)^5 = -(-3)^5(x^3)^5 = -(-243)x^{15} = 243x^{15}$

84. $(x^2)^{-5} = x^{-10} = \dfrac{1}{x^{10}}$

85. $(x^{-4})^{-5} = x^{20}$

86. $(-4x^{-5})^2 = (-4)^2x^{-10} = \dfrac{16}{x^{10}}$

87. $(-3x^{-6})^{-4} = (-3)^{-4}x^{24} = \dfrac{x^{24}}{81}$

88. $\dfrac{x^6}{x^4} = x^{6-4} = x^2$

89. $\dfrac{x^{12}}{x^7} = x^{12-7} = x^5$

90. $\dfrac{a^7}{a^{12}} = a^{7-12} = a^{-5} = \dfrac{1}{a^5}$

91. $\dfrac{a^4}{a^7} = a^{4-7} = a^{-3} = \dfrac{1}{a^3}$

92. $\dfrac{y^{-3}}{y^4} = y^{-3-4} = y^{-7} = \dfrac{1}{y^7}$

93. $\dfrac{y^5}{y^{-4}} = y^{5-(-4)} = y^9$

94. $\dfrac{x^{-5}}{x^{-4}} = x^{-5-(-4)} = x^{-1} = \dfrac{1}{x}$

95. $\dfrac{x^{-6}}{x^{-9}} = x^{-6-(-9)} = x^3$

96. $(-3a^3b^2)^{-4} = (-3)^{-4}a^{-12}b^{-8} = \dfrac{1}{81a^{12}b^8}$

97. $\left(\dfrac{4y^{-2}}{5y^{-3}}\right)^3 = \dfrac{4^3(y^{-2})^3}{5^3(y^{-3})^3} = \dfrac{64y^{-6}}{125y^{-9}} = \dfrac{64y^3}{125}$

98. $4^2 - 3(2) + 5^3 = 16 - 3(2) + 125 = 16 - 6 + 125 = 135$

99. $-2^2 - 4(5) - 3(7 - 12) = -2^2 - 4(5) - 3(-5) = -4 - 4(5) - 3(-5) = -4 - 20 + 15 = -9$

100. $19{,}300{,}000{,}000 = 1.93 \times 10^{10}$

101. $0.0000000273 = 2.73 \times 10^{-8}$

102. $8.4 \times 10^6 = 8{,}400{,}000$

103. $8.3 \times 10^{-9} = 0.0000000083$

104. $\dfrac{306{,}000{,}000 \text{ persons}}{1} \cdot \dfrac{1640 \text{ gallons}}{1 \text{ person}} = (3.06 \times 10^8)(1.64 \times 10^3) \text{ gallons} \approx 5.02 \times 10^{11} \text{ gallons}$

© 2015 Cengage Learning. All Rights Reserved. May not be scanned, copied or duplicated, or posted to a publicly accessible website, in whole or in part.

105.
$$\begin{aligned} 7x+30&=2\\ 7x+30-30&=2-30\\ 7x&=-28\\ \frac{7x}{7}&=\frac{-28}{7}\\ x&=-4 \end{aligned}$$

106.
$$\begin{aligned} -5x-8&=17\\ -5x-8+8&=17+8\\ -5x&=25\\ \frac{-5x}{-5}&=\frac{25}{-5}\\ x&=-5 \end{aligned}$$

107.
$$\begin{aligned} 4(y-1)&=28\\ 4y-4&=28\\ 4y-4+4&=28+4\\ 4y&=32\\ \frac{4y}{4}&=\frac{32}{4}\\ y&=8 \end{aligned}$$

108.
$$\begin{aligned} 6(x+2)&=-36\\ 6x+12&=-36\\ 6x+12-12&=-36-12\\ 6x&=-48\\ \frac{6x}{6}&=\frac{-48}{6}\\ x&=-8 \end{aligned}$$

109.
$$\begin{aligned} 13(x-9)-2&=7x-5\\ 13x-117-2&=7x-5\\ 13x-119&=7x-5\\ 13x-7x-119&=7x-7x-5\\ 6x-119+119&=-5+119\\ 6x&=114\\ \frac{6x}{6}&=\frac{114}{6}\\ x&=19 \end{aligned}$$

110.
$$\begin{aligned} \frac{8(x-5)}{3}&=2(x-4)\\ 3\cdot\frac{8(x-5)}{3}&=3\cdot 2(x-4)\\ 8(x-5)&=6(x-4)\\ 8x-40&=6x-24\\ 8x-6x-40&=6x-6x-24\\ 2x-40+40&=-24+40\\ 2x&=16\\ \frac{2x}{2}&=\frac{16}{2}\\ x&=8 \end{aligned}$$

111.
$$\begin{aligned} \frac{3y}{4}-13&=-\frac{y}{3}\\ 12\left(\frac{3y}{4}-13\right)&=12\left(-\frac{y}{3}\right)\\ 12\left(\frac{3y}{4}\right)-12(13)&=-4y\\ 3(3y)-156&=-4y\\ 9y-156&=-4y\\ 9y+4y-156&=-4y+4y\\ 13y-156+156&=0+156\\ 13y&=156\\ \frac{13y}{13}&=\frac{156}{13}\\ y&=12 \end{aligned}$$

112.
$$\begin{aligned} \frac{2y}{5}+5&=\frac{4y}{10}\\ 10\left(\frac{2y}{5}+5\right)&=10\left(\frac{4y}{10}\right)\\ 10\left(\frac{2y}{5}\right)+10(5)&=4y\\ 2(2y)+50&=4y\\ 4y+50&=4y\\ 4y-4y+50&=4y-4y\\ 50&\neq 0 \end{aligned}$$
no solution $\Rightarrow \emptyset$

© 2015 Cengage Learning. All Rights Reserved. May not be scanned, copied or duplicated, or posted to a publicly accessible website, in whole or in part.

113.
$$\begin{aligned}0.07a &= 0.10 - 0.04(a-3)\\ 0.07a &= 0.10 - 0.04a + 0.12\\ 0.07a &= 0.22 - 0.04a\\ 0.07a + 0.04a &= 0.22\\ 0.11a &= 0.22\\ \frac{0.11a}{0.11} &= \frac{0.22}{0.11}\\ a &= 2\end{aligned}$$

114.
$$\begin{aligned}0.12x + 0.06(50000 - 2x) &= 3000\\ 0.12x + 3000 - 0.12x &= 3000\\ 3000 &= 3000\end{aligned}$$
all real numbers $\Rightarrow \mathbb{R}$

115.
$$\begin{aligned}V &= \frac{4}{3}\pi r^3\\ 3V &= 3\left(\frac{4}{3}\pi r^3\right)\\ 3V &= 4\pi r^3\\ \frac{3V}{4\pi} &= \frac{4\pi r^3}{4\pi}\\ \frac{3V}{4\pi} &= r^3, \text{ or } r^3 = \frac{3V}{4\pi}\end{aligned}$$

116.
$$\begin{aligned}V &= \frac{1}{3}\pi r^2 h\\ 3V &= 3\left(\frac{1}{3}\pi r^2 h\right)\\ 3V &= \pi r^2 h\\ \frac{3V}{\pi r^2} &= \frac{\pi r^2 h}{\pi r^2}\\ \frac{3V}{\pi r^2} &= h, \text{ or } h = \frac{3V}{\pi r^2}\end{aligned}$$

117.
$$\begin{aligned}v &= \frac{1}{6}ab(x+y)\\ 6v &= 6\cdot\frac{1}{6}ab(x+y)\\ 6v &= ab(x+y)\\ \frac{6v}{ab} &= \frac{ab(x+y)}{ab}\\ \frac{6v}{ab} &= x+y\\ \frac{6v}{ab} - y &= x + y - y\\ \frac{6v}{ab} - y &= x, \text{ or } x = \frac{6v}{ab} - y\end{aligned}$$
$$\left[\text{also equivalent: } x = \frac{6v - aby}{ab}\right]$$

118.
$$\begin{aligned}V &= \pi h^2\left(r - \frac{h}{3}\right)\\ \frac{V}{\pi h^2} &= \frac{\pi h^2\left(r - \frac{h}{3}\right)}{\pi h^2}\\ \frac{V}{\pi h^2} &= r - \frac{h}{3}\\ \frac{V}{\pi h^2} + \frac{h}{3} &= r - \frac{h}{3} + \frac{h}{3}\\ \frac{V}{\pi h^2} + \frac{h}{3} &= r, \text{ or } r = \frac{V}{\pi h^2} + \frac{h}{3}\end{aligned}$$

119. Let $x =$ the length of the first piece.
Then $3x =$ the length of the other piece.

Sum of lengths = Total length

$$\begin{aligned}x + 3x &= 20\\ 4x &= 20\\ x &= 5\end{aligned}$$

He should cut the board 5 feet from one end.

120. Let $w =$ the width of the rectangle. Then $w + 8 =$ the length of the rectangle.

Perimeter = 60

$$\begin{aligned}w + w + w + 8 + w + 8 &= 60\\ 4w + 16 &= 60\\ 4w &= 44\\ w &= 11\end{aligned}$$

The dimensions are 11 meters by 19 meters, for an area of 209 m^2.

© 2015 Cengage Learning. All Rights Reserved. May not be scanned, copied or duplicated, or posted to a publicly accessible website, in whole or in part.

121.

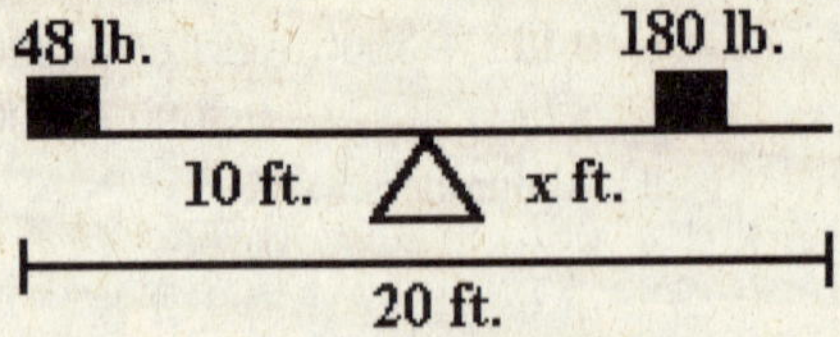

[Sue's force times distance] = [Father's force times distance]

$$\begin{aligned} 48(10) &= 180x \\ 480 &= 180x \\ \tfrac{480}{180} &= x \\ \tfrac{8}{3} &= x \\ 2\tfrac{2}{3} &= x \end{aligned}$$

The father should sit $2\frac{2}{3}$ feet from the fulcrum.

122. Let $x =$ the amount she invests at 10%. Then $25{,}000 - x =$ amount invested at 9%.

[Interest at 10%] + [Interest at 9%] = [Total interest]

$$\begin{aligned} 0.10x + 0.09(25000 - x) &= 2430 \\ 0.10x + 2250 - 0.09x &= 2430 \\ 0.01x &= 180 \\ x &= 18000 \end{aligned}$$

She invested \$18,000 at 10% interest and \$7000 at 9% interest.

123. Let $x =$ liters of water added.

[Alcohol at start] + [Alcohol added] = [Alcohol at end]

$$\begin{aligned} 0.14(25) + 0(x) &= 0.10(25 + x) \\ 3.5 &= 2.5 + 0.1x \\ 1 &= 0.1x \\ 10 &= x \end{aligned}$$

10 liters of water should be added.

124. Let $t =$ time for them to travel.

	Rate	Time	Dist.
Car	55	t	$55t$
Motorcycle	40	t	$40t$

[Car distance] − [Motorcycle distance] = [Distance between them]

$$\begin{aligned} 55t - 40t &= 5 \\ 15t &= 5 \\ t &= \tfrac{5}{15} = \tfrac{1}{3} \end{aligned}$$

It will take $\frac{1}{3}$ of an hour for the vehicles to be 5 miles apart.

Chapter 1 Test (page 87)

1. 1, 2 and 5 are natural numbers.

2. $\sqrt{7}$ is an irrational number.

3.

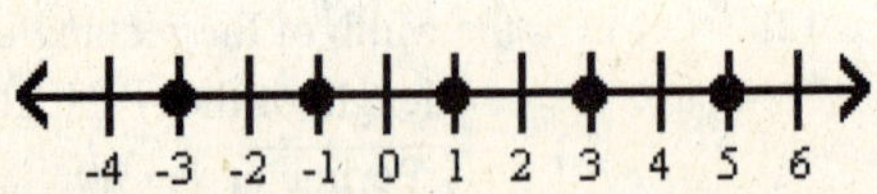

4.

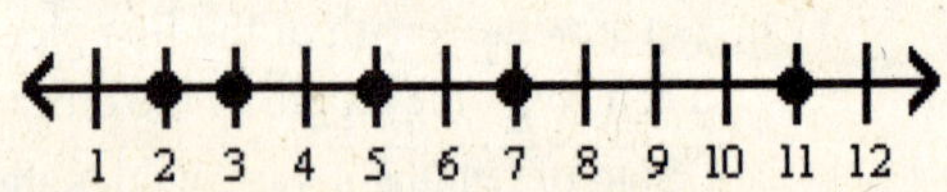

5. $\{x|x > 4\} \Rightarrow$

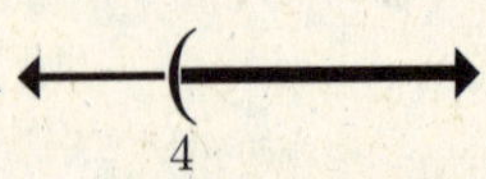

6. $[-3, \infty) \Rightarrow$

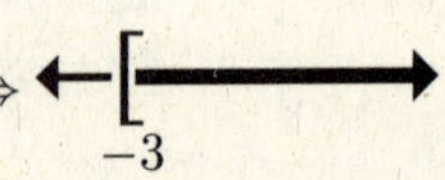

7. $\{x|-2 \le x < 4\} \Rightarrow$

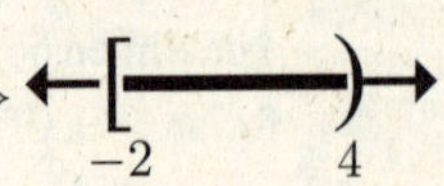

8. $(-\infty, -1] \cup [2, \infty) \Rightarrow$

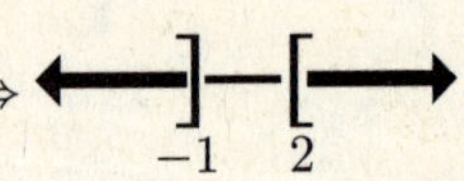

© 2015 Cengage Learning. All Rights Reserved. May not be scanned, copied or duplicated, or posted to a publicly accessible website, in whole or in part.

CHAPTER 1 TEST

9. $-|8| = -8$

10. $|-5| = 5$

11. $7 + (-5) = 2$

12. $-5(-4) = 20$

13. $\dfrac{12}{-3} = -4$

14. $-4 - \dfrac{-15}{3} = -4 - (-5) = -4 + 5 = 1$

15. $\text{mean} = \dfrac{-2+0+2+(-2)+3+(-1)+(-1)+1+1+2}{10} = \dfrac{3}{10} = 0.3$

16. $-2, -2, -1, -1, 0, 1, 1, 2, 2, 3 : \text{median} = \dfrac{0+1}{2} = \dfrac{1}{2} = 0.5$

17. $ab = (2)(-3) = -6$

18. $a + bc = 2 + (-3)(4) = 2 + (-12) = -10$

19. $$\begin{aligned} ab - bc &= 2(-3) - (-3)(4) = -6 - (-12) \\ &= -6 + 12 = 6 \end{aligned}$$

20. $$\begin{aligned} \frac{-3b+a}{ac-b} &= \frac{-3(-3)+2}{2(4)-(-3)} = \frac{9+2}{8+3} = \frac{11}{11} \\ &= 1 \end{aligned}$$

21. commutative property of addition

22. distributive property

23. $x^3x^5 = x^{3+5} = x^8$

24. $(2x^2y^3)^3 = 2^3(x^2)^3(y^3)^3 = 8x^6y^9$

25. $(m^{-4})^2 = m^{-8} = \dfrac{1}{m^8}$

26. $$\begin{aligned} \left(\frac{m^2n^3}{m^4n^{-2}}\right)^{-2} &= \left(\frac{m^4n^{-2}}{m^2n^3}\right)^2 = \left(\frac{m^2}{n^5}\right)^2 \\ &= \frac{m^4}{n^{10}} \end{aligned}$$

27. $4{,}700{,}000 = 4.7 \times 10^6$

28. $0.00000023 = 2.3 \times 10^{-7}$

29. $6.53 \times 10^5 = 653{,}000$

30. $-3 - 12 \div 6 - (-4)^2 = -3 - 12 \div 6 - 16 = -3 - 2 - 16 = -21$

31. $$\begin{aligned} 9(x+4) + 4 &= 4(x-5) \\ 9x + 36 + 4 &= 4x - 20 \\ 9x + 40 &= 4x - 20 \\ 9x - 4x + 40 &= 4x - 4x - 20 \\ 5x + 40 &= -20 \\ 5x + 40 - 40 &= -20 - 40 \\ 5x &= -60 \\ \tfrac{5x}{5} &= \tfrac{-60}{5} \\ x &= -12 \end{aligned}$$

© 2015 Cengage Learning. All Rights Reserved. May not be scanned, copied or duplicated, or posted to a publicly accessible website, in whole or in part.

CHAPTER 1 TEST

32.

$$\begin{aligned}
\frac{2y+3}{3}+\frac{3y-4}{6}&=\frac{y-2}{2}\\
6\left(\frac{2y+3}{3}+\frac{3y-4}{6}\right)&=6\left(\frac{y-2}{2}\right)\\
6\left(\frac{2y+3}{3}\right)+6\left(\frac{3y-4}{6}\right)&=3(y-2)\\
2(2y+3)+3y-4&=3y-6\\
4y+6+3y-4&=3y-6\\
7y+2&=3y-6\\
7y-3y+2&=3y-3y-6\\
4y+2&=-6\\
4y+2-2&=-6-2\\
4y&=-8\\
\frac{4y}{4}&=\frac{-8}{4}\\
y&=-2
\end{aligned}$$

33.

$$\begin{aligned}
\frac{y-1}{5}+2&=\frac{2y-3}{3}\\
15\left(\frac{y-1}{5}+2\right)&=15\left(\frac{2y-3}{3}\right)\\
15\left(\frac{y-1}{5}\right)+15(2)&=5(2y-3)\\
3(y-1)+30&=10y-15\\
3y-3+30&=10y-15\\
3y+27&=10y-15\\
3y-10y+27&=10y-10y-15\\
-7y+27-27&=-15-27\\
-7y&=-42\\
\frac{-7y}{-7}&=\frac{-42}{-7}\\
y&=6
\end{aligned}$$

34.

$$\begin{aligned}
400+1.5x&=0.5(200+3x)\\
400+1.5x&=100+1.5x\\
400+1.5x-1.5x&=100+1.5x-1.5x\\
400&\neq 100
\end{aligned}$$

no solution $\Rightarrow \emptyset$

35.

$$\begin{aligned}
P&=L+\frac{s}{f}i\\
P-L&=L-L+\frac{s}{f}i\\
P-L&=\frac{s}{f}i\\
f(P-L)&=f\cdot\frac{s}{f}i\\
f(P-L)&=si\\
\frac{f(P-L)}{s}&=\frac{si}{s}\\
\frac{f(P-L)}{s}&=i,\text{ or } i=\frac{f(P-L)}{s}
\end{aligned}$$

36. Let $w =$ the width of the rectangle. Then $w+5 =$ the length of the rectangle.

$$\begin{aligned}
\boxed{\text{Perimeter}}&=26\\
w+w+w+5+w+5&=26\\
4w+10&=26\\
4w&=16\\
w&=4
\end{aligned}$$

The dimensions are 4 cm by 9 cm, for an area of 36 cm^2.

© 2015 Cengage Learning. All Rights Reserved. May not be scanned, copied or duplicated, or posted to a publicly accessible website, in whole or in part.

37. Let $x =$ the amount invested at 9%. Then $10000 - x =$ the amount at 8%.

$$\boxed{\text{Interest at 9\%}} + \boxed{\text{Interest at 8\%}} = \boxed{\text{Total interest}}$$

$$\begin{aligned} 0.09x + 0.08(10000 - x) &= 860 \\ 0.09x + 800 - 0.08x &= 860 \\ 0.01x &= 60 \\ x &= 6000 \end{aligned}$$

He has \$6000 invested at 9% interest, and **\$4000 invested at 8% interest.**

38. Let $x =$ liters of water added.

$$\boxed{\text{Salt at start}} + \boxed{\text{Salt added}} = \boxed{\text{Salt at end}}$$

$$\begin{aligned} 0.05(20) + 0(x) &= 0.01(20 + x) \\ 1 + 0 &= 0.2 + 0.01x \\ 0.8 &= 0.01x \\ 80 &= x \end{aligned}$$

80 liters of water should be added.

Exercises 2.1 (page 102)

1.
$$\begin{aligned} -2x - 4y &= 12 \\ -2(2) - 4y &= 12 \\ -4 - 4y &= 12 \\ -4y &= 16 \\ y &= -4 \end{aligned}$$

3.
$$\begin{aligned} y &= \frac{5}{2}x - 3 \\ y &= \frac{5}{2}(2) - 3 \\ y &= 5 - 3 \\ y &= 2 \end{aligned}$$

5. $y = 3$

7.
$$\begin{aligned} 8y &= 12 \\ y &= \frac{12}{8} \\ y &= \frac{3}{2} \end{aligned}$$

9. $-5 - 2(-7) = -5 + 14 = 9$

11. $\dfrac{-3 + 5(2)}{9 + 5} = \dfrac{-3 + 10}{14} = \dfrac{7}{14} = \dfrac{1}{2}$

13.
$$\begin{aligned} -3x + 5 &= 26 \\ -3x + 5 - 5 &= 26 - 5 \\ -3x &= 21 \\ \tfrac{-3x}{-3} &= \tfrac{21}{-3} \\ x &= -7 \end{aligned}$$

15. ordered pair

17. origin

19. rectangular or Cartesian coordinate

21. linear

23. y-axis

25. vertical

27. sub 1

29-35.

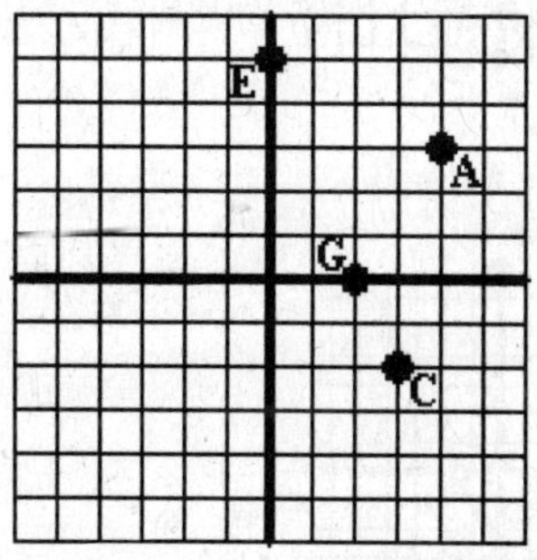

37. $(2, 4)$

39. $(-2, -1)$

41. $(4, 0)$

43. $(0, 0)$

© 2015 Cengage Learning. All Rights Reserved. May not be scanned, copied or duplicated, or posted to a publicly accessible website, in whole or in part.

45. $y = -x + 4$

$y = -(-1) + 4 = 1 + 4 = 5$

$y = -(0) + 4 = 0 + 4 = 4$

$y = -2 + 4 = 2$

x	y
-1	5
0	4
2	2

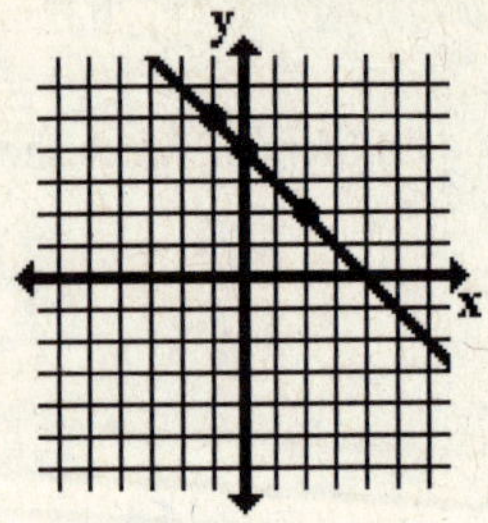

47. $y = 2x - 3$

$y = 2(-1) - 3 = -2 - 3 = -5$

$y = 2(0) - 3 = 0 - 3 = -3$

$y = 2(3) - 3 = 6 - 3 = 3$

x	y
-1	-5
0	-3
3	3

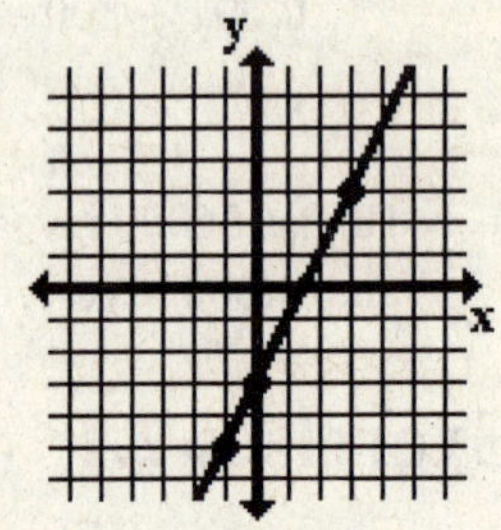

49.

$$\begin{aligned} 3x + 4y &= 12 \\ 3x + 4(0) &= 12 \\ 3x &= 12 \\ x &= 4 \end{aligned}$$

x-intercept: $(4, 0)$

$$\begin{aligned} 3x + 4y &= 12 \\ 3(0) + 4y &= 12 \\ 4y &= 12 \\ y &= 3 \end{aligned}$$

y-intercept: $(0, 3)$

$3x + 4y = 12$

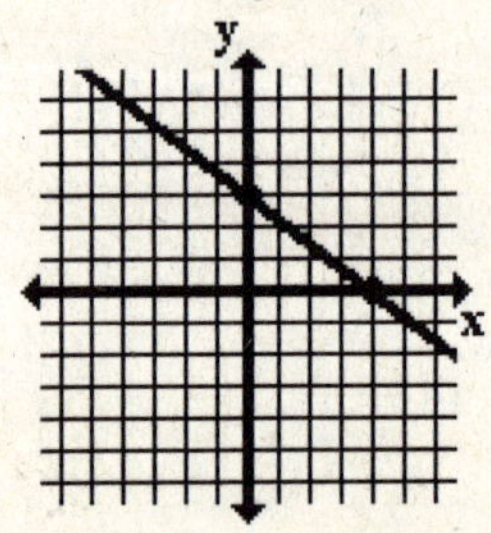

51.

$$\begin{aligned} 6x - 3y &= 12 \\ 6x - 3(0) &= 12 \\ 6x &= 12 \\ x &= 2 \end{aligned}$$

x-intercept: $(2, 0)$

$$\begin{aligned} 6x - 3y &= 12 \\ 6(0) - 3y &= 12 \\ -3y &= 12 \\ y &= -4 \end{aligned}$$

y-intercept: $(0, -4)$

$6x - 3y = 12$

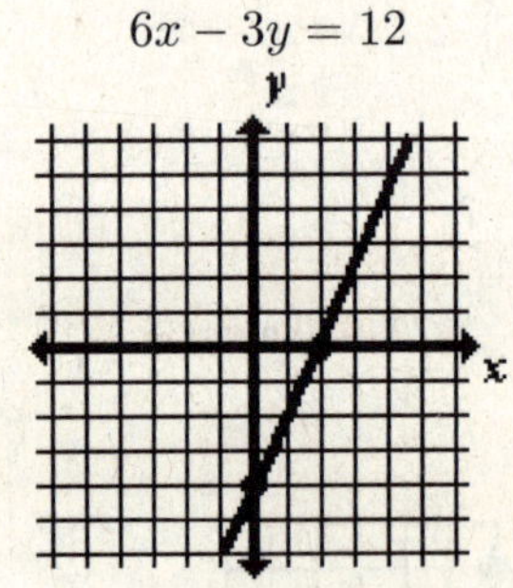

53.

$$\begin{aligned} 3y &= 6x - 9 \\ 3(0) &= 6x - 9 \\ 0 &= 6x - 9 \\ -6x &= -9 \\ x &= \tfrac{-9}{-6} = \tfrac{3}{2} \end{aligned}$$

x-intercept: $\left(\frac{3}{2}, 0\right)$

$$\begin{aligned} 3y &= 6x - 9 \\ 3y &= 6(0) - 9 \\ 3y &= 0 - 9 \\ 3y &= -9 \\ y &= \tfrac{-9}{3} = -3 \end{aligned}$$

y-intercept: $(0, -3)$

$3y = 6x - 9$

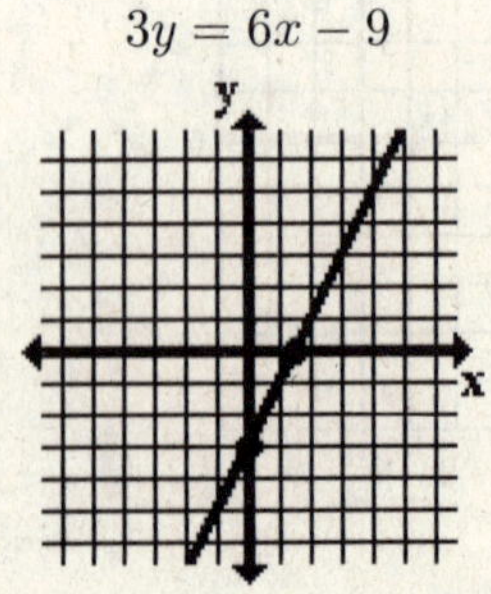

© 2015 Cengage Learning. All Rights Reserved. May not be scanned, copied or duplicated, or posted to a publicly accessible website, in whole or in part.

SECTION 2.1

55.
$$\begin{aligned} 3x+4y-8&=0 \\ 3x+4(0)-8&=0 \\ 3x-8&=0 \\ 3x&=8 \\ x&=\tfrac{8}{3} \end{aligned}$$
x-intercept: $\left(\frac{8}{3},0\right)$

$$\begin{aligned} 3x+4y-8&=0 \\ 3(0)+4y-8&=0 \\ 4y-8&=0 \\ 4y&=8 \\ y&=2 \end{aligned}$$
y-intercept: $(0,2)$

$3x+4y-8=0$

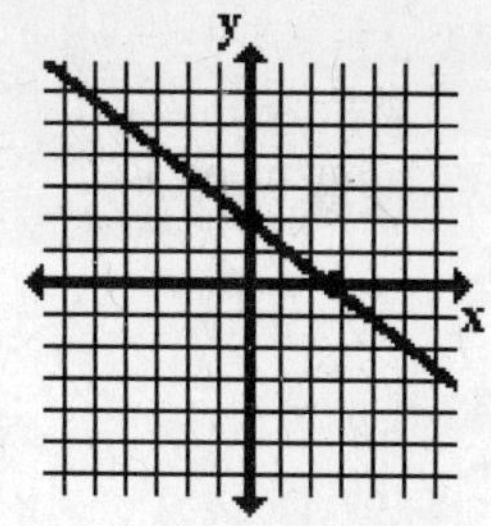

57. $x=3$
vertical line with
x-coordinate of 3

$x=3$

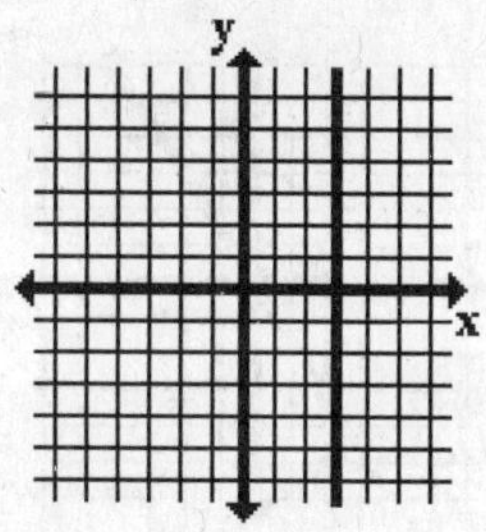

59.
$$\begin{aligned} -3y+2&=5 \\ -3y&=3 \\ y&=-1 \end{aligned}$$
horizontal line
with y-coordinate
of -1

$y=-1$

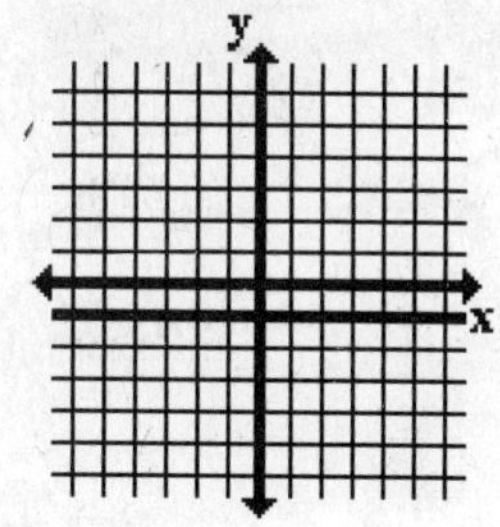

61.

x	2	4	5	6
y	12	24	30	36

Graph the ordered pairs and draw a line.

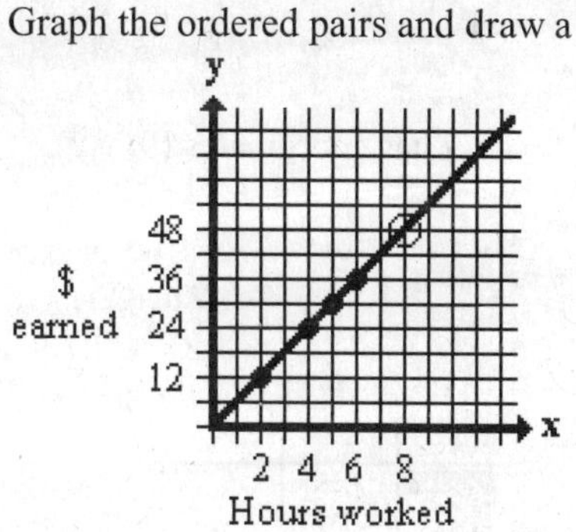

The point $(8,48)$ is on the graph. The student will earn \$48 for 8 hours of work.

63.

x	0	1	3
y	15,000	12,000	6000

Graph the ordered pairs and draw a line.

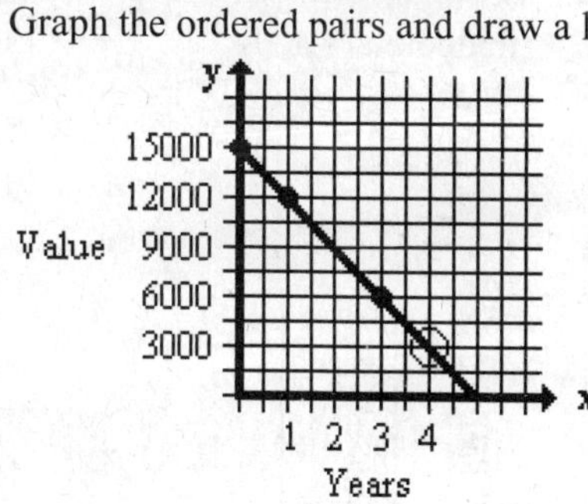

The point $(4,3000)$ is on the graph.
The car will be worth \$3000 in 4 years.

65. $x=\frac{x_1+x_2}{2}=\frac{0+2}{2}=\frac{2}{2}=1$
$y=\frac{y_1+y_2}{2}=\frac{0+4}{2}=\frac{4}{2}=2$
midpoint: $(1,2)$

67. $x=\frac{x_1+x_2}{2}=\frac{4+14}{2}=\frac{18}{2}=9$
$y=\frac{y_1+y_2}{2}=\frac{6+20}{2}=\frac{26}{2}=13$
midpoint: $(9,13)$

69. $x=\frac{x_1+x_2}{2}=\frac{6+5}{2}=\frac{11}{2}$
$y=\frac{y_1+y_2}{2}=\frac{7+1}{2}=\frac{8}{2}=4$
midpoint: $\left(\frac{11}{2},4\right)$

71. $x=\frac{x_1+x_2}{2}=\frac{-10+3}{2}=-\frac{7}{2}$
$y=\frac{y_1+y_2}{2}=\frac{-2+12}{2}=\frac{10}{2}=5$
midpoint: $\left(-\frac{7}{2},5\right)$

© 2015 Cengage Learning. All Rights Reserved. May not be scanned, copied or duplicated, or posted to a publicly accessible website, in whole or in part.

73.

$y = -3x + 2$
$0 = -3x + 2$
$3x = 2$
$x = \frac{2}{3}$
x-intercept: $\left(\frac{2}{3}, 0\right)$

$y = -3x + 2$
$y = -3(0) + 2$
$y = 2$
y-intercept: $(0, 2)$

$y = -3x + 2$

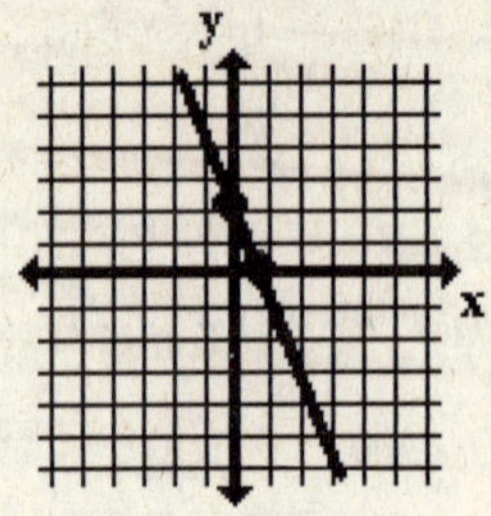

75.

$y = \frac{3}{2}x$
$0 = \frac{3}{2}x$
$2(0) = 2 \cdot \frac{3}{2}x$
$0 = 3x$
$x = \frac{0}{3} = 0$
x-intercept: $(0, 0)$

Pick a value for x such as $x = 4$.
$y = \frac{3}{2}x$
$y = \frac{3}{2}(4)$
$y = 6$
point: $(4, 6)$

$y = \frac{3}{2}x$

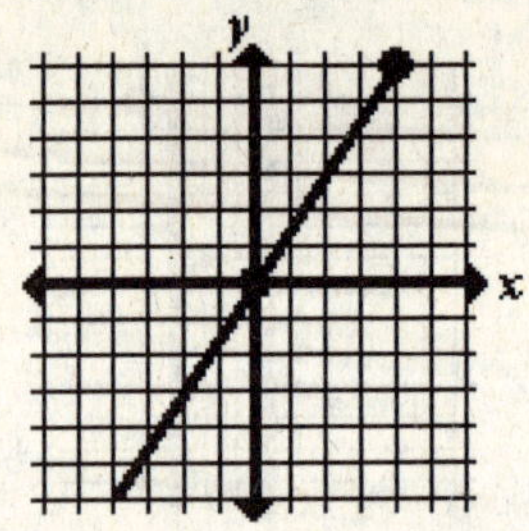

77. $x = \frac{x_1+x_2}{2} = \frac{-3+(-5)}{2} = \frac{-8}{2} = -4$
$y = \frac{y_1+y_2}{2} = \frac{5+(-5)}{2} = \frac{0}{2} = 0$
midpoint: $(-4, 0)$

79.

$x = \frac{x_1+x_2}{2}$
$7 = \frac{5+x_2}{2}$
$14 = 5 + x_2$
$9 = x_2$

$y = \frac{y_1+y_2}{2}$
$-4 = \frac{-6+y_2}{2}$
$-8 = -6 + y_2$
$-2 = y_2$

Q has coordinates $(9, -2)$.

81. $y = 2.5x + 1.5$

X	Y1
-2.5	-4.75
-2	-3.5
-1.5	-2.25
-1	-1
-.5	.25
0	1.5
.5	2.75

X=-2.5

When $x = -1$, $y = -1$.

83. $3.2x - 1.5y = 2.7$
$-1.5y = -3.2x + 2.7$
$y = \dfrac{-3.2x + 2.7}{-1.5}$

X	Y1
-2.5	-7.133
-2	-6.067
-1.5	-5
-1	-3.933
-.5	-2.867
0	-1.8
.5	-.7333

X=-2.5

When $x = -1$, $y = -3.933$.

© 2015 Cengage Learning. All Rights Reserved. May not be scanned, copied or duplicated, or posted to a publicly accessible website, in whole or in part.

85. $y = 3.7x - 4.5$; x-intercept: $(1.22, 0)$

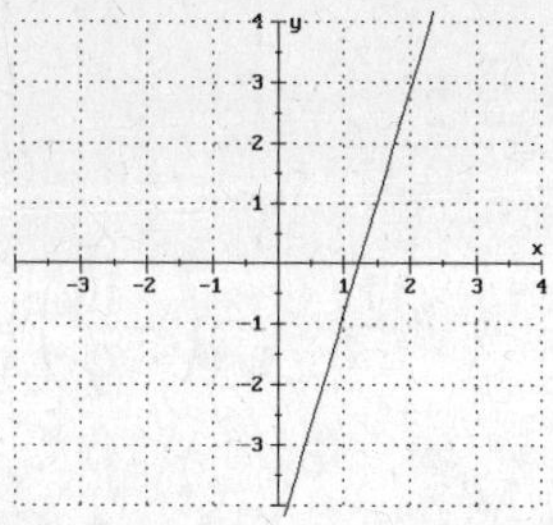

87. $1.5x - 3y = 7$; x-intercept: $(4.67, 0)$

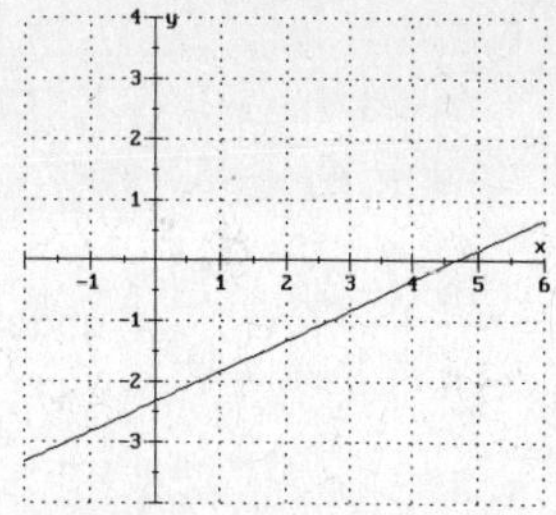

89. To find the value of the house after 5 years, let $x = 5$:

$$\begin{aligned} y &= 7500x + 125{,}000 \\ &= 7500(5) + 125{,}000 \\ &= 37{,}500 + 125{,}000 \\ &= 162{,}500 \end{aligned}$$

It will be worth \$162,500 after 5 years.

91. To find the number of TVs sold, let $p = 150$:

$$\begin{aligned} p &= -\frac{1}{10}q + 170 \\ 150 &= -\frac{1}{10}q + 170 \\ \frac{1}{10}q &= 20 \\ q &= 200 \end{aligned}$$

200 TVs will be sold at a price of \$150.

93. The shuttle is launched from the point $(3, 0)$ along the line $x = 3$. Let $x = 3$ in the equation: $y = 2x + 6 = 2(3) + 6 = 6 + 6 = 12 \Rightarrow$ The shuttle will stay in range for 12 miles.

95. $y = 0.25x + 50$

x	y
40	60
80	70
120	80
160	90

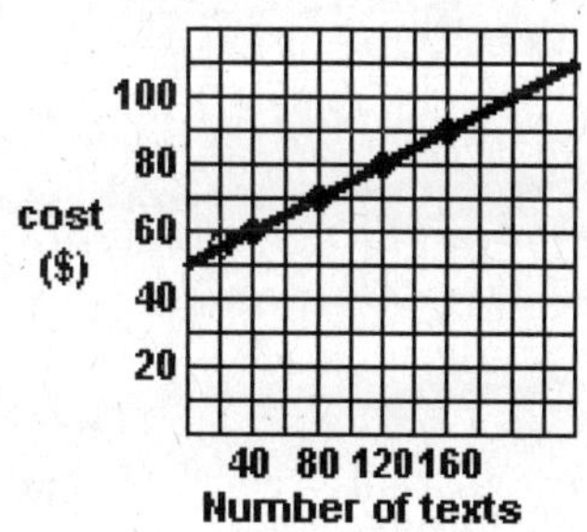

It will cost \$55 to make 20 texts.

97. **a.** The s-intercept occurs when $t = 0$. This will indicate the number of swimmers (in millions) when $t = 0$, or during 1990. Since $s = 65.5$ when $t = 0$, there were 65.5 million swimmers in 1990.

b. Find s when $t = 25$.

$$\begin{aligned} s &= -0.9t + 65.5 \\ &= -0.9(25) + 65.5 \\ &= -22.5 + 65.5 = 43 \end{aligned}$$

The estimate is 43 million swimmers in 2015.

99. **Answers may vary.**

101. If the line $y = ax + b$ passes through only quadrants I and II, then $a = 0$ and $b > 0$.

© 2015 Cengage Learning. All Rights Reserved. May not be scanned, copied or duplicated, or posted to a publicly accessible website, in whole or in part.

Exercises 2.2 (page 113)

1. $\dfrac{a-b}{c-d}=\dfrac{2-3}{-1-4}=\dfrac{-1}{-5}=\dfrac{1}{5}$

3. $\dfrac{a-b}{c-d}=\dfrac{-2-(-2)}{-3-4}=\dfrac{0}{-7}=0$

5. $\dfrac{a-b}{c-d}=\dfrac{4-(-3)}{6-6}=\dfrac{7}{0}\Rightarrow$ undefined

7. $-\dfrac{1}{-0.2}=-\dfrac{1}{-\frac{2}{10}}=-\left(-\dfrac{10}{2}\right)=5$

9. $\left(x^{-3}y^{2}\right)^{-4}=x^{12}y^{-8}=\dfrac{x^{12}}{y^{8}}$

11. $\left(\dfrac{3x^2y^3}{8}\right)^0=1$

13. change

15. rise

17. horizontal

19. positive

21. perpendicular; reciprocals

23. $m=\dfrac{\Delta y}{\Delta x}=\dfrac{5-(-3)}{2-(-2)}=\dfrac{8}{4}=2$

25. $m=\dfrac{\Delta y}{\Delta x}=\dfrac{-3-5}{7-(-2)}=\dfrac{-8}{9}=-\dfrac{8}{9}$

27. $m=\dfrac{\Delta y}{\Delta x}=\dfrac{1-8}{6-(-1)}=\dfrac{-7}{7}=-1$

29. $m=\dfrac{\Delta y}{\Delta x}=\dfrac{2-(-1)}{-6-3}=\dfrac{3}{-9}=-\dfrac{1}{3}$

31. Find two points on the line:

Let $x=0$:	Let $y=0$:
$5x+3y=15$	$5x+3y=15$
$5(0)+3y=15$	$5x+3(0)=15$
$3y=15$	$5x=15$
$y=5$	$x=3$
$(0,5)$	$(3,0)$

$$m=\frac{\Delta y}{\Delta x}=\frac{5-0}{0-3}=\frac{5}{-3}=-\frac{5}{3}$$

33. Find two points on the line:

Let $x=0$:	Let $y=0$:
$3x=4y-2$	$3x=4y-2$
$3(0)=4y-2$	$3x=4(0)-2$
$0=4y-2$	$3x=-2$
$2=4y$	$x=-\frac{2}{3}$
$\frac{1}{2}=y$	$\left(-\frac{2}{3},0\right)$
$\left(0,\frac{1}{2}\right)$	

$$m=\frac{\Delta y}{\Delta x}=\frac{\frac{1}{2}-0}{0-\left(-\frac{2}{3}\right)}=\frac{\frac{1}{2}}{\frac{2}{3}}=\frac{3}{4}$$

35. negative

37. positive

39. not defined

41. 0

43. $m_1\neq m_2\Rightarrow$ not parallel
$m_1\cdot m_2=5\left(-\frac{1}{5}\right)=-1\Rightarrow$ perpendicular

45. $m_1\neq m_2\Rightarrow$ not parallel
$m_1\cdot m_2=-(0.5)(-2)=1$
$\Rightarrow$ not perpendicular $\Rightarrow$ Neither

47. $m_{\overline{PQ}}=\dfrac{\Delta y}{\Delta x}=\dfrac{-11-5}{6-(-2)}=\dfrac{-16}{8}=-2$
same slope $\Rightarrow$ parallel

49. $m_{\overline{PQ}}=\dfrac{\Delta y}{\Delta x}=\dfrac{5-1}{6-(-2)}=\dfrac{4}{8}=\dfrac{1}{2}$
opposite reciprocal slope $\Rightarrow$ perpendicular

51. $m=\dfrac{\Delta y}{\Delta x}=\dfrac{9-9}{-8-4}=\dfrac{0}{-12}=0$

53. $m=\dfrac{\Delta y}{\Delta x}=\dfrac{-2-(-5)}{-7-(-7)}=\dfrac{3}{0}$: not defined

© 2015 Cengage Learning. All Rights Reserved. May not be scanned, copied or duplicated, or posted to a publicly accessible website, in whole or in part.

55. Find two points on the line:

Let $x = 0$:

$$y = \frac{3x-6}{2}$$
$$y = \frac{3(0)-6}{2}$$
$$y = \frac{-6}{2}$$
$$y = -3$$
$$(0,-3)$$

Let $y = 0$:

$$y = \frac{3x-6}{2}$$
$$0 = \frac{3x-6}{2}$$
$$0 = 3x-6$$
$$6 = 3x$$
$$2 = x$$
$$(2,0)$$

$$m = \frac{\Delta y}{\Delta x} = \frac{0-(-3)}{2-0} = \frac{3}{2}$$

57.
$$4y = 3(y+2)$$
$$4y = 3y+6$$
$$y = 6$$
horizontal line $\Rightarrow m = 0$

59.
$$m = \frac{\Delta y}{\Delta x} = \frac{2.5-3.7}{3.7-2.5} = \frac{-1.2}{1.2} = -1$$
$$m = \frac{\Delta y}{\Delta x} = \frac{-1.7-(-2.3)}{2.3-1.7} = \frac{0.6}{0.6} = 1$$
opposite reciprocal slopes $\Rightarrow$ perpendicular

61. $m_1 \neq m_2 \Rightarrow$ not parallel
$$m_1 \cdot m_2 = \frac{3.2}{-9.1} \cdot \frac{-9.1}{3.2} = 1$$
$\Rightarrow$ not perpendicular

63.
$$m_{\overline{PQ}} = \frac{\Delta y}{\Delta x} = \frac{-9-3}{4-(-2)} = \frac{-12}{6} = -2$$
same slope $\Rightarrow$ parallel

65.
$$m_{\overline{PQ}} = \frac{\Delta y}{\Delta x} = \frac{6-8}{1-0} = \frac{-2}{1} = -2$$
$$m_{\overline{PR}} = \frac{\Delta y}{\Delta x} = \frac{2-8}{3-0} = \frac{-6}{3} = -2$$
Since they have the same slope and a common point, the three points are on the same line.

67.
$$m_{\overline{PQ}} = \frac{\Delta y}{\Delta x} = \frac{-10-(-13)}{-8-(-10)} = \frac{3}{2}$$
$$m_{\overline{PR}} = \frac{\Delta y}{\Delta x} = \frac{-16-(-13)}{-12-(-10)} = \frac{-3}{-2} = \frac{3}{2}$$
Since they have the same slope and a common point, the three points are on the same line.

69.
$$m_{\overline{PQ}} = \frac{\Delta y}{\Delta x} = \frac{-12-(-4)}{0-8} = \frac{-8}{-8} = 1$$
$$m_{\overline{PR}} = \frac{\Delta y}{\Delta x} = \frac{-20-(-4)}{8-8} = \frac{-16}{0} \Rightarrow \text{und.}$$
Since they do not have the same slope, the three points are not on the same line.

71. On the y-axis, all x-coordinates are 0. The equation is $x = 0$, and the slope is not defined.

73. Call the points $A(0,0)$, $B(12,0)$ and $C(13,12)$. Compute these slopes:
$$m_{\overline{AB}} = \frac{\Delta y}{\Delta x} = \frac{0-0}{12-0} = \frac{0}{12} = 0$$
$$m_{\overline{AC}} = \frac{\Delta y}{\Delta x} = \frac{12-0}{13-0} = \frac{12}{13}$$
$$m_{\overline{BC}} = \frac{\Delta y}{\Delta x} = \frac{12-0}{13-12} = \frac{12}{1} = 12$$
Since none of the sides are perpendicular, it is not a right triangle.

© 2015 Cengage Learning. All Rights Reserved. May not be scanned, copied or duplicated, or posted to a publicly accessible website, in whole or in part.

SECTION 2.2

75. Call the points $A(2b, a)$, $B(b, b)$ and $C(a, 0)$. Compute these slopes:

$$m_{\overline{AB}} = \frac{\Delta y}{\Delta x} = \frac{b-a}{b-2b} = \frac{b-a}{-b} = -\frac{b-a}{b}$$
$$m_{\overline{AC}} = \frac{\Delta y}{\Delta x} = \frac{0-a}{a-2b} = \frac{-a}{a-2b} = -\frac{a}{a-2b}$$
$$m_{\overline{BC}} = \frac{\Delta y}{\Delta x} = \frac{b-0}{b-a} = \frac{b}{b-a}$$

Since $\overline{AB}$ and $\overline{BC}$ are perpendicular, it is a right triangle.

77. Call the points $A(0, 0)$, $B(0, b)$, $C(8, b+2)$ and $D(12, 3)$. Compute these slopes:

$$m_{\overline{AB}} = \frac{\Delta y}{\Delta x} = \frac{0-b}{0-0} = \frac{-b}{0} \Rightarrow \text{undefined}$$
$$m_{\overline{BC}} = \frac{\Delta y}{\Delta x} = \frac{b-(b+2)}{0-8} = \frac{b-b-2}{-8} = \frac{-2}{-8} = \frac{1}{4}$$
$$m_{\overline{CD}} = \frac{\Delta y}{\Delta x} = \frac{b+2-3}{8-12} = \frac{b-1}{-4} = -\frac{b-1}{4}$$
$$m_{\overline{DA}} = \frac{\Delta y}{\Delta x} = \frac{0-3}{0-12} = \frac{-3}{-12} = \frac{1}{4}$$

Thus, $\overline{BC} \parallel \overline{DA}$ and the figure is a trapezoid.

79. $\text{slope} = \frac{\text{rise}}{\text{run}} = \frac{3\text{ ft}}{12\text{ ft}} = \frac{1}{4}$

81. $m = \frac{\Delta y}{\Delta x} = \frac{2}{50} = \frac{1}{25}$; $m = \frac{\Delta y}{\Delta x} = \frac{5}{50} = \frac{1}{10}$; $m = \frac{\Delta y}{\Delta x} = \frac{8}{50} = \frac{4}{25}$

83. $m = \frac{\Delta y}{\Delta x} = \frac{0.7}{50} = \frac{7}{500}$; $\frac{7}{500}$ degree per year

85. **a.** $m = \frac{\Delta y}{\Delta x} = \frac{2}{25}$ **b.** $m = \frac{\Delta y}{\Delta x} = \frac{1}{20}$ (each part)

c. Design #1 has just one level, but the slope is steep. Design #2 has slopes which are less steep, but there are two levels.

87. Let x represent the number of years, and let y represent the price. Then we know two points on the line: $(-10, 5700)$ and $(-2, 1499)$. Find the slope:

$$m = \frac{\Delta y}{\Delta x} = \frac{5700-1499}{-10-(-2)} = \frac{4201}{-8} \approx -\$525.13 \Rightarrow \text{The cost is decreasing about \$525.13 per year.}$$

89. **Answers may vary.**

© 2015 Cengage Learning. All Rights Reserved. May not be scanned, copied or duplicated, or posted to a publicly accessible website, in whole or in part.

91. Find two points on the line:

Let $x = 0$:

$$y = mx + b$$
$$y = m(0) + b$$
$$y = b$$

$(0, b)$ is on the line.

Let $y = 0$:

$$y = mx + b$$
$$0 = mx + b$$
$$-b = mx$$
$$\tfrac{-b}{m} = x$$

$\left(-\frac{b}{m}, 0\right)$ is on the line.

$$m = \frac{\Delta y}{\Delta x} = \frac{b - 0}{0 - \left(-\frac{b}{m}\right)} = \frac{b}{\frac{b}{m}} = m$$

93. $m_{\overline{AB}} = \dfrac{\Delta y}{\Delta x} = \dfrac{3 - 7}{1 - (-2)} = \dfrac{-4}{3} = -\dfrac{4}{3}$. Then $m_{\overline{CD}}$ must equal $\dfrac{3}{4}$.

$$m_{\overline{CD}} = \frac{\Delta y}{\Delta x} = \frac{b - (-1)}{4 - 8} = \frac{b + 1}{-4} = \frac{3}{4}$$
$$-4 \cdot \frac{b + 1}{-4} = -4 \cdot \frac{3}{4}$$
$$b + 1 = -3$$
$$b = -4$$

Exercises 2.3 (page 126)

1.
$$2x - 3y = 6$$
$$-3y = -2x + 6$$
$$\frac{-3y}{-3} = \frac{-2x + 6}{-3}$$
$$y = \frac{2}{3}x - 2$$

3. $m = \dfrac{\Delta y}{\Delta x} = \dfrac{8 - 5}{6 - (-1)} = \dfrac{3}{7}$

5. $x = 3 \Rightarrow$ vertical $\Rightarrow$ undefined slope

7.
$$4x + 3 = 2(x - 5) + 7$$
$$4x + 3 = 2x - 10 + 7$$
$$4x + 3 = 2x - 3$$
$$2x = -6$$
$$x = -3$$

9.
$$\frac{5(2 - x)}{3} - 1 = x + 5$$
$$\frac{10 - 5x}{3} = x + 6$$
$$3 \cdot \frac{10 - 5x}{3} = 3(x + 6)$$
$$10 - 5x = 3x + 18$$
$$10 = 8x + 18$$
$$-8 = 8x$$
$$-1 = x$$

11. The alloy is to contain 25% gold. Let $x =$ the amount of copper added to the alloy.

Gold at start	+	Gold added	=	Gold at end

$$20 + 0 = 0.25(60 + x)$$
$$20 = 15 + 0.25x$$
$$5 = 0.25x$$
$$20 = x$$

Add 20 oz of copper to the alloy.

13. $y - y_1 = m(x - x_1)$

15. $Ax + By = C$

17. perpendicular

© 2015 Cengage Learning. All Rights Reserved. May not be scanned, copied or duplicated, or posted to a publicly accessible website, in whole or in part.

SECTION 2.3

19. $y - y_1 = m(x - x_1)$

$y - 5 = \frac{1}{2}(x - 1)$

21. $y - y_1 = m(x - x_1)$

$y - 0 = -3(x - 2)$

$y = -3(x - 2)$

23. Find the slope of the line:

$m = \frac{\Delta y}{\Delta x} = \frac{8-5}{4-2} = \frac{3}{2}$

Use point-slope form to find the equation:

$y - y_1 = m(x - x_1)$

$y - 5 = \frac{3}{2}(x - 2)$

$y - 5 = \frac{3}{2}x - 3$

$y = \frac{3}{2}x + 2$

25. Find the slope of the line:

$m = \frac{\Delta y}{\Delta x} = \frac{4-(-3)}{3-0} = \frac{7}{3}$

Use point-slope form to find the equation:

$y - y_1 = m(x - x_1)$

$y - 4 = \frac{7}{3}(x - 3)$

$y - 4 = \frac{7}{3}x - 7$

$y = \frac{7}{3}x - 3$

27. $y = mx + b$

$y = 5x - 12$

29. $y = mx + b$

$5 = -7(7) + b$

$5 = -49 + b$

$54 = b$

$y = -7x + 54$

31. $y = mx + b$

$-4 = 0(2) + b$

$-4 = b$

$y = 0x + (-4)$

$y = -4$

33. Find the slope of the line:

$m = \frac{\Delta y}{\Delta x} = \frac{2-(-7)}{-5-3} = \frac{9}{-8} = -\frac{9}{8}$

$y = mx + b$

$-7 = -\frac{9}{8}(3) + b$

$-\frac{56}{8} = -\frac{27}{8} + b$

$-\frac{29}{8} = b$

$y = -\frac{9}{8}x - \frac{29}{8}$

35. $5x + 2y = 10$

$2y = -5x + 10$

$y = -\frac{5}{2}x + 5$

$m = -\frac{5}{2};\, b = 5 \Rightarrow (0, 5)$

37. $-2(x + 3y) = 5$

$-2x - 6y = 5$

$-6y = 2x + 5$

$y = -\frac{1}{3}x - \frac{5}{6}$

$m = -\frac{1}{3};\, b = -\frac{5}{6} \Rightarrow (0, -\frac{5}{6})$

© 2015 Cengage Learning. All Rights Reserved. May not be scanned, copied or duplicated, or posted to a publicly accessible website, in whole or in part.

39.
$$y + 1 = x$$
$$y = x - 1$$
$$m = 1$$
$$b = -1; (0, -1)$$

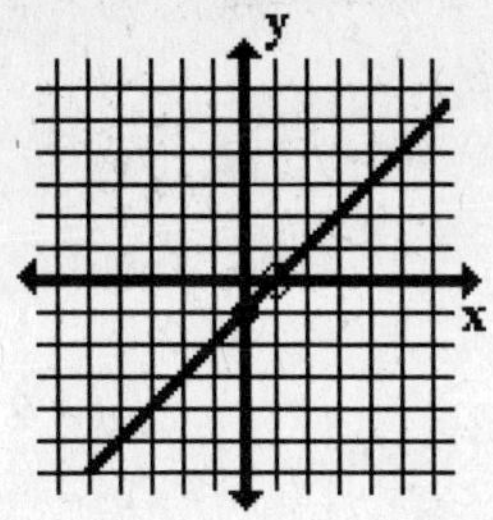

41.
$$x = \tfrac{3}{2}y - 3$$
$$2x = 3y - 6$$
$$-3y = -2x - 6$$
$$y = \tfrac{2}{3}x + 2$$
$$m = \tfrac{2}{3}$$
$$b = 2; (0, 2)$$

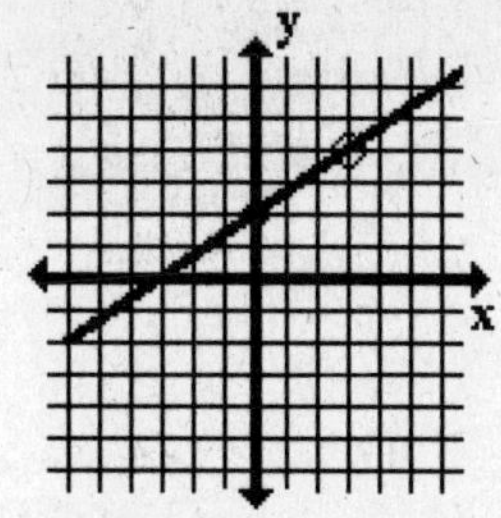

43. $y = 3x + 4$ $\quad$ $y = 3x - 7$

$m = 3$ $\quad$ $m = 3$

parallel

45. $x + y = 2$ $\quad$ $y = x + 5$

$y = -x + 2$ $\quad$ $m = 1$

$m = -1$

perpendicular

47. $y = 3x + 7$ $\quad$ $2y = 6x - 9$

$m = 3$ $\quad$ $y = 3x - \dfrac{9}{2}$

$m = 3$

parallel

49. $x = 3y + 4$ $\quad$ $y = -3x + 7$

$-3y = -x + 4$ $\quad$ $m = -3$

$y = \tfrac{1}{3}x - \tfrac{4}{3}$

$m = \tfrac{1}{3}$

perpendicular

51. Find the slope of the given line:

$y = 4x - 7 \Rightarrow m = 4$

Use the parallel slope.

$$y - y_1 = m(x - x_1)$$
$$y - 0 = 4(x - 0)$$
$$y = 4x$$

53. Find the slope of the given line:

$$4x - y = 7$$
$$-y = -4x + 7$$
$$y = 4x - 7 \Rightarrow m = 4$$

Use the parallel slope.

$$y - y_1 = m(x - x_1)$$
$$y - 5 = 4(x - 2)$$
$$y - 5 = 4x - 8$$
$$y = 4x - 3$$

55. Find the slope of the given line:

$y = 7x - 2 \Rightarrow m = 7$

Use the perpendicular slope.

$$y - y_1 = m(x - x_1)$$
$$y - (-3) = -\tfrac{1}{7}(x - 1)$$
$$y + 3 = -\tfrac{1}{7}x + \tfrac{1}{7}$$
$$y = -\tfrac{1}{7}x + \tfrac{1}{7} - 3$$
$$y = -\tfrac{1}{7}x - \tfrac{20}{7}$$

57. Find the slope of the given line:

$$4x - y = 7$$
$$-y = -4x + 7$$
$$y = 4x - 7 \Rightarrow m = 4$$

Use the perpendicular slope.

$$y - y_1 = m(x - x_1)$$
$$y - 5 = -\tfrac{1}{4}(x - 2)$$
$$y - 5 = -\tfrac{1}{4}x + \tfrac{1}{2}$$
$$y = -\tfrac{1}{4}x + \tfrac{1}{2} + 5$$
$$y = -\tfrac{1}{4}x + \tfrac{11}{2}$$

© 2015 Cengage Learning. All Rights Reserved. May not be scanned, copied or duplicated, or posted to a publicly accessible website, in whole or in part.

59.

$$\begin{aligned} 4x + 5y &= 20 \\ 5y &= -4x + 20 \\ y &= -\tfrac{4}{5}x + 4 \\ m &= -\tfrac{4}{5} \end{aligned} \qquad \begin{aligned} 5x - 4y &= 20 \\ -4y &= -5x + 20 \\ y &= \tfrac{5}{4}x - 5 \\ m &= \tfrac{5}{4} \end{aligned}$$

perpendicular

61.

$$\begin{aligned} 2x + 3y &= 12 \\ 3y &= -2x + 12 \\ y &= -\tfrac{2}{3}x + 4 \\ m &= -\tfrac{2}{3} \end{aligned} \qquad \begin{aligned} 6x + 9y &= 32 \\ 9y &= -6x + 32 \\ y &= -\tfrac{2}{3}x + \tfrac{32}{9} \\ m &= -\tfrac{2}{3} \end{aligned}$$

parallel

63.

$y = -5$ horizontal line $\qquad$ $x = 2$ vertical line

perpendicular

65.

$$\begin{aligned} x &= \frac{y-2}{3} \\ 3x &= y - 2 \\ -y &= -3x - 2 \\ y &= 3x + 2 \\ m &= 3 \end{aligned} \qquad \begin{aligned} 3(y-3) + x &= 0 \\ 3y - 9 + x &= 0 \\ 3y &= -x + 9 \\ y &= -\tfrac{1}{3}x + 3 \\ m &= -\tfrac{1}{3} \end{aligned}$$

perpendicular

67. Find the slope of the given line:

$$\begin{aligned} x &= \frac{5}{4}y - 2 \\ 4x &= 5y - 8 \\ -5y &= -4x - 8 \\ y &= \tfrac{4}{5}x + \tfrac{8}{5} \Rightarrow m = \tfrac{4}{5} \end{aligned}$$

Use the parallel slope.

$$\begin{aligned} y - y_1 &= m(x - x_1) \\ y - (-2) &= \tfrac{4}{5}(x - 4) \\ y + 2 &= \tfrac{4}{5}x - \tfrac{16}{5} \\ y &= \tfrac{4}{5}x - \tfrac{16}{5} - 2 \\ y &= \tfrac{4}{5}x - \tfrac{26}{5} \end{aligned}$$

69. Find the slope of the given line:

$$\begin{aligned} x &= \tfrac{5}{4}y - 2 \\ 4x &= 5y - 8 \\ -5y &= -4x - 8 \\ y &= \tfrac{4}{5}x + \tfrac{8}{5} \Rightarrow m = \tfrac{4}{5} \end{aligned}$$

Use the perpendicular slope.

$$\begin{aligned} y - y_1 &= m(x - x_1) \\ y - (-2) &= -\tfrac{5}{4}(x - 4) \\ y + 2 &= -\tfrac{5}{4}x + 5 \\ y &= -\tfrac{5}{4}x + 3 \end{aligned}$$

71.

$$\begin{aligned} x &= \frac{2y-4}{7} \\ 7x &= 2y - 4 \\ -2y &= -7x - 4 \\ y &= \tfrac{7}{2}x + 2 \end{aligned}$$

$m = \tfrac{7}{2};\ b = 2 \Rightarrow (0, 2)$

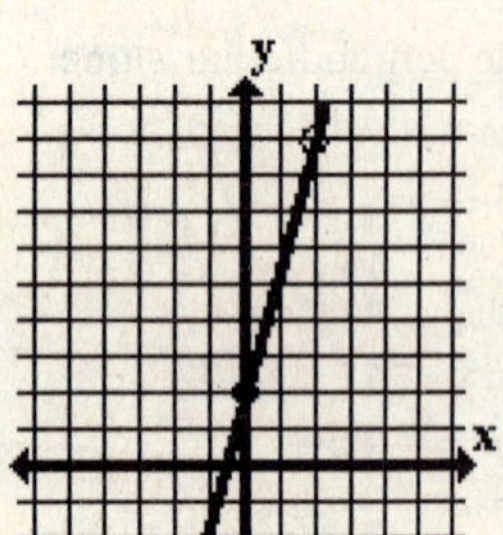

73.

$$\begin{aligned} 3(y-4) &= -2(x-3) \\ 3y - 12 &= -2x + 6 \\ 3y &= -2x + 18 \\ y &= -\frac{2}{3}x + 6 \end{aligned}$$

$m = -\tfrac{2}{3};\ b = 6 \Rightarrow (0, 6)$

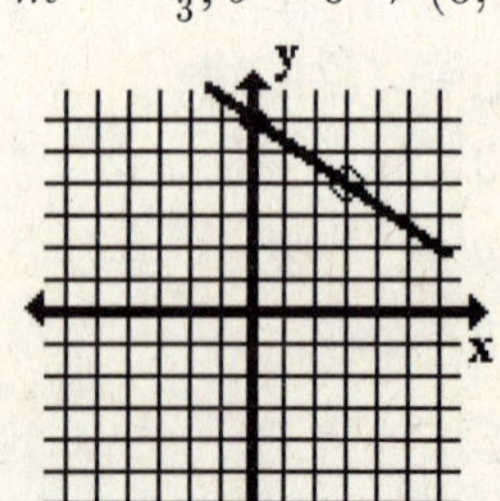

© 2015 Cengage Learning. All Rights Reserved. May not be scanned, copied or duplicated, or posted to a publicly accessible website, in whole or in part.

75. Notice that the line goes through $P(2,5)$ and the point $(-1,3)$. Find the slope:

$$m = \frac{\Delta y}{\Delta x} = \frac{5-3}{2-(-1)} = \frac{2}{3}$$

Use point-slope form to find the equation:

$$\begin{aligned} y - y_1 &= m(x - x_1) \\ y - 5 &= \tfrac{2}{3}(x-2) \\ y - 5 &= \tfrac{2}{3}x - \tfrac{4}{3} \\ y &= \tfrac{2}{3}x - \tfrac{4}{3} + \tfrac{15}{3} \\ y &= \tfrac{2}{3}x + \tfrac{11}{3} \end{aligned}$$

77. Notice that the line goes through $(2,-3)$ and the point $(-2,4)$. Find the slope:

$$m = \frac{\Delta y}{\Delta x} = \frac{4-(-3)}{-2-2} = \frac{7}{-4} = -\frac{7}{4}$$

Use point-slope form to find the equation:

$$\begin{aligned} y - y_1 &= m(x - x_1) \\ y - 4 &= -\tfrac{7}{4}(x-(-2)) \\ y - 4 &= -\tfrac{7}{4}(x+2) \\ y - 4 &= -\tfrac{7}{4}x - \tfrac{7}{2} \\ y &= -\tfrac{7}{4}x - \tfrac{7}{2} + 4 \\ y &= -\tfrac{7}{4}x + \tfrac{1}{2} \end{aligned}$$

79. The line $y = 3$ is horizontal, so any perpendicular line will be vertical.
Find the midpoint of the described segment:

$$x = \frac{x_1 + x_2}{2} = \frac{2+(-6)}{2} = \frac{-4}{2} = -2; \; y = \frac{y_1 + y_2}{2} = \frac{4+10}{2} = \frac{14}{2} = 7$$

The vertical line through the point $(-2,7)$ is $x = -2$.

81. The line $x = 3$ is vertical, so any parallel line will be vertical.
Find the midpoint of the described segment:

$$x = \frac{x_1 + x_2}{2} = \frac{2+8}{2} = \frac{10}{2} = 5; \; y = \frac{y_1 + y_2}{2} = \frac{-4+12}{2} = \frac{8}{2} = 4$$

The vertical line through the point $(5,4)$ is $x = 5$.

83.

$$\begin{aligned} Ax + By &= C \\ By &= -Ax + C \\ \frac{By}{B} &= \frac{-Ax + C}{B} \\ y &= -\frac{A}{B}x + \frac{C}{B} \Rightarrow m = -\frac{A}{B}; \; y\text{-intercept: } \left(0, \frac{C}{B}\right) \end{aligned}$$

85. Let x represent the number of years since the truck was purchased, and let y represent the value of the truck. Then we know two points on the depreciation line: $(0, 19984)$ and $(8, 1600)$. Find the slope and then use point-slope form to find the depreciation equation.

$$m = \frac{\Delta y}{\Delta x} = \frac{19984 - 1600}{0 - 8} = \frac{18384}{-8} = -2298$$

$$\begin{aligned} y - y_1 &= m(x - x_1) \\ y - 19984 &= -2298(x - 0) \\ y &= -2298x + 19984 \end{aligned}$$

© 2015 Cengage Learning. All Rights Reserved. May not be scanned, copied or duplicated, or posted to a publicly accessible website, in whole or in part.

87. Let x represent the time in years after 1987, and let y represent the value of the painting. Then we know two points on the appreciation line: $(0, 36225000)$ and $(20, 72450000)$. Find the slope and then use point-slope form to find the appreciation equation.

$$m = \frac{\Delta y}{\Delta x} = \frac{72450000 - 36225000}{20 - 0} = \frac{36225000}{20} = 1811250$$

$$y - y_1 = m(x - x_1)$$
$$y - 36225000 = 1811250(x - 0)$$
$$y = 1{,}811{,}250x + 36{,}225{,}000$$

89. Let x represent the number of years since the painting was purchased, and let y represent the value of the painting. Then we know two points on the appreciation line: $(0, 250000)$ and $(5, 500000)$. Find the slope and then use point-slope form to find the appreciation equation.

$$m = \frac{\Delta y}{\Delta x} = \frac{500000 - 250000}{5 - 0} = \frac{250000}{5} = 50000$$

$$y - y_1 = m(x - x_1)$$
$$y - 250000 = 50000(x - 0)$$
$$y = 50{,}000x + 250{,}000$$

91. Let x represent the number of years since the TV was purchased, and let y represent the value of the TV. Then we know two points on the depreciation line: $(0, 1750)$ and $(3, 800)$. Find the slope and then use point-slope form to find the depreciation equation.

$$m = \frac{\Delta y}{\Delta x} = \frac{1750 - 800}{0 - 3} = \frac{950}{-3} = -\frac{950}{3}$$

$$y - y_1 = m(x - x_1)$$
$$y - 1750 = -\tfrac{950}{3}(x - 0)$$
$$y = -\tfrac{950}{3}x + 1750$$

93. Let x represent the number of years since the copy machine was purchased, and let y represent its value. Since the value decreases \$180 per year, the slope of the depreciation line is -180. Since the copier is worth \$1750 new, the y-intercept is 1750. Thus the depreciation equation can be found:

$$y = mx + b$$
$$y = -180x + 1750$$

Now let $x = 7$: $y = -180x + 1750 = -180(7) + 1750 = -1260 + 1750 = \$490$

95. Let x represent the number of years since the house was purchased, and let y represent its value. Since the value increases \$4000 per year, the slope of the depreciation line is 4000. We also know a point on the line: $(2, 122000)$. Use point-slope form to find the appreciation equation.

$$y - y_1 = m(x - x_1)$$
$$y - 122000 = 4000(x - 2)$$
$$y - 122000 = 4000x - 8000$$
$$y = 4000x + 114000$$

Substitute $x = 10$ and find its value:

$y = 4000x + 114000 = 4000(10) + 114000 = 40000 + 114000 = \$154{,}000$

© 2015 Cengage Learning. All Rights Reserved. May not be scanned, copied or duplicated, or posted to a publicly accessible website, in whole or in part.

97. The relationship between number of copies and the total cost is linear. Let x represent the number of copies (in hundreds), and let y represent the total cost. The slope of the line will be the charge per one hundred copies, or 15. We also know a point on the line: $(3, 75)$. Use point-slope form to find the equation of the line:

$$\begin{aligned} y - y_1 &= m(x - x_1) \\ y - 75 &= 15(x - 3) \\ y - 75 &= 15x - 45 \\ y &= 15x + 30 \end{aligned}$$

Let $x = 10$ and find the cost:
$y = 15x + 30 = 15(10) + 30 = 150 + 30 = \$180$

99-105. Answers may vary.

107. To pass through II and IV, the slope must be negative. To pass through II and I, the y-intercept must be positive. Thus $a < 0$ and $b > 0$.

Exercises 2.4 (page 138)

1. $\{-2, 0, 3\}$

3. $\{(-2, 0), (0, 3), (3, 9)\}$

5.
$$\begin{aligned} 3x - 2y &= 8 \\ -2y &= -3x + 8 \\ y &= \tfrac{3}{2}x - 4 \end{aligned}$$

7.
$$\begin{aligned} \frac{y+2}{2} &= 4(y+2) \\ 2 \cdot \frac{y+2}{2} &= 2 \cdot 4(y+2) \\ y + 2 &= 8(y+2) \\ y + 2 &= 8y + 16 \\ -7y &= 14 \\ y &= -2 \end{aligned}$$

9.
$$\begin{aligned} \frac{2a}{3} + \frac{1}{2} &= \frac{6a-1}{6} \\ 6\left(\frac{2a}{3} + \frac{1}{2}\right) &= 6 \cdot \frac{6a-1}{6} \\ 2(2a) + 3 &= 6a - 1 \\ 4a + 3 &= 6a - 1 \\ -2a &= -4 \\ a &= 2 \end{aligned}$$

11. relation

13. domain

15. 0

17. cannot

19. slope; y-intercept

21. y

23. y

25. domain $= \{-2, -1, 5, 6\}$
range $= \{-4, 1, 3\}$; function

27. domain $= \{-2, 6, 5\}$
range $= \{3, 8, 5, 4\}$; not a function

29. The domain is the set of all x-coordinates on the graph. The domain is $(-\infty, 1]$.
The range is the set of all y-coordinates on the graph. The range is $(-\infty, \infty)$.
Since a vertical line can pass through the graph more than once, it is not the graph of a function.

31. The domain is the set of all x-coordinates on the graph. The domain is $(-\infty, \infty)$.
The range is the set of all y-coordinates on the graph. The range is $(-\infty, \infty)$.
Since each vertical line passes through the graph at most once, it is the graph of a function.

© 2015 Cengage Learning. All Rights Reserved. May not be scanned, copied or duplicated, or posted to a publicly accessible website, in whole or in part.

SECTION 2.4

33. $f(x) = 3x$
$f(3) = 3(3) = 9$
$f(-1) = 3(-1) = -3$

$f(x) = 0$
$3x = 0$
$x = 0$

35. $f(x) = 2x - 3$
$f(3) = 2(3) - 3 = 3$
$f(-1) = 2(-1) - 3 = -5$

$f(x) = 0$
$2x - 3 = 0$
$2x = 3$
$x = \frac{3}{2}$

37. $f(x) = x^2$
$f(2) = 2^2 = 4$
$f(3) = 3^2 = 9$

39. $f(x) = x^3 - 3$
$f(2) = 2^3 - 3 = 8 - 3 = 5$
$f(3) = 3^3 - 3 = 27 - 3 = 24$

41. $f(x) = |x| - 3$
$f(2) = |2| - 3 = 2 - 3 = -1$
$f(-2) = |-2| - 3 = 2 - 3 = -1$

43. $f(x) = x^2 - 2$
$f(2) = 2^2 - 2 = 4 - 2 = 2$
$f(-2) = (-2)^2 - 2 = 4 - 2 = 2$

45. $g(x) = 5x$
$g(w) = 5w$
$g(w+1) = 5(w+1) = 5w + 5$

47. $g(x) = 3x - 5$
$g(w) = 3w - 5$
$g(w+1) = 3(w+1) - 5 = 3w + 3 - 5 = 3w - 2$

49. $f(3) + f(2) = [2(3) + 1] + [2(2) + 1] = 6 + 1 + 4 + 1 = 12$

51. $f(b) - f(a) = [2b + 1] - [2a + 1] = 2b + 1 - 2a - 1 = 2b - 2a$

53. $y = 2x + 3$ **is a function**, since each value of x corresponds to exactly one value of y.

55. $y = 2x^2$ **is a function**, since each value of x corresponds to exactly one value of y.

57. $y^2 = x + 1$ **is not a function**, since $x = 3$ corresponds to both $y = 2$ and $y = -2$.

59. $x = |y|$ **is not a function**, since $x = 2$ corresponds to both $y = 2$ and $y = -2$.

61. domain $= (-\infty, \infty)$

63. To ensure that the denominator is not 0, x cannot equal -5. The domain is $(-\infty, -5) \cup (-5, \infty)$.

65. To ensure that the denominator is not 0, x cannot equal -3. The domain is $(-\infty, -3) \cup (-3, \infty)$.

67. The denominator will never be 0. The domain is $(-\infty, \infty)$.

© 2015 Cengage Learning. All Rights Reserved. May not be scanned, copied or duplicated, or posted to a publicly accessible website, in whole or in part.

SECTION 2.4

69. $f(x) = 2x - 1$
D = $(-\infty, \infty)$
R = $(-\infty, \infty)$

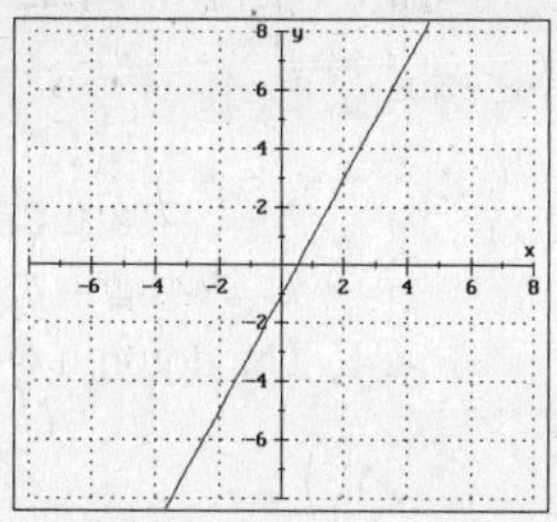

71. $2x - 3y = 6$
D = $(-\infty, \infty)$
R = $(-\infty, \infty)$

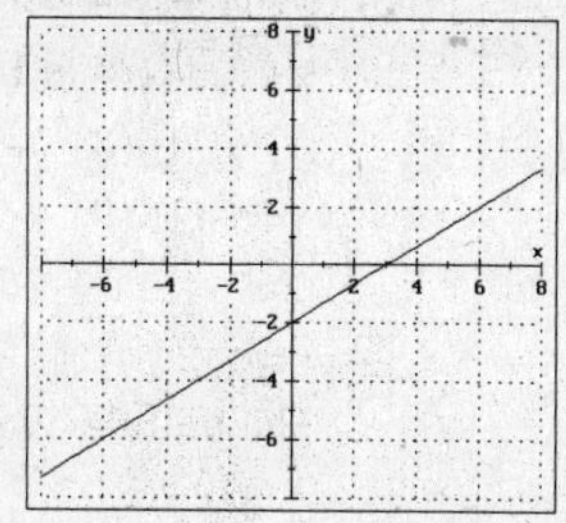

73. $y = 4x^2 + 3 \Rightarrow$ This is not a linear function because the exponent on x is 2.

75. $x = 4y - 3 \Rightarrow y = \frac{1}{4}x + \frac{3}{4} \Rightarrow$ This is a linear function $\left(m = \frac{1}{4}, b = \frac{3}{4}\right)$.

77. $f(x) = (x+1)^2$
$f(2) = (2+1)^2 = 3^2 = 9$
$f(3) = (3+1)^2 = 4^2 = 16$

79. $f(x) = 2x^2 - x$
$f(2) = 2(2)^2 - 2 = 2(4) - 2 = 8 - 2 = 6$
$f(3) = 2(3)^2 - 3 = 2(9) - 3 = 18 - 3 = 15$

81. $f(x) = 7 + 5x$
$f(3) = 7 + 5(3) = 22$
$f(-1) = 7 + 5(-1) = 2$

83. $f(x) = 9 - 2x$
$f(3) = 9 - 2(3) = 3$
$f(-1) = 9 - 2(-1) = 11$

85. $f(b) - 1 = 2b + 1 - 1 = 2b$

87. $f(0) + f(-\frac{1}{2}) = [2(0) + 1] + [2(-\frac{1}{2}) + 1] = 0 + 1 - 1 + 1 = 1$

89. $x = 1$ corresponds to both $y = 4$ and $y = -7$. $\Rightarrow$ not a function

91. Each x value corresponds to only one y value. $\Rightarrow$ function

93. domain = $\{2\}$; range = $(-\infty, \infty)$
not a function

95. domain = $(-\infty, \infty)$; range = $\{1\}$
function

97. domain = $[0, \infty)$; range = $(-\infty, \infty)$
not a function

99. Set $R(x)$ equal to $C(x)$ and solve for x.

$$R(x) = C(x)$$
$$120x = 57.50x + 12000$$
$$62.50x = 12000$$
$$x = \frac{12000}{62.50}$$
$$x = 192$$

The company must sell 192 players.

101. **a.** $I(h) = 0.10h + 50$

b. $I(115) = 0.10(115) + 50$
$= 11.5 + 50 = 61.5$
He makes \$61.50.

© 2015 Cengage Learning. All Rights Reserved. May not be scanned, copied or duplicated, or posted to a publicly accessible website, in whole or in part.

SECTION 2.4

103. Let $t = 3$ and find s.

$$\begin{aligned} s = f(t) &= -16t^2 + 256t \\ &= -16(3)^2 + 256(3) \\ &= -16(9) + 768 \\ &= -144 + 768 \\ &= 624 \end{aligned}$$

The bullet will have a height of 624 ft.

105. Let $t = 1.5$ and find h.

$$\begin{aligned} h &= -16t^2 + 32t \\ h &= -16(1.5)^2 + 32(1.5) \\ h &= -16(2.25) + 48 \\ h &= -36 + 48 = 12 \end{aligned}$$

The dolphin will be 12 feet above the water.

107. Let $C = 25$ and find F.

$$\begin{aligned} F(C) &= \frac{9}{5}C + 32 \\ &= \frac{9}{5}(25) + 32 \\ &= 45 + 32 \\ &= 77 \end{aligned}$$

The temperature is $77°F$.

109. **Answers may vary.**

111. $$\begin{aligned} f(x) + g(x) &= 2x + 1 + x^2 \\ &= x^2 + 2x + 1 \\ &= g(x) + f(x) \end{aligned}$$

Yes, they are equal.

Exercises 2.5 (page 149)

1. $$\begin{aligned} f(x) &= x^2 - 4 \\ f(3) &= (3)^2 - 4 \\ &= 9 - 4 \\ &= 5 \\ f(-2) &= (-2)^2 - 4 \\ &= 4 - 4 \\ &= 0 \end{aligned}$$

3. $$\begin{aligned} f(x) &= x^3 - 1 \\ f(3) &= (3)^3 - 1 \\ &= 27 - 1 \\ &= 26 \\ f(-2) &= (-2)^3 - 1 \\ &= -8 - 1 \\ &= -9 \end{aligned}$$

5. $$\begin{aligned} f(x) &= |x| - 2 \\ f(3) &= |3| - 2 \\ &= 3 - 2 \\ &= 1 \\ f(-2) &= |-2| - 2 \\ &= 2 - 2 \\ &= 0 \end{aligned}$$

7. $g(-3) = -(-3)^2 - 4 = -9 - 4 = -13$

9. 23, 29

11. $a \cdot b = b \cdot a$

13. 1

15. quadratic; parabola

17. absolute value

19. horizontal

21. 2; downward

23. 7; to the left

25. reflected

27. $f(x) = x^2 - 3$

x	-2	-1	0	1	2
y	1	-2	-3	-2	1

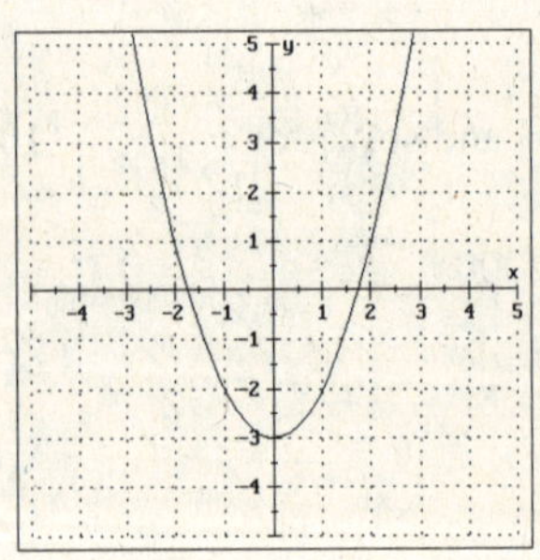

© 2015 Cengage Learning. All Rights Reserved. May not be scanned, copied or duplicated, or posted to a publicly accessible website, in whole or in part.

SECTION 2.5

29. $f(x) = (x-1)^3$

x	-1	0	1	2	3
y	-8	-1	0	1	8

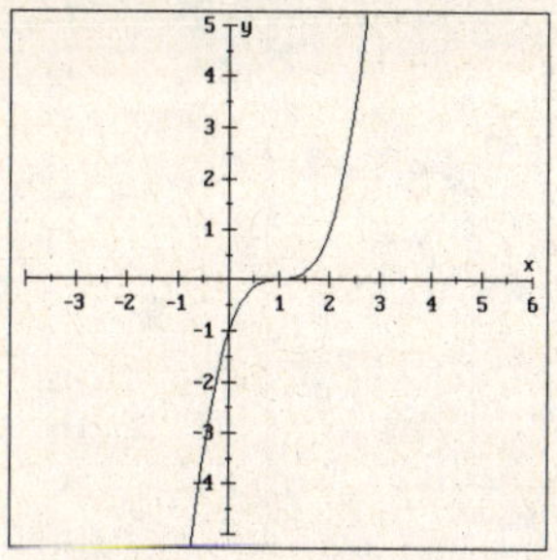

31. $f(x) = |x| - 2$

x	-2	-1	0	1	2
y	0	-1	-2	-1	0

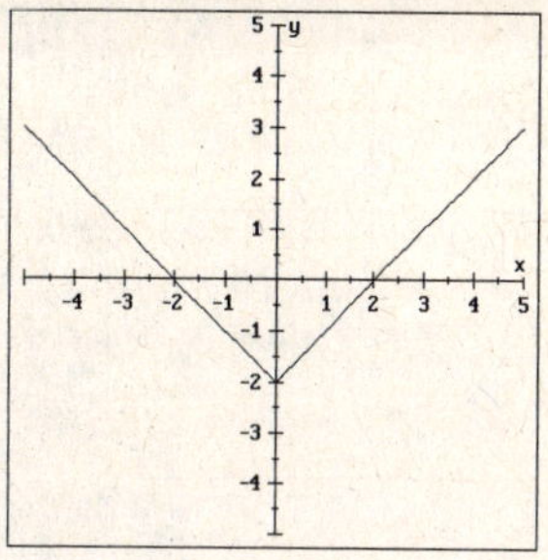

33. $f(x) = x^2 - 5$
Shift $f(x) = x^2$ D 5.

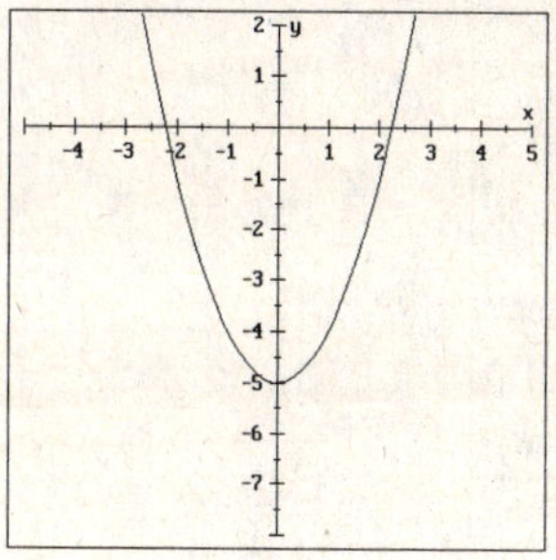

35. $f(x) = |x| + 6$
Shift $f(x) = |x|$ U 6.

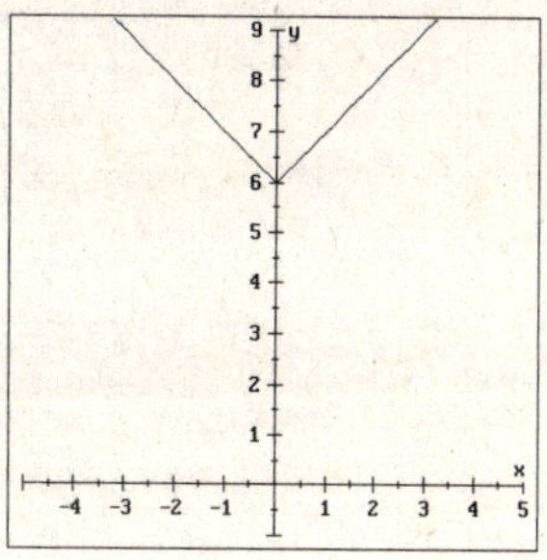

37. $f(x) = (x-1)^3$
Shift $f(x) = x^3$ R 1.

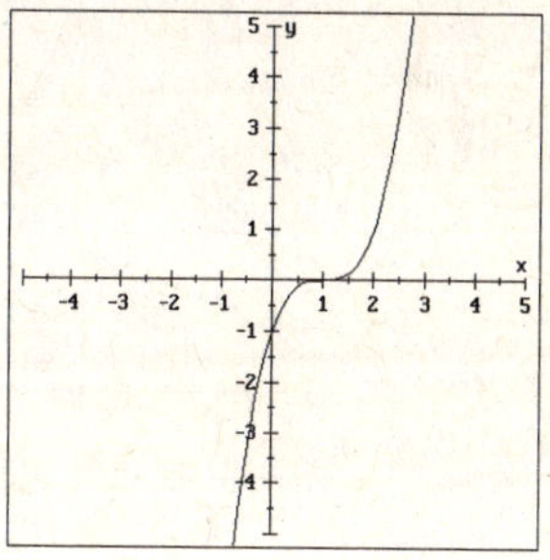

39. $f(x) = (x-3)^2$
Shift $f(x) = x^2$ R 3.

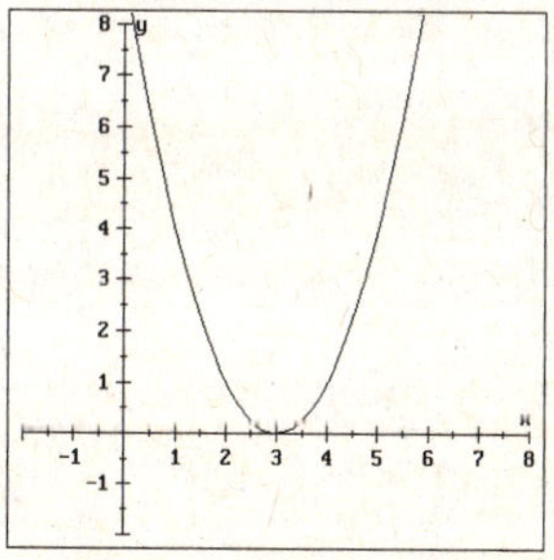

41. $f(x) = |x-2| - 1$
Shift $f(x) = |x|$ D 1, R 2.

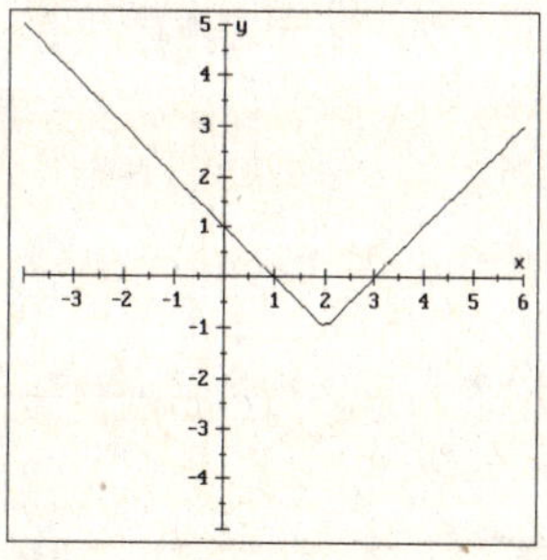

43. $f(x) = (x+1)^3 - 2$
Shift $f(x) = x^3$ D 2, L 1.

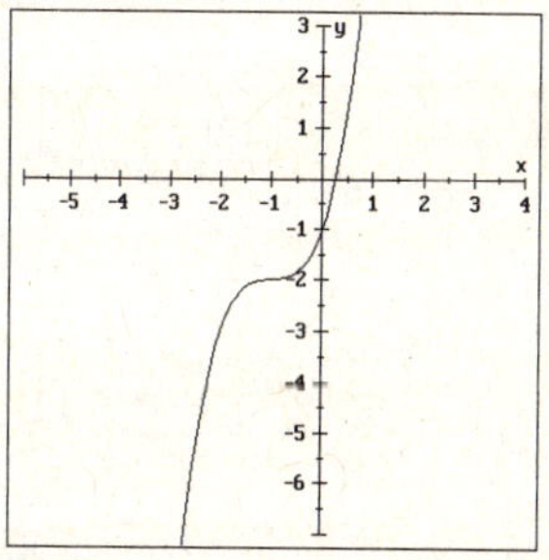

© 2015 Cengage Learning. All Rights Reserved. May not be scanned, copied or duplicated, or posted to a publicly accessible website, in whole or in part.

SECTION 2.5

45. $f(x) = -|x| + 1$
Shift $f(x) = -|x|$ U 1.

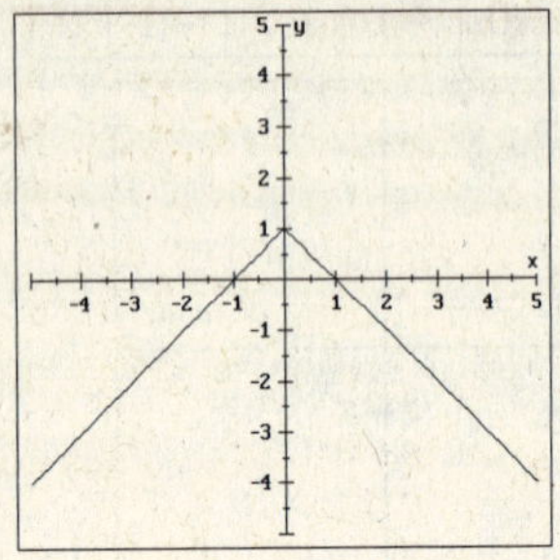

47. $f(x) = -(x-1)^2$
Shift $f(x) = -x^2$ R 1.

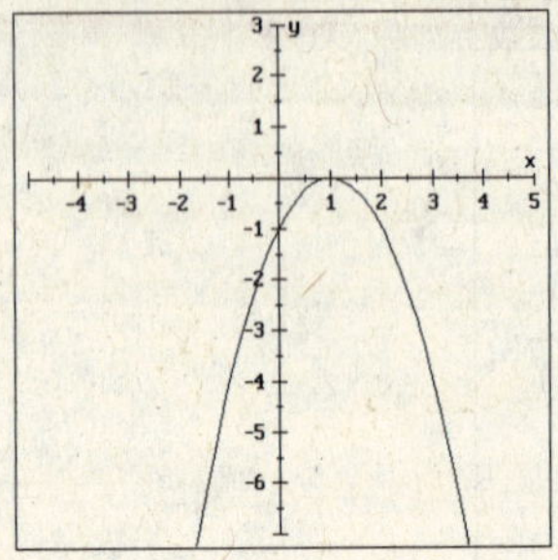

49. $f(x) = |x - 1|$

x	-1	0	1	2	3
y	2	1	0	1	2

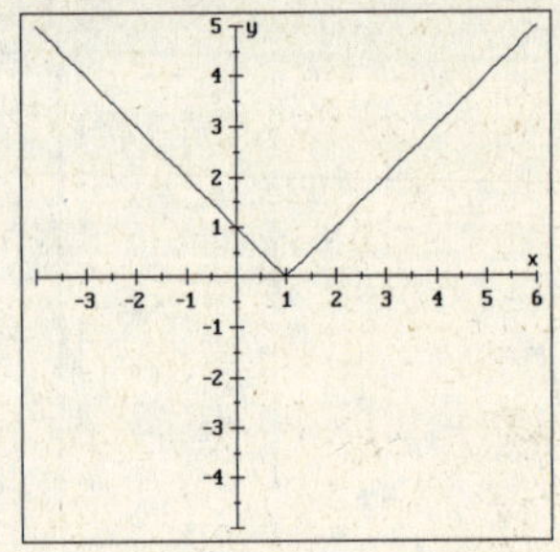

51. $f(x) = x^2 + 8$
Shift $f(x) = x^2$ U 8.

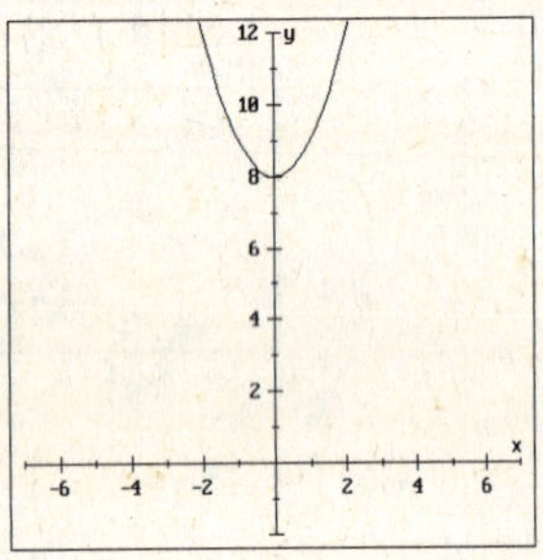

53. $f(x) = |x + 5|$
Shift $f(x) = |x|$ L 5.

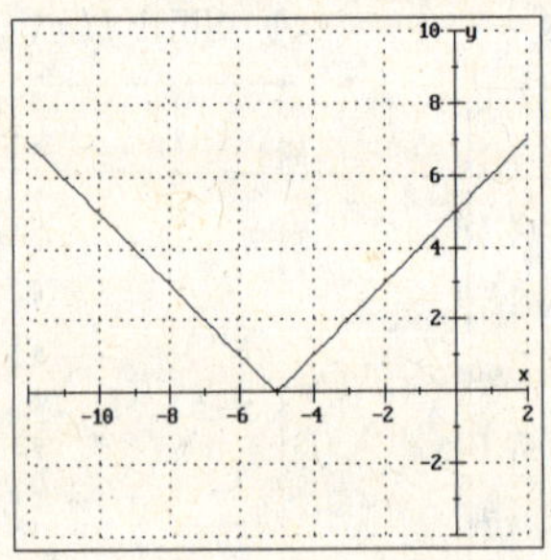

55. $f(x) = (x - 6)^2$
Shift $f(x) = x^2$ R 6.

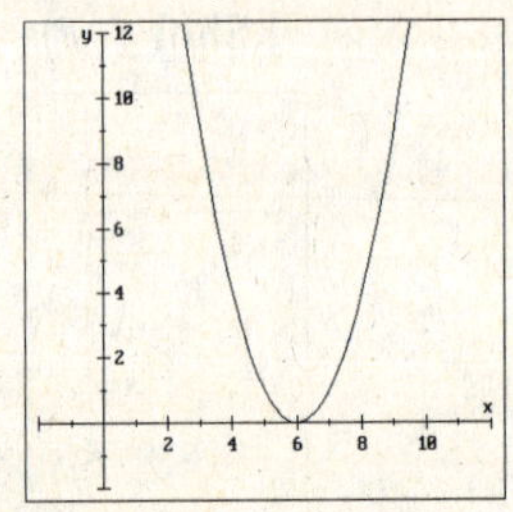

57. $f(x) = x^3 + 8$
x: $[-4, 4]$, y: $[-4, 4]$

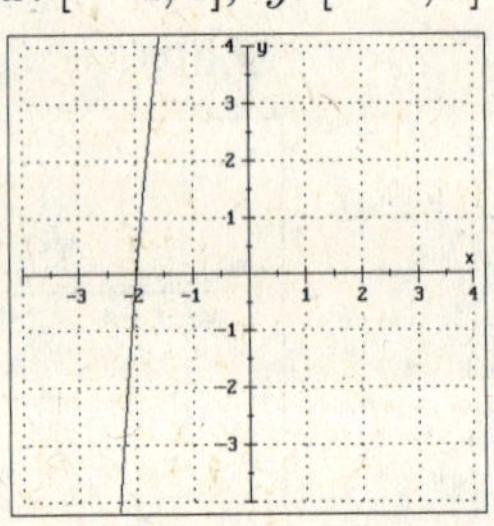

x: $[-10, 10]$, y: $[-4, 16]$

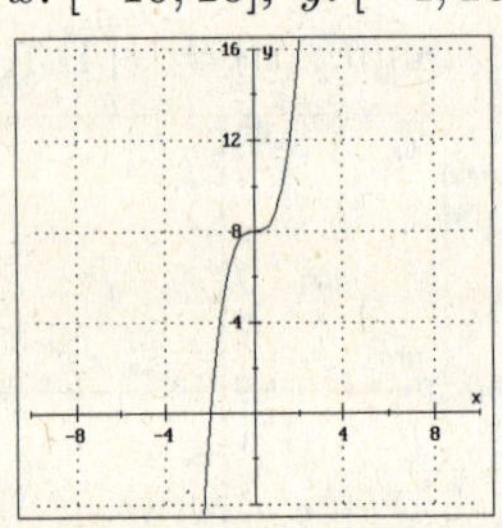

59-67. Answers may vary.

© 2015 Cengage Learning. All Rights Reserved. May not be scanned, copied or duplicated, or posted to a publicly accessible website, in whole or in part.

Chapter 2 Review (page 156)

1.

x	y
−9	[−10]
[−6]	−8
−3	[−6]
[0]	−4
3	[−2]
[6]	0
[9]	2

$$\begin{aligned} 4x-6y&=24\\ 4(-9)-6y&=24\\ -36-6y&=24\\ -6y&=60\\ y&=-10\end{aligned}\qquad \begin{aligned} 4x-6y&=24\\ 4x-6(-8)&=24\\ 4x+48&=24\\ 4x&=-24\\ x&=-6\end{aligned}\qquad \begin{aligned} 4x-6y&=24\\ 4(-3)-6y&=24\\ -12-6y&=24\\ -6y&=36\\ y&=-6\end{aligned}$$

$$\begin{aligned} 4x-6y&=24\\ 4x-6(-4)&=24\\ 4x+24&=24\\ 4x&=0\\ x&=0\end{aligned}\qquad \begin{aligned} 4x-6y&=24\\ 4(3)-6y&=24\\ 12-6y&=24\\ -6y&=12\\ y&=-2\end{aligned}\qquad \begin{aligned} 4x-6y&=24\\ 4x-6(0)&=24\\ 4x&=24\\ x&=6\end{aligned}\qquad \begin{aligned} 4x-6y&=24\\ 4x-6(2)&=24\\ 4x-12&=24\\ 4x&=36\\ x&=9\end{aligned}$$

2.

$$\begin{aligned} x+y&=4\\ 0+y&=4\\ y&=4\\ (0,&4)\end{aligned}\qquad \begin{aligned} x+y&=4\\ x+0&=4\\ x&=4\\ (4,&0)\end{aligned}$$

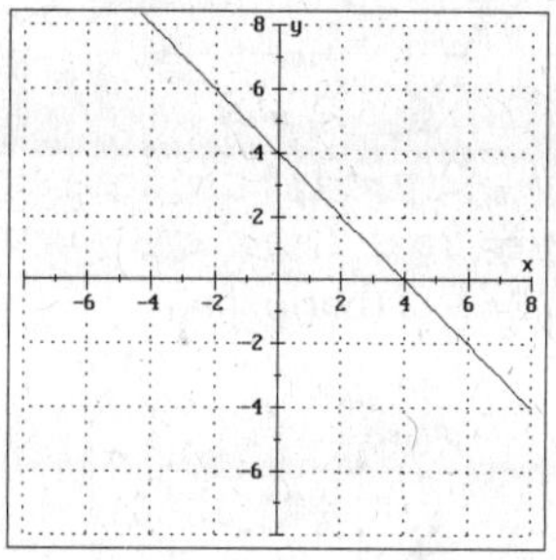

3.

$$\begin{aligned} 2x-y&=8\\ 2(0)-y&=8\\ -y&=8\\ y&=-8\\ (0,&-8)\end{aligned}\qquad \begin{aligned} 2x-y&=8\\ 2x-0&=8\\ 2x&=8\\ x&=4\\ (4,&0)\end{aligned}$$

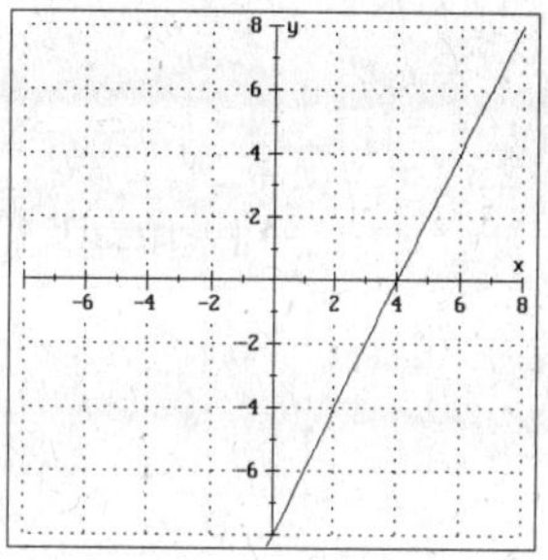

4.

$$\begin{aligned} y&=3x+4\\ y&=3(0)+4\\ y&=0+4\\ y&=4\\ (0,&4)\end{aligned}\qquad \begin{aligned} y&=3x+4\\ y&=3(1)+4\\ y&=3+4\\ y&=7\\ (1,&7)\end{aligned}$$

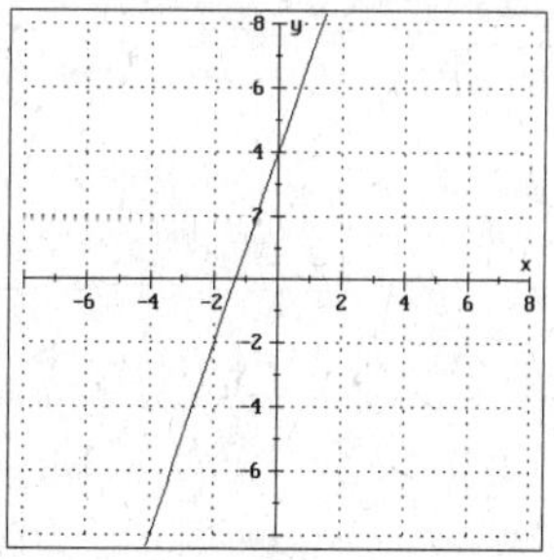

© 2015 Cengage Learning. All Rights Reserved. May not be scanned, copied or duplicated, or posted to a publicly accessible website, in whole or in part.

5. $x = 4 - 2y$ $\quad x = 4 - 2y$
$0 = 4 - 2y$ $\quad x = 4 - 2(0)$
$2y = 4$ $\quad x = 4$
$y = 2$ $\quad (4, 0)$
$(0, 2)$

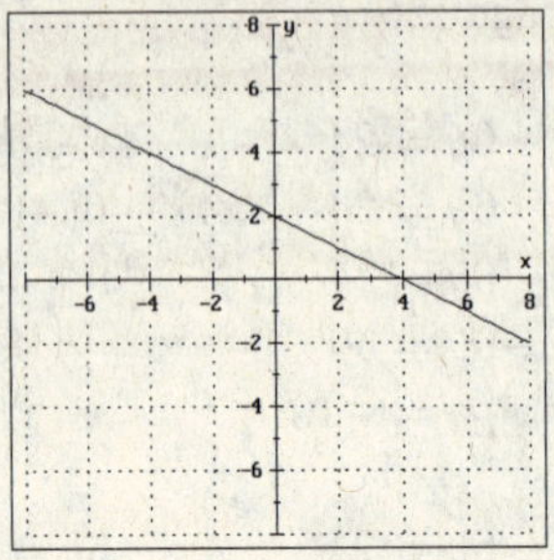

6. $y = 4$ (horizontal)

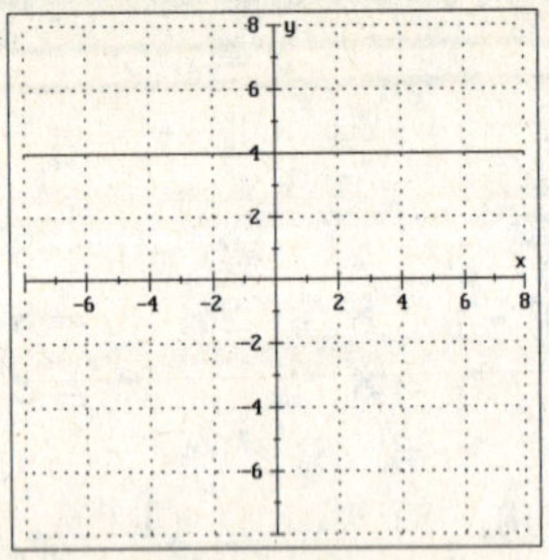

7. $x = -2$ (vertical)

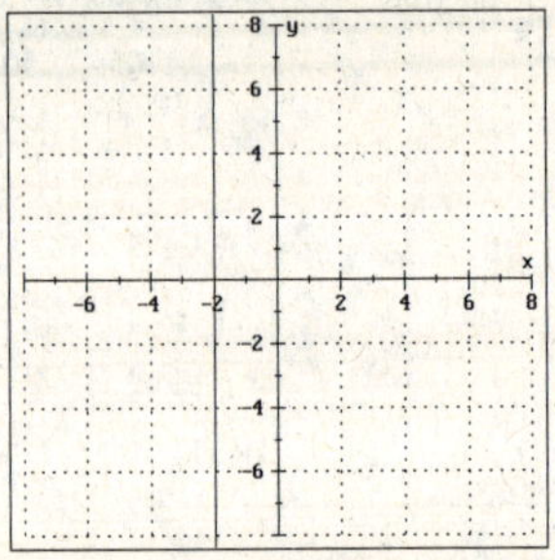

8.
$2(x+3) = x + 2$
$2x + 6 = x + 2$
$x = -4$ (vertical)

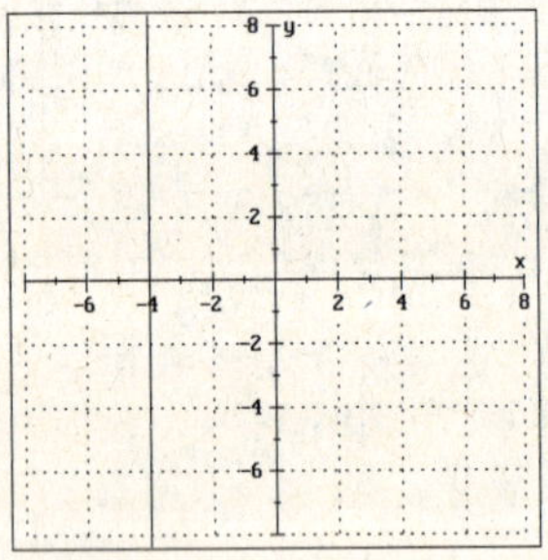

9.
$3y = 2(y-1)$
$3y = 2y - 2$
$y = -2$ (horiozntal)

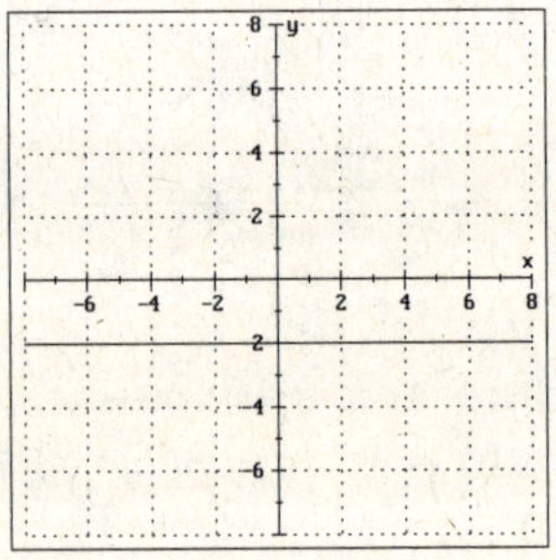

10. $x = \dfrac{x_1 + x_2}{2} = \dfrac{-4+9}{2} = \dfrac{5}{2}$; $y = \dfrac{y_1 + y_2}{2} = \dfrac{3 + (-7)}{2} = \dfrac{-4}{2} = -2$; midpoint: $\left(\frac{5}{2}, -2\right)$

11. $m = \dfrac{\Delta y}{\Delta x} = \dfrac{8-5}{5-2} = \dfrac{3}{3} = 1$

12. $m = \dfrac{\Delta y}{\Delta x} = \dfrac{12-(-2)}{6-(-3)} = \dfrac{14}{9}$

13. $m = \dfrac{\Delta y}{\Delta x} = \dfrac{-6-4}{-5-(-3)} = \dfrac{-10}{-2} = 5$

14. $m = \dfrac{\Delta y}{\Delta x} = \dfrac{-9-(-4)}{-6-5} = \dfrac{-5}{-11} = \dfrac{5}{11}$

© 2015 Cengage Learning. All Rights Reserved. May not be scanned, copied or duplicated, or posted to a publicly accessible website, in whole or in part.

15. $m = \dfrac{\Delta y}{\Delta x} = \dfrac{4-4}{8-(-2)} = \dfrac{0}{10} = 0$

16. $m = \dfrac{\Delta y}{\Delta x} = \dfrac{8-(-4)}{-5-(-5)} = \dfrac{12}{0}$

no defined slope

17.
$$\begin{aligned} 2x - 3y &= 18 \\ -3y &= -2x + 18 \\ \frac{-3y}{-3} &= \frac{-2x}{-3} + \frac{18}{-3} \\ y &= \tfrac{2}{3}x - 6 \Rightarrow m = \tfrac{2}{3} \end{aligned}$$

18.
$$\begin{aligned} 4x + 5y &= 10 \\ 5y &= -4x + 10 \\ \frac{5y}{5} &= \frac{-4x}{5} + \frac{10}{5} \\ y &= -\tfrac{4}{5}x + 2 \Rightarrow m = -\tfrac{4}{5} \end{aligned}$$

19.
$$\begin{aligned} -2(x-3) &= 10 \\ -2x + 6 &= 10 \\ -2x &= 4 \\ x &= -2 \end{aligned}$$

The slope is undefined.

20.
$$\begin{aligned} 2y - 7 &= 1 \\ 2y &= 8 \\ y &= 4 \end{aligned}$$

$m = 0$ (horizontal)

21. $m_1 \neq m_2 \Rightarrow$ not parallel

$m_1 \cdot m_2 = 4(-\frac{1}{4}) = -1 \Rightarrow$ perpendicular

22. $m_1 = m_2 \Rightarrow$ parallel

23. $m_1 \neq m_2 \Rightarrow$ not parallel

$m_1 \cdot m_2 = 0.5(-\frac{1}{2}) = -0.25$

$\Rightarrow$ not perpendicular $\Rightarrow$ Neither

24. $m_1 \neq m_2 \Rightarrow$ not parallel

$m_1 \cdot m_2 = 5(-0.2) = -1$

$\Rightarrow$ perpendicular

25. Let $y =$ sales and let $x =$ year of business.

$$m = \frac{\Delta y}{\Delta x} = \frac{y_2 - y_1}{x_1 - x_1} = \frac{130{,}000 - 65{,}000}{4-1} = \frac{65{,}000}{3} \approx \$21{,}666.67 \text{ per year}$$

26.
$$\begin{aligned} y - y_1 &= m(x - x_1) \\ y - 5 &= 3(x+8) \\ y - 5 &= 3x + 24 \\ y &= 3x + 29 \end{aligned}$$

27. $m = \dfrac{\Delta y}{\Delta x} = \dfrac{4-(-9)}{-2-6} = \dfrac{13}{-8} = -\dfrac{13}{8}$

$$\begin{aligned} y - y_1 &= m(x - x_1) \\ y - 4 &= -\tfrac{13}{8}(x+2) \\ y - 4 &= -\tfrac{13}{8}x - \tfrac{13}{4} \\ y &= -\tfrac{13}{8}x - \tfrac{13}{4} + 4 \\ y &= -\tfrac{13}{8}x + \tfrac{3}{4} \end{aligned}$$

© 2015 Cengage Learning. All Rights Reserved. May not be scanned, copied or duplicated, or posted to a publicly accessible website, in whole or in part.

28. Find the slope of the given line:

$3x - 2y = 7$

$-2y = -3x + 7$

$y = \frac{3}{2}x - \frac{7}{2} \Rightarrow m = \frac{3}{2}$

Use the parallel slope:

$y - y_1 = m(x - x_1)$

$y + 5 = \frac{3}{2}(x + 3)$

$y + 5 = \frac{3}{2}x + \frac{9}{2}$

$y = \frac{3}{2}x + \frac{9}{2} - 5$

$y = \frac{3}{2}x - \frac{1}{2}$

29. Find the slope of the given line:

$3x - 2y = 7$

$-2y = -3x + 7$

$y = \frac{3}{2}x - \frac{7}{2} \Rightarrow m = \frac{3}{2}$

Use the perpendicular slope:

$y - y_1 = m(x - x_1)$

$y + 5 = -\frac{2}{3}(x + 3)$

$y + 5 = -\frac{2}{3}x - 2$

$y = -\frac{2}{3}x - 7$

30. Let x represent the number of years since the copy machine was purchased, and let y represent its value. We know two points on the line: $(0, 8700)$ and $(5, 100)$.

$$m = \frac{\Delta y}{\Delta x} = \frac{8700 - 100}{0 - 5} = \frac{8600}{-5} = -1720$$

$y - y_1 = m(x - x_1)$

$y - 8700 = -1720(x - 0)$

$y = -1720x + 8700$

31. $y = 6x - 4$ **is a function**, since each value of x corresponds to exactly one value of y.

32. $y = 4 - x$ **is a function**, since each value of x corresponds to exactly one value of y.

33. $y^2 = x$ **is not a function**, since $x = 9$ corresponds to both $y = 3$ and $y = -3$.

34. $|y| = x$ **is not a function**, since $x = 2$ corresponds to both $y = 2$ and $y = -2$.

35. $f(x) = 3x + 2$

$f(6) = 3(6) + 2 = 18 + 2 = 20$

36. $g(x) = x^2 - 4$

$g(-5) = (-5)^2 - 4 = 25 - 4 = 21$

37. $g(x) = x^2 - 4$

$g(-2) = (-2)^2 - 4 = 4 - 4 = 0$

38. $f(x) = 3x + 2$

$f(5) = 3(5) + 2 = 15 + 2 = 17$

39. x can equal any value; D $= (-\infty, \infty)$

40. x can equal any value; D $= (-\infty, \infty)$

41. x can equal any value; D $= (-\infty, \infty)$

42. Since $x = 2$ makes the denominator equal to 0, the domain is $(-\infty, 2) \cup (2, \infty)$.

43. Since $x = 3$ makes the denominator equal to 0, the domain is $(-\infty, 3) \cup (3, \infty)$.

44. x can equal any value; D $= (-\infty, \infty)$

45. function; D $= (-\infty, \infty)$; R $= (-\infty, \infty)$

46. Since a vertical line can pass through the graph more than once, it is not a function. D $= [-4, 4]$; R $= [-2, 2]$

47. Since a vertical line can pass through the graph more than once, it is not a function. D $= (-\infty, \infty)$; R $= [-1, \infty)$

© 2015 Cengage Learning. All Rights Reserved. May not be scanned, copied or duplicated, or posted to a publicly accessible website, in whole or in part.

CHAPTER 2 REVIEW

48. function; D $= (-\infty, \infty)$; R $= [-3, \infty)$

49. $y = 4x + 2 \Rightarrow$ This is a linear function $(m = 4, b = 2)$.

50. $y = \dfrac{x+5}{4} \Rightarrow y = \dfrac{1}{4}x + \dfrac{5}{4} \Rightarrow$ This is a linear function $\left(m = \dfrac{1}{4}, b = \dfrac{5}{4}\right)$.

51. $4x - 3y = 12 \Rightarrow y = \frac{4}{3}x - 4 \Rightarrow$ This is a linear function $\left(m = \frac{4}{3}, b = -4\right)$.

52. $y = x^2 - 10 \Rightarrow$ This is not a linear function because the exponent on x is 2.

53. $f(x) = x^2 - 3$
Shift $y = x^2$ down 3.

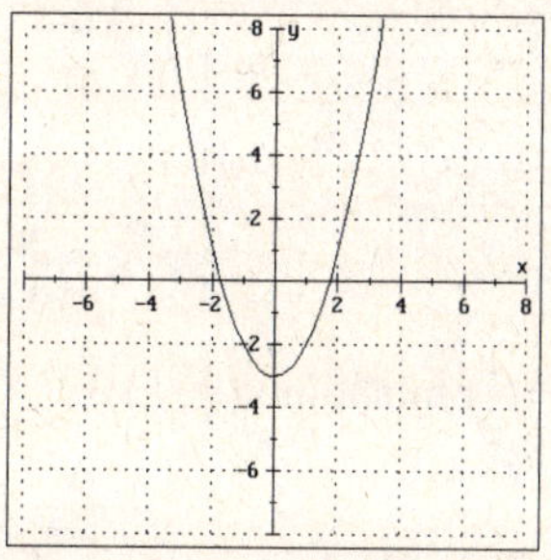

54. $f(x) = |x| - 4$
Shift $y = |x|$ down 4.

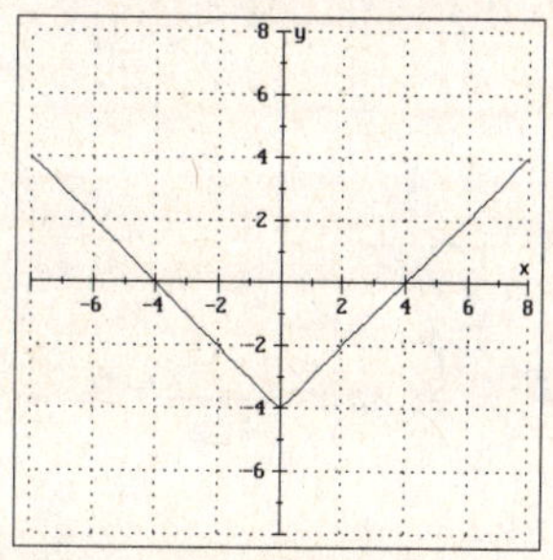

55. $f(x) = (x-2)^3$
Shift $y = x^3$ right 2.

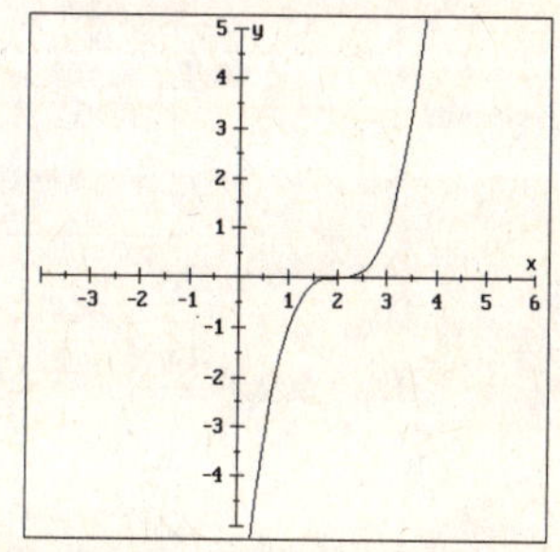

56. $f(x) = (x+4)^2 - 3$
Shift $y = x^2$ down 3, left 4.

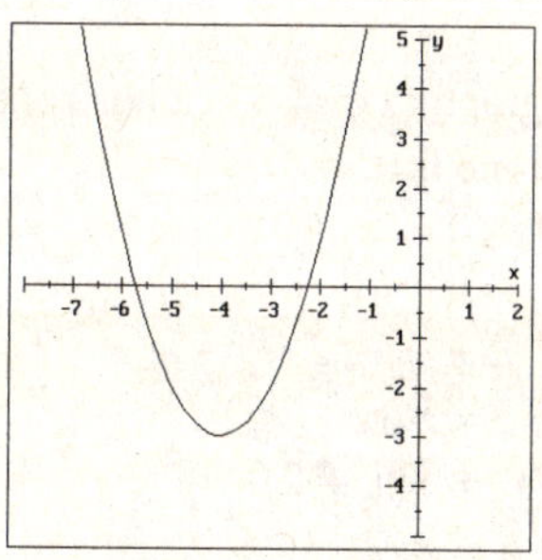

57-60. Compare your graphs to the graphs in numbers **53-56.**

61. $f(x) = -|x - 3|$

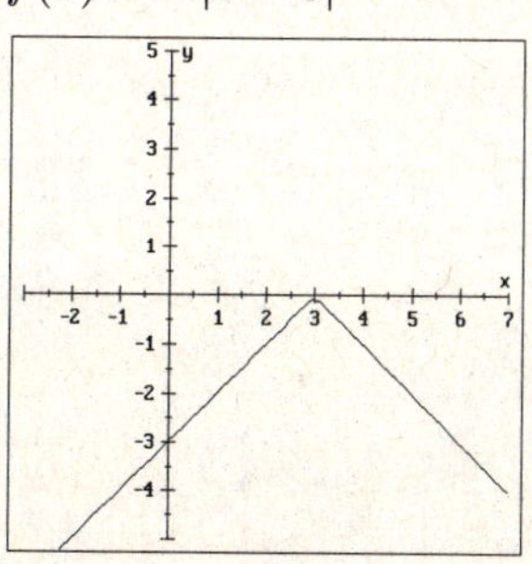

Domain $= (-\infty, \infty)$

Range $= (-\infty, 0]$

© 2015 Cengage Learning. All Rights Reserved. May not be scanned, copied or duplicated, or posted to a publicly accessible website, in whole or in part.

Chapter 2 Test (page 160)

1.
$$\begin{aligned} 2x - 5y &= 10 \\ 2(0) - 5y &= 10 \\ -5y &= 10 \\ y &= -2 \end{aligned}$$
$(0, -2)$

$$\begin{aligned} 2x - 5y &= 10 \\ 2x - 5(0) &= 10 \\ 2x &= 10 \\ x &= 5 \end{aligned}$$
$(5, 0)$

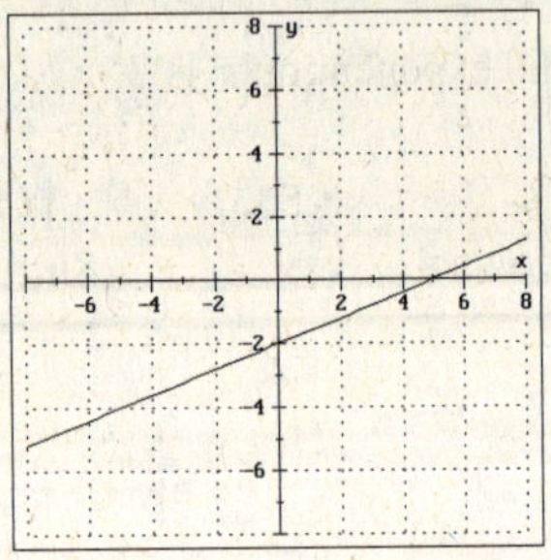

2.
$$\begin{aligned} y &= \frac{x+2}{7} \\ 0 &= \frac{x+2}{7} \\ 0 &= x+2 \\ -2 &= x \end{aligned}$$
x-intercept: $(-2, 0)$

$$\begin{aligned} y &= \frac{x+2}{7} \\ y &= \frac{0+2}{7} \\ y &= \frac{2}{7} \end{aligned}$$
y-intercept: $(0, \frac{2}{7})$

3. $x = \frac{x_1 + x_2}{2} = \frac{-3+4}{2} = \frac{1}{2}$; $y = \frac{y_1 + y_2}{2} = \frac{3+(-2)}{2} = \frac{1}{2}$; midpoint: $(\frac{1}{2}, \frac{1}{2})$

4. $m = \frac{\Delta y}{\Delta x} = \frac{4-8}{-2-6} = \frac{-4}{-8} = \frac{1}{2}$

5.
$$\begin{aligned} 2x - 3y &= 8 \\ -3y &= -2x + 8 \\ y &= \frac{2}{3}x - \frac{8}{3} \Rightarrow m = \frac{2}{3} \end{aligned}$$

6. The graph of $x = 5$ is a vertical line, so the slope is undefined.

7. The graph of $y = 5$ is a horizontal line, so the slope is 0.

8.
$$\begin{aligned} y - y_1 &= m(x - x_1) \\ y + 5 &= \frac{2}{3}(x - 4) \\ y + 5 &= \frac{2}{3}x - \frac{8}{3} \\ y &= \frac{2}{3}x - \frac{8}{3} - 5 \\ y &= \frac{2}{3}x - \frac{23}{3} \end{aligned}$$

9. $m = \frac{\Delta y}{\Delta x} = \frac{6-(-10)}{-2-(-4)} = \frac{16}{2} = 8$
$$\begin{aligned} y - y_1 &= m(x - x_1) \\ y - 6 &= 8(x + 2) \\ y - 6 &= 8x + 16 \\ y &= 8x + 22 \end{aligned}$$

© 2015 Cengage Learning. All Rights Reserved. May not be scanned, copied or duplicated, or posted to a publicly accessible website, in whole or in part.

CHAPTER 2 TEST

10.
$$\begin{aligned}-2(x-3) &= 3(2y+5)\\ -2x+6 &= 6y+15\\ 6y+15 &= -2x+6\\ 6y &= -2x-9\\ y &= \tfrac{-2}{6}x-\tfrac{9}{6}\\ y &= -\tfrac{1}{3}x-\tfrac{3}{2}\end{aligned}$$
$m=-\frac{1}{3}, b=-\frac{3}{2} \Rightarrow \left(0,-\frac{3}{2}\right)$

11.
$$\begin{aligned}x-3y &= 9\\ -3y &= -x+9\\ y &= \tfrac{1}{3}x-3 \Rightarrow m=\tfrac{1}{3}\end{aligned}$$
$y=\frac{1}{3}x+2 \Rightarrow m=\frac{1}{3}$
parallel

12. $y=-\frac{2}{3}x+4 \Rightarrow m=-\frac{2}{3}$
$2y=3x-3$
$y=\frac{3}{2}x-\frac{3}{2} \Rightarrow m=\frac{3}{2}$
perpendicular

13. $y=\frac{3}{2}x-7 \Rightarrow m=\frac{3}{2}$
Use the parallel slope:
$y-y_1=m(x-x_1)$
$y-0=\frac{3}{2}(x-0)$
$y=\frac{3}{2}x$

14. $y=-\frac{2}{3}x-7 \Rightarrow m=-\frac{2}{3}$
Use the perpendicular slope:
$y-y_1=m(x-x_1)$
$y-6=\frac{3}{2}(x+3)$
$y-6=\frac{3}{2}x+\frac{9}{2}$
$y=\frac{3}{2}x+\frac{21}{2}$

15. $|y|=x$ **is not a function**, since $x=2$ corresponds to both $y=2$ and $y=-2$.

16. Domain: $(-\infty,\infty)$; Range: $[0,\infty)$

17. Domain: $(-\infty,\infty)$; Range:$(-\infty,\infty)$

18. $f(x)=3x+1$
$f(3)=3(3)+1=9+1=10$

19. $g(x)=x^2-2$
$g(-3)=(-3)^2-2=9-2=7$

20. $f(x)=3x+1$
$f(a)=3a+1$

21. $g(x)=x^2-2$
$g(-x)=(-x)^2-2=x^2-2$

22. function

23. not a function

24. $f(x)=x^2-1$
Shift $y=x^2$ down 1.

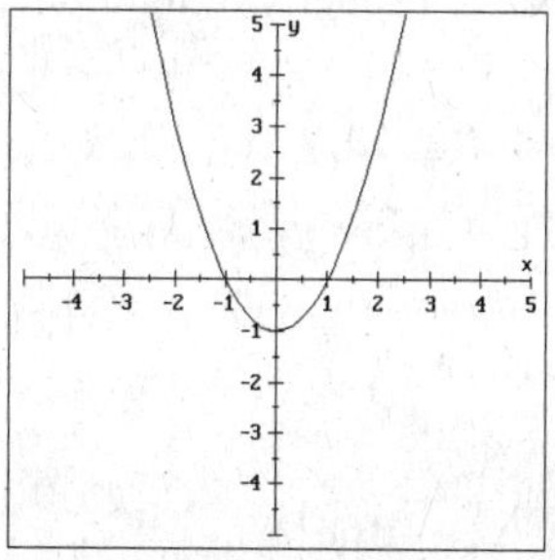

25. $f(x)=-|x+2|$
Shift $y=-|x|$ left 2.

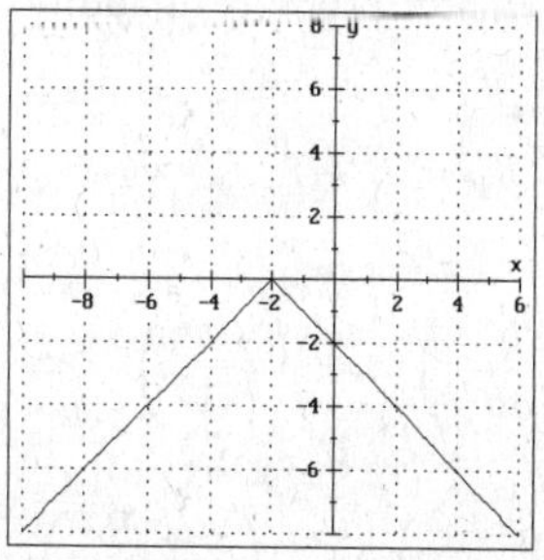

© 2015 Cengage Learning. All Rights Reserved. May not be scanned, copied or duplicated, or posted to a publicly accessible website, in whole or in part.

Cumulative Review Exercises (page 161)

1. natural numbers: $1, 2, 6, 7$

2. whole numbers: $0, 1, 2, 6, 7$

3. rational numbers: $-2, 0, 1, 2, \frac{13}{12}, 6, 7$

4. irrational numbers: $\sqrt{5}, \pi$

5. negative numbers: -2

6. real numbers: $-2, 0, 1, 2, \frac{13}{12}, 6, 7, \sqrt{5}, \pi$

7. prime numbers: $2, 7$

8. composite numbers: 6

9. even numbers: $-2, 0, 2, 6$

10. odd numbers: $1, 7$

11. $\{x| -2 < x \leq 5\} \Rightarrow$ 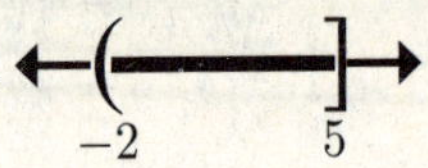

12. $[-5, 0) \cup [3, 6] \Rightarrow$ 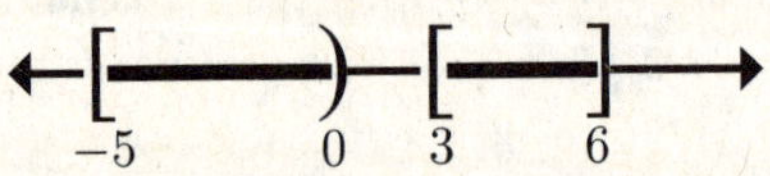

13. $-|-4| + |6| = -4 + 6 = 2$

14. $\frac{|-5| + |-3|}{-|4|} = \frac{5+3}{-4} = \frac{8}{-4} = -2$

15. $3 + 5 \cdot 4 = 3 + 20 = 23$

16. $\frac{9 - (6-3)}{6-9} = \frac{9-3}{-3} = \frac{6}{-3} = -2$

17. $20 \div (-10 \div 2) = 20 \div (-5) = -4$

18. $\frac{6 + 3(6+4)}{2(3-9)} = \frac{6 + 3(10)}{2(-6)} = \frac{6+30}{-12}$
$= \frac{36}{-12} = -3$

19. $-3x - y = -3(2) - (-3) = -6 + 3 = -3$

20. $\frac{x^2 - y^2}{2x + y} = \frac{2^2 - (-3)^2}{2(2) + (-3)} = \frac{4-9}{4-3} = \frac{-5}{1}$
$= -5$

21. associative property of addition

22. distributive property

23. commutative property of addition

24. associative property of multiplication

25. $(-2x^4y)^3 = (-2)^3(x^4)^3y^3 = -8x^{12}y^3$

26. $\frac{c^4c^8}{(c^5)^2} = \frac{c^{12}}{c^{10}} = c^2$

27. $\left(-\frac{a^3b^{-2}}{ab}\right)^{-1} = \left(-\frac{a^2}{b^3}\right)^{-1} = -\frac{b^3}{a^2}$

28. $\left(\frac{-2a^{-3}b^2}{8a^4b^{-5}}\right)^{-2} = \left(\frac{-a^{-3}b^2}{4a^4b^{-5}}\right)^{-2} = \left(\frac{4a^4b^{-5}}{-a^{-3}b^2}\right)^2 = \left(-\frac{4a^7}{b^7}\right)^2 = \frac{16a^{14}}{b^{14}}$

© 2015 Cengage Learning. All Rights Reserved. May not be scanned, copied or duplicated, or posted to a publicly accessible website, in whole or in part.

CUMULATIVE REVIEW EXERCISES

29. $0.0000382 = 3.82 \times 10^{-5}$

30. $9.32 \times 10^{8} = 932{,}000{,}000$

31.
$$\begin{aligned} 4x - 8 &= 20 \\ 4x &= 28 \\ x &= 7 \end{aligned}$$

32.
$$\begin{aligned} \frac{2x-6}{3} &= x + 7 \\ 3 \cdot \frac{2x-6}{3} &= 3(x+7) \\ 2x - 6 &= 3x + 21 \\ -x &= 27 \\ x &= -27 \end{aligned}$$

33.
$$\begin{aligned} 4(y-3) + 4 &= -3(y+5) \\ 4y - 12 + 4 &= -3y - 15 \\ 4y - 8 &= -3y - 15 \\ 7y &= -7 \\ y &= -1 \end{aligned}$$

34.
$$\begin{aligned} 2x - \frac{3(x-2)}{2} &= 7 - \frac{x-3}{3} \\ 6(2x) - 6 \cdot \frac{3x-6}{2} &= 6(7) - 6 \cdot \frac{x-3}{3} \\ 12x - 3(3x-6) &= 42 - 2(x-3) \\ 12x - 9x + 18 &= 42 - 2x + 6 \\ 3x + 18 &= 48 - 2x \\ 5x &= 30 \\ x &= 6 \end{aligned}$$

35.
$$\begin{aligned} S &= \frac{n(a+l)}{2} \\ 2S &= 2 \cdot \frac{n(a+l)}{2} \\ 2S &= n(a+l) \\ \frac{2S}{n} &= \frac{n(a+l)}{n} \\ \frac{2S}{n} &= a + l \\ \frac{2S}{n} - l &= a, \text{ or } a = \frac{2S}{n} - l \end{aligned}$$

36.
$$\begin{aligned} A &= \frac{1}{2}h(b_1 + b_2) \\ 2A &= 2 \cdot \frac{1}{2}h(b_1 + b_2) \\ 2A &= h(b_1 + b_2) \\ \frac{2A}{b_1 + b_2} &= \frac{h(b_1 + b_2)}{b_1 + b_2} \\ \frac{2A}{b_1 + b_2} &= h, \text{ or } h = \frac{2A}{b_1 + b_2} \end{aligned}$$

37. Let x represent the first even integer. Then $x + 2$ and $x + 4$ represent the others.

[1st] + [2nd] + [3rd] = [Sum of the integers]

$$\begin{aligned} x + x + 2 + x + 4 &= 90 \\ 3x + 6 &= 90 \\ 3x &= 84 \\ x &= 28 \end{aligned}$$

The integers are 28, 30 and 32.

38. Let w represent the width of the rectangle. Then $3w$ represents the length.

2 · [Length] + 2 · [Width] = [Perimeter]

$$\begin{aligned} 2(3w) + 2w &= 112 \\ 6w + 2w &= 112 \\ 8w &= 112 \\ w &= 14 \end{aligned}$$

The dimensions are 14 cm by 42 cm.

© 2015 Cengage Learning. All Rights Reserved. May not be scanned, copied or duplicated, or posted to a publicly accessible website, in whole or in part.

CUMULATIVE REVIEW EXERCISES

39. $2x - 3y = 6$

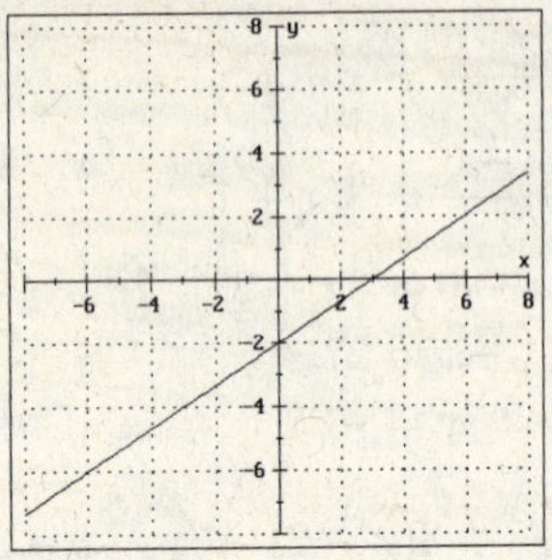

The equation defines a function.

40. $m = \dfrac{\Delta y}{\Delta x} = \dfrac{5-(-9)}{-2-8} = \dfrac{14}{-10} = -\dfrac{7}{5}$

41. Find the slope and use slope-intercept form:

$$m = \frac{\Delta y}{\Delta x} = \frac{5-(-9)}{-2-8} = \frac{14}{-10} = -\frac{7}{5}$$

$$\begin{aligned} y - y_1 &= m(x - x_1) \\ y - 5 &= -\frac{7}{5}(x+2) \\ y - 5 &= -\frac{7}{5}x - \frac{14}{5} \\ y &= -\frac{7}{5}x + \frac{11}{5} \end{aligned}$$

42. Find the slope of the given line, and use that slope to find the equation of the desired line:

$$3x + y = 8$$

$$\begin{aligned} y &= -3x + 8 \Rightarrow m = -3 \\ y - y_1 &= m(x - x_1) \\ y - 3 &= -3(x+2) \\ y - 3 &= -3x - 6 \\ y &= -3x - 3 \end{aligned}$$

43.

$$\begin{aligned} f(x) &= 3x^2 + 2 \\ f(-1) &= 3(-1)^2 + 2 \\ f(-1) &= 3(1) + 2 = 3 + 2 = 5 \end{aligned}$$

44.

$$\begin{aligned} g(x) &= 2x - 1 \\ g(8) &= 2(8) - 1 = 16 - 1 = 15 \end{aligned}$$

45.

$$\begin{aligned} g(x) &= 2x - 1 \\ g(t) &= 2t - 1 \end{aligned}$$

46.

$$\begin{aligned} f(x) &= 3x^2 + 2 \\ f(-r) &= 3(-r)^2 + 2 = 3r^2 + 2 \end{aligned}$$

47. $y = -x^2 + 1$

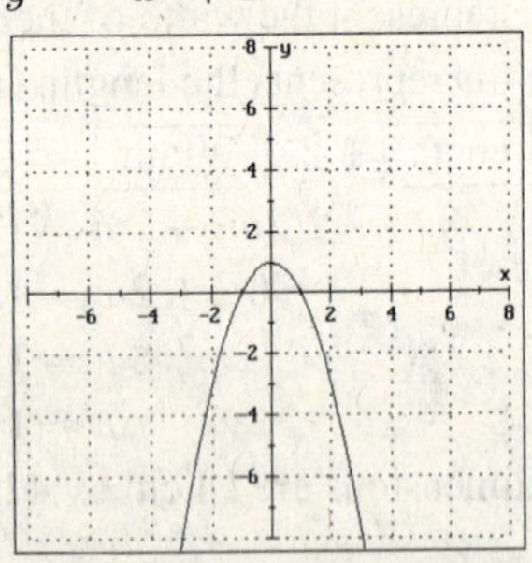

The equation describes a function.
domain: $(-\infty, \infty)$; range: $(-\infty, 1]$

48. $y = \left|\frac{1}{2}x - 3\right|$

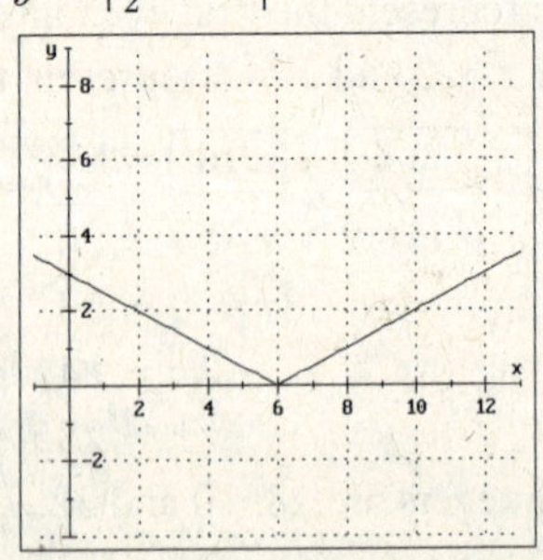

The equation describes a function.
domain: $(-\infty, \infty)$; range: $[0, \infty)$

© 2015 Cengage Learning. All Rights Reserved. May not be scanned, copied or duplicated, or posted to a publicly accessible website, in whole or in part.

Exercises 3.1 (page 172)

1.
$$y = 3x \qquad y = \frac{1}{3}x + \frac{8}{3}$$
$$3 \stackrel{?}{=} 3(1) \qquad 3 \stackrel{?}{=} \frac{1}{3}(1) + \frac{8}{3}$$
$$3 = 3 \qquad 3 \stackrel{?}{=} \frac{1}{3} + \frac{8}{3}$$
$$3 = \frac{9}{3}$$
$(1, 3)$ is a solution.

3.
$$2x - 4y = 8 \qquad 4x + y = 9$$
$$2(2) - 4(-1) \stackrel{?}{=} 8 \qquad 4(2) + (-1) \stackrel{?}{=} 9$$
$$4 + 4 \stackrel{?}{=} 8 \qquad 8 - 1 \stackrel{?}{=} 9$$
$$8 = 8 \qquad 7 \neq 9$$
$(2, -1)$ is not a solution.

5. $850{,}000{,}000 = 8.5 \times 10^8$

7. $239 \times 10^3 = 2.39 \times 10^2 \times 10^3 = 2.39 \times 10^5$

9. system

11. inconsistent

13. dependent

15. $\begin{cases} x + y = 6 \\ x - y = 2 \end{cases}$

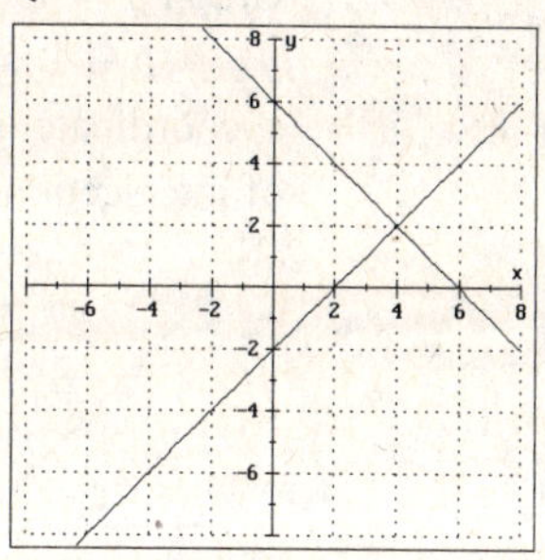

$(4, 2)$ is the solution.

17. $\begin{cases} 2x + y = 1 \\ x - 2y = -7 \end{cases}$

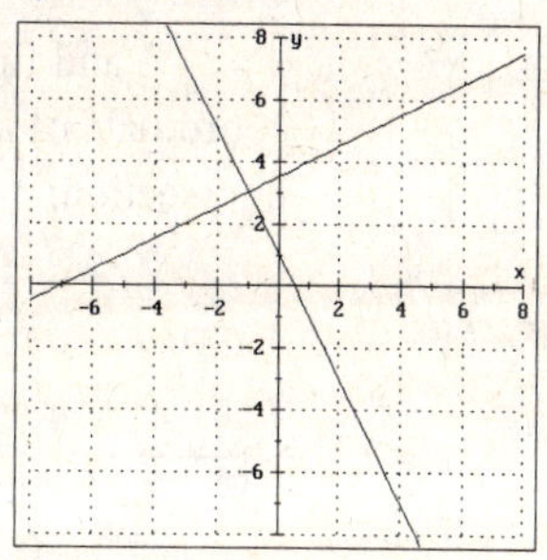

$(-1, 3)$ is the solution.

19. $\begin{cases} 2x + 3y = 0 \\ 2x + y = 4 \end{cases}$

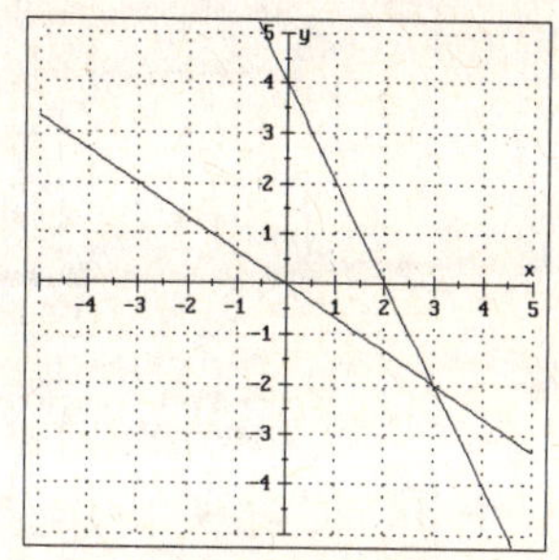

$(3, -2)$ is the solution.

21. $\begin{cases} y = 3 \\ x = 2 \end{cases}$

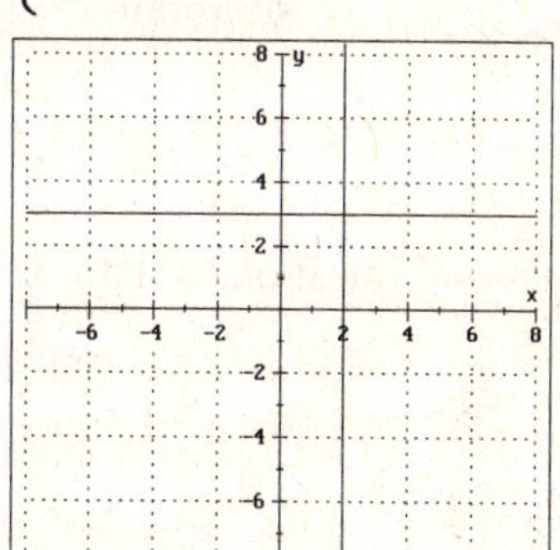

$(2, 3)$ is the solution.

23. $\begin{cases} x = 2 \\ y = \dfrac{4 - x}{2} \end{cases}$

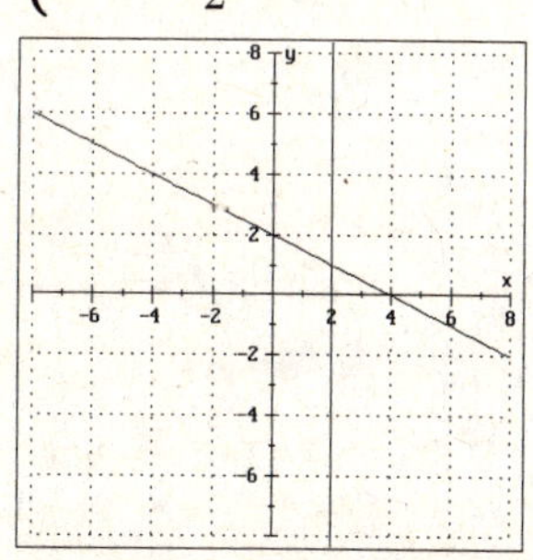

$(2, 1)$ is the solution.

25. $\begin{cases} \dfrac{5}{2}x + y = \dfrac{1}{2} \\ 2x - \dfrac{3}{2}y = 5 \end{cases}$

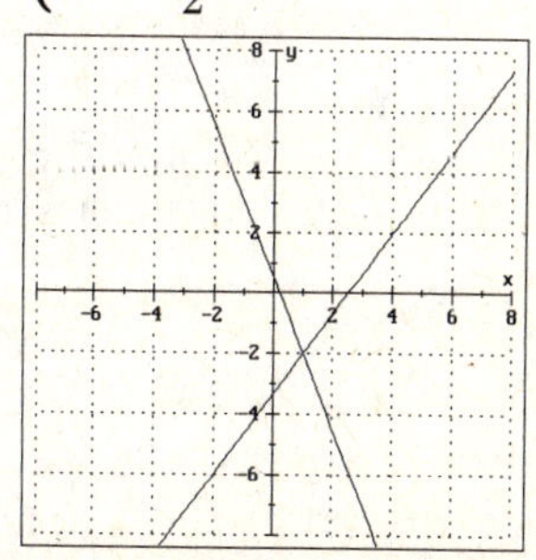

$(1, -2)$ is the solution.

© 2015 Cengage Learning. All Rights Reserved. May not be scanned, copied or duplicated, or posted to a publicly accessible website, in whole or in part.

27. $\begin{cases} 3x = 5 - 2y \\ 3x + 2y = 7 \end{cases}$

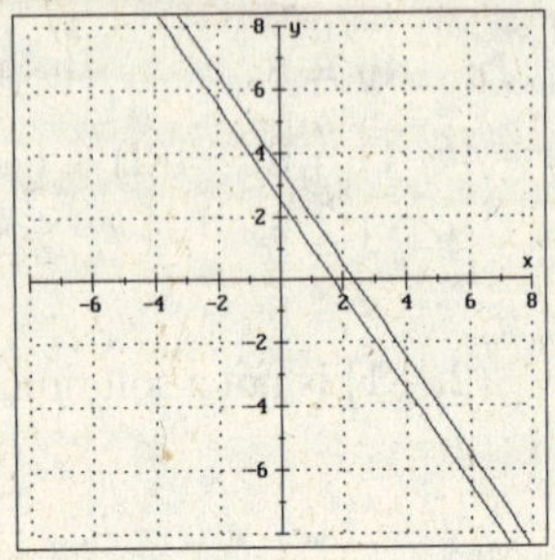

inconsistent system; $\emptyset$

29. $\begin{cases} x = 2y - 8 \\ y = \frac{1}{2}x - 5 \end{cases}$

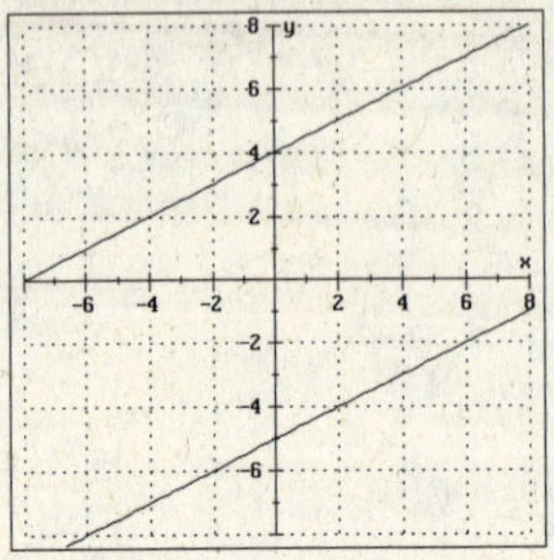

inconsistent system; $\emptyset$

31. $\begin{cases} x = 3 - 2y \\ 2x + 4y = 6 \end{cases}$

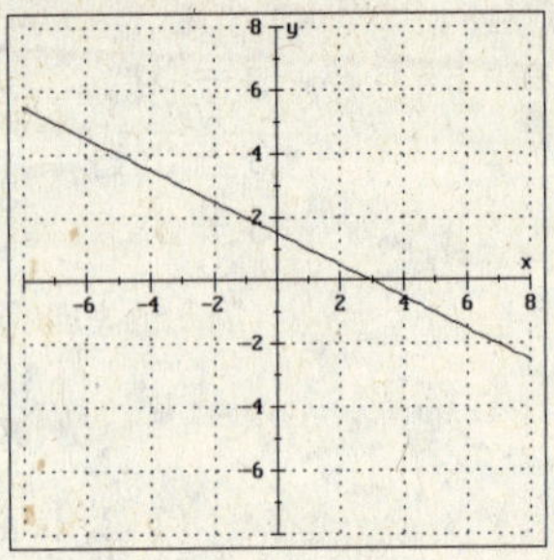

dependent system

$2x + 4y = 6$

$y = -\frac{1}{2}x + \frac{3}{2}$

$(x, -\frac{1}{2}x + \frac{3}{2})$

33. $\begin{cases} 6x + 3y = 9 \\ y + 2x = 3 \end{cases}$

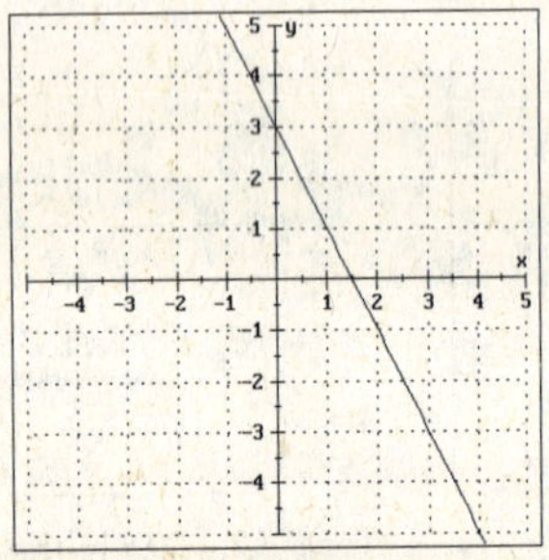

dependent system

$y + 2x = 3$

$y = -2x + 3$

$(x, -2x + 3)$

35. Graph $y = \frac{3}{2}x + 4$ and $y = -5$ and find the x-coordinate of any point of intersection.

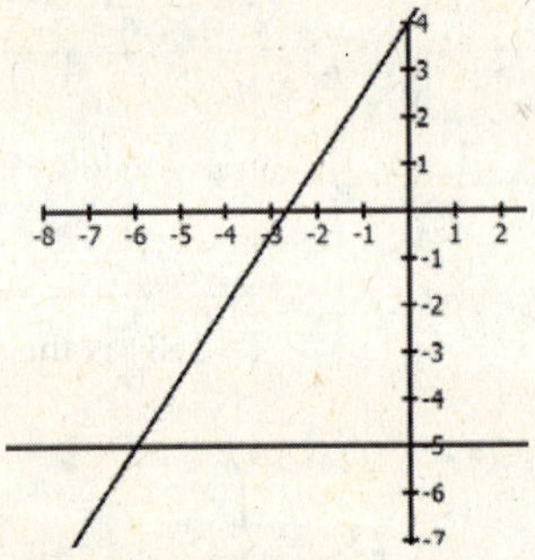

solution: $a = -6$

37. Graph $y = a + 2$ and $y = -a - 1$ and find the x-coordinate of any point of intersection.

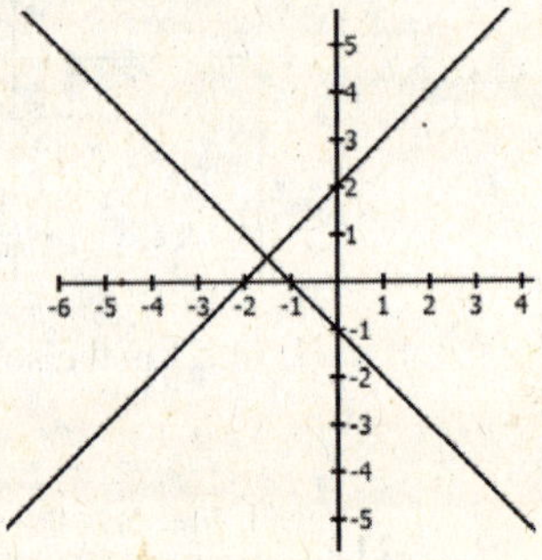

solution: $a = -\frac{3}{2}$

39. $\begin{cases} y = 5x \\ y = 5x + 3 \end{cases}$

Parallel lines: no solution

41. $\begin{cases} y = 6x + 1 \\ y = -6x + 1 \end{cases}$

Not parallel lines: one solution

© 2015 Cengage Learning. All Rights Reserved. May not be scanned, copied or duplicated, or posted to a publicly accessible website, in whole or in part.

43. $\begin{cases} x = \dfrac{11-2y}{3} \\ y = \dfrac{11-6x}{4} \end{cases}$

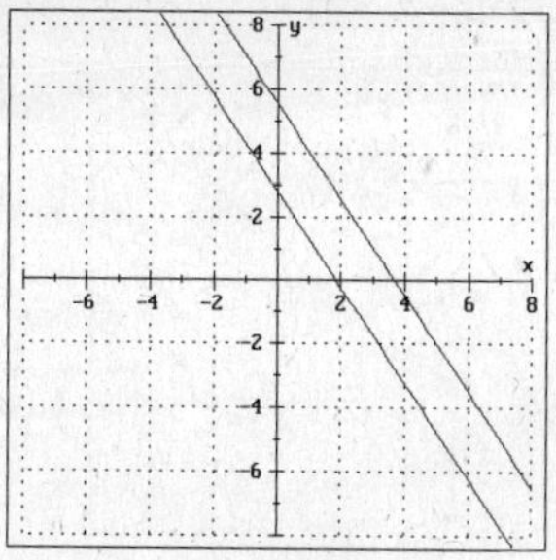

inconsistent system; $\emptyset$

45. $\begin{cases} \dfrac{5}{2}x + 3y = 6 \\ y = \dfrac{24-10x}{12} \end{cases}$

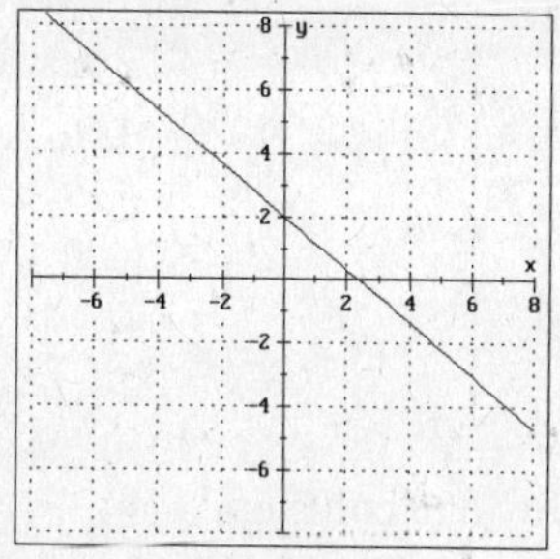

dependent system

$y = \dfrac{24-10x}{12}$

$y = 2 - \frac{5}{6}x$

$(x, 2 - \frac{5}{6}x)$

47. $\begin{cases} x = 13 - 4y \\ 3x = 4 + 2y \end{cases}$

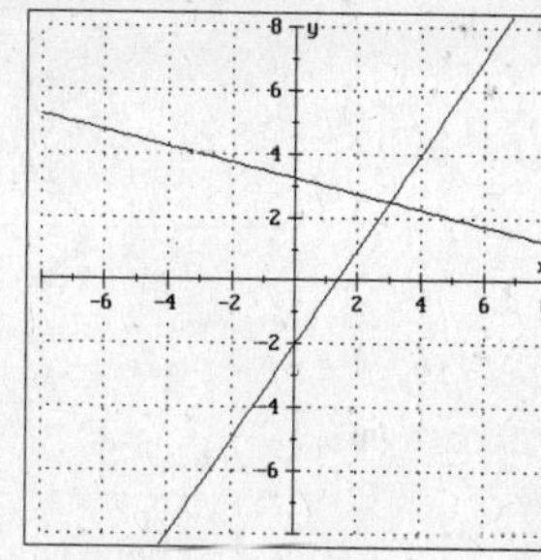

$(3, \frac{5}{2})$ is the solution.

49. $\begin{cases} x = -\dfrac{3}{2}y \\ x = \dfrac{3}{2}y - 2 \end{cases}$

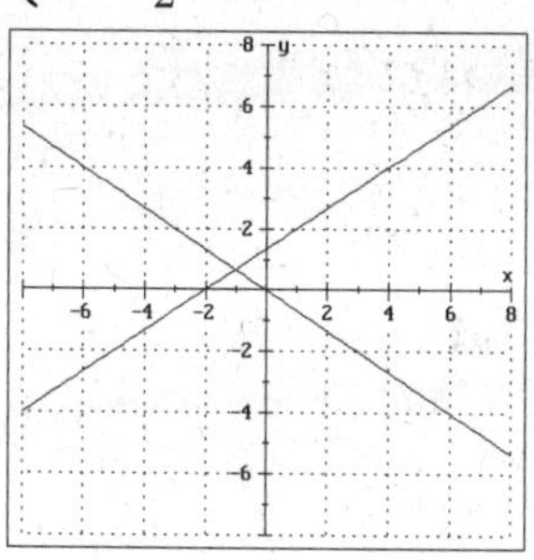

$(-1, \frac{2}{3})$ is the solution.

51. $\begin{cases} y = 3.2x - 1.5 \\ y = -2.7x - 3.7 \end{cases}$

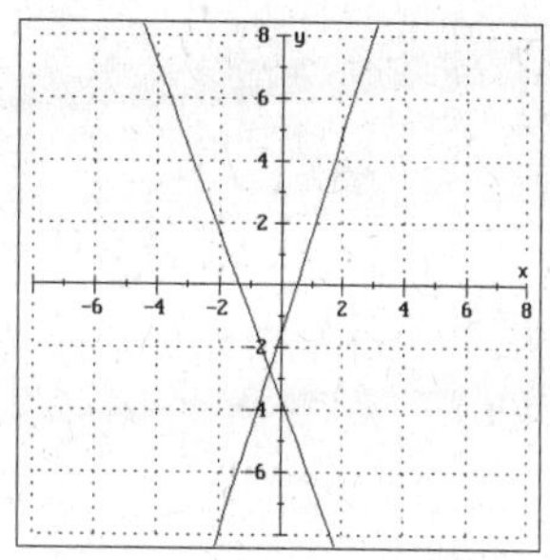

solution: $(-0.37, -2.69)$

53. $\begin{cases} 1.7x + 2.3y = 3.2 \\ y = 0.25x + 8.95 \end{cases}$

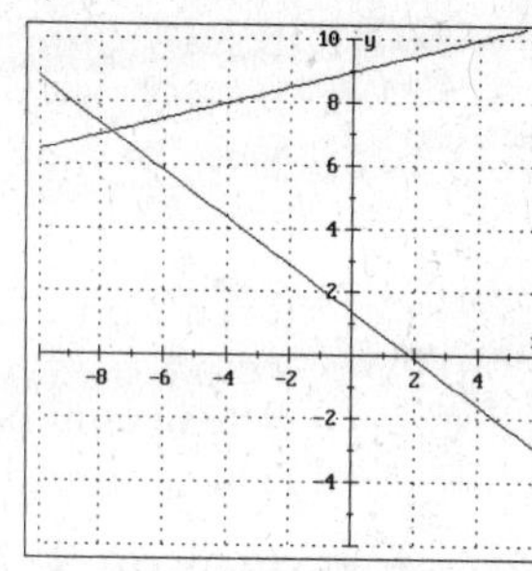

solution: $(-7.64, 7.04)$

55. **a.** The point $(15, 2.0)$ is on the graph of the cost function, so it costs \$2 million to manufacture 15,000 cameras.

b. The point $(20, 3.0)$ is on the graph of the revenue function, so there is a revenue of \$3 million when 20,000 cameras are sold.

c. The graphs of the cost function and the revenue function meet at the point $(10, 1.5)$, so the revenue and cost functions are equal for 10,000 cameras.

© 2015 Cengage Learning. All Rights Reserved. May not be scanned, copied or duplicated, or posted to a publicly accessible website, in whole or in part.

SECTION 3.1

57. $\begin{cases} 2x + 3y = 6 \\ 2x - 3y = 9 \end{cases}$

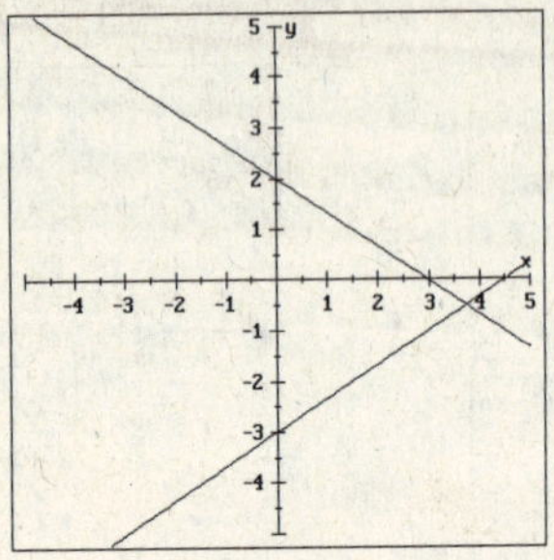

a. There is a possibility of a collision.

b. The danger point is at $(3.75, -0.5)$.

c. The collision is not certain, since the ships could be there at different times.

59. Let $x =$ the number of hours spent and let $y =$ the total cost.

$\begin{cases} y = 50x \\ y = 40x + 30 \end{cases}$

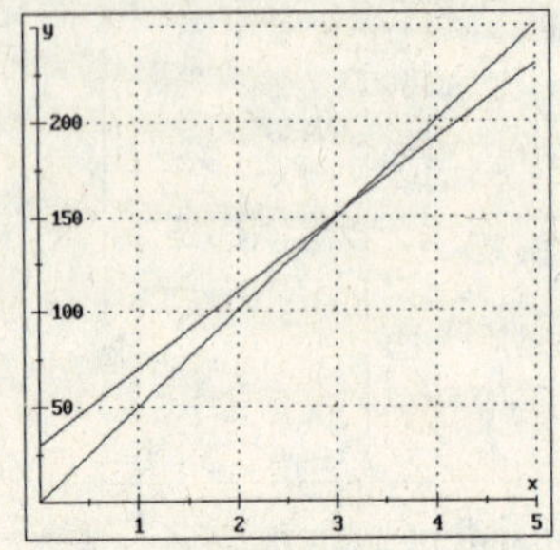

The lines meet at $(3, 150)$. Thus, the repair takes 3 hours.

61. $\begin{cases} y = 200x + 400 \\ y = 280x \end{cases}$

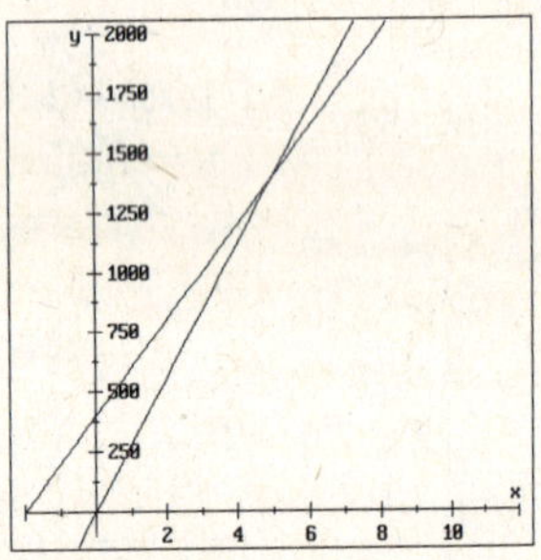

The lines meet at $(5, 1400)$.

63. **Answers may vary.**

65. One possible answer:

$\begin{cases} x + y = -3 \\ x - y = -7 \end{cases}$

Exercises 3.2 (page 185)

1.
$$\begin{aligned} 3x - 4y &= 7 \\ 3x - 4(x-2) &= 7 \\ 3x - 4x + 8 &= 7 \\ -x &= -1 \\ x &= 1 \end{aligned}$$

3.
$$\begin{array}{rcl} 4x - 2y &=& 6 \\ 3x + 2y &=& 8 \\ \hline 7x \qquad &=& 14 \end{array}$$

5.
$$\begin{aligned} (a^3b^4)^2(ab^2)^3 &= (a^3)^2(b^4)^2a^3(b^2)^3 \\ &= a^6b^8a^3b^6 = a^9b^{14} \end{aligned}$$

7.
$$\begin{aligned} \left(\frac{-3x^3y^4}{x^{-5}y^3}\right)^{-4} &= \left(\frac{x^{-5}y^3}{-3x^3y^4}\right)^4 \\ &= \left(\frac{1}{-3x^8y}\right)^4 = \frac{1}{81x^{32}y^4} \end{aligned}$$

9. setup; unit

11. parallelogram

13. Opposite

© 2015 Cengage Learning. All Rights Reserved. May not be scanned, copied or duplicated, or posted to a publicly accessible website, in whole or in part.

SECTION 3.2

15. $\begin{cases} (1) & y = 3x \\ (2) & x + y = 8 \end{cases}$

Substitute $y = 3x$ from (1) into (2):

$$\begin{aligned} x + \boldsymbol{y} &= 8 \\ x + \boldsymbol{3x} &= 8 \\ 4x &= 8 \\ x &= 2 \end{aligned}$$

Substitute this and solve for y:

$y = 3x = 3(2) = 6$

Solution: $(2, 6)$

17. $\begin{cases} (1) & x - y = 2 \\ (2) & 2x + y = 13 \end{cases}$

Substitute $x = y + 2$ from (1) into (2):

$$\begin{aligned} 2\boldsymbol{x} + y &= 13 \\ 2(\boldsymbol{y+2}) + y &= 13 \\ 2y + 4 + y &= 13 \\ 3y &= 9 \\ y &= 3 \end{aligned}$$

Substitute this and solve for x:

$x = y + 2 = 3 + 2 = 5$

Solution: $(5, 3)$

19. $\frac{x}{2} + \frac{y}{2} = \ \ 6 \Rightarrow \times 2 \ \ x + y = \ \ 12$

$\frac{x}{2} - \frac{y}{2} = -2 \Rightarrow \times 2 \ \ x - y = -4$

$\begin{cases} (1) & x + y = 12 \\ (2) & x - y = -4 \end{cases}$

Substitute $x = -y + 12$ from (1) into (2):

$$\begin{aligned} \boldsymbol{x} - y &= -4 \\ \boldsymbol{-y + 12} - y &= -4 \\ -2y + 12 &= -4 \\ -2y &= -16 \\ y &= 8 \end{aligned}$$

Substitute this and solve for x:

$x = -y + 12 = -8 + 12 = 4$

Solution: $(4, 8)$

21. $x = \frac{2}{3}y \quad \Rightarrow \times 3 \quad 3x = 2y$

$y = \ 4x + 5$

$\begin{cases} (1) & 3x = 2y \\ (2) & y = 4x + 5 \end{cases}$

Substitute $y = 4x + 5$ from (2) into (1):

$$\begin{aligned} 3x &= 2\boldsymbol{y} \\ 3x &= 2(\boldsymbol{4x + 5}) \\ 3x &= 8x + 10 \\ -5x &= 10 \\ x &= -2 \end{aligned}$$

Substitute this and solve for y:

$y = 4x + 5 = 4(-2) + 5 = -3$

Solution: $(-2, -3)$

23.

$$\begin{array}{rcl} 2x - y &=& -5 \\ \underline{2x + y} &=& \underline{-3} \\ 4x &=& -8 \\ x &=& -2 \end{array}$$

Substitute and solve for y:

$$\begin{aligned} 2x + y &= -3 \\ 2(-2) + y &= -3 \\ -4 + y &= -3 \\ y &= 1 \end{aligned}$$

The solution is $(-2, 1)$.

25.

$$\begin{array}{lll} 3x + 5y = \ \ 2 \Rightarrow \times (3) & 9x + 15y = \ \ 6 & \\ 4x - 3y = 22 \Rightarrow \times (5) & \underline{20x - 15y = 110} & \\ & 29x \qquad = 116 & \\ & x \qquad\ \ = \ \ 4 & \end{array}$$

$$\begin{aligned} 3x + 5y &= 2 \\ 3(4) + 5y &= 2 \\ 5y &= -10 \\ y &= -2 \end{aligned}$$

Solution: $\boxed{(4, -2)}$

27.

$$\begin{array}{ll} 5x - 2y = 19 \Rightarrow \times (2) & 10x - 4y = 38 \\ 3x + 4y = \ \ 1 & \underline{\ \ 3x + 4y = \ \ 1} \\ & 13x \qquad = 39 \\ & x \qquad\ \ = \ \ 3 \end{array}$$

$$\begin{aligned} 3x + 4y &= 1 \\ 3(3) + 4y &= 1 \\ 4y &= -8 \\ y &= -2 \end{aligned}$$

Solution: $\boxed{(3, -2)}$

© 2015 Cengage Learning. All Rights Reserved. May not be scanned, copied or duplicated, or posted to a publicly accessible website, in whole or in part.

SECTION 3.2

29.
$$\begin{array}{lll} 4x+6y=5 \Rightarrow \times(-2) & -8x-12y=-10 & 4x+6y=5 \\ \underline{8x-9y=3} & \underline{8x-9y=3} & 4x+6\left(\tfrac{1}{3}\right)=5 \\ & -21y=-7 & 4x+2=5 \\ & y=\tfrac{-7}{-21}=\tfrac{1}{3} & 4x=3 \\ & & x=\tfrac{3}{4} \end{array}$$

Solution: $\boxed{\left(\frac{3}{4},\frac{1}{3}\right)}$

31.
$$\begin{array}{lll} \frac{5}{6}x+\frac{2}{3}y=\frac{7}{6} \Rightarrow \times 6 & 5x+4y=7 \Rightarrow \times 7 & 35x+28y=49 \\ \underline{\frac{10}{7}x-\frac{4}{9}y=\frac{17}{21}} \Rightarrow \times 63 & \underline{90x-28y=51} \Rightarrow & \underline{90x-28y=51} \\ & & 125x=100 \\ & & x=\frac{100}{125}=\frac{4}{5} \end{array}$$

$$\begin{aligned} 5x+4y&=7 \\ 5\cdot\frac{4}{5}+4y&=7 \\ 4+4y&=7 \\ 4y&=3 \\ y&=\tfrac{3}{4} \Rightarrow \text{Solution: } \boxed{\left(\tfrac{4}{5},\tfrac{3}{4}\right)} \end{aligned}$$

33.
$$\begin{array}{llll} \frac{3}{4}x+\frac{2}{3}y=7 \Rightarrow \times 12 & 9x+8y=84 \Rightarrow \times 2 & 18x+16y=168 & 6x-5y=180 \\ \underline{\frac{3}{5}x-\frac{1}{2}y=18} \Rightarrow \times 10 & \underline{6x-5y=180} \Rightarrow \times(-3) & \underline{-18x+15y=-540} & 6x-5(-12)=180 \\ & & 31y=-372 & 6x+60=180 \\ & & y=-12 & 6x=120 \\ & & & x=20 \end{array}$$

Solution: $\boxed{(20,-12)}$

35.
$$\begin{array}{lll} 3x=2y-4 \Rightarrow 3x-2y=-4 \Rightarrow \times(-2) & -6x+4y=8 \\ \underline{6x-4y=-4} \Rightarrow \underline{6x-4y=-4} \Rightarrow & \underline{6x-4y=-4} \\ & 0\neq 4 \end{array}$$

$\boxed{\text{No solution, } \emptyset}$

37.
$$\begin{array}{ll} x=5y+2 \Rightarrow x-5y=2 \Rightarrow \times(-3) & -3x+15y=-6 \\ \underline{3x=15y+10} \Rightarrow \underline{3x-15y=10} \Rightarrow & \underline{3x-15y=10} \\ & 0\neq 4 \end{array}$$

$\boxed{\text{No solution, } \emptyset}$

39.
$$\begin{array}{ll} 12x=4y+6 \Rightarrow 12x-4y=6 \Rightarrow & 12x-4y=6 \\ \underline{6x-2y=3} \Rightarrow \underline{6x-2y=3} \Rightarrow \times(-2) & \underline{-12x+4y=-6} \\ & 0=0 \end{array}$$

$6x-2y=3 \Rightarrow y=3x-\frac{3}{2}$

$\boxed{\text{Dependent equations } \left(x,3x-\tfrac{3}{2}\right)}$

41.
$$\begin{array}{ll} y-2x=6 \Rightarrow -2x+y=6 \Rightarrow \times(2) & -4x+2y=12 \\ \underline{4x+12=2y} \Rightarrow \underline{4x-2y=-12} \Rightarrow & \underline{4x-2y=-12} \\ & 0=0 \end{array}$$

$y-2x=6 \Rightarrow y=2x+6$

$\boxed{\text{Dependent equations } (x,2x+6)}$

© 2015 Cengage Learning. All Rights Reserved. May not be scanned, copied or duplicated, or posted to a publicly accessible website, in whole or in part.

43. Let $x = 0.6666\overline{6}$. Then $10x = 6.6666\overline{6}$.

$$\begin{aligned} 10x &= 6.6666\overline{6} \\ x &= 0.6666\overline{6} \\ \hline 9x &= 6 \\ \frac{9x}{9} &= \frac{6}{9} \\ x &= \tfrac{2}{3} \end{aligned}$$

The fraction is $\frac{2}{3}$.

45. Let $x = -0.348989\overline{89}$.

Then $100x = -34.898989\overline{89}$.

$$\begin{aligned} 100x &= -34.898989\overline{89} \\ x &= -0.348989\overline{89} \\ \hline 99x &= -34.55 \\ \frac{99x}{99} &= -\frac{34.55}{99} \\ x &= -\frac{34.55(100)}{99(100)} \\ x &= -\frac{3455}{9900} = -\frac{691}{1980} \end{aligned}$$

The fraction is $-\frac{691}{1980}$.

47. $\begin{cases} (1) & y = 2x \\ (2) & x + y = 6 \end{cases}$

Substitute $y = 2x$ from (1) into (2):

$$\begin{aligned} x + \boldsymbol{y} &= 6 \\ x + \boldsymbol{2x} &= 6 \\ 3x &= 6 \\ x &= 2 \end{aligned}$$

Substitute this and solve for y:

$y = 2x = 2(2) = 4$

Solution: (2, 4)

49. $\begin{cases} (1) & 3x - 4y = 9 \\ (2) & x + 2y = 8 \end{cases}$

Substitute $x = -2y + 8$ from (2) into (1):

$$\begin{aligned} 3\boldsymbol{x} - 4y &= 9 \\ 3(\boldsymbol{-2y + 8}) - 4y &= 9 \\ -6y + 24 - 4y &= 9 \\ -10y &= -15 \\ y &= \tfrac{3}{2} \end{aligned}$$

Substitute this and solve for x:

$$x = -2y + 8 = -2\left(\frac{3}{2}\right) + 8 = 5$$

Solution: $(5, \frac{3}{2})$

51.

$$\begin{aligned} x - y &= 6 \\ x + y &= 2 \\ \hline 2x \quad &= 8 \\ x \quad &= 4 \end{aligned}$$

Substitute and solve for y:

$$\begin{aligned} x + y &= 2 \\ 4 + y &= 2 \\ y &= -2 \end{aligned}$$

The solution is $(4, -2)$.

The solution is $(4, -2)$.

53.

$$\begin{aligned} 2x + 2y &= -1 \Rightarrow \times(-2) & -4x - 4y &= 2 \\ 3x + 4y &= 0 & 3x + 4y &= 0 \\ \hline & & -x \quad &= 2 \\ & & x \quad &= -2 \end{aligned}$$

$$\begin{aligned} 2x + 2y &= -1 \\ 2(-2) + 2y &= -1 \\ -4 + 2y &= -1 \\ 2y &= 3 \\ y &= \tfrac{3}{2} \end{aligned}$$

Solution: $\boxed{(-2, \frac{3}{2})}$

© 2015 Cengage Learning. All Rights Reserved. May not be scanned, copied or duplicated, or posted to a publicly accessible website, in whole or in part.

55.
$$\begin{array}{l} \frac{2}{5}x-\frac{1}{6}y=\frac{7}{10} \Rightarrow \times 30 \quad 12x-5y=21 \Rightarrow \times 3 \quad 36x-15y=63 \\ \frac{3}{4}x-\frac{2}{3}y=\frac{19}{8} \Rightarrow \times 24 \quad 18x-16y=57 \Rightarrow \times(-2) \quad -36x+32y=-114 \\ \hline 17y=-51 \\ y=-3 \end{array}$$

$$\begin{aligned} 12x-5y&=21 \\ 12x-5(-3)&=21 \\ 12x+15&=21 \\ 12x&=6 \\ x&=\tfrac{6}{12}=\tfrac{1}{2} \Rightarrow \text{Solution: } \boxed{\left(\tfrac{1}{2},-3\right)} \end{aligned}$$

57.
$$\begin{array}{lll} 3x+5y=-14 \Rightarrow \times(3) & 9x+15y=-42 & 3x+5y=-14 \\ 2x-3y=16 \Rightarrow \times(5) & 10x-15y=80 & 3(2)+5y=-14 \\ & \overline{19x=38} & 5y=-20 \\ & x=2 & y=-4 \end{array}$$

Solution: $\boxed{(2,-4)}$

For problems #59-61, begin each problem by letting $m=\frac{1}{x}$ and $n=\frac{1}{y}$. Solve for m and n, and then solve for x and y.

59.
$$\begin{array}{l} \frac{1}{x}+\frac{1}{y}=\frac{5}{6} \Rightarrow m+n=\frac{5}{6} \\ \frac{1}{x}-\frac{1}{y}=\frac{1}{6} \Rightarrow m-n=\frac{1}{6} \\ \hline 2m=\frac{6}{6} \\ 2m=1 \\ m=\frac{1}{2} \end{array}$$

Solve for n:
$$m+n=\frac{5}{6}, \quad \frac{1}{2}+n=\frac{5}{6}, \quad n=\frac{5}{6}-\frac{1}{2}, \quad n=\frac{1}{3}$$

Solve for x:
$$m=\frac{1}{x}, \quad \frac{1}{2}=\frac{1}{x}, \quad 2=x$$

Solve for y:
$$n=\frac{1}{y}, \quad \frac{1}{3}=\frac{1}{y}, \quad 3=y$$

Solution: $\boxed{(2,3)}$

61.
$$\begin{array}{l} \frac{1}{x}+\frac{2}{y}=-1 \Rightarrow m+2n=-1 \Rightarrow \quad m+2n=-1 \\ \frac{2}{x}-\frac{1}{y}=-7 \Rightarrow 2m-n=-7 \Rightarrow \times 2 \quad 4m-2n=-14 \\ \hline 5m=-15 \\ m=-3 \end{array}$$

Solve for n:
$$m+2n=-1, \quad -3+2n=-1, \quad 2n=2, \quad n=1$$

Solve for x:
$$m=\frac{1}{x}, \quad -3=\frac{1}{x}, \quad -\frac{1}{3}=x$$

Solve for y:
$$n=\frac{1}{y}, \quad 1=\frac{1}{y}, \quad 1=y$$

Solution: $\boxed{\left(-\tfrac{1}{3},1\right)}$

© 2015 Cengage Learning. All Rights Reserved. May not be scanned, copied or duplicated, or posted to a publicly accessible website, in whole or in part.

63. Let $x =$ the cost of the pair of shoes and $y =$ the cost of the sweater.

$$\begin{array}{lll} (1)\ x+y=110 & x+\boldsymbol{y}=110 & y=x+20 \\ (2)\quad\ y=x+20 & x+\boldsymbol{x+20}=110 & y=45+20 \\ & 2x=90 & y=65 \\ & x=45 & \end{array}$$

The sweater cost \$65.

65.

$$\begin{array}{lll} (1)\ R_1+R_2=1375 & \boldsymbol{R_1}+R_2=1375 & R_1=R_2+125 \\ (2)\quad\ R_1=R_2+125 & \boldsymbol{R_2+125}+R_2=1375 & R_1=625+125 \\ & 2R_2=1250 & R_1=750 \\ & R_2=625 & \end{array}$$

The resistances are $R_1 = 750$ ohms and $R_2 = 625$ ohms.

67. Let $l =$ the length of the field and $w =$ the width of the field.

$$\begin{array}{lll} 2w+2l=72 \Rightarrow \times(-1) & -2w-2l=-72 & 2w+2l=72 \\ \underline{3w+2l=88} \Rightarrow & \underline{3w+2l=\ \ 88} & 2(16)+2l=72 \\ & w\ \ =\ \ 16 & 32+2l=72 \\ & & 2l=40 \\ & & l=20 \end{array}$$

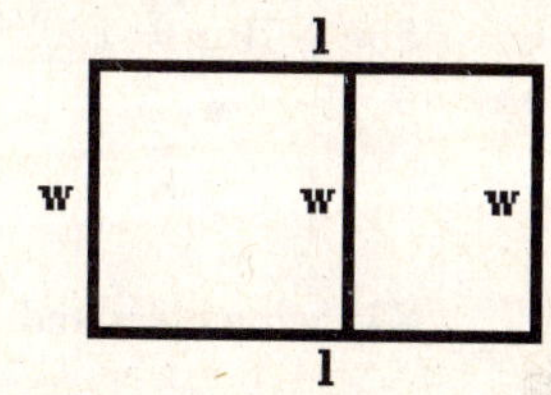

The dimensions of the field are 20 meters by 16 meters.

69. Let $x =$ the amount invested at 10% and $y =$ the amount invested at 12%.

$$\begin{array}{lll} x+\quad y=8000 \Rightarrow & x+\quad y=\ 8000 \Rightarrow \times(-10) & -10x-10y=-80000 \\ \underline{0.10x+0.12y=\ 900} \Rightarrow \times 100 & \underline{10x+12y=90000} \Rightarrow & \underline{10x+12y=\ \ 90000} \\ & & 2y=\ \ 10000 \\ & & y=\ \ 5000 \end{array}$$

$$\begin{array}{l} x+y=8000 \\ x+5000=8000 \end{array}$$

$x = 3000 \Rightarrow$ \$3000 was invested at 10%, while \$5000 was invested at 12%.

71. Let $x =$ the # of ounces of the 8% solution and $y =$ the # of ounces of the 15% solution.

$$\begin{array}{lll} x+\quad y=\ 100 \Rightarrow & x+\quad y=\ 100 \Rightarrow \times(-8) & -8x-\ 8y=-800 \\ \underline{0.08x+0.15y=12.2} \Rightarrow \times 100 & \underline{8x+15y=1220} \Rightarrow & \underline{8x+15y=\ 1220} \\ & & 7y=\ \ 420 \\ & & y=\ \ 60 \end{array}$$

$$\begin{array}{l} x+y=100 \\ x+60=100 \end{array}$$

$x = 40 \Rightarrow$ 40 oz of the 8% and 60 oz of the 15% solution should be used.

73. Let $c =$ the speed of the car and $p =$ the speed of the plane. Remember: distance $=$ rate $\cdot$ time, so time $= \dfrac{\text{distance}}{\text{rate}}$. Form one equation from the fact that the car travels 50 miles in the same time that the plane travels 180 miles:

continued on next page...

© 2015 Cengage Learning. All Rights Reserved. May not be scanned, copied or duplicated, or posted to a publicly accessible website, in whole or in part.

73. continued

Car time = Plane time $\Rightarrow \dfrac{50}{c} = \dfrac{180}{p} \Rightarrow 50p = 180c \Rightarrow 50p - 180c = 0.$

Form a second equation from the relationship given between the rates: $p = c + 143.$

$$\begin{aligned} &(1) \quad 50p - 180c = 0 \\ &(2) \quad p = c + 143 \end{aligned} \qquad \begin{aligned} 50p - 180c &= 0 \\ 50(\boldsymbol{c+143}) - 180c &= 0 \\ 50c + 7150 - 180c &= 0 \\ -130c &= -7150 \\ c &= 55 \end{aligned}$$

The car's speed is 55 mph.

75. Let r = the number of racing bikes and m = the number of mountain bikes.

$$\begin{aligned} 60r + 90m = 15900 &\Rightarrow \times(-7) \\ 55r + 70m = 13075 &\Rightarrow \times 9 \end{aligned} \qquad \begin{aligned} -420r - 630m &= -111300 \\ 495r + 630m &= 117675 \\ \hline 75r &= 6375 \\ r &= 85 \end{aligned} \qquad \begin{aligned} 60r + 90m &= 15900 \\ 60(85) + 90m &= 15900 \\ 5100 + 90m &= 15900 \\ 90m &= 10800 \\ m &= 120 \end{aligned}$$

85 racing bikes and 120 mountain bikes can be built.

77. Let x = the number of plates produced at the break-even point. Let C_1 = the cost of the first machine: $C_1 = 300 + 2x$. Let C_2 = the cost of the second machine: $C_2 = 500 + x$. To find the break point, find x such that $C_1 = C_2$:

$$\begin{aligned} C_1 &= C_2 \\ 300 + 2x &= 500 + x \\ x &= 200 \Rightarrow \text{The break point is 200 plates.} \end{aligned}$$

79. Let x = the number of computers sold at the break-even point. Let C = the costs of the store: $C = 8925 + 850x$. Let R = the revenue of the store: $R = 1275x$. To find the break-even point, find x such that $C = R$:

$$\begin{aligned} C &= R \\ 8925 + 850x &= 1275x \\ 8925 &= 425x \\ 21 &= x \Rightarrow \text{The break-even point is 21 computers.} \end{aligned}$$

81. Let x = the number of pieces of software sold at the break-even point. Let C = the costs of the business: $C = 18375 + 5.45x$. Let R = the revenue of the business: $R = 29.95x$. To find the break-even point, find x such that $C = R$:

$$\begin{aligned} C &= R \\ 18375 + 5.45x &= 29.95x \\ 18375 &= 24.50x \\ 750 &= x \Rightarrow \text{The break-even point is 750 pieces of software.} \end{aligned}$$

© 2015 Cengage Learning. All Rights Reserved. May not be scanned, copied or duplicated, or posted to a publicly accessible website, in whole or in part.

SECTION 3.2

83. Let $x =$ the number of gallons of paint A sold at the break-even point. Let $C =$ the costs of the company: $C = 32500 + 13x$. Let $R =$ the revenue of the business: $R = 18x$. To find the break-even point, find x such that $C = R$:

$$\begin{aligned} C &= R \\ 32500 + 13x &= 18x \\ 32500 &= 5x \\ 6500 &= x \end{aligned}$$

$\Rightarrow$ The break-even point is 6500 gallons of paint A (per month).

85. Calculate the profit made using each process (profit = revenue − cost)
A: revenue − cost $= 18(6000) - (32500 + 13 \cdot 6000) = 108000 - 110500 = -2500$
B: revenue − cost $= 18(6000) - (80600 + 5 \cdot 6000) = 108000 - 110600 = -2600$
Since the loss is less with process A, process A should be used.

87. Let $x =$ the number of pumps using process A sold at the break-even point. Let $C =$ the costs of the company: $C = 12390 + 29x$. Let $R =$ the revenue of the business: $R = 50x$. To find the break-even point, find x such that $C = R$:

$$\begin{aligned} C &= R \\ 12390 + 29x &= 50x \\ 12390 &= 21x \\ 590 &= x \end{aligned}$$

$\Rightarrow$ The break-even point is 590 pumps using process A (per month).

89. Calculate the profit made using each process (profit = revenue − cost)
A: revenue − cost $= 550(50) - (12390 + 550 \cdot 29) = 27500 - 28340 = -840$
B: revenue − cost $= 550(50) - (20460 + 550 \cdot 17) = 27500 - 29810 = -2310$
Since the loss is less with process A, process A should be used.

91. Calculate the profit made using each process (profit = revenue − cost)
A: revenue − cost $= 650(50) - (12390 + 650 \cdot 29) = 32500 - 31240 = 1260$
B: revenue − cost $= 650(50) - (20460 + 650 \cdot 17) = 32500 - 31510 = 990$
Since the profit is more with process A, process A should be used.

93. Let $x =$ the measure of the first angle and $y =$ the measure of the second angle.

$$\begin{aligned} x + y &= 180 \\ x - y &= 110 \\ \hline 2x &= 290 \\ x &= 145 \end{aligned} \qquad \begin{aligned} x + y &= 180 \\ 145 + y &= 180 \\ y &= 35 \end{aligned}$$

The angles have measures of 35° and 145°.

95. $\angle A$ and $\angle B$ are supplementary $\Rightarrow 2x + y + 3x = 180 \Rightarrow 5x + y = 180$.
$\angle A$ and $\angle D$ are supplementary $\Rightarrow 2x + y + y = 180 \Rightarrow 2x + 2y = 180$.

$$\begin{aligned} 5x + y &= 180 \Rightarrow \times(-2) \\ 2x + 2y &= 180 \Rightarrow \end{aligned} \qquad \begin{aligned} -10x - 2y &= -360 \\ 2x + 2y &= 180 \\ \hline -8x &= -180 \\ x &= 22.5 \end{aligned}$$

Solve for y:

$$\begin{aligned} 2x + 2y &= 180 \\ 2(22.5) + 2y &= 180 \\ 45 + 2y &= 180 \\ 2y &= 135 \\ y &= 67.5 \end{aligned}$$

© 2015 Cengage Learning. All Rights Reserved. May not be scanned, copied or duplicated, or posted to a publicly accessible website, in whole or in part.

97. Let $x =$ the measure of the angle of the range of motion and $y =$ the measure of the second angle.

$$\begin{array}{llll} (1) & x + y = 90 & \boldsymbol{x} + y = 90 & x = 4y \\ (2) & x = 4y & \boldsymbol{4y} + y = 90 & x = 4(18) \\ & & 5y = 90 & x = 72 \\ & & y = 18 & \end{array}$$

The range of motion is 72°.

99. Set $X_L = X_C$:

$$\begin{aligned} X_L &= X_C \\ 2\pi f L &= \frac{1}{2\pi f C} \\ 2\pi f L \cdot 2\pi f C &= \frac{1}{2\pi f C} \cdot 2\pi f C \\ 4\pi^2 f^2 LC &= 1 \\ \frac{4\pi^2 f^2 LC}{4\pi^2 LC} &= \frac{1}{4\pi^2 LC} \\ f^2 &= \frac{1}{4\pi^2 LC} \end{aligned}$$

101-103. Answers may vary.

Exercises 3.3 (page 196)

1.

$$\begin{array}{ccc} 2x + y - 3z = 0 & 3x - 2y + 4z = 5 & 4x + 2y - 6z = 0 \\ 2(1) + 1 - 3(1) \stackrel{?}{=} 0 & 3(1) - 2(1) + 4(1) \stackrel{?}{=} 5 & 4(1) + 2(1) - 6(1) \stackrel{?}{=} 0 \\ 2 + 1 - 3 \stackrel{?}{=} 0 & 3 - 2 + 4 \stackrel{?}{=} 5 & 4 + 2 - 6 \stackrel{?}{=} 0 \\ 0 = 0 & 5 = 5 & 0 = 0 \end{array}$$

$(1, 1, 1)$ is a solution to the system.

3. $m = \dfrac{\Delta y}{\Delta x} = \dfrac{3 - (-5)}{2 - (-1)} = \dfrac{8}{3}$

5. $f(0) = 2(0)^2 + 1 = 2(0) + 1 = 0 + 1 = 1$

7. $f(3s) = 2(3s)^2 + 1 = 2(9s^2) + 1 = 18s^2 + 1$

9. plane

11. infinitely

13.

$$\begin{array}{ccc} x + y - z = 2 & 2x + y + 3z = 7 & x - \frac{1}{3}y + \frac{2}{3}z = 1 \\ 1 + 2 - 1 \stackrel{?}{=} 2 & 2(1) + 2 + 3(1) \stackrel{?}{=} 7 & 1 - \frac{1}{3}(2) + \frac{2}{3}(1) \stackrel{?}{=} 1 \\ 2 = 2 & 2 + 2 + 3 \stackrel{?}{=} 7 & 1 - \frac{2}{3} + \frac{2}{3} \stackrel{?}{=} 1 \\ & 7 = 7 & 1 = 1 \end{array}$$

$(1, 2, 1)$ is a solution to the system.

© 2015 Cengage Learning. All Rights Reserved. May not be scanned, copied or duplicated, or posted to a publicly accessible website, in whole or in part.

SECTION 3.3

15.
$$\begin{array}{ll} (1) & x+y+z=4 \\ (2) & 2x+y-z=1 \\ (3) & 2x-3y+z=1 \end{array} \quad \begin{array}{ll} (1) & x+\ y+z=4 \\ (2) & \underline{2x+\ y-z=1} \\ (4) & 3x+2y \quad\;\; =5 \end{array} \quad \begin{array}{ll} (2) & 2x+\ y-z=1 \\ (3) & \underline{2x-3y+z=1} \\ (5) & 4x-2y \quad\;\; =2 \end{array}$$

$$\begin{array}{ll} (4) & 3x+2y=5 \\ (5) & \underline{4x-2y=2} \\ & 7x \qquad =7 \\ & x \qquad\; =1 \end{array} \quad \begin{array}{r} 3x+2y=5 \\ 3(1)+2y=5 \\ 3+2y=5 \\ 2y=2 \\ y=1 \end{array} \quad \begin{array}{r} x+y+z=4 \\ 1+1+z=4 \\ 2+z=4 \\ z=2 \end{array}$$

$\boxed{\text{The solution is } (1,1,2).}$

17.
$$\begin{array}{ll} (1) & 4x+3z=4 \\ (2) & 2y-6z=-1 \\ (3) & 8x+4y+3z=9 \end{array} \quad \begin{array}{ll} (2) & 2y-6z=-1 \\ 2\cdot(1) & \underline{8x \qquad +6z=\ \ 8} \\ (4) & 8x+2y \qquad =\ \ 7 \end{array} \quad \begin{array}{ll} (2) & 2y-6z=-1 \\ 2\cdot(3) & \underline{16x+\ 8y+6z=\ 18} \\ (5) & 16x+10y \qquad =\ 17 \end{array}$$

$$\begin{array}{l} 8x+\ 2y=\ 7 \Rightarrow \times(-2) \\ \underline{16x+10y=17 \Rightarrow} \end{array} \quad \begin{array}{r} -16x-\ 4y=-14 \\ \underline{16x+10y=\ \ 17} \\ 6y=\ \ 3 \\ y=\ \ \frac{1}{2} \end{array} \quad \begin{array}{r} 8x+2y=7 \\ 8x+2\left(\frac{1}{2}\right)=7 \\ 8x+1=7 \\ 8x=6 \\ x=\frac{3}{4} \end{array} \quad \begin{array}{r} 4x+3z=4 \\ 4\left(\frac{3}{4}\right)+3z=4 \\ 3+3z=4 \\ 3z=1 \\ z=\frac{1}{3} \end{array}$$

Solution: $\boxed{\left(\frac{3}{4},\frac{1}{2},\frac{1}{3}\right)}$

19.
$$\begin{array}{ll} (1) & 3a-2b-c=4 \\ (2) & 6a-4b-2c=10 \\ (3) & a+3b+c=2 \end{array} \quad \begin{array}{ll} -2\cdot(1) & -6a+4b+2c=-8 \\ (2) & \underline{\ 6a-4b-2c=\ 10} \\ (4) & 0=\ \ 2 \end{array}$$

Since equation (4) is always false, there is no solution. The system is inconsistent. $\boxed{\emptyset}$

21.
$$\begin{array}{ll} (1) & 7x-2y-z=1 \\ (2) & 9x-6y+z=7 \\ (3) & x-2y+z=3 \end{array} \quad \begin{array}{ll} -3\cdot(1) & -21x+6y+3z=-3 \\ (2) & \underline{\ 9x-6y+\ z=\ \ 7} \\ (4) & -12x \qquad +4z=\ \ 4 \end{array} \quad \begin{array}{ll} (2) & 9x-6y+\ z=\ \ 7 \\ -3\cdot(3) & \underline{-3x+6y-3z=-9} \\ (5) & 6x \qquad -2z=-2 \end{array}$$

$$\begin{array}{l} -12x+4z=\ \ 4 \Rightarrow \\ \underline{\ 6x-2z=-2 \Rightarrow \times 2} \end{array} \quad \begin{array}{r} -12x+4z\ =\ \ 4 \\ \underline{12x-4z\ =-4} \\ 0=\ \ 0 \end{array}$$

Dependent equations

If $x=x$, then

$$\begin{array}{r} 6x \quad 2z - -2 \\ -2z=-6x-2 \\ z=3x+1 \end{array} \quad \begin{array}{r} x-2y+z=3 \\ x-2y+3x+1=3 \\ -2y=-4x+2 \\ y=2x-1 \end{array}$$

$\boxed{(x,2x-1,3x+1)}$

© 2015 Cengage Learning. All Rights Reserved. May not be scanned, copied or duplicated, or posted to a publicly accessible website, in whole or in part.

SECTION 3.3

23.

$$\begin{array}{lrcl} (1) & x + 2y + z = 1 & 2\cdot(1) & 2x + 4y + 2z = \ \ 2 \\ (2) & 3x + y + 3z = 3 & (3) & \underline{-2x + \ \ y - 2z = -2} \\ (3) & -2x + y - 2z = -2 & (4) & 5y \qquad = \ \ 0 \\ & & & y \qquad = \ \ 0 \end{array}$$

$$\begin{array}{llllll} (1) & x + 2y + z = 1 & (1) & x + 2(0) + z = 1 & (1) & x + z = 1 \\ (2) & 3x + y + 3z = 3 & (2) & 3x + 0 + 3z = 3 & (2) & 3x + 3z = 3 \\ (3) & -2x + y - 2z = -2 & (3) & -2x + 0 - 2z = -2 & (3) & -2x - 2z = -2 \end{array}$$

$$\begin{array}{lr} 2\cdot(1) & 2x + 2z = \ \ 2 \\ (3) & \underline{-2x - 2z = -2} \\ (4) & 0 = \ \ 0 \end{array}$$

Dependent equations: $x + z = 1 \Rightarrow z = 1 - x$

Solution: $\boxed{(x, 0, 1 - x)}$

25.

$$\begin{array}{llllll} (1) & a + 3b + c = 10 & -3\cdot(1) & -3a - 9b - 3c = -30 & -3\cdot(2) & -9a - 3b - 3c = -36 \\ (2) & 3a + b + c = 12 & (3) & \underline{a + \ \ b + 3c = \ \ \ 8} & (3) & \underline{a + \ \ b + 3c = \ \ \ 8} \\ (3) & a + b + 3c = 8 & (4) & -2a - 8b \qquad = -22 & (5) & -8a - 2b \qquad = -28 \end{array}$$

$$\begin{array}{lrrr} -2a - 8b = -22 \Rightarrow \times(-4) & 8a + 32b = \ \ 88 & -2a - 8b = -22 & a + 3b + c = 10 \\ \underline{-8a - 2b = -28} \Rightarrow & \underline{-8a - \ \ 2b = -28} & -2a - 8(2) = -22 & 3 + 3(2) + c = 10 \\ & 30b = \ \ 60 & -2a - 16 = -22 & 3 + 6 + c = 10 \\ & b = \ \ \ 2 & -2a = -6 & c = 1 \\ & & a = 3 & \end{array}$$

Solution: $\boxed{(3, 2, 1)}$

27.

$$\begin{array}{lll} x + \frac{1}{3}y + z = 13 & \Rightarrow \times 3 & (1)\ 3x + y + 3z = 39 \\ \frac{1}{2}x - y + \frac{1}{3}z = -2 & \Rightarrow \times 6 & (2)\ 3x - 6y + 2z = -12 \\ x + \frac{1}{2}y - \frac{1}{3}z = 2 & \Rightarrow \times 6 & (3)\ 6x + 3y - 2z = 12 \end{array}$$

$$\begin{array}{llll} (2) & 3x - 6y + \ \ 2z = -12 & (2) & 3x - 6y + 2z = -12 \\ 6\cdot(1) & \underline{18x + 6y + 18z = \ \ 234} & 2\cdot(3) & \underline{12x + 6y - 4z = \ \ \ 24} \\ (4) & 21x \qquad + 20z = \ \ 222 & (5) & 15x \qquad - 2z = \ \ \ 12 \end{array}$$

$$\begin{array}{lrrr} 21x + 20z = 222 \Rightarrow & 21x + 20z = 222 & 15x - 2z = 12 & 3x + y + 3z = 39 \\ \underline{15x - \ \ 2z = \ \ 12} \Rightarrow \times 10 & \underline{150x - 20z = 120} & 15(2) - 2z = 12 & 3(2) + y + 3(9) = 39 \\ & 171x \qquad = 342 & 30 - 2z = 12 & 6 + y + 27 = 39 \\ & x \qquad = \ \ \ 2 & -2z = -18 & y = 6 \\ & & z = 9 & \end{array}$$

Solution: $\boxed{(2, 6, 9)}$

© 2015 Cengage Learning. All Rights Reserved. May not be scanned, copied or duplicated, or posted to a publicly accessible website, in whole or in part.

SECTION 3.3

29.
$$\begin{array}{ll} (1) & x-3y+z=1 \\ (2) & 2x-y-2z=2 \\ (3) & x+2y-3z=-1 \end{array} \quad \begin{array}{rr} -2\cdot(1) & -2x+6y-2z=-2 \\ (2) & \underline{2x-\ \ y-2z=\ \ 2} \\ (4) & 5y-4z=\ \ 0 \end{array} \quad \begin{array}{rr} -2\cdot(3) & -2x-4y+6z=2 \\ (2) & \underline{2x-\ \ y-2z=2} \\ (5) & -5y+4z=4 \end{array}$$

$$\begin{array}{r} 5y-4z\ =0 \\ \underline{-5y+4z\ =4} \\ 0\neq 4 \end{array}$$
Since this equation is always false, there is no solution. **The system is inconsistent.**

31.
$$\begin{array}{ll} (1) & 2x+3y+4z=6 \\ (2) & 2x-3y-4z=-4 \\ (3) & 4x+6y+8z=12 \end{array} \quad \begin{array}{l} (1)\ 2x+3y+4z=\ \ 6 \\ (2)\ \underline{2x-3y-4z=-4} \\ (4)\ 4x \qquad\quad =\ 2 \\ \qquad x \qquad\quad =\ \tfrac{1}{2} \end{array}$$

Substitute $x=\frac{1}{2}$ into (1):
$$\begin{aligned} 2\left(\tfrac{1}{2}\right)+3y+4z&=6 \\ 1+3y+4z&=6 \\ y&=\tfrac{5}{3}-\tfrac{4}{3}z \end{aligned}$$

Since (3) is a multiple of (1), the equations are dependent: $\boxed{\left(\frac{1}{2},\frac{5}{3}-\frac{4}{3}z,z\right)}$

33.
$$\begin{array}{ll} (1) & 2x+2y+3z=10 \\ (2) & 3x+y-z=0 \\ (3) & x+y+2z=6 \end{array} \quad \begin{array}{rr} (1) & 2x+2y+3z=10 \\ 3\cdot(2) & \underline{9x+3y-3z=\ \ 0} \\ (4) & 11x+5y \qquad =10 \end{array} \quad \begin{array}{rr} (3) & x+\ \ y+2z=6 \\ 2\cdot(2) & \underline{6x+2y-2z=0} \\ (5) & 7x+3y \qquad =6 \end{array}$$

$$\begin{array}{l} 11x+5y=10\Rightarrow\times 3 \\ \underline{7x+3y=\ \ 6\Rightarrow\times(-5)} \end{array} \quad \begin{array}{rl} 33x+15y= & 30 \\ \underline{-35x-15y=} & \underline{-30} \\ -2x \quad = & 0 \\ x \quad = & 0 \end{array} \quad \begin{aligned} 11x+5y&=10 \\ 11(0)+5y&=10 \\ 0+5y&=10 \\ 5y&=10 \\ y&=2 \end{aligned} \quad \begin{aligned} x+y+2z&=6 \\ 0+2+2z&=6 \\ 2+2z&=6 \\ 2z&=4 \\ z&=2 \end{aligned}$$

Solution: $\boxed{(0,2,2)}$

35. Let $x=$ the number of \$5 statues, $y=$ the number of \$4 statues and $z=$ the number of \$3 statues.

$$\begin{array}{lrl} (1) & x+y+z=180 & \text{(total number made)} \\ (2) & 5x+4y+3z=650 & \text{(total cost)} \\ (3) & 20x+12y+9z=2100 & \text{(total revenue)} \end{array} \quad \begin{array}{rr} -3\cdot(1) & -3x-3y-3z=-540 \\ (2) & \underline{5x+4y+3z=\ \ 650} \\ (4) & 2x+\ y \qquad =\ \ 110 \end{array}$$

$$\begin{array}{rr} -9\cdot(1) & -9x-\ \ 9y-9z=-1620 \\ (3) & \underline{20x+12y+9z=\ \ 2100} \\ (5) & 11x+\ \ 3y \qquad =\ \ 480 \end{array} \quad \begin{array}{l} 2x+\ \ y=110\Rightarrow\times(-3) \\ \underline{11x+3y=480\Rightarrow} \end{array} \quad \begin{array}{rl} -6x-3y= & -330 \\ \underline{11x+3y=} & \underline{\ \ 480} \\ 5x \quad = & 150 \\ x \quad = & 30 \end{array}$$

$$\begin{aligned} 2x+y&=110 \\ 2(30)+y&=110 \\ 60+y&=110 \\ y&=50 \end{aligned} \quad \begin{aligned} x+y+z&=180 \\ 30+50+z&=180 \\ z&=100 \end{aligned}$$

30 of the \$5, 50 of the \$4 and 100 of the \$3 statues should be made.

© 2015 Cengage Learning. All Rights Reserved. May not be scanned, copied or duplicated, or posted to a publicly accessible website, in whole or in part.

SECTION 3.3

37. Substitute the coordinates of each point for x and y in the equation $y = ax^2 + bx + c$.

$$\begin{array}{lll} y = ax^2 + bx + c & y = ax^2 + bx + c & y = ax^2 + bx + c \\ 0 = a(0)^2 + b(0) + c & -4 = a(2)^2 + b(2) + c & 0 = a(4)^2 + b(4) + c \\ 0 = c & -4 = 4a + 2b + c & 0 = 16a + 4b + c \end{array}$$

Solve the system of equations formed from the three equations:

$$\begin{array}{lllll} (1) \quad c = 0 & 4a + 2b = -4 \Rightarrow \times(-2) & -8a - 4b = 8 & 16a + 4b = 0 \\ (2) \quad 4a + 2b + c = -4 & \underline{16a + 4b = \ \ 0} \Rightarrow & \underline{16a + 4b = 0} & 16(1) + 4b = 0 \\ (3) \quad 16a + 4b + c = 0 & & 8a \quad = 8 & 16 + 4b = 0 \\ & & a \quad = 1 & 4b = -16 \\ & & & b = -4 \end{array}$$

The equation is $y = x^2 - 4x$.

39. Let $x =$ the first integer, $y =$ the second integer and $z =$ the third integer.

$$\begin{array}{llll} x + y + z = 19 & \Rightarrow \ (1) \ x + y + z = 19 & (1) & x + y + z = 19 \\ z = 2y & \Rightarrow \ (2) \ -2y + z = 0 & (3) & \underline{-x + y \quad\ = \ 5} \\ y = x + 5 & \Rightarrow \ (3) \ -x + y = 5 & (4) & 2y + z = 24 \end{array}$$

$$\begin{array}{llll} 2y + z = 24 & 2y + z = 24 & x + y + z = 19 & \text{The integers} \\ \underline{-2y + z = \ 0} & 2y + 12 = 24 & x + 6 + 12 = 19 & \text{are } 1, 6 \text{ and } 12. \\ 2z = 24 & 2y = 12 & x = 1 & \\ z = 12 & y = 6 & & \end{array}$$

41. Let A, B and C represent the measures of the three angles.

$$\begin{array}{llll} A + B + C = 180 & \Rightarrow \ (1) \ A + B + C = 180 & (1) & A + B + C = 180 \\ A = B + C - 80 & \Rightarrow \ (2) \ A - B - C = -80 & (2) & \underline{A - B - C = -80} \\ C = 2B - 50 & \Rightarrow \ (3) \ -2B + C = -50 & (4) & 2A \quad = 100 \\ & & & A \quad = 50 \end{array}$$

$$\begin{array}{llll} A + B + C = 180 & -1 \cdot (3) \quad 2B - C = \ 50 & B + C = 130 & \text{The angles have measures} \\ 50 + B + C = 180 & \underline{(5) \qquad B + C = 130} & 60 + C = 130 & \text{of } A = 50°, B = 60° \text{ and} \\ (5) \quad B + C = 130 & 3B \quad = 180 & C = 70 & C = 70°. \\ & B \quad = \ 60 & & \end{array}$$

43. Let $A =$ the units of food A, $B =$ the units of food B and $C =$ the units of food C.

$$\begin{array}{llll} (1) \quad A + 2B + 2C = 11 \quad \text{(fat)} & (1) & A + 2B + 2C = \ 11 \\ (2) \quad A + B + C = 6 \quad \text{(carbohydrate)} & -2 \cdot (2) & \underline{-2A - 2B - 2C = -12} \\ (3) \quad 2A + B + 2C = 10 \quad \text{(protein)} & (4) & -A \quad = -1 \\ & & A \quad = \ 1 \end{array}$$

$$\begin{array}{llll} (1) & A + 2B + 2C = \ 11 & -3A - 2C = -9 & A + B + C = 6 \\ -2 \cdot (3) & \underline{-4A - 2B - 4C = -20} & -3(1) - 2C = -9 & 1 + B + 3 = 6 \\ (5) & -3A \qquad - 2C = \ -9 & -3 - 2C = -9 & B + 4 = 6 \\ & & -2C = -6 & B = 2 \\ & & C = 3 & \end{array}$$

1 unit of food A, 2 units of food B and 3 units of food C should be used.

© 2015 Cengage Learning. All Rights Reserved. May not be scanned, copied or duplicated, or posted to a publicly accessible website, in whole or in part.

SECTION 3.3

45. Let $x =$ the number of \$5 tickets, $y =$ the number of \$3 tickets and $z =$ the number of \$2 tickets.

$$\begin{array}{lrl} (1) & x+y+z=750 & \text{(total sold)} \\ (2) & x=2z \Rightarrow x-2z=0 & \text{(twice as many)} \\ (3) & 5x+3y+2z=2625 & \text{(total revenue)} \end{array} \qquad \begin{array}{lrl} 2\cdot(1) & 2x+2y+2z= & 1500 \\ (2) & x \qquad -2z= & 0 \\ \hline (4) & 3x+2y \qquad = & 1500 \end{array}$$

$$\begin{array}{lrl} (2) & x \qquad -2z= & 0 \\ (3) & 5x+3y+2z= & 2625 \\ \hline (5) & 6x+3y \qquad = & 2625 \end{array} \qquad \begin{array}{r} 3x+2y=1500 \Rightarrow \times(-2) \\ \underline{6x+3y=2625 \Rightarrow} \end{array} \qquad \begin{array}{rl} -6x-4y= & -3000 \\ 6x+3y= & 2625 \\ \hline -y= & -375 \\ y= & 375 \end{array}$$

$$\begin{array}{rl} 3x+2y= & 1500 \\ 3x+2(375)= & 1500 \\ 3x+750= & 1500 \\ 3x= & 750 \\ x= & 250 \end{array} \qquad \begin{array}{rl} x+y+z= & 750 \\ 250+375+z= & 750 \\ 625+z= & 750 \\ z= & 125 \end{array}$$

250 of the \$5, 375 of the \$3 and 125 of the \$2 tickets were sold.

47. Let $x =$ the number of totem poles, $y =$ the number of bears and $z =$ the number of deer.

$$\begin{array}{lrl} (1) & 2x+2y+z=14 & \text{(carving)} \\ (2) & x+2y+2z=15 & \text{(sanding)} \\ (3) & 3x+2y+2z=21 & \text{(painting)} \end{array} \qquad \begin{array}{lrl} -2\cdot(1) & -4x-4y-2z= & -28 \\ (2) & x+2y+2z= & 15 \\ \hline (4) & -3x-2y \qquad = & -13 \end{array}$$

$$\begin{array}{lrl} -2\cdot(1) & -4x-4y-2z= & -28 \\ (3) & 3x+2y+2z= & 21 \\ \hline (5) & -x-2y \qquad = & -7 \end{array} \qquad \begin{array}{r} -3x-2y=-13 \Rightarrow \times(-1) \\ \underline{-x-2y=-7 \Rightarrow} \end{array} \qquad \begin{array}{rl} 3x+2y= & 13 \\ -x-2y= & -7 \\ \hline 2x= & 6 \\ x= & 3 \end{array}$$

$$\begin{array}{rl} 3x+2y= & 13 \\ 3(3)+2y= & 13 \\ 9+2y= & 13 \\ 2y= & 4 \\ y= & 2 \end{array} \qquad \begin{array}{rl} 2x+2y+z= & 14 \\ 2(3)+2(2)+z= & 14 \\ 6+4+z= & 14 \\ z= & 4 \end{array}$$

3 totem poles, 2 bears and 4 deer should be made.

49. Let $x =$ the % of nitrogen, $y =$ the % of oxygen and $z =$ the % of other gases.

$$\begin{array}{lrr} (1) & x+y+z=100 & \\ (2) & x=3(y+z)+12 \Rightarrow & x-3y-3z=12 \\ (3) & z=y-20 \Rightarrow & -y+z=-20 \end{array} \qquad \begin{array}{lrl} (1) & x+y+\ z= & 100 \\ (3) & -y+\ z= & -20 \\ \hline (4) & x \qquad +2z= & 80 \end{array}$$

$$\begin{array}{lrl} 3\cdot(1) & 3x+3y+3z= & 300 \\ (2) & x-3y-3z= & 12 \\ \hline (5) & 4x \qquad = & 312 \\ & x \qquad = & 78 \end{array} \qquad \begin{array}{rl} x+2z= & 80 \\ 78+2z= & 80 \\ 2z= & 2 \\ z= & 1 \end{array} \qquad \begin{array}{rl} x+y+z= & 100 \\ 78+y+1= & 100 \\ y+79= & 100 \\ y= & 21 \end{array}$$

The composition is 78% nitrogen, 21% oxygen, and 1% other gases.

© 2015 Cengage Learning. All Rights Reserved. May not be scanned, copied or duplicated, or posted to a publicly accessible website, in whole or in part.

51. Substitute the coordinates of each point for x and y in the equation $x^2+y^2+cx+dy+e=0$.

$$\begin{aligned} x^2+y^2+cx+dy+e&=0 \\ (1)^2+(3)^2+c(1)+d(3)+e&=0 \\ 1+9+c+3d+e&=0 \\ c+3d+e&=-10 \end{aligned} \qquad \begin{aligned} x^2+y^2+cx+dy+e&=0 \\ (3)^2+(1)^2+c(3)+d(1)+e&=0 \\ 9+1+3c+d+e&=0 \\ 3c+d+e&=-10 \end{aligned}$$

$$\begin{aligned} x^2+y^2+cx+dy+e&=0 \\ (1)^2+(-1)^2+c(1)+d(-1)+e&=0 \\ 1+1+c-d+e&=0 \\ c-d+e&=-2 \end{aligned} \qquad \begin{aligned} (1)\quad c+3d+e&=-10 \\ (2)\quad 3c+d+e&=-10 \\ (3)\quad c-d+e&=-2 \end{aligned}$$

$$\begin{array}{rl} (1) & c+3d+e=-10 \\ -1\cdot(2) & -3c-d-e=10 \\ \hline (4) & -2c+2d=0 \end{array} \qquad \begin{array}{rl} (1) & c+3d+e=-10 \\ -1\cdot(3) & -c+d-e=2 \\ \hline (5) & 4d=-8 \\ & d=-2 \end{array} \qquad \begin{aligned} -2c+2d&=0 \\ -2c+2(-2)&=0 \\ -2c-4&=0 \\ -2c&=4 \\ c&=-2 \end{aligned}$$

$$\begin{aligned} c+3d+e&=-10 \\ -2+3(-2)+e&=-10 \\ -2-6+e&=-10 \\ -8+e&=-10 \\ e&=-2 \end{aligned}$$

The equation is $x^2+y^2-2x-2y-2=0$.

53. **Answers may vary.**

55.

$$\begin{aligned} (1)\quad x+y+z+w&=3 \\ (2)\quad x-y-z-w&=-1 \\ (3)\quad x+y-z-w&=1 \\ (4)\quad x+y-z+w&=3 \end{aligned} \qquad \begin{array}{rl} (1) & x+y+z+w=3 \\ (2) & x-y-z-w=-1 \\ \hline & 2x=2 \\ & x=1 \end{array} \qquad \begin{array}{rl} (1) & x+y+z+w=3 \\ (3) & x+y-z-w=1 \\ \hline & 2x+2y=4 \end{array}$$

$$\begin{array}{rl} (1) & x+y+z+w=3 \\ (4) & x+y-z+w=3 \\ \hline & 2x+2y+2w=6 \end{array} \qquad \begin{aligned} 2x+2y&=4 \\ 2(1)+2y&=4 \\ 2y&=2 \\ y&=1 \end{aligned} \qquad \begin{aligned} 2x+2y+2w&=6 \\ 2(1)+2(1)+2w&=6 \\ 4+2w&=6 \\ 2w&=2 \\ w&=1 \end{aligned} \qquad \begin{aligned} x+y+z+w&=3 \\ 1+1+z+1&=3 \\ 3+z&=3 \\ z&=0 \end{aligned}$$

The solution is $(1,1,0,1)$.

Exercises 3.4 (page 205)

1. $\begin{cases} 2x-y=3 \\ x+5y=7 \end{cases}$

3. $\begin{cases} 3x-2y+4z=1 \\ 5x+2y-3z=-7 \\ -x+9y+8z=0 \end{cases}$

5. $470{,}000{,}000=4.7\times10^8$

7. $75\times10^4=7.5\times10^1\times10^4=7.5\times10^5$

9. matrix

11. 3; columns

© 2015 Cengage Learning. All Rights Reserved. May not be scanned, copied or duplicated, or posted to a publicly accessible website, in whole or in part.

13. augmented; coefficient

15. $\begin{bmatrix} 2 & -3 \\ 4 & 2 \end{bmatrix}$

17. yes

Note: The notation $3R_1 + R_3 \Rightarrow R_2$ means to multiply Row #1 of the previous matrix by 3, add that result to Row #3 of the previous matrix, and write the final result in Row #2 of the current matrix.

19. $\begin{bmatrix} 3 & 1 & 2 \\ 4 & 5 & 7 \end{bmatrix} \Rightarrow \overset{-R_1 + R_2 \Rightarrow R_2}{\begin{bmatrix} 3 & 1 & 2 \\ 1 & 4 & \boxed{5} \end{bmatrix}}$

21. $\begin{bmatrix} -5 & -1 & 3 \\ -2 & 3 & 1 \end{bmatrix} \Rightarrow \overset{2R_2 \Rightarrow R_2}{\begin{bmatrix} -5 & -1 & 3 \\ -4 & 6 & \boxed{2} \end{bmatrix}}$

23. $\left[\begin{array}{cc|c} 2 & -3 & 3 \\ 2 & -1 & 5 \end{array}\right] \Rightarrow \overset{R_1 + (-R_2) \Rightarrow R_2}{\left[\begin{array}{cc|c} 2 & -3 & 3 \\ 0 & -2 & -2 \end{array}\right]} \Rightarrow \overset{-\frac{1}{2}R_2 \Rightarrow R_2}{\left[\begin{array}{cc|c} 2 & -3 & 3 \\ 0 & 1 & 1 \end{array}\right]}$

From R_2, $y = 1$. From R_1:

$$\begin{aligned} 2x - 3y &= 3 \\ 2x - 3(1) &= 3 \\ 2x - 3 &= 3 \\ 2x &= 6 \Rightarrow x = 3 \end{aligned}$$

The solution is $(3, 1)$.

25. $\left[\begin{array}{cc|c} 1 & 2 & -2 \\ 3 & -1 & 8 \end{array}\right] \Rightarrow \overset{-3R_1 + R_2 \Rightarrow R_2}{\left[\begin{array}{cc|c} 1 & 2 & -2 \\ 0 & -7 & 14 \end{array}\right]} \Rightarrow \overset{-\frac{1}{7}R_2 \Rightarrow R_2}{\left[\begin{array}{cc|c} 1 & 2 & -2 \\ 0 & 1 & -2 \end{array}\right]}$

From R_2, $y = -2$. From R_1:

$$\begin{aligned} x + 2y &= -2 \\ x + 2(-2) &= -2 \\ x - 4 &= -2 \Rightarrow x = 2 \end{aligned}$$

The solution is $(2, -2)$.

27. $\left[\begin{array}{ccc|c} 1 & 1 & 1 & 6 \\ 1 & 2 & 1 & 8 \\ 1 & 1 & 2 & 9 \end{array}\right] \Rightarrow \overset{\substack{-R_1 + R_2 \Rightarrow R_2 \\ -R_1 + R_3 \Rightarrow R_3}}{\left[\begin{array}{ccc|c} 1 & 1 & 1 & 6 \\ 0 & 1 & 0 & 2 \\ 0 & 0 & 1 & 3 \end{array}\right]}$

From R_3, $z = 3$. From R_2, $y = 2$. From R_1:

$$\begin{aligned} x + y + z &= 6 \\ x + 2 + 3 &= 6 \\ x + 5 &= 6 \\ x &= 1 \end{aligned}$$

The solution is $(1, 2, 3)$.

© 2015 Cengage Learning. All Rights Reserved. May not be scanned, copied or duplicated, or posted to a publicly accessible website, in whole or in part.

SECTION 3.4

29. $$\begin{bmatrix} 1 & -1 & 0 & | & 1 \\ 0 & 1 & 1 & | & 1 \\ 1 & 0 & 1 & | & 2 \end{bmatrix} \overset{-R_1+R_3 \Rightarrow R_3}{\Rightarrow} \begin{bmatrix} 1 & -1 & 0 & | & 1 \\ 0 & 1 & 1 & | & 1 \\ 0 & 1 & 1 & | & 1 \end{bmatrix} \overset{-R_2+R_3 \Rightarrow R_3}{\Rightarrow} \begin{bmatrix} 1 & -1 & 0 & | & 1 \\ 0 & 1 & 1 & | & 1 \\ 0 & 0 & 0 & | & 0 \end{bmatrix}$$

From R_2:
$$y+z=1$$
$$y=1-z$$

From R_1:
$$x-y=1$$
$$x-(1-z)=1$$
$$x-1+z=1$$
$$x+z=2$$
$$x=2-z$$

The solution is $(2-z, 1-z, z)$.

31. $$\begin{bmatrix} 1 & 1 & | & 3 \\ 3 & -1 & | & 1 \\ 2 & 1 & | & 4 \end{bmatrix} \overset{\substack{-3R_1+R_2 \Rightarrow R_2 \\ -2R_1+R_3 \Rightarrow R_3}}{\Rightarrow} \begin{bmatrix} 1 & 1 & | & 3 \\ 0 & -4 & | & -8 \\ 0 & -1 & | & -2 \end{bmatrix} \overset{\substack{-\frac{1}{4}R_2 \Rightarrow R_2 \\ -4R_3+R_2 \Rightarrow R_3}}{\Rightarrow} \begin{bmatrix} 1 & 1 & | & 3 \\ 0 & 1 & | & 2 \\ 0 & 0 & | & 0 \end{bmatrix}$$

From R_2, $y=2$.

From R_1:
$$x+y=3$$
$$x+2=3$$
$$x=1$$

The solution is $(1, 2)$.

33. $$\begin{bmatrix} 2 & 1 & | & 7 \\ 1 & -1 & | & 2 \\ -1 & 3 & | & -2 \end{bmatrix} \overset{\substack{-2R_2+R_1 \Rightarrow R_2 \\ 2R_3+R_1 \Rightarrow R_3}}{\Rightarrow} \begin{bmatrix} 2 & 1 & | & 7 \\ 0 & 3 & | & 3 \\ 0 & 7 & | & 3 \end{bmatrix} \overset{\substack{\frac{1}{3}R_2 \Rightarrow R_2 \\ -\frac{7}{3}R_2+R_3 \Rightarrow R_3}}{\Rightarrow} \begin{bmatrix} 2 & 1 & | & 7 \\ 0 & 1 & | & 1 \\ 0 & 0 & | & -4 \end{bmatrix}$$

R_3: $0x+0y=-4$, or $0=-4$, which is an impossible equation. $\emptyset$

35. $$\begin{bmatrix} 1 & 3 & 2 & | & 4 \\ -1 & -2 & -1 & | & -2 \end{bmatrix} \overset{R_1+R_2 \Rightarrow R_2}{\Rightarrow} \begin{bmatrix} 1 & 3 & 2 & | & 4 \\ 0 & 1 & 1 & | & 2 \end{bmatrix}$$

From R_2:
$$y+z=2$$
$$y=2-z$$

From R_1:
$$x+3y+2z=4$$
$$x+3(2-z)+2z=4$$
$$x+6-3z+2z=4$$
$$x-z=-2$$
$$x=z-2$$

The solution is $(z-2, 2-z, z)$.

© 2015 Cengage Learning. All Rights Reserved. May not be scanned, copied or duplicated, or posted to a publicly accessible website, in whole or in part.

37. $\left[\begin{array}{ccc|c} -3 & -2 & 2 & 0 \\ -1 & 1 & -1 & -5 \end{array}\right] \overset{-3R_2+R_1 \Rightarrow R_2}{\Rightarrow} \left[\begin{array}{ccc|c} -3 & -2 & 2 & 0 \\ 0 & -5 & 5 & 15 \end{array}\right] \overset{-\frac{1}{5}R_2 \Rightarrow R_2}{\Rightarrow} \left[\begin{array}{ccc|c} -3 & -2 & 2 & 0 \\ 0 & 1 & -1 & -3 \end{array}\right]$

From R_2:

$$\begin{aligned} y - z &= -3 \\ y &= z - 3 \end{aligned}$$

From R_1:

$$\begin{aligned} -3x - 2y + 2z &= 0 \\ -3x - 2(z-3) + 2z &= 0 \\ -3x - 2z + 6 + 2z &= 0 \\ -3x &= -6 \\ x &= 2 \end{aligned}$$

The solution is $(2, z-3, z)$.

39. $\left[\begin{array}{cc|c} 5 & -4 & 8 \\ 10 & 3 & -6 \end{array}\right] \overset{-2R_1+R_2 \Rightarrow R_2}{\Rightarrow} \left[\begin{array}{cc|c} 5 & -4 & 8 \\ 0 & 11 & -22 \end{array}\right] \overset{\frac{1}{11}R_2 \Rightarrow R_2}{\Rightarrow} \left[\begin{array}{cc|c} 5 & -4 & 8 \\ 0 & 1 & -2 \end{array}\right]$

From R_2, $y = -2$.

From R_1:

$$\begin{aligned} 5x - 4y &= 8 \\ 5x - 4(-2) &= 8 \\ 5x + 8 &= 8 \\ 5x = 0 &\Rightarrow x = 0 \end{aligned}$$

The solution is $(0, -2)$.

41. $\begin{cases} 5a = 24 + 2b \\ 5b = 3a + 16 \end{cases} \Rightarrow \begin{cases} 5a - 2b = 24 \\ -3a + 5b = 16 \end{cases}$

$\left[\begin{array}{cc|c} 5 & -2 & 24 \\ -3 & 5 & 16 \end{array}\right] \overset{2R_2+R_1 \Rightarrow R_1}{\Rightarrow} \left[\begin{array}{cc|c} -1 & 8 & 56 \\ -3 & 5 & 16 \end{array}\right] \overset{-R_1 \Rightarrow R_1}{\Rightarrow} \left[\begin{array}{cc|c} 1 & -8 & -56 \\ -3 & 5 & 16 \end{array}\right] \overset{3R_1+R_2 \Rightarrow R_2}{\Rightarrow} \left[\begin{array}{cc|c} 1 & -8 & -56 \\ 0 & -19 & -152 \end{array}\right]$

$\overset{-\frac{1}{19}R_2 \Rightarrow R_2}{\Rightarrow} \left[\begin{array}{cc|c} 1 & -8 & -56 \\ 0 & 1 & 8 \end{array}\right]$

From R_2, $b = 8$.

From R_1:

$$\begin{aligned} a - 8b &= -56 \\ a - 8(8) &= -56 \\ a - 64 = -56 &\Rightarrow a = 8 \end{aligned}$$

The solution is $(8, 8)$.

43. $\left[\begin{array}{ccc|c} 3 & 1 & -3 & 5 \\ 1 & -2 & 4 & 10 \\ 1 & 1 & 1 & 13 \end{array}\right] \overset{\substack{-3R_2+R_1 \Rightarrow R_2 \\ -3R_3+R_1 \Rightarrow R_3}}{\Rightarrow} \left[\begin{array}{ccc|c} 3 & 1 & -3 & 5 \\ 0 & 7 & -15 & -25 \\ 0 & -2 & -6 & -34 \end{array}\right] \overset{\frac{2}{7}R_2+R_3 \Rightarrow R_3}{\Rightarrow} \left[\begin{array}{ccc|c} 3 & 1 & -3 & 5 \\ 0 & 7 & -15 & -25 \\ 0 & 0 & -\frac{72}{7} & -\frac{288}{7} \end{array}\right]$

From R_3:

$$\begin{aligned} -\frac{72}{7}c &= -\frac{288}{7} \\ -\frac{7}{72}\left(-\frac{72}{7}c\right) &= -\frac{7}{72}\left(-\frac{288}{7}\right) \\ c &= 4 \end{aligned}$$

From R_2:

$$\begin{aligned} 7b - 15c &= -25 \\ 7b - 15(4) &= -25 \\ 7b &= 35 \\ b &= 5 \end{aligned}$$

From R_1:

$$\begin{aligned} 3a + b - 3c &= 5 \\ 3a + 5 - 3(4) &= 5 \\ 3a - 7 &= 5 \\ 3a &= 12 \\ a &= 4 \end{aligned}$$

The solution is $(4, 5, 4)$.

© 2015 Cengage Learning. All Rights Reserved. May not be scanned, copied or duplicated, or posted to a publicly accessible website, in whole or in part.

45. $$\begin{bmatrix} 1 & 2 & 2 & | & 2 \\ 2 & 1 & -1 & | & 1 \\ 4 & 5 & 3 & | & 3 \end{bmatrix} \overset{\substack{-2R_1+R_2 \Rightarrow R_2 \\ -4R_1+R_3 \Rightarrow R_3}}{\Rightarrow} \begin{bmatrix} 1 & 2 & 2 & | & 2 \\ 0 & -3 & -5 & | & -3 \\ 0 & -3 & -5 & | & -5 \end{bmatrix} \overset{-R_2+R_3 \Rightarrow R_3}{\Rightarrow} \begin{bmatrix} 1 & 2 & 2 & | & 2 \\ 0 & -3 & -5 & | & -3 \\ 0 & 0 & 0 & | & -2 \end{bmatrix}$$

The last row indicates that $0x + 0y + 0z = -2$. This is false, so there is no solution. $\emptyset$

47. $$\begin{bmatrix} 2 & -1 & | & 4 \\ 1 & 3 & | & 2 \\ -1 & -4 & | & -2 \end{bmatrix} \overset{\substack{-2R_2+R_1 \Rightarrow R_2 \\ 2R_3+R_1 \Rightarrow R_3}}{\Rightarrow} \begin{bmatrix} 2 & -1 & | & 4 \\ 0 & -7 & | & 0 \\ 0 & -9 & | & 0 \end{bmatrix} \overset{\substack{-\frac{1}{7}R_2 \Rightarrow R_2 \\ -\frac{9}{7}R_2+R_3 \Rightarrow R_3}}{\Rightarrow} \begin{bmatrix} 2 & -1 & | & 4 \\ 0 & 1 & | & 0 \\ 0 & 0 & | & 0 \end{bmatrix}$$

From R_2, $y = 0$. From R_1:

$$\begin{aligned} 2x - y &= 4 \\ 2x - 0 &= 4 \\ 2x &= 4 \\ x &= 2 \end{aligned}$$

The solution is $(2, 0)$.

49. $$\begin{bmatrix} 1 & 3 & | & 7 \\ 1 & 1 & | & 3 \\ 3 & 1 & | & 5 \end{bmatrix} \overset{\substack{-R_2+R_1 \Rightarrow R_2 \\ -3R_1+R_3 \Rightarrow R_3}}{\Rightarrow} \begin{bmatrix} 1 & 3 & | & 7 \\ 0 & 2 & | & 4 \\ 0 & -8 & | & -16 \end{bmatrix} \overset{\substack{\frac{1}{2}R_2 \Rightarrow R_2 \\ 4R_2+R_3 \Rightarrow R_3}}{\Rightarrow} \begin{bmatrix} 1 & 3 & | & 7 \\ 0 & 1 & | & 2 \\ 0 & 0 & | & 0 \end{bmatrix}$$

From R_2, $y = 2$. From R_1:

$$\begin{aligned} x + 3y &= 7 \\ x + 3(2) &= 7 \\ x + 6 = 7 &\Rightarrow x = 1 \end{aligned}$$

The solution is $(1, 2)$.

51. $$\begin{bmatrix} 5 & -2 & | & 4 \\ 2 & -4 & | & -8 \end{bmatrix} \overset{\frac{1}{2}R_2 \Rightarrow R_2}{\Rightarrow} \begin{bmatrix} 5 & -2 & | & 4 \\ 1 & -2 & | & -4 \end{bmatrix} \overset{R_1+(-5R_2) \Rightarrow R_2}{\Rightarrow} \begin{bmatrix} 5 & -2 & | & 4 \\ 0 & 8 & | & 24 \end{bmatrix} \overset{\frac{1}{8}R_2 \Rightarrow R_2}{\Rightarrow} \begin{bmatrix} 5 & -2 & | & 4 \\ 0 & 1 & | & 3 \end{bmatrix}$$

From R_2, $y = 3$. From R_1:

$$\begin{aligned} 5x - 2y &= 4 \\ 5x - 2(3) &= 4 \\ 5x = 10 &\Rightarrow x = 2 \end{aligned}$$

The solution is $(2, 3)$.

53. $$\begin{bmatrix} 2 & 1 & | & -4 \\ 6 & 3 & | & 1 \end{bmatrix} \overset{-3R_1+R_2 \Rightarrow R_2}{\Rightarrow} \begin{bmatrix} 2 & 1 & | & -4 \\ 0 & 0 & | & 13 \end{bmatrix}$$ The last equation is impossible. $\emptyset$

© 2015 Cengage Learning. All Rights Reserved. May not be scanned, copied or duplicated, or posted to a publicly accessible website, in whole or in part.

55.
$$\begin{bmatrix} 3 & -2 & 4 & | & 4 \\ 1 & 1 & 1 & | & 3 \\ 6 & -2 & -3 & | & 10 \end{bmatrix} \overset{\substack{-3R_2+R_1 \Rightarrow R_2 \\ -2R_1+R_3 \Rightarrow R_3}}{\Rightarrow} \begin{bmatrix} 3 & -2 & 4 & | & 4 \\ 0 & -5 & 1 & | & -5 \\ 0 & 2 & -11 & | & 2 \end{bmatrix} \overset{\frac{2}{5}R_2+R_3 \Rightarrow R_3}{\Rightarrow} \begin{bmatrix} 3 & -2 & 4 & | & 4 \\ 0 & -5 & 1 & | & -5 \\ 0 & 0 & -\frac{53}{5} & | & 0 \end{bmatrix}$$

From R_3, $z = 0$.

From R_2:
$$\begin{aligned} -5y + z &= -5 \\ -5y + 0 &= -5 \\ -5y &= -5 \\ y &= 1 \end{aligned}$$

From R_1:
$$\begin{aligned} 3x - 2y + 4z &= 4 \\ 3x - 2(1) + 4(0) &= 4 \\ 3x - 2 &= 4 \\ 3x &= 6 \\ x &= 2 \end{aligned}$$

The solution is $(2, 1, 0)$.

57.
$$\begin{bmatrix} 3 & -1 & | & 9 \\ -6 & 2 & | & -18 \end{bmatrix} \overset{2R_1+R_2 \Rightarrow R_2}{\Rightarrow} \begin{bmatrix} 3 & -1 & | & 9 \\ 0 & 0 & | & 0 \end{bmatrix}$$

From R_2:
$$\begin{aligned} 3x - y &= 9 \\ y &= 3x - 9 \end{aligned}$$

The solution is $(x, 3x - 9)$.

59.
$$\begin{bmatrix} 1 & 1 & 1 & | & 6 \\ 1 & -1 & 1 & | & 2 \end{bmatrix} \overset{-R_1+R_2 \Rightarrow R_2}{\Rightarrow} \begin{bmatrix} 1 & 1 & 1 & | & 6 \\ 0 & -2 & 0 & | & -4 \end{bmatrix} \Rightarrow -2y = -4 \Rightarrow y = 2$$

From R_1:
$$\begin{aligned} x + y + z &= 6 \\ x + 2 + z &= 6 \\ x &= 4 - z \end{aligned}$$

The solution is $(4 - z, 2, z)$.

61.
$$\begin{bmatrix} 1 & 2 & 1 & | & 1 \\ 2 & -1 & 2 & | & 2 \\ 3 & 1 & 3 & | & 3 \end{bmatrix} \overset{\substack{-2R_1+R_2 \Rightarrow R_2 \\ -3R_1+R_3 \Rightarrow R_3}}{\Rightarrow} \begin{bmatrix} 1 & 2 & 1 & | & 1 \\ 0 & -5 & 0 & | & 0 \\ 0 & -5 & 0 & | & 0 \end{bmatrix} \overset{R_2+(-R_3) \Rightarrow R_3}{\Rightarrow} \begin{bmatrix} 1 & 2 & 1 & | & 1 \\ 0 & -5 & 0 & | & 0 \\ 0 & 0 & 0 & | & 0 \end{bmatrix}$$

From R_2:
$$\begin{aligned} -5y &= 0 \\ y &= 0 \end{aligned}$$

From R_1:
$$\begin{aligned} x + 2y + z &= 1 \\ x + 2(0) + z &= 1 \\ z &= 1 - x \end{aligned}$$

The solution is $(x, 0, 1 - x)$.

© 2015 Cengage Learning. All Rights Reserved. May not be scanned, copied or duplicated, or posted to a publicly accessible website, in whole or in part.

63. $$\left[\begin{array}{ccc|c} 2 & 1 & 3 & 3 \\ -2 & -1 & 1 & 5 \\ 4 & -2 & 2 & 2 \end{array}\right] \xrightarrow{\substack{R_1+R_2\Rightarrow R_2 \\ -2R_1+R_3\Rightarrow R_3}} \left[\begin{array}{ccc|c} 2 & 1 & 3 & 3 \\ 0 & 0 & 4 & 8 \\ 0 & -4 & -4 & -4 \end{array}\right] \xrightarrow{R_2\Leftrightarrow R_3} \left[\begin{array}{ccc|c} 2 & 1 & 3 & 3 \\ 0 & -4 & -4 & -4 \\ 0 & 0 & 4 & 8 \end{array}\right] \xrightarrow{\substack{-\frac{1}{4}R_2\Rightarrow R_2 \\ \frac{1}{4}R_3\Rightarrow R_3}} \left[\begin{array}{ccc|c} 2 & 1 & 3 & 3 \\ 0 & 1 & 1 & 1 \\ 0 & 0 & 1 & 2 \end{array}\right].$$

From R_3, $z = 2$.

From R_2:
$$\begin{aligned} y+z&=1 \\ y+2&=1 \\ y&=-1 \end{aligned}$$

From R_1:
$$\begin{aligned} 2x+y+3z&=3 \\ 2x+(-1)+3(2)&=3 \\ 2x-1+6&=3 \\ 2x+5&=3 \\ 2x&=-2 \\ x&=-1 \end{aligned}$$

The solution is $(-1,-1,2)$.

65. Let $x =$ the number of nickels, $y =$ the number of dimes, and $z =$ the number of quarters.
$$\begin{cases} x+y+z=64 \\ 5x+10y+25z=600 \\ 10x+5y+25z=500 \end{cases}$$

$$\left[\begin{array}{ccc|c} 1 & 1 & 1 & 64 \\ 5 & 10 & 25 & 600 \\ 10 & 5 & 25 & 500 \end{array}\right] \xrightarrow{\substack{-5R_1+R_2\Rightarrow R_2 \\ -10R_1+R_3\Rightarrow R_3}} \left[\begin{array}{ccc|c} 1 & 1 & 1 & 64 \\ 0 & 5 & 20 & 280 \\ 0 & -5 & 15 & -140 \end{array}\right] \xrightarrow{R_2+R_3\Rightarrow R_3} \left[\begin{array}{ccc|c} 1 & 1 & 1 & 64 \\ 0 & 5 & 20 & 280 \\ 0 & 0 & 35 & 140 \end{array}\right] \xrightarrow{\substack{\frac{1}{5}R_2\Rightarrow R_2 \\ \frac{1}{35}R_3\Rightarrow R_3}} \left[\begin{array}{ccc|c} 1 & 1 & 1 & 64 \\ 0 & 1 & 4 & 56 \\ 0 & 0 & 1 & 4 \end{array}\right]$$

From R_3, $z = 4$.

From R_2,
$$\begin{aligned} y+4z&=56 \\ y+4(4)&=56 \\ y+16&=56 \\ y&=40 \end{aligned}$$

From R_3,
$$\begin{aligned} x+y+z&=64 \\ x+40+4&=64 \\ x+44&=64 \\ x&=20 \end{aligned}$$

There are 20 nickels, 40 dimes, and 4 quarters.

67. Plug the coordinates of the points into the general equation to form and solve a system of equations.

$$\begin{aligned} y&=ax^2+bx+c \\ 1&=a(0)^2+b(0)+c \\ 1&=c \end{aligned} \qquad \begin{aligned} y&=ax^2+bx+c \\ 2&=a(1)^2+b(1)+c \\ 2&=a+b+c \end{aligned} \qquad \begin{aligned} y&=ax^2+bx+c \\ 4&=a(-1)^2+b(-1)+c \\ 4&=a-b+c \end{aligned}$$

$$\begin{cases} a+b+c=2 \\ a-b+c=4 \\ c=1 \end{cases} \Rightarrow \left[\begin{array}{ccc|c} 1 & 1 & 1 & 2 \\ 1 & -1 & 1 & 4 \\ 0 & 0 & 1 & 1 \end{array}\right] \xrightarrow{-R_2+R_1\Rightarrow R_2} \left[\begin{array}{ccc|c} 1 & 1 & 1 & 2 \\ 0 & 2 & 0 & -2 \\ 0 & 0 & 1 & 1 \end{array}\right] \xrightarrow{\frac{1}{2}R_2\Rightarrow R_2} \left[\begin{array}{ccc|c} 1 & 1 & 1 & 2 \\ 0 & 1 & 0 & -1 \\ 0 & 0 & 1 & 1 \end{array}\right]$$

From R_3, $c = 1$.

From R_2, $b = -1$. From R_1:
$$\begin{aligned} a+b+c&=2 \\ a+(-1)+1&=2 \\ a&=2 \end{aligned}$$

© 2015 Cengage Learning. All Rights Reserved. May not be scanned, copied or duplicated, or posted to a publicly accessible website, in whole or in part.

69. Let $x =$ the measure of the first angle and $y =$ the measure of the second angle. Form and solve this system of equations: $\begin{cases} x + y = 180 \\ y = x + 28 \end{cases} \Rightarrow \begin{cases} x + y = 180 \\ -x + y = 28 \end{cases}$

$$\left[\begin{array}{cc|c} 1 & 1 & 180 \\ -1 & 1 & 28 \end{array}\right] \overset{R_1 + R_2 \Rightarrow R_2}{\Rightarrow} \left[\begin{array}{cc|c} 1 & 1 & 180 \\ 0 & 2 & 208 \end{array}\right] \overset{\frac{1}{2}R_2 \Rightarrow R_2}{\Rightarrow} \left[\begin{array}{cc|c} 1 & 1 & 180 \\ 0 & 1 & 104 \end{array}\right]$$

From R_2, $y = 104$.

From R_1:

$$\begin{aligned} x + y &= 180 \\ x + 104 &= 180 \\ x &= 76 \end{aligned}$$

The angles have measures of 76° and 104°.

71. Let $x =$ the measure of the first angle and $y =$ the measure of the second angle. Form and solve this system of equations: $\begin{cases} x + y = 90 \\ y = x + 28 \end{cases} \Rightarrow \begin{cases} x + y = 90 \\ -x + y = 28 \end{cases}$

$$\left[\begin{array}{cc|c} 1 & 1 & 90 \\ -1 & 1 & 28 \end{array}\right] \overset{R_1 + R_2 \Rightarrow R_2}{\Rightarrow} \left[\begin{array}{cc|c} 1 & 1 & 90 \\ 0 & 2 & 118 \end{array}\right] \overset{\frac{1}{2}R_2 \Rightarrow R_2}{\Rightarrow} \left[\begin{array}{cc|c} 1 & 1 & 90 \\ 0 & 1 & 59 \end{array}\right]$$

From R_2, $y = 59$.

From R_1:

$$\begin{aligned} x + y &= 90 \\ x + 59 &= 90 \\ x &= 31 \end{aligned}$$

The angles have measures of 31° and 59°.

73. Let A, B and C represent the measures of the three angles.

$$\begin{cases} A + B + C = 180 \\ B = A + 25 \\ C = 2A - 5 \end{cases} \Rightarrow \begin{cases} A + B + C = 180 \\ -A + B = 25 \\ -2A + C = -5 \end{cases}$$

$$\left[\begin{array}{ccc|c} 1 & 1 & 1 & 180 \\ -1 & 1 & 0 & 25 \\ -2 & 0 & 1 & -5 \end{array}\right] \overset{\substack{R_1 + R_2 \Rightarrow R_2 \\ 2R_1 + R_3 \Rightarrow R_3}}{\Rightarrow} \left[\begin{array}{ccc|c} 1 & 1 & 1 & 180 \\ 0 & 2 & 1 & 205 \\ 0 & 2 & 3 & 355 \end{array}\right] \overset{-R_2 + R_3 \Rightarrow R_3}{\Rightarrow} \left[\begin{array}{ccc|c} 1 & 1 & 1 & 180 \\ 0 & 2 & 1 & 205 \\ 0 & 0 & 2 & 150 \end{array}\right] \overset{\frac{1}{2}R_3 \Rightarrow R_3}{\Rightarrow} \left[\begin{array}{ccc|c} 1 & 1 & 1 & 180 \\ 0 & 2 & 1 & 205 \\ 0 & 0 & 1 & 75 \end{array}\right]$$

From R_3, $C = 75$.

From R_2:

$$\begin{aligned} 2B + C &= 205 \\ 2B + 75 &= 205 \\ 2B &= 130 \\ B &= 65 \end{aligned}$$

From R_1:

$$\begin{aligned} A + B + C &= 180 \\ A + 65 + 75 &= 180 \\ A + 140 &= 180 \\ A &= 40 \end{aligned}$$

The angles have measures of 40°, 65° and 75°.

75. **Answers may vary.**

77. The last equation represents the equation $0x + 0y + 0z = k$, or $0 = k$. If $k = 0$, then the system can be solved. However, if $k \neq 0$, the system will have no solution.

© 2015 Cengage Learning. All Rights Reserved. May not be scanned, copied or duplicated, or posted to a publicly accessible website, in whole or in part.

Exercises 3.5 (page 215)

1. $ad - bc = 4(-1) - 3(2)$
$= -4 - 6 = -10$

3. $ad - bc = 0(5) - 8(2)$
$= 0 - 16 = -16$

5. $\begin{vmatrix} -21 & -1 \\ 7 & 5 \end{vmatrix}$

7.
$$\begin{aligned} 5(x-2) - (3-x) &= x+2 \\ 5x - 10 - 3 + x &= x+2 \\ 6x - 13 &= x+2 \\ 5x &= 15 \\ x &= 3 \end{aligned}$$

9.
$$\begin{aligned} \frac{5}{3}(5x+6) - 10 &= 0 \\ 3 \cdot \frac{5}{3}(5x+6) - 3 \cdot 10 &= 3 \cdot 0 \\ 5(5x+6) - 30 &= 0 \\ 25x + 30 - 30 &= 0 \\ 25x &= 0 \\ x &= 0 \end{aligned}$$

11. number; square

13. $\begin{vmatrix} a_2 & c_2 \\ a_3 & c_3 \end{vmatrix}$

15. Cramer's rule

17. consistent; independent

19. $\begin{vmatrix} 4 & 2 \\ -3 & 2 \end{vmatrix} = 4(2) - 2(-3)$
$= 8 + 6 = 14$

21. $\begin{vmatrix} -2 & 5 \\ 1 & -3 \end{vmatrix} = -2(-3) - 5(1)$
$= 6 - 5 = 1$

23. $\begin{vmatrix} 1 & 1 & 1 \\ 1 & 0 & 2 \\ 0 & 2 & 0 \end{vmatrix} = 1\begin{vmatrix} 0 & 2 \\ 2 & 0 \end{vmatrix} - 1\begin{vmatrix} 1 & 2 \\ 0 & 0 \end{vmatrix} + 1\begin{vmatrix} 1 & 0 \\ 0 & 2 \end{vmatrix} = 1(-4) - 1(0) + 1(2) = -4 - 0 + 2 = -2$

25. $\begin{vmatrix} -1 & 2 & 1 \\ 2 & 1 & -3 \\ 1 & 1 & 1 \end{vmatrix} = -1\begin{vmatrix} 1 & -3 \\ 1 & 1 \end{vmatrix} - 2\begin{vmatrix} 2 & -3 \\ 1 & 1 \end{vmatrix} + 1\begin{vmatrix} 2 & 1 \\ 1 & 1 \end{vmatrix} = -1(4) - 2(5) + 1(1) = -4 - 10 + 1$
$= -13$

27. $\begin{vmatrix} 1 & -2 & 3 \\ -2 & 1 & 1 \\ -3 & -2 & 1 \end{vmatrix} = 1\begin{vmatrix} 1 & 1 \\ -2 & 1 \end{vmatrix} - (-2)\begin{vmatrix} -2 & 1 \\ -3 & 1 \end{vmatrix} + 3\begin{vmatrix} -2 & 1 \\ -3 & -2 \end{vmatrix} = 1(3) + 2(1) + 3(7) = 3 + 2 + 21$
$= 26$

29. $\begin{vmatrix} 2 & 4 & 6 \\ 1 & 3 & 5 \\ 9 & 8 & 7 \end{vmatrix} = 2\begin{vmatrix} 3 & 5 \\ 8 & 7 \end{vmatrix} - 4\begin{vmatrix} 1 & 5 \\ 9 & 7 \end{vmatrix} + 6\begin{vmatrix} 1 & 3 \\ 9 & 8 \end{vmatrix} = 2(-19) - 4(-38) + 6(-19)$
$= -38 + 152 - 114 = 0$

© 2015 Cengage Learning. All Rights Reserved. May not be scanned, copied or duplicated, or posted to a publicly accessible website, in whole or in part.

SECTION 3.5

31. $x = \dfrac{\begin{vmatrix} 5 & 3 \\ -9 & -2 \end{vmatrix}}{\begin{vmatrix} 4 & 3 \\ 3 & -2 \end{vmatrix}} = \dfrac{-10-(-27)}{-8-9} = \dfrac{17}{-17} = -1;\ y = \dfrac{\begin{vmatrix} 4 & 5 \\ 3 & -9 \end{vmatrix}}{\begin{vmatrix} 4 & 3 \\ 3 & -2 \end{vmatrix}} = \dfrac{-36-15}{-17} = \dfrac{-51}{-17} = 3$

solution: $(-1, 3)$

33. $x = \dfrac{\begin{vmatrix} 6 & 1 \\ 2 & -1 \end{vmatrix}}{\begin{vmatrix} 1 & 1 \\ 1 & -1 \end{vmatrix}} = \dfrac{-6-2}{-1-1} = \dfrac{-8}{-2} = 4;\ y = \dfrac{\begin{vmatrix} 1 & 6 \\ 1 & 2 \end{vmatrix}}{\begin{vmatrix} 1 & 1 \\ 1 & -1 \end{vmatrix}} = \dfrac{2-6}{-2} = \dfrac{-4}{-2} = 2$; solution: $(4, 2)$

35. $\begin{cases} 4x = 2y - 5 \\ y = \dfrac{4x-5}{2} \end{cases} \Rightarrow \begin{cases} 4x = 2y - 5 \\ 2y = 4x - 5 \end{cases} \Rightarrow \begin{cases} 4x - 2y = -5 \\ -4x + 2y = -5 \end{cases}$

$x = \dfrac{\begin{vmatrix} -5 & -2 \\ -5 & 2 \end{vmatrix}}{\begin{vmatrix} 4 & -2 \\ -4 & 2 \end{vmatrix}} = \dfrac{-10-10}{8-8} = \dfrac{-20}{0} \Rightarrow$ denominator $= 0$, numerator $\neq 0 \Rightarrow \emptyset$

37. $\begin{cases} 2x = \dfrac{y+5}{2} \\ y = 4x - 6 \end{cases} \Rightarrow \begin{cases} 4x = y + 5 \\ y = 4x - 6 \end{cases} \Rightarrow \begin{cases} 4x - y = 5 \\ -4x + y = -6 \end{cases}$

$x = \dfrac{\begin{vmatrix} 5 & -1 \\ -6 & 1 \end{vmatrix}}{\begin{vmatrix} 4 & -1 \\ -4 & 1 \end{vmatrix}} = \dfrac{5-6}{4-4} = \dfrac{-1}{0} \Rightarrow$ denominator $= 0$, numerator $\neq 0 \Rightarrow \emptyset$

39. $\begin{cases} 2x + 3y = 9 \\ y = -\dfrac{2}{3}x + 3 \end{cases} \Rightarrow \begin{cases} 2x + 3y = 9 \\ 3y = -2x + 9 \end{cases} \Rightarrow \begin{cases} 2x + 3y = 9 \\ 2x + 3y = 9 \end{cases}$

$x = \dfrac{\begin{vmatrix} 9 & 3 \\ 9 & 3 \end{vmatrix}}{\begin{vmatrix} 2 & 3 \\ 2 & 3 \end{vmatrix}} = \dfrac{27-27}{6-6} = \dfrac{0}{0} \Rightarrow$ denominator $= 0$, numerator $= 0 \Rightarrow$ dependent equations

solution: $\left(x, -\frac{2}{3}x + 3\right)$

41. $\begin{cases} 4x - 3y = 6 \\ y = \dfrac{4x-6}{3} \end{cases} \Rightarrow \begin{cases} 4x - 3y = 6 \\ 3y = 4x - 6 \end{cases} \Rightarrow \begin{cases} 4x - 3y = 6 \\ -4x + 3y = -6 \end{cases}$

$x = \dfrac{\begin{vmatrix} 6 & -3 \\ -6 & 3 \end{vmatrix}}{\begin{vmatrix} 4 & -3 \\ -4 & 3 \end{vmatrix}} = \dfrac{18-18}{12-12} = \dfrac{0}{0} \Rightarrow$ denominator $= 0$, numerator $= 0 \Rightarrow$ dependent equations

$y = \frac{4x-6}{3} = \frac{4}{3}x - 2$; solution: $\left(x, \frac{4}{3}x - 2\right)$

© 2015 Cengage Learning. All Rights Reserved. May not be scanned, copied or duplicated, or posted to a publicly accessible website, in whole or in part.

SECTION 3.5

Note: In the following problems, D stands for the denominator determinant, while N_x, N_y and N_z stand for the numerator determinants for x, y and z, respectively.

43. $D = \begin{vmatrix} 1 & 1 & 1 \\ 1 & 1 & -1 \\ 1 & -1 & 1 \end{vmatrix} = 1\begin{vmatrix} 1 & -1 \\ -1 & 1 \end{vmatrix} - 1\begin{vmatrix} 1 & -1 \\ 1 & 1 \end{vmatrix} + 1\begin{vmatrix} 1 & 1 \\ 1 & -1 \end{vmatrix} = 1(0) - 1(2) + 1(-2) = -4$

$N_x = \begin{vmatrix} 4 & 1 & 1 \\ 0 & 1 & -1 \\ 2 & -1 & 1 \end{vmatrix} = 4\begin{vmatrix} 1 & -1 \\ -1 & 1 \end{vmatrix} - 1\begin{vmatrix} 0 & -1 \\ 2 & 1 \end{vmatrix} + 1\begin{vmatrix} 0 & 1 \\ 2 & -1 \end{vmatrix} = 4(0) - 1(2) + 1(-2) = -4$

$N_y = \begin{vmatrix} 1 & 4 & 1 \\ 1 & 0 & -1 \\ 1 & 2 & 1 \end{vmatrix} = 1\begin{vmatrix} 0 & -1 \\ 2 & 1 \end{vmatrix} - 4\begin{vmatrix} 1 & -1 \\ 1 & 1 \end{vmatrix} + 1\begin{vmatrix} 1 & 0 \\ 1 & 2 \end{vmatrix} = 1(2) - 4(2) + 1(2) = -4$

$N_z = \begin{vmatrix} 1 & 1 & 4 \\ 1 & 1 & 0 \\ 1 & -1 & 2 \end{vmatrix} = 1\begin{vmatrix} 1 & 0 \\ -1 & 2 \end{vmatrix} - 1\begin{vmatrix} 1 & 0 \\ 1 & 2 \end{vmatrix} + 4\begin{vmatrix} 1 & 1 \\ 1 & -1 \end{vmatrix} = 1(2) - 1(2) + 4(-2) = -8$

$x = \frac{N_x}{D} = \frac{-4}{-4} = 1;\ y = \frac{N_y}{D} = \frac{-4}{-4} = 1;\ z = \frac{N_z}{D} = \frac{-8}{-4} = 2 \Rightarrow$ solution: $(1, 1, 2)$

45. $D = \begin{vmatrix} 1 & 1 & 2 \\ 1 & 2 & 1 \\ 2 & 1 & 1 \end{vmatrix} = 1\begin{vmatrix} 2 & 1 \\ 1 & 1 \end{vmatrix} - 1\begin{vmatrix} 1 & 1 \\ 2 & 1 \end{vmatrix} + 2\begin{vmatrix} 1 & 2 \\ 2 & 1 \end{vmatrix} = 1(1) - 1(-1) + 2(-3) = -4$

$N_x = \begin{vmatrix} 7 & 1 & 2 \\ 8 & 2 & 1 \\ 9 & 1 & 1 \end{vmatrix} = 7\begin{vmatrix} 2 & 1 \\ 1 & 1 \end{vmatrix} - 1\begin{vmatrix} 8 & 1 \\ 9 & 1 \end{vmatrix} + 2\begin{vmatrix} 8 & 2 \\ 9 & 1 \end{vmatrix} = 7(1) - 1(-1) + 2(-10) = -12$

$N_y = \begin{vmatrix} 1 & 7 & 2 \\ 1 & 8 & 1 \\ 2 & 9 & 1 \end{vmatrix} = 1\begin{vmatrix} 8 & 1 \\ 9 & 1 \end{vmatrix} - 7\begin{vmatrix} 1 & 1 \\ 2 & 1 \end{vmatrix} + 2\begin{vmatrix} 1 & 8 \\ 2 & 9 \end{vmatrix} = 1(-1) - 7(-1) + 2(-7) = -8$

$N_z = \begin{vmatrix} 1 & 1 & 7 \\ 1 & 2 & 8 \\ 2 & 1 & 9 \end{vmatrix} = 1\begin{vmatrix} 2 & 8 \\ 1 & 9 \end{vmatrix} - 1\begin{vmatrix} 1 & 8 \\ 2 & 9 \end{vmatrix} + 7\begin{vmatrix} 1 & 2 \\ 2 & 1 \end{vmatrix} = 1(10) - 1(-7) + 7(-3) = -4$

$x = \frac{N_x}{D} = \frac{-12}{-4} = 3;\ y = \frac{N_y}{D} = \frac{-8}{-4} = 2;\ z = \frac{N_z}{D} = \frac{-4}{-4} = 1 \Rightarrow$ solution: $(3, 2, 1)$

47. $D = \begin{vmatrix} 2 & 1 & -1 \\ 1 & 2 & 2 \\ 4 & 5 & 3 \end{vmatrix} = 2\begin{vmatrix} 2 & 2 \\ 5 & 3 \end{vmatrix} - 1\begin{vmatrix} 1 & 2 \\ 4 & 3 \end{vmatrix} + (-1)\begin{vmatrix} 1 & 2 \\ 4 & 5 \end{vmatrix} = 2(-4) - 1(-5) - 1(-3) = 0$

$N_x = \begin{vmatrix} 1 & 1 & -1 \\ 2 & 2 & 2 \\ 3 & 5 & 3 \end{vmatrix} = 1\begin{vmatrix} 2 & 2 \\ 5 & 3 \end{vmatrix} - 1\begin{vmatrix} 2 & 2 \\ 3 & 3 \end{vmatrix} + (-1)\begin{vmatrix} 2 & 2 \\ 3 & 5 \end{vmatrix} = 1(-4) - 1(0) - 1(4) = -8$

$x = \frac{N_x}{D} = \frac{-8}{0} \Rightarrow$ denominator $= 0$, numerator $\neq 0 \Rightarrow \emptyset$

© 2015 Cengage Learning. All Rights Reserved. May not be scanned, copied or duplicated, or posted to a publicly accessible website, in whole or in part.

SECTION 3.5

49. $D = \begin{vmatrix} 2 & 3 & 4 \\ 2 & -3 & -4 \\ 4 & 6 & 8 \end{vmatrix} = 2\begin{vmatrix} -3 & -4 \\ 6 & 8 \end{vmatrix} - 3\begin{vmatrix} 2 & -4 \\ 4 & 8 \end{vmatrix} + 4\begin{vmatrix} 2 & -3 \\ 4 & 6 \end{vmatrix} = 2(0) - 3(32) + 4(24) = 0$

$N_x = \begin{vmatrix} 6 & 3 & 4 \\ -4 & -3 & -4 \\ 12 & 6 & 8 \end{vmatrix} = 6\begin{vmatrix} -3 & -4 \\ 6 & 8 \end{vmatrix} - 3\begin{vmatrix} -4 & -4 \\ 12 & 8 \end{vmatrix} + 4\begin{vmatrix} -4 & -3 \\ 12 & 6 \end{vmatrix} = 6(0) - 3(16) + 4(12) = 0$

$x = \frac{N_x}{D} = \frac{0}{0} \Rightarrow$ denominator $= 0$, numerator $= 0 \Rightarrow$ dependent equations.

$$\begin{array}{c} -R_1 + R_2 \Rightarrow R_2 \\ -2R_1 + R_3 \Rightarrow R_3 \end{array}$$

$$\left[\begin{array}{ccc|c} 2 & 3 & 4 & 6 \\ 2 & -3 & -4 & -4 \\ 4 & 6 & 8 & 12 \end{array}\right] \Rightarrow \left[\begin{array}{ccc|c} 2 & 3 & 4 & 6 \\ 0 & -6 & -8 & -10 \\ 0 & 0 & 0 & 0 \end{array}\right]$$

From R_2:

$-6y - 8z = -10$

$y = \frac{5}{3} - \frac{4}{3}z$

From R_1:

$2x + 3y + 4z = 6$

$2x + 3\left(\frac{5}{3} - \frac{4}{3}z\right) + 4z = 6$

$2x + 5 - 4z + 4z = 6$

$x = \frac{1}{2}$

The solution is $\left(\frac{1}{2}, \frac{5}{3} - \frac{4}{3}z, z\right)$.

51. $\begin{vmatrix} 2a & b \\ b & 2a \end{vmatrix} = 2a(2a) - b(b) = 4a^2 - b^2$

53. $\begin{vmatrix} a & 2a & -a \\ 2 & -1 & 3 \\ 1 & 2 & -3 \end{vmatrix} = a\begin{vmatrix} -1 & 3 \\ 2 & -3 \end{vmatrix} - 2a\begin{vmatrix} 2 & 3 \\ 1 & -3 \end{vmatrix} + (-a)\begin{vmatrix} 2 & -1 \\ 1 & 2 \end{vmatrix} = a(-3) - 2a(-9) - a(5)$

$= -3a + 18a - 5a = 10a$

55. $\begin{vmatrix} 1 & a & b \\ 1 & 2a & 2b \\ 1 & 3a & 3b \end{vmatrix} = 1\begin{vmatrix} 2a & 2b \\ 3a & 3b \end{vmatrix} - a\begin{vmatrix} 1 & 2b \\ 1 & 3b \end{vmatrix} + b\begin{vmatrix} 1 & 2a \\ 1 & 3a \end{vmatrix} = 1(0) - a(b) + b(a) = 0 - ab + ab = 0$

57. $\begin{vmatrix} 2 & -3 & 4 \\ -1 & 2 & 4 \\ 3 & -3 & 1 \end{vmatrix} = -23$

59. $\begin{vmatrix} 2 & 1 & -3 \\ -2 & 2 & 4 \\ 1 & -2 & 2 \end{vmatrix} = 26$

61. $x = \dfrac{\begin{vmatrix} 0 & 3 \\ -4 & 6 \end{vmatrix}}{\begin{vmatrix} 2 & 3 \\ 4 & -6 \end{vmatrix}} = \dfrac{0 - (-12)}{-12 - 12} = \dfrac{12}{-24} = -\dfrac{1}{2}$; $y = \dfrac{\begin{vmatrix} 2 & 0 \\ 4 & -4 \end{vmatrix}}{\begin{vmatrix} 2 & 3 \\ 4 & -6 \end{vmatrix}} = \dfrac{-8 - 0}{-24} = \dfrac{-8}{-24} = \dfrac{1}{3}$

solution: $\left(-\frac{1}{2}, \frac{1}{3}\right)$

© 2015 Cengage Learning. All Rights Reserved. May not be scanned, copied or duplicated, or posted to a publicly accessible website, in whole or in part.

63. $\begin{cases} y = \dfrac{-2x+1}{3} \\ 3x - 2y = 8 \end{cases} \Rightarrow \begin{cases} 3y = -2x+1 \\ 3x - 2y = 8 \end{cases} \Rightarrow \begin{cases} 2x + 3y = 1 \\ 3x - 2y = 8 \end{cases}$

$$x = \frac{\begin{vmatrix} 1 & 3 \\ 8 & -2 \end{vmatrix}}{\begin{vmatrix} 2 & 3 \\ 3 & -2 \end{vmatrix}} = \frac{-2-24}{-4-9} = \frac{-26}{-13} = 2; \; y = \frac{\begin{vmatrix} 2 & 1 \\ 3 & 8 \end{vmatrix}}{\begin{vmatrix} 2 & 3 \\ 3 & -2 \end{vmatrix}} = \frac{16-3}{-13} = \frac{13}{-13} = -1; \text{ solution: } (2, -1)$$

65. $\begin{cases} x = \dfrac{5y-4}{2} \\ y = \dfrac{3x-1}{5} \end{cases} \Rightarrow \begin{cases} 2x = 5y - 4 \\ 5y = 3x - 1 \end{cases} \Rightarrow \begin{cases} 2x - 5y = -4 \\ -3x + 5y = -1 \end{cases}$

$$x = \frac{\begin{vmatrix} -4 & -5 \\ -1 & 5 \end{vmatrix}}{\begin{vmatrix} 2 & -5 \\ -3 & 5 \end{vmatrix}} = \frac{-20-5}{10-15} = \frac{-25}{-5} = 5; \; y = \frac{\begin{vmatrix} 2 & -4 \\ -3 & -1 \end{vmatrix}}{\begin{vmatrix} 2 & -5 \\ -3 & 5 \end{vmatrix}} = \frac{-2-12}{-5} = \frac{-14}{-5} = \frac{14}{5}$$

solution: $\left(5, \frac{14}{5}\right)$

67. $D = \begin{vmatrix} 2 & 1 & 1 \\ 1 & -2 & 3 \\ 1 & 1 & -4 \end{vmatrix} = 2\begin{vmatrix} -2 & 3 \\ 1 & -4 \end{vmatrix} - 1\begin{vmatrix} 1 & 3 \\ 1 & -4 \end{vmatrix} + 1\begin{vmatrix} 1 & -2 \\ 1 & 1 \end{vmatrix} = 2(5) - 1(-7) + 1(3) = 20$

$$N_x = \begin{vmatrix} 5 & 1 & 1 \\ 10 & -2 & 3 \\ -3 & 1 & -4 \end{vmatrix} = 5\begin{vmatrix} -2 & 3 \\ 1 & -4 \end{vmatrix} - 1\begin{vmatrix} 10 & 3 \\ -3 & -4 \end{vmatrix} + 1\begin{vmatrix} 10 & -2 \\ -3 & 1 \end{vmatrix}$$

$$= 5(5) - 1(-31) + 1(4) = 60$$

$$N_y = \begin{vmatrix} 2 & 5 & 1 \\ 1 & 10 & 3 \\ 1 & -3 & -4 \end{vmatrix} = 2\begin{vmatrix} 10 & 3 \\ -3 & -4 \end{vmatrix} - 5\begin{vmatrix} 1 & 3 \\ 1 & -4 \end{vmatrix} + 1\begin{vmatrix} 1 & 10 \\ 1 & -3 \end{vmatrix}$$

$$= 2(-31) - 5(-7) + 1(-13) = -40$$

$$N_z = \begin{vmatrix} 2 & 1 & 5 \\ 1 & -2 & 10 \\ 1 & 1 & -3 \end{vmatrix} = 2\begin{vmatrix} -2 & 10 \\ 1 & -3 \end{vmatrix} - 1\begin{vmatrix} 1 & 10 \\ 1 & -3 \end{vmatrix} + 5\begin{vmatrix} 1 & -2 \\ 1 & 1 \end{vmatrix} = 2(-4) - 1(-13) + 5(3) = 20$$

$$x = \frac{N_x}{D} = \frac{60}{20} = 3; \; y = \frac{N_y}{D} = \frac{-40}{20} = -2; \; z = \frac{N_z}{D} = \frac{20}{20} = 1 \Rightarrow \text{solution: } (3, -2, 1)$$

69. $D = \begin{vmatrix} 4 & 0 & 3 \\ 0 & 2 & -6 \\ 8 & 4 & 3 \end{vmatrix} = 4\begin{vmatrix} 2 & -6 \\ 4 & 3 \end{vmatrix} - 0\begin{vmatrix} 0 & -6 \\ 8 & 3 \end{vmatrix} + 3\begin{vmatrix} 0 & 2 \\ 8 & 4 \end{vmatrix} = 4(30) - 0(48) + 3(-16) = 72$

$$N_x = \begin{vmatrix} 4 & 0 & 3 \\ -1 & 2 & -6 \\ 9 & 4 & 3 \end{vmatrix} = 4\begin{vmatrix} 2 & -6 \\ 4 & 3 \end{vmatrix} - 0\begin{vmatrix} -1 & -6 \\ 9 & 3 \end{vmatrix} + 3\begin{vmatrix} -1 & 2 \\ 9 & 4 \end{vmatrix} = 4(30) - 0(51) + 3(-22) = 54$$

$$N_y = \begin{vmatrix} 4 & 4 & 3 \\ 0 & -1 & -6 \\ 8 & 9 & 3 \end{vmatrix} = 4\begin{vmatrix} -1 & -6 \\ 9 & 3 \end{vmatrix} - 4\begin{vmatrix} 0 & -6 \\ 8 & 3 \end{vmatrix} + 3\begin{vmatrix} 0 & -1 \\ 8 & 9 \end{vmatrix} = 4(51) - 4(48) + 3(8) = 36$$

continued on next page...

© 2015 Cengage Learning. All Rights Reserved. May not be scanned, copied or duplicated, or posted to a publicly accessible website, in whole or in part.

69. continued

$$N_z = \begin{vmatrix} 4 & 0 & 4 \\ 0 & 2 & -1 \\ 8 & 4 & 9 \end{vmatrix} = 4\begin{vmatrix} 2 & -1 \\ 4 & 9 \end{vmatrix} - 0\begin{vmatrix} 0 & -1 \\ 8 & 9 \end{vmatrix} + 4\begin{vmatrix} 0 & 2 \\ 8 & 4 \end{vmatrix} = 4(22) - 0(8) + 4(-16) = 24$$

$$x = \frac{N_x}{D} = \frac{54}{72} = \frac{3}{4};\ y = \frac{N_y}{D} = \frac{36}{72} = \frac{1}{2};\ z = \frac{N_z}{D} = \frac{24}{72} = \frac{1}{3} \Rightarrow \text{solution: } \left(\frac{3}{4}, \frac{1}{2}, \frac{1}{3}\right)$$

71. $\begin{cases} x + y = 1 \Rightarrow & x + y = 1 \\ \frac{1}{2}y + z = \frac{5}{2} \Rightarrow & y + 2z = 5 \\ x - z = -3 \Rightarrow & x - z = -3 \end{cases}$

$$D = \begin{vmatrix} 1 & 1 & 0 \\ 0 & 1 & 2 \\ 1 & 0 & -1 \end{vmatrix} = 1\begin{vmatrix} 1 & 2 \\ 0 & -1 \end{vmatrix} - 1\begin{vmatrix} 0 & 2 \\ 1 & -1 \end{vmatrix} + 0\begin{vmatrix} 0 & 1 \\ 1 & 0 \end{vmatrix} = 1(-1) - 1(-2) + 0(-1) = 1$$

$$N_x = \begin{vmatrix} 1 & 1 & 0 \\ 5 & 1 & 2 \\ -3 & 0 & -1 \end{vmatrix} = 1\begin{vmatrix} 1 & 2 \\ 0 & -1 \end{vmatrix} - 1\begin{vmatrix} 5 & 2 \\ -3 & -1 \end{vmatrix} + 0\begin{vmatrix} 5 & 1 \\ -3 & 0 \end{vmatrix} = 1(-1) - 1(1) + 0(3) = -2$$

$$N_y = \begin{vmatrix} 1 & 1 & 0 \\ 0 & 5 & 2 \\ 1 & -3 & -1 \end{vmatrix} = 1\begin{vmatrix} 5 & 2 \\ -3 & -1 \end{vmatrix} - 1\begin{vmatrix} 0 & 2 \\ 1 & -1 \end{vmatrix} + 0\begin{vmatrix} 0 & 5 \\ 1 & -3 \end{vmatrix} = 1(1) - 1(-2) + 0(-5) = 3$$

$$N_z = \begin{vmatrix} 1 & 1 & 1 \\ 0 & 1 & 5 \\ 1 & 0 & -3 \end{vmatrix} = 1\begin{vmatrix} 1 & 5 \\ 0 & -3 \end{vmatrix} - 1\begin{vmatrix} 0 & 5 \\ 1 & -3 \end{vmatrix} + 1\begin{vmatrix} 0 & 1 \\ 1 & 0 \end{vmatrix} = 1(-3) - 1(-5) + 1(-1) = 1$$

$$x = \frac{N_x}{D} = \frac{-2}{1} = -2;\ y = \frac{N_y}{D} = \frac{3}{1} = 3;\ z = \frac{N_z}{D} = \frac{1}{1} = 1 \Rightarrow \text{solution: } (-2, 3, 1)$$

73.
$$\begin{vmatrix} x & 2 \\ -3 & 1 \end{vmatrix} = 2$$
$$x + 6 = 2$$
$$x = -4$$

75.
$$\begin{vmatrix} x & -2 \\ 3 & 1 \end{vmatrix} = \begin{vmatrix} 4 & 2 \\ x & 3 \end{vmatrix}$$
$$x - (-6) = 12 - 2x$$
$$x + 6 = 12 - 2x$$
$$3x = 6$$
$$x = 2$$

77. $\begin{cases} 2x + y = 180 \\ y = x + 30 \end{cases} \Rightarrow \begin{cases} 2x + y = 180 \\ -x + y = 30 \end{cases}$

$$x = \frac{\begin{vmatrix} 180 & 1 \\ 30 & 1 \end{vmatrix}}{\begin{vmatrix} 2 & 1 \\ -1 & 1 \end{vmatrix}} = \frac{180 - 30}{2 - (-1)} = \frac{150}{3} = 50°;\ y = \frac{\begin{vmatrix} 2 & 180 \\ -1 & 30 \end{vmatrix}}{\begin{vmatrix} 2 & 1 \\ -1 & 1 \end{vmatrix}} = \frac{60 - (-180)}{3} = \frac{240}{3} = 80°$$

© 2015 Cengage Learning. All Rights Reserved. May not be scanned, copied or duplicated, or posted to a publicly accessible website, in whole or in part.

SECTION 3.5

79. Let $x =$ the amount invested in HiTech, $y =$ the amount invested in SaveTel and $z =$ the amount invested in HiGas. Form and solve the following system of equations:

$$\begin{cases} x+y+z=20000 \Rightarrow & x+y+z=20000 \\ 0.10x+0.05y+0.06z=0.066(20000) \Rightarrow & 10x+5y+6x=132000 \\ y+z=3x \Rightarrow & -3x+y+z=0 \end{cases}$$

$$D=\begin{vmatrix} 1 & 1 & 1 \\ 10 & 5 & 6 \\ -3 & 1 & 1 \end{vmatrix}=1\begin{vmatrix} 5 & 6 \\ 1 & 1 \end{vmatrix}-1\begin{vmatrix} 10 & 6 \\ -3 & 1 \end{vmatrix}+1\begin{vmatrix} 10 & 5 \\ -3 & 1 \end{vmatrix}=1(-1)-1(28)+1(25)=-4$$

$$N_x=\begin{vmatrix} 20000 & 1 & 1 \\ 132000 & 5 & 6 \\ 0 & 1 & 1 \end{vmatrix}=20000\begin{vmatrix} 5 & 6 \\ 1 & 1 \end{vmatrix}-1\begin{vmatrix} 132000 & 6 \\ 0 & 1 \end{vmatrix}+1\begin{vmatrix} 132000 & 5 \\ 0 & 1 \end{vmatrix}$$

$$=20000(-1)-1(132000)+1(132000)=-20000$$

$$N_y=\begin{vmatrix} 1 & 20000 & 1 \\ 10 & 132000 & 6 \\ -3 & 0 & 1 \end{vmatrix}=1\begin{vmatrix} 132000 & 6 \\ 0 & 1 \end{vmatrix}-20000\begin{vmatrix} 10 & 6 \\ -3 & 1 \end{vmatrix}+1\begin{vmatrix} 10 & 132000 \\ -3 & 0 \end{vmatrix}$$

$$=1(132000)-20000(28)+1(396000)=-32000$$

$$N_z=\begin{vmatrix} 1 & 1 & 20000 \\ 10 & 5 & 132000 \\ -3 & 1 & 0 \end{vmatrix}=1\begin{vmatrix} 5 & 132000 \\ 1 & 0 \end{vmatrix}-1\begin{vmatrix} 10 & 132000 \\ -3 & 0 \end{vmatrix}+20000\begin{vmatrix} 10 & 5 \\ -3 & 1 \end{vmatrix}$$

$$=1(-132000)-1(396000)+20000(25)=-28000$$

$$x=\frac{N_x}{D}=\frac{-20000}{-4}=5000;\ y=\frac{N_y}{D}=\frac{-32000}{-4}=8000;\ z=\frac{N_z}{D}=\frac{-28000}{-4}=7000$$

He should invest \$5000 in HiTech, \$8000 in SaveTel and \$7000 in HiGas.

81. **Answers may vary.**

83.
$$\begin{vmatrix} x & y & 1 \\ -2 & 3 & 1 \\ 3 & 5 & 1 \end{vmatrix}=0$$

$$x\begin{vmatrix} 3 & 1 \\ 5 & 1 \end{vmatrix}-y\begin{vmatrix} -2 & 1 \\ 3 & 1 \end{vmatrix}+1\begin{vmatrix} -2 & 3 \\ 3 & 5 \end{vmatrix}=0$$

$$x(3-5)-y(-2-3)+1(-10-9)=0$$

$$-2x+5y-19=0$$

$$2x-5y=-19$$ Verify that both points satisfy this equation.

85.
$$\begin{vmatrix} 1 & 0 & 2 & 1 \\ 2 & 1 & 1 & 3 \\ 1 & 1 & 1 & 1 \\ 2 & 1 & 1 & 1 \end{vmatrix}=1\begin{vmatrix} 1 & 1 & 3 \\ 1 & 1 & 1 \\ 1 & 1 & 1 \end{vmatrix}-0\begin{vmatrix} 2 & 1 & 3 \\ 1 & 1 & 1 \\ 2 & 1 & 1 \end{vmatrix}+2\begin{vmatrix} 2 & 1 & 3 \\ 1 & 1 & 1 \\ 2 & 1 & 1 \end{vmatrix}-1\begin{vmatrix} 2 & 1 & 1 \\ 1 & 1 & 1 \\ 2 & 1 & 1 \end{vmatrix}$$

$$=1(0)-0(???)+2(-2)-1(0)=-4$$

© 2015 Cengage Learning. All Rights Reserved. May not be scanned, copied or duplicated, or posted to a publicly accessible website, in whole or in part.

Chapter 3 Review (page 221)

1. $\begin{cases} 2x + y = 11 \\ -x + 2y = 7 \end{cases}$

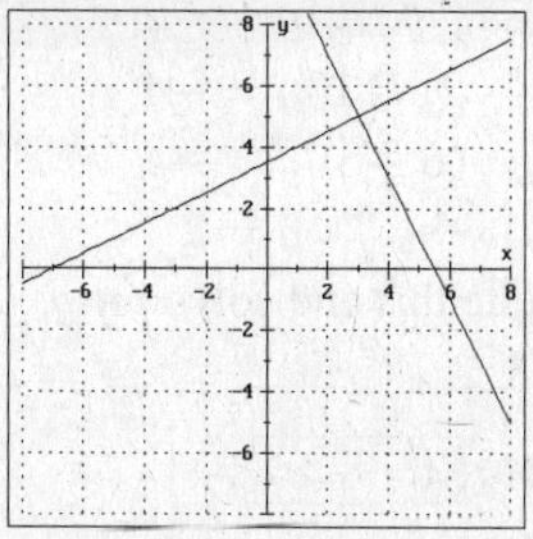

$(3, 5)$ is the solution.

2. $\begin{cases} 3x + 2y = 0 \\ 2x - 3y = -13 \end{cases}$

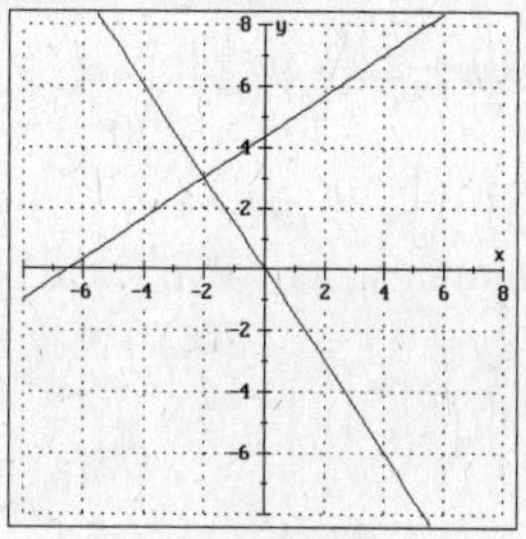

$(-2, 3)$ is the solution.

3. $\begin{cases} \frac{1}{2}x + \frac{1}{3}y = 2 \\ y = 6 - \frac{3}{2}x \end{cases}$

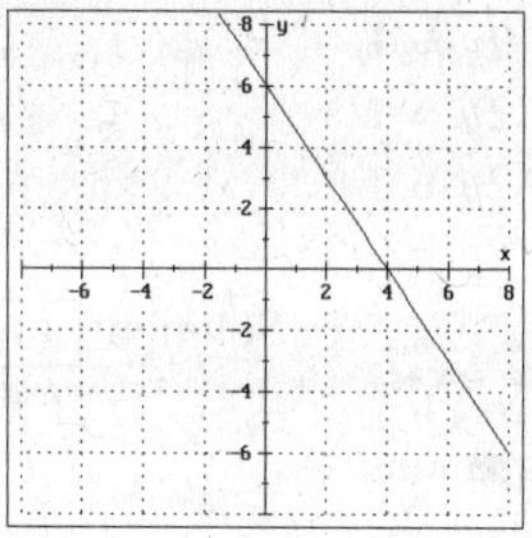

dependent equations; $\left(x, 6 - \frac{3}{2}x\right)$

4. $\begin{cases} \frac{1}{3}x - \frac{1}{2}y = 1 \\ 6x - 9y = 2 \end{cases}$

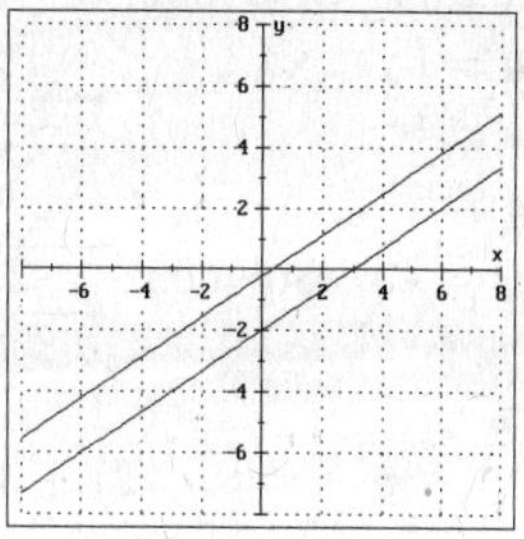

inconsistent system; $\emptyset$

5. $\begin{cases} (1) \quad y = 2x + 5 \\ (2) \quad 3x + 4y = 9 \end{cases}$

Substitute $y = 2x + 5$ from (1) into (2):

$$3x + 4\boldsymbol{y} = 9$$
$$3x + 4(\mathbf{2x + 5}) = 9$$
$$3x + 8x + 20 = 9$$
$$11x = -11$$
$$x = -1$$

Substitute this and solve for y:

$y = 2x + 5 = 2(-1) + 5 = 3$

Solution: $\boxed{(-1, 3)}$

6. $\begin{cases} (1) \quad y = 3x + 5 \\ (2) \quad 3x - y = -5 \end{cases}$

Substitute $y = 3x + 5$ from (1) into (2):

$$3x - \boldsymbol{y} = -5$$
$$3x - (\mathbf{3x + 5}) = -5$$
$$3x - 3x - 5 = -5$$
$$-5 = -5$$

Dependent equations

$\boxed{(x, 3x + 5)}$

© 2015 Cengage Learning. All Rights Reserved. May not be scanned, copied or duplicated, or posted to a publicly accessible website, in whole or in part.

7. $\begin{cases} (1) & x + 2y = 11 \\ (2) & 2x - y = 2 \end{cases}$

Substitute $x = -2y + 11$ from (1) into (2):

$$\begin{aligned} 2\boldsymbol{x} - y &= 2 \\ 2(\boldsymbol{-2y + 11}) - y &= 2 \\ -4y + 22 - y &= 2 \\ -5y &= -20 \\ y &= 4 \end{aligned}$$

Substitute this and solve for x:

$x = -2y + 11 = -2(4) + 11 = 3$

Solution: $\boxed{(3, 4)}$

8. $\begin{cases} (1) & 2x + 3y = -2 \\ (2) & 3x + 5y = -2 \end{cases}$

Substitute $x = \frac{-3y-2}{2}$ from (1) into (2):

$$\begin{aligned} 3\boldsymbol{x} + 5y &= -2 \\ 3 \cdot \boldsymbol{\tfrac{-3y-2}{2}} + 5y &= -2 \\ 3(-3y - 2) + 10y &= -4 \\ -9y - 6 + 10y &= -4 \\ y &= 2 \end{aligned}$$

Substitute this and solve for y:

$x = \frac{-3y-2}{2} = \frac{-3(2)-2}{2} = \frac{-8}{2} = -4$

Solution: $\boxed{(-4, 2)}$

9.

$$\begin{array}{rcl} x - y = 1 & \Rightarrow \times(2) & 2x - 2y = 2 \\ \underline{5x + 2y = -16} & \Rightarrow & \underline{5x + 2y = -16} \\ & & 7x \quad = -14 \\ & & x \quad = -2 \end{array}$$

Substitute and solve for y:

$$\begin{aligned} x - y &= 1 \\ -2 - y &= 1 \\ -y &= 3 \\ y &= -3 \end{aligned}$$

Solution: $\boxed{(-2, -3)}$

10.

$$\begin{array}{rcl} 3x + 2y = 1 & \Rightarrow \times 3 & 9x + 6y = 3 \\ \underline{2x - 3y = 5} & \Rightarrow \times 2 & \underline{4x - 6y = 10} \\ & & 13x \quad = 13 \\ & & x \quad = 1 \end{array}$$

Substitute and solve for y:

$$\begin{aligned} 3x + 2y &= 1 \\ 3(1) + 2y &= 1 \\ 3 + 2y &= 1 \\ 2y &= -2 \\ y &= -1 \end{aligned}$$

Solution: $\boxed{(1, -1)}$

11.

$$\begin{array}{rcl} 2x + 6 = -3y & \Rightarrow & 2x + 3y = -6 \Rightarrow \\ \underline{-2x = 3y + 6} & \Rightarrow & \underline{-2x - 3y = 6} \Rightarrow \\ & & 0 = 0 \end{array}$$

Dependent equations

$$\begin{aligned} 2x + 3y &= -6 \\ y &= \tfrac{-2x-6}{3} \\ y &= -\tfrac{2}{3}x - 2 \Rightarrow \boxed{\left(x, -\tfrac{2}{3}x - 2\right)} \end{aligned}$$

12.

$$\begin{array}{rcccl} y = \dfrac{2x-3}{2} & \Rightarrow \times 2 & 2y = 2x - 3 & \Rightarrow & -2x + 2y = -3 \\ \underline{x = \dfrac{2y+7}{2}} & \Rightarrow \times 2 & \underline{2x = 2y + 7} & \Rightarrow & \underline{2x - 2y = 7} \\ & & & & 0 \neq 4 \end{array}$$

Inconsistent system

No solutions: $\boxed{\emptyset}$

13.

$$\begin{array}{llllll} (1) & x + y + z = 6 & (1) & x + y + z = 6 & (1) & x + y + z = 6 \\ (2) & x - y - z = -4 & (2) & \underline{x - y - z = -4} & (3) & \underline{-x + y - z = -2} \\ (3) & -x + y - z = -2 & (4) & 2x \quad = 2 & (5) & 2y \quad = 4 \\ & & & x \quad = 1 & & y \quad = 2 \end{array}$$

$$\begin{aligned} x + y + z &= 6 \\ 1 + 2 + z &= 6 \\ 3 + z &= 6 \\ z &= 3 \end{aligned}$$

Solution: $\boxed{(1, 2, 3)}$

© 2015 Cengage Learning. All Rights Reserved. May not be scanned, copied or duplicated, or posted to a publicly accessible website, in whole or in part.

14.
$$\begin{array}{ll} (1) & 2x+3y+z=-5 \\ (2) & -x+2y-z=-6 \\ (3) & 3x+y+2z=4 \end{array} \quad \begin{array}{ll} (1) & 2x+3y+z=\ -5 \\ (2) & \underline{-x+2y-z=\ -6} \\ (4) & x+5y \qquad =-11 \end{array} \quad \begin{array}{ll} (3) & 3x+\ y+2z=\ \ 4 \\ 2\cdot(2) & \underline{-2x+4y-2z=-12} \\ (5) & x+5y \qquad =\ -8 \end{array}$$

$$\begin{array}{l} x+5y=-11 \Rightarrow \times(-1) \\ \underline{x+5y=\ -8 \Rightarrow} \end{array} \quad \begin{array}{r} -x-5y\ =\ 11 \\ \underline{x+5y\ =-8} \\ 0=\ \ 3 \end{array}$$

Since this is an impossible equation, there is no solution. It is an inconsistent system. $\boxed{\emptyset}$

15.
$$\begin{array}{ll} (1) & -x+5y+2z=-6 \\ (2) & x-5y-z=6 \\ (3) & 3x-15y-6z=18 \end{array} \quad \begin{array}{ll} (1) & -x+5y+2z=-6 \\ (2) & \underline{x-5y-\ z=\ \ 6} \\ (4) & z=\ \ 0 \end{array} \quad \begin{array}{ll} (1) & -x+\ 5y+2(0)=-6 \\ (3) & \underline{3x-15y-6(0)=\ 18} \end{array}$$

$$\begin{array}{l} -x+\ 5y=-6 \Rightarrow \times(3) \\ \underline{3x-15y=\ 18 \Rightarrow} \end{array} \quad \begin{array}{r} -3x+15y\ =-18 \\ \underline{3x-15y\ =\ \ 18} \\ 0=\ \ \ 0 \end{array} \quad \text{Dependent equations} \quad \begin{array}{r} -x+5y=-6 \\ y=\frac{x-6}{5} \\ y=\frac{1}{5}x-\frac{6}{5} \end{array}$$

$$\boxed{\left(x,\tfrac{1}{5}x-\tfrac{6}{5},0\right)}$$

16.
$$\left[\begin{array}{ccc|c} 1 & 1 & 1 & 6 \\ 2 & -1 & 1 & 1 \\ 4 & 1 & -1 & 5 \end{array}\right] \xRightarrow[-4R_1+R_3 \Rightarrow R_3]{-2R_1+R_2 \Rightarrow R_2} \left[\begin{array}{ccc|c} 1 & 1 & 1 & 6 \\ 0 & -3 & -1 & -11 \\ 0 & -3 & -5 & -19 \end{array}\right] \xRightarrow[-R_3+R_2 \Rightarrow R_3]{-R_2 \Rightarrow R_2} \left[\begin{array}{ccc|c} 1 & 1 & 1 & 6 \\ 0 & 3 & 1 & 11 \\ 0 & 0 & 4 & 8 \end{array}\right] \xRightarrow{\frac{1}{4}R_3 \Rightarrow R_3} \left[\begin{array}{ccc|c} 1 & 1 & 1 & 6 \\ 0 & 3 & 1 & 11 \\ 0 & 0 & 1 & 2 \end{array}\right]$$

From R_3, $z=2$. From R_2: From R_1: Solution: $\boxed{(1,3,2)}$

$$\begin{array}{r} 3y+z=11 \\ 3y+2=11 \\ 3y=9 \\ y=3 \end{array} \qquad \begin{array}{r} x+y+z=6 \\ x+3+2=6 \\ x=1 \end{array}$$

17.
$$\left[\begin{array}{cc|c} 1 & 1 & 3 \\ 1 & -2 & -3 \\ 2 & 1 & 4 \end{array}\right] \xRightarrow[-2R_1+R_3 \Rightarrow R_3]{-R_1+R_2 \Rightarrow R_2} \left[\begin{array}{cc|c} 1 & 1 & 3 \\ 0 & -3 & -6 \\ 0 & -1 & -2 \end{array}\right] \xRightarrow[-3R_3+R_2 \Rightarrow R_3]{-\frac{1}{3}R_2 \Rightarrow R_2} \left[\begin{array}{cc|c} 1 & 1 & 3 \\ 0 & 1 & 2 \\ 0 & 0 & 0 \end{array}\right]$$

From R_2, $y=2$. From R_1: Solution: $\boxed{(1,2)}$

$$\begin{array}{r} x+y=3 \\ x+2=3 \\ x=1 \end{array}$$

© 2015 Cengage Learning. All Rights Reserved. May not be scanned, copied or duplicated, or posted to a publicly accessible website, in whole or in part.

18. $\left[\begin{array}{ccc|c} 1 & -3 & 1 & 4 \\ 2 & -5 & 3 & 6 \end{array}\right] \Rightarrow \begin{array}{c} -2R_1 + R_2 \Rightarrow R_2 \\ \left[\begin{array}{ccc|c} 1 & -3 & 1 & 4 \\ 0 & 1 & 1 & -2 \end{array}\right] \end{array}$

From R_2:
$$y + z = -2$$
$$y = -2 - z$$

From R_1:
$$x - 3y + z = 4$$
$$x - 3(-2 - z) + z = 4$$
$$x + 6 + 3z + z = 4$$
$$x + 4z = -2$$
$$x = -2 - 4z$$

Solution: $\boxed{(-2 - 4z, -2 - z, z)}$

19. $\begin{vmatrix} 3 & 1 \\ -2 & 4 \end{vmatrix} = 3(4) - 1(-2) = 12 + 2 = 14$

20. $\begin{vmatrix} -4 & 5 \\ 6 & -2 \end{vmatrix} = -4(-2) - 5(6) = 8 - 30 = -22$

21. $\begin{vmatrix} -1 & 2 & -1 \\ 2 & -1 & 3 \\ 1 & -2 & 2 \end{vmatrix} = -1\begin{vmatrix} -1 & 3 \\ -2 & 2 \end{vmatrix} - 2\begin{vmatrix} 2 & 3 \\ 1 & 2 \end{vmatrix} + (-1)\begin{vmatrix} 2 & -1 \\ 1 & -2 \end{vmatrix} = -1(4) - 2(1) - 1(-3) = -3$

22. $\begin{vmatrix} 3 & -2 & 2 \\ 1 & -2 & -2 \\ 2 & 1 & -1 \end{vmatrix} = 3\begin{vmatrix} -2 & -2 \\ 1 & -1 \end{vmatrix} - (-2)\begin{vmatrix} 1 & -2 \\ 2 & -1 \end{vmatrix} + 2\begin{vmatrix} 1 & -2 \\ 2 & 1 \end{vmatrix} = 3(4) + 2(3) + 2(5) = 28$

23. $x = \dfrac{\begin{vmatrix} 10 & 4 \\ 1 & -3 \end{vmatrix}}{\begin{vmatrix} 3 & 4 \\ 2 & -3 \end{vmatrix}} = \dfrac{-30 - 4}{-9 - 8} = \dfrac{-34}{-17} = 2; \; y = \dfrac{\begin{vmatrix} 3 & 10 \\ 2 & 1 \end{vmatrix}}{\begin{vmatrix} 3 & 4 \\ 2 & -3 \end{vmatrix}} = \dfrac{3 - 20}{-17} = \dfrac{-17}{-17} = 1$

Solution: $\boxed{(2, 1)}$

24. $x = \dfrac{\begin{vmatrix} -17 & -5 \\ 3 & 2 \end{vmatrix}}{\begin{vmatrix} 2 & -5 \\ 3 & 2 \end{vmatrix}} = \dfrac{-34 + 15}{4 + 15} = \dfrac{-19}{19} = -1; \; y = \dfrac{\begin{vmatrix} 2 & -17 \\ 3 & 3 \end{vmatrix}}{\begin{vmatrix} 2 & -5 \\ 3 & 2 \end{vmatrix}} = \dfrac{6 + 51}{19} = \dfrac{57}{19} = 3$

Solution: $\boxed{(-1, 3)}$

25. $D = \begin{vmatrix} 1 & 2 & 1 \\ 2 & 1 & 1 \\ 1 & 1 & 2 \end{vmatrix} = 1\begin{vmatrix} 1 & 1 \\ 1 & 2 \end{vmatrix} - 2\begin{vmatrix} 2 & 1 \\ 1 & 2 \end{vmatrix} + 1\begin{vmatrix} 2 & 1 \\ 1 & 1 \end{vmatrix} = 1(1) - 2(3) + 1(1) = -4$

$N_x = \begin{vmatrix} 0 & 2 & 1 \\ 3 & 1 & 1 \\ 5 & 1 & 2 \end{vmatrix} = 0\begin{vmatrix} 1 & 1 \\ 1 & 2 \end{vmatrix} - 2\begin{vmatrix} 3 & 1 \\ 5 & 2 \end{vmatrix} + 1\begin{vmatrix} 3 & 1 \\ 5 & 1 \end{vmatrix} = 0(1) - 2(1) + 1(-2) = -4$

$N_y = \begin{vmatrix} 1 & 0 & 1 \\ 2 & 3 & 1 \\ 1 & 5 & 2 \end{vmatrix} = 1\begin{vmatrix} 3 & 1 \\ 5 & 2 \end{vmatrix} - 0\begin{vmatrix} 2 & 1 \\ 1 & 2 \end{vmatrix} + 1\begin{vmatrix} 2 & 3 \\ 1 & 5 \end{vmatrix} = 1(1) - 0(3) + 1(7) = 8$

continued on next page...

© 2015 Cengage Learning. All Rights Reserved. May not be scanned, copied or duplicated, or posted to a publicly accessible website, in whole or in part.

25. continued

$$N_z = \begin{vmatrix} 1 & 2 & 0 \\ 2 & 1 & 3 \\ 1 & 1 & 5 \end{vmatrix} = 1\begin{vmatrix} 1 & 3 \\ 1 & 5 \end{vmatrix} - 2\begin{vmatrix} 2 & 3 \\ 1 & 5 \end{vmatrix} + 0\begin{vmatrix} 2 & 1 \\ 1 & 1 \end{vmatrix} = 1(2) - 2(7) + 0(1) = -12$$

$$x = \frac{N_x}{D} = \frac{-4}{-4} = 1;\ y = \frac{N_y}{D} = \frac{8}{-4} = -2;\ z = \frac{N_z}{D} = \frac{-12}{-4} = 3 \Rightarrow \text{solution: } (1, -2, 3)$$

26. $$D = \begin{vmatrix} 1 & 1 & 1 \\ -1 & 1 & -1 \\ -1 & -1 & 1 \end{vmatrix} = 1\begin{vmatrix} 1 & -1 \\ -1 & 1 \end{vmatrix} - 1\begin{vmatrix} -1 & -1 \\ -1 & 1 \end{vmatrix} + 1\begin{vmatrix} -1 & 1 \\ -1 & -1 \end{vmatrix} = 1(0) - 1(-2) + 1(2) = 0$$

$$N_x = \begin{vmatrix} 1 & 1 & 1 \\ 3 & 1 & -1 \\ 5 & -1 & 1 \end{vmatrix} = 1\begin{vmatrix} 1 & -1 \\ -1 & 1 \end{vmatrix} - 1\begin{vmatrix} 3 & -1 \\ 5 & 1 \end{vmatrix} + 1\begin{vmatrix} 3 & 1 \\ 5 & -1 \end{vmatrix} = 1(0) - 1(8) + 1(-8) = -16$$

$$x = \frac{N_x}{D} = \frac{-16}{0} \Rightarrow \text{denominator} = 0, \text{ numerator} \neq 0 \Rightarrow \emptyset$$

Chapter 3 Test (page 226)

1. $\begin{cases} 2x + y = 5 \\ y = 2x - 3 \end{cases}$

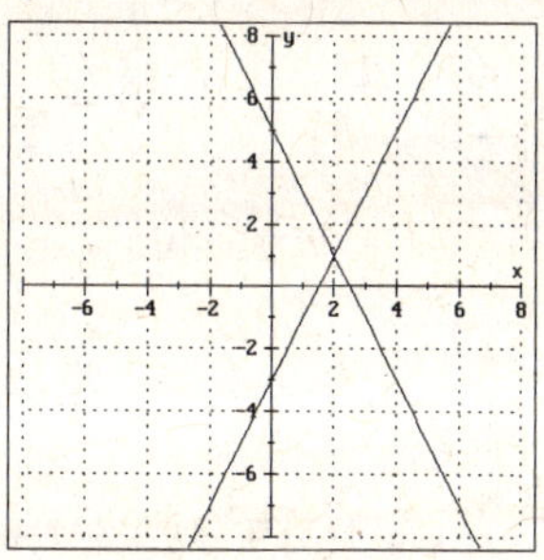

(2, 1) is the solution.

2. $\begin{cases} (1)\ \ 2x - 4y = 14 \\ (2)\ \ x = -2y + 7 \end{cases}$

Substitute $x = -2y + 7$ from (2) into (1):

$$\begin{aligned} 2\boldsymbol{x} - 4y &= 14 \\ 2(\boldsymbol{-2y + 7}) - 4y &= 14 \\ -4y + 14 - 4y &= 14 \\ -8y &= 0 \\ y &= 0 \end{aligned}$$

Substitute this and solve for x:

$$\begin{aligned} x &= -2y + 7 \\ x &= -2(0) + 7 = 7 \end{aligned}$$

Solution: $(7, 0)$

3. $\begin{aligned} 2x + 3y &= -5 \Rightarrow \times 2 \\ 3x - 2y &= \ \ 12 \Rightarrow \times 3 \end{aligned}$ $\quad \begin{aligned} 4x + 6y &= -10 \\ 9x - 6y &= 36 \\ \hline 13x &= 26 \\ x &= 2 \end{aligned}$ $\quad \begin{aligned} 2x + 3y &= -5 \\ 2(2) + 3y &= -5 \\ 4 + 3y &= -5 \\ 3y &= -9 \\ y &= -3 \end{aligned}$

Solution: $\boxed{(2, -3)}$

4. $\begin{aligned} \frac{x}{2} - \frac{y}{4} &= -4 \Rightarrow \times 4 \\ x + y &= -2 \Rightarrow \end{aligned}$ $\quad \begin{aligned} 2x - y &= -16 \\ x + y &= -2 \\ \hline 3x &= -18 \\ x &= -6 \end{aligned}$ $\quad \begin{aligned} x + y &= -2 \\ -6 + y &= -2 \\ y &= 4 \end{aligned}$

Solution: $\boxed{(-6, 4)}$

© 2015 Cengage Learning. All Rights Reserved. May not be scanned, copied or duplicated, or posted to a publicly accessible website, in whole or in part.

5. $3(x+y) = x - 3 \Rightarrow \quad 2x + 3y = -3 \qquad 2x + 3y = -3 \Rightarrow \qquad 2x + 3y = -3$

$-y = \dfrac{2x+3}{3} \Rightarrow \quad 2x + 3y = -3 \qquad 2x + 3y = -3 \Rightarrow \times(-1) \; -2x - 3y = 3$

$0 = 0$

The equation is an identity, so the system has infinitely many solutions. $\Rightarrow$ dependent equations

6. See #5. Since the system has at least one solution, it is a consistent system.

7. $\begin{bmatrix} 1 & 3 & -2 \\ 4 & -2 & -1 \end{bmatrix} \Rightarrow \begin{bmatrix} 1 & 3 & -2 \\ -1 & -17 & \boxed{9} \end{bmatrix}$ ($-5R_1 + R_2 \Rightarrow R_2$)

8. $\begin{bmatrix} -1 & 3 & 6 \\ 3 & -2 & 4 \end{bmatrix} \Rightarrow \begin{bmatrix} -1 & 3 & 6 \\ 5 & -8 & \boxed{-8} \end{bmatrix}$ ($-2R_1 + R_2 \Rightarrow R_2$)

9. $\left[\begin{array}{ccc|c} 1 & 1 & 1 & 4 \\ 1 & 1 & -1 & 6 \\ 2 & -3 & 1 & -1 \end{array}\right]$

For #10-12, find the denominator determinant:

$$D = \begin{vmatrix} 1 & 1 & 1 \\ 1 & 1 & -1 \\ 2 & -3 & 1 \end{vmatrix} = 1\begin{vmatrix} 1 & -1 \\ -3 & 1 \end{vmatrix} - 1\begin{vmatrix} 1 & -1 \\ 2 & 1 \end{vmatrix} + 1\begin{vmatrix} 1 & 1 \\ 2 & -3 \end{vmatrix} = 1(-2) - 1(3) + 1(-5) = -10$$

10. $$N_x = \begin{vmatrix} 4 & 1 & 1 \\ 6 & 1 & -1 \\ -1 & -3 & 1 \end{vmatrix} = 4\begin{vmatrix} 1 & -1 \\ -3 & 1 \end{vmatrix} - 1\begin{vmatrix} 6 & -1 \\ -1 & 1 \end{vmatrix} + 1\begin{vmatrix} 6 & 1 \\ -1 & -3 \end{vmatrix} = 4(-2) - 1(5) + 1(-17)$$

$$= -30$$

$$x = \frac{N_x}{D} = \frac{-30}{-10} = 3$$

11. $$N_y = \begin{vmatrix} 1 & 4 & 1 \\ 1 & 6 & -1 \\ 2 & -1 & 1 \end{vmatrix} = 1\begin{vmatrix} 6 & -1 \\ -1 & 1 \end{vmatrix} - 4\begin{vmatrix} 1 & -1 \\ 2 & 1 \end{vmatrix} + 1\begin{vmatrix} 1 & 6 \\ 2 & -1 \end{vmatrix} = 1(5) - 4(3) + 1(-13) = -20$$

$$y = \frac{N_y}{D} = \frac{-20}{-10} = 2$$

12. $$N_z = \begin{vmatrix} 1 & 1 & 4 \\ 1 & 1 & 6 \\ 2 & -3 & -1 \end{vmatrix} = 1\begin{vmatrix} 1 & 6 \\ -3 & -1 \end{vmatrix} - 1\begin{vmatrix} 1 & 6 \\ 2 & -1 \end{vmatrix} + 4\begin{vmatrix} 1 & 1 \\ 2 & -3 \end{vmatrix} = 1(17) - 1(-13) + 4(-5) = 10$$

$$z = \frac{N_z}{D} = \frac{10}{-10} = -1$$

© 2015 Cengage Learning. All Rights Reserved. May not be scanned, copied or duplicated, or posted to a publicly accessible website, in whole or in part.

CHAPTER 3 TEST

13. $\left[\begin{array}{cc|c} 1 & 1 & 4 \\ 2 & -1 & 2 \end{array}\right] \xRightarrow{-2R_1+R_2 \Rightarrow R_2} \left[\begin{array}{cc|c} 1 & 1 & 4 \\ 0 & -3 & -6 \end{array}\right] \xRightarrow{-\frac{1}{3}R_2 \Rightarrow R_2} \left[\begin{array}{cc|c} 1 & 1 & 4 \\ 0 & 1 & 2 \end{array}\right]$

From R_2, $y = 2$. From R_1:

$$\begin{aligned} x + y &= 4 \\ x + 2 &= 4 \\ x &= 2 \end{aligned}$$

Solution: $\boxed{(2, 2)}$

14. $\left[\begin{array}{cc|c} 1 & 1 & 2 \\ 1 & -1 & -4 \\ 2 & 1 & 1 \end{array}\right] \xRightarrow[-2R_1+R_3 \Rightarrow R_3]{-R_2+R_1 \Rightarrow R_2} \left[\begin{array}{cc|c} 1 & 1 & 2 \\ 0 & 2 & 6 \\ 0 & -1 & -3 \end{array}\right] \xRightarrow[2R_3+R_2 \Rightarrow R_3]{\frac{1}{2}R_2 \Rightarrow R_2} \left[\begin{array}{cc|c} 1 & 1 & 2 \\ 0 & 1 & 3 \\ 0 & 0 & 0 \end{array}\right]$

From R_2, $y = 3$. From R_1:

$$\begin{aligned} x + y &= 2 \\ x + 3 &= 2 \\ x &= -1 \end{aligned}$$

Solution: $\boxed{(-1, 3)}$

15. $\begin{vmatrix} 2 & -3 \\ 4 & 5 \end{vmatrix} = 2(5) - (-3)(4) = 10 - (-12) = 22$

16. $\begin{vmatrix} -3 & -4 \\ -2 & 3 \end{vmatrix} = -3(3) - (-4)(-2) = -9 - 8 = -17$

17. $\begin{vmatrix} 1 & 2 & 0 \\ 2 & 0 & 3 \\ 1 & -2 & 2 \end{vmatrix} = 1\begin{vmatrix} 0 & 3 \\ -2 & 2 \end{vmatrix} - 2\begin{vmatrix} 2 & 3 \\ 1 & 2 \end{vmatrix} + 0\begin{vmatrix} 2 & 0 \\ 1 & -2 \end{vmatrix} = 1(6) - 2(1) + 0(-4) = 4$

18. $\begin{vmatrix} 1 & -2 & 3 \\ 3 & 1 & -2 \\ 2 & -4 & 6 \end{vmatrix} = 1\begin{vmatrix} 1 & -2 \\ -4 & 6 \end{vmatrix} - (-2)\begin{vmatrix} 3 & -2 \\ 2 & 6 \end{vmatrix} + 3\begin{vmatrix} 3 & 1 \\ 2 & -4 \end{vmatrix} = 1(-2) + 2(22) + 3(-14) = 0$

19. $\begin{vmatrix} -6 & -1 \\ -6 & 1 \end{vmatrix}$

20. $\begin{vmatrix} 1 & -1 \\ 3 & 1 \end{vmatrix}$

21. $x = \dfrac{\begin{vmatrix} -6 & -1 \\ -6 & 1 \end{vmatrix}}{\begin{vmatrix} 1 & -1 \\ 3 & 1 \end{vmatrix}} = \dfrac{-6-6}{1-(-3)} = \dfrac{-12}{4} = -3$

22. $y = \dfrac{\begin{vmatrix} 1 & -6 \\ 3 & -6 \end{vmatrix}}{\begin{vmatrix} 1 & -1 \\ 3 & 1 \end{vmatrix}} = \dfrac{-6-(-18)}{1-(-3)} = \dfrac{12}{4} = 3$

23. $\left[\begin{array}{ccc|c} 2 & -2 & -2 & 2 \\ 2 & 3 & 1 & 2 \\ 3 & 2 & 0 & 0 \end{array}\right] \xRightarrow{\frac{1}{2}R_1 \Rightarrow R_1} \left[\begin{array}{ccc|c} 1 & -1 & -1 & 1 \\ 2 & 3 & 1 & 2 \\ 3 & 2 & 0 & 0 \end{array}\right] \xRightarrow[-3R_1+R_3 \Rightarrow R_3]{-2R_1+R_2 \Rightarrow R_2} \left[\begin{array}{ccc|c} 1 & -1 & -1 & 1 \\ 0 & 5 & 3 & 0 \\ 0 & 5 & 3 & -3 \end{array}\right] \xRightarrow{-R_2+R_3 \Rightarrow R_3} \left[\begin{array}{ccc|c} 1 & -1 & -1 & 1 \\ 0 & 5 & 3 & 0 \\ 0 & 0 & 0 & -3 \end{array}\right]$

The last row indicates that $0x + 0y + 0z = -3$. This is false, so there is no solution. $\emptyset$

© 2015 Cengage Learning. All Rights Reserved. May not be scanned, copied or duplicated, or posted to a publicly accessible website, in whole or in part.

24. From R_2:
$$y + 3z = 2$$
$$y = 2 - 3z$$
From R_1:
$$x + 4z = 4$$
$$x = 4 - 4z$$
The solution is $(4 - 4z, 2 - 3z, z)$.

25.
$$\left[\begin{array}{ccc|c} -2 & -1 & 1 & -4 \\ 1 & -1 & -3 & -1 \end{array}\right] \overset{R_1 \Leftrightarrow R_2}{\Rightarrow} \left[\begin{array}{ccc|c} 1 & -1 & -3 & -1 \\ -2 & -1 & 1 & -4 \end{array}\right] \overset{2R_1 + R_2 \Rightarrow R_2}{\Rightarrow} \left[\begin{array}{ccc|c} 1 & -1 & -3 & -1 \\ 0 & -3 & -5 & -6 \end{array}\right]$$
$$\overset{-\frac{1}{3}R_2 \Rightarrow R_2}{\Rightarrow} \left[\begin{array}{ccc|c} 1 & -1 & -3 & -1 \\ 0 & 1 & \frac{5}{3} & 2 \end{array}\right]$$
From R_2:
$$y + \tfrac{5}{3}z = 2$$
$$y = -\tfrac{5}{3}z + 2$$
From R_1:
$$x - y - 3z = -1$$
$$x - (-\tfrac{5}{3}z + 2) - 3z = -1$$
$$x + \tfrac{5}{3}z - 2 - 3z = -1$$
$$x - \tfrac{4}{3}z = 1$$
$$x = \tfrac{4}{3}z + 1$$
The solution is $\left(\tfrac{4}{3}z + 1, -\tfrac{5}{3}z + 2, z\right)$

Exercise 4.1 (page 238)

1. $x < 2$
$\{x|x < 2\}$

3. $x \geq -5$
$\{x|x \geq -5\}$

5. $x > -6$
$(-6, \infty)$

7. $x \leq 2$
$(-\infty, 2]$

9.
$$\left(\frac{t^4t^2t^{-3}}{t^5t^{-7}}\right)^{-2} = \left(\frac{t^5t^{-7}}{t^4t^2t^{-3}}\right)^2 = \left(\frac{t^{-2}}{t^3}\right)^2 = \left(\frac{1}{t^5}\right)^2 = \frac{1}{t^{10}}$$

11. Let $x =$ the number of pies made.
[Expenses] = [Income]
$$1200 + 4.20x = 7.95x$$
$$1200 = 3.75x$$
$$320 = x$$
He must sell at least 320 pies.

13. $\neq$

15. $<$

17. $\geq$

19. $a < c$

21. reversed

23. $c < x; x < d$

25. open

27.
$$x + 4 < 5$$
$$x + 4 - 4 < 5 - 4$$
$$x < 1$$
solution set: $\{x|x < 1\}$

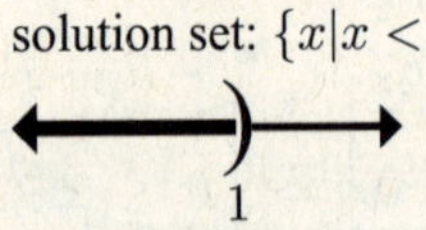

29.
$$2(x + 4) \geq -2$$
$$2x + 8 \geq -2$$
$$2x + 8 - 8 \geq -2 - 8$$
$$2x \geq -10$$
$$x \geq -5$$
solution set: $\{x|x \geq -5\}$

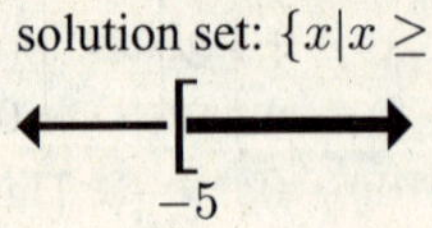

31.
$$2x + 3 < 9$$
$$2x + 3 - 3 < 9 - 3$$
$$2x < 6$$
$$\tfrac{2x}{2} < \tfrac{6}{2}$$
$$x < 3$$
solution set: $\{x|x < 3\}$

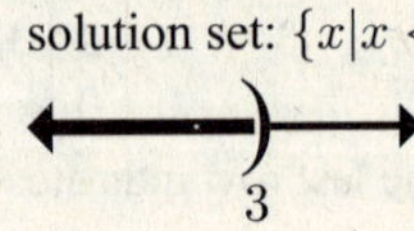

© 2015 Cengage Learning. All Rights Reserved. May not be scanned, copied or duplicated, or posted to a publicly accessible website, in whole or in part.

33.
$$\begin{aligned} -2(4x+1) &\le 14 \\ -8x-2 &\le 14 \\ -8x-2+2 &\le 14+2 \\ -8x &\le 16 \\ \tfrac{-8x}{-8} &\ge \tfrac{16}{-8} \\ x &\ge -2 \end{aligned}$$
solution set: $\{x|x \ge -2\}$

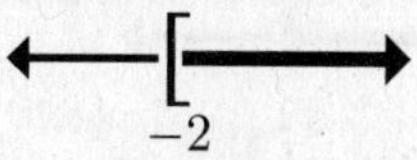

35.
$$\begin{aligned} 3(z-2) &\le 2(z+7) \\ 3z-6 &\le 2z+14 \\ 3z-6+6 &\le 2z+14+6 \\ 3x-2z &\le 2z-2z+20 \\ z &\le 20 \end{aligned}$$
solution set: $(-\infty, 20]$

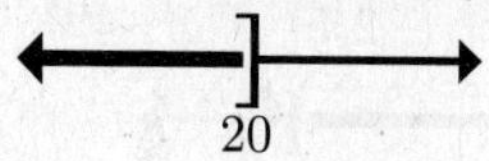

37.
$$\begin{aligned} -11(2-b) &< 4(2b+2) \\ -22+11b &< 8b+8 \\ -22+22+11b &< 8b+8+22 \\ 11b-8b &< 8b-8b+30 \\ 3b &< 30 \\ \frac{3b}{3} &< \frac{30}{3} \\ b &< 10 \end{aligned}$$
solution set: $(-\infty, 10)$

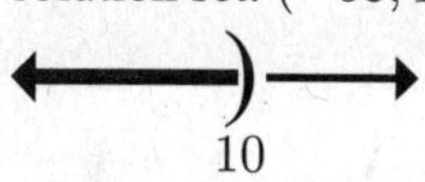

39.
$$\begin{aligned} \frac{1}{2}y+2 &> \frac{1}{3}y-4 \\ 6\left(\frac{1}{2}y+2\right) &> 6\left(\frac{1}{3}y-4\right) \\ 3y+12 &> 2y-24 \\ 3y+12-12 &> 2y-24-12 \\ 3y-2y &> 2y-2y-36 \\ y &> -36 \end{aligned}$$
solution set: $(-36, \infty)$

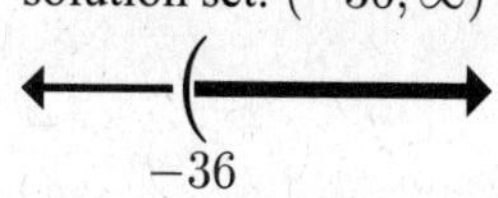

41.
$$\begin{aligned} \frac{2}{3}x+\frac{3}{2}(x-5) &\le x \\ 6\left[\frac{2}{3}x+\frac{3}{2}(x-5)\right] &\le 6x \\ 4x+9(x-5) &\le 6x \\ 4x+9x-45 &\le 6x \\ 13x-45+45 &\le 6x+45 \\ 13x-6x &\le 6x-6x+45 \\ 7x &\le 45 \\ \tfrac{7x}{7} &\le \tfrac{45}{7} \\ x &\le \tfrac{45}{7} \end{aligned}$$
solution set: $\left(-\infty, \frac{45}{7}\right]$

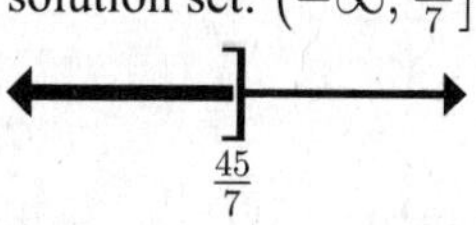

43.
$$\begin{aligned} &-4 < 2x < 8 \\ &\frac{-4}{2} < \frac{2x}{2} < \frac{8}{2} \\ &-2 < x < 4 \end{aligned}$$
solution set: $(-2, 4)$

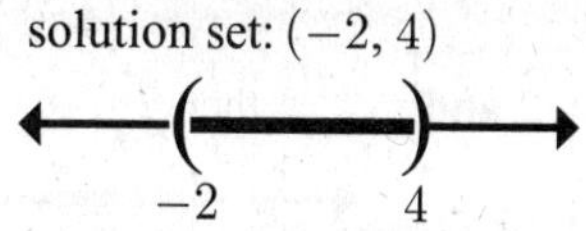

© 2015 Cengage Learning. All Rights Reserved. May not be scanned, copied or duplicated, or posted to a publicly accessible website, in whole or in part.

45.
$$\begin{aligned} -2 &< -b + 3 < 5 \\ -2 - 3 &< -b + 3 - 3 < 5 - 3 \\ -5 &< -b < 2 \\ \frac{-5}{-1} &> \frac{-b}{-1} > \frac{2}{-1} \\ 5 &> b > -2, \text{ or } -2 < b < 5 \end{aligned}$$
solution set: $(-2, 5)$

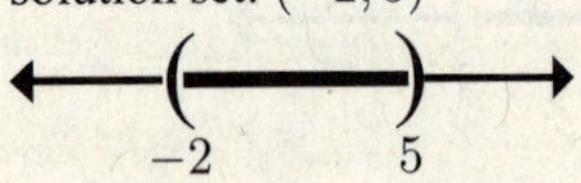

47.
$$\begin{aligned} 15 &> 2x - 7 > 9 \\ 15 + 7 &> 2x - 7 + 7 > 9 + 7 \\ 22 &> 2x > 16 \\ \frac{22}{2} &> \frac{2x}{2} > \frac{16}{2} \\ 11 &> x > 8, \text{ or } 8 < x < 11 \end{aligned}$$
solution set: $(8, 11)$

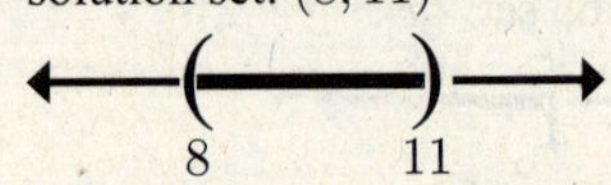

49.
$$\begin{aligned} -6 &< -3(x - 4) \le 24 \\ \frac{-6}{-3} &> \frac{-3(x-4)}{-3} \ge \frac{24}{-3} \\ 2 &> x - 4 \ge -8 \\ 2 + 4 &> x - 4 + 4 \ge -8 + 4 \\ 6 &> x \ge -4, \text{ or } -4 \le x < 6 \end{aligned}$$
solution set: $[-4, 6)$

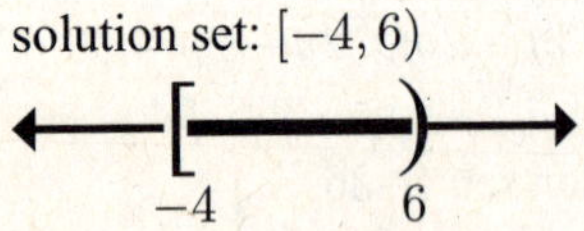

51.
$$x + 3 < 3x - 1 < 2x + 2$$
$$\begin{aligned} x + 3 &< 3x - 1 & &\text{and} & 3x - 1 &< 2x + 2 \\ x + 3 - 3 &< 3x - 1 - 3 & & & 3x - 1 + 1 &< 2x + 2 + 1 \\ x - 3x &< 3x - 3x - 4 & & & 3x - 2x &< 2x - 2x + 3 \\ -2x &< -4 & & & x &< 3 \\ \frac{-2x}{-2} &> \frac{-4}{-2} & & & & \\ x &> 2 & & & & \end{aligned}$$
If $x > 2$ **and** $x < 3$, then $2 < x < 3$. The solution set is $(2, 3)$.

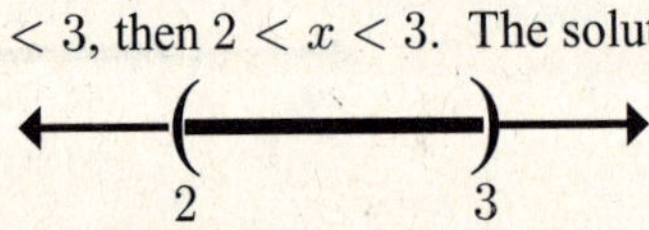

53.
$$5(x + 1) \le 4(x + 3) < 3(x - 1)$$
$$\begin{aligned} 5(x + 1) &\le 4(x + 3) & &\text{and} & 4(x + 3) &< 3(x - 1) \\ 5x + 5 - 5 &\le 4x + 12 - 5 & & & 4x + 12 - 12 &< 3x - 3 - 12 \\ 5x - 4x &\le 4x - 4x + 7 & & & 4x - 3x &< 3x - 3x - 15 \\ x &\le 7 & & & x &< -15 \end{aligned}$$
If $x \le 7$ **and** $x < -15$, then $x < -15$. The solution set is $(-\infty, -15)$.

© 2015 Cengage Learning. All Rights Reserved. May not be scanned, copied or duplicated, or posted to a publicly accessible website, in whole or in part.

55.

$$\begin{aligned} 3x+2 &< 8 & \text{or} \quad 2x-3 &> 11 \\ 3x+2-2 &< 8-2 & 2x-3+3 &> 11+3 \\ 3x &< 6 & 2x &> 14 \\ x &< 2 & x &> 7 \end{aligned}$$

If $x < 2$ **or** $x > 7$, then the solution set is $(-\infty, 2) \cup (7, \infty)$.

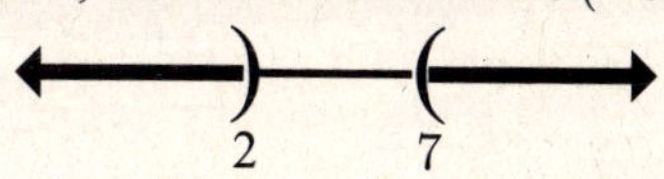

57.

$$\begin{aligned} -4(x+2) &\geq 12 & \text{or} \quad 3x+8 &< 11 \\ -4x-8+8 &\geq 12+8 & 3x+8-8 &< 11-8 \\ -4x &\geq 20 & 3x &< 3 \\ \frac{-4x}{-4} &\leq \frac{20}{-4} & x &< 1 \\ x &\leq -5 \end{aligned}$$

If $x \leq -5$ **or** $x < 1$, then $x < 1$. The solution set is $(-\infty, 1)$.

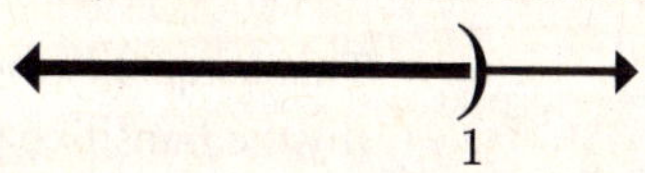

59.

$x < -3$ and $x > 3$

It is impossible for x to be both less than -3 and greater than 3. There is no solution. $\emptyset$

61.

$$\begin{aligned} 5(x-2) &\geq 0 & \text{and} \quad -3x &< 9 \\ 5x-10+10 &\geq 0+10 & \frac{-3x}{-3} &> \frac{9}{-3} \\ 5x &\geq 10 & x &> -3 \\ x &\geq 2 \end{aligned}$$

If $x \geq 2$ **and** $x > -3$, then $x \geq 2$. The solution set is $[2, \infty)$.

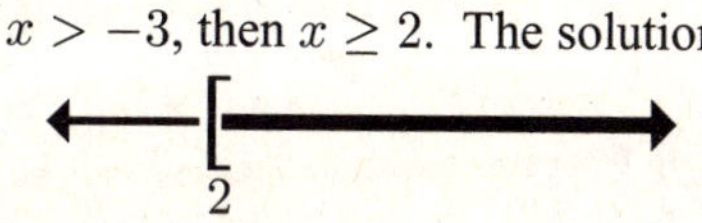

63.

$$\begin{aligned} 6x+7 &> 19 \\ 6x+7-7 &> 19-7 \\ 6x &> 12 \\ \frac{6x}{6} &> \frac{12}{6} \\ x &> 2 \end{aligned}$$

solution set: $(2, \infty)$

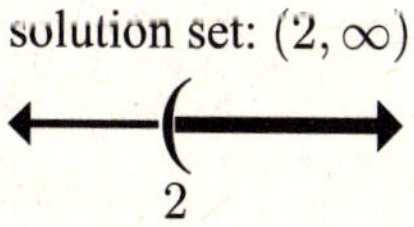

65.

$$\begin{aligned} 6x+12 &> 2x \\ 6x-2x+12 &> 2x-2x \\ 4x+12-12 &> 0-12 \\ 4x &> -12 \\ \frac{4x}{4} &> \frac{-12}{4} \\ x &> -3 \end{aligned}$$

solution set: $(-3, \infty)$

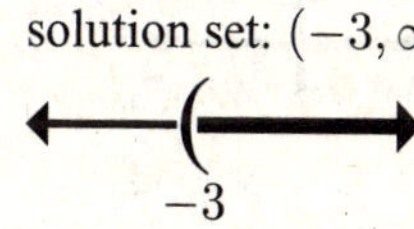

© 2015 Cengage Learning. All Rights Reserved. May not be scanned, copied or duplicated, or posted to a publicly accessible website, in whole or in part.

67.
$$\begin{aligned} -\frac{1}{3}x + 2 &\geq 1 \\ 3\left(-\frac{1}{3}x + 2\right) &\geq 3(1) \\ -x + 6 &\geq 3 \\ -x + 6 - 6 &\geq 3 - 6 \\ -x &\geq -3 \\ \frac{-x}{-1} &\leq \frac{-3}{-1} \\ x &\leq 3 \end{aligned}$$
solution set: $(-\infty, 3]$

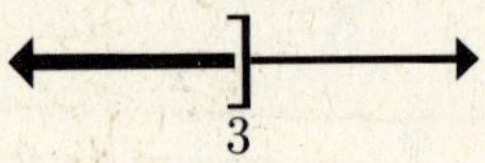

69.
$$\begin{aligned} 4(x + 5) &< 12 \\ 4x + 20 &< 12 \\ 4x + 20 - 20 &< 12 - 20 \\ 4x &< -8 \\ \frac{4x}{4} &< \frac{-8}{4} \\ x &< -2 \end{aligned}$$
solution set: $(-\infty, -2)$

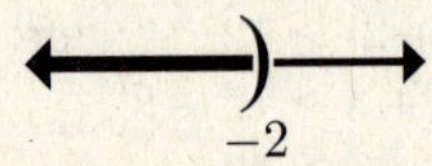

71.
$$\begin{aligned} 0.4x + 0.4 &\leq 0.1x + 0.85 \\ 0.4x - 0.1x + 0.4 &\leq 0.1x - 0.1x + 0.85 \\ 0.3x + 0.4 - 0.4 &\leq 0.85 - 0.4 \\ 0.3x &\leq 0.45 \\ \frac{0.3x}{0.3} &\leq \frac{0.45}{0.3} \\ x &\leq 1.5 \end{aligned}$$
solution set: $(-\infty, 1.5]$

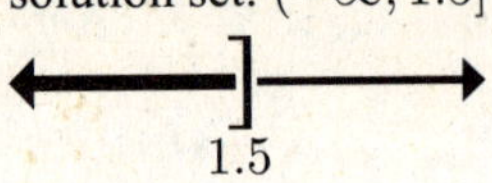

73. $0 \geq \frac{1}{2}x - 4 > 6$

This inequality indicates that $0 \geq 6$ (by the transitive property). Since this is not possible, there is no solution to the inequality.

75.
$$\begin{aligned} 0 \leq \frac{4 - x}{3} &\leq 2 \\ 3(0) \leq 3\left(\frac{4 - x}{3}\right) &\leq 3(2) \\ 0 \leq 4 - x &\leq 6 \\ 0 - 4 \leq 4 - 4 - x &\leq 6 - 4 \\ -4 \leq -x &\leq 2 \\ \frac{-4}{-1} \geq \frac{-x}{-1} &\geq \frac{2}{-1} \\ 4 \geq x \geq -2, &\text{ or } -2 \leq x \leq 4 \end{aligned}$$
solution set: $[-2, 4]$

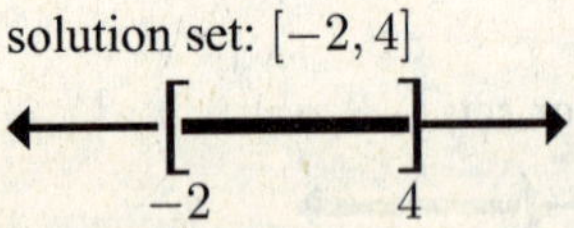

© 2015 Cengage Learning. All Rights Reserved. May not be scanned, copied or duplicated, or posted to a publicly accessible website, in whole or in part.

77.

$$\begin{aligned} x - 2 &\le 5x + 2 \\ x - 5x - 2 &\le 5x - 5x + 2 \\ -4x - 2 + 2 &\le 2 + 2 \\ -4x &\le 4 \\ \frac{-4x}{-4} &\ge \frac{4}{-4} \\ x &\ge -1 \end{aligned} \quad \text{or} \quad \begin{aligned} 5x + 2 &\le 2x - 4 \\ 5x - 2x + 2 &\le 2x - 2x - 4 \\ 3x + 2 - 2 &\le -4 - 2 \\ 3x &\le -6 \\ \frac{3x}{3} &\le \frac{-6}{3} \\ x &\le -2 \end{aligned}$$

If $x \ge -1$ **or** $x \le -2$, then the solution set is $(-\infty, -2] \cup [-1, \infty)$.

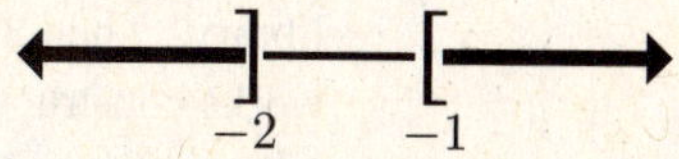

79.

$$\begin{aligned} 3x - 2 &< 4 \\ 3x - 2 + 2 &< 4 + 2 \\ 3x &< 6 \\ x &< 2 \end{aligned} \quad \text{or} \quad \begin{aligned} -6x - 1 &\le 5 \\ -6x - 1 + 1 &\le 5 + 1 \\ -6x &\le 6 \\ \frac{-6x}{-6} &\ge \frac{6}{-6} \\ x &\ge -1 \end{aligned}$$

Every real number satisfies one of these conditions. The solution set is $(-\infty, \infty)$.

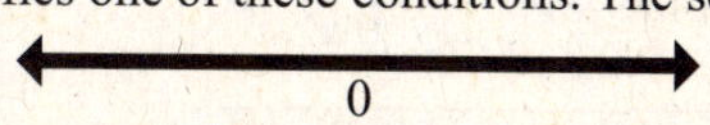

81. $2x + 3 < 5$; Graph $y = 2x + 3$ and $y = 5$ and find the x-coordinates when the first graph is below the second:

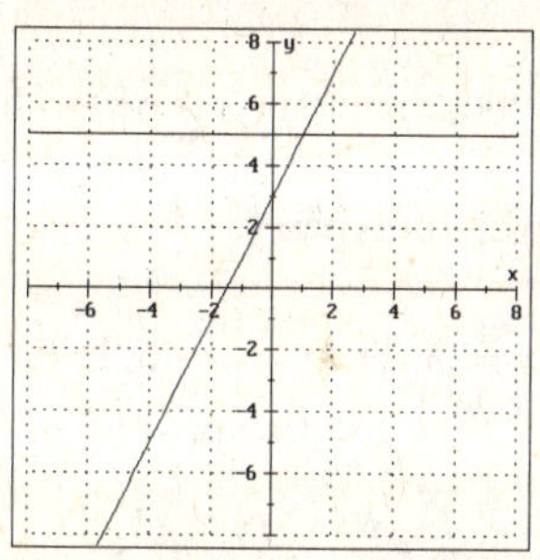

Solution: $x < 1$

83. $5x + 2 \ge -18$; Graph $y = 5x + 2$ and $y = -18$ and find the x-coordinates when the first graph is on or above the second:

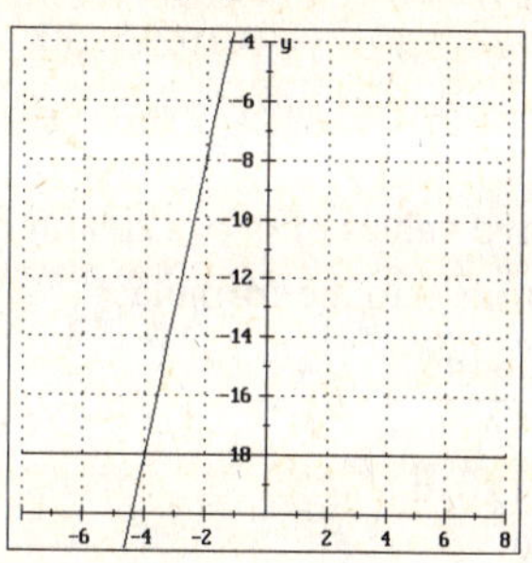

Solution: $x \ge -4$

85. Let $x =$ the number of additional hours the person rents the rototiller.

$$\begin{aligned} 15.50 + 7.95x &< 50 \\ 7.95x &< 34.50 \\ x &< 4.34 \end{aligned}$$

The person may rent the rototiller for 4 additional hours, or for 5 total hours.

87. Let $x =$ the number of children.

$$\begin{aligned} 205 + 175 + 90x &\le 750 \\ 90x &\le 370 \\ x &\le 4.11 \end{aligned}$$

At most 4 children can ride along.

© 2015 Cengage Learning. All Rights Reserved. May not be scanned, copied or duplicated, or posted to a publicly accessible website, in whole or in part.

89. Let $x =$ the amount invested at 9%.
$$\begin{aligned} 0.08(10000) + 0.09x &> 1250 \\ 800 + 0.09x &> 1250 \\ 0.09x &> 450 \\ x &> 5000 \end{aligned}$$
She must invest more than \$5000.

91. Let $x =$ the # of compact discs bought.
$$\begin{aligned} 175 + 8.50x &\leq 330 \\ 8.50x &\leq 155 \\ x &\leq 18.24 \end{aligned}$$
He can buy at most 18 discs.

93. Let $x =$ the student's score on 4th exam.
$$\begin{aligned} \frac{70 + 77 + 85 + x}{4} &\geq 80 \\ 232 + x &\geq 320 \\ x &\geq 88 \end{aligned}$$
The student needs a score of at least 88.

95. Let $x =$ the # of hours he works at the library. Then $20 - x =$ the # of hours he works construction.
$$\begin{aligned} 5x + 9(20 - x) &> 125 \\ 5x + 180 - 9x &> 125 \\ -4x &> -55 \\ x &< 13.75 \end{aligned}$$
He can work up to 13 full hours at the library.

97. Let $x =$ the total bill. Under Plan 1, the employee will pay \$100, plus 30% of the rest (the amount over \$100). This amount could be represented by $100 + 0.30(x - 100)$. Similarly, under Plan 2 the employee would pay $200 + 0.20(x - 200)$.
$$\boxed{\text{Amount paid in Plan 1}} > \boxed{\text{Amount paid in Plan 2}}$$
$$\begin{aligned} 100 + 0.30(x - 100) &> 200 + 0.20(x - 200) \\ 100 + 0.30x - 30 &> 200 + 0.20x - 40 \\ 70 + 0.30x &> 160 + 0.20x \\ 0.10x &> 90 \\ x &> 900 \end{aligned}$$
$\Rightarrow$ If the total bill is over \$900, then Plan 2 is better than Plan 1.

99. We have values of $\sigma = 120$ and $E = 20$. Substitute into the formula:
$$\begin{aligned} \frac{3.84\sigma^2}{N} &< E^2 \\ \frac{3.84(120)^2}{N} &< 20^2 \\ \frac{55296}{N} &< 400 \\ 55296 &< 400N \\ 138.24 &< N \end{aligned}$$
The sample size must be at least 139.

101. **Answers may vary.**

103. Let $x = 2$. Then $x > -3$, but $x^2 = 2^2 = 4 < 9$. No.

105. Transitive: $=$, $\leq$ and $\not>$

Exercise 4.2 (page 251)

1. $-5, 5$

3. $-12, 8$

5. $(-8, 8)$

7. $(-\infty, -2] \cup [2, \infty)$

9. $(-\infty, \infty)$

11. $(-\infty, 7) \cup (9, \infty)$

© 2015 Cengage Learning. All Rights Reserved. May not be scanned, copied or duplicated, or posted to a publicly accessible website, in whole or in part.

SECTION 4.2

13.
$$\begin{aligned} 3(2a-1) &= 2a \\ 6a-3 &= 2a \\ 4a &= 3 \\ a &= \frac{3}{4} \end{aligned}$$

15.
$$\begin{aligned} \frac{5x}{2} - 1 &= \frac{x}{3} + 12 \\ 6\left(\frac{5x}{2} - 1\right) &= 6\left(\frac{x}{3} + 12\right) \\ 15x - 6 &= 2x + 72 \\ 13x &= 78 \\ x &= 6 \end{aligned}$$

17.
$$\begin{aligned} A &= p + prt \\ A - p &= prt \\ \frac{A-p}{pr} &= t, \text{ or } t = \frac{A-p}{pr} \end{aligned}$$

19. x

21. 0

23. reflected

25. $a = b$ or $a = -b$

27. $x \le -k$ or $x \ge k$

29. $f(x) = |x| - 2$; Shift $f(x) = |x|$ down 2.

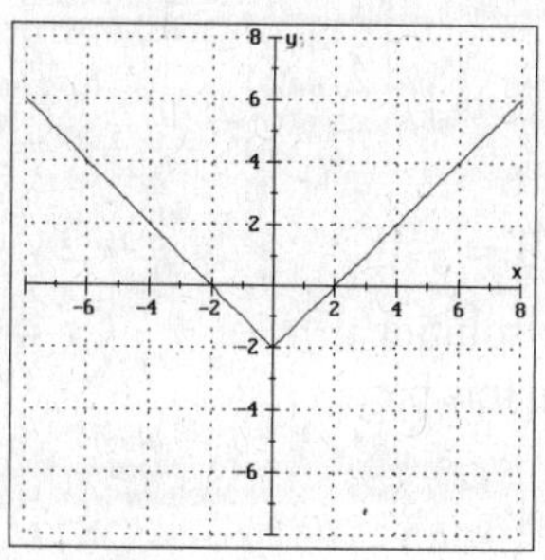

31. $f(x) = -|x+4|$; Reflect $f(x) = |x|$ about the x-axis and shift left 4.

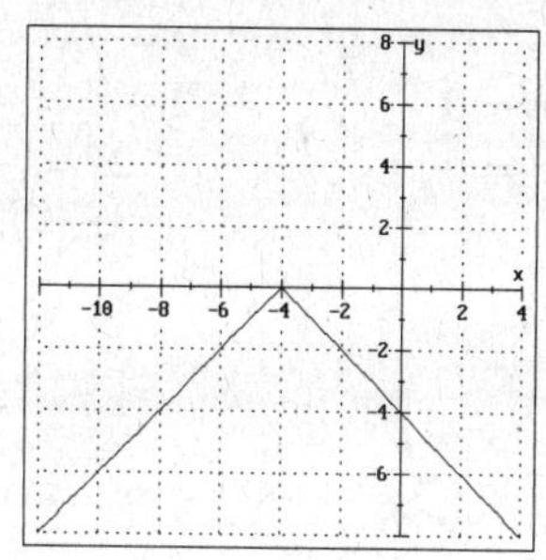

33.
$$\begin{aligned} |x| &= 4 \\ x = 4 \quad &\textbf{or} \quad x = -4 \end{aligned}$$

35.
$$\begin{aligned} |x-3| &= 6 \\ x - 3 = 6 \quad &\textbf{or} \quad x - 3 = -6 \\ x = 9 \quad & \quad\quad x = -3 \end{aligned}$$

37.
$$\begin{aligned} |2x-3| &= 5 \\ 2x - 3 = 5 \quad &\textbf{or} \quad 2x - 3 = -5 \\ 2x = 8 \quad & \quad\quad 2x = -2 \\ x = 4 \quad & \quad\quad x = -1 \end{aligned}$$

39.
$$\begin{aligned} |3-4x| &= 5 \\ 3 - 4x = 5 \quad &\textbf{or} \quad 3 - 4x = -5 \\ -4x = 2 \quad & \quad\quad -4x = -8 \\ x = -\tfrac{1}{2} \quad & \quad\quad x = 2 \end{aligned}$$

41.
$$\begin{aligned} |2x+1| - 3 &= 12 \\ |2x+1| &= 15 \\ 2x + 1 = 15 \quad &\textbf{or} \quad 2x + 1 = -15 \\ 2x = 14 \quad & \quad\quad 2x = -16 \\ x = 7 \quad & \quad\quad x = -8 \end{aligned}$$

43.
$$\begin{aligned} |x+3| + 7 &= 10 \\ |x+3| &= 3 \\ x + 3 = 3 \quad &\textbf{or} \quad x + 3 = -3 \\ x = 0 \quad & \quad\quad x = -6 \end{aligned}$$

© 2015 Cengage Learning. All Rights Reserved. May not be scanned, copied or duplicated, or posted to a publicly accessible website, in whole or in part.

SECTION 4.2

45.

$$\left|\frac{5}{2}x-3\right|+2=4$$
$$\left|\frac{5}{2}x-3\right|=2$$
$$\frac{5}{2}x-3=2 \quad \textbf{or} \quad \frac{5}{2}x-3=-2$$
$$\frac{5}{2}x=5 \qquad\qquad \frac{5}{2}x=1$$
$$x=2 \qquad\qquad x=\tfrac{2}{5}$$

47.

$$\left|\frac{2}{3}x+4\right|+3=7$$
$$\left|\frac{2}{3}x+4\right|=4$$
$$\frac{2}{3}x+4=4 \quad \textbf{or} \quad \frac{2}{3}x+4=-4$$
$$\frac{2}{3}x=0 \qquad\qquad \frac{2}{3}x=-8$$
$$x=0 \qquad\qquad x=-12$$

49.

$$|3x+24|=0$$
$$3x+24=0 \quad \textbf{or} \quad 3x+24=-0$$
$$3x=-24 \qquad\qquad 3x=-24$$
$$x=-8 \qquad\qquad x=-8$$

51.

$$|5-x|+7=7$$
$$|5-x|=0$$
$$5-x=0 \quad \textbf{or} \quad 5-x=-0$$
$$-x=-5 \qquad\qquad -x=-5$$
$$x=5 \qquad\qquad x=5$$

53.

$$|2-x|=|3x+2|$$
$$2-x=3x+2 \quad \textbf{or} \quad 2-x=-(3x+2)$$
$$-4x=0 \qquad\qquad 2-x=-3x-2$$
$$x=0 \qquad\qquad 2x=-4$$
$$x=-2$$

55.

$$\left|\frac{x}{2}+2\right|=\left|\frac{x}{2}-2\right|$$
$$\frac{x}{2}+2=\frac{x}{2}-2 \quad \textbf{or} \quad \frac{x}{2}+2=-\left(\frac{x}{2}-2\right)$$
$$0=-4 \qquad\qquad \frac{x}{2}+2=-\frac{x}{2}+2$$

No solution from this part. $\qquad x=0$

57. $|2x|<8$

$$-8<2x<8$$
$$\frac{-8}{2}<\frac{2x}{2}<\frac{8}{2}$$
$$-4<x<4$$

solution set: $(-4,4)$

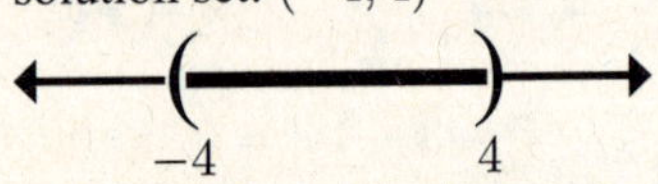

59. $|x+9|\le 12$

$$-12\le x+9\le 12$$
$$-12-9\le x+9-9\le 12-9$$
$$-21\le x\le 3$$

solution set: $[-21,3]$

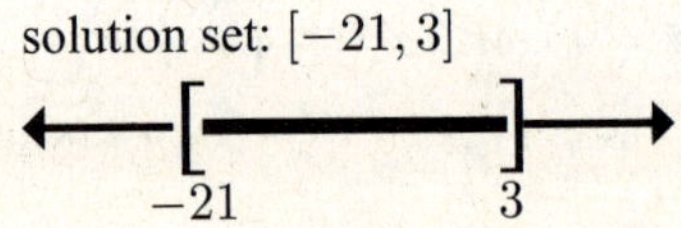

61. $\left|\frac{1}{2}x-3\right|-4<2$

$$\left|\tfrac{1}{2}x-3\right|<6$$
$$-6<\tfrac{1}{2}x-3<6$$
$$-3<\tfrac{1}{2}x<9$$
$$-6<x<18$$

solution set: $(-6,18)$

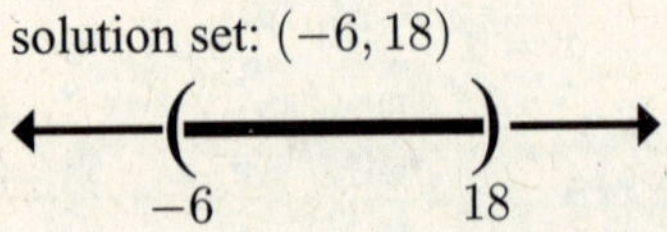

63.

$$-|5x-1|+2<0$$
$$-|5x-1|<-2$$
$$|5x-1|>2$$
$$5x-1<-2 \quad \textbf{or} \quad 5x-1>2$$
$$5x<-1 \qquad\qquad 5x>3$$
$$x<-\tfrac{1}{5} \qquad\qquad x>\tfrac{3}{5}$$

solution set: $\left(-\infty,-\tfrac{1}{5}\right)\cup\left(\tfrac{3}{5},\infty\right)$

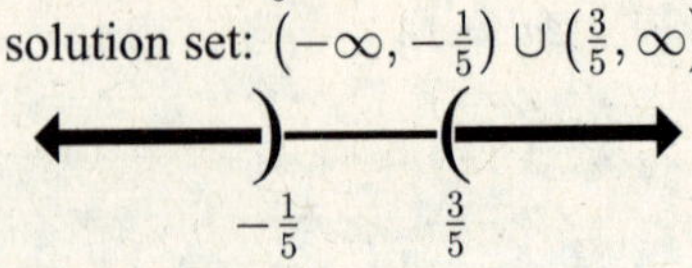

© 2015 Cengage Learning. All Rights Reserved. May not be scanned, copied or duplicated, or posted to a publicly accessible website, in whole or in part.

SECTION 4.2

65.

$$|5x| + 2 > 7$$
$$|5x| > 5$$
$$5x < -5 \quad \textbf{or} \quad 5x > 5$$
$$x < -1 \qquad\qquad x > 1$$

solution set: $(-\infty, -1) \cup (1, \infty)$

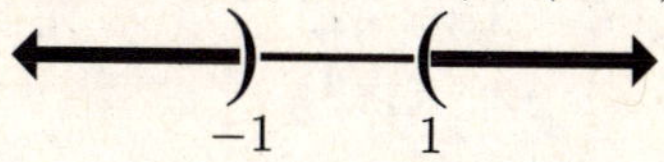

67.

$$|2 - 3x| \geq 8$$
$$2 - 3x \leq -8 \quad \textbf{or} \quad 2 - 3x \geq 8$$
$$-3x \leq -10 \qquad\qquad -3x \geq 6$$
$$x \geq \tfrac{10}{3} \qquad\qquad x \leq -2$$

solution set: $(-\infty, -2] \cup [\frac{10}{3}, \infty)$

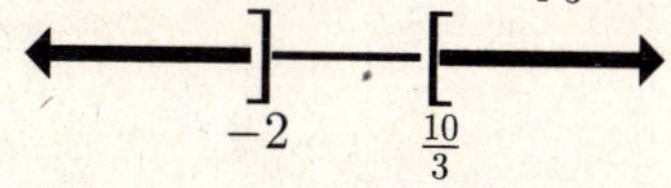

69.

$$\left|\frac{x-2}{3}\right| > 4$$
$$\frac{x-2}{3} < -4 \quad \textbf{or} \quad \frac{x-2}{3} > 4$$
$$x - 2 < -12 \qquad\qquad x - 2 > 12$$
$$x < -10 \qquad\qquad x > 14$$

solution set: $(-\infty, -10) \cup (14, \infty)$

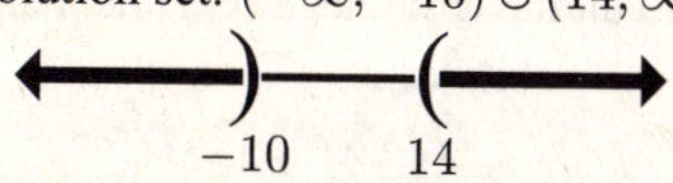

71.

$$\left|\frac{x-7}{2}\right| > 4$$
$$\frac{x-7}{2} < -4 \quad \textbf{or} \quad \frac{x-7}{2} > 4$$
$$x - 7 < -8 \qquad\qquad x - 7 > 8$$
$$x < -1 \qquad\qquad x > 15$$

solution set: $(-\infty, -1) \cup (15, \infty)$

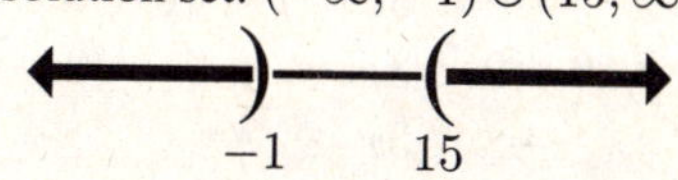

73.

$$\left|\frac{5-x}{2}\right| - \frac{3}{2} > 3$$
$$\left|\frac{5-x}{2}\right| > \frac{9}{2}$$
$$\frac{5-x}{2} < -\frac{9}{2} \quad \textbf{or} \quad \frac{5-x}{2} > \frac{9}{2}$$
$$5 - x < -9 \qquad\qquad 5 - x > 9$$
$$-x < -14 \qquad\qquad -x > 4$$
$$x > 14 \qquad\qquad x < -4$$

solution set: $(-\infty, -4) \cup (14, \infty)$

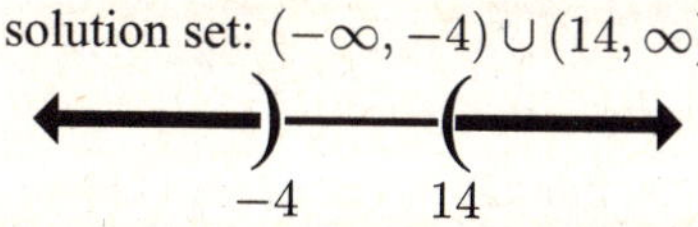

75.

$$\left|\frac{1}{3}x + 7\right| + 5 > 6$$
$$\left|\frac{1}{3}x + 7\right| > 1$$
$$\frac{1}{3}x + 7 < -1 \quad \textbf{or} \quad \frac{1}{3}x + 7 > 1$$
$$\tfrac{1}{3}x < -8 \qquad\qquad \tfrac{1}{3}x > -6$$
$$x < -24 \qquad\qquad x > -18$$

solution set: $(-\infty, -24) \cup (-18, \infty)$

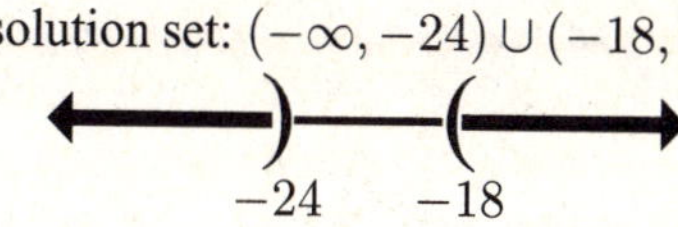

77.

$$|x - 7| \geq -3$$

Since an absolute value is always at least 0, this inequality is true for all real numbers.

solution set: $(-\infty, \infty)$

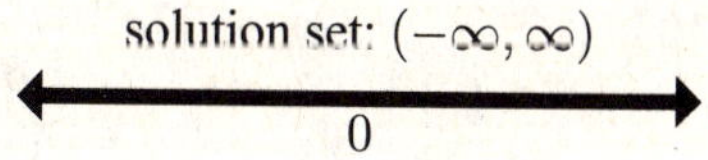

79.

$$|4x + 3| > -5$$

Since an absolute value is always at least 0, this inequality is true for all real numbers.

solution set: $(-\infty, \infty)$

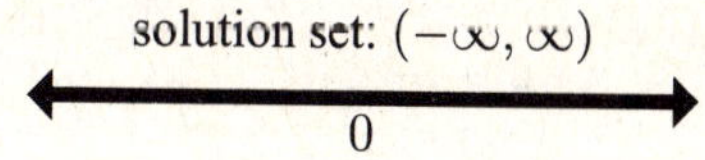

81.

$$|3x - 30| + 10 \leq 4$$
$$|3x - 30| \leq -6$$

Since an absolute value can never be negative, this inequality has no solution. $\emptyset$

83.

$$\left|\tfrac{2}{7}x - 3\right| + 5 < 3$$
$$\left|\tfrac{2}{7}x - 3\right| < -2$$

Since an absolute value can never be negative, this inequality has no solution. $\emptyset$

© 2015 Cengage Learning. All Rights Reserved. May not be scanned, copied or duplicated, or posted to a publicly accessible website, in whole or in part.

SECTION 4.2

85.
$$\left|\frac{x}{2}-1\right|=3$$
$$\frac{x}{2}-1=3 \quad \textbf{or} \quad \frac{x}{2}-1=-3$$
$$\frac{x}{2}=4 \qquad \frac{x}{2}=-2$$
$$x=8 \qquad x=-4$$

87.
$$\left|\frac{3x+48}{3}\right|=12$$
$$\frac{3x+48}{3}=12 \quad \textbf{or} \quad \frac{3x+48}{3}=-12$$
$$3x+48=36 \qquad 3x+48=-36$$
$$3x=-12 \qquad 3x=-84$$
$$x=-4 \qquad x=-28$$

89.
$$|3x+2|=|5x-6|$$
$$3x+2=5x-6 \quad \textbf{or} \quad 3x+2=-(5x-6)$$
$$-2x=-8 \qquad 3x+2=-5x+6$$
$$x=4 \qquad 8x=4$$
$$x=\tfrac{1}{2}$$

91.
$$|3x+7|=-|8x-2|$$
Since an absolute value cannot be negative, there is no solution. $\emptyset$

93. $|3x-2|<10$
$$-10<3x-2<10$$
$$-10+2<3x-2+2<10+2$$
$$-8<3x<12$$
$$\frac{-8}{3}<\frac{3x}{3}<\frac{12}{3}$$
$$-\tfrac{8}{3}<x<4$$
solution set: $(-\tfrac{8}{3},4)$

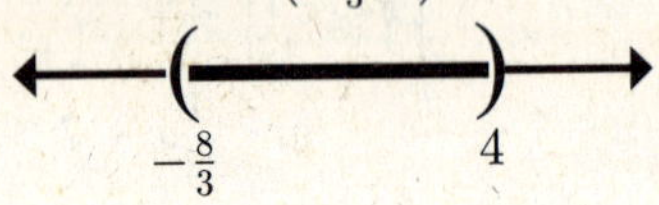

95. $|5x-12|<-5$; Since an absolute value can never be negative, this inequality has no solution. $\emptyset$

97.
$$|3x+2|>14$$
$$3x+2<-14 \quad \textbf{or} \quad 3x+2>14$$
$$3x<-16 \qquad 3x>12$$
$$x<-\tfrac{16}{3} \qquad x>4$$
solution set: $(-\infty,-\tfrac{16}{3})\cup(4,\infty)$

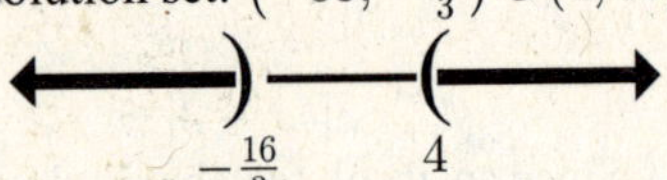

99.
$$|4x+3|>0$$
$$4x+3<-0 \quad \textbf{or} \quad 4x+3>0$$
$$4x<-3 \qquad 4x>-3$$
$$x<-\tfrac{3}{4} \qquad x>-\tfrac{3}{4}$$
solution set: $(-\infty,-\tfrac{3}{4})\cup(-\tfrac{3}{4},\infty)$

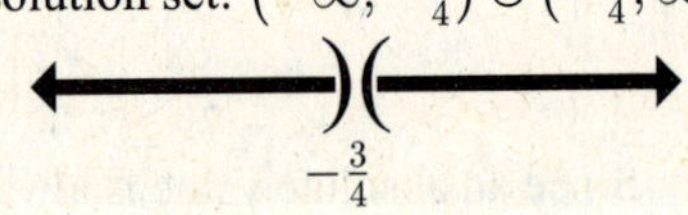

101.
$$|2x+10|=0$$
$$2x+10=0 \quad \textbf{or} \quad 2x+10=-0$$
$$2x=-10 \qquad 2x=-10$$
$$x=-5 \qquad x=-5$$

103.
$$|3x+2|=16$$
$$3x+2=16 \quad \textbf{or} \quad 3x+2=-16$$
$$3x=14 \qquad 3x=-18$$
$$x=\tfrac{14}{3} \qquad x=-6$$

105.
$$|3x-1|=|x+5|$$
$$3x-1=x+5 \quad \textbf{or} \quad 3x-1=-(x+5)$$
$$2x=6 \qquad 3x-1=-x-5$$
$$x=3 \qquad 4x=-4$$
$$x=-1$$

© 2015 Cengage Learning. All Rights Reserved. May not be scanned, copied or duplicated, or posted to a publicly accessible website, in whole or in part.

SECTION 4.2

107.

$$|x + \tfrac{1}{3}| = |x - 3|$$

$$x + \tfrac{1}{3} = x - 3 \quad \textbf{or} \quad x + \tfrac{1}{3} = -(x - 3)$$

$$3x + 1 = 3x - 9 \qquad x + \tfrac{1}{3} = -x + 3$$

$$0 = -10 \qquad 3x + 1 = -3x + 9$$

(no solution from this part) $\qquad 6x = 8$

$$x = \tfrac{4}{3}$$

109. $|4x - 1| \le 7$

$$-7 \le 4x - 1 \le 7$$

$$-7 + 1 \le 4x - 1 + 1 \le 7 + 1$$

$$-6 \le 4x \le 8$$

$$\frac{-6}{4} \le \frac{4x}{4} \le \frac{8}{4}$$

$$-\tfrac{3}{2} \le x \le 2$$

solution set: $\left[-\tfrac{3}{2}, 2\right]$

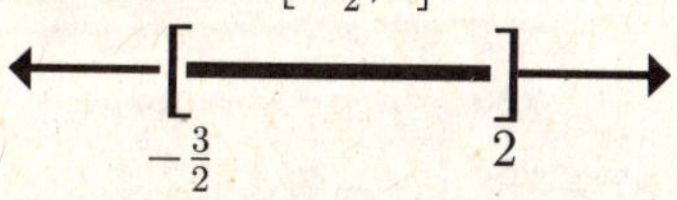

111.

$$3|2x + 5| \ge 9$$

$$|2x + 5| \ge 3$$

$$2x + 5 \le -3 \quad \textbf{or} \quad 2x + 5 \ge 3$$

$$2x \le -8 \qquad 2x \ge -2$$

$$x \le -4 \qquad x \ge -1$$

solution set: $(-\infty, -4] \cup [-1, \infty)$

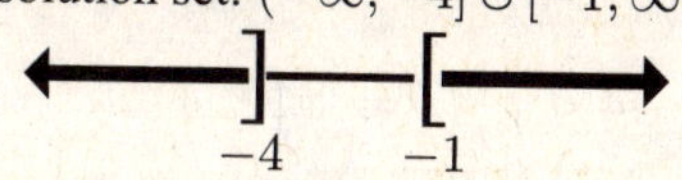

113.

$$-|2x - 3| < -7$$

$$|2x - 3| > 7$$

$$2x - 3 < -7 \quad \textbf{or} \quad 2x - 3 > 7$$

$$2x < -4 \qquad 2x > 10$$

$$x < -2 \qquad x > 5$$

solution set: $(-\infty, -2) \cup (5, \infty)$

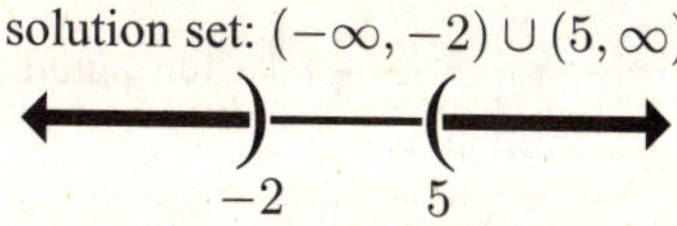

115.

$$|3x - 2| + 2 \ge 0$$

$$|3x - 2| \ge -2$$

Since an absolute value is always at least 0, this inequality is true for all real numbers.

solution set: $(-\infty, \infty)$

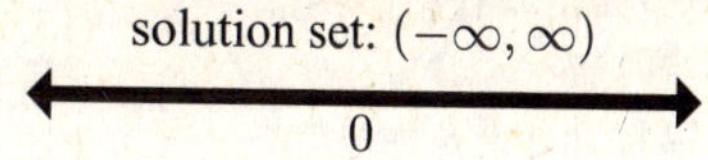

117.

$$|x - 12| > 24$$

$$x - 12 < -24 \quad \textbf{or} \quad x - 12 > 24$$

$$x < -12 \qquad x > 36$$

solution set: $(-\infty, -12) \cup (36, \infty)$

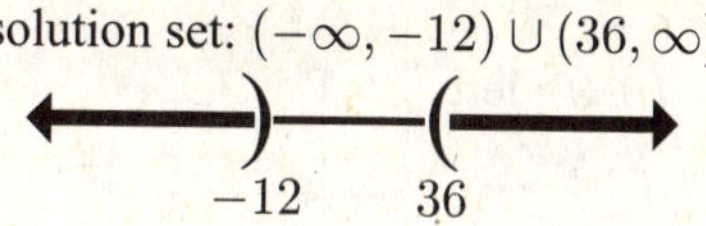

119. $|3x + 2| \le -3$

Since an absolute value can never be negative, this inequality has no solution. $\emptyset$

121.

$$\left|\frac{3}{5}x - 4\right| - 2 = -2$$

$$\left|\frac{3}{5}x - 4\right| = 0$$

$$\frac{3}{5}x - 4 = 0 \quad \textbf{or} \quad \frac{3}{5}x - 4 = -0$$

$$\frac{3}{5}x = 4 \qquad \frac{3}{5}x = 4$$

$$x = \tfrac{20}{3} \qquad x = \tfrac{20}{3}$$

© 2015 Cengage Learning. All Rights Reserved. May not be scanned, copied or duplicated, or posted to a publicly accessible website, in whole or in part.

SECTION 4.2

123.
$$\left|\tfrac{x-5}{10}\right| + 5 \le 5$$
$$\left|\tfrac{x-5}{10}\right| \le 0$$
Since an absolute value can never be less than zero, the only solution for this inequality is when the absolute value is equal to 0.
$$\left|\tfrac{x-5}{10}\right| = 0$$
$\frac{x-5}{10} = 0$ **or** $\frac{x-5}{10} = -0$
$x - 5 = 0 \qquad x - 5 = 0$
$x = 5 \qquad x = 5$
solution set: $[5, 5]$

125. $|x - 21| = -8$
Since an absolute value can never be negative, this equation has no solution. $\emptyset$

127. $-4 < x < 4 \Rightarrow |x| < 4$

129. $x + 3 < -6$ or $x + 3 > 6 \Rightarrow |x + 3| > 6$

131. $|d - 5| \le 1$
$$-1 \le d - 5 \le 1$$
$$-1 + 5 \le d - 5 + 5 \le 1 + 5$$
$$4 \text{ ft} \le d \le 6 \text{ ft}$$

133. $|t - 78°| \le 8°$
$$-8° \le t - 78° \le 8°$$
$$-8° + 78° \le t - 78° + 78° \le 8° + 78°$$
$$70° \le t \le 86°$$

135. $0.6° - 0.5° \le c \le 0.6° + 0.5°$
$$-0.5° \le c - 0.6° \le 0.5°$$
$$|c - 0.6°| \le 0.5°$$

137-139. Answers may vary.

141. $|x| + k = 0$
$$|x| = -k$$
If this equation has exactly two solutions, then $-k > 0$, or $k < 0$.

143. $|x| + |y| > |x + y|$ when x and y have different signs.

Exercise 4.3 (page 259)

1. (0, 0): below

3. (6, 1): above

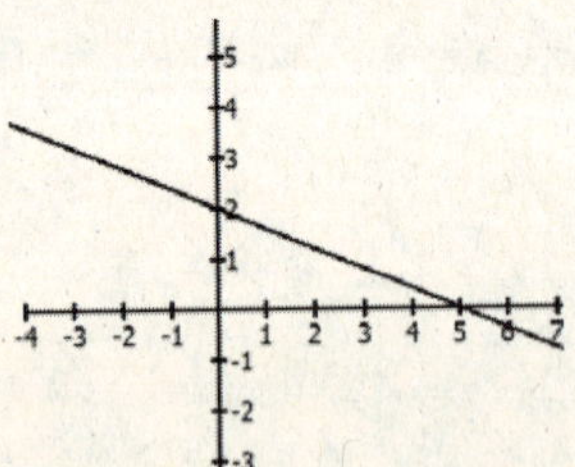

5. (0, 0): left

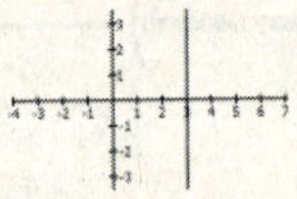

7. (0, 0): above

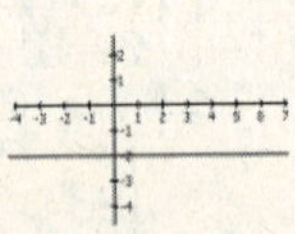

© 2015 Cengage Learning. All Rights Reserved. May not be scanned, copied or duplicated, or posted to a publicly accessible website, in whole or in part.

SECTION 4.3

9.

$$\begin{aligned} x+y&=3\\ x-y&=5\\ \hline 2x\quad&=8\\ x\quad&=4 \end{aligned}$$

Substitute and solve for y:

$$\begin{aligned} x+y&=3\\ 4+y&=3\\ y&=-1 \end{aligned}$$

The solution is $(4,-1)$.

11.

$$\begin{aligned} 4x+3y=12 &\Rightarrow \times 2 & 8x+6y&=24\\ 5x-2y=15 &\Rightarrow \times 3 & 15x-6y&=45\\ & & \hline 23x\quad&=69\\ & & x\quad&=3 \end{aligned}$$

Substitute and solve for y:

$$\begin{aligned} 4x+3y&=12\\ 4(3)+3y&=12\\ 3y&=0\\ y&=0 \end{aligned}$$

The solution is $(3,0)$.

13. linear

15. edge

17. dashed

19. $y \geq x$

Boundary (solid) $y = x$

x	y
0	0
2	2

Test point: $(1,2)$

$$\begin{aligned} y &\geq x\\ 2 &\stackrel{?}{\geq} 1\\ 2 &\geq 1 \end{aligned}$$

same half-plane

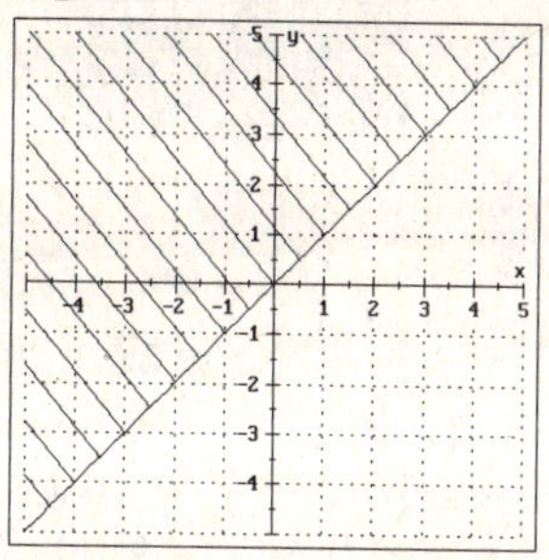

21. $y \geq -2$

The boundary will be a solid line.
Shade all points above $y=-2$.

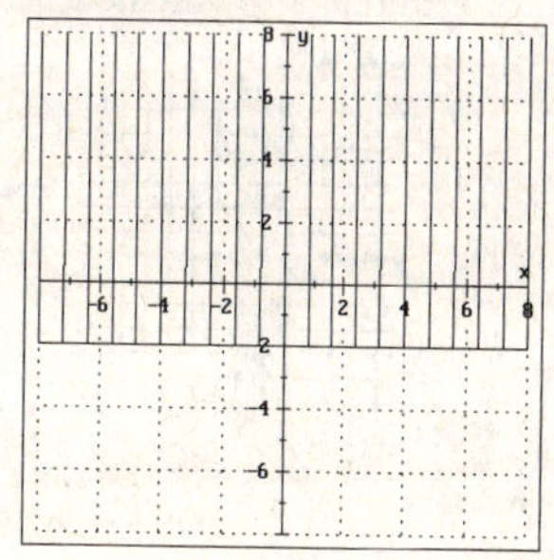

23. $y > x+1$

Boundary (dashed) $y = x+1$

x	y
0	1
2	3

Test point: $(0,0)$

$$\begin{aligned} y &> x+1\\ 0 &\stackrel{?}{>} 0+1\\ 0 &\not> 1 \end{aligned}$$

opposite half-plane

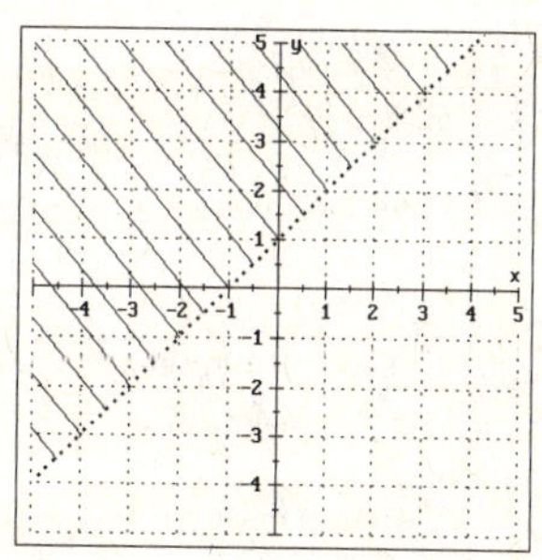

© 2015 Cengage Learning. All Rights Reserved. May not be scanned, copied or duplicated, or posted to a publicly accessible website, in whole or in part.

SECTION 4.3

25. $x < 4$

The boundary will be a dashed line.

Shade all points left of $x = 4$.

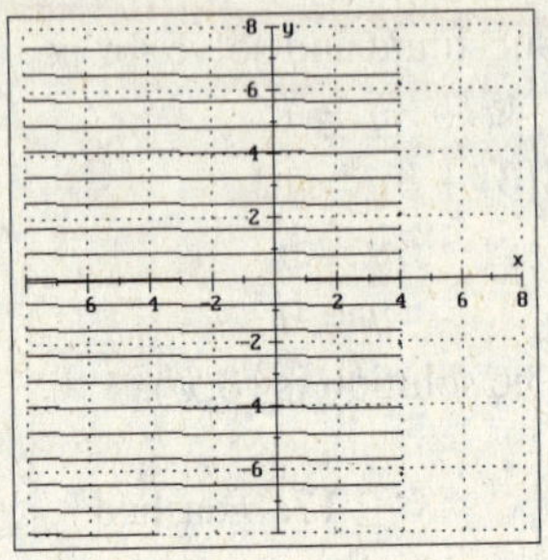

27. $-2 \leq x < 0$

Shade points between $x = -2$ (solid) and $x = 0$ (dashed).

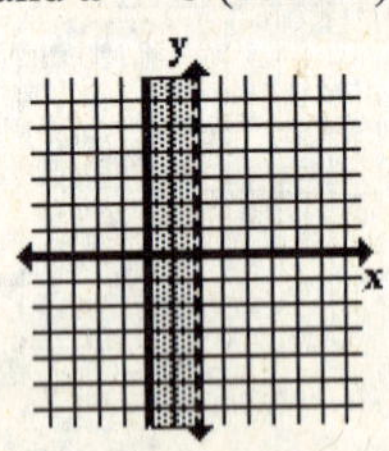

29. $y < -2$ or $y > 3$

Shade points below $y = -2$ (dashed) and points above $y = 3$ (dashed).

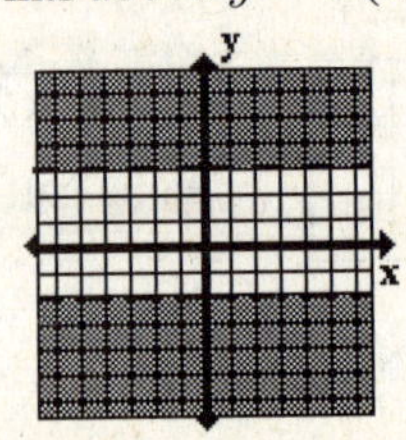

31. $y \geq 1 - \dfrac{3}{2}x$

Boundary (solid)

$y = 1 - \frac{3}{2}x$

x	y
0	1
2	-2

Test point: $(0, 0)$

$y \geq 1 - \frac{3}{2}x$

$0 \overset{?}{\geq} 1 - \frac{3}{2}(0)$

$0 \not\geq 1$

opposite half-plane

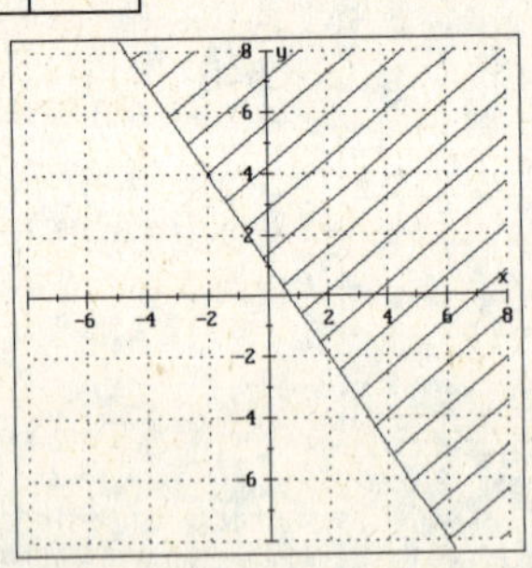

33. $0.5x + y > 1.5 + x$

$y > 0.5x + 1.5$

Boundary (dashed)

$y = 0.5x + 1.5$

x	y
1	2
3	3

Test point: $(0, 0)$

$y > 0.5x + 1.5$

$0 \overset{?}{>} 0.5(0) + 1.5$

$0 \not> 1.5$

opposite half-plane

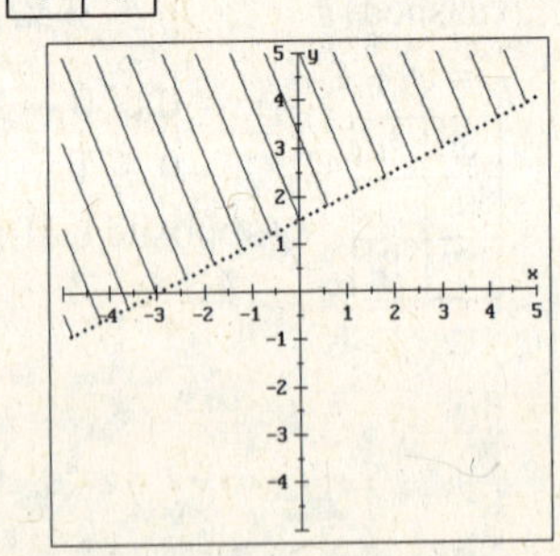

© 2015 Cengage Learning. All Rights Reserved. May not be scanned, copied or duplicated, or posted to a publicly accessible website, in whole or in part.

SECTION 4.3

35. $y < \dfrac{1}{3}x - 1$

Boundary (dashed)

$y = \frac{1}{3}x - 1$

x	y
0	−1
3	0

Test point: $(0,0)$

$y < \frac{1}{3}x - 1$

$0 \overset{?}{<} \frac{1}{3}(0) - 1$

$0 \not< -1$

opposite half-plane

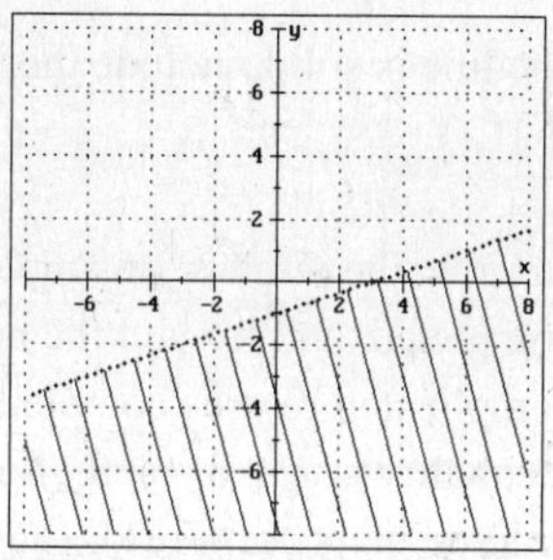

37. $3x \geq -y + 3$

Boundary (solid)

$3x = -y + 3$

x	y
0	3
1	0

Test point: $(0,0)$

$3x \geq -y + 3$

$3(0) \overset{?}{\geq} -(0) + 3$

$0 \not\geq 3$

opposite half-plane

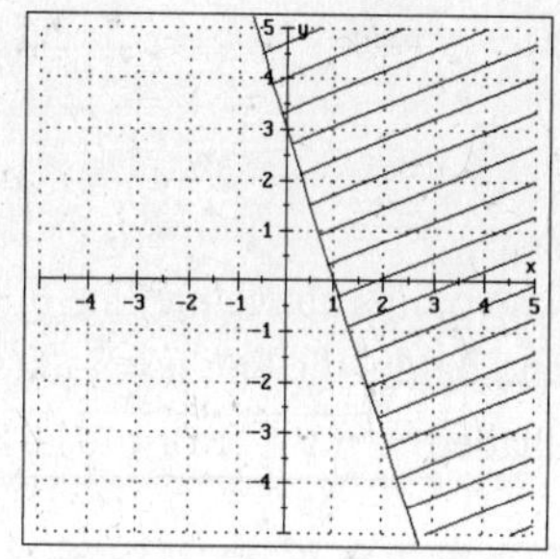

39. $y < 0.27x - 1$

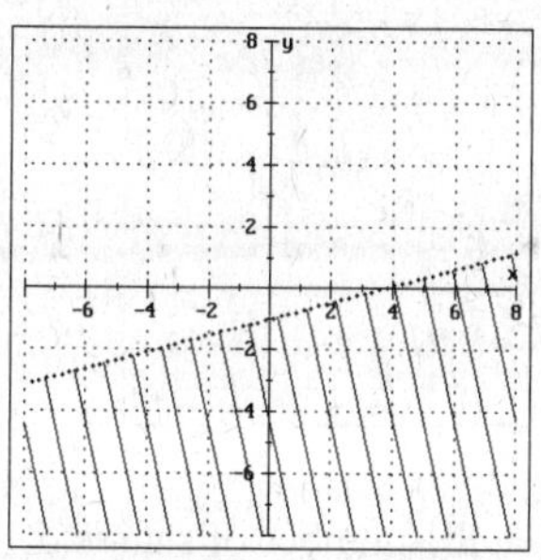

41. $y \geq -2.37x + 1.5$

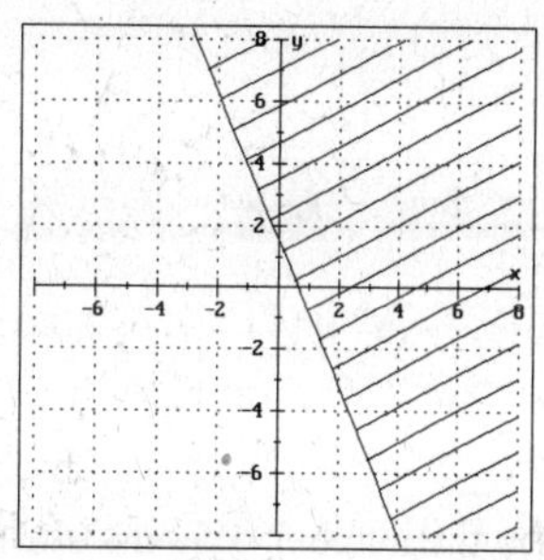

43. The boundary line is $x = 3$. Since points with x-coordinates less than 3 are shaded, and since the line is solid, the inequality is $\boxed{x \leq 3}$.

45. The boundary line goes through $(2,0)$ and $(0,3)$. Use point-slope form to find the equation:

$$m = \frac{\Delta y}{\Delta x} = \frac{3-0}{0-2} = \frac{3}{-2} = -\frac{3}{2}$$

$$y - y_1 = m(x - x_1)$$

$$y - 0 = -\frac{3}{2}(x - 2)$$

$$y = -\frac{3}{2}x + 3$$

Pick a point which is shaded, like (2, 1), and substitute its coordinates for x and y:

$y \,\square\, -\frac{3}{2}x + 3$

$1 \,\square\, -\frac{3}{2}(2) + 3$

$1 > -3 + 3$

Since the line is dashed, do not include the line:

$\boxed{y > -\frac{3}{2}x + 3}$

© 2015 Cengage Learning. All Rights Reserved. May not be scanned, copied or duplicated, or posted to a publicly accessible website, in whole or in part.

SECTION 4.3

47. The points between the lines $x = -2$ and $x = 3$ are shaded. Both lines are solid, so they should be included. $\boxed{-2 \leq x \leq 3}$

49. The boundary line goes through $(0, 0)$ and $(1, 1)$. Use point-slope form to find the equation:

$$m = \frac{\Delta y}{\Delta x} = \frac{1-0}{1-0} = \frac{1}{1} = 1$$

$$y - y_1 = m(x - x_1)$$

$$y - 0 = 1(x - 0)$$

$$y = x$$

Pick a point which is shaded, and substitute the coordinates for x and y:

$y \,\square\, x$

$0 < 1$

Since the line is solid, include the line:

$\boxed{y \leq x}$

51. The points above the line $y = -1$ (dashed) and below the line $y = -3$ (solid) are shaded. $\boxed{y > -1 \text{ or } y \leq -3}$

53. Let $x =$ the number of simple returns completed, and let $y =$ the number of complicated returns completed. The inequality is $x + 3y \leq 9$. Some ordered pairs are: $(1, 1)$, $(2, 1)$, $(2, 2)$

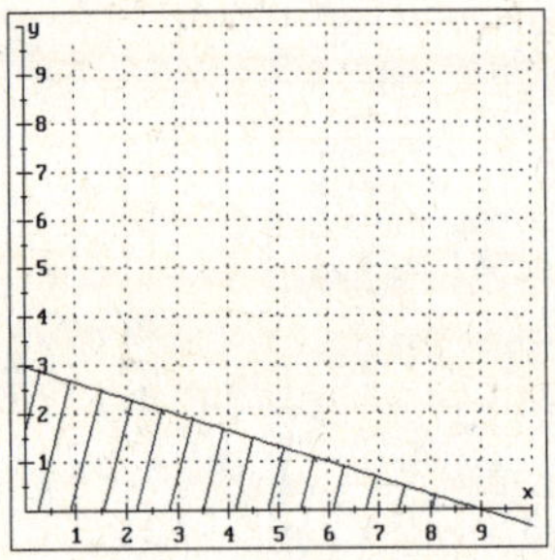

55. Let $x =$ the number of hours she uses the first, and let $y =$ the number of hours she uses the second. The inequality is $6x + 7y \leq 42$. Some ordered pairs are: $(2, 2)$, $(3, 3)$, $(5, 1)$

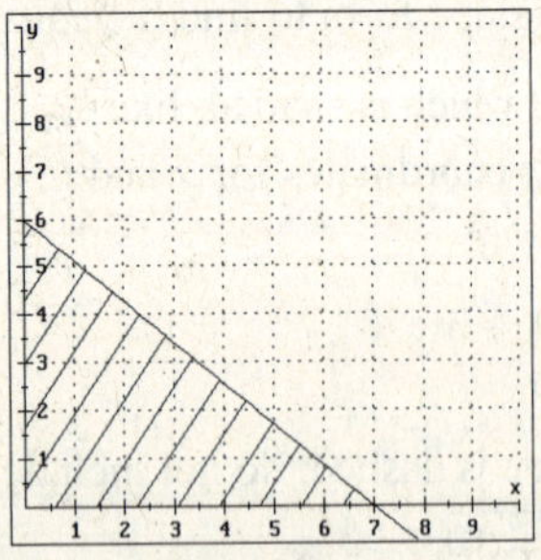

57. Let $x =$ the number of shares of Traffico, and let $y =$ the number of shares of Cleanco. The inequality is $50x + 60y \leq 6000$. Some ordered pairs are: $(40, 20)$, $(60, 40)$, $(80, 20)$

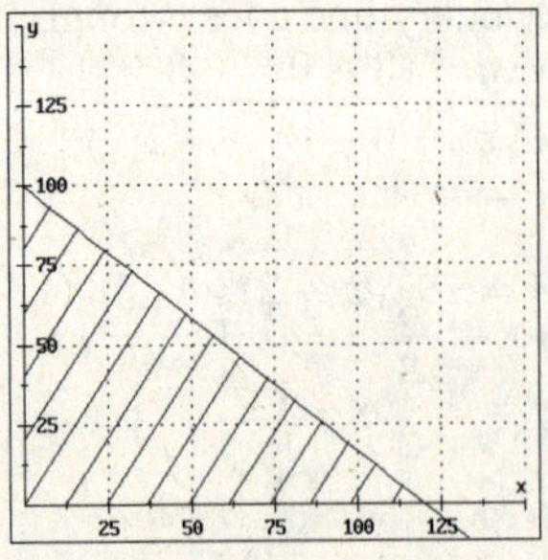

59. **Answers may vary.**

61. An inequality such as $x + (-x) < 10$ is an identity.

© 2015 Cengage Learning. All Rights Reserved. May not be scanned, copied or duplicated, or posted to a publicly accessible website, in whole or in part.

Exercise 4.4 (page 267)

1.

$$y \;\boxed{\phantom{<}}\; x+2 \qquad y \;\boxed{\phantom{<}}\; x-2$$
$$0 \;\boxed{\phantom{<}}\; 1+2 \qquad 0 \;\boxed{\phantom{<}}\; 1-2$$
$$0 \;\boxed{<}\; 3 \qquad 0 \;\boxed{>}\; -1$$
$$y < x+2 \qquad y > x-2$$

3.

$$A = p + prt$$
$$A - p = prt$$
$$\frac{A-p}{pr} = t, \text{ or } t = \frac{A-p}{pr}$$

5.

$$z = \frac{x-\mu}{\sigma}$$
$$z\sigma = x - \mu$$
$$z\sigma + \mu = x, \text{ or } x = z\sigma + \mu$$

7.

$$l = a + (n-1)d$$
$$l - a = (n-1)d$$
$$\frac{l-a}{n-1} = d, \text{ or } d = \frac{l-a}{n-1}$$

9. intersect

11. $\begin{cases} y < 3x + 2 \\ y < -2x + 3 \end{cases}$

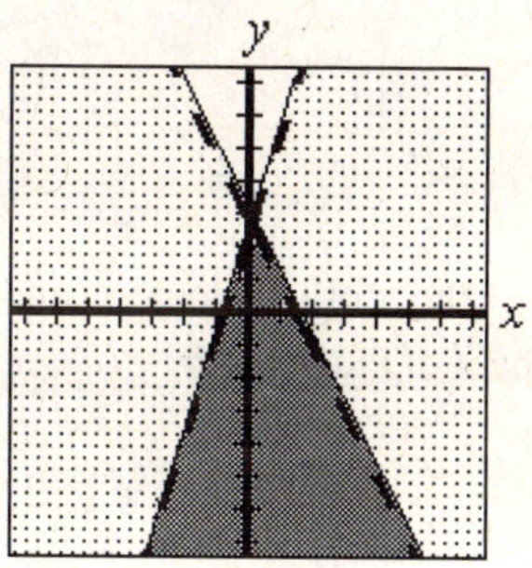

13. $\begin{cases} 3x + 2y > 6 \\ x + 3y \le 2 \end{cases}$

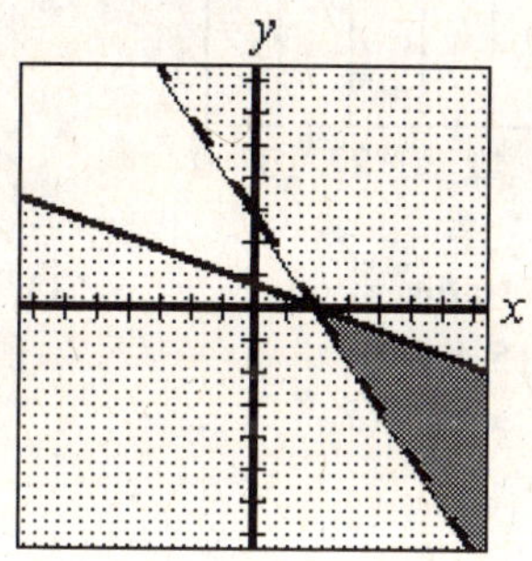

15. $\begin{cases} 2x + 3y \le 6 \\ 3x + y \le 1 \\ x \le 0 \end{cases}$

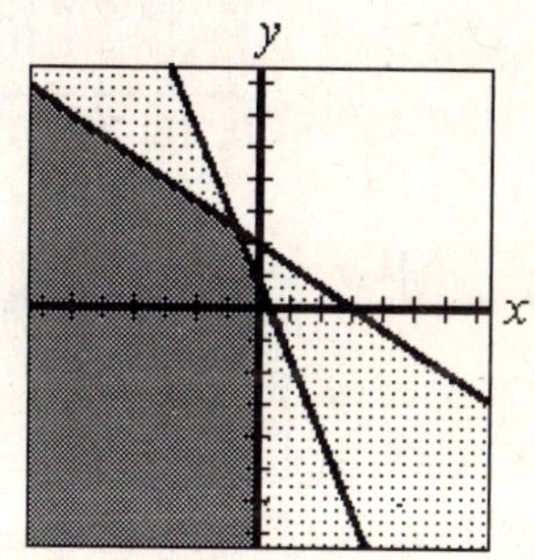

17. $\begin{cases} x - y < 4 \\ y \le 0 \\ x \ge 0 \end{cases}$

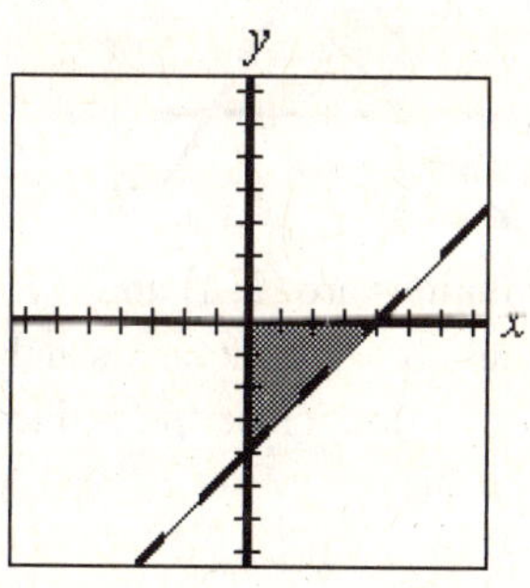

19. $\begin{cases} 2x - y > 4 \\ y < -x^2 + 2 \end{cases}$

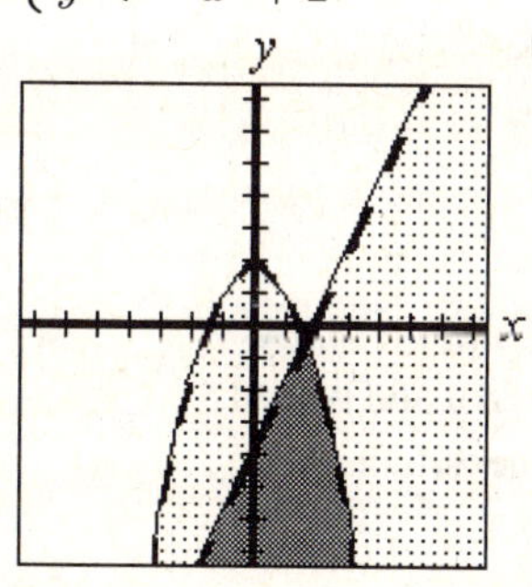

21. $\begin{cases} y > x^2 - 4 \\ y < -x^2 + 4 \end{cases}$

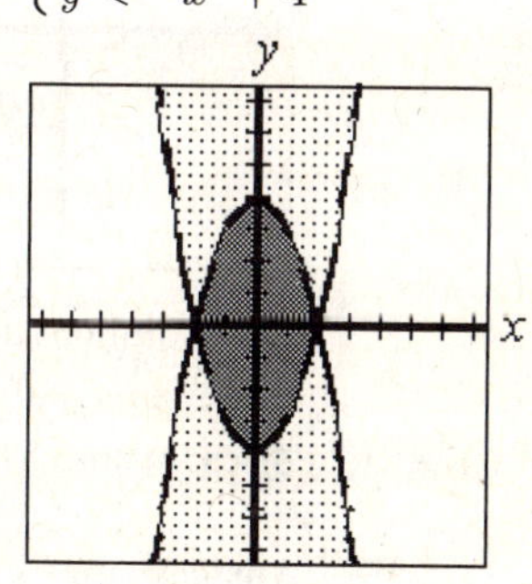

© 2015 Cengage Learning. All Rights Reserved. May not be scanned, copied or duplicated, or posted to a publicly accessible website, in whole or in part.

SECTION 4.4

23. $\begin{cases} 3x + y \leq 1 \\ -x + 2y \geq 6 \end{cases}$

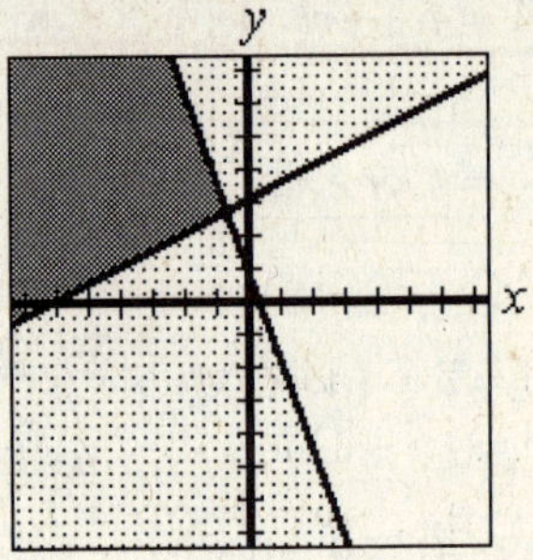

25. $\begin{cases} x \geq 0, y \geq 0 \\ 9x + 3y \leq 18 \\ 3x + 6y \leq 18 \end{cases}$

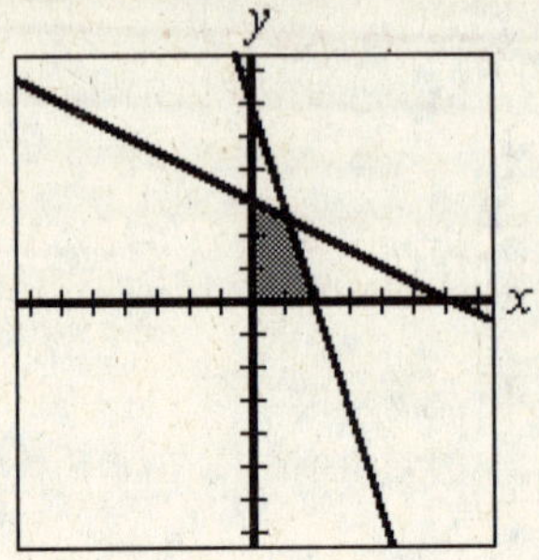

27. See #11.

29.

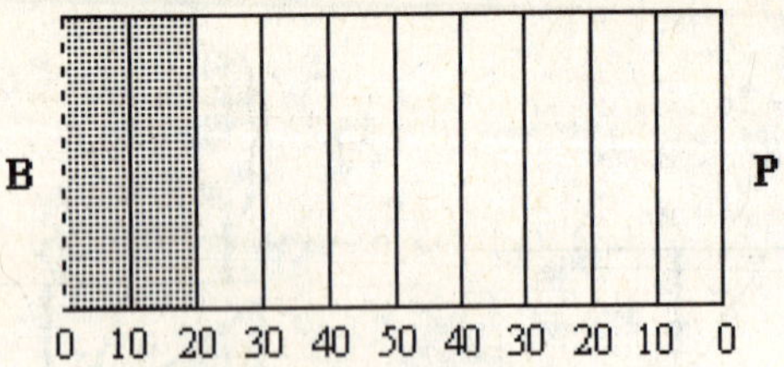

31. Let $x =$ the number of \$10 discs and let $y =$ the number of \$15 discs.

$\begin{cases} 10x + 15y \geq 30, & x \geq 0 \\ 10x + 15y \leq 60, & y \geq 0 \end{cases}$

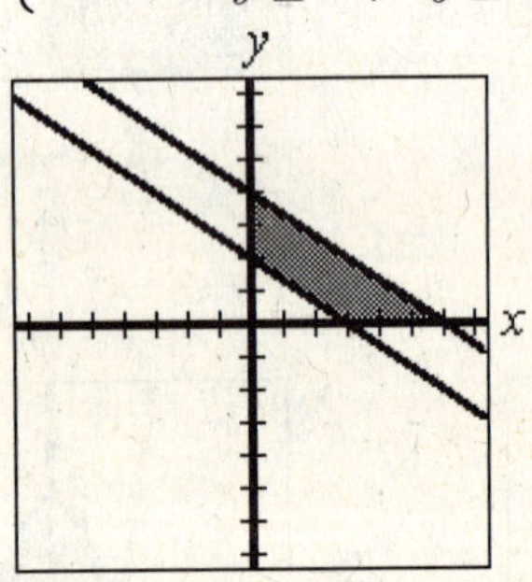

Two solutions are $(1, 2)$ and $(4, 1)$. A customer could buy one \$10 and two \$15 discs, or four \$10 and one \$15 disc.

33. Let $x =$ the # of desk chairs, and let $y =$ the # of side chairs.

$\begin{cases} 150x + 100y \leq 900, & y > x \\ x \geq 0, & y \geq 0 \end{cases}$

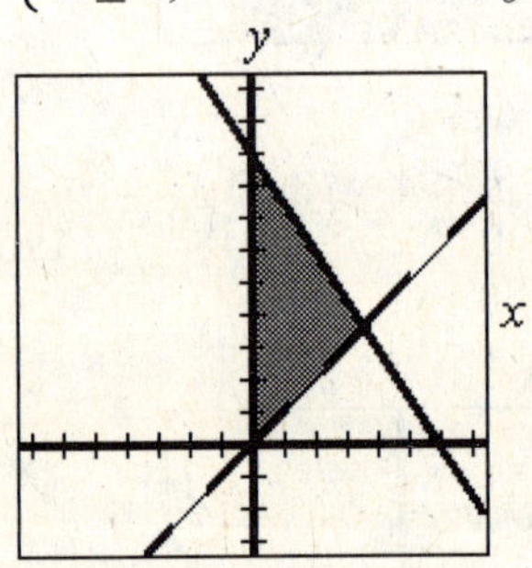

Two solutions are $(2, 4)$ and $(1, 5)$. Best could buy 2 desk chairs and 4 side chairs, or 1 desk chair and 5 side chairs.

35. **Answers may vary.**

37. No.

Exercise 4.5 (page 277)

1. $P = 2x + 5y = 2(0) + 5(4) = 20$

© 2015 Cengage Learning. All Rights Reserved. May not be scanned, copied or duplicated, or posted to a publicly accessible website, in whole or in part.

SECTION 4.5

3. $\begin{array}{l} x + y = 5 \Rightarrow \times(-2) \\ 3x + 2y = 12 \Rightarrow \end{array} \begin{array}{rcr} -2x - 2y &=& -10 \\ 3x + 2y &=& 12 \\ \hline x &=& 2 \end{array}$

Substitute and solve for y:

$x + y = 5$

$2 + y = 5$

$y = 3$

The solution is $(2, 3)$.

5. $m = \dfrac{\Delta y}{\Delta x} = \dfrac{-4-5}{-1-3} = \dfrac{-9}{-4} = \dfrac{9}{4}$

7.
$$\begin{aligned} y - y_1 &= m(x - x_1) \\ y - 5 &= \tfrac{9}{4}(x-3) \\ y - 5 &= \tfrac{9}{4}x - \tfrac{27}{4} \\ y &= \tfrac{9}{4}x - \tfrac{7}{4} \end{aligned}$$

9. constraints

11. objective

13.

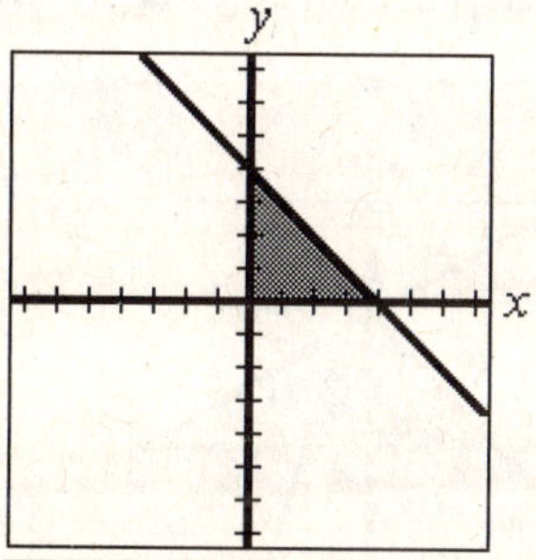

Vertex	$P = 2x + 3y$	Maximum?
$(0,0)$	$= 0$	No
$(4,0)$	$= 8$	No
$(0,4)$	$= 12$	YES

P has a maximum value of 12 at $(0, 4)$.

15.

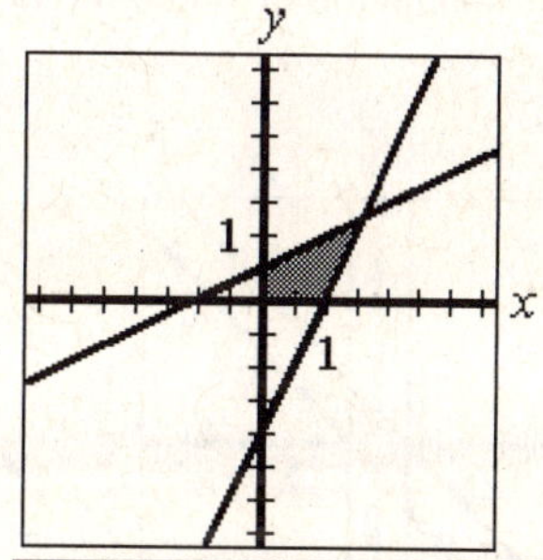

Vertex	$P = y + \frac{1}{2}x$	Maximum?
$(0,0)$	$= 0$	No
$(1,0)$	$= \frac{1}{2}$	No
$\left(\frac{5}{3}, \frac{4}{3}\right)$	$= \frac{13}{6}$	YES
$\left(0, \frac{1}{2}\right)$	$= \frac{1}{2}$	No

P has a maximum value of $\frac{13}{6}$ at $\left(\frac{5}{3}, \frac{4}{3}\right)$.

17.

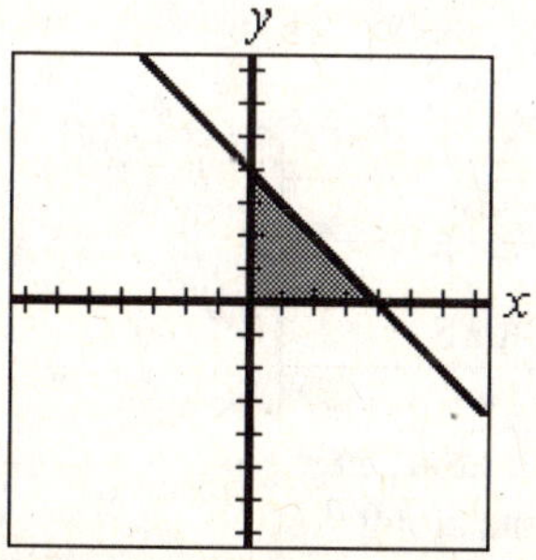

Vertex	$P = 5x + 12y$	Minimum?
$(4,0)$	$= 20$	No
$(0,4)$	$= 48$	No
$(0,0)$	$= 0$	YES

P has a minimum value of 0 at $(0, 0)$.

© 2015 Cengage Learning. All Rights Reserved. May not be scanned, copied or duplicated, or posted to a publicly accessible website, in whole or in part.

19.

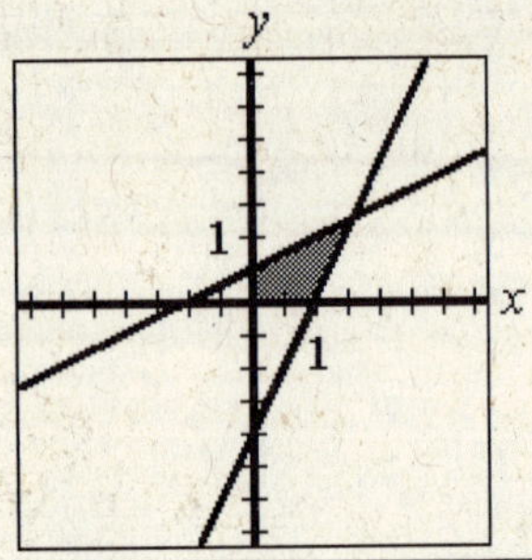

Vertex	$P = 3y + x$	Minimum?
$(0,0)$	$= 0$	YES
$(1,0)$	$= 1$	No
$\left(\frac{5}{3}, \frac{4}{3}\right)$	$= \frac{17}{3}$	No
$\left(0, \frac{1}{2}\right)$	$= \frac{3}{2}$	No

P has a minimum value of 0 at $(0,0)$.

21.

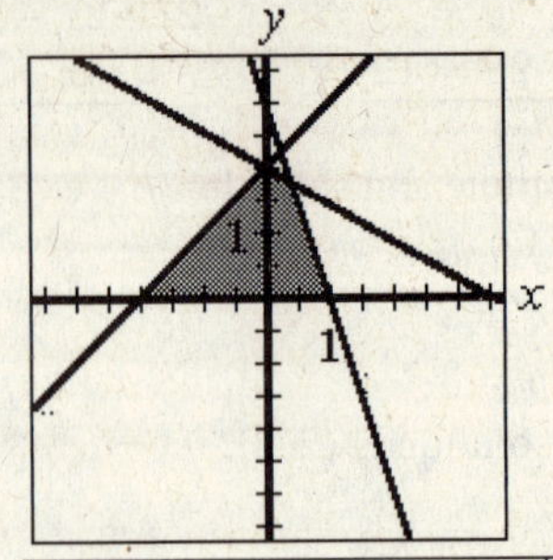

Vertex	$P = 2x + y$	Maximum?
$(1,0)$	$= 2$	No
$\left(\frac{3}{7}, \frac{12}{7}\right)$	$= \frac{18}{7}$	YES
$(0,2)$	$= 2$	No
$(-2,0)$	$= -4$	No

P has a maximum value of $\frac{18}{7}$ at $\left(\frac{3}{7}, \frac{12}{7}\right)$.

23.

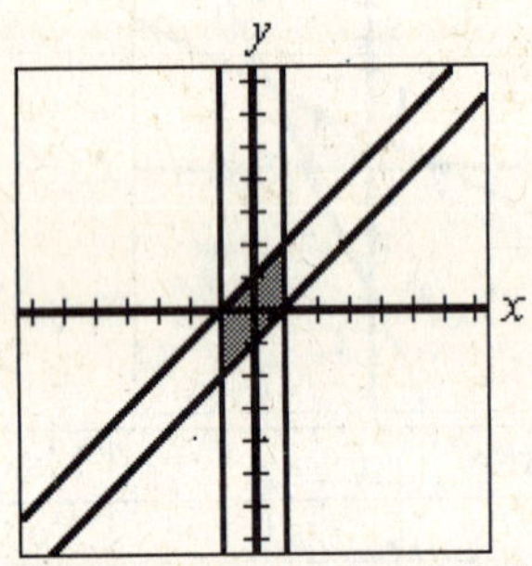

Vertex	$P = 3x - 2y$	Maximum?
$(1,0)$	$= 3$	YES
$(1,2)$	$= -1$	No
$(-1,0)$	$= -3$	No
$(-1,-2)$	$= 1$	No

P has a maximum value of 3 at $(1,0)$.

25.

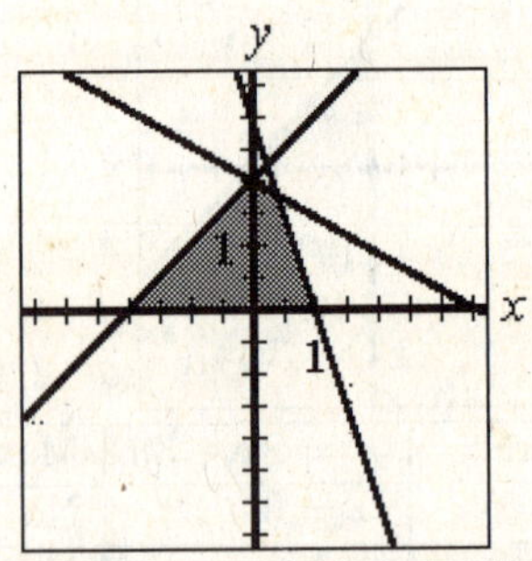

Vertex	$P = 6x + 2y$	Minimum?
$(1,0)$	$= 6$	No
$\left(\frac{3}{7}, \frac{12}{7}\right)$	$= 6$	No
$(0,2)$	$= 4$	No
$(-2,0)$	$= -12$	YES

P has a minimum value of -12 at $(-2,0)$.

27.

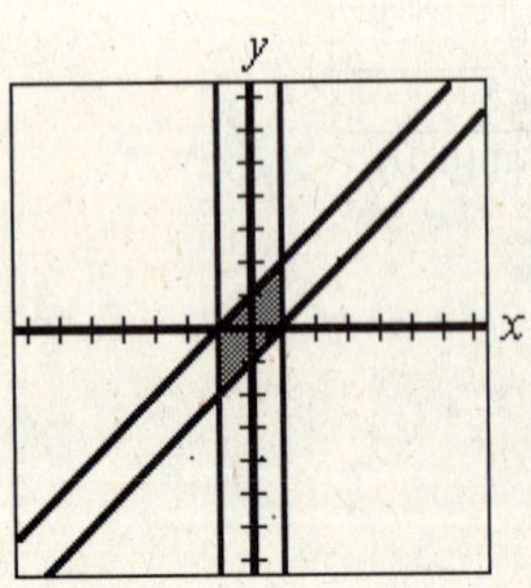

Vertex	$P = 2x - 2y$	Minimum?
$(1,0)$	$= 2$	No
$(1,2)$	$= -2$	YES
$(-1,0)$	$= -2$	YES
$(-1,-2)$	$= 2$	No

P has a minimum value of -2 at $(1,2)$ and at $(-1,0)$ (and at all points along the boundary between them).

© 2015 Cengage Learning. All Rights Reserved. May not be scanned, copied or duplicated, or posted to a publicly accessible website, in whole or in part.

29. Let $x =$ the number of tables made, and let $y =$ the number of chairs made.

$$\begin{cases} x \geq 0 & y \geq 0 \\ 2x + 3y \leq 42 & \text{(Tom's hours)} \\ 6x + 2y \leq 42 & \text{(Carlos' hours)} \end{cases}$$

Profit $= P = 100x + 80y$.

Vertex	$P = 100x + 80y$
$(0, 0)$	$= 0$
$(7, 0)$	$= 700$
$(3, 12)$	$= 1260$
$(0, 14)$	$= 1120$

The maximum income of \$1260 results when they make 3 tables and 12 chairs.

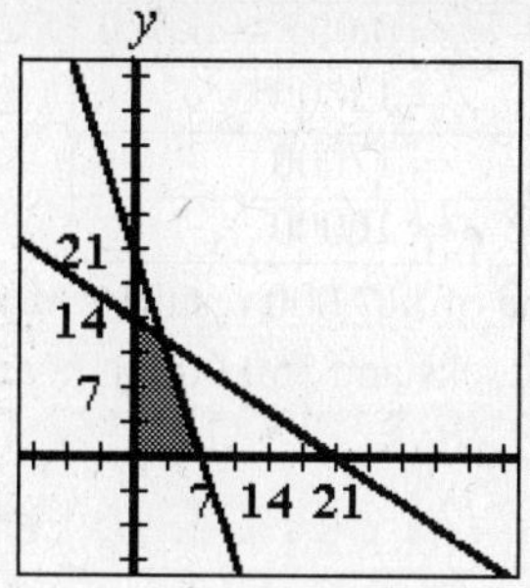

31. Let $x =$ the number of IBMs stocked, and let $y =$ the number of Macintoshes stocked.

$$\begin{cases} 20 \leq x \leq 30 \\ 30 \leq y \leq 50 \\ x + y \leq 60 & \text{(Total stock)} \end{cases}$$

Let the objective function be the total commissions: $P = 50x + 40y$.

Vertex	$P = 50x + 40y$
$(20, 30)$	$= 2200$
$(30, 30)$	$= 2700$
$(20, 40)$	$= 2600$

The maximum amount of commissions is \$2700, which results from stocking 30 of each type.

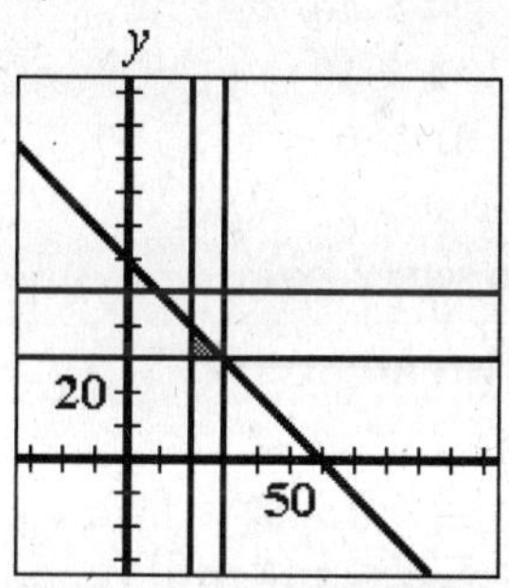

33. Let $x =$ the number of DVD players made, and let $y =$ the number of TVs made.

$$\begin{cases} x \geq 0 & y \geq 0 \\ 3x + 4y \leq 180 & \text{(electronics hours)} \\ 2x + 3y \leq 120 & \text{(assembly hours)} \\ 2x + y \leq 60 & \text{(finishing hours)} \end{cases}$$

Let the objective function be the profit: $P = 40x + 32y$.

Vertex	$P = 40x + 32y$
$(0, 0)$	$= 0$
$(30, 0)$	$= 1200$
$(15, 30)$	$= 1560$
$(0, 40)$	$= 1280$

The maximum profit of \$1560 results when they make 15 DVD players and 30 TVs.

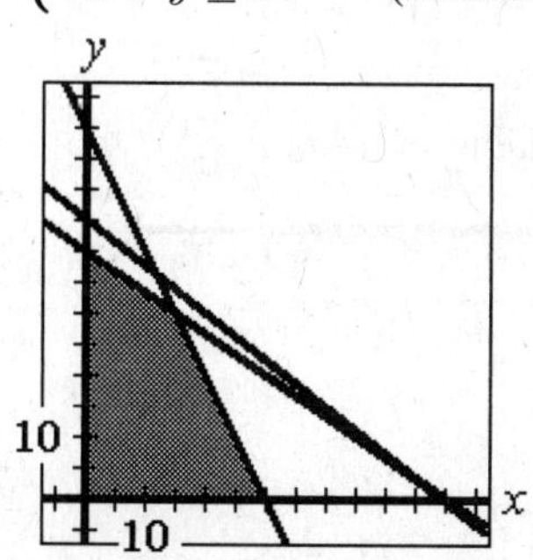

© 2015 Cengage Learning. All Rights Reserved. May not be scanned, copied or duplicated, or posted to a publicly accessible website, in whole or in part.

SECTION 4.5

35. Let $x =$ the amount in stocks, and let $y =$ the amount in bonds.

$$\begin{cases} x \geq 100000 \\ y \geq 50000 \\ x + y \leq 200000 \quad \text{(total amount)} \end{cases}$$

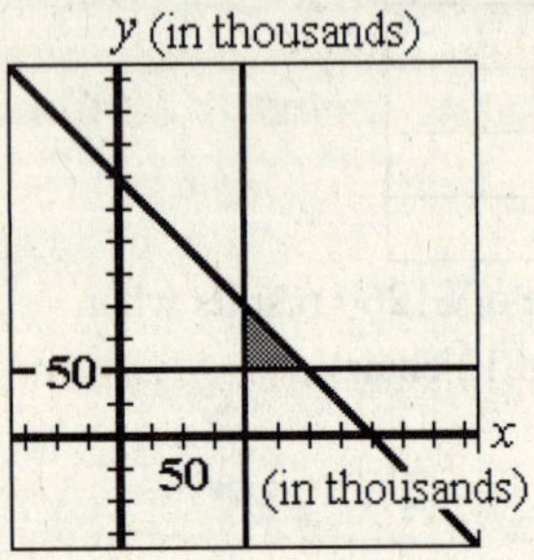

Let the objective function be the income:
$P = 0.09x + 0.07y$.

Vertex	$P = 0.09x + 0.07y$
$(100000, 50000)$	$= 12500$
$(150000, 50000)$	$= 17000$
$(100000, 100000)$	$= 16000$

The maximum income of \$17,000 results when she invests \$150,000 in stocks and \$50,000 in bonds.

37-39. Answers may vary.

Chapter 4 Review (page 282)

1.
$$\begin{aligned} 4(x-2) &\leq 4 \\ 4x - 8 + 8 &\leq 4 + 8 \\ 4x &\leq 12 \\ x &\leq 3 \end{aligned}$$
solution set: $(-\infty, 3]$

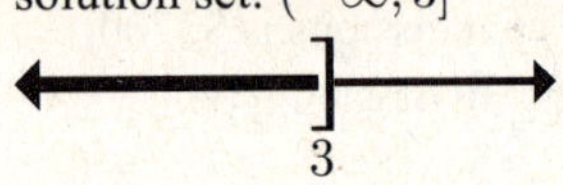

2.
$$\begin{aligned} 3x + 4 &> 10 \\ 3x + 4 - 4 &> 10 - 4 \\ 3x &> 6 \\ x &> 2 \end{aligned}$$
solution set: $(2, \infty)$

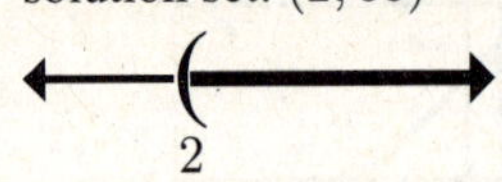

3.
$$\begin{aligned} \frac{1}{3}x - 2 &\geq \frac{1}{2}x + 2 \\ 6\left(\frac{1}{3}x - 2\right) &\geq 6\left(\frac{1}{2}x + 2\right) \\ 2x - 12 + 12 &\geq 3x + 12 + 12 \\ 2x - 3x &\geq 3x - 3x + 24 \\ -x &\geq 24 \\ x &\leq -24 \end{aligned}$$
solution set: $(-\infty, -24]$

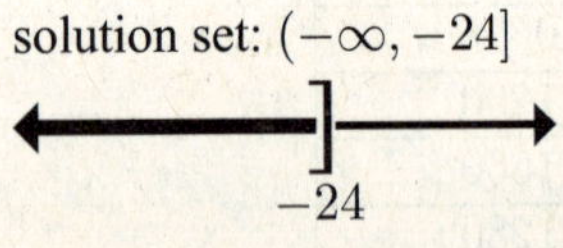

4.
$$\begin{aligned} \frac{7}{4}(x+3) &< \frac{3}{8}(x-3) \\ 8 \cdot \frac{7}{4}(x+3) &< 8 \cdot \frac{3}{8}(x-3) \\ 14(x+3) &< 3(x-3) \\ 14x + 42 &< 3x - 9 \\ 14x &< 3x - 51 \\ 11x &< -51 \\ x &< -\tfrac{51}{11} \end{aligned}$$
solution set: $(-\infty, -\frac{51}{11})$

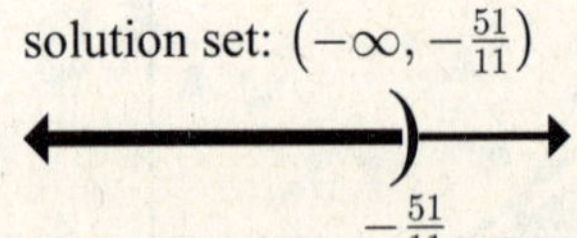

© 2015 Cengage Learning. All Rights Reserved. May not be scanned, copied or duplicated, or posted to a publicly accessible website, in whole or in part.

5.
$$3 < 3x + 4 < 10$$
$$3 - 4 < 3x + 4 - 4 < 10 - 4$$
$$-1 < 3x < 6$$
$$-\tfrac{1}{3} < \tfrac{3x}{3} < \tfrac{6}{3}$$
$-\frac{1}{3} < x < 2 \Rightarrow$ solution set: $(-\frac{1}{3}, 2) \Rightarrow$ 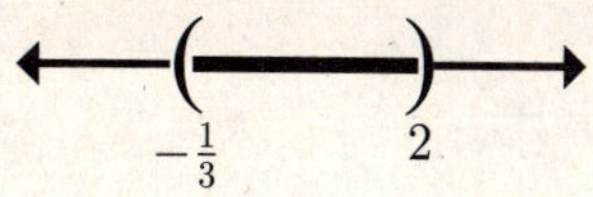

6.
$$4x > 3x + 2 > x - 3$$

$4x > 3x + 2$ and $3x + 2 > x - 3$

$4x > 3x + 2$	and	$3x + 2 > x - 3$
$4x - 3x > 3x - 3x + 2$		$3x + 2 - 2 > x - 3 - 2$
$x > 2$		$3x - x > x - x - 5$
$x > 2$		$2x > -5$
		$x > -\frac{5}{2}$

If $x > 2$ **and** $x > -\frac{5}{2}$, $x > 2$. The solution set is $(2, \infty)$.

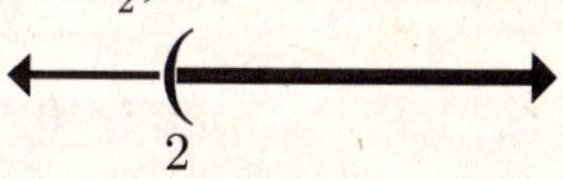

7.

$2x - 3 \geq -5$	or	$2x - 3 > 5$
$2x - 3 + 3 \geq -5 + 3$		$2x - 3 + 3 > 5 + 3$
$2x \geq -2$		$2x > 8$
$x \geq -1$		$x > 4$

If $x \geq -1$ **or** $x > 4$, $x \geq -1$. The solution set is $[-1, \infty)$.

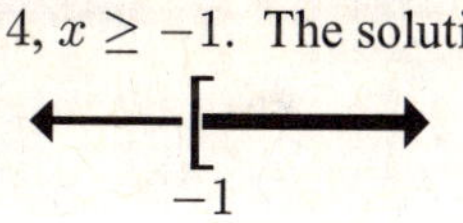

8. Let $x =$ the amount invested at 5%.
$$0.03(25000) + 0.05x \geq 3000$$
$$750 + 0.05x \geq 3000$$
$$0.05x \geq 2250$$
$x \geq 45000 \Rightarrow$ She must invest at least \$45,000.

9. $|-9| = 9$ **10.** $|8| = 8$ **11.** $-|7| = -7$ **12.** $-|-6| = -6$

13. $f(x) = |x + 1| - 3$; Shift $f(x) = |x|$ left 1 and down 3.

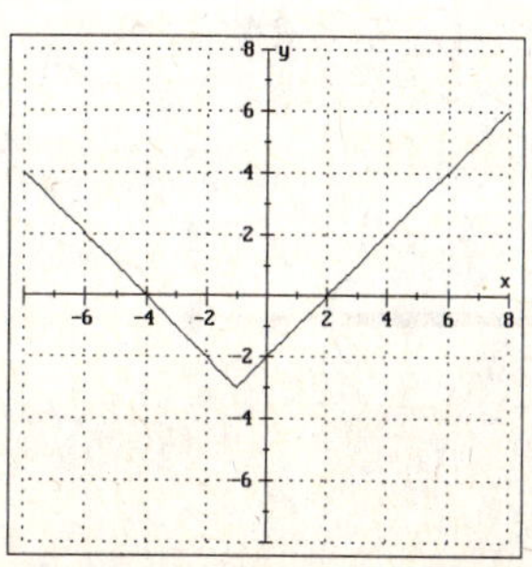

© 2015 Cengage Learning. All Rights Reserved. May not be scanned, copied or duplicated, or posted to a publicly accessible website, in whole or in part.

14. $f(x) = |x - 2| + 1$; Shift $f(x) = |x|$ right 2 and up 1.

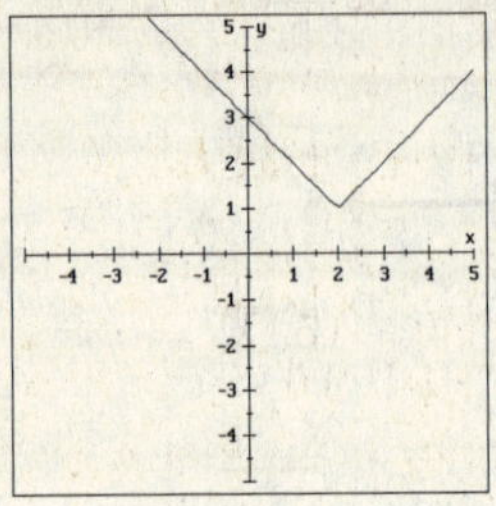

15.
$$|4x + 3| = 7$$
$$4x + 3 = 7 \quad \textbf{or} \quad 4x + 3 = -7$$
$$4x = 4 \qquad\qquad 4x = -10$$
$$x = 1 \qquad\qquad x = -\tfrac{10}{4} = -\tfrac{5}{2}$$

16.
$$\left|\tfrac{3}{2}x - 4\right| + 2 = 11$$
$$\left|\tfrac{3}{2}x - 4\right| = 9$$
$$\tfrac{3}{2}x - 4 = 9 \quad \textbf{or} \quad \tfrac{3}{2}x - 4 = -9$$
$$\tfrac{3}{2}x = 13 \qquad\qquad \tfrac{3}{2}x = -5$$
$$x = \tfrac{26}{3} \qquad\qquad x = -\tfrac{10}{3}$$

17.
$$\left|\frac{2 - x}{3}\right| = 4$$
$$\frac{2 - x}{3} = 4 \quad \textbf{or} \quad \frac{2 - x}{3} = -4$$
$$2 - x = 12 \qquad\qquad 2 - x = -12$$
$$-x = 10 \qquad\qquad -x = -14$$
$$x = -10 \qquad\qquad x = 14$$

18.
$$|6x + 1| = |4x - 3|$$
$$6x + 1 = 4x - 3 \quad \textbf{or} \quad 6x + 1 = -(4x - 3)$$
$$2x = -4 \qquad\qquad 6x + 1 = -4x + 3$$
$$x = -2 \qquad\qquad 10x = 2$$
$$x = \tfrac{2}{10} = \tfrac{1}{5}$$

19.
$$|5x - 4| = |4x - 5|$$
$$5x - 4 = 4x - 5 \quad \textbf{or} \quad 5x - 4 = -(4x - 5)$$
$$x = -1 \qquad\qquad 5x - 4 = -4x + 5$$
$$9x = 9$$
$$x = 1$$

20.
$$\left|\frac{3 - 2x}{2}\right| = \left|\frac{3x - 2}{3}\right|$$
$$\frac{3 - 2x}{2} = \frac{3x - 2}{3} \quad \textbf{or} \quad \frac{3 - 2x}{2} = -\frac{3x - 2}{3}$$
$$3(3 - 2x) = 2(3x - 2) \qquad\qquad 3(3 - 2x) = -2(3x - 2)$$
$$9 - 6x = 6x - 4 \qquad\qquad 9 - 6x = -6x + 4$$
$$-12x = -13 \qquad\qquad 0 = -5$$
$$x = \tfrac{13}{12}$$
(no solution from this part)

21.
$$|2x + 7| < 3$$
$$-3 < 2x + 7 < 3$$
$$-3 - 7 < 2x + 7 - 7 < 3 - 7$$
$$-10 < 2x < -4$$
$$-5 < x < -2$$
solution set: $(-5, -2)$

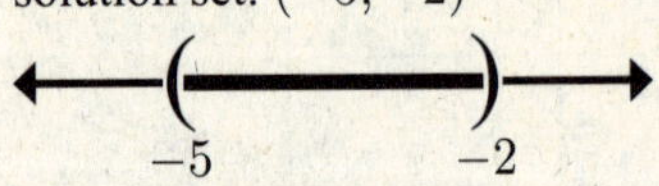

22.
$$|5 - 3x| \le 14$$
$$-14 \le 5 - 3x \le 14$$
$$-14 - 5 \le 5 - 5 - 3x \le 14 - 5$$
$$-19 \le -3x \le 9$$
$$\tfrac{-19}{-3} \ge \tfrac{-3x}{-3} \ge \tfrac{9}{-3}$$
$$\tfrac{19}{3} \ge x \ge -3, \text{ or } -3 \le x \le \tfrac{19}{3}$$
solution set: $\left[-3, \tfrac{19}{3}\right]$

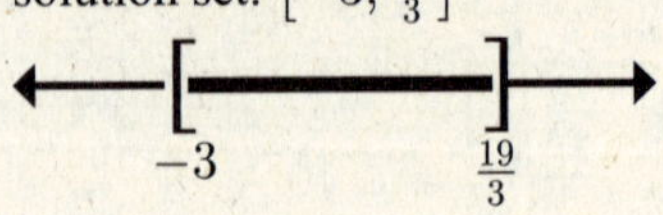

© 2015 Cengage Learning. All Rights Reserved. May not be scanned, copied or duplicated, or posted to a publicly accessible website, in whole or in part.

23. $\left|\frac{2}{3}x + 14\right| < 0$

Since an absolute value can never be negative, there is no solution. $\emptyset$

24.

$$\left|\frac{1-5x}{3}\right| > 7$$

$$\frac{1-5x}{3} < -7 \quad \textbf{or} \quad \frac{1-5x}{3} > 7$$

$$1 - 5x < -21 \qquad 1 - 5x > 21$$

$$-5x < -22 \qquad -5x > 20$$

$$x > \tfrac{22}{5} \qquad x < -4$$

solution set: $(-\infty, -4) \cup \left(\frac{22}{5}, \infty\right)$

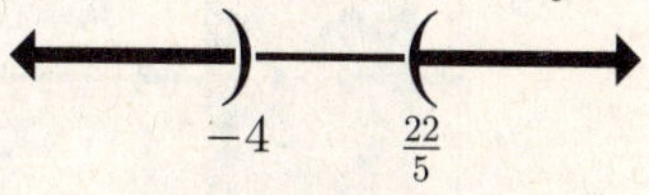

25.

$$|3x - 8| - 3 \geq 1$$

$$|3x - 8| \geq 4$$

$$3x - 8 \leq -4 \quad \textbf{or} \quad 3x - 8 \geq 4$$

$$3x \leq 4 \qquad 3x \geq 12$$

$$x \leq \tfrac{4}{3} \qquad x \geq 4$$

solution set: $\left(-\infty, \frac{4}{3}\right] \cup [4, \infty)$

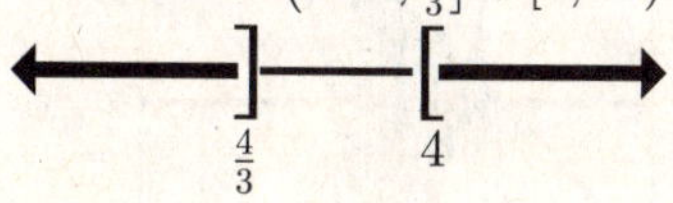

26. $\left|\frac{3}{2}x - 14\right| \geq 0$

Since an absolute value is always at least 0, this inequality is true for all real #s. Solution set: $(-\infty, \infty)$

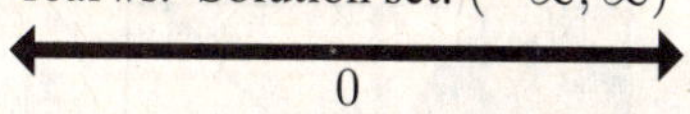

27. $2x + 3y > 6$

Boundary (dashed)

$2x + 3y = 6$

x	y
0	2
3	0

Test point: $(0, 0)$

$$2x + 3y > 6$$

$$2(0) + 3(0) \overset{?}{>} 6$$

$$0 \not> 6$$

opposite half-plane

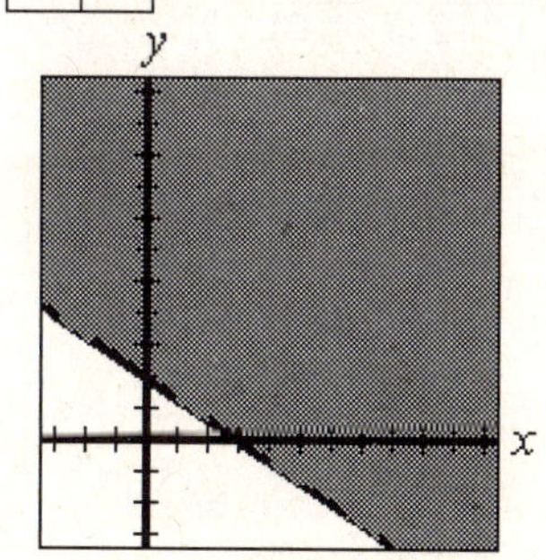

28. $y \leq 4 - x$

Boundary (solid)

$y = 4 - x$

x	y
0	4
2	2

Test point: $(0, 0)$

$$y \leq 4 - x$$

$$0 \overset{?}{\leq} 4 - 0$$

$$0 \leq 4$$

same half-plane

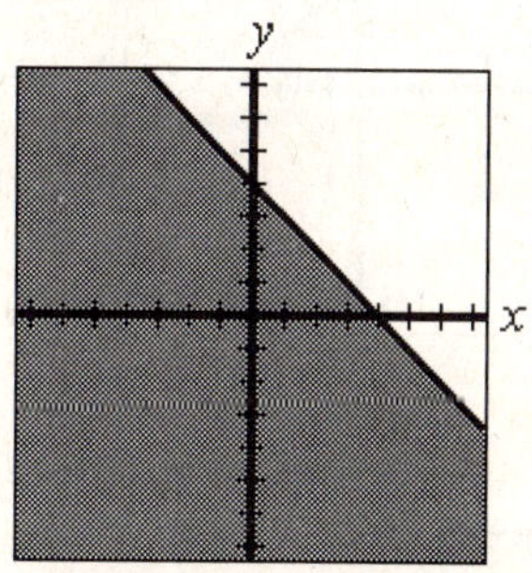

© 2015 Cengage Learning. All Rights Reserved. May not be scanned, copied or duplicated, or posted to a publicly accessible website, in whole or in part.

29. $-2 < x < 4$

Shade points between $x = -2$ and $x = 4$.

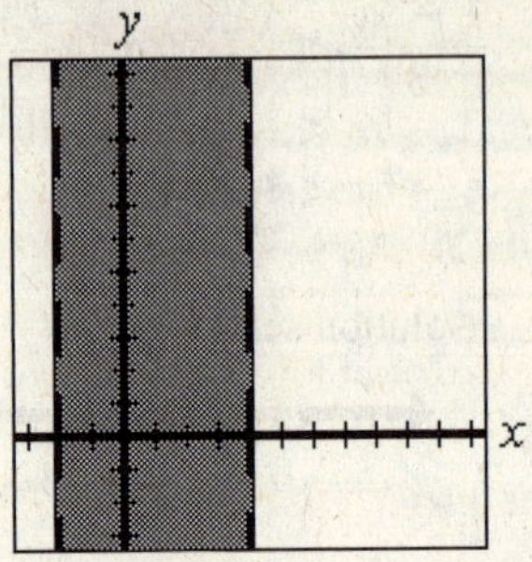

30. $y \leq -2$ or $y > 1$

Shade points below $y = -2$ or above $y = 1$.

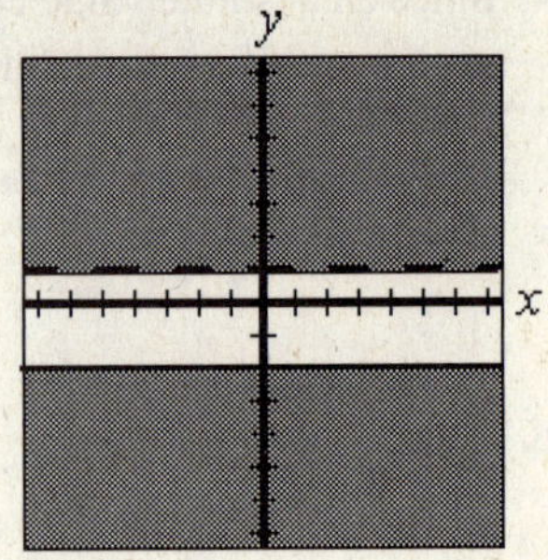

31. $\begin{cases} y \geq x + 1 \\ 3x + 2y < 6 \end{cases}$

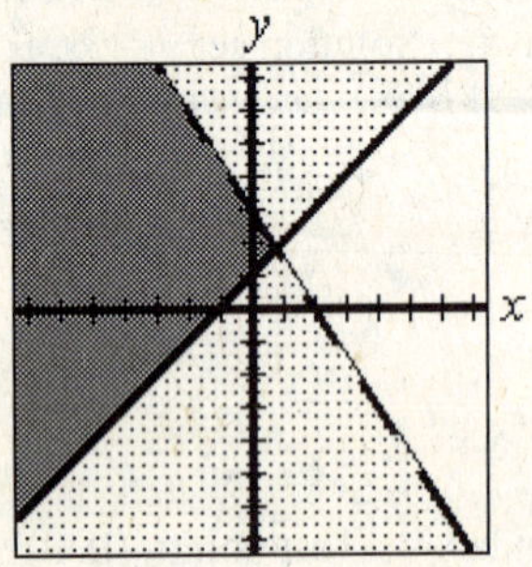

32. $\begin{cases} y \geq x^2 - 4 \\ y < x + 3 \end{cases}$

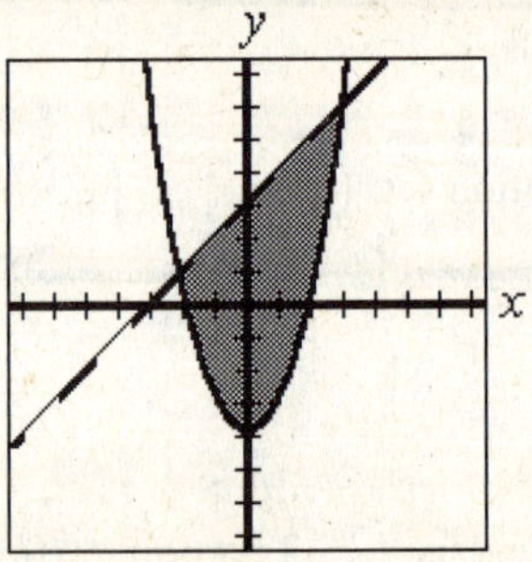

33.

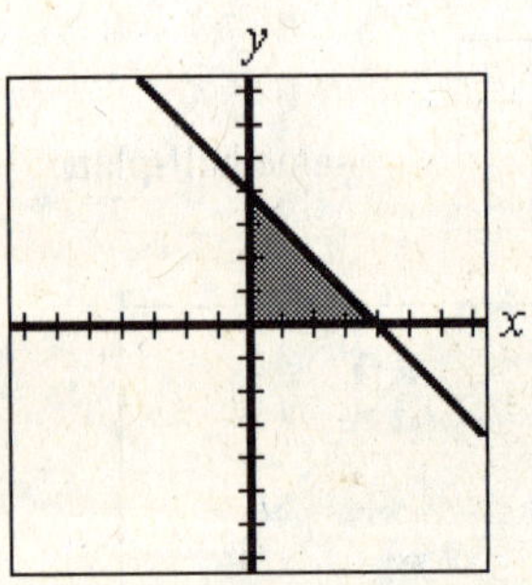

Vertex	$P = 3x + y$	Maximum?
$(4, 0)$	$= 12$	YES
$(0, 4)$	$= 4$	No
$(0, 0)$	$= 0$	No

P has a maximum value of 12 at $(4, 0)$.

© 2015 Cengage Learning. All Rights Reserved. May not be scanned, copied or duplicated, or posted to a publicly accessible website, in whole or in part.

34. Let $x =$ the number of bags of fertilizer X, and let $y =$ the number of bags of fertilizer Y.

$$\begin{cases} x \geq 0 & y \geq 0 \\ 6x + 10y \leq 20000 & \text{(Nitrogen)} \\ 8x + 6y \leq 16400 & \text{(Phosphorus)} \\ 6x + 4y \leq 12000 & \text{(Potash)} \end{cases}$$

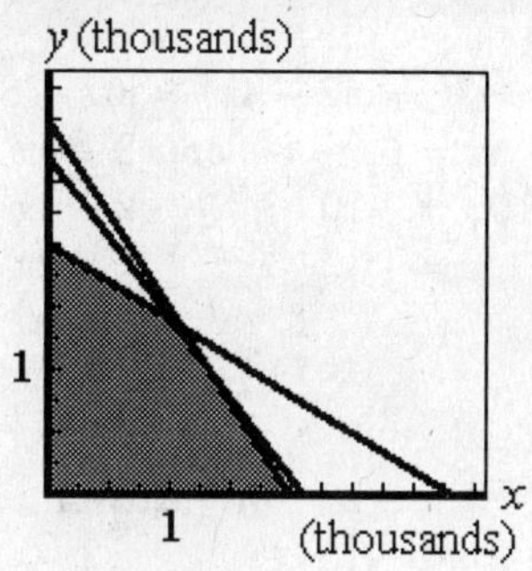

Let the objective function be the total profit: $P = 6x + 5y$.

Vertex	$P = 6x + 5y$
$(2000, 0)$	$= 12000$
$(1600, 600)$	$= 12600$
$(1000, 1400)$	$= 13000$
$(0, 2000)$	$= 10000$

The maximum profit is \$13,000, which results from using 1000 bags of fertilizer X and 1400 bags of fertilizer Y.

Chapter 4 Test (page 286)

1.
$$\begin{aligned} -2(2x+3) &\geq 14 \\ -4x - 6 + 6 &\geq 14 + 6 \\ -4x &\geq 20 \\ \frac{-4x}{-4} &\leq \frac{20}{-4} \\ x &\leq -5 \end{aligned}$$
solution set: $(-\infty, -5]$

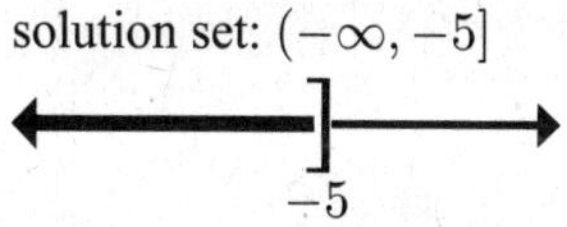

2.
$$\begin{aligned} -2 < &\frac{x-4}{3} < 4 \\ -6 < &\, x - 4 < 12 \\ -6 + 4 < &\, x - 4 + 4 \leq 12 + 4 \\ -2 < &\, x < 16 \end{aligned}$$
solution set: $(-2, 16)$

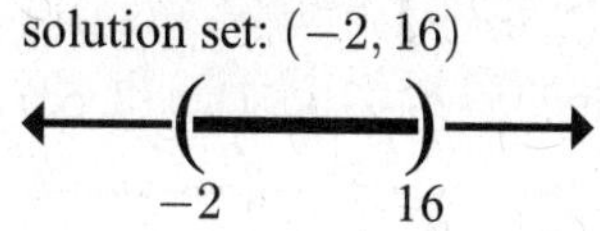

3. $|6 - 10| = |-4| = 4$

4. $|4\pi - 4| = 4\pi - 4 \ (4\pi - 4 > 0)$

5. $f(x) = |x + 1| - 4$; Shift $f(x) = |x|$ left 1 and down 4.

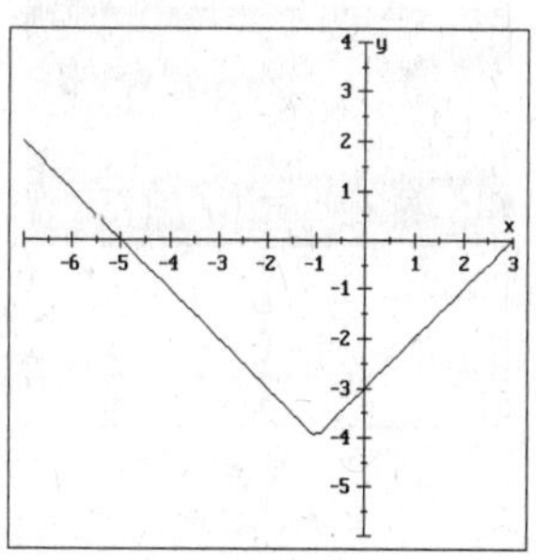

6. $f(x) = |x - 2| + 3$; Shift $f(x) = |x|$ right 2 and up 3.

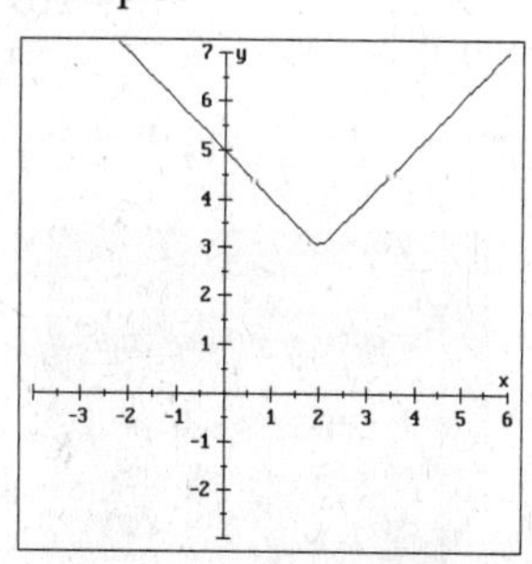

© 2015 Cengage Learning. All Rights Reserved. May not be scanned, copied or duplicated, or posted to a publicly accessible website, in whole or in part.

CHAPTER 4 TEST

7.

$$|5x - 6| = 9$$
$$5x - 6 = 9 \quad \textbf{or} \quad 5x - 6 = -9$$
$$5x = 15 \qquad 5x = -3$$
$$x = 3 \qquad x = -\tfrac{3}{5}$$

8.

$$|4 - 3x| - 2 = 17$$
$$|4 - 3x| = 19$$
$$4 - 3x = 19 \quad \textbf{or} \quad 4 - 3x = -19$$
$$-3x = 15 \qquad -3x = -23$$
$$x = -5 \qquad x = \tfrac{23}{3}$$

9.

$$|3x + 7| + 8 = 5$$
$$|3x + 7| = -3$$

Since an absolute value can never be negative, there is no solution. $\emptyset$

10.

$$|2 - 4x| = |6x - 8|$$
$$2 - 4x = 6x - 8 \quad \textbf{or} \quad 2 - 4x = -(6x - 8)$$
$$-10x = -10 \qquad 2 - 4x = -6x + 8$$
$$x = 1 \qquad 2x = 6$$
$$x = 3$$

11. $|x + 3| \le 4$

$$-4 \le x + 3 \le 4$$
$$-4 - 3 \le x + 3 - 3 \le 4 - 3$$
$$-7 \le x \le 1$$

solution set: $[-7, 1]$

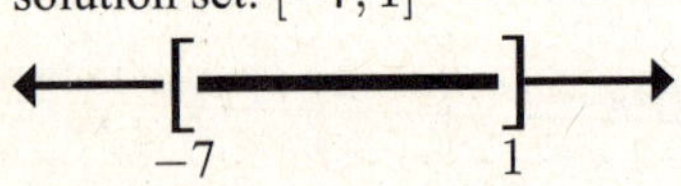

12.

$$|2x - 4| > 22$$
$$2x - 4 < -22 \quad \textbf{or} \quad 2x - 4 > 22$$
$$2x < -18 \qquad 2x > 26$$
$$x < -9 \qquad x > 13$$

solution set: $(-\infty, -9) \cup (13, \infty)$

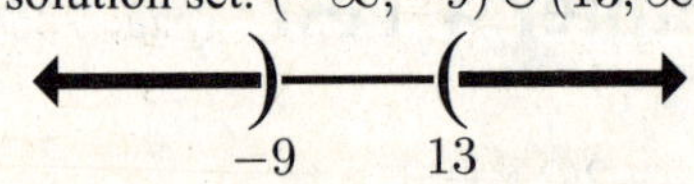

13.

$$|4 - 2x| > 2$$
$$4 - 2x < -2 \quad \textbf{or} \quad 4 - 2x > 2$$
$$-2x < -6 \qquad -2x > -2$$
$$x > 3 \qquad x < 1$$

solution set: $(-\infty, 1) \cup (3, \infty)$

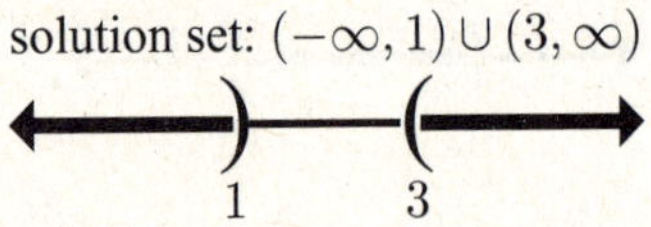

14. $|2x - 4| \le -2$

Since an absolute value can never be negative, there is no solution. $\emptyset$

15.

$$3x + 2y \ge 6$$

Boundary (solid)

$$3x + 2y = 6$$

x	y
0	3
2	0

Test point: $(0, 0)$

$$3x + 2y \ge 6$$
$$3(0) + 2(0) \overset{?}{\ge} 6$$
$$0 \not\ge 6$$

opposite half-plane

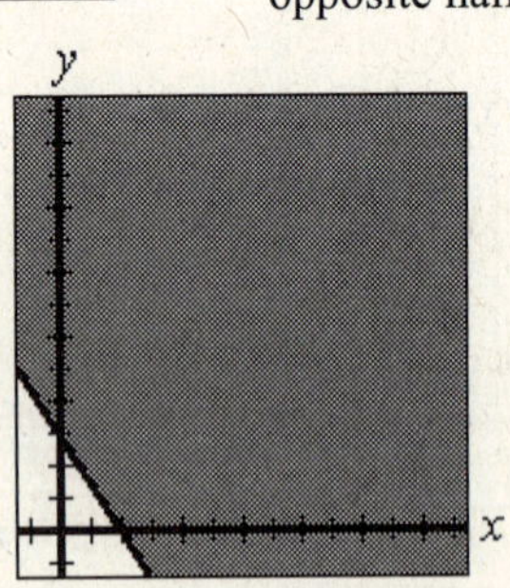

16.

$$-2 \le y < 5$$

Shade points between $y = -2$ and $y = 5$.

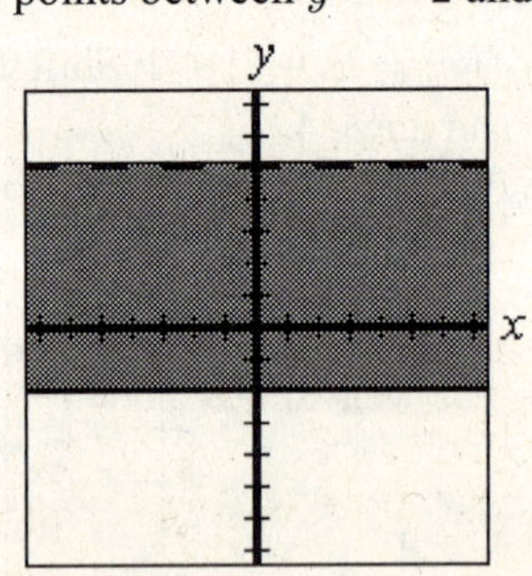

© 2015 Cengage Learning. All Rights Reserved. May not be scanned, copied or duplicated, or posted to a publicly accessible website, in whole or in part.

CHAPTER 4 TEST

17. $\begin{cases} 2x - 3y \geq 6 \\ y \leq -x + 1 \end{cases}$

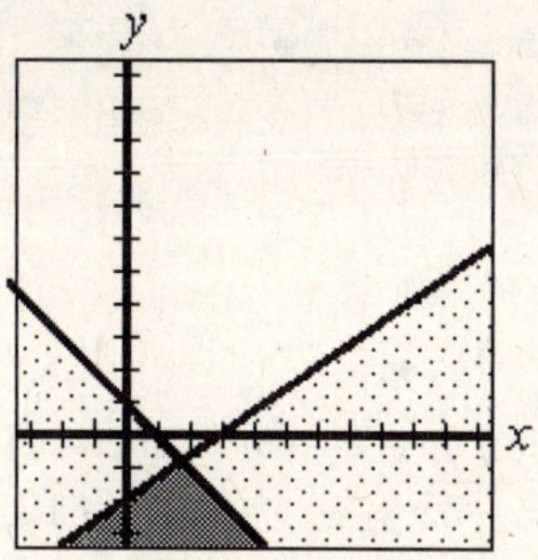

18. $\begin{cases} y \geq x^2 \\ y < x + 3 \end{cases}$

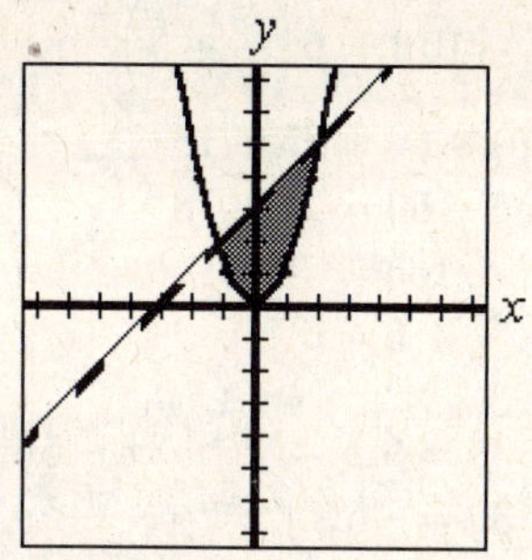

19.

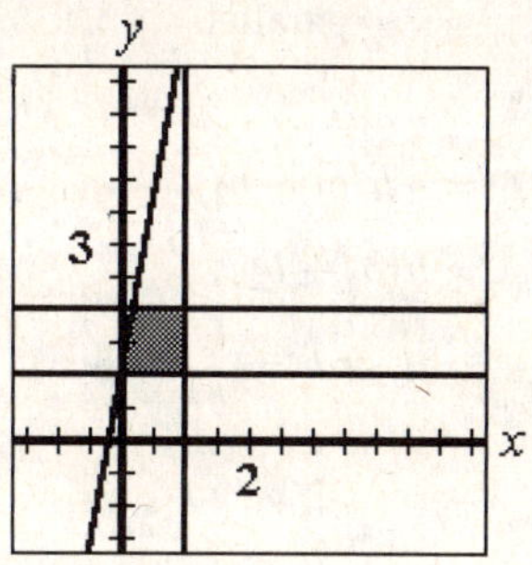

Vertex	$P = 3x - y$	Maximum?
$(0, 1)$	$= -1$	No
$(1, 1)$	$= 2$	YES
$(1, 2)$	$= 1$	No
$(\frac{1}{3}, 2)$	$= -1$	No

P has a maximum value of 2 at $(1, 1)$.

Cumulative Review Exercises (page 287)

1.

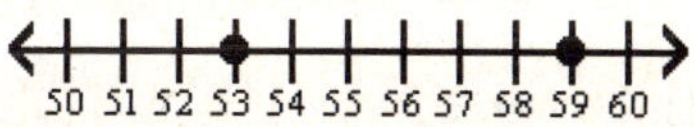

2. The additive inverse of -7 is 7.

3. $x^2 - y = 2^2 - (-4) = 4 + 4 = 8$

4. $\dfrac{x^2 - y^2}{3x + y} = \dfrac{2^2 - (-4)^2}{3(2) + (-4)} = \dfrac{4 - 16}{6 - 4}$
$= \frac{-12}{2} = -6$

5. $(x^3x^5)^3 = (x^8)^3 = x^{24}$

6. $(x^2)^3(x^4)^2 = x^6x^8 = x^{14}$

7. $\left(\dfrac{x^3}{x^5}\right)^{-2} = \left(\dfrac{x^5}{x^3}\right)^2 = (x^2)^2 = x^4$

8. $\dfrac{a^2b^n}{a^nb^2} = a^{2-n}b^{n-2}$

9. $455{,}000{,}000 = 4.55 \times 10^8$

10. $0.000012 = 1.2 \times 10^{-5}$

11.
$$\begin{aligned} 4x - 10 &= 12 \\ 4x &= 22 \\ x &= \frac{22}{4} = \frac{11}{2} \end{aligned}$$

12.
$$\begin{aligned} 6(x - 1) &= 2(x + 3) \\ 6x - 6 &= 2x + 6 \\ 4x &= 12 \\ x &= 3 \end{aligned}$$

© 2015 Cengage Learning. All Rights Reserved. May not be scanned, copied or duplicated, or posted to a publicly accessible website, in whole or in part.

CUMULATIVE REVIEW EXERCISES

13.
$$\frac{5b}{2} - 10 = \frac{b}{3} + 3$$
$$6\left(\frac{5b}{2} - 10\right) = 6\left(\frac{b}{3} + 3\right)$$
$$3(5b) - 60 = 2b + 18$$
$$15b - 60 = 2b + 18$$
$$13b = 78$$
$$b = 6$$

14.
$$2a - 5 = -2a + 4(a - 2) + 1$$
$$2a - 5 = -2a + 4a - 8 + 1$$
$$2a - 5 = 2a - 7$$
$$-5 = -7$$
$$\emptyset$$

15.
$$5x + 3y = 9 \qquad 3x - 5y = 10$$
$$3y = -5x + 9 \qquad -5y = -3x + 10$$
$$y = -\tfrac{5}{3}x + 3 \qquad y = \tfrac{3}{5}x - 2$$
$$m = -\tfrac{5}{3} \qquad m = \tfrac{3}{5}$$
perpendicular

16.
$$3x = y + 4 \qquad y = 3(x - 4) - 1$$
$$-y = -3x + 4 \qquad y = 3x - 12 - 1$$
$$y = 3x - 4 \qquad y = 3x - 13$$
$$m = 3 \qquad m = 3$$
parallel

17.
$$3x + y = 8$$
$$y = -3x + 8$$
$$m = -3$$
Use $m = \frac{1}{3}$ (the perpendicular slope):
$$y - y_1 = m(x - x_1)$$
$$y - 3 = \tfrac{1}{3}(x + 2)$$
$$y = \tfrac{1}{3}x + \tfrac{11}{3}$$

18.
$$A = \frac{1}{2}h(b_1 + b_2)$$
$$2A = h(b_1 + b_2)$$
$$\frac{2A}{b_1 + b_2} = h, \text{ or } h = \frac{2A}{b_1 + b_2}$$

19. $f(-1) = 3(-1)^2 - (-1) = 3(1) + 1$
$= 3 + 1 = 4$

20. $f(-3) = 3(-3)^2 - (-3) = 3(9) + 3$
$= 27 + 3 = 30$

21.
$$\begin{cases} 2x + y = 5 \\ x - 2y = 0 \end{cases}$$

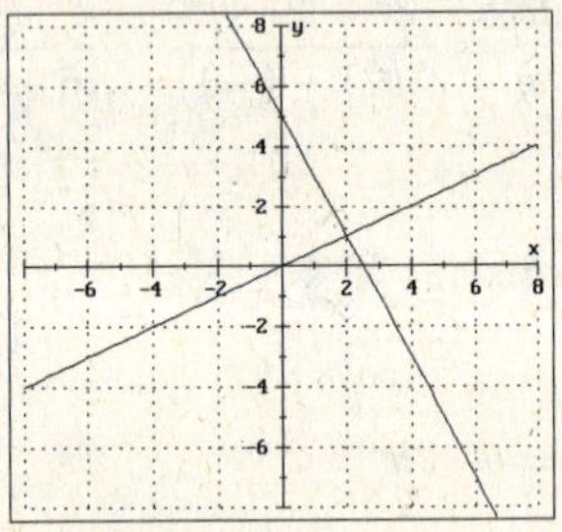

solution: $(2, 1)$

22.
$$\begin{cases} 3x + y = 4 & (1) \\ 2x - 3y = -1 & (2) \end{cases}$$
Substitute $y = -3x + 4$ from (1) into (2):
$$2x - 3\boldsymbol{y} = -1$$
$$2x - 3\boldsymbol{(-3x + 4)} = -1$$
$$2x + 9x - 12 = -1$$
$$11x = 11$$
$$x = 1$$
Substitute this and solve for y:
$$y = -3x + 4$$
$$y = -3(1) + 4$$
$$y = 1$$
Solution: $(1, 1)$

© 2015 Cengage Learning. All Rights Reserved. May not be scanned, copied or duplicated, or posted to a publicly accessible website, in whole or in part.

CUMULATIVE REVIEW EXERCISES

23. $\begin{array}{l} x+2y=-2 \\ 2x-y=6 \Rightarrow \times 2 \end{array} \quad \begin{array}{rl} x+2y &= -2 \\ 4x-2y &= 12 \\ \hline 5x &= 10 \\ x &= 2 \end{array}$

Substitute and solve for y:

$x+2y=-2$
$2+2y=-2$
$2y=-4$
$y=-2$

The solution is $(2,-2)$.

24. $\begin{array}{l} \dfrac{x}{10}+\dfrac{y}{5}=\dfrac{1}{2} \Rightarrow \times 10 \\ \dfrac{x}{2}-\dfrac{y}{5}=\dfrac{13}{10} \Rightarrow \times 10 \end{array} \quad \begin{array}{rl} x+2y &= 5 \\ 5x-2y &= 13 \\ \hline 6x &= 18 \\ x &= 3 \end{array}$

Substitute and solve for y:

$x+2y=5$
$3+2y=5$
$2y=2$
$y=1$

The solution is $(3,1)$.

25. $\begin{array}{ll} (1) & x+y+z=1 \\ (2) & 2x-y-z=-4 \\ (3) & x-2y+z=4 \end{array} \quad \begin{array}{ll} (1) & x+y+z=1 \\ (2) & 2x-y-z=-4 \\ \hline (4) & 3x=-3 \\ & x=-1 \end{array} \quad \begin{array}{ll} (2) & 2x-y-z=-4 \\ (3) & x-2y+z=4 \\ \hline (5) & 3x-3y=0 \end{array}$

$\begin{array}{r} 3x-3y=0 \\ 3(-1)-3y=0 \\ -3-3y=0 \\ -3y=3 \\ y=-1 \end{array} \quad \begin{array}{r} x+y+z=1 \\ -1+(-1)+z=1 \\ -2+z=1 \\ z=3 \end{array}$

Solution: $\boxed{(-1,-1,3)}$

26. $\begin{array}{ll} (1) & x+2y+3z=1 \\ (2) & 3x-2y+z=-1 \\ (3) & 2y+2z=1 \end{array} \quad \begin{array}{ll} (1) & x+2y+3z=1 \\ (2) & 3x-2y+z=-1 \\ \hline (4) & 4x+4z=0 \end{array} \quad \begin{array}{ll} (1) & x+2y+3z=1 \\ -1\cdot(3) & -2y-2z=-1 \\ \hline (5) & x+z=0 \end{array}$

$\begin{array}{l} 4x+4z=0 \Rightarrow \\ x+z=0 \Rightarrow \times(-4) \end{array} \quad \begin{array}{r} 4x+4z=0 \\ -4x-4z=0 \\ \hline 0=0 \end{array}$

Dependent equations. $x+z=0 \Rightarrow x=-z$
$2y+2z=1 \Rightarrow 2y=1-2z \Rightarrow y=\frac{1}{2}-z$

Solution: $\boxed{\left(-z, \frac{1}{2}-z, z\right)}$

27. $\begin{vmatrix} 3 & -2 \\ 1 & -1 \end{vmatrix} = 3(-1)-(-2)(1) = -3-(-2) = -3+2 = -1$

28. $\begin{vmatrix} 2 & 3 & -1 \\ -1 & -1 & 2 \\ 4 & 1 & -1 \end{vmatrix} = 2\begin{vmatrix} -1 & 2 \\ 1 & -1 \end{vmatrix} - 3\begin{vmatrix} -1 & 2 \\ 4 & -1 \end{vmatrix} + (-1)\begin{vmatrix} -1 & -1 \\ 4 & 1 \end{vmatrix} = 2(-1)-3(-7)-1(3) = 16$

29. $x = \dfrac{\begin{vmatrix} -1 & -3 \\ -7 & 4 \end{vmatrix}}{\begin{vmatrix} 4 & -3 \\ 3 & 4 \end{vmatrix}} = \dfrac{-4-(21)}{16-(-9)} = \dfrac{-25}{25} = -1;\ y = \dfrac{\begin{vmatrix} 4 & -1 \\ 3 & -7 \end{vmatrix}}{\begin{vmatrix} 4 & -3 \\ 3 & 4 \end{vmatrix}} = \dfrac{-28-(-3)}{25} = \dfrac{-25}{25} = -1$

solution: $(-1,-1)$

© 2015 Cengage Learning. All Rights Reserved. May not be scanned, copied or duplicated, or posted to a publicly accessible website, in whole or in part.

30. $D = \begin{vmatrix} 1 & -2 & -1 \\ 3 & 1 & -1 \\ 2 & -1 & 1 \end{vmatrix} = 1\begin{vmatrix} 1 & -1 \\ -1 & 1 \end{vmatrix} - (-2)\begin{vmatrix} 3 & -1 \\ 2 & 1 \end{vmatrix} + (-1)\begin{vmatrix} 3 & 1 \\ 2 & -1 \end{vmatrix} = 1(0) + 2(5) - 1(-5)$

$= 15$

$N_x = \begin{vmatrix} -2 & -2 & -1 \\ 6 & 1 & -1 \\ -1 & -1 & 1 \end{vmatrix} = -2\begin{vmatrix} 1 & -1 \\ -1 & 1 \end{vmatrix} - (-2)\begin{vmatrix} 6 & -1 \\ -1 & 1 \end{vmatrix} + (-1)\begin{vmatrix} 6 & 1 \\ -1 & -1 \end{vmatrix} = 15$

$N_y = \begin{vmatrix} 1 & -2 & -1 \\ 3 & 6 & -1 \\ 2 & -1 & 1 \end{vmatrix} = 1\begin{vmatrix} 6 & -1 \\ -1 & 1 \end{vmatrix} - (-2)\begin{vmatrix} 3 & -1 \\ 2 & 1 \end{vmatrix} + (-1)\begin{vmatrix} 3 & 6 \\ 2 & -1 \end{vmatrix} = 30$

$N_z = \begin{vmatrix} 1 & -2 & -2 \\ 3 & 1 & 6 \\ 2 & -1 & -1 \end{vmatrix} = 1\begin{vmatrix} 1 & 6 \\ -1 & -1 \end{vmatrix} - (-2)\begin{vmatrix} 3 & 6 \\ 2 & -1 \end{vmatrix} + (-2)\begin{vmatrix} 3 & 1 \\ 2 & -1 \end{vmatrix} = -15$

$x = \frac{N_x}{D} = \frac{15}{15} = 1;\ y = \frac{N_y}{D} = \frac{30}{15} = 2;\ z = \frac{N_z}{D} = \frac{-15}{15} = -1 \Rightarrow$ solution: $(1, 2, -1)$

31.
$$-3(x-4) \geq x - 32$$
$$-3x + 12 \geq x - 32$$
$$-4x \geq -44$$
$$x \leq 11 \quad (-\infty, 11]$$

32.
$$-8 < -3x + 1 < 10$$
$$-9 < -3x < 9$$
$3 > x > -3$, or $-3 < x < 3 \quad (-3, 3)$

33.
$$|5x - 7| = 3$$
$5x - 7 = 3$ **or** $5x - 7 = -3$

$5x = 10 \qquad 5x = 4$

$x = 2 \qquad x = \frac{4}{5}$

34.
$$|2x - 1| = |3x + 4|$$
$2x - 1 = 3x + 4$ **or** $2x - 1 = -(3x + 4)$

$-x = 5 \qquad 2x - 1 = -3x - 4$

$x = -5 \qquad 5x = -3$

$x = -\frac{3}{5}$

35.
$$|3x - 2| \leq 4$$
$$-4 \leq 3x - 2 \leq 4$$
$$-2 \leq 3x \leq 6$$
$$-\frac{2}{3} \leq x \leq 2 \quad \left[-\frac{2}{3}, 2\right]$$

36.
$$|2x + 3| - 1 > 4$$
$$|2x + 3| > 5$$
$2x + 3 < -5$ **or** $2x + 3 > 5$

$2x < -8 \qquad 2x > 2$

$x < -4 \qquad x > 1$

$(-\infty, -4) \cup (1, \infty)$

37.
$$2x - 3y \leq 12$$

Boundary (solid)

$2x - 3y = 12$

x	y
0	−4
6	0

Test point: $(0, 0)$

$2x - 3y \leq 12$

$2(0) - 3(0) \overset{?}{\leq} 12$

$0 \leq 12$

same half-plane

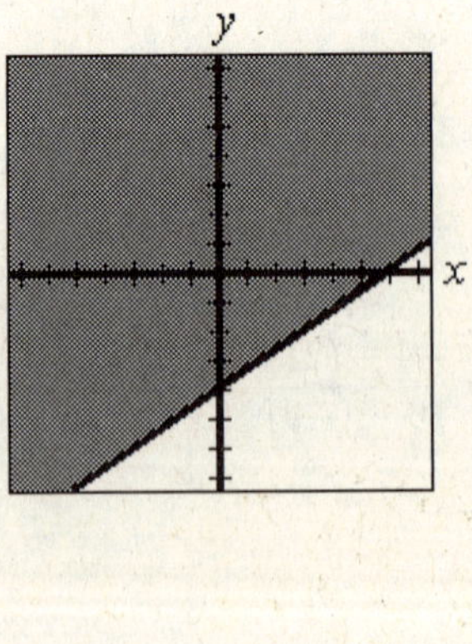

© 2015 Cengage Learning. All Rights Reserved. May not be scanned, copied or duplicated, or posted to a publicly accessible website, in whole or in part.

CUMULATIVE REVIEW EXERCISES

38. $3 > x \geq -2$

$-2 \leq x < 3$

Shade all points between $x = -2$ and $x = 3$.

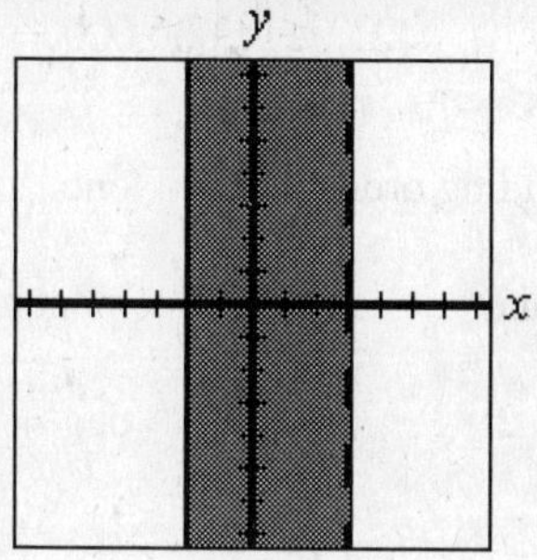

39. $\begin{cases} 3x - 2y < 6 \\ y < -x + 2 \end{cases}$

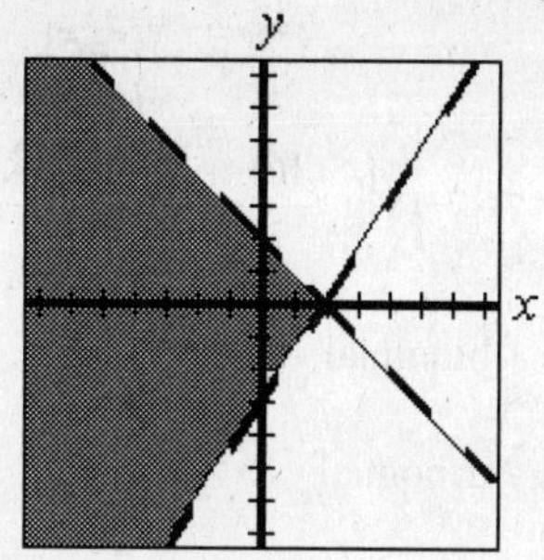

40. $\begin{cases} y < x + 2 \\ 3x + y \leq 6 \end{cases}$

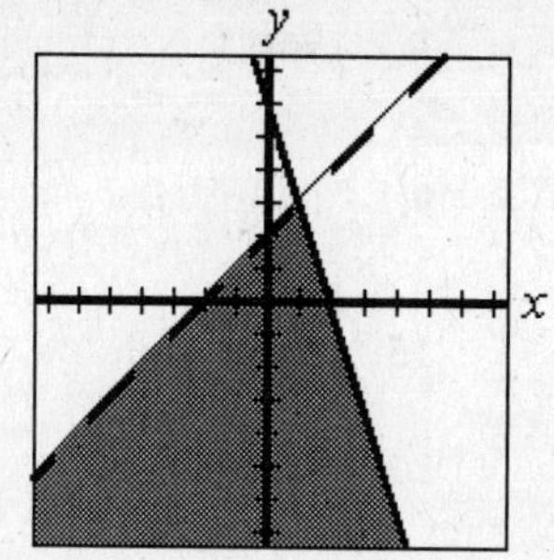

41. Let $x =$ the number of mice used, and let $y =$ the number of rats used.

Form the following system of constraints:

$$\begin{cases} x \geq 0 \\ y \geq 0 \\ 12x + 8y \leq 240 & \text{(Maze 1)} \\ 10x + 15y \leq 300 & \text{(Maze 2)} \end{cases}$$

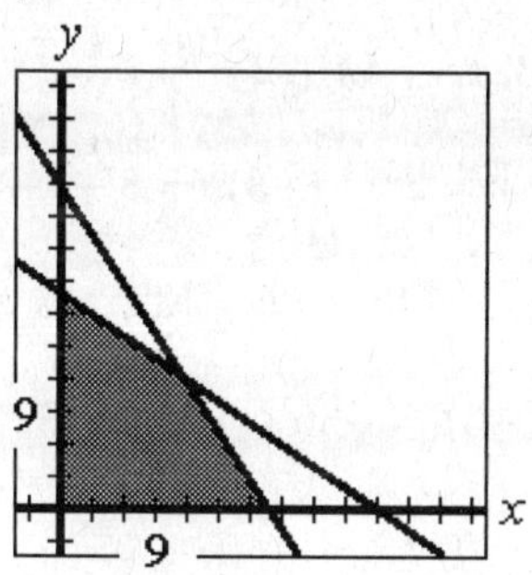

Let the objective function be the total number of animals: $P = x + y$.

Vertex	$P = x + y$
$(0, 0)$	$= 0$
$(20, 0)$	$= 20$
$(12, 12)$	$= 24$
$(0, 20)$	$= 20$

The maximum number of animals that can be used in both mazes is 24, resulting from using 12 mice and 12 rats.

© 2015 Cengage Learning. All Rights Reserved. May not be scanned, copied or duplicated, or posted to a publicly accessible website, in whole or in part.

Exercises 5.1 (page 298)

1. yes **3.** no **5.** yes **7.** no

9. $(y^4y^3)^2 = (y^7)^2 = y^{14}$ **11.** $\frac{3(y^3)^{10}}{y^3y^4} = \frac{3y^{30}}{y^7} = 3y^{23}$ **13.** $114{,}000{,}000 = 1.14 \times 10^8$

15. sum; whole **17.** binomial **19.** descending order **21.** one

23. monomial **25.** trinomial **27.** binomial **29.** monomial

31. deg = 7 **33.** deg = 0 **35.** deg = 2 **37.** deg = 8

39. deg $= 3 + 1 + 6 = 10$ **41.** $3x - 2x^4 + 7 - 5x^2 = -2x^4 - 5x^2 + 3x + 7$

43. $4y^2 - 2y^5 + 7y - 5y^3 = 7y + 4y^2 - 5y^3 - 2y^5$

45. $P(0) = 2(0)^2 + 0 + 2 = 2(0) + 2 = 2$ **47.** $P(-2) = 2(-2)^2 + (-2) + 2 = 2(4) + 0 = 8$

49. $P(5a) = 2(5a)^2 + 5a + 2 = 2(25a^2) + 5a + 2 = 50a^2 + 5a + 2$

51. $P(-4x) = 2(-4x)^2 + (-4x) + 2 = 2(16x^2) - 4x + 2 = 32x^2 - 4x + 2$

53. $x^2 + y^2 = 2^2 + (-3)^2 = 4 + 9 = 13$ **55.** $x^3 - y^3 = 2^3 - (-3)^3 = 8 - (-27) = 35$

57. $4x^2y^2 + 2xy = 4(2)^2(-3)^2 + 2(2)(-3) = 4(4)(9) + (-12) = 144 - 12 = 132$

59. $-2xy^2 + x^2y = -2(2)(-3)^2 + (2)^2(-3) = -2(2)(9) + 4(-3) = -36 - 12 = -48$

61. $h = f(t) = f(0) = -16(0)^2 + 64(0) = 0 + 0 = 0$ ft

63. $h = f(t) = f(2) = -16(2)^2 + 64(2) = -16(4) + 128 = -64 + 128 = 64$ f

65. $f(x) = x^2 + 2$
Shift $y = x^2$ up 2.

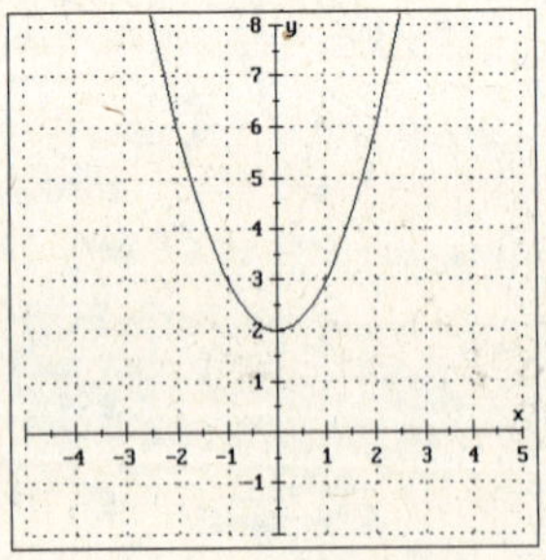

67. $f(x) = -x^3$
Reflect $y = x^3$ about x-axis.

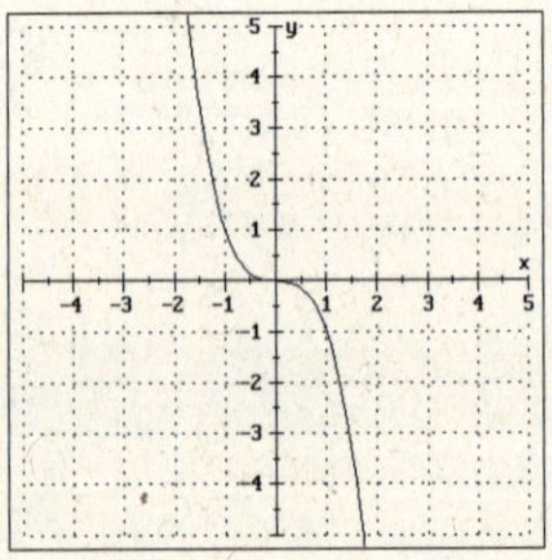

© 2015 Cengage Learning. All Rights Reserved. May not be scanned, copied or duplicated, or posted to a publicly accessible website, in whole or in part.

69. $f(x) = -x^3 + x$

x	y
-2	6
-1	0
$-\frac{1}{2}$	$-\frac{3}{8}$
0	0

x	y
$\frac{1}{2}$	$\frac{3}{8}$
1	0
2	-6

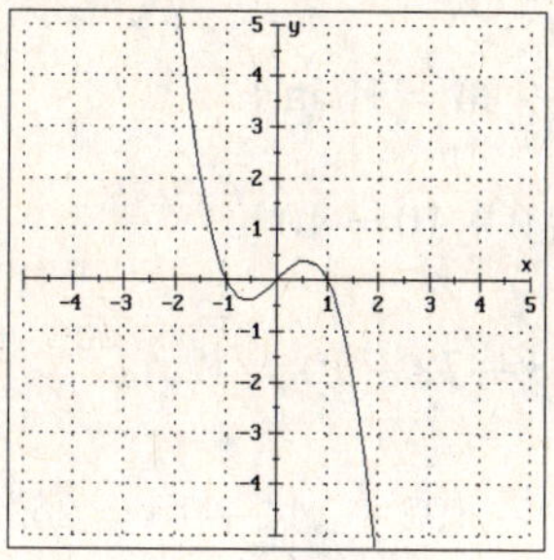

71. $f(x) = x^2 - 2x + 1$

x	y
-1	4
0	1
1	0

x	y
2	1
3	4

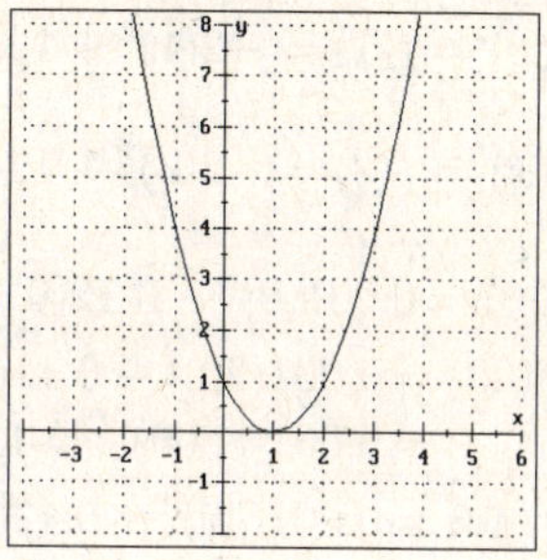

73. $P(0) = -3(0) + 2 = 0 + 2 = 2$

75. $P(-1) = -3(-1) + 2 = 3 + 2 = 5$

77. $ax - ax^3 + 2a^4x^5 - 4a^3x^2 = 2a^4x^5 - ax^3 - 4a^3x^2 + ax$; $\text{deg} = 4 + 5 = 9$

79. $5x^3y^6 + 2x^4y - 5x^3y^3 + x^5y^7 - 2y^4 = 2x^4y - 5x^3y^3 - 2y^4 + 5x^3y^6 + x^5y^7$; $\text{deg} = 5 + 7 = 12$

Problems 81-85 are to be solved using a calculator. The keystrokes needed to solve each problem using a TI-84 graphing calculator appear in each solution. There may be other solutions. Keystrokes for other calculators may be slightly different.

81. $x^2y = (3.7)^2(-2.5) \Rightarrow$ [3] [.] [7] [x^2] [×] [(–)] [2] [.] [5] [ENTER] $\{-34.225\}$

83. $\dfrac{x^2}{z^2} = \dfrac{(3.7)^2}{(8.9)^2} \Rightarrow$ [3] [.] [7] [x^2] [÷] [8] [.] [9] [x^2] [ENTER] $\{0.1728317132\}$

85. $\dfrac{x + y + z}{xyz} = \dfrac{3.7 + (-2.5) + 8.9}{(3.7)(-2.5)(8.9)} \Rightarrow$ [(] [3] [.] [7] [+] [(–)] [2] [.] [5] [+] [8] [.] [9] [)] [÷] [(] [3] [.] [7] [×] [(–)] [2] [.] [5] [×] [8] [.] [9] [)] [ENTER] $\{-0.1226844822\}$

87. $f(x) = 2.75x^2 - 4.7x + 1.5$

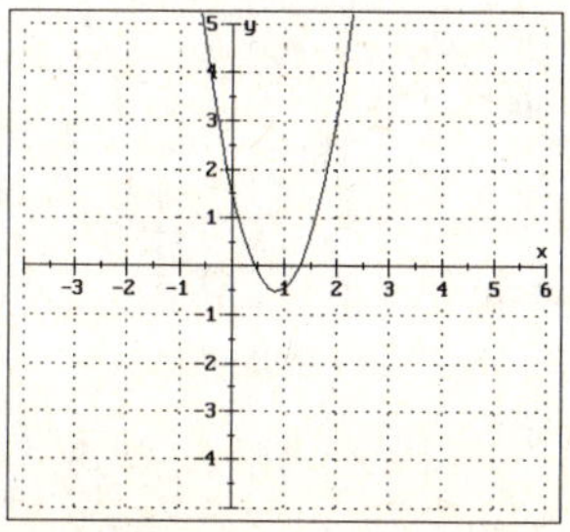

89. $f(x) = 0.25x^2 - 0.5x - 2.5$

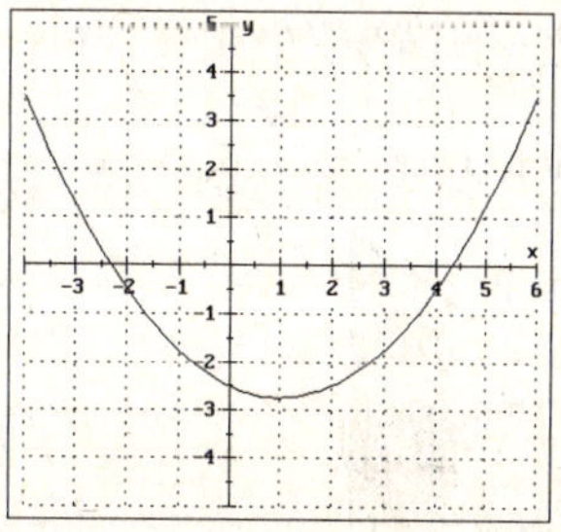

© 2015 Cengage Learning. All Rights Reserved. May not be scanned, copied or duplicated, or posted to a publicly accessible website, in whole or in part.

91. $f(1) = -16(1)^2 + 32(1) + 4 = -16(1) + 32 + 4 = -16 + 32 + 4 = 20$; 20 feet

93. $d = f(v) = f(30) = 0.04(30)^2 + 0.9(30) = 0.04(900) + 27 = 36 + 27 = 63$ ft

95. $d = f(v) = f(60) = 0.04(60)^2 + 0.9(60) = 0.04(3600) + 54 = 144 + 54 = 198$ ft

97. $c = f(x) = f(1) = -2(1)^2 + 12(1) = -2(1) + 12 = -2 + 12 = 10$ in.^{2}

99. $c = f(x) = f(3) = -2(3)^2 + 12(3) = -2(9) + 36 = -18 + 36 = 18$ in.^{2}

101. **a.** $f(0) = 0.001(0)^3 - 0.12(0)^2 + 3.6(0) + 10 = 0 - 0 + 0 + 10 = 10$ m

b. $f(20) = 0.001(20)^3 - 0.12(20)^2 + 3.6(20) + 10$
$= 0.001(8000) - 0.12(400) + 72 + 10 = 8 - 48 + 72 + 10 = 42$ m

c. $f(40) = 0.001(40)^3 - 0.12(40)^2 + 3.6(40) + 10$
$= 0.001(64000) - 0.12(1600) + 144 + 10 = 64 - 192 + 144 + 10 = 26$ m

d. $f(60) = 0.001(60)^3 - 0.12(60)^2 + 3.6(60) + 10$
$= 0.001(216000) - 0.12(3600) + 216 + 10 = 216 - 432 + 216 + 10 = 10$ m

103. Answers may vary.

105. $P(2) + P(3) = [2^2 - 5(2)] + [3^2 - 5(3)] = [4 - 10] + [9 - 15] = -6 + (-6) = -12$
$P(2+3) = P(5) = 5^2 - 5(5) = 0 \Rightarrow$ They are not equal.

107. $P(P(0)) = P(0^2 - 2(0) - 3) = P(-3) = (-3)^2 - 2(-3) - 3 = 9 + 6 - 3 = 12$

109. Answers may vary.

Exercises 5.2 (page 303)

1. $2 \cdot 3 + 4 \cdot 5 = 6 + 20 = 26$

3. $4(5+5) - 6(1+2) = 4(10) - 6(3)$
$= 40 - 18 = 22$

5. $-(6+2) = -(8) = -8$

7. $5x + 6 \le -9$
$5x \le -15$
$x \le -3$
solution set: $(-\infty, -3]$

9. $|x - 4| < 5$
$-5 < x - 4 < 5$
$-1 < x < 9$
solution set: $(-1, 9)$

11. exponents

13. coefficients

15. like terms, $5x + 20x = 25x$

17. unlike terms

19. like terms, $3r^2t^3 - 8r^2t^3 = -5r^2t^3$

21. unlike terms

© 2015 Cengage Learning. All Rights Reserved. May not be scanned, copied or duplicated, or posted to a publicly accessible website, in whole or in part.

SECTION 5.2

23. $9y + 7y = 16y$

25. $5x^3y^2z - 3x^3y^2z = 2x^3y^2z$

27. $4xy^3 + 6x^3y - 5xy^3 = 6x^3y - xy^3$

29. $(3x^2y)^2 + 2x^4y^2 - x^4y^2 = 9x^4y^2 + 2x^4y^2 - x^4y^2 = 10x^4y^2$

31. $(5a + 8) + (2a + 3) = 5a + 8 + 2a + 3 = 7a + 11$

33. $(3x^2 + 2x + 1) + (-2x^2 - 7x + 5) = 3x^2 + 2x + 1 - 2x^2 - 7x + 5 = x^2 - 5x + 6$

35. $(10ab^2 + 6) + (3ab^2 - 11) = 10ab^2 + 6 + 3ab^2 - 11 = 13ab^2 - 5$

37. $(-4a^2b^4 + 2a^2b^2) + (8a^2b^4 - 2a^2b^2) = -4a^2b^4 + 2a^2b^2 + 8a^2b^4 - 2a^2b^2 = 4a^2b^4$

39. $14x^2 - 5x^2 = 9x^2$

41. $7x^3y^2 - 4x^3y^2 = 3x^3y^2$

43. $(5t + 7) - (3t + 6) = 5t + 7 - 3t - 6 = 2t + 1$

45. $(-a^2 + 2a + 3) - (4a^2 - 2a - 1) = -a^2 + 2a + 3 - 4a^2 + 2a + 1 = -5a^2 + 4a + 4$

47. $(7pq^2 + 3pq) - (2pq^2 - 2pq) = 7pq^2 + 3pq - 2pq^2 + 2pq = 5pq^2 + 5pq$

49. $(10xy - 3x^2y^2) - (-3xy - 5x^2y^2) = 10xy - 3x^2y^2 + 3xy + 5x^2y^2 = 13xy + 2x^2y^2$

51. $$\begin{aligned} 7(x+1) - 4(x+4) &= 7x + 7 - 4x - 16 \\ &= 3x - 9 \end{aligned}$$

53. $$\begin{aligned} -3(t-5) - 6(t-2) &= -3t + 15 - 6t + 12 \\ &= -9t + 27 \end{aligned}$$

55. $2(x^3 + x^2) + 3(2x^3 - x^2) = 2x^3 + 2x^2 + 6x^3 - 3x^2 = 8x^3 - x^2$

57. $$\begin{aligned} &5(2a^2 + 4a - 2) - 2(-3a^2 - a + 12) - 2(a^2 + 3a - 5) \\ &\qquad = 10a^2 + 20a - 10 + 6a^2 + 2a - 24 - 2a^2 - 6a + 10 \\ &\qquad = 14a^2 + 16a - 24 \end{aligned}$$

59. $(x^2 + 2x + 1) + (2x^2 - 2x + 1) = x^2 + 2x + 1 + 2x^2 - 2x + 1 = 3x^2 + 2$

61. $(2x^2 - x - 3) + (x^2 - 3x - 1) = 2x^2 - x - 3 + x^2 - 3x - 1 = 3x^2 - 4x - 4$

63. $(7y^3 + 4y^2 + y + 3) + (-8y^3 - y + 3) = 7y^3 + 4y^2 + y + 3 - 8y^3 - y + 3 = -y^3 + 4y^2 + 6$

65. $$\begin{aligned} &(3x^3 - 2x + 3) + (4x^3 + 3x^2 - 2) + (-4x^3 - 3x^2 + x + 12) \\ &\qquad = 3x^3 - 2x + 3 + 4x^3 + 3x^2 - 2 - 4x^3 - 3x^2 + x + 12 \\ &\qquad = 3x^3 - x + 13 \end{aligned}$$

67. $(4x^3 + 2x - 3) - (5x^3 + 2x^2 + 1) = 4x^3 + 2x - 3 - 5x^3 - 2x^2 - 1 = -x^3 - 2x^2 + 2x - 4$

© 2015 Cengage Learning. All Rights Reserved. May not be scanned, copied or duplicated, or posted to a publicly accessible website, in whole or in part.

SECTION 5.2

69. $(3y^2 - 2y + 4) + \left[(2y^2 - 3y + 2) - (y^2 + 4y + 3)\right]$

$$\begin{aligned} &= 3y^2 - 2y + 4 + \left[2y^2 - 3y + 2 - y^2 - 4y - 3\right] \\ &= 3y^2 - 2y + 4 + \left[y^2 - 7y - 1\right] \\ &= 3y^2 - 2y + 4 + y^2 - 7y - 1 = 4y^2 - 9y + 3 \end{aligned}$$

71. $2(y^2 + 5y) - 7(y^2 - 3) = 2y^2 + 10y - 7y^2 + 21 = -5y^2 + 10y + 21$

73. $5(p - 2q) - 4(2p + q) = 5p - 10q - 8p - 4q = -3p - 14q$

75. $-3(3a^2 + 4b^3 + 7) + 4(5a^2 - 2b^3 + 3) = -9a^2 - 12b^3 - 21 + 20a^2 - 8b^3 + 12$

$$= 11a^2 - 20b^3 - 9$$

77.

$$\begin{array}{rr} & 3x^3 - 2x^2 + 4x - 3 \\ & -2x^3 + 3x^2 + 3x - 2 \\ + & 5x^3 - 7x^2 + 7x - 12 \\ \hline & 6x^3 - 6x^2 + 14x - 17 \end{array}$$

79.

$$\begin{array}{rr} & -2y^4 - 2y^3 + 4y^2 - 3y + 10 \\ & -3y^4 + 7y^3 - y^2 + 14y - 3 \\ & -3y^3 - 5y^2 - 5y + 7 \\ + & -4y^4 + y^3 - 13y^2 + 14y - 2 \\ \hline & -9y^4 + 3y^3 - 15y^2 + 20y + 12 \end{array}$$

81.

$$\begin{array}{rr} & 3x^2 - 4x + 17 \\ - & 2x^2 + 4x - 5 \\ \hline & 3x^2 - 4x + 17 \\ + & -2x^2 - 4x + 5 \\ \hline & x^2 - 8x + 22 \end{array}$$

83.

$$\begin{array}{rr} & -5y^3 + 4y^2 - 11y + 3 \\ - & -2y^3 - 14y^2 + 17y - 32 \\ \hline & -5y^3 + 4y^2 - 11y + 3 \\ + & 2y^3 + 14y^2 - 17y + 32 \\ \hline & -3y^3 + 18y^2 - 28y + 35 \end{array}$$

85. $x + (x + 1) + (x + 2) = x + x + 1 + x + 2 = 3x + 3$

87. $(2x^2 + 5x + 1) - (x^2 - 5) = 2x^2 + 5x + 1 - x^2 + 5 = x^2 + 5x + 6$ meters

89. $f(x) = 1100x + 125000$; $f(10) = 1100(10) + 125000 = 11000 + 125000 = \$136{,}000$

91. $f(x) = (1100x + 125000) + (1400x + 150000) = 2500x + 275000$

93. $f(x) = -2100x + 16600$

95. $f(x) = (-2100x + 16600) + (-2700x + 19200) = -4800x + 35800$

97. **Answers may vary.**

99. $\left[(-2x^2 - x + 7) + (5x^2 + 3x - 1)\right] - (3x^2 + 4x - 3)$

$$\begin{aligned} &= \left[-2x^2 - x + 7 + 5x^2 + 3x - 1\right] - 3x^2 - 4x + 3 \\ &= -2x^2 - x + 7 + 5x^2 + 3x - 1 - 3x^2 - 4x + 3 \\ &= -2x + 9 \end{aligned}$$

© 2015 Cengage Learning. All Rights Reserved. May not be scanned, copied or duplicated, or posted to a publicly accessible website, in whole or in part.

101. $\left[(2x^2 - 4x + 3) - (8x^2 + 5x - 3)\right] + (-2x^2 + 7x - 4)$
$= \left[2x^2 - 4x + 3 - 8x^2 - 5x + 3\right] - 2x^2 + 7x - 4$
$= 2x^2 - 4x + 3 - 8x^2 - 5x + 3 - 2x^2 + 7x - 4$
$= -8x^2 - 2x + 2$

Exercises 5.3 (page 313)

1. distributive property

3. associative property of multiplication

5. commutative property of multiplication

7. $|3a - b| = |3(-2) - 4| = |-6 - 4|$
$= |-10| = 10$

9. $-|a^2b - b^0| = -|(-2)^2(4) - 1| = -|4(4) - 1| = -|16 - 1| = -|15| = -15$

11. The current value of ABC = \$126.5 per share, and the current value of WD = \$73.5 per share. Total value = 200(126.5) + 350(73.5) = 25300 + 25725 = \$51,025.

13. variable

15. term

17. $x^2 + 2xy + y^2$

19. $x^2 - y^2$; conjugate; difference

21. $(-6a^2b)(4ab^2) = -24a^3b^3$

23. $(-3ab^2c)(5ac^2) = -15a^2b^2c^3$

25. $(4a^2b)(-5a^3b^2)(6a^4) = -120a^9b^3$

27. $(5x^3y^2)^4\left(\frac{1}{5}x^2\right)^2 = (625x^{12}y^8)\left(\frac{1}{25}x^4\right)$
$= 25x^{16}y^8$

29. $7(x + 4) = 7x + 7(4) = 7x + 28$

31. $-a(a - b) = -a(a) + (-a)(-b) = -a^2 + ab$

33. $5x(3x^2 + 4) = 5x(3x^2) + 5x(4) = 15x^3 + 20x$

35. $-2x(3x^2 - 3x + 2) = -2x(3x^2) + (-2x)(-3x) + (-2x)(2) = -6x^3 + 6x^2 - 4x$

37. $5a^2b^3(2a^4b - 5b^3) = 5a^2b^3(2a^4b) + 5a^2b^3(-5b^3) = 10a^6b^4 - 25a^2b^6$

39. $7rst(r^2 + s^2 - t^2) = 7rst(r^2) + 7rst(s^2) + 7rst(-t^2) = 7r^3st + 7rs^3t - 7rst^3$

41. $(x + 4)(x + 2) = x^2 + 2x + 4x + 8$
$= x^2 + 6x + 8$

43. $(z - 3)(z - 8) = z^2 - 3z - 8z + 24$
$= z^2 - 11z + 24$

45. $(3y + 1)(2y^2 + 3y + 2) = 3y(2y^2 + 3y + 2) + 1(2y^2 + 3y + 2)$
$= 6y^3 + 9y^2 + 6y + 2y^2 + 3y + 2 = 6y^3 + 11y^2 + 9y + 2$

47. $(3x + 4)(2x^2 - 3x + 5) = 3x(2x^2 - 3x + 5) + 4(2x^2 - 3x + 5)$
$= 6x^3 - 9x^2 + 15x + 8x^2 - 12x + 20 = 6x^3 - x^2 + 3x + 20$

© 2015 Cengage Learning. All Rights Reserved. May not be scanned, copied or duplicated, or posted to a publicly accessible website, in whole or in part.

SECTION 5.3

49. $(2a+1)(a-2) = 2a^2 - 4a + a - 2 = 2a^2 - 3a - 2$

51. $(2x-3y)(x+2y) = 2x^2 + 4xy - 3xy - 6y^2 = 2x^2 + xy - 6y^2$

53. $(3x+y)(3x-3y) = 9x^2 - 9xy + 3xy - 3y^2 = 9x^2 - 6xy - 3y^2$

55. $(4a-3b)(2a+5b) = 8a^2 + 20ab - 6ab - 15b^2 = 8a^2 + 14ab - 15b^2$

57. $(x+2)^2 = (x+2)(x+2) = x^2 + 2x + 2x + 4 = x^2 + 4x + 4$

59. $(a-4)^2 = (a-4)(a-4) = a^2 - 4a - 4a + 16 = a^2 - 8a + 16$

61. $(2a+b)^2 = (2a+b)(2a+b) = 4a^2 + 2ab + 2ab + b^2 = 4a^2 + 4ab + b^2$

63. $(2x-y)^2 = (2x-y)(2x-y) = 4x^2 - 2xy - 2xy + y^2 = 4x^2 - 4xy + y^2$

65. $$\begin{aligned}(x+2)(x-2) &= x^2 - 2x + 2x - 4\\ &= x^2 - 4\end{aligned}$$

67. $$\begin{aligned}(a+b)(a-b) &= a^2 - ab + ab - b^2\\ &= a^2 - b^2\end{aligned}$$

69. $$\begin{aligned}(2p+q)(p-2q)^2 = (2p+q)(p-2q)(p-2q) &= (2p+q)\left(p^2 - 4pq + 4q^2\right)\\ &= 2p\left(p^2 - 4pq + 4q^2\right) + q\left(p^2 - 4pq + 4q^2\right)\\ &= 2p^3 - 8p^2q + 8pq^2 + p^2q - 4pq^2 + 4q^3\\ &= 2p^3 - 7p^2q + 4pq^2 + 4q^3\end{aligned}$$

71. $$\begin{aligned}(a+b)^3 = (a+b)(a+b)(a+b) &= \left(a^2 + 2ab + b^2\right)(a+b)\\ &= a^2(a+b) + 2ab(a+b) + b^2(a+b)\\ &= a^3 + a^2b + 2a^2b + 2ab^2 + ab^2 + b^3\\ &= a^3 + 3a^2b + 3ab^2 + b^3\end{aligned}$$

73. $(a+b)(a-b)(a-3b) = (a^2 - b^2)(a-3b) = a^3 - 3a^2b - ab^2 + 3b^3$

75. $(x+3)(2x+1)(x-3) = (x+3)(x-3)(2x+1) = (x^2-9)(2x+1) = 2x^3 + x^2 - 18x - 9$

77. $x^3(2x^2 + x^{-2}) = 2x^5 + x$

79. $x^3y^{-6}z^{-2}(3x^{-2}y^2z - x^3y^{-4}) = 3xy^{-4}z^{-1} - x^6y^{-10}z^{-2} = \dfrac{3x}{y^4z} - \dfrac{x^6}{y^{10}z^2}$

81. $(x^{-1}+y)(x^{-1}-y) = x^{-2} - x^{-1}y + x^{-1}y - y^2 = \dfrac{1}{x^2} - y^2$

83. $$\begin{aligned}(2x^{-3}+y^3)(2x^3-y^{-3}) = 4x^0 - 2x^{-3}y^{-3} + 2x^3y^3 - y^0 &= 4 - \frac{2}{x^3y^3} + 2x^3y^3 - 1\\ &= 2x^3y^3 - \frac{2}{x^3y^3} + 3\end{aligned}$$

© 2015 Cengage Learning. All Rights Reserved. May not be scanned, copied or duplicated, or posted to a publicly accessible website, in whole or in part.

SECTION 5.3

85. $x^n(x^{2n} - x^n) = x^{3n} - x^{2n}$

87. $a^{2n}(a^n + a^{2n}) = a^{3n} + a^{4n}$

89. $3x(2x + 4) - 3x^2 = 6x^2 + 12x - 3x^2 = 3x^2 + 12x$

91.
$$\begin{aligned}(3y + 1)^2 + (2y - 4)^2 &= (3y + 1)(3y + 1) + (2y - 4)(2y - 4)\\ &= 9y^2 + 3y + 3y + 1 + 4y^2 - 8y - 8y + 16\\ &= 13y^2 - 10y + 17\end{aligned}$$

93. $(2x + 3y)(2x - 3y) = 4x^2 - 6xy + 6xy - 9y^2 = 4x^2 - 9y^2$

95.
$$\begin{aligned}(x - y)(x^2 + xy + y^2) &= x(x^2 + xy + y^2) - y(x^2 + xy + y^2)\\ &= x^3 + x^2y + xy^2 - x^2y - xy^2 - y^3 = x^3 - y^3\end{aligned}$$

97. $(-5xx^2)(-3xy)^4 = (-5x^3)(81x^4y^4) = -405x^7y^4$

99. $4m^2n(-3mn)(m + n) = -12m^3n^2(m + n) = -12m^4n^2 - 12m^3n^3$

101.
$$\begin{aligned}(2x - y)(5x^2 + 3xy + 4y^2) &= 2x(5x^2 + 3xy + 4y^2) - y(5x^2 + 3xy + 4y^2)\\ &= 10x^3 + 6x^2y + 8xy^2 - 5x^2y - 3xy^2 - 4y^3\\ &= 10x^3 + x^2y + 5xy^2 - 4y^3\end{aligned}$$

103.
$$\begin{aligned}(2a - b)(4a^2 + 2ab + b^2) &= 2a(4a^2 + 2ab + b^2) - b(4a^2 + 2ab + b^2)\\ &= 8a^3 + 4a^2b + 2ab^2 - 4a^2b - 2ab^2 - b^3 = 8a^3 - b^3\end{aligned}$$

105.
$$\begin{aligned}(2x - 1)[2x^2 - 3(x + 2)] = (2x - 1)(2x^2 - 3x - 6) &= 2x(2x^2 - 3x - 6) - 1(2x^2 - 3x - 6)\\ &= 4x^3 - 6x^2 - 12x - 2x^2 + 3x + 6\\ &= 4x^3 - 8x^2 - 9x + 6\end{aligned}$$

107.
$$\begin{aligned}(x + 2)(x^2 - 2x + 4) &= x(x^2 - 2x + 4) + 2(x^2 - 2x + 4)\\ &= x^3 - 2x^2 + 4x + 2x^2 - 4x + 8 = x^3 + 8\end{aligned}$$

109.
$$\begin{aligned}2m(m - n) - (m + n)(m - 2n) &= 2m^2 - 2mn - (m^2 - 2mn + mn - 2n^2)\\ &= 2m^2 - 2mn - m^2 + 2mn - mn + 2n^2\\ &= m^2 - mn + 2n^2\end{aligned}$$

111.
$$\begin{aligned}(3x - 4)^2 - (2x + 3)^2 &= (3x - 4)(3x - 4) - (2x + 3)(2x + 3)\\ &= 9x^2 - 12x - 12x + 16 - (4x^2 + 6x + 6x + 9)\\ &= 9x^2 - 24x + 16 - 4x^2 - 6x - 6x - 9\\ &= 5x^2 - 36x + 7\end{aligned}$$

© 2015 Cengage Learning. All Rights Reserved. May not be scanned, copied or duplicated, or posted to a publicly accessible website, in whole or in part.

113. $5(2y-z)^2+4(y+2z)^2 = 5(2y-z)(2y-z)+4(y+2z)(y+2z)$
$= 5(4y^2-2yz-2yz+z^2)+4(y^2+2yz+2yz+4z^2)$
$= 20y^2-10yz-10yz+5z^2+4y^2+8yz+8yz+16z^2$
$= 24y^2-4yz+21z^2$

115. $(x^n-y^n)(x^n-y^{-n}) = x^{2n}-x^ny^{-n}-x^ny^n+y^0 = x^{2n}-\dfrac{x^n}{y^n}-x^ny^n+1$

117. $(x^{2n}+y^{2n})(x^{2n}-y^{2n}) = x^{4n}-x^{2n}y^{2n}+x^{2n}y^{2n}-y^{4n} = x^{4n}-y^{4n}$

119. $(x^n+y^n)(x^n+1) = x^{2n}+x^n+x^ny^n+y^n$

121. $(3.21x-7.85)(2.87x+4.59) = 9.2127x^2+14.7339x-22.5295x-36.0315$
$= 9.2127x^2-7.7956x-36.0315$

123. $(-17.3y+4.35)^2 = (-17.3y+4.35)(-17.3y+4.35)$
$= 299.29y^2-75.255y-75.255y+18.9225$
$= 299.29y^2-150.51y+18.9225$

125. **a.** Area of large square $= x^2$
b. Area I $= (x-y)^2 = x^2-2xy+y^2$
c. Area II $= y(x-y) = xy-y^2$
d. Area III $= y(x-y) = xy-y^2$
e. Area IV $= y^2$
Area I = Area of large square − Area II − Area III − Area IV
$(x-y)^2 = x^2-(xy-y^2)-(xy-y^2)-y^2$
$(x-y)^2 = x^2-xy+y^2-xy+y^2-y^2$
$(x-y)^2 = x^2-2xy+y^2$

127. $x = -\frac{1}{5}(375)+90 = -75+90 = 15$

129. $r = p\left(-\frac{1}{5}p+90\right) = -\frac{1}{5}p^2+90p$

131. $A = lw = (2x-3)(2x-3) = 4x^2-6x-6x+9 = (4x^2-12x+9)$ ft^2

133. $A = \frac{1}{2}bh = \frac{1}{2}(b+5)(b-2) = \frac{1}{2}(b^2-2b+5b-10) = \frac{1}{2}(b^2+3b-10)$ in.^{2}

135. Answers may vary.

137. $0.35\times10^7+1.96\times10^7 = (0.35+1.96)\times10^7 = 2.31\times10^7$

Exercises 5.4 (page 322)

1. 2 terms

3. 4 terms

5. $13 = 1\cdot13$, prime

7. $(a+9)(a-9) = a^2-81$

9. $(4r^2+3s)(4r^2-3s) = 16r^4-9s^2$

11. $(n-5)(n^2+5n+25) = n^3+5n^2+25n-5n^2-25n-125 = n^3-125$

© 2015 Cengage Learning. All Rights Reserved. May not be scanned, copied or duplicated, or posted to a publicly accessible website, in whole or in part.

SECTION 5.4

13. factoring

15. greatest common factor

17. $24 = 8 \cdot 3 = 2^3 \cdot 3$

19. $135 = 27 \cdot 5 = 3^3 \cdot 5$

21. $128 = 2^7$

23. $325 = 5^2 \cdot 13$

25. $36 = 2^2 \cdot 3^2$; $48 = 2^4 \cdot 3$; $\text{gcf} = 2^2 \cdot 3 = 12$

27. $42 = 2 \cdot 3 \cdot 7$; $36 = 2^2 \cdot 3^2$; $98 = 2 \cdot 7^2$
$\text{gcf} = 2$

29. $4a^2b = 2^2 \cdot a^2b$; $8a^3c = 2^3 \cdot a^3c$; $\text{gcf} = 2^2 \cdot a^2 = 4a^2$

31. $18x^4y^3z^2 = 2 \cdot 3^2 \cdot x^4y^3z^2$; $-12xy^2z^3 = -1 \cdot 2^2 \cdot 3xy^2z^2$; $\text{gcf} = 2 \cdot 3 \cdot x \cdot y^2 \cdot z^2 = 6xy^2z^2$

33. $6a - 24 = 6(a - \underline{4})$

35. $8z^2 + 2z = 2z(4z + \underline{1})$

37. $2x + 8 = \mathbf{2} \cdot x + \mathbf{2} \cdot 4 = \mathbf{2}(x + 4)$

39. $2x^2 - 6x = \mathbf{2x} \cdot x - \mathbf{2x} \cdot 3 = \mathbf{2x}(x - 3)$

41. $7x^2 + 14x = \mathbf{7x} \cdot x + \mathbf{7x} \cdot 2 = \mathbf{7x}(x + 2)$

43. $63x^3y^2 + 81x^2y^4 = \mathbf{9x^2y^2} \cdot 7x + \mathbf{9x^2y^2} \cdot 9y^2 = \mathbf{9x^2y^2}(7x + 9y^2)$

45. $9a^3b^3 - 15a^2b^2 + 3ab = \mathbf{3ab}(3a^2b^2) - \mathbf{3ab}(5ab) + \mathbf{3ab}(1) = \mathbf{3ab}(3a^2b^2 - 5ab + 1)$

47. $45x^{10}y^3 - 63x^7y^7 + 81x^{10}y^{10} = \mathbf{9x^7y^3}(5x^3) - \mathbf{9x^7y^3}(7y^4) + \mathbf{9x^7y^3}(9x^3y^7)$
$= \mathbf{9x^7y^3}(5x^3 - 7y^4 + 9x^3y^7)$

49. $16x^2y^2 + 8ab \Rightarrow$ prime

51. $11m^3n^2 - 12x^2y \Rightarrow$ prime

53. $-5a - 10 = (\mathbf{-5})(a) + (\mathbf{-5})(2)$
$= \mathbf{-5}(a + 2)$

55. $-2x^4 + x^3 = (\mathbf{-x^3})(2x) + (\mathbf{-x^3})(-1)$
$= \mathbf{-x^3}(2x - 1)$

57. $-6x^2 - 3xy = (\mathbf{-3x})(2x) + (\mathbf{-3x})(y) = \mathbf{-3x}(2x + y)$

59. $-36a^2b^2 + 48ab^3 = (\mathbf{-12ab^2})(3a) + (\mathbf{-12ab^2})(-4b) = \mathbf{-12ab^2}(3a - 4b)$

61. $x^{9n} + x^{6n} + x^{3n} = x^{3n}(x^{9n-3n} + x^{6n-3n} + x^{3n-3n}) = x^{3n}(x^{6n} + x^{3n} + x^0)$
$= x^{3n}(x^{6n} + x^{3n} + 1)$

63. $2y^{n+2} - 3y^{n+3} = y^n(2y^{n+2-n} - 3y^{n+3-n}) = y^n(2y^2 - 3y^3)$

65. $4(\mathbf{x + y}) + t(\mathbf{x + y}) = (\mathbf{x + y})(4 + t)$

67. $(\mathbf{a - b})r - (\mathbf{a - b})s = (\mathbf{a - b})(r - s)$

69. $(u + v)^2 - (u + v) = (u + v)(\mathbf{u + v}) - 1(\mathbf{u + v}) = (\mathbf{u + v})[(u + v) - 1] = (u + v)(u + v - 1)$

71. $-a(\mathbf{x + y}) + b(\mathbf{x + y}) = (\mathbf{x + y})(-a + b) = -(x + y)(a - b)$

73. $(\mathbf{a + b})x - (\mathbf{a + b})y + (\mathbf{a + b})z = (\mathbf{a + b})(x - y + z)$

© 2015 Cengage Learning. All Rights Reserved. May not be scanned, copied or duplicated, or posted to a publicly accessible website, in whole or in part.

SECTION 5.4

75. $x(\boldsymbol{a}-\boldsymbol{b}+\boldsymbol{c})-y(\boldsymbol{a}-\boldsymbol{b}+\boldsymbol{c})+z(\boldsymbol{a}-\boldsymbol{b}+\boldsymbol{c})=(\boldsymbol{a}-\boldsymbol{b}+\boldsymbol{c})(x-y+z)$

77. $3c-cd+3d-c^2=3c+3d-c^2-cd=3(c+d)-c(c+d)=(c+d)(3-c)$

79. $a^2-4b+ab-4a=a^2+ab-4a-4b=a(a+b)-4(a+b)=(a+b)(a-4)$

81. $ax+bx-a-b=x(a+b)-1(a+b)=(a+b)(x-1)$

83. $x^2+xy+xz+ax+ay+az=x(x+y+z)+a(x+y+z)=(x+y+z)(x+a)$

85.
$$\begin{aligned} mpx+mqx+npx+nqx &= x(mp+mq+np+nq)=x[m(p+q)+n(p+q)] \\ &= x(p+q)(m+n) \end{aligned}$$

87.
$$\begin{aligned} x^2y+xy^2+2xyz+xy^2+y^3+2y^2z &= y(x^2+xy+2xz+xy+y^2+2yz) \\ &= y[x(x+y+2z)+y(x+y+2z)] \\ &= y(x+y+2z)(x+y) \end{aligned}$$

89.
$$\begin{aligned} 2n^4p-2n^2-n^3p^2+np+2mn^3p-2mn &= n(2n^3p-2n-n^2p^2+p+2mn^2p-2m) \\ &= n(2n^3p-n^2p^2+2mn^2p-2n+p-2m) \\ &= n[n^2p(2n-p+2m)-1(2n-p+2m)] \\ &= n(2n-p+2m)(n^2p-1) \end{aligned}$$

91.
$$\begin{aligned} 2a^3b-5a^2b^2-6a^2b+15ab^2 = ab\left[2a^2-5ab-6a+15b\right] &= ab[a(2a-5b)-3(2a-5b)] \\ &= ab(2a-5b)(a-3) \end{aligned}$$

93.
$$\begin{aligned} r_1r_2 &= rr_2+rr_1 \\ r_1r_2-rr_1 &= rr_2 \\ r_1(r_2-r) &= rr_2 \\ r_1 &= \frac{rr_2}{r_2-r} \end{aligned}$$

95.
$$\begin{aligned} d_1d_2 &= fd_2+fd_1 \\ d_1d_2 &= f(d_2+d_1) \\ \frac{d_1d_2}{d_2+d_1} &= f, \text{ or } f=\frac{d_1d_2}{d_2+d_1} \end{aligned}$$

97.
$$\begin{aligned} b^2x^2+a^2y^2 &= a^2b^2 \\ b^2x^2 &= a^2b^2-a^2y^2 \\ b^2x^2 &= a^2(b^2-y^2) \\ \frac{b^2x^2}{b^2-y^2} &= a^2, \text{ or } a^2=\frac{b^2x^2}{b^2-y^2} \end{aligned}$$

99.
$$\begin{aligned} S(1-r) &= a-lr \\ S-Sr &= a-lr \\ S-a &= Sr-lr \\ S-a &= r(S-l) \\ \frac{S-a}{S-l} &= r, \text{ or } r=\frac{S-a}{S-l} \end{aligned}$$

101. $27z^3+12z^2+3z=\boldsymbol{3z}\cdot 9z^2+\boldsymbol{3z}\cdot 4z+\boldsymbol{3z}\cdot 1=\boldsymbol{3z}(9z^2+4z+1)$

103. $25x^3-14y^3+36x^3y^3 \Rightarrow$ prime

105. $-63u^3v^6z^9+28u^2v^7z^2-21u^3v^3z^4=-7u^2v^3z^2(9uv^3z^7-4v^4+3uz^2)$

107. $8y^{2n}+12+16y^{-2n}=4y^{-2n}\left(2y^{2n-(-2n)}+3y^{0-(-2n)}+4y^{-2n-(-2n)}\right)=4y^{-2n}(2y^{4n}+3y^{2n}+4)$

© 2015 Cengage Learning. All Rights Reserved. May not be scanned, copied or duplicated, or posted to a publicly accessible website, in whole or in part.

109. $3(\boldsymbol{m+n+p}) + x(\boldsymbol{m+n+p}) = (\boldsymbol{m+n+p})(3+x)$

111. $(x+y)(\boldsymbol{x+y}) + z(\boldsymbol{x+y}) = (\boldsymbol{x+y})[(x+y)+z] = (x+y)(x+y+z)$

113. $ax + ay + bx + by = a(x+y) + b(x+y) = (x+y)(a+b)$

115. $x^2 + 2x + xy + 2y = x(x+2) + y(x+2) = (x+2)(x+y)$

117. $x^4 - 5x^6 = x^{-2}\left(x^{4-(-2)} - 5x^{6-(-2)}\right) = x^{-2}(x^6 - 5x^8)$

119. $a^{-3}b - 5a^{-6}b^{-1} = a^{-3}b^{-1}\left(a^{-3-(-3)}b^{1-(-1)} - 5a^{-6-(-3)}b^{-1-(-1)}\right) = a^{-3}b^{-1}\left(a^0b^2 - 5a^{-3}b^0\right) = a^{-3}b^{-1}\left(b^2 - 5a^{-3}\right)$

121.
$$\begin{aligned} H(a+b) &= 2ab \\ Ha + Hb &= 2ab \\ Hb &= 2ab - Ha \\ Hb &= a(2b - H) \\ \frac{Hb}{2b-H} &= a, \text{ or } a = \frac{Hb}{2b-H} \end{aligned}$$

123.
$$\begin{aligned} 3xy - x &= 2y + 3 \\ 3xy - 2y &= x + 3 \\ y(3x-2) &= x + 3 \\ y &= \frac{x+3}{3x-2} \end{aligned}$$

125. **a.** $(2x^2)(6x) = 12x^3$ in.^{2}

b. $(5x)(4x) = 20x^2$ in.^{2}

c. $12x^3 - 20x^2 = 4x^2(3x - 5)$ in.^{2}

127. **Answers may vary.**

129. **Answers may vary.**

131. $14 = 2 \cdot 7$; $45 = 3^2 \cdot 5$; gcf $= 1 \Rightarrow$ relatively prime

133. $60 = 2^2 \cdot 3 \cdot 5$; $28 = 2^2 \cdot 7$; $36 = 2^2 \cdot 3^2$; gcf $= 2^2 \Rightarrow$ not relatively prime

135. $12x^2y = 2^2 \cdot 3 \cdot x^2y$; $5ab^3 = 5 \cdot ab^3$; $35x^2b^3 = 5 \cdot 7 \cdot x^2b^3$; gcf $= 1 \Rightarrow$ relatively prime

Exercises 5.5 (page 329)

1. $36a^4 = (6a^2)^2$

3. $y^6 = (y^3)^2$ or $(y^2)^3$

5. $8b^6 = (2b^2)^3$

7. $(x+1)(x+1) = x^2 + x + x + 1 = x^2 + 2x + 1$

9. $(2m+n)(2m+n) = 4m^2 + 2mn + 2mn + n^2 = 4m^2 + 4mn + n^2$

11. $(a+4)(a+3) = a^2 + 3a + 4a + 12 = a^2 + 7a + 12$

13. $(4r - 3s)(2r - s) = 8r^2 - 4rs - 6rs + 3s^2 = 8r^2 - 10rs + 3s^2$

15. $1, 4, 9, 16, 25, 36, 49, 64, 81, 100$

17. cannot

19. $(p^2 - pq + q^2)$

21. $x^2 - 4 = x^2 - 2^2 = (x+2)(x-2)$

23. $t^2 - 225 = t^2 - 15^2 = (t+15)(t-15)$

25. $b^2 - 81 = b^2 - 9^2 = (b+9)(b-9)$

© 2015 Cengage Learning. All Rights Reserved. May not be scanned, copied or duplicated, or posted to a publicly accessible website, in whole or in part.

SECTION 5.5

27. $16x^4 - 81y^2 = (4x^2)^2 - (9y)^2$
$= (4x^2 + 9y)(4x^2 - 9y)$

29. $625a^2 - 169b^4 = (25a)^2 - (13b^2)^2$
$= (25a + 13b^2)(25a - 13b^2)$

31. $x^4 - y^4 = (x^2 + y^2)(x^2 - y^2) = (x^2 + y^2)(x + y)(x - y)$

33. $16a^4 - 81b^4 = (4a^2 + 9b^2)(4a^2 - 9b^2) = (4a^2 + 9b^2)(2a + 3b)(2a - 3b)$

35. $(x + y)^2 - z^2 = [(x + y) + z][(x + y) - z]$
$= (x + y + z)(x + y - z)$

37. $(a - b)^2 - c^2 = [(a - b) + c][(a - b) - c]$
$= (a - b + c)(a - b - c)$

39. $2x^2 - 288 = 2(x^2 - 144)$
$= 2(x + 12)(x - 12)$

41. $2x^3 - 32x = 2x(x^2 - 16)$
$= 2x(x + 4)(x - 4)$

43. $5x^3 - 125x = 5x(x^2 - 25)$
$= 5x(x + 5)(x - 5)$

45. $r^2s^2t^2 - t^2x^4y^2 = t^2(r^2s^2 - x^4y^2)$
$= t^2(rs + x^2y)(rs - x^2y)$

47. $x^3 + 27 = x^3 + 3^3 = (x + 3)(x^2 - 3x + 3^2)$
$= (x + 3)(x^2 - 3x + 9)$

49. $p^3 + 64 = p^3 + 4^3 = (p + 4)(p^2 - 4p + 4^2)$
$= (p + 4)(p^2 - 4p + 16)$

51. $t^3 - v^3 = (t - v)(t^2 + tv + v^2)$

53. $x^3 - 8y^3 = x^3 - (2y)^3 = (x - 2y)[x^2 + x(2y) + (2y)^2] = (x - 2y)(x^2 + 2xy + 4y^2)$

55. $5x^3 + 625 = 5(x^3 + 125) = 5(x^3 + 5^3) = 5(x + 5)(x^2 - 5x + 25)$

57. $4x^5 - 256x^2 = 4x^2(x^3 - 64) = 4x^2(x^3 - 4^3) = 4x^2(x - 4)(x^2 + 4x + 16)$

59. $128u^2v^3 - 2t^3u^2 = 2u^2(64v^3 - t^3) = 2u^2[(4v)^3 - t^3] = 2u^2(4v - t)[(4v)^2 + 4vt + t^2]$
$= 2u^2(4v - t)(16v^2 + 4vt + t^2)$

61. $(a + b)x^3 + 27(a + b) = (a + b)(x^3 + 27) = (a + b)(x^3 + 3^3) = (a + b)(x + 3)(x^2 - 3x + 9)$

63. $x^6 - y^6 = (x^3)^2 - (y^3)^2 = (x^3 + y^3)(x^3 - y^3)$
$= (x + y)(x^2 - xy + y^2)(x - y)(x^2 + xy + y^2)$

65. $1 - a^6 = (1)^2 - (a^3)^2 = (1 + a^3)(1 - a^3)$
$= (1 + a)(1^2 - 1a + a^2)(1 - a)(1^2 + 1a + a^2)$
$= (1 + a)(1 - a + a^2)(1 - a)(1 + a + a^2)$

67. $x^3 - (y + z)^3 = [x - (y + z)][x^2 + x(y + z) + (y + z)^2]$
$= (x - y - z)(x^2 + xy + xz + y^2 + 2yz + z^2)$

69. $(a - b)^3 - c^3 = [(a - b) - c][(a - b)^2 + (a - b)c + c^2]$
$= (a - b - c)(a^2 - 2ab + b^2 + ac - bc + c^2)$

© 2015 Cengage Learning. All Rights Reserved. May not be scanned, copied or duplicated, or posted to a publicly accessible website, in whole or in part.

SECTION 5.5

71. $a^2 - b^2 + a + b = (a^2 - b^2) + (a + b) = (a + b)(a - b) + 1(a + b)$
$= (a + b)(a - b + 1)$

73. $a^2 - b^2 + 2a - 2b = (a^2 - b^2) + 2(a - b) = (a + b)(a - b) + 2(a - b)$
$= (a - b)(a + b + 2)$

75. $x^2 + 25 \Rightarrow$ prime (sum of two squares)

77. $4y^4 - 1 = (2y^2) - 1^2 = (2y^2 + 1)(2y^2 - 1)$

79. $m^3 + n^3 = (m + n)(m^2 - mn + n^2)$

81. $81x^3 + y^3 \Rightarrow$ prime (81 is not a perfect cube)

83. $x^3 + (y - z)^3 = [x + (y - z)][x^2 - x(y - z) + (y - z)^2]$
$= (x + y - z)(x^2 - xy + xz + y^2 - 2yz + z^2)$

85. $4a^2b^4c^6 - 9d^8 = (2ab^2c^3)^2 - (3d^4)^2 = (2ab^2c^3 + 3d^4)(2ab^2c^3 - 3d^4)$

87. $8x^6 + 125y^3 = (2x^2)^3 + (5y)^3 = (2x^2 + 5y)[(2x^2)^2 - (2x^2)(5y) + (5y)^2]$
$= (2x^2 + 5y)(4x^4 - 10x^2y + 25y^2)$

89. $1000a^6 - 343b^3c^6 = (10a^2)^3 - (7bc^2)^3 = (10a^2 - 7bc^2)[(10a^2)^2 + (10a^2)(7bc^2) + (7bc^2)^2]$
$= (10a^2 - 7bc^2)(100a^4 + 70a^2bc^2 + 49b^2c^4)$

91. $2x + y + 4x^2 - y^2 = (2x + y) + (4x^2 - y^2) = (2x + y)(1) + (2x + y)(2x - y)$
$= (2x + y)(1 + 2x - y)$

93. $x^{3n} - 8 = (x^n)^3 - 2^3 = (x^n - 2)[(x^n)^2 + 2x^n + 2^2] = (x^n - 2)(x^{2n} + 2x^n + 4)$

95. $a^{3m} + b^{3n} = (a^m)^3 + (b^n)^3 = (a^m + b^n)[(a^m)^2 - a^mb^n + (b^n)^2] = (a^m + b^n)(a^{2m} - a^mn^n + b^{2n})$

97. $2x^{6m} + 16y^{3m} = 2(x^{6m} + 8y^{3m}) = 2[(x^{2m})^3 + (2y^m)^3]$
$= 2(x^{2m} + 2y^m)[(x^{2m})^2 - 2x^{2m}y^m + (2y^m)^2]$
$= 2(x^{2m} + 2y^m)(x^{4m} - 2x^{2m}y^m + 4y^{2m})$

99. $0.5gt_1^2 - 0.5gt_2^2 = 0.5g(t_1^2 - t_2^2) = 0.5g(t_1 + t_2)(t_1 - t_2)$

101. $V = \frac{4}{3}\pi r_1^3 - \frac{4}{3}\pi r_2^3 = \frac{4}{3}\pi(r_1^3 - r_2^3) = \frac{4}{3}\pi(r_1 - r_2)(r_1^2 + r_1r_2 + r_2^2)$

103. Answers may vary.

105. $x^{32} - y^{32} = (x^{16} + y^{16})(x^{16} - y^{16}) = (x^{16} + y^{16})(x^8 + y^8)(x^8 - y^8)$
$= (x^{16} + y^{16})(x^8 + y^8)(x^4 + y^4)(x^4 - y^4)$
$= (x^{16} + y^{16})(x^8 + y^8)(x^4 + y^4)(x^2 + y^2)(x^2 - y^2)$
$= (x^{16} + y^{16})(x^8 + y^8)(x^4 + y^4)(x^2 + y^2)(x + y)(x - y)$

© 2015 Cengage Learning. All Rights Reserved. May not be scanned, copied or duplicated, or posted to a publicly accessible website, in whole or in part.

Exercises 5.6 (page 341)

1. 20, 3

3. 2, −10

5. 15, −2

7. $\frac{2+x}{11} = 3$
$2 + x = 33$
$x = 31$

9. $\frac{2}{3}(5t-3) = 38$
$2(5t-3) = 114$
$10t - 6 = 114$
$10t = 120$
$t = 12$

11. $11r + 6(3-r) = 3$
$11r + 18 - 6r = 3$
$5r = -15$
$r = -3$

13. $2xy + y^2$; $2xy + y^2$

15. key number

17. $x + 2$

19. $x - 3$

21. $2a + 1$

23. $2m + 3n$

25. $x^2 + 9x + 8 = (x+1)(x+8)$

27. $x^2 - 7x + 10 = (x-5)(x-2)$

29. $b^2 + 8b + 18 \Rightarrow$ prime

31. $x^2 - x - 30 = (x-6)(x+5)$

33. $x^2 + 2x + 1 = (x+1)(x+1) = (x+1)^2$

35. $a^2 - 18a + 81 = (a-9)(a-9) = (a-9)^2$

37. $4y^2 + 4y + 1 = (2y+1)(2y+1)$
$= (2y+1)^2$

39. $9b^2 - 12b + 4 = (3b-2)(3b-2)$
$= (3b-2)^2$

41. $-4a + a^2 - 32 = a^2 - 4a - 32$
$= (a-8)(a+4)$

43. $-5x + x^2 + 6 = x^2 - 5x + 6$
$= (x-2)(x-3)$

45. $a^2b^2 - 13ab^2 + 22b^2 = b^2(a^2 - 13a + 22)$
$= b^2(a-11)(a-2)$

47. $b^2x^2 - 12bx^2 + 35x^2 = x^2(b^2 - 12b + 35)$
$= x^2(b-5)(b-7)$

49. $6y^2 + 7y + 2$; $a = 6$, $b = 7$, $c = 2$
$b^2 - 4ac = 7^2 - 4(6)(2) = 1$
$1 = 1^2 \Rightarrow$ factorable
$6y^2 + 7y + 2 = (2y+1)(3y+2)$

51. $6x^2 - 5x - 4$; $a = 6$, $b = -5$, $c = -4$
$b^2 - 4ac = (-5)^2 - 4(6)(-4) = 121$
$121 = 11^2 \Rightarrow$ factorable
$6x^2 - 5x - 4 = (3x-4)(2x+1)$

53. $8a^2 + 6a - 9$; $a = 8$, $b = 6$, $c = -9$
$b^2 - 4ac = 6^2 - 4(8)(-9) = 324$
$324 = 18^2 \Rightarrow$ factorable
$8a^2 + 6a - 9 = (4a-3)(2a+3)$

55. $6z^2 + 17z + 12$; $a = 6$, $b = 17$, $c = 12$
$b^2 - 4ac = 17^2 - 4(6)(12) = 1$
$1 = 1^2 \Rightarrow$ factorable
$6z^2 + 17z + 12 = (3z+4)(2z+3)$

57. $5x^2 + 4x + 1$; $a = 5$, $b = 4$, $c = 1$
$b^2 - 4ac = 4^2 - 4(5)(1) = -4$
$-4 < 0 \Rightarrow$ not factorable
$5x^2 + 4x + 1 \Rightarrow$ prime

59. $9x^2 + 12x + 16$; $a = 9$, $b = 12$, $c = 16$
$b^2 - 4ac = 12^2 - 4(9)(16) = -432$
$-432 < 0 \Rightarrow$ not factorable
$9x^2 + 12x + 16 \Rightarrow$ prime

© 2015 Cengage Learning. All Rights Reserved. May not be scanned, copied or duplicated, or posted to a publicly accessible website, in whole or in part.

61. $-12z - 14yz - 4y^2z = -2z\left(6 + 7y + 2y^2\right) = -2z\left(2y^2 + 7y + 6\right)$
$a = 2, b = 7, c = 6; b^2 - 4ac = 7^2 - 4(2)(6) = 1 = 1^2 \Rightarrow$ factorable
$-2z\left(2y^2 + 7y + 6\right) = -2z(2y + 3)(y + 2)$

63. $-5z + 27y^2z - 10y^4z = -z\left(5 - 27y^2 + 10y^4\right) = -z\left(10y^4 - 27y^2 + 5\right)$
$a = 10, b = -27, c = 5; b^2 - 4ac = (-14)^2 - 4(8)(3) = 529 = 23^2 \Rightarrow$ factorable
$-z\left(10y^4 - 27y^2 + 5\right) = -z(5y^2 - 1)(2y^2 - 5)$

65. $x^3 - 60xy^2 + 7x^2y = x(x^2 - 60y^2 + 7xy) = x(x^2 + 7xy - 60y^2)$
$a = 1, b = 7, c = -60; b^2 - 4ac = 7^2 - 4(1)(-60) = 289 = 17^2 \Rightarrow$ factorable
$x(x^2 + 7xy - 60y^2) = x(x + 12y)(x - 5y)$

67. $14x + 9x^3 + x^5 = x\left(14 + 9x^2 + x^4\right) = x\left(x^4 + 9x^2 + 14\right)$
$a = 1, b = 9, c = 14; b^2 - 4ac = 9^2 - 4(1)(14) = 25 = 5^2 \Rightarrow$ factorable
$x\left(x^4 + 9x^2 + 14\right) = x\left(x^2 + 7\right)\left(x^2 + 2\right)$

69. $x^{2n} + 2x^n + 1 = (x^n + 1)(x^n + 1)$
$= (x^n + 1)^2$

71. $2a^{6n} - 3a^{3n} - 2 = (2a^{3n} + 1)(a^{3n} - 2)$

73. Let $A = x + 1$.
$(x + 1)^2 + 2(x + 1) + 1 = A^2 + 2A + 1 = (A + 1)(A + 1) = (A + 1)^2 = (x + 1 + 1)^2$
$= (x + 2)^2$

75. Let $A = a + b$.
$(a + b)^2 - 2(a + b) - 24 = A^2 - 2A - 24 = (A - 6)(A + 4) = (a + b - 6)(a + b + 4)$

77. $a^2 - 17a + 16$: $a = 1, b = -17, c = 16 \Rightarrow$ key # $= ac = 1(16) = 16$.
Find two factors of 16 which add to equal -17: -1 and -16.
Rewrite and factor: $a^2 - 17a + 16 = a^2 - a - 16a + 16$
$= a(a - 1) - 16(a - 1) = (a - 1)(a - 16)$

79. $2u^2 + 5u + 3$: $a = 2, b = 5, c = 3 \Rightarrow$ key # $= ac = 2(3) = 6$.
Find two factors of 6 which add to equal 5: $+2$ and $+3$.
Rewrite and factor: $2u^2 + 5u + 3 = 2u^2 + 2u + 3u + 3$
$= 2u(u + 1) + 3(u + 1) = (u + 1)(2u + 3)$

81. $20r^2 - 7rs - 6s^2$: $a = 20, b = -7, c = -6 \Rightarrow$ key # $= ac = 20(-6) = -120$.
Find two factors of -120 which add to equal -7: -15 and $+8$.
Rewrite and factor: $20r^2 - 7rs - 6s^2 = 20r^2 - 15rs + 8rs - 6s^2$
$= 5r(4r - 3s) + 2s(4r - 3s) = (4r - 3s)(5r + 2s)$

© 2015 Cengage Learning. All Rights Reserved. May not be scanned, copied or duplicated, or posted to a publicly accessible website, in whole or in part.

SECTION 5.6

83. $20u^2 + 19uv + 3v^2$: $a = 20, b = 19, c = 3 \Rightarrow$ key # $= ac = 20(3) = 60$.
Find two factors of 60 which add to equal 19: +15 and +4.
Rewrite and factor: $20u^2 + 19uv + 3v^2 = 20u^2 + 15uv + 4uv + 3v^2$
$= 5u(4u + 3v) + v(4u + 3v) = (4u + 3v)(5u + v)$

85. $x^2 + 4x + 4 - y^2 = (x^2 + 4x + 4) - y^2 = (x + 2)(x + 2) - y^2 = (x + 2)^2 - y^2$
$= [(x + 2) + y][(x + 2) - y]$
$= (x + 2 + y)(x + 2 - y)$

87. $x^2 + 2x + 1 - 9z^2 = (x^2 + 2x + 1) - 9z^2 = (x + 1)(x + 1) - 9z^2$
$= (x + 1)^2 - (3z)^2$
$= [(x + 1) + 3z][(x + 1) - 3z]$
$= (x + 1 + 3z)(x + 1 - 3z)$

89. $c^2 - 4a^2 + 4ab - b^2 = c^2 - (4a^2 - 4ab + b^2) = c^2 - (2a - b)(2a - b)$
$= c^2 - (2a - b)^2$
$= [c + (2a - b)][c - (2a - b)]$
$= (c + 2a - b)(c - 2a + b)$

91. $a^2 - b^2 + 8a + 16 = (a^2 + 8a + 16) - b^2 = (a + 4)(a + 4) - b^2 = (a + 4)^2 - b^2$
$= [(a + 4) + b][(a + 4) - b]$
$= (a + 4 + b)(a + 4 - b)$

93. $a^2 - 3ab - 4b^2 = (a - 4b)(a + b)$

95. $9z^2 + 24z + 16 = (3z + 4)(3z + 4)$
$= (3z + 4)^2$

97. $a^2 + 5a - 50 = (a + 10)(a - 5)$

99. $y^2 - 4y - 21 = (y - 7)(y + 3)$

101. $x^2 + 4x - 28 \Rightarrow$ prime

103. $-3x^2 + 15x - 18 = -3(x^2 - 5x + 6)$
$= -3(x - 2)(x - 3)$

105. $3x^2 + 12x - 63 = 3(x^2 + 4x - 21)$
$= 3(x + 7)(x - 3)$

107. $-4x^2 + 4x + 80 = -4(x^2 - x - 20)$
$= -4(x - 5)(x + 4)$

109. $-4x^2 - 9 + 12x = -4x^2 + 12x - 9 = -(4x^2 - 12x + 9) = -(2x - 3)(2x - 3) = -(2x - 3)^2$

111. $-3a^2 + ab + 2b^2 = -(3a^2 - ab - 2b^2) = -(3a + 2b)(a - b)$

113. $2y^2 + yt - 6t^2 = (2y - 3t)(y + 2t)$

115. $3t^3 - 3t^2 + t = t(3t^2 - 3t + 1)$

117. $-90x^2 + 2 - 8x = -90x^2 - 8x + 2 = -2(45x^2 + 4x - 1) = -2(9x - 1)(5x + 1)$

119. $x^{4n} + 2x^{2n}y^{2n} + y^{4n} = (x^{2n} + y^{2n})(x^{2n} + y^{2n}) = (x^{2n} + y^{2n})^2$

121. $6x^{2n} + 7x^n - 3 = (3x^n - 1)(2x^n + 3)$

© 2015 Cengage Learning. All Rights Reserved. May not be scanned, copied or duplicated, or posted to a publicly accessible website, in whole or in part.

SECTION 5.6

123. $21x^4 - 10x^3 - 16x^2 = x^2(21x^2 - 10x - 16) = x^2(7x - 8)(3x + 2)$

125. $x^4 + 8x^2 + 15 = (x^2 + 5)(x^2 + 3)$

127. $y^4 - 13y^2 + 30 = (y^2 - 10)(y^2 - 3)$

129. $a^4 - 13a^2 + 36 = (a^2 - 4)(a^2 - 9) = (a + 2)(a - 2)(a + 3)(a - 3)$

131. $z^4 - z^2 - 12 = (z^2 - 4)(z^2 + 3) = (z + 2)(z - 2)(z^2 + 3)$

133. $4x^3 + x^6 + 3 = x^6 + 4x^3 + 3 = (x^3 + 1)(x^3 + 3) = (x + 1)(x^2 - x + 1)(x^3 + 3)$

135. $6(x + y)^2 - 7(x + y) - 20 = [(3(x + y) + 4][2(x + y) - 5] = (3x + 3y + 4)(2x + 2y - 5)$

137.
$$\begin{aligned} 4x^2 - z^2 + 4xy + y^2 = (4x^2 + 4xy + y^2) - z^2 &= (2x + y)(2x + y) - z^2 \\ &= (2x + y)^2 - z^2 \\ &= [(2x + y) + z][(2x + y) - z] \\ &= (2x + y + z)(2x + y - z) \end{aligned}$$

139. $x^2 + 6x + 9 = (x + 3)(x + 3)$
One side is $x + 3$.

141. $4x^2 + 20x - 11 = (2x + 11)(2x - 1)$
length $= 2x + 11$ in., width $= 2x - 1$ in.
$$\begin{aligned} \text{difference} &= (2x + 11) - (2x - 1) \\ &= 2x + 11 - 2x + 1 = 12 \end{aligned}$$
The difference is 12 inches.

143. Answers may vary.

145. $x^2 - q^2 = x^2 + 0x - q^2$: $a = 1, b = 0, c = -q^2 \Rightarrow b^2 - 4ac = 0^2 - 4(1)(-q^2) = 4q^2 = (2q)^2$.
Since $4q^2$ is a perfect square, the binomial is factorable, and the test works.

Exercises 5.7 (page 346)

1. 2; GCF = 3

3. 3; GCF = 2

5. 2; no GCF

7. $(7a^2 + 6a - 3) + (5a^2 - 2a - 4) = 7a^2 + 6a - 3 + 5a^2 - 2a - 4 = 12a^2 + 4a - 7$

9. $5(2y^2 - 3y + 3) - 2(3y^2 - 2y + 6) = 10y^2 - 15y + 15 - 6y^2 + 4y - 12 = 4y^2 - 11y + 3$

11. $(n + 5)(n - 9) = n^2 - 9n + 5n - 45 = n^2 - 4n - 45$

13. common factors

15. trinomial

17. $x^2 - 10x + 25 = (x - 5)(x - 5) = (x - 5)^2$

19.
$$\begin{aligned} 64x^3 - 125y^3 = (4x)^3 - (5y)^3 &= (4x - 5y)[(4x)^2 + (4x)(5y) + (5y)^2] \\ &= (4x - 5y)(16x^2 + 20xy + 25y^2) \end{aligned}$$

© 2015 Cengage Learning. All Rights Reserved. May not be scanned, copied or duplicated, or posted to a publicly accessible website, in whole or in part.

SECTION 5.7

21. $xy - ty + xs - ts = y(x-t) + s(x-t)$
$= (x-t)(y+s)$

23. $36a^2 - 49b^2 = (6a)^2 - (7b)^2$
$= (6a+7b)(6a-7b)$

25. $12x^2 + 52x + 35 = (6x+5)(2x+7)$

27. $6x^2 - 14x + 8 = 2(3x^2 - 7x + 4)$
$= 2(3x-4)(x-1)$

29. $16y^2 + 81 \Rightarrow$ prime

31. $7x^2 - 57x + 8 = (7x-1)(x-8)$

33. $100z^2 - 81t^2 = (10z)^2 - (9t)^2 = (10z+9t)(10z-9t)$

35. $4x^2y^2z^2 - 26x^2y^2z^3 = 2x^2y^2z^2(2-13z)$

37. $12x^2 - 9x + 8 \Rightarrow$ prime

39. $ae + bf + af + be = ae + af + be + bf = a(e+f) + b(e+f) = (e+f)(a+b)$

41. $5(x+y)^2 + (x+y) - 4 = [5(x+y) - 4][(x+y)+1] = (5x+5y-4)(x+y+1)$

43. $625x^4 - 256y^4 = (25x^2)^2 - (16y^2)^2 = (25x^2+16y^2)(25x^2-16y^2)$
$= (25x^2+16y^2)(5x+4y)(5x-4y)$

45. $8x^4 - 6x^3 + 12x^2 + 24x = 2x(4x^3 - 3x^2 + 6x + 12)$

47. $6x^2 - 63 - 13x = 6x^2 - 13x - 63 = (3x+7)(2x-9)$

49. $x^4 - x^4y^4 = x^4(1-y^4) = x^4(1+y^2)(1-y^2) = x^4(1+y^2)(1+y)(1-y)$

51. $x^4 - 17x^2 + 16 = (x^2-16)(x^2-1) = (x+4)(x-4)(x+1)(x-1)$

53. $x^2 + 10x + 25 - y^8 = (x+5)^2 - (y^4)^2 = (x+5+y^4)(x+5-y^4)$

55. $9x^2 - 6x + 1 - 25y^2 = (3x-1)^2 - (5y)^2 = (3x-1+5y)(3x-1-5y)$

57. $64x^3 + 81 \Rightarrow$ prime

59. $x^5 + x^2 - x^3 - 1 = x^2(x^3+1) - 1(x^3+1) = (x^3+1)(x^2-1)$
$= (x+1)(x^2-x+1)(x+1)(x-1)$
$= (x+1)^2\big(x^2-x+1\big)(x-1)$

61. $x^5 - 9x^3 + 8x^2 - 72 = x^3(x^2-9) + 8(x^2-9) = (x^2-9)(x^3+8)$
$= (x+3)(x-3)(x+2)(x^2-2x+4)$

63. $2x^5z - 2x^2y^3z - 2x^3y^2z + 2y^5z = 2z(x^5 - x^2y^3 - x^3y^2 + y^5)$
$= 2z[x^2(x^3-y^3) - y^2(x^3-y^3)]$
$= 2z(x^3-y^3)(x^2-y^2)$
$= 2z(x-y)(x^2+xy+y^2)(x+y)(x-y)$
$= 2z(x+y)(x-y)^2\big(x^2+xy+y^2\big)$

© 2015 Cengage Learning. All Rights Reserved. May not be scanned, copied or duplicated, or posted to a publicly accessible website, in whole or in part.

SECTION 5.7

65. $x^{2m} - x^m - 6 = (x^m - 3)(x^m + 2)$

67. $$\begin{aligned} a^{3n} - b^{3n} &= (a^n)^3 - (b^n)^3 \\ &= (a^n - b^n)(a^{2n} + a^n b^n + b^{2n}) \end{aligned}$$

69. $$\begin{aligned} x^3 - 3x^2 + 2x - x^2y + 3xy - 2y &= x(x^2 - 3x + 2) - y(x^2 - 3x + 2) \\ &= (x^2 - 3x + 2)(x - y) = (x - 2)(x - 1)(x - y) \end{aligned}$$

71. **Answers may vary.**

73. $6x^{-2} - 5x^{-1} - 6 = (3x^{-1} + 2)(2x^{-1} - 3) = \left(\frac{3}{x} + 2\right)\left(\frac{2}{x} - 3\right)$

75. $$\begin{aligned} x^4 + x^2 + 1 = x^4 + x^2 + x^2 + 1 - x^2 = x^4 + 2x^2 + 1 - x^2 &= (x^2 + 1)^2 - x^2 \\ &= (x^2 + 1 + x)(x^2 + 1 - x) \end{aligned}$$

Exercises 5.8 (page 353)

1. $$\begin{aligned} (x - 2)(x - 3) &= 0 \\ x - 2 = 0 \quad &\textbf{or} \quad x - 3 = 0 \\ x = 2 \qquad & \qquad\quad x = 3 \end{aligned}$$

3. $$\begin{aligned} (x - 2)(x - 3)(x + 1) &= 0 \\ x - 2 = 0 \quad \textbf{or} \quad x - 3 = 0 \quad &\textbf{or} \quad x + 1 = 0 \\ x = 2 \qquad\qquad x = 3 \qquad & \qquad\quad x = -1 \end{aligned}$$

5. 2, 3, 5, 7

7. $V = \frac{4}{3}\pi r^3 = \frac{4}{3}\pi(21.23)^3 = 40{,}081.00 \text{ cm}^3$

9. $ax^2 + bx + c = 0$

11. $$\begin{aligned} z^2 + 9z + 20 &= 0 \\ (z + 5)(z + 4) &= 0 \\ z + 5 = 0 \quad &\textbf{or} \quad z + 4 = 0 \\ z = -5 \qquad & \qquad\quad z = -4 \end{aligned}$$

13. $$\begin{aligned} y^2 - 7y + 12 &= 0 \\ (y - 4)(y - 3) &= 0 \\ y - 4 = 0 \quad &\textbf{or} \quad y - 3 = 0 \\ y = 4 \qquad & \qquad\quad y = 3 \end{aligned}$$

15. $$\begin{aligned} 6m^2 + 24m &= 0 \\ 6m(m + 4) &= 0 \\ 6m = 0 \quad &\textbf{or} \quad m + 4 = 0 \\ m = 0 \qquad & \qquad\quad m = -4 \end{aligned}$$

17. $$\begin{aligned} x^2 + x &= 0 \\ x(x + 1) &= 0 \\ x = 0 \quad &\textbf{or} \quad x + 1 = 0 \\ & \qquad\quad x = -1 \end{aligned}$$

19. $$\begin{aligned} x^2 - 9 &= 0 \\ (x + 3)(x - 3) &= 0 \\ x + 3 = 0 \quad &\textbf{or} \quad x - 3 = 0 \\ x = -3 \qquad & \qquad\quad x = 3 \end{aligned}$$

21. $$\begin{aligned} 4a^2 - 25 &- 0 \\ (2a + 5)(2a - 5) &= 0 \\ 2a + 5 = 0 \quad &\textbf{or} \quad 2a - 5 = 0 \\ 2a = -5 \qquad & \qquad\quad 2a = 5 \\ a = -\tfrac{5}{2} \qquad & \qquad\quad a = \tfrac{5}{2} \end{aligned}$$

© 2015 Cengage Learning. All Rights Reserved. May not be scanned, copied or duplicated, or posted to a publicly accessible website, in whole or in part.

SECTION 5.8

23.
$$\begin{aligned}\frac{3a^2}{2} &= \frac{1}{2} - a\\ 3a^2 &= 1 - 2a\\ 3a^2 + 2a - 1 &= 0\\ (3a-1)(a+1) &= 0\end{aligned}$$
$3a - 1 = 0$ **or** $a + 1 = 0$
$3a = 1$ $\quad a = -1$
$a = \frac{1}{3}$

25.
$$\begin{aligned}\frac{1}{2}x^2 - \frac{5}{4}x &= -\frac{1}{2}\\ 2x^2 - 5x &= -2\\ 2x^2 - 5x + 2 &= 0\\ (2x-1)(x-2) &= 0\end{aligned}$$
$2x - 1 = 0$ **or** $x - 2 = 0$
$2x = 1$ $\quad x = 2$
$x = \frac{1}{2}$

27.
$$\begin{aligned}x^3 - 4x^2 - 21x &= 0\\ x(x^2 - 4x - 21) &= 0\\ x(x+3)(x-7) &= 0\end{aligned}$$
$x = 0$ **or** $x + 3 = 0$ **or** $x - 7 = 0$
$x = -3$ $\quad x = 7$

29.
$$\begin{aligned}x^3 + x^2 &= 0\\ x^2(x+1) &= 0\end{aligned}$$
$x^2 = 0$ **or** $x + 1 = 0$
$x = 0$ $\quad x = -1$

31.
$$\begin{aligned}x^4 - 26x^2 + 25 &= 0\\ (x^2 - 25)(x^2 - 1) &= 0\\ (x+5)(x-5)(x+1)(x-1) &= 0\end{aligned}$$
$x + 5 = 0$ **or** $x - 5 = 0$ **or** $x + 1 = 0$ **or** $x - 1 = 0$
$x = -5$ $\quad x = 5$ $\quad x = -1$ $\quad x = 1$

33.
$$\begin{aligned}x^4 - 20x^2 + 64 &= 0\\ (x^2 - 16)(x^2 - 4) &= 0\\ (x+4)(x-4)(x+2)(x-2) &= 0\end{aligned}$$
$x + 4 = 0$ **or** $x - 4 = 0$ **or** $x + 2 = 0$ **or** $x - 2 = 0$
$x = -4$ $\quad x = 4$ $\quad x = -2$ $\quad x = 2$

35.
$$\begin{aligned}f(x) &= 0\\ x^2 - 49 &= 0\\ (x+7)(x-7) &= 0\end{aligned}$$
$x + 7 = 0$ **or** $x - 7 = 0$
$x = -7$ $\quad x = 7$

37.
$$\begin{aligned}f(x) &= 0\\ 2x^2 + 5x - 3 &= 0\\ (2x-1)(x+3) &= 0\end{aligned}$$
$2x - 1 = 0$ **or** $x + 3 = 0$
$2x = 1$ $\quad x = -3$
$x = \frac{1}{2}$

39.
$$\begin{aligned}y^2 - 7y + 6 &= 0\\ (y-6)(y-1) &= 0\end{aligned}$$
$y - 6 = 0$ **or** $y - 1 = 0$
$y = 6$ $\quad y = 1$

41.
$$\begin{aligned}3m^2 + 10m + 3 &= 0\\ (3m+1)(m+3) &= 0\end{aligned}$$
$3m + 1 = 0$ **or** $m + 3 = 0$
$3m = -1$ $\quad m = -3$
$m = -\frac{1}{3}$

© 2015 Cengage Learning. All Rights Reserved. May not be scanned, copied or duplicated, or posted to a publicly accessible website, in whole or in part.

SECTION 5.8

43.
$$\begin{aligned} b(6b-7) &= 10 \\ 6b^2 - 7b - 10 &= 0 \\ (6b+5)(b-2) &= 0 \end{aligned}$$
$6b+5=0$ **or** $b-2=0$
$6b=-5$ $\quad b=2$
$b=-\frac{5}{6}$

45.
$$\begin{aligned} 3a(a^2+5a) &= -18a \\ 3a^3+15a^2+18a &= 0 \\ 3a(a^2+5a+6) &= 0 \\ 3a(a+2)(a+3) &= 0 \end{aligned}$$
$3a=0$ **or** $a+2=0$ **or** $a+3=0$
$a=0$ $\quad a=-2$ $\quad a=-3$

47.
$$\begin{aligned} x^3+3x^2-x-3 &= 0 \\ x^2(x+3)-1(x+3) &= 0 \\ (x+3)(x^2-1) &= 0 \\ (x+3)(x+1)(x-1) &= 0 \end{aligned}$$
$x+3=0$ **or** $x+1=0$ **or** $x-1=0$
$x=-3$ $\quad x=-1$ $\quad x=1$

49.
$$\begin{aligned} 2r^3+3r^2-18r-27 &= 0 \\ r^2(2r+3)-9(2r+3) &= 0 \\ (2r+3)(r^2-9) &= 0 \\ (2r+3)(r+3)(r-3) &= 0 \end{aligned}$$
$2r+3=0$ **or** $r+3=0$ **or** $r-3=0$
$2r=-3$ $\quad r=-3$ $\quad r=3$
$r=-\frac{3}{2}$

51.
$$\begin{aligned} 5y^3-45y &= 0 \\ 5y(y^2-9) &= 0 \\ 5y(y+3)(y-3) &= 0 \end{aligned}$$
$5y=0$ **or** $y+3=0$ **or** $y-3=0$
$y=0$ $\quad y=-3$ $\quad y=3$

53.
$$\begin{aligned} 2y^2-5y+2 &= 0 \\ (2y-1)(y-2) &= 0 \end{aligned}$$
$2y-1=0$ **or** $y-2=0$
$2y=1$ $\quad y=2$
$y=\frac{1}{2}$

55.
$$\begin{aligned} 2x^2-x-1 &= 0 \\ (2x+1)(x-1) &= 0 \end{aligned}$$
$2x+1=0$ **or** $x-1=0$
$2x=-1$ $\quad x=1$
$x=-\frac{1}{2}$

57.
$$\begin{aligned} 3s^2-5s-2 &= 0 \\ (3s+1)(s-2) &= 0 \end{aligned}$$
$3s+1=0$ **or** $s-2=0$
$3s=-1$ $\quad s=2$
$s=-\frac{1}{3}$

59.
$$\begin{aligned} 10x^2+9x &= 7 \\ 10x^2+9x-7 &= 0 \\ (5x+7)(2x-1) &= 0 \end{aligned}$$
$5x+7=0$ **or** $2x-1=0$
$5x=-7$ $\quad 2x=1$
$x=-\frac{7}{5}$ $\quad x=\frac{1}{2}$

61.
$$\begin{aligned} 5y^2-25y &= 0 \\ 5y(y-5) &= 0 \end{aligned}$$
$5y=0$ **or** $y-5=0$
$y=0$ $\quad y=5$

© 2015 Cengage Learning. All Rights Reserved. May not be scanned, copied or duplicated, or posted to a publicly accessible website, in whole or in part.

SECTION 5.8

63.
$$\begin{aligned} x(x-6)+9&=0\\ x^2-6x+9&=0\\ (x-3)(x-3)&=0 \end{aligned}$$
$$x-3=0 \quad \textbf{or} \quad x-3=0$$
$$x=3 \qquad\qquad x=3$$

65.
$$\begin{aligned} x^2+6x+8&=0\\ (x+4)(x+2)&=0 \end{aligned}$$
$$x+4=0 \quad \textbf{or} \quad x+2=0$$
$$x=-4 \qquad\qquad x=-2$$

67.
$$\begin{aligned} x\left(3x+\frac{22}{5}\right)&=1\\ 3x^2+\frac{22}{5}x&=1\\ 15x^2+22x&=5\\ 15x^2+22x-5&=0 \end{aligned}$$
$$(5x-1)(3x+5)=0$$
$$5x-1=0 \quad \textbf{or} \quad 3x+5=0$$
$$5x=1 \qquad\qquad 3x=-5$$
$$x=\tfrac{1}{5} \qquad\qquad x=-\tfrac{5}{3}$$

69.
$$\begin{aligned} \frac{x^2(6x+37)}{35}&=x\\ x^2(6x+37)&=35x\\ 6x^3+37x^2&=35x\\ 6x^3+37x^2-35x&=0\\ x(6x^2+37x-35)&=0\\ x(6x-5)(x+7)&=0 \end{aligned}$$
$$x(6x-5)(x+7)=0$$
$$x=0 \quad \textbf{or} \quad 6x-5=0 \quad \textbf{or} \quad x+7=0$$
$$6x=5 \qquad\qquad x=-7$$
$$x=\tfrac{5}{6}$$

71.
$$\begin{aligned} 3y^3+y^2&=4(3y+1)\\ 3y^3+y^2&=12y+4\\ 3y^3+y^2-12y-4&=0\\ y^2(3y+1)-4(3y+1)&=0\\ (3y+1)(y^2-4)&=0\\ (3y+1)(y+2)(y-2)&=0 \end{aligned}$$
$$(3y+1)(y+2)(y-2)=0$$
$$3y+1=0 \quad \textbf{or} \quad y+2=0 \quad \textbf{or} \quad y-2=0$$
$$3y=-1 \qquad\qquad y=-2 \qquad\qquad y=2$$
$$y=-\tfrac{1}{3}$$

73.
$$\begin{aligned} f(x)&=0\\ 5x^3+3x^2-2x&=0\\ x(5x^2+3x-2)&=0\\ x(5x-2)(x+1)&=0 \end{aligned}$$
$$x=0 \ \textbf{or}\ 5x-2=0 \ \textbf{or}\ x+1=0$$
$$5x=2 \qquad\qquad x=-1$$
$$x=\tfrac{2}{5}$$

© 2015 Cengage Learning. All Rights Reserved. May not be scanned, copied or duplicated, or posted to a publicly accessible website, in whole or in part.

SECTION 5.8

75. $x^2 - 4x + 7 = 0$

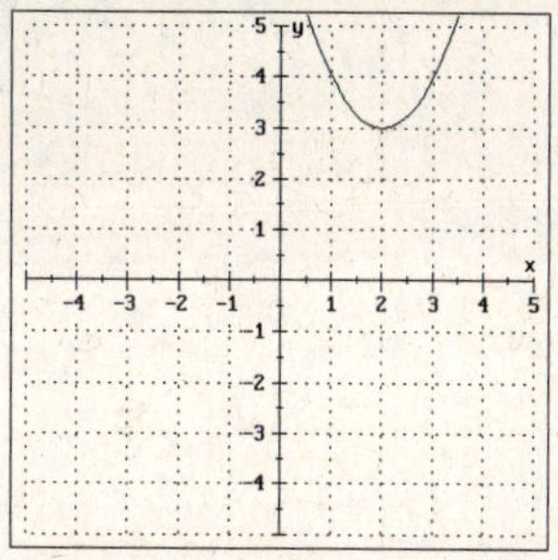

x-intercepts: none $\Rightarrow$ no solution

77. $-3x^3 - 2x^2 + 5 = 0$

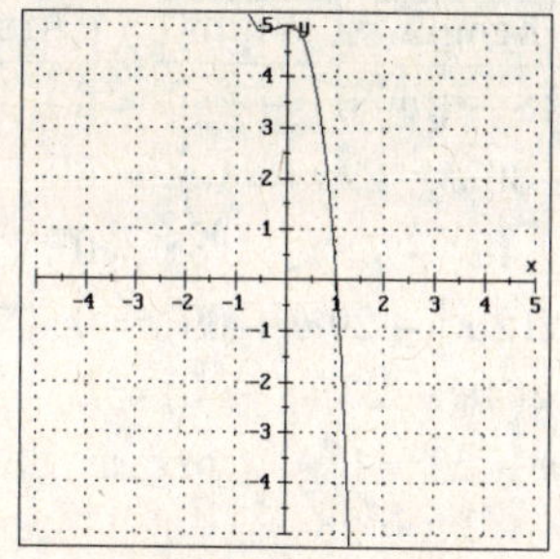

x-intercept: 1.00

79. Let $x =$ the first even integer and $x + 2 =$ the second.

$$\boxed{\text{First}} \cdot \boxed{\text{Second}} = 224$$
$$x(x+2) = 224$$
$$x^2 + 2x - 224 = 0$$
$$(x+16)(x-14) = 0$$
$$x + 16 = 0 \quad \textbf{or} \quad x - 14 = 0$$
$$x = -16 \qquad\qquad x = 14$$

The integers are -16 and -14, or 14 and 16.

81. Let $x =$ the first positive integer and $x + 1 =$ the second.

$$\boxed{\text{First}^2} + \boxed{\text{Second}^2} = 85$$
$$x^2 + (x+1)^2 = 85$$
$$x^2 + (x+1)(x+1) = 85$$
$$x^2 + x^2 + 2x + 1 = 85$$
$$2x^2 + 2x - 84 = 0$$
$$2(x^2 + x - 42) = 0$$
$$2(x+7)(x-6) = 0$$
$$x + 7 = 0 \quad \textbf{or} \quad x - 6 = 0$$
$$x = -7 \qquad\qquad x = 6$$

(not positive)

The integers are 6 and 7.

83. Let $x =$ one side and $3x =$ the other side.

$$\boxed{\text{Length}} \cdot \boxed{\text{Width}} = \boxed{\text{Area}}$$
$$3x \cdot x = 147$$
$$3x^2 - 147 = 0$$
$$3(x^2 - 49) = 0$$
$$3(x+7)(x-7) = 0$$
$$x + 7 = 0 \quad \textbf{or} \quad x - 7 = 0$$
$$x = -7 \qquad\qquad x = 7$$

(not positive)

The dimensions are 7 cm by 21cm.

85. $$\frac{1}{2} \cdot \boxed{\text{Base}} \cdot \boxed{\text{Height}} = \boxed{\text{Area}}$$
$$\frac{1}{2}(2h+3)h = 162$$
$$(2h+3)h = 324$$
$$2h^2 + 3h - 324 = 0$$
$$(2h+27)(h-12) = 0$$
$$2h + 27 = 0 \quad \textbf{or} \quad h - 12 = 0$$
$$2h = -27 \qquad\qquad h = 12$$

(h is not positive)

The height is 12 cm.

© 2015 Cengage Learning. All Rights Reserved. May not be scanned, copied or duplicated, or posted to a publicly accessible website, in whole or in part.

87. The length of the mural is $18 - 2w$, while the width is $11 - 2w$.

$$\boxed{\text{Length}} \cdot \boxed{\text{Width}} = \boxed{\text{Area}}$$
$$(18 - 2w)(11 - 2w) = 60$$
$$198 - 36w - 22w + 4w^2 = 60$$
$$4w^2 - 58w + 138 = 0$$
$$2(2w^2 - 29w + 69) = 0$$
$$2(2w - 23)(w - 3) = 0$$
$$2w - 23 = 0 \quad \textbf{or} \quad w - 3 = 0$$
$$2w = 23 \qquad w = 3$$
$$w = \frac{23}{2}$$

(makes width < 0)

The dimensions are 12 ft by 5 ft.

89. Let $w =$ the width of the room and let $2w =$ the length of the room.

$$\boxed{\text{Length}} \cdot \boxed{\text{Width}} = \boxed{\text{Area}}$$
$$(2w - 12)w = 560$$
$$2w^2 - 12w - 560 = 0$$
$$2(w^2 - 6w - 280) = 0$$
$$2(w + 14)(w - 20) = 0$$
$$w + 14 = 0 \quad \textbf{or} \quad w - 20 = 0$$
$$w = -14 \qquad w = 20$$

(impossible)

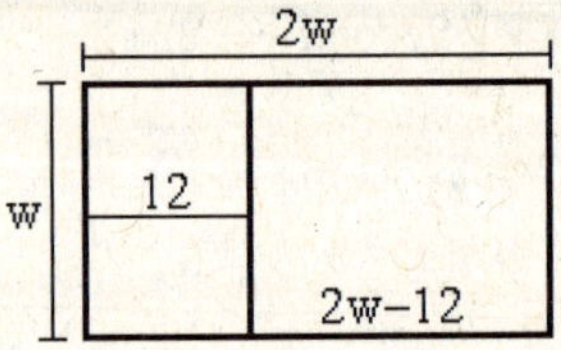

The dimensions are 20 feet by 40 feet.

91.

$$h = vt - 16t^2$$
$$0 = 160t - 16t^2$$
$$0 = 16t(10 - t)$$
$$16t = 0 \quad \textbf{or} \quad 10 - t = 0$$
$$t = 0 \qquad 10 = t$$

The object will hit the ground after 10 sec.

93.

$$h = vt - 16t^2$$
$$3344 = 480t - 16t^2$$
$$16t^2 - 480t + 3344 = 0$$
$$16(t^2 - 30t + 209) = 0$$
$$16(t - 11)(t - 19) = 0$$
$$t - 11 = 0 \quad \textbf{or} \quad t - 19 = 0$$
$$t = 11 \qquad t = 19$$

The cannonball will be at that height after 11 and 19 seconds.

95. Let $w =$ the width and $2w + 20 =$ the length.

$$\boxed{\text{Length}} \cdot \boxed{\text{Width}} = \boxed{\text{Area}}$$
$$(2w + 20)w = 6000$$
$$2w^2 + 20w - 6000 = 0$$
$$2(w^2 + 10w - 3000) = 0$$
$$2(w - 50)(w + 60) = 0$$
$$w - 50 = 0 \quad \textbf{or} \quad w + 60 = 0$$
$$w = 50 \qquad w = -60$$

(impossible)

The width is 50 meters.

© 2015 Cengage Learning. All Rights Reserved. May not be scanned, copied or duplicated, or posted to a publicly accessible website, in whole or in part.

97. Let $x =$ the width of the pool and $2x - 10 =$ the length of the pool.

$$\boxed{\text{Length}} \cdot \boxed{\text{Width}} = \boxed{\text{Area of pool}}$$

$$(2x - 10)x = 1500$$
$$2x^2 - 10x - 1500 = 0$$
$$2(x^2 - 5x - 750) = 0$$
$$2(x + 25)(x - 30) = 0$$
$$x + 25 = 0 \quad \textbf{or} \quad x - 30 = 0$$
$$x = -25 \qquad\qquad x = 30$$
(impossible)

The dimensions of the pool are 30 feet by 50 feet. Then let $w =$ the width of the walkway.

$$30w + 30w + 50w + 50w + 4w^2 = \boxed{\text{Area of walkway}}$$
$$4w^2 + 160w = 516$$
$$4w^2 + 160w - 516 = 0$$
$$4(w^2 + 40w - 129) = 0$$
$$4(w + 43)(w - 3) = 0$$
$$w + 43 = 0 \quad \textbf{or} \quad w - 3 = 0$$
$$w = -43 \qquad\qquad w = 3$$
(impossible)

The walkway should be 3 feet wide.

99. **Answers may vary.**

101. $(x - 3)(x - 5) = 0$
$x^2 - 8x + 15 = 0$

103. $(x - 0)(x + 5) = 0$
$x^2 + 5x = 0$

Chapter 5 Review (page 358)

1. degree $= 5$

2. degree $= 4 + 4 = 8$; $8x^4y^4 + 13x^3y^2 + 9x^2y$

3. $P(-1) = -(-1)^2 + 4(-1) + 6$
$= -(1) - 4 + 6 = 1$

4. $P(-2) = -(-2)^2 + 4(-2) + 6$
$= -4 - 8 + 6 = -6$

5. $P(-t) = -(-t)^2 + 4(-t) + 6$
$= -t^2 - 4t + 6$

6. $P(z) = -z^2 + 4z + 6$

7. $f(x) = x^3 - 1$
Shift $f(x) = x^3$ down 1.

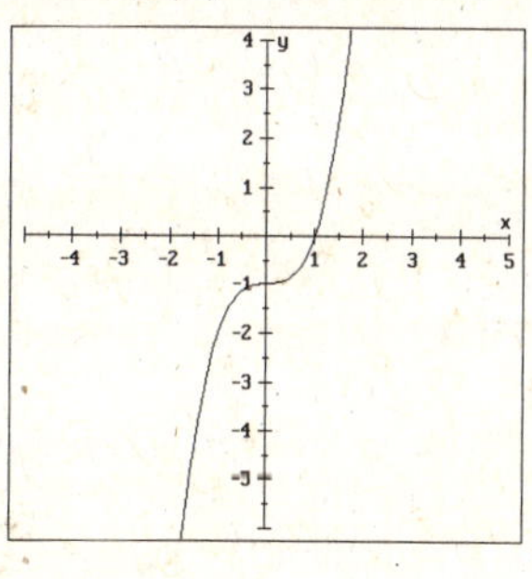

8. $f(x) = x^2 - 2x$

x	y
-1	3
0	0

x	y
1	-1
2	0

x	y
3	3

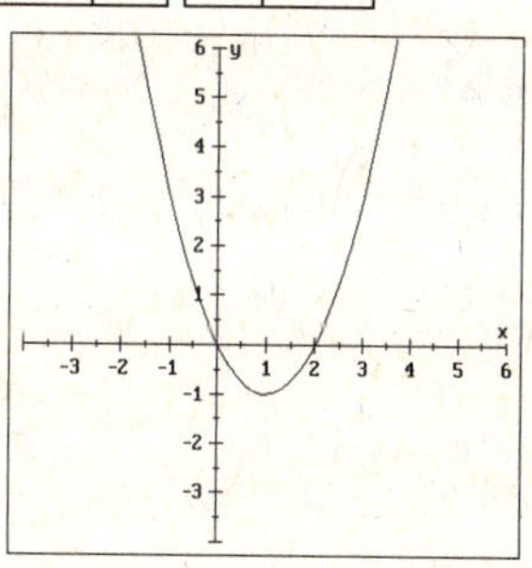

9. $(5x^2 + 7x - 2) + (4x^2 - 10x - 3) = 5x^2 + 7x - 2 + 4x^2 - 10x - 3 = 9x^2 - 3x - 5$

10. $(4x^3 + 4x^2 + 7) - (-2x^3 - x - 2) = 4x^3 + 4x^2 + 7 + 2x^3 + x + 2 = 6x^3 + 4x^2 + x + 9$

© 2015 Cengage Learning. All Rights Reserved. May not be scanned, copied or duplicated, or posted to a publicly accessible website, in whole or in part.

CHAPTER 5 REVIEW

11. $(2x^2 - 5x + 9) - (x^2 - 3) - (-3x^2 + 4x - 7)$
$= 2x^2 - 5x + 9 - x^2 + 3 + 3x^2 - 4x + 7 = 4x^2 - 9x + 19$

12. $2(7x^3 - 6x^2 + 4x - 3) - 3(7x^3 + 6x^2 + 4x - 3)$
$= 14x^3 - 12x^2 + 8x - 6 - 21x^3 - 18x^2 - 12x + 9 = -7x^3 - 30x^2 - 4x + 3$

13. $(8a^2b^2)(-2abc) = -16a^3b^3c$

14. $(-3xy^2z)(2xz^3) = -6x^2y^2z^4$

15. $2xy^2(x^3y - 4xy^5) = 2x^4y^3 - 8x^2y^7$

16. $a^2b(a^2 + 2ab + b^2) = a^4b + 2a^3b^2 + a^2b^3$

17. $(6x + 7)(3x - 2) = 18x^2 - 12x + 21x - 14 = 18x^2 + 9x - 14$

18. $(3x + 2)(2x - 4) = 6x^2 - 12x + 4x - 8 = 6x^2 - 8x - 8$

19. $(3x - 8y)^2 = (3x - 8y)(3x - 8y) = 9x^2 - 24xy - 24xy + 64y^2 = 9x^2 - 48xy + 64y^2$

20. $(3x + 2)(3x - 2) = 9x^2 - 6x + 6x - 4 = 9x^2 - 4$

21. $(3x^2 - 2)(x^2 - x + 2) = 3x^4 - 3x^3 + 6x^2 - 2x^2 + 2x - 4 = 3x^4 - 3x^3 + 4x^2 + 2x - 4$

22. $(2a - b)(a + b)(a - 2b) = \left(2a^2 + ab - b^2\right)(a - 2b) = \left(2a^2 + ab - b^2\right)a + \left(2a^2 + ab - b^2\right)(-2b)$
$= 2a^3 + a^2b - ab^2 - 4a^2b - 2ab^2 + 2b^3$
$= 2a^3 - 3a^2b - 3ab^2 + 2b^3$

23. $-6x - 18 = -6(x + 3)$

24. $8x^3 + 32x^2 = 8x^2(x + 4)$

25. $5x^2y^3 - 10xy^2 = 5xy^2(xy - 2)$

26. $7a^4b^2 + 49a^3b = 7a^3b(ab + 7)$

27. $-8x^2y^3z^4 - 12x^4y^3z^2 = -4x^2y^3z^2(2z^2 + 3x^2)$

28. $12a^6b^4c^2 + 15a^2b^4c^6 = 3a^2b^4c^2(4a^4 + 5c^4)$

29. $27x^3y^3z^3 + 81x^4y^5z^2 - 90x^2y^3z^7 = 9x^2y^3z^2(3xz + 9x^2y^2 - 10z^5)$

30. $-36a^5b^4c^2 + 60a^7b^5c^3 - 24a^2b^3c^7 = -12a^2b^3c^2(3a^3b - 5a^5b^2c + 2c^5)$

31. $x^{2n} + x^n = x^n(x^{2n-n} + 1) = x^n(x^n + 1)$

32. $y^{2n} - y^{4n} = y^{2n}(1 - y^{4n-2n}) = y^{2n}(1 - y^{2n})$

33. $x^{-4} - x^{-2} = x^{-2}(x^{-4-(-2)} - 1)$
$= x^{-2}(x^{-2} - 1)$

34. $a^6 + 1 = a^{-3}(a^{6-(-3)} + a^{0-(-3)})$
$= a^{-3}(a^9 + a^3)$

35. $5x^2(x + y)^3 - 15x^3(x + y)^4 = 5x^2(x + y)^3[1 - 3x(x + y)] = 5x^2(x + y)^3(1 - 3x^2 - 3xy)$

36. $-49a^3b^2(a - b)^4 + 63a^2b^4(a - b)^3 = -7a^2b^2(a - b)^3[7a(a - b) - 9b^2]$
$= -7a^2b^2(a - b)^3(7a^2 - 7ab - 9b^2)$

© 2015 Cengage Learning. All Rights Reserved. May not be scanned, copied or duplicated, or posted to a publicly accessible website, in whole or in part.

37. $mn + 5n + 6m + 30 = n(m + 5) + 6(m + 5) = (m + 5)(n + 6)$

38. $ac + bc - 3a - 3b = c(a + b) - 3(a + b) = (a + b)(c - 3)$

39. $x^4 + 4y + 4x^2 + x^2y = x^4 + x^2y + 4x^2 + 4y = x^2(x^2 + y) + 4(x^2 + y) = (x^2 + y)(x^2 + 4)$

40. $a^5 + b^2c + a^2c + a^3b^2 = a^5 + a^3b^2 + a^2c + b^2c = a^3(a^2 + b^2) + c(a^2 + b^2) = (a^2 + b^2)(a^3 + c)$

41.
$$\begin{aligned} S &= 2wh + 2wl + 2lh \\ S - 2wl &= 2wh + 2lh \\ S - 2wl &= h(2w + 2l) \\ \frac{S - 2wl}{2w + 2l} &= h, \text{ or } h = \frac{S - 2wl}{2w + 2l} \end{aligned}$$

42.
$$\begin{aligned} S &= 2wh + 2wl + 2lh \\ S - 2wh &= 2wl + 2lh \\ S - 2wh &= l(2w + 2h) \\ \frac{S - 2wh}{2w + 2h} &= l, \text{ or } l = \frac{S - 2wh}{2w + 2h} \end{aligned}$$

43. $z^2 - 36 = z^2 - 6^2 = (z + 6)(z - 6)$

44. $y^4 - 16 = (y^2)^2 - 4^2 = (y^2 + 4)(y^2 - 4) = (y^2 + 4)(y + 2)(y - 2)$

45.
$$\begin{aligned} x^2y^4 - 64z^6 &= (xy^2)^2 - (8z^3)^2 \\ &= (xy^2 + 8z^3)(xy^2 - 8z^3) \end{aligned}$$

46. $a^2b^2 + c^2 \Rightarrow$ prime

47.
$$\begin{aligned} (x + z)^2 - t^2 &= [(x + z) + t][(x + z) - t] \\ &= (x + z + t)(x + z - t) \end{aligned}$$

48.
$$\begin{aligned} c^2 - (a + b)^2 &= [c + (a + b)][c - (a + b)] \\ &= (c + a + b)(c - a - b) \end{aligned}$$

49.
$$\begin{aligned} 2x^4 - 98 &= 2(x^4 - 49) \\ &= 2(x^2 + 7)(x^2 - 7) \end{aligned}$$

50.
$$\begin{aligned} 3x^6 - 300x^2 &= 3x^2(x^4 - 100) \\ &= 3x^2(x^2 + 10)(x^2 - 10) \end{aligned}$$

51. $x^3 + 343 = x^3 + 7^3 = (x + 7)(x^2 - 7x + 49)$

52. $3p^3 + 81 = 3(p^3 + 27) = 3(p^3 + 3^3) = 3(p + 3)(p^2 - 3p + 3^2) = 3(p + 3)(p^2 - 3p + 9)$

53. $8y^3 - 512 = 8(y^3 - 64) = 8(y^3 - 4^3) = 8(y - 4)(y^2 + 4y + 16)$

54. $4x^3y + 108yz^3 = 4y(x^3 + 27z^3) = 4y\left[x^3 + (3z)^3\right] = 4y(x + 3z)(x^2 - 3xz + 9z^2)$

55. $x^2 + 18x + 81 = (x + 9)(x + 9)$

56. $a^2 - 14a + 49 = (a - 7)(a - 7)$

57. $y^2 + 21y + 20 = (y + 20)(y + 1)$

58. $z^2 - 11z + 30 = (z - 5)(z - 6)$

59.
$$\begin{aligned} -x^2 - 3x + 28 &= -(x^2 + 3x - 28) \\ &= -(x + 7)(x - 4) \end{aligned}$$

60. $y^2 - 5y - 36 = (y - 9)(y + 4)$

61. $4a^2 - 5a + 1 = (4a - 1)(a - 1)$

62. $3b^2 + 2b + 1 \Rightarrow$ prime

© 2015 Cengage Learning. All Rights Reserved. May not be scanned, copied or duplicated, or posted to a publicly accessible website, in whole or in part.

63. $7x^2 + x + 2 \Rightarrow$ prime

64. $-15x^2 + 14x + 8 = -(15x^2 - 14x - 8)$
$= -(5x + 2)(3x - 4)$

65. $y^3 + y^2 - 2y = y(y^2 + y - 2)$
$= y(y + 2)(y - 1)$

66. $2a^4 + 4a^3 - 6a^2 = 2a^2(a^2 + 2a - 3)$
$= 2a^2(a + 3)(a - 1)$

67. $-8x^2 + 22x - 12 = -2(4x^2 - 11x + 6)$
$= -2(4x - 3)(x - 2)$

68. $8x^2 - 4x - 24 = 4(2x^2 - x - 6)$
$= 4(2x + 3)(x - 2)$

69. $15x^2 - 57xy - 12y^2 = 3(5x^2 - 19xy - 4y^2)$
$= 3(5x + y)(x - 4y)$

70. $30x^2 + 65xy + 10y^2 = 5(6x^2 + 13xy + 2y^2)$
$= 5(6x + y)(x + 2y)$

71. $24x^2 - 23xy - 12y^2 = (8x + 3y)(3x - 4y)$

72. $14x^2 + 13xy - 12y^2 = (2x + 3y)(7x - 4y)$

73. $10x^3 + 5x^2 - 15x = 5x(2x^2 + x - 3)$
$= 5x(2x + 3)(x - 1)$

74. $3x^2y - 12xy - 63y = 3y(x^2 - 4x - 21)$
$= 3y(x - 7)(x + 3)$

75. $z^2 - 4 + zx - 2x = (z^2 - 4) + zx - 2x = (z + 2)(z - 2) + x(z - 2) = (z - 2)(z + 2 + x)$

76. $x^2 + 4x + 4 - y^2 = (x^2 + 4x + 4) - y^2 = (x + 2)^2 - y^2 = (x + 2 + y)(x + 2 - y)$

77. $x^2 + 4x + 4 - 4p^4 = (x^2 + 4x + 4) - 4p^4 = (x + 2)^2 - (2p^2)^2$
$= (x + 2 + 2p^2)(x + 2 - 2p^2)$

78. $8y^3 - 125 = (2y)^3 - 5^3 = (2y - 5)\left[(2y)^2 + (2y)(5) + 5^2\right] = (2y - 5)(4y^2 + 10y + 25)$

79. $x^{2m} + 2x^m - 3 = (x^m + 3)(x^m - 1)$

80. $x^4 - 5x^2 - 36 = (x^2 + 4)(x^2 - 9)$
$= (x^2 + 4)(x + 3)(x - 3)$

81.
$5x^2 - 7x = 0$
$x(5x - 7) = 0$
$x = 0$ **or** $5x - 7 = 0$
$5x = 7$
$x = \frac{7}{5}$

82.
$2x^2 - 18 = 0$
$2(x^2 - 9) = 0$
$2(x + 3)(x - 3) = 0$
$x + 3 = 0$ **or** $x - 3 = 0$
$x = -3$ $x = 3$

83.
$12x^2 + 4x - 5 = 0$
$(2x - 1)(6x + 5) = 0$
$2x - 1 = 0$ **or** $6x + 5 = 0$
$2x = 1$ $6x = -5$
$x = \frac{1}{2}$ $x = -\frac{5}{6}$

84.
$7y^2 - 37y + 10 = 0$
$(7y - 2)(y - 5) = 0$
$7y - 2 = 0$ **or** $y - 5 = 0$
$7y = 2$ $y = 5$
$y = \frac{2}{7}$

© 2015 Cengage Learning. All Rights Reserved. May not be scanned, copied or duplicated, or posted to a publicly accessible website, in whole or in part.

85.
$$t^2(15t-2)=8t$$
$$15t^3-2t^2=8t$$
$$15t^3-2t^2-8t=0$$
$$t(15t^2-2t-8)=0$$
$$t(3t+2)(5t-4)=0$$
$$t=0 \quad \textbf{or} \quad 3t+2=0 \quad \textbf{or} \quad 5t-4=0$$
$$3t=-2 \qquad\qquad 5t=4$$
$$t=-\tfrac{2}{3} \qquad\qquad t=\tfrac{4}{5}$$

86.
$$3u^3=u(19u+14)$$
$$3u^3=19u^2+14u$$
$$3u^3-19u^2-14u=0$$
$$u(3u^2-19u-14)=0$$
$$u(3u+2)(u-7)=0$$
$$u=0 \quad \textbf{or} \quad 3u+2=0 \quad \textbf{or} \quad u-7=0$$
$$3u=-2 \qquad\qquad u=7$$
$$u=-\tfrac{2}{3}$$

87. Let h = the height and $h+3$ = the width
$$lwh=V$$
$$12(h+3)h=840$$
$$12h^2+36h-840=0$$
$$3(4h^2+12h-280)=0$$
$$3(4h+40)(h-7)=0$$
$$4h+40=0 \quad \textbf{or} \quad h-7=0$$
$$4h=-40 \qquad\qquad h=7$$
$$h=-10$$
(impossible) The height is 7 cm.

88. Let x = one side of the base and $x+3$ = the other side of the base.
$$\tfrac{Bh}{3}=V$$
$$\tfrac{x(x+3)9}{3}=1020$$
$$\tfrac{9x^2+27x}{3}=1020$$
$$9x^2+27x=3060$$
$$9x^2+27x-3060=0$$
$$9(x^2+3x-340)=0$$
$$9(x+20)(x-17)=0$$
$$x+20=0 \quad \textbf{or} \quad x-17=0$$
$$x=-20 \qquad\qquad x=17$$
(impossible) The base is 17 m by 20 m.

Chapter 5 Test (page 363)

1. degree = 6

2. degree = $9+4=13$

3. $P(-3)=-3(-3)^2+2(-3)-1=-3(9)-6-1=-27-6-1=-34$

4. $P(-1)=-3(-1)^2+2(-1)-1=-3(1)-2-1=-3-2-1=-6$

5. $f(x)=x^2+2x$

x	y
-3	3
-2	0
-1	-1

x	y
0	0
1	3

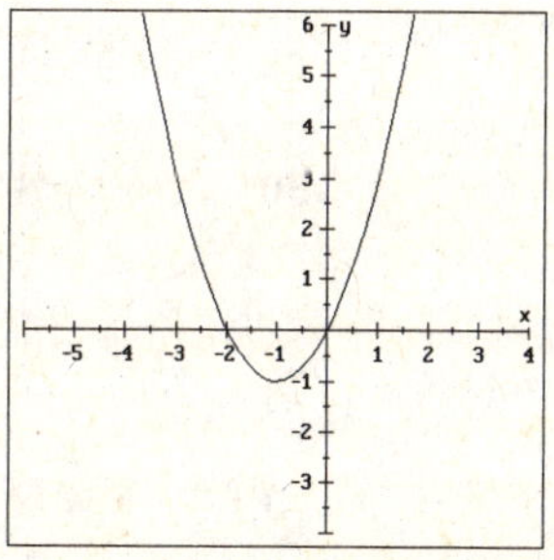

6. $(2y^2+4y+3)+(3y^2-3y-4)=2y^2+4y+3+3y^2-3y-4=5y^2+y-1$

© 2015 Cengage Learning. All Rights Reserved. May not be scanned, copied or duplicated, or posted to a publicly accessible website, in whole or in part.

7. $(-3u^2+2u-7)-(u^2+7)=-3u^2+2u-7-u^2-7=-4u^2+2u-14$

8. $3(2a^2-4a+2)-4(-a^2-3a-4)=6a^2-12a+6+4a^2+12a+16=10a^2+22$

9. $-2(2x^2-2)+3(x^2+5x-2)=-4x^2+4+3x^2+15x-6=-x^2+15x-2$

10. $(3x^3y^2z)(-2xy^{-1}z^3)=-6x^4yz^4$

11. $-5a^2b(3ab^3-2ab^4)=-15a^3b^4+10a^3b^5$

12. $(z+9)(z-9)=z^2-9z+9z-81$
$=z^2-81$

13. $(5x+3)(6x-7)=30x^2-35x+18x-21$
$=30x^2-17x-21$

14. $(u-v)^2(2u+v)=\left(u^2-2uv+v^2\right)(2u+v)=\left(u^2-2uv+v^2\right)(2u)+\left(u^2-2uv+v^2\right)(v)$
$=2u^3-4u^2v+2uv^2+u^2v-2uv^2+v^3$
$=2u^3-3u^2v+v^3$

15. $4x^2y-12xy^2=4xy(x-3y)$

16. $12a^3b^2c-3a^2b^2c^2+6abc^3=3abc(4a^2b-abc+2c^2)$

17. $x^2y^{n+2}+y^n=y^n(x^2y^{n+2-n}+y^{n-n})$
$=y^n(x^2y^2+1)$

18. $4x^2-12x+9=(2x-3)(2x-3)$
$=(2x-3)^2$

19. $(u-v)r+(u-v)s=(u-v)(r+s)$

20. $ax-xy+ay-y^2=x(a-y)+y(a-y)$
$=(a-y)(x+y)$

21. $x^2-64=x^2-8^2=(x+8)(x-8)$

22. $2x^2-32=2(x^2-16)=2(x+4)(x-4)$

23. $4y^4-64=4(y^4-16)=4(y^2+4)(y^2-4)=4(y^2+4)(y+2)(y-2)$

24. $b^3+64=b^3+4^3$
$=(b+4)(b^2-4b+16)$

25. $5x^2-8x-6 \Rightarrow$ prime

26. $3u^3-24=3(u^3-8)=3(u^3-2^3)=3(u-2)(u^2+2u+4)$

27. $a^2-9a-10=(a-10)(a+1)$

28. $6b^2+b-2=(3b+2)(2b-1)$

29. $6u^2+9u-6=3(2u^2+3u-2)$
$=3(2u-1)(u+2)$

30. $20r^2-15r-5=5(4r^2-3r-1)$
$=5(4r+1)(r-1)$

31. $x^{2n}+2x^n+1=(x^n+1)(x^n+1)=(x^n+1)^2$

32. $x^2+6x+9-y^2=(x^2+6x+9)-y^2=(x+3)^2-y^2=(x+3+y)(x+3-y)$

© 2015 Cengage Learning. All Rights Reserved. May not be scanned, copied or duplicated, or posted to a publicly accessible website, in whole or in part.

CHAPTER 5 TEST

33.
$$\begin{aligned} x^2 + x - 6 &= 0 \\ (x-2)(x+3) &= 0 \end{aligned}$$
$x - 2 = 0$ **or** $x + 3 = 0$
$x = 2$ $\quad$ $x = -3$

34.
$$\begin{aligned} 5x^2 + 9x - 18 &= 0 \\ (5x-6)(x+3) &= 0 \end{aligned}$$
$5x - 6 = 0$ **or** $x + 3 = 0$
$5x = 6$ $\quad$ $x = -3$
$x = \frac{6}{5}$

35.
$$\begin{aligned} x^2 &= 9x + 10 \\ x^2 - 9x - 10 &= 0 \\ (x-10)(x+1) &= 0 \end{aligned}$$
$x - 10 = 0$ **or** $x + 1 = 0$
$x = 10$ $\quad$ $x = -1$

36.
$$\begin{aligned} 4x^3 + 24x^2 &= -35x \\ 4x^3 + 24x^2 + 35x &= 0 \\ x(4x^2 + 24x + 35) &= 0 \\ x(2x+7)(2x+5) &= 0 \end{aligned}$$
$x = 0$ **or** $2x + 7 = 0$ **or** $2x + 5 = 0$
$2x = -7$ $\quad$ $2x = -5$
$x = -\frac{7}{2}$ $\quad$ $x = -\frac{5}{2}$

37.
$$\begin{aligned} r_1r_2 - r_2r &= r_1r \\ r_1r_2 &= r_1r + r_2r \\ r_1r_2 &= r(r_1 + r_2) \\ \frac{r_1r_2}{r_1 + r_2} &= r, \text{ or } r = \frac{r_1r_2}{r_1 + r_2} \end{aligned}$$

38.
$$\begin{aligned} x^2 - 5x - 6 &= 0 \\ (x+1)(x-6) &= 0 \end{aligned}$$
$x + 1 = 0$ **or** $x - 6 = 0$
$x = -1$ $\quad$ $x = 6$

39. Let x and $x + 1$ represent the integers.
$$\begin{aligned} x(x+1) &= 156 \\ x^2 + x - 156 &= 0 \\ (x+13)(x-12) &= 0 \end{aligned}$$
$x + 13 = 0$ **or** $x - 12 = 0$
$x = -13$ $\quad$ $x = 12$
(impossible)
The integers are 12 and 13 (sum of 25).

40.
$$\begin{aligned} \boxed{\text{Area of border}} &= 70 \\ 2w(1) + 2w(1) + w(1) + w(1) + 4(1) &= 70 \\ 6w + 4 &= 70 \\ 6w &= 66 \\ w &= 11 \end{aligned}$$
The dimensions are 11 ft by 22 ft.

© 2015 Cengage Learning. All Rights Reserved. May not be scanned, copied or duplicated, or posted to a publicly accessible website, in whole or in part.

Exercises 6.1 (page 374)

1. $\frac{x}{x-2} = \frac{0}{0-2} = \frac{0}{-2} = 0$

3. $\frac{y+1}{y^2-9} = \frac{-3+1}{(-3)^2-9} = \frac{-2}{0}$; undefined

5. $\frac{x-1}{1-x} = \frac{2-1}{1-2} = \frac{1}{-1} = -1$

7. $3x^2 - 9x = 3x(x-3)$

9. $27x^6 + 64y^3 = (3x^2)^3 + (4y)^3 = (3x^2 + 4y)(9x^4 - 12x^2y + 16y^2)$

11. rational

13. asymptote

15. a; 1; 0

17. $\frac{a}{b}$; 0

19. $x + 6 \neq 0$
$x \neq -6$

21. $3x + 4 \neq 0$
$3x \neq -4$
$x \neq -\frac{4}{3}$

23. $x + 6 \neq 0$
$x \neq -6$; $(-\infty, -6) \cup (-6, \infty)$

25. $3x + 4 \neq 0$
$3x \neq -4$
$x \neq -\frac{4}{3}$; $(-\infty, -\frac{4}{3}) \cup (-\frac{4}{3}, \infty)$

27. $f(x) = \frac{x}{x-2}$

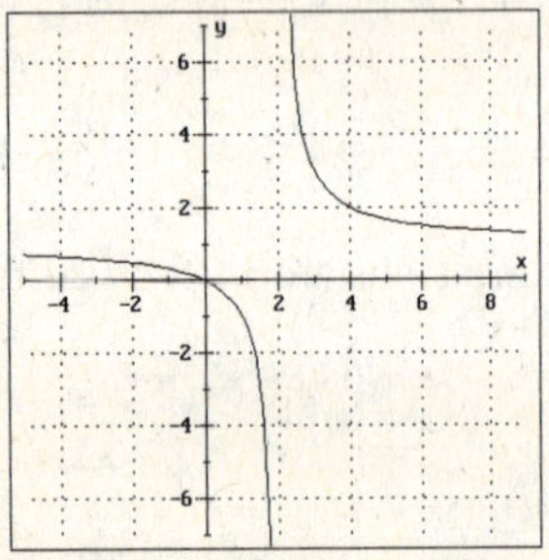

domain: $(-\infty, 2) \cup (2, \infty)$

29. $f(x) = \frac{x+1}{x^2-4}$

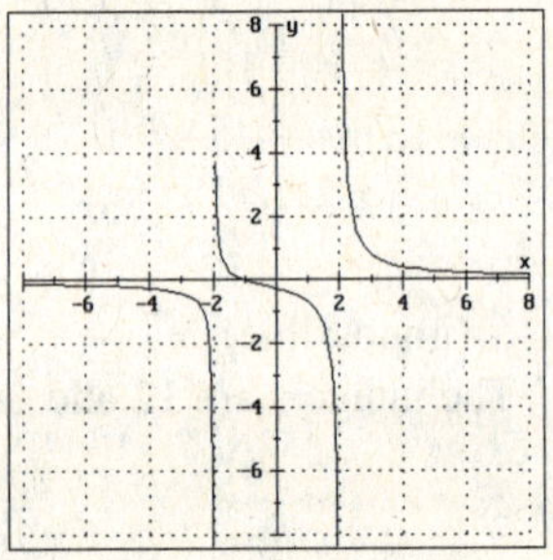

domain: $(-\infty, -2) \cup (-2, 2) \cup (2, \infty)$

31. $-\frac{18}{24} = -\frac{\cancel{6} \cdot 3}{\cancel{6} \cdot 4} = -\frac{3}{4}$

33. $-\frac{112}{36} = -\frac{\cancel{4} \cdot 28}{\cancel{4} \cdot 9} = -\frac{28}{9}$

35. $\frac{12x^3}{3x} = \frac{12}{3} \cdot \frac{x^3}{x} = 4x^2$

37. $\frac{-24x^3y^4}{18x^4y^3} = -\frac{24}{18} \cdot \frac{x^3y^4}{x^4y^3} = -\frac{4y}{3x}$

39. $\frac{x^2-81}{x+9} = \frac{(x+9)(x-9)}{x+9} = x - 9$

41. $\frac{x+y}{x^2-y^2} = \frac{x+y}{(x+y)(x-y)} = \frac{1}{x-y}$

43. $\frac{x^2+2x-15}{x^2-25} = \frac{(x+5)(x-3)}{(x+5)(x-5)} = \frac{x-3}{x-5}$

45. $\frac{2a^2+5a+3}{6a+9} = \frac{(2a+3)(a+1)}{3(2a+3)} = \frac{a+1}{3}$

© 2015 Cengage Learning. All Rights Reserved. May not be scanned, copied or duplicated, or posted to a publicly accessible website, in whole or in part.

SECTION 6.1

47. $\dfrac{x^2+2x+1}{x^2+4x+3} = \dfrac{(x+1)(x+1)}{(x+1)(x+3)} = \dfrac{x+1}{x+3}$

49. $\dfrac{4x^2+24x+32}{16x^2+8x-48} = \dfrac{4(x^2+6x+8)}{8(2x^2+x-6)} = \dfrac{4(x+4)(x+2)}{8(2x-3)(x+2)} = \dfrac{x+4}{2(2x-3)}$

51. $\dfrac{(x^2-1)(x+1)}{(x^2-2x+1)^2} = \dfrac{(x+1)(x-1)(x+1)}{[(x-1)^2]^2} = \dfrac{(x+1)^2(x-1)}{(x-1)^4} = \dfrac{(x+1)^2}{(x-1)^3}$

53. $$\dfrac{(2x^2+3xy+y^2)(3a+b)}{(x+y)(2xy+2bx+y^2+by)} = \dfrac{(2x+y)(x+y)(3a+b)}{(x+y)[2x(y+b)+y(y+b)]} = \dfrac{(2x+y)(x+y)(3a+b)}{(x+y)(y+b)(2x+y)} = \dfrac{3a+b}{y+b}$$

55. $\dfrac{(p+q)(p-r)}{(r-p)(p+q)} = \dfrac{(p-r)}{-(p-r)} = -1$

57. $\dfrac{-3ab^2(a-b)}{9ab(b-a)} = \dfrac{-3ab^2(a-b)}{-9ab(a-b)} = \dfrac{b}{3}$

59. $t = f(r) = \dfrac{600}{r}$
$f(30) = \dfrac{600}{30}$
$= 20;\ t = 20$ hours

61. $t = f(r) = \dfrac{600}{r}$
$f(50) = \dfrac{600}{50}$
$= 12;\ t = 12$ hours

63. $c = f(p) = \dfrac{50{,}000p}{100-p}$
$f(10) = \dfrac{50{,}000(10)}{100-10}$
$= \dfrac{500{,}000}{90}$
$\approx 5555.56;\ c = \$5555.56$

65. $c = f(p) = \dfrac{50{,}000p}{100-p}$
$f(50) = \dfrac{50{,}000(50)}{100-50}$
$= \dfrac{2{,}500{,}000}{50}$
$= 50{,}000.00;\ c = \$50{,}000$

67. $\dfrac{288}{312} = \dfrac{\cancel{24}\cdot 12}{\cancel{24}\cdot 13} = \dfrac{12}{13}$

69. $-\dfrac{96a^5}{112a^3} = -\dfrac{96}{112}\cdot\dfrac{a^5}{a^3} = -\dfrac{6a^2}{7}$

71. $\dfrac{(5x^4)^2}{15x^5} = \dfrac{25x^8}{15x^5} = \dfrac{5x^3}{3}$

73. $\dfrac{-14x^2(x+y)^2}{49x^3(x+y)} = -\dfrac{2(x+y)}{7x}$

75. $\dfrac{9y^2(y-z)}{21y(y-z)^2} = \dfrac{3y}{7(y-z)}$

77. $\dfrac{4x^2+8x+3}{6+x-2x^2} = \dfrac{4x^2+8x+3}{-(2x^2-x-6)} = \dfrac{(2x+3)(2x+1)}{-(2x+3)(x-2)} = \dfrac{2x+1}{-(x-2)} = \dfrac{2x+1}{2-x}$

79. $\dfrac{y^3-8}{20-5y^2} = \dfrac{(y-2)(y^2+2y+4)}{5(4-y^2)} = \dfrac{(y-2)(y^2+2y+4)}{5(2+y)(2-y)} = -\dfrac{y^2+2y+4}{5(2+y)}$

© 2015 Cengage Learning. All Rights Reserved. May not be scanned, copied or duplicated, or posted to a publicly accessible website, in whole or in part.

SECTION 6.1

81. $\dfrac{x^2+x-30}{x^2-x-20}=\dfrac{(x+6)(x-5)}{(x-5)(x+4)}=\dfrac{x+6}{x+4}$

83. $\dfrac{x^2+y^2}{x+y} \Rightarrow$ lowest terms

85. $\dfrac{3m-6n}{3n-6m}=\dfrac{3(m-2n)}{3(n-2m)}=\dfrac{m-2n}{n-2m}$

87. $\dfrac{x^3+8}{x^2-2x+4}=\dfrac{(x+2)(x^2-2x+4)}{x^2-2x+4}$
$=x+2$

89. $\dfrac{x^2+3x+9}{x^3-27}=\dfrac{x^2+3x+9}{(x-3)(x^2+3x+9)}$
$=\dfrac{1}{x-3}$

91. $\dfrac{3x+6y}{x+2y}=\dfrac{3(x+2y)}{x+2y}=3$

93. $\dfrac{4a^2-9b^2}{2a^2-ab-6b^2}=\dfrac{(2a+3b)(2a-3b)}{(2a+3b)(a-2b)}=\dfrac{2a-3b}{a-2b}$

95. $\dfrac{x-y}{x^3-y^3-x+y}=\dfrac{x-y}{(x^3-y^3)-(x-y)}=\dfrac{x-y}{(x-y)(x^2+xy+y^2)-1(x-y)}$
$=\dfrac{x-y}{(x-y)(x^2+xy+y^2-1)}$
$=\dfrac{1}{x^2+xy+y^2-1}$

97. $\dfrac{px-py+qx-qy}{px+qx+py+qy}=\dfrac{p(x-y)+q(x-y)}{x(p+q)+y(p+q)}=\dfrac{(x-y)(p+q)}{(p+q)(x+y)}=\dfrac{x-y}{x+y}$

99. $\dfrac{m^3-mn^2}{mn^2+m^2n-2m^3}=\dfrac{m(m^2-n^2)}{m(n^2+mn-2m^2)}=\dfrac{m(m+n)(m-n)}{m(n+2m)(n-m)}=\dfrac{m(m+n)(m-n)}{-m(n+2m)(m-n)}$
$=-\dfrac{m+n}{2m+n}$

101. $\dfrac{3x^2-3y^2}{x^2+2y+2x+yx}=\dfrac{3(x^2-y^2)}{x^2+2x+yx+2y}=\dfrac{3(x+y)(x-y)}{x(x+2)+y(x+2)}=\dfrac{3(x+y)(x-y)}{(x+2)(x+y)}=\dfrac{3(x-y)}{x+2}$

103. $f(t)=\dfrac{t^2+2t}{2t+2}$

$f(15)=\dfrac{15^2+2(15)}{2(15)+2}$
$=\dfrac{225+30}{30+2}=\dfrac{255}{32}\approx 7.96875$

It will take them almost 8 days.

105. $f(t)=\dfrac{t^2+3t}{2t+3}$

Small pipe: 7 hrs $\Rightarrow$ Large pipe: 4 hrs

$f(4)=\dfrac{4^2+3(4)}{2(4)+3}$
$=\dfrac{16+12}{8+3}=\dfrac{28}{11}=2.5455$

It will take the pipes about 2.55 hours.

© 2015 Cengage Learning. All Rights Reserved. May not be scanned, copied or duplicated, or posted to a publicly accessible website, in whole or in part.

107. **a.** $f(p) = \dfrac{50{,}000p}{100 - p}$
$f(40) = \dfrac{50{,}000(40)}{100 - 40}$
$= \dfrac{2{,}000{,}000}{60} \approx 33333.33$
\$33,333.33

b. $f(p) = \dfrac{50{,}000p}{100 - p}$
$f(70) = \dfrac{50{,}000(70)}{100 - 70}$
$= \dfrac{3{,}500{,}000}{30} \approx 116{,}666.67$
\$116,666.67

109. $c = f(x) = 1.25x + 700$

111. Refer to **#109**.
$c = f(500) = 1.25(500) + 700 = \$1325$

113. Refer to **#110**. $\overline{c} = f(1000) = \dfrac{1.25(1000) + 700}{1000} = \dfrac{1950}{1000} = \$1.95$

115. $c = f(n) = 0.09n + 7.50$

117. Refer to **#115**. $c = f(775) = 0.09(775) + 7.50 = 69.75 + 7.50 = \$77.25$

119. Refer to **#116**. $\overline{c} = f(1000) = \dfrac{0.09(1000) + 7.50}{1000} = \dfrac{97.50}{1000} = \$0.0975 = 9.75¢$

121. **Answers may vary.**

123. $\dfrac{a - 3b}{2b - a} = \dfrac{-(3b - a)}{-(a - 2b)} = \dfrac{3b - a}{a - 2b}$.
The two answers are the same.

125. You can divide out the 4's in parts a and d.

Exercises 6.2 (page 383)

1. $\dfrac{2 \cdot 7}{7 \cdot 2^2} = \dfrac{1}{2}$

3. $\dfrac{3(5 - 1)}{3^2(1 - 5)} = -\dfrac{1}{3}$

5. $\dfrac{2^2 \cdot 3^3}{2 \cdot 3^2} = \dfrac{2 \cdot 3}{1} = 6$

7. $-7x^3(4x^4 - x^2) = -28x^7 + 7x^5$

9. $(m^n + 2)(m^n - 2) = m^{2n} - 4$

11. $\dfrac{ac}{bd}$

13. 0

15. $\dfrac{10}{9} \cdot \dfrac{3}{5} \cdot \dfrac{7}{8} = \dfrac{\cancel{2} \cdot \cancel{5}}{\cancel{3} \cdot 3} \cdot \dfrac{\cancel{3}}{\cancel{5}} \cdot \dfrac{7}{\cancel{2} \cdot 4} = \dfrac{7}{12}$

17. $-\dfrac{8}{13} \div \dfrac{12}{26} = -\dfrac{8}{13} \cdot \dfrac{26}{12} = -\dfrac{\cancel{4} \cdot 2}{\cancel{13}} \cdot \dfrac{\cancel{13} \cdot 2}{\cancel{4} \cdot 3}$
$= -\dfrac{4}{3}$

19. $\dfrac{x^2y^2}{cd} \cdot \dfrac{c^{-2}d^2}{x} = x^{2-1}y^2c^{-2-1}d^{2-1}$
$= xy^2c^{-3}d = \dfrac{xy^2d}{c^3}$

© 2015 Cengage Learning. All Rights Reserved. May not be scanned, copied or duplicated, or posted to a publicly accessible website, in whole or in part.

SECTION 6.2

21. $\dfrac{-x^2y^{-2}}{x^{-1}y^{-3}} \div \dfrac{x^{-3}y^2}{x^4y^{-1}} = \dfrac{-x^2y^{-2}}{x^{-1}y^{-3}} \cdot \dfrac{x^4y^{-1}}{x^{-3}y^2} = \dfrac{-x^6y^{-3}}{x^{-4}y^{-1}} = -x^{6-(-4)}y^{-3-(-1)} = -x^{10}y^{-2} = -\dfrac{x^{10}}{y^2}$

23. $\dfrac{x^2-4}{x^3} \cdot \dfrac{x^2}{x+2} = \dfrac{(x+2)(x-2)}{x^3} \cdot \dfrac{x^2}{x+2} = \dfrac{x-2}{x}$

25. $\dfrac{x^2-x-6}{x^2-4} \cdot \dfrac{x^2-x-2}{9-x^2} = \dfrac{(x-3)(x+2)}{(x+2)(x-2)} \cdot \dfrac{(x-2)(x+1)}{(3+x)(3-x)} = -\dfrac{x+1}{x+3}$

27. $\dfrac{x^3+2x^2+4x+8}{y^2-1} \cdot \dfrac{y^2+2y+1}{x^4-16} = \dfrac{x^2(x+2)+4(x+2)}{(y+1)(y-1)} \cdot \dfrac{(y+1)(y+1)}{(x^2+4)(x^2-4)}$

$= \dfrac{(x+2)(x^2+4)}{(y+1)(y-1)} \cdot \dfrac{(y+1)(y+1)}{(x^2+4)(x+2)(x-2)} = \dfrac{y+1}{(y-1)(x-2)}$

29. $\dfrac{m^2-n^2}{2x^2+3x-2} \cdot \dfrac{2x^2+5x-3}{n^2-m^2} = \dfrac{(m+n)(m-n)}{(2x-1)(x+2)} \cdot \dfrac{(2x-1)(x+3)}{(n+m)(n-m)} = -\dfrac{x+3}{x+2}$

31. $\dfrac{2x^2-x-3}{x^2-1} \cdot \dfrac{x^2+x-2}{2x^2+x-6} = \dfrac{(2x-3)(x+1)}{(x+1)(x-1)} \cdot \dfrac{(x+2)(x-1)}{(2x-3)(x+2)} = 1$

33. $\dfrac{2p^2-5p-3}{p^2-9} \cdot \dfrac{2p^2+5p-3}{2p^2+5p+2} = \dfrac{(2p+1)(p-3)}{(p+3)(p-3)} \cdot \dfrac{(2p-1)(p+3)}{(2p+1)(p+2)} = \dfrac{2p-1}{p+2}$

35. $(9-x^2) \cdot \dfrac{x}{x^3-4x^2+3x} = (9-x^2) \cdot \dfrac{x}{x(x^2-4x+3)} = \dfrac{(3+x)(3-x)}{1} \cdot \dfrac{x}{x(x-3)(x-1)}$

$= -\dfrac{x+3}{x-1}$

37. $\dfrac{x}{2x^2+x} \cdot (2x^2-9x-5) = \dfrac{x}{2x^2+x} \cdot \dfrac{2x^2-9x-5}{1} = \dfrac{x}{x(2x+1)} \cdot \dfrac{(2x+1)(x-5)}{1} = x-5$

39. $\dfrac{x-7}{x^2+3}$: lowest terms. $\left(\dfrac{x-7}{x^2+3}\right)^2 = \dfrac{(x-7)(x-7)}{(x^2+3)(x^2+3)} = \dfrac{x^2-14x+49}{x^4+6x^2+9}$

41. $\dfrac{2m^2-m-3}{x^2-1} = \dfrac{(2m-3)(m+1)}{(x+1)(x-1)}$ is in lowest terms.

$\left(\dfrac{2m^2-m-3}{x^2-1}\right)^2 = \dfrac{(2m^2-m-3)(2m^2-m-3)}{(x^2-1)(x^2-1)}$

$= \dfrac{4m^4-2m^3-6m^2-2m^3+m^2+3m-6m^2+3m+9}{x^4-2x^2+1}$

$= \dfrac{4m^4-4m^3-11m^2+6m+9}{x^4-2x^2+1}$

43. $\dfrac{x^2-36}{x^2-81} \div \dfrac{x+6}{x-9} = \dfrac{x^2-36}{x^2-81} \cdot \dfrac{x-9}{x+6} = \dfrac{(x+6)(x-6)}{(x+9)(x-9)} \cdot \dfrac{x-9}{x+6} = \dfrac{x-6}{x+9}$

© 2015 Cengage Learning. All Rights Reserved. May not be scanned, copied or duplicated, or posted to a publicly accessible website, in whole or in part.

SECTION 6.2

45. $$\frac{3n^2+5n-2}{12n^2-13n+3} \div \frac{n^2+3n+2}{4n^2+5n-6} = \frac{3n^2+5n-2}{12n^2-13n+3} \cdot \frac{4n^2+5n-6}{n^2+3n+2} = \frac{(3n-1)(n+2)}{(3n-1)(4n-3)} \cdot \frac{(4n-3)(n+2)}{(n+2)(n+1)} = \frac{n+2}{n+1}$$

47. $$\frac{a^2+2a-35}{12x} \div \frac{ax-3x}{a^2+4a-21} = \frac{a^2+2a-35}{12x} \cdot \frac{a^2+4a-21}{ax-3x} = \frac{(a+7)(a-5)}{12x} \cdot \frac{(a+7)(a-3)}{x(a-3)} = \frac{(a+7)^2(a-5)}{12x^2}$$

49. $$\frac{x^2-6x+9}{4-x^2} \div \frac{x^2-9}{x^2-8x+12} = \frac{x^2-6x+9}{4-x^2} \cdot \frac{x^2-8x+12}{x^2-9} = \frac{(x-3)(x-3)}{(2+x)(2-x)} \cdot \frac{(x-6)(x-2)}{(x+3)(x-3)} = -\frac{(x-3)(x-6)}{(x+2)(x+3)}$$

51. $$\frac{x^2-4}{x} \div (x+2) = \frac{x^2-4}{x} \div \frac{x+2}{1} = \frac{x^2-4}{x} \cdot \frac{1}{x+2} = \frac{(x+2)(x-2)}{x} \cdot \frac{1}{x+2} = \frac{x-2}{x}$$

53. $$\frac{x^3-25x}{x^2} \div \left(x^2+10x+25\right) = \frac{x(x^2-25)}{x^2} \cdot \frac{1}{x^2+10x+25} = \frac{x(x+5)(x-5)}{x^2} \cdot \frac{1}{(x+5)(x+5)} = \frac{x-5}{x(x+5)}$$

55. $$\frac{3x^2y^2}{6x^3y} \cdot \frac{-4x^7y^{-2}}{18x^{-2}y} \div \frac{36x}{18y^{-2}} = \frac{3x^2y^2}{6x^3y} \cdot \frac{-4x^7y^{-2}}{18x^{-2}y} \cdot \frac{18y^{-2}}{36x} = \frac{-2^3 \cdot 3^3x^9y^{-2}}{2^4 \cdot 3^5x^2y^2} = -\frac{x^7}{18y^4}$$

57. $$(5x+10) \cdot \frac{x^3}{10x-20} \div \frac{2x+4}{4} = \frac{5x+10}{1} \cdot \frac{x^3}{10x-20} \cdot \frac{4}{2x+4} = \frac{5(x+2)}{1} \cdot \frac{x^3}{10(x-2)} \cdot \frac{4}{2(x+2)} = \frac{x^3}{x-2}$$

59. $$\frac{2x^2-2x-4}{x^2+2x-8} \cdot \frac{3x^2+15x}{x+1} \div \frac{4x^2-100}{x^2-x-20} = \frac{2x^2-2x-4}{x^2+2x-8} \cdot \frac{3x^2+15x}{x+1} \cdot \frac{x^2-x-20}{4x^2-100} = \frac{2(x-2)(x+1)}{(x+4)(x-2)} \cdot \frac{3x(x+5)}{x+1} \cdot \frac{(x-5)(x+4)}{4(x+5)(x-5)} = \frac{3x}{2}$$

61. $$\frac{2t^2+5t+2}{t^2-4t+16} \div \frac{t+2}{t^3+64} \div \frac{2t^3+9t^2+4t}{t+1} = \frac{2t^2+5t+2}{t^2-4t+16} \cdot \frac{t^3+64}{t+2} \cdot \frac{t+1}{2t^3+9t^2+4t} = \frac{(2t+1)(t+2)}{t^2-4t+16} \cdot \frac{(t+4)(t^2-4t+16)}{t+2} \cdot \frac{t+1}{t(2t+1)(t+4)} = \frac{t+1}{t}$$

© 2015 Cengage Learning. All Rights Reserved. May not be scanned, copied or duplicated, or posted to a publicly accessible website, in whole or in part.

SECTION 6.2

63. $$\frac{x^2-y^2}{2x^2+2xy+x+y}\cdot\frac{2x^2-5x-3}{yx-3y-x^2+3x}=\frac{(x+y)(x-y)}{2x(x+y)+1(x+y)}\cdot\frac{(2x+1)(x-3)}{y(x-3)-x(x-3)}$$
$$=\frac{(x+y)(x-y)}{(x+y)(2x+1)}\cdot\frac{(2x+1)(x-3)}{(x-3)(y-x)}=-1$$

65. $$\frac{x^3+y^3}{x^3-y^3}\div\frac{x^2-xy+y^2}{x^2+xy+y^2}=\frac{x^3+y^3}{x^3-y^3}\cdot\frac{x^2+xy+y^2}{x^2-xy+y^2}=\frac{(x+y)(x^2-xy+y^2)}{(x-y)(x^2+xy+y^2)}\cdot\frac{x^2+xy+y^2}{x^2-xy+y^2}$$
$$=\frac{x+y}{x-y}$$

67. $$(x^2-7x+10)\div(x-2)\div(x+5)=\frac{x^2-7x+10}{1}\div\frac{x-2}{1}\div\frac{x+5}{1}$$
$$=\frac{(x-5)(x-2)}{1}\cdot\frac{1}{x-2}\cdot\frac{1}{x+5}=\frac{x-5}{x+5}$$

69. $$\frac{3x^2-2x}{3x+2}\div(3x-2)\div\frac{3x}{3x-3}=\frac{3x^2-2x}{3x+2}\cdot\frac{1}{3x-2}\cdot\frac{3x-3}{3x}$$
$$=\frac{x(3x-2)}{3x+2}\cdot\frac{1}{3x-2}\cdot\frac{3(x-1)}{3x}=\frac{x-1}{3x+2}$$

71. $$\frac{2x^2+5x-3}{x^2+2x-3}\div\left(\frac{x^2+2x-35}{x^2-6x+5}\div\frac{x^2-9x+14}{2x^2-5x+2}\right)=$$
$$=\frac{2x^2+5x-3}{x^2+2x-3}\div\left(\frac{x^2+2x-35}{x^2-6x+5}\cdot\frac{2x^2-5x+2}{x^2-9x+14}\right)$$
$$=\frac{2x^2+5x-3}{x^2+2x-3}\div\left(\frac{(x+7)(x-5)}{(x-5)(x-1)}\cdot\frac{(2x-1)(x-2)}{(x-7)(x-2)}\right)$$
$$=\frac{2x^2+5x-3}{x^2+2x-3}\div\frac{(x+7)(2x-1)}{(x-1)(x-7)}=\frac{(2x-1)(x+3)}{(x+3)(x-1)}\cdot\frac{(x-1)(x-7)}{(x+7)(2x-1)}=\frac{x-7}{x+7}$$

73. $$\frac{x^2-x-12}{x^2+x-2}\div\frac{x^2-6x+8}{x^2-3x-10}\cdot\frac{x^2-3x+2}{x^2-2x-15}=\frac{x^2-x-12}{x^2+x-2}\cdot\frac{x^2-3x-10}{x^2-6x+8}\cdot\frac{x^2-3x+2}{x^2-2x-15}$$
$$=\frac{(x-4)(x+3)}{(x+2)(x-1)}\cdot\frac{(x-5)(x+2)}{(x-4)(x-2)}\cdot\frac{(x-2)(x-1)}{(x-5)(x+3)}=1$$

75. $$A=\frac{1}{2}bh=\frac{1}{2}\cdot\frac{b^2-4}{b+3}\cdot\frac{b^2-9}{b+2}=\frac{1}{2}\cdot\frac{(b+2)(b-2)}{b+3}\cdot\frac{(b+3)(b-3)}{b+2}=\frac{(b-2)(b-3)}{2}\text{ cm}^2$$

77. $$d=rt=\frac{k^2-k-6}{k-4}\cdot\frac{k^2-16}{k^2-2k-3}=\frac{(k-3)(k+2)}{k-4}\cdot\frac{(k+4)(k-4)}{(k-3)(k+1)}=\frac{(k+2)(k+4)}{k+1}\text{ miles}$$

79. **Answers may vary.**

81. $$\frac{x^2}{y}\boxed{\div}\frac{x}{y^2}\boxed{\times}\frac{x^2}{y^2}=\frac{x^3}{y}$$

© 2015 Cengage Learning. All Rights Reserved. May not be scanned, copied or duplicated, or posted to a publicly accessible website, in whole or in part.

Exercises 6.3 (page 392)

1. $x + x = 2x$

3. $5x + (4x + 3) = 5x + 4x + 3 = 9x + 3$

5. $(3y + 1) - (y + 3) = 3y + 1 - y - 3$
$= 2y - 2$

7. $(-1, 4] \Rightarrow$ 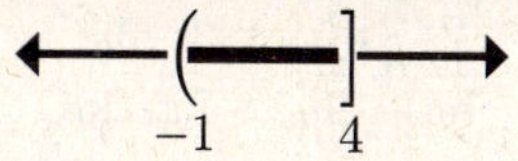

9. $V = \pi r^2 h$
$\frac{V}{\pi r^2} = \frac{\pi r^2 h}{\pi r^2}$
$\frac{V}{\pi r^2} = h$, or $h = \frac{V}{\pi r^2}$

11. $\frac{a + c}{b}$

13. subtract; keep

15. LCD

17. $\frac{4}{9} + \frac{2}{9} = \frac{6}{9} = \frac{2}{3}$

19. $\frac{10}{33} - \frac{21}{33} = \frac{-11}{33} = -\frac{1}{3}$

21. $\frac{5}{6x} + \frac{3}{6x} = \frac{8}{6x} = \frac{4}{3x}$

23. $\frac{a}{a + b} - \frac{7}{a + b} = \frac{a - 7}{a + b}$

25. $\frac{3x}{2x + 2} + \frac{x + 4}{2x + 2} = \frac{4x + 4}{2x + 2} = \frac{4(x + 1)}{2(x + 1)} = 2$

27. $\frac{2x}{x - 5} - \frac{10}{x - 5} = \frac{2x - 10}{x - 5} = \frac{2(x - 5)}{x - 5} = 2$

29. $\frac{5x}{x + 1} + \frac{3}{x + 1} - \frac{2x}{x + 1} = \frac{3x + 3}{x + 1}$
$= \frac{3(x + 1)}{x + 1} = 3$

31. $\frac{3(x^2 + x)}{x^2 - 5x + 6} + \frac{-3(x^2 - x)}{x^2 - 5x + 6} = \frac{3x^2 + 3x}{x^2 - 5x + 6} + \frac{-3x^2 + 3x}{x^2 - 5x + 6} = \frac{6x}{(x - 3)(x - 2)}$

33. $8 = 2^3$; $12 = 2^2 \cdot 3$; $18 = 2 \cdot 3^2 \Rightarrow \text{LCD} = 2^3 \cdot 3^2 = 72$

35. $x^2 - 4 = (x + 2)(x - 2)$; $x^2 + 2x = x(x + 2) \Rightarrow \text{LCD} = x(x + 2)(x - 2)$

37. $x^3 + 27 = (x + 3)(x^2 - 3x + 9)$; $x^2 + 6x + 9 = (x + 3)^2 \Rightarrow \text{LCD} = (x + 3)^2(x^2 - 3x + 9)$

39. $2x^2 + 5x + 3 = (2x + 3)(x + 1)$; $4x^2 + 12x + 9 = (2x + 3)^2$; $x^2 + 2x + 1 = (x + 1)^2$
$\text{LCD} = (2x + 3)^2(x + 1)^2$

41. $\frac{x + 3}{x - 4} - \frac{x - 11}{4 - x} = \frac{x + 3}{x - 4} - \frac{-x + 11}{x - 4} = \frac{2x - 8}{x - 4} = \frac{2(x - 4)}{x - 4} = 2$

43. $\frac{2a + 1}{3a - 2} - \frac{a - 4}{2 - 3a} = \frac{2a + 1}{3a - 2} - \frac{-a + 4}{3a - 2} = \frac{3a - 3}{3a - 2} = \frac{3(a - 1)}{3a - 2}$

© 2015 Cengage Learning. All Rights Reserved. May not be scanned, copied or duplicated, or posted to a publicly accessible website, in whole or in part.

SECTION 6.3

45. $\frac{1}{5}+\frac{1}{4}=\frac{1\cdot 4}{5\cdot 4}+\frac{1\cdot 5}{4\cdot 5}=\frac{4}{20}+\frac{5}{20}=\frac{9}{20}$

47. $\frac{a}{2}+\frac{2a}{5}=\frac{a\cdot 5}{2\cdot 5}+\frac{2a\cdot 2}{5\cdot 2}=\frac{5a}{10}+\frac{4a}{10}=\frac{9a}{10}$

49. $\frac{2}{7x}+\frac{5}{2y}=\frac{2\cdot 2y}{7x\cdot 2y}+\frac{5\cdot 7x}{2y\cdot 7x}=\frac{4y}{14xy}+\frac{35x}{14xy}=\frac{4y+35x}{14xy}$

51. $\frac{3a}{2b}-\frac{2b}{3a}=\frac{3a\cdot 3a}{2b\cdot 3a}-\frac{2b\cdot 2b}{3a\cdot 2b}=\frac{9a^2}{6ab}-\frac{4b^2}{6ab}=\frac{9a^2-4b^2}{6ab}$

53. $4+\frac{7}{x-3}=\frac{4}{1}+\frac{7}{x-3}=\frac{4(x-3)}{1(x-3)}+\frac{7}{x-3}=\frac{4x-12}{x-3}+\frac{7}{x-3}=\frac{4x-5}{x-3}$

55. $\frac{9}{y-3}+6=\frac{9}{y-3}+\frac{6}{1}=\frac{9}{y-3}+\frac{6(y-3)}{1(y-3)}=\frac{9}{y-3}+\frac{6y-18}{y-3}=\frac{6y-9}{y-3}$, or $\frac{3(2y-3)}{y-3}$

57. $\frac{3}{x+2}+\frac{5}{x-4}=\frac{3(x-4)}{(x+2)(x-4)}+\frac{5(x+2)}{(x-4)(x+2)}=\frac{3x-12}{(x+2)(x-4)}+\frac{5x+10}{(x+2)(x-4)}$
$=\frac{8x-2}{(x+2)(x-4)}$, or $\frac{2(4x-1)}{(x+2)(x-4)}$

59. $\frac{x+2}{x+5}-\frac{x-3}{x+7}=\frac{(x+2)(x+7)}{(x+5)(x+7)}-\frac{(x-3)(x+5)}{(x+7)(x+5)}=\frac{x^2+9x+14}{(x+5)(x+7)}-\frac{x^2+2x-15}{(x+5)(x+7)}$
$=\frac{7x+29}{(x+5)(x+7)}$

61. $\frac{x}{x^2+5x+6}+\frac{x}{x^2-4}=\frac{x}{(x+2)(x+3)}+\frac{x}{(x+2)(x-2)}$
$=\frac{x(x-2)}{(x+2)(x+3)(x-2)}+\frac{x(x+3)}{(x+2)(x-2)(x+3)}$
$=\frac{x^2-2x}{(x+2)(x+3)(x-2)}+\frac{x^2+3x}{(x+2)(x+3)(x-2)}$
$=\frac{2x^2+x}{(x+2)(x+3)(x-2)}$

63. $\frac{4}{x^2-2x-3}-\frac{x}{3x^2-7x-6}=\frac{4}{(x-3)(x+1)}-\frac{x}{(3x+2)(x-3)}$
$=\frac{4(3x+2)}{(x-3)(x+1)(3x+2)}-\frac{x(x+1)}{(3x+2)(x-3)(x+1)}$
$=\frac{12x+8}{(x-3)(x+1)(3x+2)}-\frac{x^2+x}{(x-3)(x+1)(3x+2)}$
$=\frac{-x^2+11x+8}{(x-3)(x+1)(3x+2)}$

© 2015 Cengage Learning. All Rights Reserved. May not be scanned, copied or duplicated, or posted to a publicly accessible website, in whole or in part.

SECTION 6.3

65. $$\frac{2x+y}{y-2x}+\frac{3x^2+y}{y^2-4xy+4x^2}=\frac{2x+y}{y-2x}+\frac{3x^2+y}{(y-2x)^2}=\frac{(2x+y)(y-2x)}{(y-2x)^2}+\frac{3x^2+y}{(y-2x)^2}$$
$$=\frac{2xy-4x^2+y^2-2xy}{(y-2x)^2}+\frac{3x^2+y}{(y-2x)^2}$$
$$=\frac{y^2+y-x^2}{(y-2x)^2}$$

67. $$\frac{x}{1-x}-\frac{2x^2-1}{x^2-2x+1}=\frac{-x}{x-1}-\frac{2x^2-1}{(x-1)^2}=\frac{-x(x-1)}{(x-1)^2}-\frac{2x^2-1}{(x-1)^2}=\frac{-x^2+x}{(x-1)^2}-\frac{2x^2-1}{(x-1)^2}$$
$$=\frac{-3x^2+x+1}{(x-1)^2}$$

69. $$\frac{8}{x^2-9}+\frac{2}{x-3}-\frac{6}{x}=\frac{8}{(x+3)(x-3)}+\frac{2}{x-3}-\frac{6}{x}$$
$$=\frac{8x}{x(x+3)(x-3)}+\frac{2x(x+3)}{x(x+3)(x-3)}-\frac{6(x+3)(x-3)}{x(x+3)(x-3)}$$
$$=\frac{8x}{x(x+3)(x-3)}+\frac{2x^2+6x}{x(x+3)(x-3)}-\frac{6(x^2-9)}{x(x+3)(x-3)}$$
$$=\frac{8x}{x(x+3)(x-3)}+\frac{2x^2+6x}{x(x+3)(x-3)}-\frac{6x^2-54}{x(x+3)(x-3)}$$
$$=\frac{-4x^2+14x+54}{x(x+3)(x-3)}$$

71. $$\frac{x}{x+1}-\frac{x}{1-x^2}+\frac{1}{x}=\frac{x}{x+1}+\frac{x}{x^2-1}+\frac{1}{x}$$
$$=\frac{x}{x+1}+\frac{x}{(x+1)(x-1)}+\frac{1}{x}$$
$$=\frac{x(x)(x-1)}{x(x+1)(x-1)}+\frac{x(x)}{x(x+1)(x-1)}+\frac{1(x+1)(x-1)}{x(x+1)(x-1)}$$
$$=\frac{x^2(x-1)}{x(x+1)(x-1)}+\frac{x^2}{x(x+1)(x-1)}+\frac{x^2-1}{x(x+1)(x-1)}$$
$$=\frac{x^3-x^2}{x(x+1)(x-1)}+\frac{x^2}{x(x+1)(x-1)}+\frac{x^2-1}{x(x+1)(x-1)}$$
$$=\frac{x^3+x^2-1}{x(x+1)(x-1)}$$

© 2015 Cengage Learning. All Rights Reserved. May not be scanned, copied or duplicated, or posted to a publicly accessible website, in whole or in part.

SECTION 6.3

73. $$\frac{y+4}{y^2+7y+12}-\frac{y-4}{y+3}+\frac{47}{y+4}=\frac{y+4}{(y+4)(y+3)}-\frac{y-4}{y+3}+\frac{47}{y+4}$$
$$=\frac{y+4}{(y+4)(y+3)}-\frac{(y-4)(y+4)}{(y+4)(y+3)}+\frac{47(y+3)}{(y+4)(y+3)}$$
$$=\frac{y+4}{(y+4)(y+3)}-\frac{y^2-16}{(y+4)(y+3)}+\frac{47y+141}{(y+4)(y+3)}$$
$$=\frac{-y^2+48y+161}{(y+4)(y+3)}$$

75. $$\frac{2}{x-2}+\frac{3}{x+2}-\frac{x-1}{x^2-4}=\frac{2}{x-2}+\frac{3}{x+2}-\frac{x-1}{(x+2)(x-2)}$$
$$=\frac{2(x+2)}{(x+2)(x-2)}+\frac{3(x-2)}{(x+2)(x-2)}-\frac{x-1}{(x+2)(x-2)}$$
$$=\frac{2x+4}{(x+2)(x-2)}+\frac{3x-6}{(x+2)(x-2)}-\frac{x-1}{(x+2)(x-2)}$$
$$=\frac{4x-1}{(x+2)(x-2)}$$

77. $$\left(\frac{1}{x-1}+\frac{1}{1-x}\right)^2=\left(\frac{1}{x-1}+\frac{-1}{x-1}\right)^2=\left(\frac{0}{x-1}\right)^2=0^2=0$$

79. $$\left(\frac{x}{x-3}+\frac{3}{3-x}\right)^3=\left(\frac{x}{x-3}+\frac{-3}{x-3}\right)^3=\left(\frac{x-3}{x-3}\right)^3=1^3=1$$

81. $$\frac{2}{x}-\frac{5}{y}=\frac{2\cdot y}{x\cdot y}-\frac{5\cdot x}{y\cdot x}=\frac{2y}{xy}-\frac{5x}{xy}=\frac{2y-5x}{xy}$$

83. $$2x+3+\frac{1}{x+1}=\frac{2x+3}{1}+\frac{1}{x+1}=\frac{(2x+3)(x+1)}{1(x+1)}+\frac{1}{x+1}$$
$$=\frac{2x^2+5x+3}{x+1}+\frac{1}{x+1}=\frac{2x^2+5x+4}{x+1}$$

85. $$\frac{2}{x-1}-\frac{2x}{x^2-1}-\frac{x}{x^2+2x+1}=\frac{2}{x-1}-\frac{2x}{(x+1)(x-1)}-\frac{x}{(x+1)^2}$$
$$=\frac{2(x+1)^2}{(x-1)(x+1)^2}-\frac{2x(x+1)}{(x-1)(x+1)^2}-\frac{x(x-1)}{(x-1)(x+1)^2}$$
$$=\frac{2x^2+4x+2}{(x-1)(x+1)^2}-\frac{2x^2+2x}{(x-1)(x+1)^2}-\frac{x^2-x}{(x-1)(x+1)^2}$$
$$=\frac{-x^2+3x+2}{(x-1)(x+1)^2}$$

87. $$a+\frac{1}{a+3}=\frac{a}{1}+\frac{1}{a+3}=\frac{a(a+3)}{1(a+3)}+\frac{1}{a+3}=\frac{a^2+3a}{a+3}+\frac{1}{a+3}=\frac{a^2+3a+1}{a+3}$$

© 2015 Cengage Learning. All Rights Reserved. May not be scanned, copied or duplicated, or posted to a publicly accessible website, in whole or in part.

89. $$\frac{x+3}{2x^2-5x+2}-\frac{3x-1}{x^2-x-2}=\frac{x+3}{(2x-1)(x-2)}-\frac{3x-1}{(x-2)(x+1)}$$
$$=\frac{(x+3)(x+1)}{(2x-1)(x-2)(x+1)}-\frac{(3x-1)(2x-1)}{(2x-1)(x-2)(x+1)}$$
$$=\frac{x^2+4x+3}{(2x-1)(x-2)(x+1)}-\frac{6x^2-5x+1}{(2x-1)(x-2)(x+1)}$$
$$=\frac{-5x^2+9x+2}{(2x-1)(x-2)(x+1)}$$
$$=\frac{-(5x^2-9x-2)}{(2x-1)(x-2)(x+1)}$$
$$=\frac{-(5x+1)(x-2)}{(2x-1)(x-2)(x+1)}=-\frac{5x+1}{(2x-1)(x+1)}$$

91. $$\frac{x-y}{2}+\frac{x+y}{3}=\frac{(x-y)3}{2(3)}+\frac{(x+y)2}{3(2)}=\frac{3x-3y}{6}+\frac{2x+2y}{6}=\frac{5x-y}{6}$$

93. $$2-x+\frac{3}{x-9}=\frac{2-x}{1}+\frac{3}{x-9}=\frac{(2-x)(x-9)}{1(x-9)}+\frac{3}{x-9}$$
$$=\frac{-x^2+11x-18}{x-9}+\frac{3}{x-9}=\frac{-x^2+11x-15}{x-9}$$

95. $$\frac{23}{x-1}+4x-5x^2=\frac{23}{x-1}+\frac{4x-5x^2}{1}=\frac{23}{x-1}+\frac{(4x-5x^2)(x-1)}{1(x-1)}$$
$$=\frac{23}{x-1}+\frac{-5x^3+9x^2-4x}{x-1}=\frac{-5x^3+9x^2-4x+23}{x-1}$$

97. $$\frac{3x}{x-3}+\frac{4}{x-2}-\frac{5x}{x^3-5x^2+6x}=\frac{3x}{x-3}+\frac{4}{x-2}-\frac{5x}{x(x-3)(x-2)}$$
$$=\frac{3x(x)(x-2)}{x(x-3)(x-2)}+\frac{4x(x-3)}{x(x-3)(x-2)}-\frac{5x}{x(x-3)(x-2)}$$
$$=\frac{3x^3-6x^2}{x(x-3)(x-2)}+\frac{4x^2-12x}{x(x-3)(x-2)}-\frac{5x}{x(x-3)(x-2)}$$
$$=\frac{3x^3-2x^2-17x}{x(x-3)(x-2)}=\frac{x(3x^2-2x-17)}{x(x-3)(x-2)}=\frac{3x^2-2x-17}{(x-3)(x-2)}$$

99. $$2+\frac{4a}{a^2-1}-\frac{2}{a+1}=\frac{2}{1}+\frac{4a}{(a+1)(a-1)}-\frac{2}{(a+1)}$$
$$=\frac{2(a+1)(a-1)}{(a+1)(a-1)}+\frac{4a}{(a+1)(a-1)}-\frac{2(a-1)}{(a+1)(a-1)}$$
$$=\frac{2a^2-2}{(a+1)(a-1)}+\frac{4a}{(a+1)(a-1)}-\frac{2a-2}{(a+1)(a-1)}$$
$$=\frac{2a^2+2a}{(a+1)(a-1)}=\frac{2a(a+1)}{(a+1)(a-1)}=\frac{2a}{a-1}$$

© 2015 Cengage Learning. All Rights Reserved. May not be scanned, copied or duplicated, or posted to a publicly accessible website, in whole or in part.

SECTION 6.3

101. $$\frac{x+5}{2x^2-2}+\frac{x}{2x+2}-\frac{3}{x-1}=\frac{x+5}{2(x+1)(x-1)}+\frac{x}{2(x+1)}-\frac{3}{x-1}$$
$$=\frac{x+5}{2(x+1)(x-1)}+\frac{x(x-1)}{2(x+1)(x-1)}-\frac{3(2)(x+1)}{2(x+1)(x-1)}$$
$$=\frac{x+5}{2(x+1)(x-1)}+\frac{x^2-x}{2(x+1)(x-1)}-\frac{6x+6}{2(x+1)(x-1)}$$
$$=\frac{x^2-6x-1}{2(x+1)(x-1)}$$

103. $$\frac{a}{a-b}+\frac{b}{a+b}+\frac{a^2+b^2}{b^2-a^2}=\frac{-a}{b-a}+\frac{b}{b+a}+\frac{a^2+b^2}{(b+a)(b-a)}$$
$$=\frac{-a(b+a)}{(b+a)(b-a)}+\frac{b(b-a)}{(b+a)(b-a)}+\frac{a^2+b^2}{(b+a)(b-a)}$$
$$=\frac{-ab-a^2}{(b+a)(b-a)}+\frac{b^2-ab}{(b+a)(b-a)}+\frac{a^2+b^2}{(b+a)(b-a)}$$
$$=\frac{2b^2-2ab}{(b+a)(b-a)}=\frac{2b(b-a)}{(b+a)(b-a)}=\frac{2b}{b+a}$$

105. $$\frac{7n^2}{m-n}+\frac{3m}{n-m}-\frac{3m^2-n}{m^2-2mn+n^2}=\frac{7n^2}{m-n}-\frac{3m}{m-n}-\frac{3m^2-n}{(m-n)^2}$$
$$=\frac{7n^2(m-n)}{(m-n)^2}-\frac{3m(m-n)}{(m-n)^2}-\frac{3m^2-n}{(m-n)^2}$$
$$=\frac{7mn^2-7n^3}{(m-n)^2}-\frac{3m^2-3mn}{(m-n)^2}-\frac{3m^2-n}{(m-n)^2}$$
$$=\frac{7mn^2-7n^3-6m^2+3mn+n}{(m-n)^2}$$

107. $$\frac{m+1}{m^2+2m+1}+\frac{m-1}{m^2-2m+1}+\frac{2}{m^2-1}=\frac{m+1}{(m+1)^2}+\frac{m-1}{(m-1)^2}+\frac{2}{(m+1)(m-1)}$$
$$=\frac{1}{m+1}+\frac{1}{m-1}+\frac{2}{(m+1)(m-1)}$$
$$=\frac{1(m-1)}{(m+1)(m-1)}+\frac{1(m+1)}{(m+1)(m-1)}+\frac{2}{(m+1)(m-1)}$$
$$=\frac{2m+2}{(m+1)(m-1)}=\frac{2(m+1)}{(m+1)(m-1)}=\frac{2}{m-1}$$

109. $$\frac{a}{b}+\frac{c}{d}=\frac{ad}{bd}+\frac{cb}{db}=\frac{ad+bc}{bd}$$

111. $$h=\frac{9x-2}{2}+x+\frac{3x+6}{x}=\frac{(9x-2)x}{2x}+\frac{x(2x)}{1(2x)}+\frac{(3x+6)2}{x(2)}$$
$$=\frac{9x^2-2x}{2x}+\frac{2x^2}{2x}+\frac{6x+12}{12}=\frac{11x^2+4x+12}{2x}\text{ ft}$$

© 2015 Cengage Learning. All Rights Reserved. May not be scanned, copied or duplicated, or posted to a publicly accessible website, in whole or in part.

SECTION 6.3

113. $2w = \dfrac{7x+5}{2} - \dfrac{8x-6}{x} = \dfrac{(7x+5)x}{2x} - \dfrac{(8x-6)2}{2x} = \dfrac{7x^2+5x}{2x} - \dfrac{16x-12}{2x} = \dfrac{7x^2-11x+12}{2x}$

$w = \dfrac{1}{2}\cdot 2w = \dfrac{1}{2}\cdot\dfrac{7x^2-11x+12}{2x} = \dfrac{7x^2-11x+12}{4x}$ cm

115. Answers may vary.

117. In the second line, the **whole** numerator must be subtracted.
The second line should have $\dfrac{8x+2-3x-8}{5}$.

Exercises 6.4 (page 401)

1. $\dfrac{2}{5} \div \dfrac{3}{5} = \dfrac{2}{5}\cdot\dfrac{5}{3} = \dfrac{2}{3}$

3. $\dfrac{a+b}{a} \div \dfrac{a-b}{a} = \dfrac{a+b}{a}\cdot\dfrac{a}{a-b} = \dfrac{a+b}{a-b}$

5. $\dfrac{6x}{y} \div \dfrac{4x^2}{y^2} = \dfrac{6x}{y}\cdot\dfrac{y^2}{4x^2} = \dfrac{3y}{2x}$

7.
$$\begin{aligned} \frac{8(a-5)}{3} &= 2(a-4) \\ 8(a-5) &= 6(a-4) \\ 8a-40 &= 6a-24 \\ 2a &= 16 \\ a &= 8 \end{aligned}$$

9.
$$\begin{aligned} a^4 - 20a^2 + 64 &= 0 \\ (a^2-4)(a^2-16) &= 0 \\ (a+2)(a-2)(a+4)(a-4) &= 0 \end{aligned}$$

$a+2=0$ **or** $a-2=0$ **or** $a+4=0$ **or** $a-4=0$

$a=-2$ $\quad a=2$ $\quad a=-4$ $\quad a=4$

11. complex

13. $\dfrac{\frac{1}{3}}{\frac{4}{9}} = \dfrac{1}{3}\div\dfrac{4}{9} = \dfrac{1}{3}\cdot\dfrac{9}{4} = \dfrac{3}{4}$

15. $\dfrac{\frac{8y}{x^2}}{\frac{12yz}{x}} = \dfrac{8y}{x^2}\div\dfrac{12yz}{x} = \dfrac{8y}{x^2}\cdot\dfrac{x}{12yz} = \dfrac{2}{3xz}$

17.
$$\begin{aligned} \frac{\frac{x^2+5x+6}{3xy}}{\frac{x^2-9}{6xy}} &= \frac{x^2+5x+6}{3xy} : \frac{x^2-9}{6xy} = \frac{x^2+5x+6}{3xy}\cdot\frac{6xy}{x^2-9} \\ &= \frac{(x+3)(x+2)}{3xy}\cdot\frac{6xy}{(x+3)(x-3)} = \frac{2(x+2)}{x-3} \end{aligned}$$

19. $\dfrac{\frac{1}{2}-\frac{2}{3}}{\frac{2}{3}+\frac{1}{2}} = \dfrac{\frac{3}{6}-\frac{4}{6}}{\frac{4}{6}+\frac{3}{6}} = \dfrac{-\frac{1}{6}}{\frac{7}{6}} = -\dfrac{1}{6}\div\dfrac{7}{6} = -\dfrac{1}{6}\cdot\dfrac{6}{7} = -\dfrac{1}{7}$

© 2015 Cengage Learning. All Rights Reserved. May not be scanned, copied or duplicated, or posted to a publicly accessible website, in whole or in part.

SECTION 6.4

21. $\dfrac{\frac{3}{x}+1}{x+2} = \dfrac{\left(\frac{3}{x}+1\right)\cdot x}{(x+2)\cdot x} = \dfrac{\frac{3}{x}\cdot x + 1\cdot x}{x\cdot x + 2\cdot x} = \dfrac{3+x}{x^2+2x}$

23. $\dfrac{\frac{1}{x}-\frac{1}{y}}{xy} = \dfrac{\left(\frac{1}{x}-\frac{1}{y}\right)\cdot xy}{(xy)\cdot xy} = \dfrac{\frac{1}{x}\cdot xy - \frac{1}{y}\cdot xy}{x^2y^2} = \dfrac{y-x}{x^2y^2}$

25. $\dfrac{\frac{5}{a}+\frac{2}{b}}{ab} = \dfrac{\left(\frac{5}{a}+\frac{2}{b}\right)\cdot ab}{ab\cdot ab} = \dfrac{\frac{5}{a}\cdot ab + \frac{2}{b}\cdot ab}{a^2b^2} = \dfrac{5b+2a}{a^2b^2}$

27. $\dfrac{5xy}{1+\frac{1}{xy}} = \dfrac{(5xy)\cdot xy}{\left(1+\frac{1}{xy}\right)\cdot xy} = \dfrac{5x^2y^2}{xy+1}$

29. $\dfrac{\frac{p}{q}-\frac{q}{p}}{\frac{1}{q}-\frac{1}{p}} = \dfrac{\left(\frac{p}{q}-\frac{q}{p}\right)\cdot pq}{\left(\frac{1}{q}-\frac{1}{p}\right)\cdot pq} = \dfrac{\frac{p}{q}\cdot pq - \frac{q}{p}\cdot pq}{\frac{1}{q}\cdot pq - \frac{1}{p}\cdot pq} = \dfrac{p^2-q^2}{p-q} = \dfrac{(p+q)(p-q)}{p-q} = p+q$

31. $\dfrac{\frac{1}{a}-\frac{1}{b}}{\frac{a}{b}-\frac{b}{a}} = \dfrac{\left(\frac{1}{a}-\frac{1}{b}\right)\cdot ab}{\left(\frac{a}{b}-\frac{b}{a}\right)\cdot ab} = \dfrac{\frac{1}{a}\cdot ab - \frac{1}{b}\cdot ab}{\frac{a}{b}\cdot ab - \frac{b}{a}\cdot ab} = \dfrac{b-a}{a^2-b^2} = \dfrac{b-a}{(a+b)(a-b)} = -\dfrac{1}{a+b}$

33. $\dfrac{x^{-1}+y^{-1}}{x^{-1}-y^{-1}} = \dfrac{\frac{1}{x}+\frac{1}{y}}{\frac{1}{x}-\frac{1}{y}} = \dfrac{\left(\frac{1}{x}+\frac{1}{y}\right)(xy)}{\left(\frac{1}{x}-\frac{1}{y}\right)(xy)} = \dfrac{y+x}{y-x}$

35. $\dfrac{x-y^{-2}}{y-x^{-2}} = \dfrac{x-\frac{1}{y^2}}{y-\frac{1}{x^2}} = \dfrac{\left(x-\frac{1}{y^2}\right)(x^2y^2)}{\left(y-\frac{1}{x^2}\right)(x^2y^2)} = \dfrac{x^3y^2-x^2}{x^2y^3-y^2} = \dfrac{x^2(xy^2-1)}{y^2(x^2y-1)}$

37. $\dfrac{1+\frac{a}{b}}{1-\frac{a}{1-\frac{a}{b}}} = \dfrac{1+\frac{a}{b}}{1-\frac{a(b)}{\left(1-\frac{a}{b}\right)(b)}} = \dfrac{1+\frac{a}{b}}{1-\frac{ab}{b-a}} = \dfrac{\left(1+\frac{a}{b}\right)(b)(b-a)}{\left(1-\frac{ab}{b-a}\right)(b)(b-a)} = \dfrac{b(b-a)+a(b-a)}{b(b-a)-ab(b)}$

$= \dfrac{b^2-ab+ab-a^2}{b^2-ab-ab^2}$

$= \dfrac{b^2-a^2}{b(b-a-ab)}$

$= \dfrac{(b+a)(b-a)}{b(b-a-ab)}$

39. $\dfrac{x-\frac{1}{x}}{1+\frac{1}{\frac{1}{x}}} = \dfrac{x-\frac{1}{x}}{1+\frac{1(x)}{\frac{1}{x}(x)}} = \dfrac{x-\frac{1}{x}}{1+\frac{x}{1}} = \dfrac{x-\frac{1}{x}}{1+x} = \dfrac{\left(x-\frac{1}{x}\right)x}{(1+x)x} = \dfrac{x^2-1}{x(x+1)} = \dfrac{(x+1)(x-1)}{x(x+1)}$

$= \dfrac{x-1}{x}$

© 2015 Cengage Learning. All Rights Reserved. May not be scanned, copied or duplicated, or posted to a publicly accessible website, in whole or in part.

SECTION 6.4

41. $$\frac{x+1-\frac{6}{x}}{\frac{1}{x}} = \frac{\left(x+1-\frac{6}{x}\right)\cdot x}{\frac{1}{x}\cdot x} = \frac{x^2+x-6}{1} = x^2+x-6$$

43. $$\frac{1+\frac{x}{y}}{1-\frac{x}{y}} = \frac{\left(1+\frac{x}{y}\right)\cdot y}{\left(1-\frac{x}{y}\right)\cdot y} = \frac{1\cdot y+\frac{x}{y}\cdot y}{1\cdot y-\frac{x}{y}\cdot y} = \frac{y+x}{y-x}$$

45. $$\frac{1+\frac{6}{x}+\frac{8}{x^2}}{1+\frac{1}{x}-\frac{12}{x^2}} = \frac{\left(1+\frac{6}{x}+\frac{8}{x^2}\right)\cdot x^2}{\left(1+\frac{1}{x}-\frac{12}{x^2}\right)\cdot x^2} = \frac{x^2+6x+8}{x^2+x-12} = \frac{(x+4)(x+2)}{(x+4)(x-3)} = \frac{x+2}{x-3}$$

47. $$\frac{\frac{1}{a+1}+1}{\frac{3}{a-1}+1} = \frac{\left(\frac{1}{a+1}+1\right)(a+1)(a-1)}{\left(\frac{3}{a-1}+1\right)(a+1)(a-1)} = \frac{1(a-1)+1(a+1)(a-1)}{3(a+1)+1(a+1)(a-1)} = \frac{a-1+a^2-1}{3a+3+a^2-1}$$
$$= \frac{a^2+a-2}{a^2+3a+2} = \frac{(a+2)(a-1)}{(a+2)(a+1)} = \frac{a-1}{a+1}$$

49. $$\frac{x^{-1}+y^{-1}}{x} = \frac{\frac{1}{x}+\frac{1}{y}}{x} = \frac{\left(\frac{1}{x}+\frac{1}{y}\right)(xy)}{x(xy)} = \frac{y+x}{x^2y}$$

51. $$\frac{y}{x^{-1}-y^{-1}} = \frac{y}{\frac{1}{x}-\frac{1}{y}} = \frac{y(xy)}{\left(\frac{1}{x}-\frac{1}{y}\right)(xy)} = \frac{xy^2}{y-x}$$

53. $$\frac{b}{b+\frac{2}{2+\frac{1}{2}}} = \frac{b}{b+\frac{2(2)}{\left(2+\frac{1}{2}\right)(2)}} = \frac{b}{b+\frac{4}{4+1}} = \frac{b}{b+\frac{4}{5}} = \frac{b(5)}{\left(b+\frac{4}{5}\right)(5)} = \frac{5b}{5b+4}$$

55. $$a+\frac{a}{1+\frac{a}{a+1}} = a+\frac{a(a+1)}{\left(1+\frac{a}{a+1}\right)(a+1)} = a+\frac{a(a+1)}{a+1+a} = a+\frac{a^2+a}{2a+1} = \frac{a(2a+1)}{2a+1}+\frac{a^2+a}{2a+1}$$
$$= \frac{2a^2+a+a^2+a}{2a+1} = \frac{3a^2+2a}{2a+1}$$

57. $$\frac{-\frac{2}{3}}{\frac{6}{9}} = -\frac{2}{3}\div\frac{6}{9} = -\frac{2}{3}\cdot\frac{9}{6} = -1$$

59. $$\frac{\frac{1}{2}+\frac{1}{3}}{\frac{1}{4}} = \frac{\frac{3}{6}+\frac{2}{6}}{\frac{1}{4}} = \frac{\frac{5}{6}}{\frac{1}{4}} = \frac{5}{6}\div\frac{1}{4} = \frac{5}{6}\cdot\frac{4}{1} = \frac{10}{3}$$

© 2015 Cengage Learning. All Rights Reserved. May not be scanned, copied or duplicated, or posted to a publicly accessible website, in whole or in part.

SECTION 6.4

61. $\dfrac{x+y}{x^{-1}+y^{-1}} = \dfrac{x+y}{\frac{1}{x}+\frac{1}{y}} = \dfrac{(x+y)(xy)}{\left(\frac{1}{x}+\frac{1}{y}\right)(xy)} = \dfrac{(x+y)xy}{y+x} = xy$

63. $\dfrac{x-\frac{1}{1-\frac{x}{2}}}{\frac{3}{x+\frac{2}{3}}-x} = \dfrac{x-\frac{1(2)}{(1-\frac{x}{2})2}}{\frac{3(3)}{\left(x+\frac{2}{3}\right)3}-x} = \dfrac{x-\frac{2}{2-x}}{\frac{9}{3x+2}-x} = \dfrac{\left(x-\frac{2}{2-x}\right)(2-x)(3x+2)}{\left(\frac{9}{3x+2}-x\right)(2-x)(3x+2)}$

$$= \frac{x(2-x)(3x+2)-2(3x+2)}{9(2-x)-x(2-x)(3x+2)}$$

$$= \frac{(2x-x^2)(3x+2)-2(3x+2)}{9(2-x)-(3x^2+2x)(2-x)}$$

$$= \frac{(3x+2)(-x^2+2x-2)}{(2-x)(-3x^2-2x+9)}$$

65. $\dfrac{2x+\frac{1}{2-\frac{x}{2}}}{\frac{4}{\frac{x}{2}-2}-x} = \dfrac{2x+\frac{1(2)}{(2-\frac{x}{2})2}}{\frac{4(2)}{(\frac{x}{2}-2)2}-x} = \dfrac{2x+\frac{2}{4-x}}{\frac{8}{x-4}-x} = \dfrac{\left(2x+\frac{2}{4-x}\right)(x-4)}{\left(\frac{8}{x-4}-x\right)(x-4)} = \dfrac{2x^2-8x-2}{8-x^2+4x}$

$$= \frac{2(x^2-4x-1)}{-x^2+4x+8}$$

67. $\dfrac{\frac{1}{x^2+3x+2}+\frac{1}{x^2+x-2}}{\frac{3x}{x^2-1}-\frac{x}{x+2}} = \dfrac{\frac{1}{(x+2)(x+1)}+\frac{1}{(x+2)(x-1)}}{\frac{3x}{(x+1)(x-1)}-\frac{x}{x+2}}$

$$= \frac{\left(\frac{1}{(x+2)(x+1)}+\frac{1}{(x+2)(x-1)}\right)(x+2)(x+1)(x-1)}{\left(\frac{3x}{(x+1)(x-1)}-\frac{x}{x+2}\right)(x+2)(x+1)(x-1)}$$

$$= \frac{x-1+x+1}{3x(x+2)-x(x+1)(x-1)}$$

$$= \frac{2x}{3x^2+6x-x^3+x}$$

$$= \frac{2x}{-x^3+3x^2+7x} = \frac{2x}{-x(x^2-3x-7)} = \frac{-2}{x^2-3x-7}$$

69. $\dfrac{1}{\frac{1}{k_1}+\frac{1}{k_2}} = \dfrac{1k_1k_2}{\left(\frac{1}{k_1}+\frac{1}{k_2}\right)k_1k_2} = \dfrac{k_1k_2}{k_2+k_1}$

71. average $= \dfrac{\frac{k}{2}+\frac{k}{3}+\frac{k}{2}}{3} = \dfrac{\frac{3k}{6}+\frac{2k}{6}+\frac{3k}{6}}{3} = \dfrac{\frac{8k}{6}}{\frac{3}{1}} = \dfrac{8k}{6}\div\dfrac{3}{1} = \dfrac{8k}{6}\cdot\dfrac{1}{3} = \dfrac{8k}{18} = \dfrac{4k}{9}$

73. **Answers may vary.**

© 2015 Cengage Learning. All Rights Reserved. May not be scanned, copied or duplicated, or posted to a publicly accessible website, in whole or in part.

75. $(x^{-1}y^{-1})(x^{-1}+y^{-1})^{-1} = \frac{1}{x}\cdot\frac{1}{y}\cdot\frac{1}{\frac{1}{x}+\frac{1}{y}} = \frac{1}{xy}\cdot\frac{1(xy)}{\left(\frac{1}{x}+\frac{1}{y}\right)xy} = \frac{1}{xy}\cdot\frac{xy}{y+x} = \frac{1}{y+x}$

Exercises 6.5 (page 411)

1. $4 = 2^2$; $x+3 = x+3$; $12 = 2^2\cdot 3 \Rightarrow \text{LCD} = 2^2\cdot 3(x+3) = 12(x+3)$

3. $a = a$; $a-5 = a-5 \Rightarrow \text{LCD} = a(a-5)$

5. $x^2-9 = (x+3)(x-3)$; $x-3 = x-3 \Rightarrow \text{LCD} = (x+3)(x-3)$

7. $(m^4n^{-2})^{-3} = m^{-12}n^6 = \frac{n^6}{m^{12}}$

9. $\frac{a^0+2a^0-3a^0}{(a-b)^0} = \frac{1+2-3}{1} = 0$

11. rational

13.
$$\begin{aligned}\frac{1}{5}+\frac{8}{x} &= 1\\ \left(\frac{1}{5}+\frac{8}{x}\right)5x &= 1(5x)\\ x+40 &= 5x\\ -4x &= -40\\ x &= 10\end{aligned}$$
The answer checks.

15.
$$\begin{aligned}\frac{34}{x}-\frac{3}{2} &= -\frac{13}{20}\\ \left(\frac{34}{x}-\frac{3}{2}\right)20x &= -\left(\frac{13}{20}\right)20x\\ 680-30x &= -13x\\ -17x &= -680\\ x &= 40\end{aligned}$$
The answer checks.

17.
$$\begin{aligned}\frac{3}{4y}+\frac{4}{y} &= 9\\ \left(\frac{3}{4y}+\frac{4}{y}\right)4y &= (9)4y\\ 3+16 &= 36y\\ -36y &= -19\\ y &= \frac{-19}{-36} = \frac{19}{36}\end{aligned}$$
The answer checks.

19.
$$\begin{aligned}\frac{7}{5x}-\frac{1}{2} &= \frac{5}{6x}+\frac{1}{3}\\ \left(\frac{7}{5x}-\frac{1}{2}\right)30x &= \left(\frac{5}{6x}+\frac{1}{3}\right)30x\\ 42-15x &= 25+10x\\ -25x &= -17\\ x &= \frac{-17}{-25} = \frac{17}{25}\end{aligned}$$
The answer checks.

21.
$$\begin{aligned}\frac{5}{x+4}+\frac{1}{x+4} &= x-1\\ \left(\frac{5}{x+4}+\frac{1}{x+4}\right)(x+4) &= (x-1)(x+4)\\ 5+1 &= x^2+3x-4\\ 0 &= x^2+3x-10\\ 0 &= (x+5)(x-2)\end{aligned}$$
$x+5=0$ **or** $x-2=0$

$x=-5$ $\quad x=2$ $\quad$ Both answers check.

© 2015 Cengage Learning. All Rights Reserved. May not be scanned, copied or duplicated, or posted to a publicly accessible website, in whole or in part.

23.
$$\begin{aligned}\frac{3}{x+1}-\frac{x-2}{2}&=\frac{x-2}{x+1}\\ \left(\frac{3}{x+1}-\frac{x-2}{2}\right)(2)(x+1)&=\left(\frac{x-2}{x+1}\right)(2)(x+1)\\ 6-(x-2)(x+1)&=2x-4\\ 6-(x^2-x-2)&=2x-4\\ 6-x^2+x+2&=2x-4\\ 0&=x^2+x-12\\ 0&=(x+4)(x-3)\end{aligned}$$

$x+4=0$ **or** $x-3=0$

$x=-4$ $\quad x=3$ Both answers check.

25.
$$\begin{aligned}\frac{x+1}{x}-\frac{x-1}{x}&=0\\ \left(\frac{x+1}{x}-\frac{x-1}{x}\right)x&=(0)x\\ x+1-(x-1)&=0\\ x+1-x+1&=0\\ 2&\neq 0\end{aligned}$$

No solution: $\emptyset$

27.
$$\begin{aligned}\frac{2x+1}{x-3}-\frac{x+4}{x-3}&=0\\ \left(\frac{2x+1}{x-3}-\frac{x+4}{x-3}\right)(x-3)&=0(x-3)\\ 2x+1-(x+4)&=0\\ 2x+1-x-4&=0\\ x-3&=0\\ x&=3\end{aligned}$$

The answer does not check.

No solution: $\emptyset$; 3 is extraneous.

29.
$$\begin{aligned}\frac{x}{x+2}&=1-\frac{3x+2}{x^2+4x+4}\\ \frac{x}{x+2}&=1-\frac{3x+2}{(x+2)(x+2)}\\ \left(\frac{x}{x+2}\right)(x+2)(x+2)&=\left(1-\frac{3x+2}{(x+2)(x+2)}\right)(x+2)(x+2)\\ x(x+2)&=(x+2)(x+2)-(3x+2)\\ x^2+2x&=x^2+4x+4-3x-2\\ x^2+2x&=x^2+x+2\\ x&=2\end{aligned}$$

The answer checks.

31.
$$\begin{aligned}\frac{5}{x+3}+\frac{1}{x-1}&=\frac{1}{x^2+2x-3}\\ \left(\frac{5}{x+3}+\frac{1}{x-1}\right)(x-1)(x+3)&=\left(\frac{1}{(x-1)(x+3)}\right)(x-1)(x+3)\\ 5(x-1)+1(x+3)&=1\\ 5x-5+x+3&=1\\ 6x&=3\\ x&=\tfrac{1}{2}\end{aligned}$$

The answer checks.

© 2015 Cengage Learning. All Rights Reserved. May not be scanned, copied or duplicated, or posted to a publicly accessible website, in whole or in part.

SECTION 6.5

33.
$$\begin{aligned}\frac{1}{R} &= \frac{1}{r_1} + \frac{1}{r_2}\\ \left(\frac{1}{R}\right)Rr_1r_2 &= \left(\frac{1}{r_1} + \frac{1}{r_2}\right)Rr_1r_2\\ r_1r_2 &= Rr_2 + Rr_1\\ r_1r_2 - Rr_1 &= Rr_2\\ r_1(r_2 - R) &= Rr_2\\ r_1 &= \frac{Rr_2}{r_2 - R}\end{aligned}$$

35.
$$\begin{aligned}\frac{1}{p} + \frac{1}{q} &= \frac{1}{f}\\ \left(\frac{1}{p} + \frac{1}{q}\right)pqf &= \left(\frac{1}{f}\right)pqf\\ qf + pf &= pq\\ f(q + p) &= pq\\ f &= \frac{pq}{q + p}\end{aligned}$$

37.
$$\begin{aligned}\frac{3 - 5y}{2 + y} &= \frac{3 + 5y}{2 - y}\\ \left(\frac{3 - 5y}{2 + y}\right)(2 + y)(2 - y) &= \left(\frac{3 + 5y}{2 - y}\right)(2 + y)(2 - y)\\ (3 - 5y)(2 - y) &= (3 + 5y)(2 + y)\\ 5y^2 - 13y + 6 &= 5y^2 + 13y + 6\\ -26y &= 0\\ y &= 0\end{aligned}$$

The answer checks.

39.
$$\begin{aligned}\frac{30}{y - 2} + \frac{24}{y - 5} &= 13\\ \left(\frac{30}{y - 2} + \frac{24}{y - 5}\right)(y - 2)(y - 5) &= 13(y - 2)(y - 5)\\ 30(y - 5) + 24(y - 2) &= 13(y^2 - 7y + 10)\\ 30y - 150 + 24y - 48 &= 13y^2 - 91y + 130\\ 0 &= 13y^2 - 145y + 328\\ 0 &= (13y - 41)(y - 8)\end{aligned}$$

$13y - 41 = 0$ **or** $y - 8 = 0$

$13y = 41$ $\quad$ $y = 8$

$y = \frac{41}{13}$

Both answers check.

41.
$$\begin{aligned}\frac{x - 4}{x - 3} + \frac{x - 2}{x - 3} &= x - 3\\ \left(\frac{x - 4}{x - 3} + \frac{x - 2}{x - 3}\right)(x - 3) &= (x - 3)(x - 3)\\ x - 4 + x - 2 &= x^2 - 6x + 9\\ 2x - 6 &= x^2 - 6x + 9\\ 0 &= x^2 - 8x + 15\\ 0 &= (x - 5)(x - 3)\end{aligned}$$

$x - 5 = 0$ **or** $x - 3 = 0$

$x = 5$ $\quad$ $x = 3$ $\quad$ $x = 5$ checks, but $x = 3$ does not and is extraneous.

© 2015 Cengage Learning. All Rights Reserved. May not be scanned, copied or duplicated, or posted to a publicly accessible website, in whole or in part.

SECTION 6.5

43.
$$\frac{4-3a}{5+a}=\frac{4-3a}{5-a}$$
$$\left(\frac{4-3a}{5+a}\right)(5+a)(5-a)=\left(\frac{4-3a}{5-a}\right)(5+a)(5-a)$$
$$(4-3a)(5-a)=(4-3a)(5+a)$$
$$20-19a+3a^2=20-11a-3a^2$$
$$6a^2-8a=0$$
$$2a(3a-4)=0$$
$2a=0$ **or** $3a-4=0$

$a=0$ $\quad a=\frac{4}{3}$ Both answers check.

45.
$$\frac{y-3}{y+2}=3-\frac{1-2y}{y+2}$$
$$\frac{y-3}{y+2}\cdot(y+2)=\left(3-\frac{1-2y}{y+2}\right)(y+2)$$
$$y-3=3(y+2)-(1-2y)$$
$$y-3=3y+6-1+2y$$
$$y-3=5y+5$$
$$-4y=8$$
$y=-2$; The answer does not check. -2 is extraneous. No solution: $\emptyset$

47.
$$\frac{x+2}{x+3}-1=\frac{1}{3-2x-x^2}$$
$$\frac{x+2}{x+3}-1=\frac{-1}{x^2+2x-3}$$
$$\frac{x+2}{x+3}-1=\frac{-1}{(x+3)(x-1)}$$
$$\left(\frac{x+2}{x+3}-1\right)(x+3)(x-1)=\left(\frac{-1}{(x+3)(x-1)}\right)(x+3)(x-1)$$
$$(x+2)(x-1)-(x+3)(x-1)=-1$$
$$x^2+x-2-(x^2+2x-3)=-1$$
$$x^2+x-2-x^2-2x+3=-1$$
$$-x=-2$$
$x=2$ The answer checks.

© 2015 Cengage Learning. All Rights Reserved. May not be scanned, copied or duplicated, or posted to a publicly accessible website, in whole or in part.

49.

$$\frac{y^2}{y+2}+4=\frac{y+6}{y+2}+3$$
$$\left(\frac{y^2}{y+2}+4\right)(y+2)=\left(\frac{y+6}{y+2}+3\right)(y+2)$$
$$y^2+4(y+2)=y+6+3(y+2)$$
$$y^2+4y+8=y+6+3y+6$$
$$y^2+4y+8=4y+12$$
$$y^2-4=0$$
$$(y+2)(y-2)=0$$
$$y+2=0 \quad \textbf{or} \quad y-2=0$$
$$y=-2 \qquad y=2$$

$y=2$ checks, but $y=-2$ does not and is extraneous.

51.

$$\frac{5}{x+4}-\frac{1}{3}=\frac{x-1}{x}$$
$$\left(\frac{5}{x+4}-\frac{1}{3}\right)(3x)(x+4)=\left(\frac{x-1}{x}\right)(3x)(x+4)$$
$$15x-x(x+4)=3(x-1)(x+4)$$
$$15x-x^2-4x=3(x^2+3x-4)$$
$$-x^2+11x=3x^2+9x-12$$
$$0=4x^2-2x-12$$
$$0=2(2x^2-x-6)$$
$$0=2(2x+3)(x-2)$$
$$2x+3=0 \quad \textbf{or} \quad x-2=0$$
$$2x=-3 \qquad x=2$$
$$x=-\tfrac{3}{2}$$

Both answers check.

53.

$$\frac{5}{x+6}-\frac{3}{x-4}=\frac{2}{x+3}$$
$$\left(\frac{5}{x+6}-\frac{3}{x-4}\right)(x+6)(x-4)(x+3)=\left(\frac{2}{x+3}\right)(x+6)(x-4)(x+3)$$
$$5(x-4)(x+3)-3(x+6)(x+3)=2(x+6)(x-4)$$
$$5(x^2-x-12)-3(x^2+9x+18)=2(x^2+2x-24)$$
$$5x^2-5x-60-3x^2-27x-54=2x^2+4x-48$$
$$2x^2-32x-114=2x^2+4x-48$$
$$-36x=66$$
$$x=\frac{66}{-36}=-\frac{11}{6}$$

The answer checks.

© 2015 Cengage Learning. All Rights Reserved. May not be scanned, copied or duplicated, or posted to a publicly accessible website, in whole or in part.

55.

$$S = \frac{a - lr}{1 - r}$$
$$S(1 - r) = \left(\frac{a - lr}{1 - r}\right)(1 - r)$$
$$S - Sr = a - lr$$
$$S - a = Sr - lr$$
$$S - a = r(S - l)$$
$$\frac{S - a}{S - l} = r, \text{ or } r = \frac{S - a}{S - l}$$

57.

$$\frac{1}{R} = \frac{1}{r_1} + \frac{1}{r_2} + \frac{1}{r_3}$$
$$\left(\frac{1}{R}\right)Rr_1r_2r_3 = \left(\frac{1}{r_1} + \frac{1}{r_2} + \frac{1}{r_3}\right)Rr_1r_2r_3$$
$$r_1r_2r_3 = Rr_2r_3 + Rr_1r_3 + Rr_1r_2$$
$$r_1r_2r_3 = R(r_2r_3 + r_1r_3 + r_1r_2)$$
$$\frac{r_1r_2r_3}{r_2r_3 + r_1r_3 + r_1r_2} = R, \text{ or } R = \frac{r_1r_2r_3}{r_2r_3 + r_1r_3 + r_1r_2}$$

59. The first painter paints $\frac{1}{5}$ of a house in 1 day, while the second paints $\frac{1}{3}$ of a house in 1 day. Let $x =$ the number of days it takes them to paint the house together. Then they can paint $\frac{1}{x}$ of a house together in 1 day.

Amount 1st paints in 1 day	+	Amount 2nd paints in 1 day	=	Amount painted by both in 1 day

$$\frac{1}{5} + \frac{1}{3} = \frac{1}{x}$$
$$\left(\frac{1}{5} + \frac{1}{3}\right)15x = \frac{1}{x} \cdot 15x$$
$$3x + 5x = 15$$
$$8x = 15$$
$$x = \frac{15}{8} = 1\frac{7}{8}$$

They can paint the house together in $1\frac{7}{8}$ days.

61. The first belt can move $\frac{1}{10}$ of the 1000 bushels in 1 minute, while the second can move $\frac{1}{14}$ of the 1000 bushels in 1 minute. Let $x =$ the number of minutes for them to move the 1000 bushels together. Then $\frac{1}{x} =$ the amount of the 1000 bushels they can move together in 1 minute.

Amount 1st moves in 1 minute	+	Amount 2nd moves in 1 minute	=	Amount moved together in 1 minute

$$\frac{1}{10} + \frac{1}{14} = \frac{1}{x}$$
$$\left(\frac{1}{10} + \frac{1}{14}\right)(70x) = \frac{1}{x} \cdot 70x$$
$$7x + 5x = 70$$

continued on next page...

© 2015 Cengage Learning. All Rights Reserved. May not be scanned, copied or duplicated, or posted to a publicly accessible website, in whole or in part.

61. continued

$$7x + 5x = 70$$
$$12x = 70$$
$$x = \frac{70}{12} = \frac{35}{6} = 5\frac{5}{6}$$

It will take them $5\frac{5}{6}$ minutes to move 1000 bushels working together.

63. The first drain can drain $\frac{1}{3}$ of the pool in 1 day, while the second can drain $\frac{1}{2}$ of the pool in 1 day. Let $x =$ the number of days needed for them to drain the pool together. Then $\frac{1}{x} =$ the amount of the pool they can drain together in 1 day.

[Amount 1st drains in 1 day] + [Amount 2nd drains in 1 day] = [Amount drained together in 1 day]

$$\frac{1}{3} + \frac{1}{2} = \frac{1}{x}$$
$$\left(\frac{1}{3} + \frac{1}{2}\right)6x = \frac{1}{x} \cdot 6x$$
$$2x + 3x = 6$$
$$5x = 6$$
$$x = \frac{6}{5} = 1\frac{1}{5}$$

It will take the two drains $1\frac{1}{5}$ days to drain the pool.

65. The first pipe fills $\frac{1}{3}$ of the pond in 1 week, while the second fills $\frac{1}{5}$ of the pond in a week and evaporation empties $\frac{1}{10}$ of the pond in a week. Let $x =$ the number of weeks needed to fill the pond with both pipes, considering evaporation. Then $\frac{1}{x}$ of the pond is filled in 1 week.

[Amount 1st pipe fills in 1 week] + [Amount 2nd pipe fills in 1 week] − [Amount emptied in 1 week] = [Total amount filled in 1 week]

$$\frac{1}{3} + \frac{1}{5} - \frac{1}{10} = \frac{1}{x}$$
$$\left(\frac{1}{3} + \frac{1}{5} - \frac{1}{10}\right)30x = \frac{1}{x} \cdot 30x$$
$$10x + 6x - 3x = 30$$
$$13x = 30$$
$$x = \frac{30}{13} = 2\frac{4}{13}$$

It will take $2\frac{4}{13}$ weeks to fill the pond.

© 2015 Cengage Learning. All Rights Reserved. May not be scanned, copied or duplicated, or posted to a publicly accessible website, in whole or in part.

67. Let w = the rate at which he walks. Then $w+5$ = the rate at which he bicycles.

	Rate	Time	Dist.
Walk	w	$\frac{24}{w}$	24
Bicycle	$w+5$	$\frac{24}{w+5}$	24

Time walking + Time bicycling = Total time

$$\frac{24}{w}+\frac{24}{w+5}=11$$
$$\left(\frac{24}{w}+\frac{24}{w+5}\right)(w)(w+5)=11w(w+5)$$
$$24(w+5)+24w=11w^2+55w$$
$$0=11w^2+7w-120$$
$$0=(11w+40)(w-3)$$
$$11w+40=0 \quad \textbf{or} \quad w-3=0$$
$$w=-\tfrac{40}{11} \qquad w=3$$

$w=3$ is the only answer that makes sense.
He walks at a rate of 3 miles per hour.

69. Let r = the first rate. Then $r-20$ = the second rate.

	Rate	Time	Dist.
First	r	$\frac{120}{r}$	120
Second	$r-20$	$\frac{120}{r-20}$	120

First time + Second time = 5

$$\frac{120}{r}+\frac{120}{r-20}=5$$
$$\left(\frac{120}{r}+\frac{120}{r-20}\right)(r)(r-20)=5r(r-20)$$
$$120(r-20)+120r=5r^2-100r$$
$$0=5r^2-340r+2400$$
$$0=5(r-60)(r-8)$$
$$r-60=0 \quad \textbf{or} \quad r-8=0$$
$$r=60 \qquad r=8$$

$r=60$ is the only answer that makes sense.
The train traveled at rates of 60 and 40 miles per hour.

© 2015 Cengage Learning. All Rights Reserved. May not be scanned, copied or duplicated, or posted to a publicly accessible website, in whole or in part.

SECTION 6.5

71. Let $c =$ the speed of the current.

	Rate	Time	Dist.
Downstream	$12+c$	$\frac{45}{12+c}$	45
Upstream	$12-c$	$\frac{27}{12-c}$	27

Time downstream = Time upstream

$$\frac{45}{12+c} = \frac{27}{12-c}$$
$$\frac{45}{12+c}(12+c)(12-c) = \frac{27}{12-c}(12+c)(12-c)$$
$$45(12-c) = 27(12+c)$$
$$540 - 45c = 324 + 27c$$
$$-72c = -216$$
$$c = \frac{-216}{-72} = 3$$

The speed of the current is 3 miles per hour.

73. Let $w =$ the speed of the wind.

	Rate	Time	Dist.
Downwind	$340+w$	$\frac{200}{340+w}$	200
Upwind	$340-w$	$\frac{140}{340-w}$	140

Time downwind = Time upwind

$$\frac{200}{340+w} = \frac{140}{340-w}$$
$$\left(\frac{200}{340+w}\right)(340+w)(340-w) = \left(\frac{140}{340-w}\right)(340+w)(340-w)$$
$$200(340-w) = 140(340+w)$$
$$68000 - 200w = 47600 + 140w$$
$$-340w = -20400$$
$$w = 60$$

The speed of the wind is 60 miles per hour.

75. Let $x =$ the number purchased.
Then the unit cost $= \frac{224}{x}$.

New unit cost $\cdot$ New # motors $= 224$

$$\left(\frac{224}{x} - 4\right)(x+1) = 224$$
$$\left(\frac{224}{x} - 4\right)(x+1)x = 224x$$
$$224(x+1) - 4x(x+1) = 224x$$
$$224x + 224 - 4x^2 - 4x = 224x$$
$$-4x^2 - 4x + 224 = 0$$
$$-4(x+8)(x-7) = 0$$
$$x + 8 = 0 \quad \textbf{or} \quad x - 7 = 0$$
$$x = -8 \qquad\qquad x = 7$$

Does not make sense.
He originally bought 7 motors.

© 2015 Cengage Learning. All Rights Reserved. May not be scanned, copied or duplicated, or posted to a publicly accessible website, in whole or in part.

77. Let x = the number of days.

Then the daily cost = $\frac{1200}{x}$.

$$\boxed{\text{New daily cost}} \cdot \boxed{\text{New \# days}} = 1200$$

$$\left(\frac{1200}{x} - 20\right)(x+3) = 1200$$

$$\left(\frac{1200}{x} - 20\right)(x+3)x = 1200x$$

$$1200(x+3) - 20x(x+3) = 1200x$$

$$1200x + 3600 - 20x^2 - 60x = 1200x$$

$$-20x^2 - 60x + 3600 = 0$$

$$-20(x+15)(x-12) = 0$$

$$x + 15 = 0 \quad \textbf{or} \quad x - 12 = 0$$

$$x = -15 \qquad x = 12$$

Does not make sense.

Her original vacation was 12 days.

79.

$$\frac{1}{f} = 0.6\left(\frac{1}{r_1} + \frac{1}{r_2}\right)$$

$$\left(\frac{1}{f}\right) fr_1r_2 = 0.6\left(\frac{1}{r_1} + \frac{1}{r_2}\right) fr_1r_2$$

$$r_1r_2 = 0.6fr_2 + 0.6fr_1$$

$$r_1r_2 = f(0.6r_2 + 0.6r_1)$$

$$\frac{r_1r_2}{0.6r_2 + 0.6r_1} = f, \text{ or } f = \frac{r_1r_2}{0.6(r_2 + r_1)}$$

$$f = \frac{r_1r_2}{0.6(r_2+r_1)} = \frac{(8\text{ cm})(8\text{ cm})}{0.6(8\text{ cm} + 8\text{ cm})} = \frac{64\text{ cm}^2}{0.6(16\text{ cm})} = \frac{64\text{ cm}^2}{9.6\text{ cm}} = \frac{640\text{ cm}^2}{96\text{ cm}} = \frac{20}{3}\text{ cm}$$

81. **Answers may vary.**

83. $\frac{x}{x-3} + \frac{3}{x-3} = \frac{6}{x-3}$

Exercises 6.6 (page 419)

1. $x - 5 + x^2 = x^2 + x - 5$

3. $5x^2 - x + 2x^3 - 6 = 2x^3 + 5x^2 - x - 6$

5. $2(x^2 + 4x - 1) + 3(2x^2 - 2x + 2) = 2x^2 + 8x - 2 + 6x^2 - 6x + 6 = 8x^2 + 2x + 4$

7. $-2(3y^3 - 2y + 7) - 3(y^2 + 2y - 4) + 4(y^3 + 2y - 1)$

$= -6y^3 + 4y - 14 - 3y^2 - 6y + 12 + 4y^3 + 8y - 4 = -2y^3 - 3y^2 + 6y - 6$

9. $\frac{1}{b}$

11. quotient

13. $\frac{2x^4y^5}{12x^2y^8} = \frac{x^2}{6y^3}$

15. $\frac{33a^{-2}b^2}{44a^2b^{-2}} = \frac{3b^4}{4a^4}$

17. $\frac{45x^{-2}y^{-3}t^0}{-63x^{-1}y^4t^2} = -\frac{5}{7xy^7t^2}$

19. $\frac{-65a^{2n}b^nc^{3n}}{-15a^nb^{-n}c} = \frac{13a^nb^{2n}c^{3n-1}}{3}$

21. $\frac{4x^2 - x^3}{6x} = \frac{4x^2}{6x} - \frac{x^3}{6x} = \frac{2x}{3} - \frac{x^2}{6}$

23.

$$\frac{6x^5y^2 + 16x^3y}{4x^2y^2} = \frac{6x^5y^2}{4x^2y^2} + \frac{16x^3y}{4x^2y^2}$$

$$= \frac{3x^3}{2} + \frac{4x}{y}$$

© 2015 Cengage Learning. All Rights Reserved. May not be scanned, copied or duplicated, or posted to a publicly accessible website, in whole or in part.

SECTION 6.6

25. $$\frac{28x^2y^3+4xy-20x^3y^2}{4xy}=\frac{28x^2y^3}{4xy}+\frac{4xy}{4xy}-\frac{20x^3y^2}{4xy}=7xy^2+1-5x^2y$$

27. $$\frac{24x^6y^7-12x^5y^{12}+36xy}{48x^2y^3}=\frac{24x^6y^7}{48x^2y^3}-\frac{12x^5y^{12}}{48x^2y^3}+\frac{36xy}{48x^2y^3}=\frac{x^4y^4}{2}-\frac{x^3y^9}{4}+\frac{3}{4xy^2}$$

29. $$\begin{array}{r} x+2 \\ x+5\,\overline{)\,x^2+7x+10} \\ \underline{x^2+5x} \\ 2x+10 \\ \underline{2x+10} \\ 0 \end{array}$$

31. $$\begin{array}{r} x-6 \\ x-2\,\overline{)\,x^2-8x+12} \\ \underline{x^2-2x} \\ -6x+12 \\ \underline{-6x+12} \\ 0 \end{array}$$

33. $$\begin{array}{r} 3x+4 \\ 2x-3\,\overline{)\,6x^2-x-12} \\ \underline{6x^2-9x} \\ 8x-12 \\ \underline{8x-12} \\ 0 \end{array}$$

35. $$\begin{array}{r} 5x^2+2x-3 \\ 3x-2\,\overline{)\,15x^3-4x^2-13x+6} \\ \underline{15x^3-10x^2} \\ 6x^2-13x \\ \underline{6x^2-4x} \\ -9x+6 \\ \underline{-9x+6} \\ 0 \end{array}$$

37. $$\begin{array}{r} 4a^2-3a+0+\frac{7}{a+1} \\ a+1\,\overline{)\,4a^3+a^2-3a+7} \\ \underline{4a^3+4a^2} \\ -3a^2-3a \\ \underline{-3a^2-3a} \\ 0a+7 \\ \underline{0a+0} \\ 7 \end{array}$$

39. $$\begin{array}{r} 3x^2+x-2+\frac{16}{2x+3} \\ 2x+3\,\overline{)\,6x^3+11x^2-x+10} \\ \underline{6x^3+9x^2} \\ 2x^2-x \\ \underline{2x^2+3x} \\ -4x+10 \\ \underline{-4x-6} \\ 16 \end{array}$$

41. $$\begin{array}{r} x-7 \\ x+2\,\overline{)\,x^2-5x-14} \\ \underline{x^2+2x} \\ -7x-14 \\ \underline{-7x-14} \\ 0 \end{array}$$

43. $$\begin{array}{r} 3x^2-x+2 \\ 3x-2\,\overline{)\,9x^3-9x^2+8x-4} \\ \underline{9x^3-6x^2} \\ -3x^2+8x \\ \underline{-3x^2+2x} \\ 6x-4 \\ \underline{6x-4} \\ 0 \end{array}$$

45. $$\begin{array}{r} 9a^2+6ab+4b^2 \\ 3a-2b\,\overline{)\,27a^3+0a^2b+0ab^2-8b^3} \\ \underline{27a^3-18a^2b} \\ 18a^2b+0ab^2 \\ \underline{18a^2b-12ab^2} \\ 12ab^2-8b^3 \\ \underline{12ab^2-8b^3} \\ 0 \end{array}$$

47. $$\begin{array}{r} a^2+a+1+\frac{2}{a-1} \\ a-1\,\overline{)\,a^3+0a^2+0a+1} \\ \underline{a^3-a^2} \\ a^2+0a \\ \underline{a^2-a} \\ a+1 \\ \underline{a-1} \\ 2 \end{array}$$

© 2015 Cengage Learning. All Rights Reserved. May not be scanned, copied or duplicated, or posted to a publicly accessible website, in whole or in part.

SECTION 6.6

49.

$$\begin{array}{r} 3x - 5 + \frac{3}{2x+3} \\ 2x+3\,\overline{\big)\,6x^2 - x - 12} \\ \underline{6x^2 + 9x} \\ -10x - 12 \\ \underline{-10x - 15} \\ 3 \end{array}$$

51.

$$\begin{array}{r} x^2 + x + 1 \\ x^2+x+2\,\overline{\big)\,x^4 + 2x^3 + 4x^2 + 3x + 2} \\ \underline{x^4 + x^3 + 2x^2} \\ x^3 + 2x^2 + 3x \\ \underline{x^3 + x^2 + 2x} \\ x^2 + x + 2 \\ \underline{x^2 + x + 2} \\ 0 \end{array}$$

53.

$$\begin{array}{r} 2y + 2 \\ 5y-2\,\overline{\big)\,10y^2 + 6y - 4} \\ \underline{10y^2 - 4y} \\ 10y - 4 \\ \underline{10y - 4} \\ 0 \end{array}$$

55.

$$\begin{array}{r} x^2 + x + 2 \\ x^2+0x+3\,\overline{\big)\,x^4 + x^3 + 5x^2 + 3x + 6} \\ \underline{x^4 + 0x^3 + 3x^2} \\ x^3 + 2x^2 + 3x \\ \underline{x^3 + 0x^2 + 3x} \\ 2x^2 + 0x + 6 \\ \underline{2x^2 + 0x + 6} \\ 0 \end{array}$$

57.

$$\begin{array}{r} x^4 + x^2 + 4 \\ x^2-2\,\overline{\big)\,x^6 - x^4 + 2x^2 - 8} \\ \underline{x^6 - 2x^4} \\ x^4 + 2x^2 \\ \underline{x^4 - 2x^2} \\ 4x^2 - 8 \\ \underline{4x^2 - 8} \\ 0 \end{array}$$

59. $$\frac{5x^ny^n - 4x^{5n}y^{5n} - 3x^{2n}y^{2n}}{x^ny^n} = \frac{5x^ny^n}{x^ny^n} - \frac{4x^{5n}y^{5n}}{x^ny^n} - \frac{3x^{2n}y^{2n}}{x^ny^n} = 5 - 4x^{4n}y^{4n} - 3x^ny^n$$

61. $$\frac{3a^{-2}b^3 - 6a^2b^{-3} + 9a^{-2}}{12a^{-1}b} = \frac{3a^{-2}b^3}{12a^{-1}b} - \frac{6a^2b^{-3}}{12a^{-1}b} + \frac{9a^{-2}}{12a^{-1}b} = \frac{b^2}{4a} - \frac{a^3}{2b^4} + \frac{3}{4ab}$$

63.

$$\begin{array}{r} 3x^2 - x + 2 \\ 3x+2\,\overline{\big)\,9x^3 + 3x^2 + 4x + 4} \\ \underline{9x^3 + 6x^2} \\ -3x^2 + 4x \\ \underline{-3x^2 - 2x} \\ 6x + 4 \\ \underline{6x + 4} \\ 0 \end{array}$$

65.

$$\begin{array}{r} 5a^2 - 3a - 4 \\ 3a-4\,\overline{\big)\,15a^3 - 29a^2 + 0a + 16} \\ \underline{15a^3 - 20a^2} \\ -9a^2 + 0a \\ \underline{-9a^2 + 12a} \\ -12a + 16 \\ \underline{-12a + 16} \\ 0 \end{array}$$

67.

$$\begin{array}{r} x - 8y \\ x-2y\,\overline{\big)\,x^2 - 10xy + 16y^2} \\ \underline{x^2 - 2xy} \\ -8xy + 16y^2 \\ \underline{-8xy + 16y^2} \\ 0 \end{array}$$

© 2015 Cengage Learning. All Rights Reserved. May not be scanned, copied or duplicated, or posted to a publicly accessible website, in whole or in part.

69.

$$\begin{array}{r|l} & 16x^4 - 8x^3y + 4x^2y^2 - 2xy^3 + y^4 \\ \hline 2x+y & 32x^5 + 0x^4y + 0x^3y^2 + 0x^2y^3 + 0xy^4 + y^5 \\ & 32x^5 + 16x^4y \\ \hline & -16x^4y + 0x^3y^2 \\ & -16x^4y - 8x^3y^2 \\ \hline & 8x^3y^2 + 0x^2y^3 \\ & 8x^3y^2 + 4x^2y^3 \\ \hline & -4x^2y^3 + 0xy^4 \\ & -4x^2y^3 - 2xy^4 \\ \hline & 2xy^4 + y^5 \\ & 2xy^4 + y^5 \\ \hline & 0 \end{array}$$

71.

$$A = lw$$
$$\frac{A}{w} = l$$
$$\frac{9x^2 + 21x + 10}{3x + 2} = l$$

$$\begin{array}{r|l} & 3x + 5 \\ \hline 3x+2 & 9x^2 + 21x + 10 \\ & 9x^2 + 6x \\ \hline & 15x + 10 \\ & 15x + 10 \\ \hline & 0 \end{array}$$

The length is $(3x + 5)$ ft.

73.

Dog sled

$$rt = d$$
$$r = \frac{d}{t} = \frac{12x^2 + 13x - 14}{4x + 7}$$

$$\begin{array}{r|l} & 3x - 2 \\ \hline 4x+7 & 12x^2 + 13x - 14 \\ & 12x^2 + 21x \\ \hline & -8x - 14 \\ & -8x - 14 \\ \hline & 0 \end{array}$$

Snowshoes

$$rt = d$$
$$t = \frac{d}{r} = \frac{3x^2 + 19x + 20}{3x + 4}$$

$$\begin{array}{r|l} & x + 5 \\ \hline 3x+4 & 3x^2 + 19x + 20 \\ & 3x^2 + 4x \\ \hline & 15x + 20 \\ & 15x + 20 \\ \hline & 0 \end{array}$$

$r = (3x - 2)$ mph

$t = (x + 5)$ hr

75. **Answers may vary.**

77.

$$\begin{array}{r|l} & 5x + 7 \\ \hline 2x-3 & 10x^2 - x - 21 \\ & 10x^2 - 15x \\ \hline & 14x - 21 \\ & 14x - 21 \\ \hline & 0 \end{array}$$

It is a factor.

Exercises 6.7 (page 426)

1.

$$\begin{array}{r|l} & 2x - 5 \\ \hline x+4 & 2x^2 + 3x - 18 \\ & 2x^2 + 8x \\ \hline & -5x - 18 \\ & -5x - 20 \\ \hline & 2 \end{array}$$

remainder = 2

3. $2(-4)^2 + 3(-4) - 18 = 2(16) - 12 - 18$
$= 32 - 12 - 18 = 2$

5. $f(-1) = 3(-1)^2 + 2(-1) - 1 = 0$

7. $f(2a) = 3(2a)^2 + 2(2a) - 1$
$= 12a^2 + 4a - 1$

© 2015 Cengage Learning. All Rights Reserved. May not be scanned, copied or duplicated, or posted to a publicly accessible website, in whole or in part.

SECTION 6.7

9. $5(x^2 - 3x + 2) + 4(3x^2 - x + 1) = 5x^2 - 15x + 10 + 12x^2 - 4x + 4 = 17x^2 - 19x + 14$

11. $x - r$

13. $P(r)$

15.
$$\begin{array}{r|rrr} 9 & 1 & -13 & 30 \\ & & 9 & -36 \\ \hline & 1 & -4 & -6 \end{array}$$
$\Rightarrow \boxed{x - 4 + \frac{-6}{x-9}}$

17.
$$\begin{array}{r|rrr} -5 & 1 & 4 & -10 \\ & & -5 & 5 \\ \hline & 1 & -1 & -5 \end{array}$$
$\Rightarrow \boxed{x - 1 + \frac{-5}{x+5}}$

19.
$$\begin{array}{r|rrrr} 2 & 2 & 0 & -5 & -6 \\ & & 4 & 8 & 6 \\ \hline & 2 & 4 & 3 & 0 \end{array}$$
$\Rightarrow \boxed{2x^2 + 4x + 3}$

21.
$$\begin{array}{r|rrrr} -2 & 4 & 0 & -2 & 36 \\ & & -8 & 16 & -28 \\ \hline & 4 & -8 & 14 & 8 \end{array}$$
$\Rightarrow \boxed{4x^2 - 8x + 14 + \frac{8}{x+2}}$

23.
$$\begin{array}{r|rrrr} -1 & 6 & 5 & 0 & 4 \\ & & -6 & 1 & -1 \\ \hline & 6 & -1 & 1 & 3 \end{array}$$
$\Rightarrow \boxed{6x^2 - x + 1 + \frac{3}{x+1}}$

25.
$$\begin{array}{r|rrrr} -2 & 3 & 4 & 0 & 8 \\ & & -6 & 4 & -8 \\ \hline & 3 & -2 & 4 & 0 \end{array}$$
$\Rightarrow \boxed{3x^2 - 2x + 4}$

27. $P(1) = 2(1)^3 - 4(1)^2 + 2(1) - 1 = \boxed{-1}$
$$\begin{array}{r|rrrr} 1 & 2 & -4 & 2 & -1 \\ & & 2 & -2 & 0 \\ \hline & 2 & -2 & 0 & \boxed{-1} \end{array}$$

29. $P(-3) = 2(-3)^3 - 4(-3)^2 + 2(-3) - 1$
$= \boxed{-97}$
$$\begin{array}{r|rrrr} -3 & 2 & -4 & 2 & -1 \\ & & -6 & 30 & -96 \\ \hline & 2 & -10 & 32 & \boxed{-97} \end{array}$$

31. $P(3) = 2(3)^3 - 4(3)^2 + 2(3) - 1 = \boxed{23}$
$$\begin{array}{r|rrrr} 3 & 2 & -4 & 2 & -1 \\ & & 6 & 6 & 24 \\ \hline & 2 & 2 & 8 & \boxed{23} \end{array}$$

33. $P(-5) = 2(-5)^3 - 4(-5)^2 + 2(-5) - 1$
$= \boxed{-361}$
$$\begin{array}{r|rrrr} -5 & 2 & -4 & 2 & -1 \\ & & -10 & 70 & -360 \\ \hline & 2 & -14 & 72 & \boxed{-361} \end{array}$$

35.
$$\begin{array}{r|rrrr} 2 & 1 & -4 & 1 & -2 \\ & & 2 & -4 & -6 \\ \hline & 1 & -2 & -3 & \boxed{-8} \end{array}$$

37.
$$\begin{array}{r|rrrrr} -2 & 1 & -2 & 1 & -3 & 2 \\ & & -2 & 8 & -18 & 42 \\ \hline & 1 & -4 & 9 & -21 & \boxed{44} \end{array}$$

39.
$$\begin{array}{r|rrrr} 3 & 1 & -3 & 5 & -15 \\ & & 3 & 0 & 15 \\ \hline & 1 & 0 & 5 & \boxed{0} \end{array}$$
factor; $P(x) = (x - 3)(x^2 + 5)$

41.
$$\begin{array}{r|rrrr} 6 & 2 & -8 & 1 & -36 \\ & & 12 & 24 & 150 \\ \hline & 2 & 4 & 25 & \boxed{114} \end{array}$$
not a factor

© 2015 Cengage Learning. All Rights Reserved. May not be scanned, copied or duplicated, or posted to a publicly accessible website, in whole or in part.

43. $$\begin{array}{r|rrrr} 3 & 3 & -10 & 5 & -6 \\ & & 9 & -3 & 6 \\ \hline & 3 & -1 & 2 & 0 \end{array}$$
$\Rightarrow \boxed{3x^2 - x + 2}$

45. $$\begin{array}{r|rrr} -2 & 1 & -5 & 14 \\ & & -2 & 14 \\ \hline & 1 & -7 & 28 \end{array}$$
$\Rightarrow \boxed{x - 7 + \frac{28}{x+2}}$

47. $$\begin{array}{r|rrr} -4 & 1 & 6 & 8 \\ & & -4 & -8 \\ \hline & 1 & 2 & 0 \end{array}$$
$\Rightarrow \boxed{x + 2}$

49. $$\begin{array}{r|rrr} -6 & 1 & 13 & 42 \\ & & -6 & -42 \\ \hline & 1 & 7 & 0 \end{array}$$
$\Rightarrow \boxed{x + 7}$

51. $$\begin{array}{r|rrrr} -2 & 1 & 1 & 0 & 1 \\ & & -2 & 2 & -4 \\ \hline & 1 & -1 & 2 & \boxed{-3} \end{array}$$

53. $$\begin{array}{r|rrrr} \frac{1}{2} & 2 & -1 & 4 & -5 \\ & & 1 & 0 & 2 \\ \hline & 2 & 0 & 4 & \boxed{-3} \end{array}$$

55. $$\begin{array}{r|rrrr} 0 & 7 & -5 & -8 & 0 \\ & & 0 & 0 & 0 \\ \hline & 7 & -5 & -8 & \boxed{0} \end{array}$$
factor; $P(x) = x(7x^2 - 5x - 8)$

57. $$\begin{array}{r|rrrr} 5 & 1 & 0 & 0 & -125 \\ & & 5 & 25 & 125 \\ \hline & 1 & 5 & 25 & \boxed{0} \end{array}$$
factor; $P(x) = (x - 5)(x^2 + 5x + 25)$

59. $$\begin{array}{r|rrrrrr} -\frac{1}{2} & 3 & 0 & 0 & 0 & 0 & 1 \\ & & -\frac{3}{2} & \frac{3}{4} & -\frac{3}{8} & \frac{3}{16} & -\frac{3}{32} \\ \hline & 3 & -\frac{3}{2} & \frac{3}{4} & -\frac{3}{8} & \frac{3}{16} & \boxed{\frac{29}{32}} \end{array}$$

61. $Q(-1) = (-1)^4 - 3(-1)^3 + 2(-1)^2 + (-1) - 3 = \boxed{2}$
$$\begin{array}{r|rrrrr} -1 & 1 & -3 & 2 & 1 & -3 \\ & & -1 & 4 & -6 & 5 \\ \hline & 1 & -4 & 6 & -5 & \boxed{2} \end{array}$$

63. $Q(2) = (2)^4 - 3(2)^3 + 2(2)^2 + (2) - 3$
$= \boxed{-1}$
$$\begin{array}{r|rrrrr} 2 & 1 & -3 & 2 & 1 & -3 \\ & & 2 & -2 & 0 & 2 \\ \hline & 1 & -1 & 0 & 1 & \boxed{-1} \end{array}$$

65. $$\begin{array}{r|rrrrrrr} 2 & 1 & 0 & 0 & 0 & 0 & 0 & 0 \\ & & 2 & 4 & 8 & 16 & 32 & 64 \\ \hline & 1 & 2 & 4 & 8 & 16 & 32 & \boxed{64} \end{array}$$

67. **Answers may vary.**

69. remainder $= P(1) = 1^{100} - 1^{99} + 1^{98} - 1^{97} + \cdots + 1^2 - 1 + 1 = 1$

© 2015 Cengage Learning. All Rights Reserved. May not be scanned, copied or duplicated, or posted to a publicly accessible website, in whole or in part.

Exercises 6.8 (page 437)

1. $\frac{15}{4} \square \frac{105}{28}$
$15 \cdot 28 \square 4 \cdot 105$
$420 \boxed{=} 420$

3. $\frac{30}{49} \square \frac{120}{245}$
$30 \cdot 245 \square 49 \cdot 120$
$7350 \boxed{\neq} 5880$

5. $t = \frac{kl}{w}$
$20 = \frac{k(4)}{40}$
$20 = \frac{4k}{40}$
$800 = 4k \Rightarrow \boxed{200 = k}$

7. $(x^4x^{-5})^2 = (x^{-1})^2$
$= x^{-2} = \frac{1}{x^2}$

9. $\frac{3y^0 - 5y^0}{y^0} = \frac{3-5}{1}$
$= \frac{-2}{1} = -2$

11. $470{,}000 = 4.7 \times 10^5$

13. $2.5 \times 10^{-3} = 0.0025$

15. unit costs; rates

17. extremes; means

19. direct

21. rational

23. joint; constant of proportionality

25. direct

27. neither

29. $\frac{x}{8} = \frac{28}{32}$
$32x = 224$
$x = 7$

31. $\frac{r-2}{3} = \frac{r}{5}$
$5(r-2) = 3r$
$5r - 10 = 3r$
$2r = 10$
$r = 5$

33. $\frac{x+2}{x-2} = \frac{7}{3}$
$3(x+2) = 7(x-2)$
$3x + 6 = 7x - 14$
$20 = 4x$
$5 = x$

35. $\frac{5}{5z+3} = \frac{2z}{2z^2+6}$
$5(2z^2+6) = 2z(5z+3)$
$10z^2 + 30 = 10z^2 + 6z$
$30 = 6z$
$5 = z$

37. $\frac{1}{x+3} = \frac{-2x}{x+5}$
$x + 5 = -2x^2 - 6x$
$2x^2 + 7x + 5 = 0$
$(2x+5)(x+1)$
$2x + 5 = 0$ **or** $x + 1 = 0$
$x = -\frac{5}{2}$ $\quad x = -1$

39. $\frac{a-4}{a+2} = \frac{a-5}{a+1}$
$(a-4)(a+1) = (a+2)(a-5)$
$a^2 - 3a - 4 = a^2 - 3a - 10$
$-3a - 4 = -3a - 10$
$-4 \neq -10$
no solution

41. $A = kp^2$

43. $a = \frac{k}{b^2}$

45. $B = kmn$

47. $X = \frac{kw}{q}$

© 2015 Cengage Learning. All Rights Reserved. May not be scanned, copied or duplicated, or posted to a publicly accessible website, in whole or in part.

SECTION 6.8

49. $\frac{2}{c} = \frac{c-3}{2}$
$4 = c^2 - 3c$
$0 = c^2 - 3c - 4$
$0 = (c+1)(c-4)$
$c + 1 = 0$ **or** $c - 4 = 0$
$c = -1$ $\quad$ $c = 4$

51. $\frac{9}{5x} = \frac{3x}{15}$
$135 = 15x^2$
$0 = 15x^2 - 135$
$0 = 15(x+3)(x-3)$
$x + 3 = 0$ **or** $x - 3 = 0$
$x = -3$ $\quad$ $x = 3$

53. $\frac{a+6}{a+4} = \frac{a-3}{a-4}$
$(a+6)(a-4) = (a+4)(a-3)$
$a^2 + 2a - 24 = a^2 + a - 12$
$a = 12$

55. $P = \frac{ka^2}{j^3}$

57. L varies jointly with m and n.

59. E varies jointly with a and the square of b.

61. X varies directly with x^2 and inversely with y^2.

63. R varies directly with L and inversely with d^2.

65. Let $c =$ the cost of 5 shirts.
$\frac{2}{25} = \frac{5}{c}$
$2c = 125$
$c = 62.5$
5 shirts will cost \$62.50.

67. Let $g =$ gallons of gas for 323 miles.
$\frac{38}{1} = \frac{323}{g}$
$38g = 323$
$g = \frac{323}{38} = 8.5$
8.5 gallons of gas are needed.

69. Let $w =$ the width if it were a real house.
$\frac{1\text{in.}}{1\text{ ft}} = \frac{32\text{ in.}}{w\text{ ft}}$
$w = 32$
The width would be 32 feet.

71. Let $h =$ the actual height of the building.
$\frac{7\text{ in.}}{280\text{ ft}} = \frac{2\text{ in.}}{h\text{ ft}}$
$7h = 560$
$h = 80$
The building is 80 feet tall.

73. Let $d =$ the dosage required.
$\frac{0.006}{1} = \frac{d}{30}$
$0.18 = d$
The dosage is 0.18 grams.

75. Let $h =$ the height of the tree.
$\frac{6}{4} = \frac{h}{28}$
$168 = 4h$
$42 = h$
The tree is 42 feet tall.

© 2015 Cengage Learning. All Rights Reserved. May not be scanned, copied or duplicated, or posted to a publicly accessible website, in whole or in part.

SECTION 6.8

77. $\frac{20}{32} = \frac{w}{75}$

$1500 = 32w$

$w = \frac{1500}{32} = 46\frac{7}{8}$

The width of the river is $46\frac{7}{8}$ feet.

79. $\frac{1350}{1} = \frac{x}{5}$

$6750 = x$

The plane will descend 6750 feet.

81. $A = kr^2$

$A = \pi r^2$

$A = \pi(8 \text{ in.})^2$

$\boxed{A = 64\pi \text{ in.}^2}$

83. $d = kg$

$288 = k(12)$

$24 = k$

$d = 24g$

$d = 24(18)$

$\boxed{d = 432 \text{ mi}}$

85. $t = \frac{k}{x}$

$10 = \frac{k}{25}$

$250 = k$

$t = \frac{250}{x}$

$t = \frac{250}{10}$

$\boxed{t = 25 \text{ days}}$

87. $V = \frac{k}{P}$

$20 = \frac{k}{6}$

$120 = k$

$V = \frac{120}{P}$

$V = \frac{120}{10}$

$\boxed{V = 12 \text{ in.}^3}$

89. $A_1 = klw$

$A_2 = k(3l)(3w) = 9klw = 9A_1$

The area is multiplied by 9.

91. $c = kth$

$1800 = k(4)(6)$

$1800 = 24k$

$75 = k$

$c = 75th$

$c = 75(10)(12)$

$\boxed{c = \$9000}$

93. $D = \frac{k}{wd^3}$

$1.1 = \frac{k}{4(4)^3}$

$1.1 = \frac{k}{256}$

$281.6 = k$

$D = \frac{281.6}{wd^3}$

$D = \frac{281.6}{2(8)^3}$

$D = \frac{281.6}{1024}$

$\boxed{D = 0.275 \text{ in.}}$

95. $P = \frac{kT}{V}$

$1 = \frac{k(273)}{1}$

$\frac{1}{273} = k$

$P = \frac{\frac{1}{273}T}{V}$

$1 = \frac{\frac{1}{273}T}{2}$

$2 = \frac{1}{273}T$

$\boxed{546 \text{ K} = T}$

97. $f = \frac{k}{l}$

$256 = \frac{k}{2}$

$512 = k$

$f = \frac{512}{l}$

$f = \frac{512}{6}$

$\boxed{f = 85\frac{1}{3}}$

99. $V = kC$

$6 = k(2)$

$3 = k \Rightarrow$ The resistance is 3 ohms.

101-103. Answers may vary.

105. This is not direct variation. For this to be direct variation, one temperature would have to be a constant multiple of the other.

© 2015 Cengage Learning. All Rights Reserved. May not be scanned, copied or duplicated, or posted to a publicly accessible website, in whole or in part.

Chapter 6 Review (page 444)

1. $f(3) = \frac{4(3)+3}{3} = \frac{15}{3} = 5$

2. $2x^2 - x - 21 \neq 0$
$(2x-7)(x+3) \neq 0$
$x \neq \frac{7}{2}, x \neq -3$

3. $f(x) = \frac{3x+2}{x}$

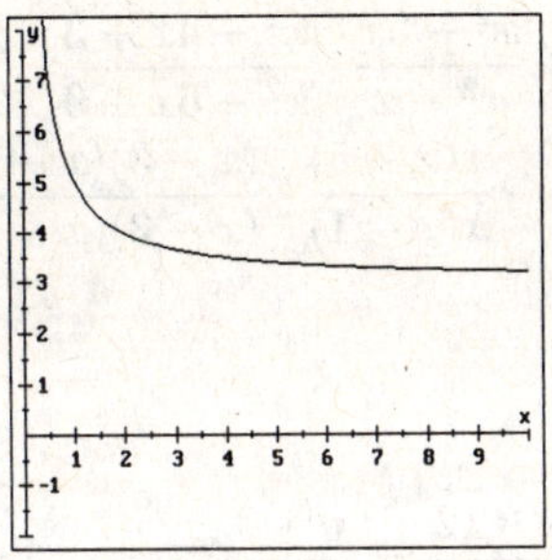

Vertical: $x = 0$

4. $x - 1 \neq 0$
$x \neq 1 \Rightarrow (-\infty, 1) \cup (1, \infty)$

5. $\frac{212m^3n}{588m^2n^3} = \frac{4 \cdot 53m^3n}{4 \cdot 147m^2n^3} = \frac{53m}{147n^2}$

6. $\frac{x^2-81}{x^2+18x+81} = \frac{(x+9)(x-9)}{(x+9)(x+9)} = \frac{x-9}{x+9}$

7. $\frac{x^2+6x+36}{x^3-216} = \frac{x^2+6x+36}{(x-6)(x^2+6x+36)} = \frac{1}{x-6}$

8. $\frac{x^2-2x+4}{2x^3+16} = \frac{x^2-2x+4}{2(x^3+8)} = \frac{x^2-2x+4}{2(x+2)(x^2-2x+4)} = \frac{1}{2(x+2)} = \frac{1}{2x+4}$

9. $\frac{5x-5y}{4y-4x} = \frac{5(x-y)}{-4(x-y)} = -\frac{5}{4}$

10. $\frac{2m-2n}{n-m} = \frac{2(m-n)}{-1(m-n)} = -2$

11. $\frac{m^3+m^2n-2mn^2}{2m^3-mn^2-m^2n} = \frac{m(m^2+mn-2n^2)}{m(2m^2-mn-n^2)} = \frac{m(m+2n)(m-n)}{m(2m+n)(m-n)} = \frac{m+2n}{2m+n}$

12. $\frac{ac-ad+bc-bd}{d^2-c^2} = \frac{a(c-d)+b(c-d)}{(d+c)(d-c)} = \frac{(c-d)(a+b)}{(d+c)(d-c)} = -\frac{a+b}{c+d}$

13. $\frac{x^2+4x+4}{x^2-x-6} \cdot \frac{x^2-9}{x^2+5x+6} = \frac{(x+2)(x+2)}{(x+2)(x-3)} \cdot \frac{(x+3)(x-3)}{(x+2)(x+3)} = 1$

14. $\frac{x^3-8}{x^2+2x+4} \div \frac{x^2-4}{x+2} = \frac{x^3-8}{x^2+2x+4} \cdot \frac{x+2}{x^2-4} = \frac{(x-2)(x^2+2x+4)}{x^2+2x+4} \cdot \frac{x+2}{(x+2)(x-2)} = 1$

© 2015 Cengage Learning. All Rights Reserved. May not be scanned, copied or duplicated, or posted to a publicly accessible website, in whole or in part.

15. $$\frac{x^2+3x+2}{x^2-x-6}\cdot\frac{3x^2-3x}{x^2-3x-4}\div\frac{x^2+3x+2}{x^2-2x-8}$$
$$=\frac{x^2+3x+2}{x^2-x-6}\cdot\frac{3x^2-3x}{x^2-3x-4}\cdot\frac{x^2-2x-8}{x^2+3x+2}$$
$$=\frac{(x+2)(x+1)}{(x-3)(x+2)}\cdot\frac{3x(x-1)}{(x-4)(x+1)}\cdot\frac{(x-4)(x+2)}{(x+2)(x+1)}$$
$$=\frac{3x(x-1)}{(x-3)(x+1)}$$

16. $$\frac{x^2-x-6}{x^2-3x-10}\div\frac{x^2-x}{x^2-5x}\cdot\frac{x^2-4x+3}{x^2-6x+9}=\frac{x^2-x-6}{x^2-3x-10}\cdot\frac{x^2-5x}{x^2-x}\cdot\frac{x^2-4x+3}{x^2-6x+9}$$
$$=\frac{(x-3)(x+2)}{(x-5)(x+2)}\cdot\frac{x(x-5)}{x(x-1)}\cdot\frac{(x-3)(x-1)}{(x-3)(x-3)}=1$$

17. $$\frac{3x}{x-y}+\frac{7}{x-y}=\frac{3x+7}{x-y}$$

18. $$\frac{5y-3}{y^2+3}-\frac{4(y-3)}{y^2+3}=\frac{5y-3-(4y-12)}{y^2+3}=\frac{5y-3-4y+12}{y^2+3}=\frac{y+9}{y^2+3}$$

19. $$\frac{3}{x+2}+\frac{2}{x+3}=\frac{3(x+3)}{(x+2)(x+3)}+\frac{2(x+2)}{(x+2)(x+3)}=\frac{3x+9+2x+4}{(x+2)(x+3)}=\frac{5x+13}{(x+2)(x+3)}$$

20. $$\frac{4x}{x-4}-\frac{3}{x+3}=\frac{4x(x+3)}{(x-4)(x+3)}-\frac{3(x-4)}{(x-4)(x+3)}=\frac{4x^2+12x-3x+12}{(x-4)(x+3)}=\frac{4x^2+9x+12}{(x-4)(x+3)}$$

21. $$\frac{2x}{x+1}+\frac{3x}{x+2}+\frac{4x}{x^2+3x+2}=\frac{2x}{x+1}+\frac{3x}{x+2}+\frac{4x}{(x+2)(x+1)}$$
$$=\frac{2x(x+2)}{(x+1)(x+2)}+\frac{3x(x+1)}{(x+1)(x+2)}+\frac{4x}{(x+2)(x+1)}$$
$$=\frac{2x^2+4x+3x^2+3x+4x}{(x+1)(x+2)}=\frac{5x^2+11x}{(x+1)(x+2)}$$

22. $$\frac{5x}{x-3}+\frac{5}{x^2-5x+6}+\frac{x+3}{x-2}=\frac{5x}{x-3}+\frac{5}{(x-3)(x-2)}+\frac{x+3}{x-2}$$
$$=\frac{5x(x-2)}{(x-3)(x-2)}+\frac{5}{(x-3)(x-2)}+\frac{(x+3)(x-3)}{(x-3)(x-2)}$$
$$=\frac{5x^2-10x+5+x^2-9}{(x-3)(x-2)}$$
$$=\frac{6x^2-10x-4}{(x-3)(x-2)}=\frac{2(3x+1)(x-2)}{(x-3)(x-2)}=\frac{2(3x+1)}{x-3}$$

© 2015 Cengage Learning. All Rights Reserved. May not be scanned, copied or duplicated, or posted to a publicly accessible website, in whole or in part.

23. $$\frac{3(x+2)}{x^2-1}-\frac{2}{x+1}+\frac{4(x+3)}{x^2-2x+1}=\frac{3(x+2)}{(x+1)(x-1)}-\frac{2}{(x+1)}+\frac{4(x+3)}{(x-1)(x-1)}$$
$$=\frac{3(x+2)(x-1)}{(x+1)(x-1)(x-1)}-\frac{2(x-1)(x-1)}{(x+1)(x-1)(x-1)}+\frac{4(x+3)(x+1)}{(x+1)(x-1)(x-1)}$$
$$=\frac{3(x^2+x-2)-2(x^2-2x+1)+4(x^2+4x+3)}{(x+1)(x-1)(x-1)}$$
$$=\frac{3x^2+3x-6-2x^2+4x-2+4x^2+16x+12}{(x+1)(x-1)(x-1)}=\frac{5x^2+23x+4}{(x+1)(x-1)^2}$$

24. $$\frac{-2(3+x)}{x^2+6x+9}+\frac{3(x+2)}{x^2-6x+9}-\frac{1}{x^2-9}$$
$$=\frac{-2(3+x)}{(x+3)(x+3)}+\frac{3(x+2)}{(x-3)(x-3)}-\frac{1}{(x+3)(x-3)}$$
$$=\frac{-2}{x+3}+\frac{3(x+2)}{(x-3)(x-3)}-\frac{1}{(x+3)(x-3)}$$
$$=\frac{-2(x-3)(x-3)}{(x+3)(x-3)(x-3)}+\frac{3(x+2)(x+3)}{(x+3)(x-3)(x-3)}-\frac{1(x-3)}{(x+3)(x-3)(x-3)}$$
$$=\frac{-2(x^2-6x+9)+3(x^2+5x+6)-x+3}{(x+3)(x-3)(x-3)}=\frac{x^2+26x+3}{(x+3)(x-3)^2}$$

25. $$\frac{\frac{5}{a}-\frac{4}{b}}{ab}=\frac{\left(\frac{5}{a}-\frac{4}{b}\right)ab}{ab(ab)}=\frac{5b-4a}{a^2b^2}$$

26. $$\frac{\frac{3}{p}+\frac{4}{q}}{\frac{4}{p}-\frac{3}{q}}=\frac{\left(\frac{3}{p}+\frac{4}{q}\right)pq}{\left(\frac{4}{p}-\frac{3}{q}\right)pq}=\frac{3q+4p}{4q-3p}$$

27. $$\frac{2x+3+\frac{1}{x}}{x+2+\frac{1}{x}}=\frac{\left(2x+3+\frac{1}{x}\right)x}{\left(x+2+\frac{1}{x}\right)x}=\frac{2x^2+3x+1}{x^2+2x+1}=\frac{(2x+1)(x+1)}{(x+1)(x+1)}=\frac{2x+1}{x+1}$$

28. $$\frac{6x+13+\frac{6}{x}}{6x+5-\frac{6}{x}}=\frac{\left(6x+13+\frac{6}{x}\right)x}{\left(6x+5-\frac{6}{x}\right)x}=\frac{6x^2+13x+6}{6x^2+5x-6}=\frac{(2x+3)(3x+2)}{(2x+3)(3x-2)}=\frac{3x+2}{3x-2}$$

29. $$\frac{1-\frac{1}{x}-\frac{2}{x^2}}{1+\frac{4}{x}+\frac{3}{x^2}}=\frac{\left(1-\frac{1}{x}-\frac{2}{x^2}\right)x^2}{\left(1+\frac{4}{x}+\frac{3}{x^2}\right)x^2}=\frac{x^2-x-2}{x^2+4x+3}=\frac{(x-2)(x+1)}{(x+3)(x+1)}=\frac{x-2}{x+3}$$

30. $$\frac{x^{-1}+1}{x+1}=\frac{\frac{1}{x}+1}{x+1}=\frac{\left(\frac{1}{x}+1\right)x}{(x+1)x}=\frac{1+x}{(x+1)x}=\frac{1}{x}$$

31. $$\frac{x^{-1}-y^{-1}}{x^{-1}+y^{-1}}=\frac{\frac{1}{x}-\frac{1}{y}}{\frac{1}{x}+\frac{1}{y}}=\frac{\left(\frac{1}{x}-\frac{1}{y}\right)xy}{\left(\frac{1}{x}+\frac{1}{y}\right)xy}=\frac{y-x}{y+x}$$

© 2015 Cengage Learning. All Rights Reserved. May not be scanned, copied or duplicated, or posted to a publicly accessible website, in whole or in part.

32. $$\frac{(x-y)^{-2}}{x^{-2}-y^{-2}}=\frac{\frac{1}{(x-y)^2}}{\frac{1}{x^2}-\frac{1}{y^2}}=\frac{\left(\frac{1}{(x-y)^2}\right)x^2y^2(x-y)^2}{\left(\frac{1}{x^2}-\frac{1}{y^2}\right)x^2y^2(x-y)^2}=\frac{x^2y^2}{y^2(x-y)^2-x^2(x-y)^2}$$
$$=\frac{x^2y^2}{(x-y)^2(y^2-x^2)}$$

33. $$\frac{3}{y}-\frac{1}{12}=\frac{7}{4y}$$
$$\left(\frac{3}{y}-\frac{1}{12}\right)12y=\frac{7}{4y}\cdot 12y$$
$$36-y=21$$
$$15=y$$ The answer checks.

34. $$\frac{2}{x+5}-\frac{1}{6}=\frac{1}{x+4}$$
$$\left(\frac{2}{x+5}-\frac{1}{6}\right)6(x+5)(x+4)=\left(\frac{1}{x+4}\right)6(x+5)(x+4)$$
$$12(x+4)-(x+5)(x+4)=6(x+5)$$
$$12x+48-x^2-9x-20=6x+30$$
$$0=x^2+3x+2$$
$$0=(x+2)(x+1)$$
$x+2=0$ **or** $x+1=0$
$x=-2$ $x=-1$ Both answers check.

35. $$\frac{2(x-5)}{x-2}=\frac{6x+12}{4-x^2}$$
$$\frac{2(x-5)}{x-2}=\frac{6(x+2)}{(2+x)(2-x)}$$
$$\frac{2(x-5)}{x-2}=\frac{-6(x+2)}{(x+2)(x-2)}$$
$$2(x-5)(x+2)=-6(x+2)$$
$$2x^2-6x-20=-6x-12$$
$$2x^2-8=0$$
$$2(x+2)(x-2)=0$$
$x=2$, or $x=-2$ Both answers are extraneous. $\emptyset$

© 2015 Cengage Learning. All Rights Reserved. May not be scanned, copied or duplicated, or posted to a publicly accessible website, in whole or in part.

36.
$$\frac{7}{x+9}-\frac{x+2}{2}=\frac{x+4}{x+9}$$
$$\left(\frac{7}{x+9}-\frac{x+2}{2}\right)2(x+9)=\left(\frac{x+4}{x+9}\right)2(x+9)$$
$$14-(x+2)(x+9)=2(x+4)$$
$$14-x^2-11x-18=2x+8$$
$$0=x^2+13x+12$$
$$0=(x+12)(x+1)$$
$x+12=0$ **or** $x+1=0$

$x=-12$ $\quad x=-1$ Both answers check.

37.
$$\frac{x^2}{a^2}-\frac{y^2}{b^2}=1$$
$$\left(\frac{x^2}{a^2}-\frac{y^2}{b^2}\right)a^2b^2=a^2b^2$$
$$x^2b^2-y^2a^2=a^2b^2$$
$$x^2b^2=a^2b^2+a^2y^2$$
$$x^2b^2=a^2\left(b^2+y^2\right)$$
$$\frac{x^2b^2}{b^2+y^2}=a^2$$

38.
$$H=\frac{2ab}{a+b}$$
$$H(a+b)=2ab$$
$$Ha+Hb=2ab$$
$$Ha=2ab-Hb$$
$$Ha=b(2a-H)$$
$$\frac{Ha}{2a-H}=b$$

39. Let r = the usual rate. Then $r-10$ = the slower rate.

	Rate	Time	Dist.
Usual	r	$\frac{200}{r}$	200
Slower	$r-10$	$\frac{200}{r-10}$	200

$$\boxed{\text{Usual time}}+1=\boxed{\text{Slower time}}$$
$$\frac{200}{r}+1=\frac{200}{r-10}$$
$$\left(\frac{200}{r}+1\right)(r)(r-10)=\frac{200}{r-10}\cdot r(r-10)$$
$$200(r-10)+r(r-10)=200r$$
$$r^2-10r-2000=0$$
$$(r+40)(r-50)=0$$
$r+40=0$ **or** $r-50=0$

$r=-40$ $\quad r=50$ $\quad r=50$ is the only answer that makes sense.

The usual rate is 50 miles per hour.

© 2015 Cengage Learning. All Rights Reserved. May not be scanned, copied or duplicated, or posted to a publicly accessible website, in whole or in part.

40. Let $r =$ the usual rate. Then $r + 40 =$ the faster rate.

	Rate	Time	Dist.
Usual	r	$\frac{600}{r}$	600
Faster	$r+40$	$\frac{600}{r+40}$	600

$$\boxed{\text{Usual time}} - \frac{1}{2} = \boxed{\text{Faster time}}$$

$$\frac{600}{r} - \frac{1}{2} = \frac{600}{r+40}$$

$$\left(\frac{600}{r} - \frac{1}{2}\right)(2r)(r+40) = \frac{600}{r+40} \cdot 2r(r+40)$$

$$1200(r+40) + r(r+40) = 1200r$$

$$r^2 + 40r - 48000 = 0$$

$$(r+240)(r-200) = 0$$

$r + 240 = 0$ **or** $r - 200 = 0$

$r = -240$ $\quad r = 200$ $\quad r = 200$ is the only answer that makes sense.

The usual rate is 200 miles per hour.

41. The first pipe can drain $\frac{1}{24}$ of the tank in 1 hour, while the second can drain $\frac{1}{36}$ of the pool in 1 hour. Let $x =$ the number of hours needed for them to drain the tank together. Then $\frac{1}{x} =$ the amount of the tank they can drain together in 1 hour.

$$\boxed{\text{Amount 1st drains in 1 hour}} + \boxed{\text{Amount 2nd drains in 1 hour}} = \boxed{\text{Amount drained together in 1 hour}}$$

$$\frac{1}{24} + \frac{1}{36} = \frac{1}{x}$$

$$\left(\frac{1}{24} + \frac{1}{36}\right)72x = \frac{1}{x} \cdot 72x$$

$$3x + 2x = 72$$

$$5x = 72$$

$$x = \frac{72}{5} = 14\frac{2}{5}$$

It will take the two pipes $14\frac{2}{5}$ hours to drain the tank.

© 2015 Cengage Learning. All Rights Reserved. May not be scanned, copied or duplicated, or posted to a publicly accessible website, in whole or in part.

CHAPTER 6 REVIEW

42. The first man can side $\frac{1}{14}$ of the house in 1 day. Let $x =$ the number of days needed for the second man to side the house alone. Then he can side $\frac{1}{x}$ of the house alone. Since both men can side the house in 8 days, they side $\frac{1}{8}$ of the house in 1 day.

Amount 1st sides in 1 day	+	Amount 2nd sides in 1 day	=	Amount sided together in 1 day

$$\frac{1}{14} + \frac{1}{x} = \frac{1}{8}$$
$$\left(\frac{1}{14} + \frac{1}{x}\right)(56x) = \frac{1}{8} \cdot 56x$$
$$4x + 56 = 7x$$
$$56 = 3x, \text{ or } x = \frac{56}{3} = 18\frac{2}{3}$$

It takes the second man $18\frac{2}{3}$ days to side the house alone.

43. $\dfrac{-8x^7y^5}{24x^5y^8} = -\dfrac{x^2}{3y^3}$

44. $\dfrac{30x^3y^2 - 15x^2y - 10xy^2}{-10xy} = \dfrac{30x^3y^2}{-10xy} + \dfrac{-15x^2y}{-10xy} + \dfrac{-10xy^2}{-10xy} = -3x^2y + \dfrac{3x}{2} + y$

45.
$$\begin{array}{r|l} & x + 5y \\ \hline 3x - 2y & 3x^2 + 13xy - 10y^2 \\ & \underline{3x^2 - 2xy} \\ & 15xy - 10y^2 \\ & \underline{15xy - 10y^2} \\ & 0 \end{array}$$

46.
$$\begin{array}{r|l} & x^2 + 2x - 1 + \frac{6}{2x+3} \\ \hline 2x + 3 & 2x^3 + 7x^2 + 4x + 3 \\ & \underline{2x^3 + 3x^2} \\ & 4x^2 + 4x \\ & \underline{4x^2 + 6x} \\ & -2x + 3 \\ & \underline{-2x - 3} \\ & 6 \end{array}$$

47.
$$\begin{array}{r|rrrr} 3 & 1 & -7 & 9 & 9 \\ & & 3 & -12 & -9 \\ \hline & 1 & -4 & -3 & \boxed{0} \end{array}$$

factor; $P(x) = (x - 3)(x^2 - 4x - 3)$

48.
$$\begin{array}{r|rrrr} -5 & 1 & 4 & -5 & 5 \\ & & -5 & 5 & 0 \\ \hline & 1 & -1 & 0 & \boxed{5} \end{array} \Rightarrow \text{not a factor}$$

49.
$$\frac{x+1}{8} = \frac{4x-2}{24}$$
$$24(x+1) = 8(4x-2)$$
$$24x + 24 = 32x - 16$$
$$-8x = -40$$
$$x = 5$$

50.
$$\frac{1}{x+6} = \frac{x+10}{12}$$
$$12 = (x+6)(x+10)$$
$$12 = x^2 + 16x + 60$$
$$0 = x^2 + 16x + 48$$
$$0 = (x+12)(x+4)$$
$$x + 12 = 0 \quad \textbf{or} \quad x + 4 = 0$$
$$x = -12 \qquad\qquad x = -4$$

© 2015 Cengage Learning. All Rights Reserved. May not be scanned, copied or duplicated, or posted to a publicly accessible website, in whole or in part.

51. Let h = the height of the tree.
$$\frac{44}{2.5} = \frac{h}{4}$$
$$176 = 2.5h$$
$70.4 = h \Rightarrow$ The tree is 70.4 feet tall.

52. $$x = ky \qquad x = 6y$$
$$18 = k(3) \qquad x = 6(9)$$
$$6 = k \qquad \boxed{x = 54}$$

53. $$x = \frac{k}{y} \qquad x = \frac{72}{y}$$
$$24 = \frac{k}{3} \qquad 12 = \frac{72}{y}$$
$$72 = k \qquad 12y = 72$$
$$\boxed{y = 6}$$

54. $$x = kyz$$
$$24 = k(3)(4)$$
$$24 = 12k$$
$$\boxed{2 = k}$$

55. $$x = \frac{kt}{y}$$
$$2 = \frac{k(8)}{64}$$
$$128 = 8k$$
$$\boxed{16 = k}$$

56. $$T = kv \qquad T = kv$$
$$1575 = k(90{,}000) \qquad T = 0.0175(312{,}000)$$
$$\frac{1575}{90{,}000} = k \qquad \boxed{T = \$5460}$$
$$0.0175 = k$$

Chapter 6 Test (page 450)

1. $$\frac{-12x^2y^3z^2}{18x^3y^4z^2} = -\frac{2}{3xy}$$

2. $$\frac{5x+15}{x^2-9} = \frac{5(x+3)}{(x+3)(x-3)} = \frac{5}{x-3}$$

3. $$\frac{7a-28b}{4b-a} = \frac{7(a-4b)}{4b-a} = -7$$

4. a. $x - 3 = 0$, $x = 3$ b. $(-\infty, 3) \cup (3, \infty)$

5. $$f(6) = \frac{7(6)-4}{6} = \frac{38}{6} = \frac{19}{3}$$

6. $$\frac{x^2y^{-2}}{x^3z^2} \cdot \frac{x^2z^4}{y^2z} = \frac{x^4y^{-2}z^4}{x^3y^2z^3} = \frac{xz}{y^4}$$

7. $$\frac{x^2-3x-4}{12x-3} \cdot \frac{3}{x-4} = \frac{(x-4)(x+1)}{3(4x-1)} \cdot \frac{3}{x-4} = \frac{x+1}{4x-1}$$

8. $$\frac{u^2+5u+6}{u^2-4} \cdot \frac{u^2-5u+6}{u^2-9} = \frac{(u+2)(u+3)}{(u+2)(u-2)} \cdot \frac{(u-2)(u-3)}{(u+3)(u-3)} = 1$$

9. $$\frac{x^3+y^3}{4} \div \frac{x^2-xy+y^2}{2x+2y} = \frac{x^3+y^3}{4} \cdot \frac{2x+2y}{x^2-xy+y^2} = \frac{(x+y)(x^2-xy+y^2)}{4} \cdot \frac{2(x+y)}{x^2-xy+y^2}$$
$$= \frac{(x+y)^2}{2}$$

10. $$\frac{x^2}{x+1} - \frac{x^2-2}{x+1} = \frac{2}{x+1}$$

© 2015 Cengage Learning. All Rights Reserved. May not be scanned, copied or duplicated, or posted to a publicly accessible website, in whole or in part.

CHAPTER 6 TEST

11. $\dfrac{a^2+7a+12}{a+3} \div \dfrac{16-a^2}{a-4} = \dfrac{a^2+7a+12}{a+3} \cdot \dfrac{a-4}{16-a^2} = \dfrac{(a+4)(a+3)}{a+3} \cdot \dfrac{a-4}{(4+a)(4-a)} = -1$

12. $\dfrac{6y}{y+6} + \dfrac{36}{y+6} = \dfrac{6y+36}{y+6} = \dfrac{6(y+6)}{y+6} = 6$

13. $\dfrac{3w}{w-5} + \dfrac{w+10}{5-w} = \dfrac{3w}{w-5} - \dfrac{w+10}{w-5} = \dfrac{2w-10}{w-5} = \dfrac{2(w-5)}{w-5} = 2$

14. $\dfrac{2}{r} + \dfrac{r}{s} = \dfrac{2s}{rs} + \dfrac{rr}{rs} = \dfrac{2s+r^2}{rs}$

15.
$$\begin{aligned}\frac{x+2}{x+1} - \frac{x+1}{x+2} = \frac{(x+2)(x+2)}{(x+1)(x+2)} - \frac{(x+1)(x+1)}{(x+1)(x+2)} &= \frac{x^2+4x+4-(x^2+2x+1)}{(x+1)(x+2)} \\ &= \frac{x^2+4x+4-x^2-2x-1}{(x+1)(x+2)} \\ &= \frac{2x+3}{(x+1)(x+2)}\end{aligned}$$

16. $\dfrac{\frac{2u^2w^3}{v^2}}{\frac{4uw^4}{uv}} = \dfrac{2u^2w^3}{v^2} \div \dfrac{4uw^4}{uv} = \dfrac{2u^2w^3}{v^2} \cdot \dfrac{uv}{4uw^4} = \dfrac{2u^3w^3v}{4uv^2w^4} = \dfrac{u^2}{2vw}$

17. $\dfrac{\frac{x}{y}+\frac{1}{2}}{\frac{x}{2}-\frac{1}{y}} = \dfrac{\left(\frac{x}{y}+\frac{1}{2}\right)2y}{\left(\frac{x}{2}-\frac{1}{y}\right)2y} = \dfrac{2x+y}{xy-2}$

18.
$$\begin{aligned}\frac{2}{x-1} + \frac{5}{x+2} &= \frac{11}{x+2} \\ \left(\frac{2}{x-1} + \frac{5}{x+2}\right)(x-1)(x+2) &= \left(\frac{11}{x+2}\right)(x-1)(x+2) \\ 2(x+2)+5(x-1) &= 11(x-1) \\ 2x+4+5x-5 &= 11x-11 \\ 7x-1 &= 11x-11 \\ -4x &= -10 \\ x &= \frac{-10}{-4} = \frac{5}{2}\end{aligned}$$

The answer checks.

© 2015 Cengage Learning. All Rights Reserved. May not be scanned, copied or duplicated, or posted to a publicly accessible website, in whole or in part.

CHAPTER 6 TEST

19.
$$\frac{u-2}{u-3}+3=u+\frac{u-4}{3-u}$$
$$\frac{u-2}{u-3}+3=u+\frac{4-u}{u-3}$$
$$\left(\frac{u-2}{u-3}+3\right)(u-3)=\left(u+\frac{4-u}{u-3}\right)(u-3)$$
$$u-2+3(u-3)=u(u-3)+4-u$$
$$4u-11=u^2-4u+4$$
$$0=u^2-8u+15$$
$$0=(u-5)(u-3)$$
$u-5=0$ **or** $u-3=0$

$u=5$ $\quad u=3$ $\quad$ $u=3$ is extraneous. $u=5$ is a solution.

20.
$$\frac{x^2}{a^2}+\frac{y^2}{b^2}=1$$
$$\left(\frac{x^2}{a^2}+\frac{y^2}{b^2}\right)a^2b^2=a^2b^2$$
$$x^2b^2+y^2a^2=a^2b^2$$
$$x^2b^2=a^2b^2-y^2a^2$$
$$x^2b^2=a^2(b^2-y^2)$$
$$x^2=\frac{a^2(b^2-y^2)}{b^2}$$

21.
$$\frac{1}{r}=\frac{1}{r_1}+\frac{1}{r_2}$$
$$\frac{1}{r}\cdot rr_1r_2=\left(\frac{1}{r_1}+\frac{1}{r_2}\right)rr_1r_2$$
$$r_1r_2=rr_2+rr_1$$
$$r_1r_2-rr_2=rr_1$$
$$r_2(r_1-r)=rr_1$$
$$r_2=\frac{rr_1}{r_1-r}$$

22. Let r = the usual rate (in nautical miles per day). Then $r+11$ = the faster rate.

	Rate	Time	Dist.
Usual	r	$\frac{440}{r}$	440
Faster	$r+11$	$\frac{440}{r+11}$	440

$$\boxed{\text{Usual time}}-2=\boxed{\text{Faster time}}$$
$$\frac{440}{r}-2=\frac{440}{r+11}$$
$$\left(\frac{440}{r}-2\right)(r)(r+11)=\frac{440}{r+11}\cdot r(r+11)$$
$$440(r+11)-2r(r+11)=440r$$
$$-2r^2-22r+4840=0$$
$$-2(r+55)(r-44)=0$$
$r+55=0$ **or** $r-44=0$

$r=-55$ $\quad r=44$ $\quad$ $r=44$ is the only answer that makes sense.

The usual rate is 44 nautical miles per day.

The usual time is $\frac{440}{44}=10$ days.

© 2015 Cengage Learning. All Rights Reserved. May not be scanned, copied or duplicated, or posted to a publicly accessible website, in whole or in part.

23. Let r = the usual rate. Then $r + 0.04$ = the higher rate.

Interest = Principal · Rate · Time (with Time = 1) ⇒ Principal = $\frac{\text{Interest}}{\text{Rate}}$

$$\boxed{\text{Original principal}} - 2000 = \boxed{\text{New principal}}$$

$$\frac{300}{r} - 2000 = \frac{300}{r+0.04}$$

$$\left(\frac{300}{r} - 2000\right)r(r+0.04) = \left(\frac{300}{r+0.04}\right)r(r+0.04)$$

$$300(r+0.04) - 2000r(r+0.04) = 300r$$

$$-2000r^2 - 80r + 12 = 0$$

$$-4(500r^2 + 20r - 3) = 0$$

$$-4(50r - 3)(10r + 1) = 0$$

$50r - 3 = 0$ **or** $10r + 1 = 0$

$r = 0.06$ $\qquad r = -0.10$ $\qquad r = 0.06$ is the only answer that makes sense.

She would invest \$5000 at 6% or \$3000 at 10%.

24. $$\frac{18x^2y^3 - 12x^3y^2 + 9xy}{-3xy^4} = \frac{18x^2y^3}{-3xy^4} + \frac{-12x^3y^2}{-3xy^4} + \frac{9xy}{-3xy^4} = -\frac{6x}{y} + \frac{4x^2}{y^2} - \frac{3}{y^3}$$

25.
$$\begin{array}{r|l} & 3x^2 + 4x + 2 \\ 2x - 1 & 6x^3 + 5x^2 + 0x - 2 \\ & \underline{6x^3 - 3x^2} \\ & \quad 8x^2 + 0x \\ & \quad \underline{8x^2 - 4x} \\ & \qquad 4x - 2 \\ & \qquad \underline{4x - 2} \\ & \qquad\qquad 0 \end{array}$$

26.
$$\begin{array}{r|l} & x^2 - 5x + 10 \\ x + 1 & x^3 - 4x^2 + 5x + 3 \\ & \underline{x^3 + x^2} \\ & \quad -5x^2 + 5x \\ & \quad \underline{-5x^2 - 5x} \\ & \qquad 10x + 3 \\ & \qquad \underline{10x + 10} \\ & \qquad\qquad -7 \end{array}$$

remainder = -7

27.
$$\begin{array}{r|rrrr} 2 & 4 & 3 & 2 & -1 \\ & & 8 & 22 & 48 \\ \hline & 4 & 11 & 24 & \boxed{47} \end{array}$$

28. Let h = the height of the tree.

$$\frac{12}{2} = \frac{h}{3}$$

$$36 = 2h$$

$$18 = h$$

The tree is 18 feet tall.

29. $$\frac{3}{x-2} = \frac{x+3}{2x}$$

$$6x = (x+3)(x-2)$$

$$6x = x^2 + x - 6$$

$$0 = x^2 - 5x - 6$$

$$0 = (x-6)(x+1)$$

$x - 6 = 0$ **or** $x + 1 = 0$

$x = 6$ $\qquad x = -1$

30. $V = \frac{k}{t}$ $\qquad V = \frac{1100}{t}$

$55 = \frac{k}{20}$ $\qquad 75 = \frac{1100}{t}$

$1100 = k$ $\qquad 75t = 1100$

$$t = \frac{1100}{75}$$

$$\boxed{t = \frac{44}{3}}$$

© 2015 Cengage Learning. All Rights Reserved. May not be scanned, copied or duplicated, or posted to a publicly accessible website, in whole or in part.

Cumulative Review Exercises (page 451)

1. $x^4y^3x^7y^{-5} = x^{11}y^{-2} = \dfrac{x^{11}}{y^2}$

2. $\dfrac{a^3b^6}{a^7b^2} = \dfrac{b^4}{a^4}$

3. $\left(\dfrac{2a^2}{3b^4}\right)^{-4} = \left(\dfrac{3b^4}{2a^2}\right)^4 = \dfrac{81b^{16}}{16a^8}$

4. $\left(\dfrac{x^{-2}y^3}{x^2x^3y^4}\right)^{-3} = \left(\dfrac{x^2x^3y^4}{x^{-2}y^3}\right)^3$
$= (x^7y)^3 = x^{21}y^3$

5. $7.23 \times 10^6 = 7{,}230{,}000$

6. $7.12 \times 10^{-4} = 0.000712$

7.
$$\frac{a+2}{5} - \frac{8}{5} = 4a - \frac{a+9}{2}$$
$$\left(\frac{a+2}{5} - \frac{8}{5}\right)10 = \left(4a - \frac{a+9}{2}\right)10$$
$$2(a+2) - 16 = 40a - 5(a+9)$$
$$2a + 4 - 16 = 40a - 5a - 45$$
$$33 = 33a$$
$$1 = a$$

8.
$$\frac{3x-4}{6} - \frac{x-2}{2} = \frac{-2x-3}{3}$$
$$\left(\frac{3x-4}{6} - \frac{x-2}{2}\right)6 = \left(\frac{-2x-3}{3}\right)6$$
$$3x - 4 - 3(x-2) = 2(-2x-3)$$
$$3x - 4 - 3x + 6 = -4x - 6$$
$$4x = -8$$
$$x = -2$$

9. $m = \dfrac{\Delta y}{\Delta x} = \dfrac{5-10}{-2-4} = \dfrac{-5}{-6} = \dfrac{5}{6}$

10.
$$3x + 4y = 13$$
$$4y = -3x + 13$$
$$y = -\tfrac{3}{4}x + \tfrac{13}{4}$$
$$m = -\tfrac{3}{4}$$

11. $y = -5x + 2 \Rightarrow m = -5$
A parallel line will also have $m = -5$.

12. $y = 3x + 2 \Rightarrow m = 3$
A perpendicular line will have $m = -\frac{1}{3}$.

13. $f(3) = 3^2 - 2(3) = 3$

14. $f(-2) = (-2)^2 - 2(-2) = 8$

15. $f\left(\frac{2}{5}\right) = \left(\frac{2}{5}\right)^2 - 2\left(\frac{2}{5}\right) = \frac{4}{25} - \frac{4}{5} = -\frac{16}{25}$

16. $f(t-1) = (t-1)^2 - 2(t-1) = t^2 - 2t + 1 - 2t + 2 = t^2 - 4t + 3$

17. $y = \dfrac{kxz}{r}$

18. Since the graph does not pass the vertical line test, it is not the graph of a function.

© 2015 Cengage Learning. All Rights Reserved. May not be scanned, copied or duplicated, or posted to a publicly accessible website, in whole or in part.

CUMULATIVE REVIEW EXERCISES

19.
$$x - 2 \le 3x + 1 \le 5x - 4$$
$$\begin{array}{rcl} x - 2 \le 3x + 1 & \text{and} & 3x + 1 \le 5x - 4 \\ x - 2 - 1 \le 3x + 1 - 1 & & 3x + 1 + 4 \le 5x - 4 + 4 \\ x - x - 3 \le 3x - x & & 3x - 3x + 5 \le 5x - 3x \\ -3 \le 2x & & 5 \le 2x \\ \frac{-3}{2} \le x & & \frac{5}{2} \le x \\ x \ge -\frac{3}{2} & & x \ge \frac{5}{2} \end{array}$$
If $x \ge -\frac{3}{2}$ **and** $x \ge \frac{5}{2}$, then $x \ge \frac{5}{2}$. The solution set is $\left[\frac{5}{2}, \infty\right)$

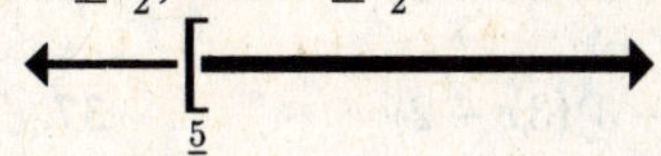

20.
$$\left|\frac{3a}{5} - 2\right| + 1 \ge \frac{6}{5}$$
$$\left|\frac{3a}{5} - 2\right| \ge \frac{1}{5}$$
$$\begin{array}{rcl} \frac{3a}{5} - 2 \le -\frac{1}{5} & \text{or} & \frac{3a}{5} - 2 \ge \frac{1}{5} \\ \frac{3a}{5} \le \frac{9}{5} & & \frac{3a}{5} \ge \frac{11}{5} \\ 3a \le 9 & & 3a \ge 11 \\ a \le 3 & & a \ge \frac{11}{3} \end{array}$$
solution set: $(-\infty, 3] \cup \left[\frac{11}{3}, \infty\right)$

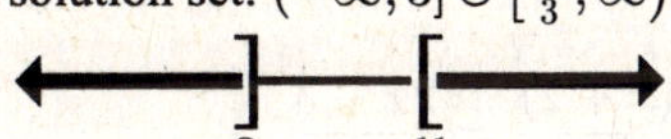

21. binomial

22. degree $= 3 + 4 = 7$

23. $f(-2) = -(-2)^3 + 7(-2) - 2 = -8$

24. $y = f(x) = 2x^2 - 3$

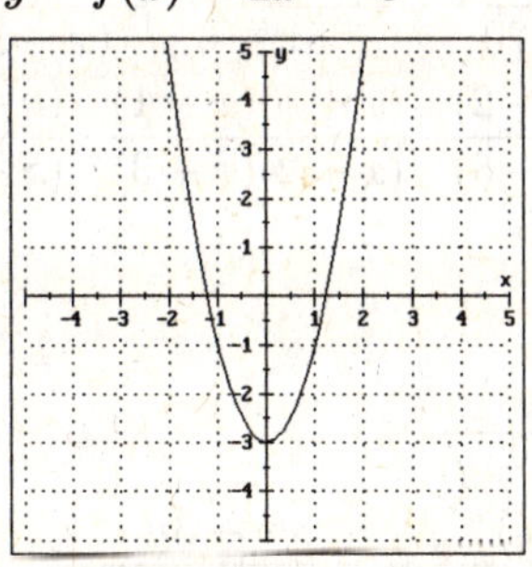

25. $(3x^2 - 2x + 7) + (-2x^2 + 2x + 5) + (3x^2 - 4x + 2)$
$$= 3x^2 - 2x + 7 - 2x^2 + 2x + 5 + 3x^2 - 4x + 2 = 4x^2 - 4x + 14$$

26. $(-5x^2 + 3x + 4) - (-2x^2 + 3x + 7) = -5x^2 + 3x + 4 + 2x^2 - 3x - 7 = -3x^2 - 3$

27. $(4x - 3)(2x + 7) = 8x^2 + 28x - 6x - 21$
$$= 8x^2 + 22x - 21$$

28. $(2x^n - 1)(x^n + 2) = 2x^{2n} + 4x^n - x^n - 2$
$$= 2x^{2n} + 3x^n - 2$$

© 2015 Cengage Learning. All Rights Reserved. May not be scanned, copied or duplicated, or posted to a publicly accessible website, in whole or in part.

CUMULATIVE REVIEW EXERCISES

29. $3r^2s^3 - 6rs^4 = 3rs^3(r - 2s)$

30. $x(a-b) - 4(a-b) = (a-b)(x-4)$

31. $xu + yv + xv + yu = xu + xv + yu + yv = x(u+v) + y(u+v) = (u+v)(x+y)$

32. $81x^4 - 16y^4 = (9x^2+4y^2)(9x^2-4y^2) = (9x^2+4y^2)(3x+2y)(3x-2y)$

33. $27y^3 - 64x^3 = (3y)^3 - (4x)^3 = (3y-4x)(9y^2+12xy+16x^2)$

34. $6x^2 + 5x - 6 = (2x+3)(3x-2)$

35. $9x^2 - 30x + 25 = (3x-5)(3x-5)$

36. $15x^2 - x - 6 = (5x+3)(3x-2)$

37. $27a^3 + 8b^3 = (3a+2b)(9a^2 - 6ab + 4b^2)$

38. $6x^2 + x - 35 = (3x-7)(2x+5)$

39. $x^2 + 10x + 25 - y^4 = (x+5)^2 - y^4 = (x+5+y^2)(x+5-y^2)$

40. $y^2 - x^2 + 4x - 4 = y^2 - (x^2 - 4x + 4) = y^2 - (x-2)^2 = (y+x-2)(y-x+2)$

41.
$$\begin{aligned} x^3 - 9x &= 0 \\ x(x^2-9) &= 0 \\ x(x+3)(x-3) &= 0 \end{aligned}$$
$$x = 0 \quad \textbf{or} \quad x+3=0 \quad \textbf{or} \quad x-3=0$$
$$x=-3 \qquad\qquad x=3$$

42.
$$\begin{aligned} 6x^2 + 7 &= -23x \\ 6x^2 + 23x + 7 &= 0 \\ (2x+7)(3x+1) &= 0 \end{aligned}$$
$$2x+7=0 \quad \textbf{or} \quad 3x+1=0$$
$$x=-\tfrac{7}{2} \qquad\qquad x=-\tfrac{1}{3}$$

43. $$\frac{2x^2y + xy - 6y}{3x^2y + 5xy - 2y} = \frac{y(2x^2+x-6)}{y(3x^2+5x-2)} = \frac{y(2x-3)(x+2)}{y(3x-1)(x+2)} = \frac{2x-3}{3x-1}$$

44. $$\begin{aligned} \frac{x^2-4}{x^2+9x+20} \div \frac{x^2+5x+6}{x^2+4x-5} \cdot \frac{x^2+3x-4}{(x-1)^2} &= \frac{x^2-4}{x^2+9x+20} \cdot \frac{x^2+4x-5}{x^2+5x+6} \cdot \frac{x^2+3x-4}{(x-1)^2} \\ &= \frac{(x+2)(x-2)}{(x+4)(x+5)} \cdot \frac{(x+5)(x-1)}{(x+2)(x+3)} \cdot \frac{(x+4)(x-1)}{(x-1)(x-1)} \\ &= \frac{x-2}{x+3} \end{aligned}$$

45. $$\begin{aligned} \frac{2}{x+y} + \frac{3}{x-y} - \frac{x-3y}{x^2-y^2} &= \frac{2}{x+y} + \frac{3}{x-y} - \frac{x-3y}{(x+y)(x-y)} \\ &= \frac{2(x-y)}{(x+y)(x-y)} + \frac{3(x+y)}{(x+y)(x-y)} - \frac{x-3y}{(x+y)(x-y)} \\ &= \frac{2x-2y+3x+3y-x+3y}{(x+y)(x-y)} \\ &= \frac{4x+4y}{(x+y)(x-y)} = \frac{4(x+y)}{(x+y)(x-y)} = \frac{4}{x-y} \end{aligned}$$

© 2015 Cengage Learning. All Rights Reserved. May not be scanned, copied or duplicated, or posted to a publicly accessible website, in whole or in part.

46. $\dfrac{\frac{a}{b}+b}{a-\frac{b}{a}} = \dfrac{\left(\frac{a}{b}+b\right)ab}{\left(a-\frac{b}{a}\right)ab} = \dfrac{a^2+ab^2}{a^2b-b^2}$

47.
$$\begin{aligned} \frac{5x-3}{x+2} &= \frac{5x+3}{x-2} \\ (5x-3)(x-2) &= (5x+3)(x+2) \\ 5x^2-13x+6 &= 5x^2 = 13x+6 \\ -26x &= 0 \\ x &= 0 \quad \text{The answer checks.} \end{aligned}$$

48.
$$\begin{aligned} \frac{3}{x-2}+\frac{x^2}{(x+3)(x-2)} &= \frac{x+4}{x+3} \\ \left(\frac{3}{x-2}+\frac{x^2}{(x+3)(x-2)}\right)(x+3)(x-2) &= \left(\frac{x+4}{x+3}\right)(x+3)(x-2) \\ 3(x+3)+x^2 &= (x+4)(x-2) \\ x^2+3x+9 &= x^2+2x-8 \\ x &= -17 \quad \text{The answer checks.} \end{aligned}$$

49.
$$\begin{array}{r|l} & x+4 \\ \hline x+5 & x^2+9x+20 \\ & \underline{x^2+5x} \\ & \quad 4x+20 \\ & \quad \underline{4x+20} \\ & \qquad 0 \end{array}$$

50.
$$\begin{array}{r|l} & -x^2+x+5+\frac{8}{x-1} \\ \hline x-1 & -x^3+2x^2+4x+3 \\ & \underline{-x^3+x^2} \\ & \quad x^2+4x \\ & \quad \underline{x^2-x} \\ & \qquad 5x+3 \\ & \qquad \underline{5x-5} \\ & \qquad\quad 8 \end{array}$$

Exercises 7.1 (page 464)

1. $(5)^2 = 25, (-5)^2 = 25$

3. $(2x)^2 = 4x^2, (-2x)^2 = 4x^2$

5. $(2)^3 = 8$

7. $(3)^4 = 81, (-3)^4 = 81$

9. $\dfrac{x^2+7x+10}{x^2-4} = \dfrac{(x+2)(x+5)}{(x+2)(x-2)} = \dfrac{x+5}{x-2}$

11. $\dfrac{x^2-x-6}{x^2-2x-3}\cdot\dfrac{x^2-1}{x^2+x-2} = \dfrac{(x-3)(x+2)}{(x-3)(x+1)}\cdot\dfrac{(x+1)(x-1)}{(x+2)(x-1)} = 1$

13. $\dfrac{3}{m+1}+\dfrac{3m}{m-1} = \dfrac{3(m-1)}{(m+1)(m-1)}+\dfrac{3m(m+1)}{(m+1)(m-1)} = \dfrac{3m^2+6m-3}{(m+1)(m-1)} = \dfrac{3(m^2+2m-1)}{(m+1)(m-1)}$

15. $(5x^2)^2; 6^2 = 36$

17. positive

19. 5; left

21. radical; index; radicand

23. $|x|$

25. x

27. even

29. $7y^2$

31. a^2+b^3

33. $10^2 = 100; 10, -10$

35. $7^2 = 49; 7, -7$

© 2015 Cengage Learning. All Rights Reserved. May not be scanned, copied or duplicated, or posted to a publicly accessible website, in whole or in part.

SECTION 7.1

37. $\sqrt{121} = \sqrt{11^2} = 11$

39. $-\sqrt{64} = -\sqrt{8^2} = -8$

41. $-\sqrt{\frac{25}{49}} = -\sqrt{\left(\frac{5}{7}\right)^2} = -\frac{5}{7}$

43. $\sqrt{-4}$: not a real number

45. $\sqrt{0.16} = \sqrt{(0.4)^2} = 0.4$

47. $\sqrt{25x^2} = \sqrt{(5x)^2} = |5x| = 5|x|$

49. $\sqrt{64y^4} = \sqrt{(8y^2)^2} = |8y^2| = 8y^2$

51. $\sqrt{(x+3)^2} = |x+3|$

53. $\sqrt{a^2+6a+9} = \sqrt{(a+3)^2} = |a+3|$

55. $\sqrt[3]{1} = \sqrt[3]{1^3} = 1$

57. $\sqrt[3]{-27} = \sqrt[3]{(-3)^3} = -3$

59. $\sqrt[3]{-\frac{64}{27}} = \sqrt[3]{\left(-\frac{4}{3}\right)^3} = -\frac{4}{3}$

61. $\sqrt[3]{0.064} = \sqrt[3]{(0.4)^3} = 0.4$

63. $\sqrt[3]{125y^3} = \sqrt[3]{(5y)^3} = 5y$

65. $\sqrt[3]{-1000p^3q^3} = \sqrt[3]{(-10pq)^3} = -10pq$

67. $-\sqrt[5]{243} = -\sqrt[5]{3^5} = -3$

69. $-\sqrt[4]{16} = -\sqrt[4]{2^4} = -2$

71. $\sqrt[6]{64} = \sqrt[6]{2^6} = 2$

73. $\sqrt[4]{\frac{16}{625}} = \sqrt[4]{\left(\frac{2}{5}\right)^4} = \frac{2}{5}$

75. $\sqrt[4]{-256} \Rightarrow$ not a real number

77. $-\sqrt[5]{-\frac{1}{32}} = -\sqrt[5]{\left(-\frac{1}{2}\right)^5} = -\left(-\frac{1}{2}\right) = \frac{1}{2}$

79. $\sqrt[4]{81x^4} = \sqrt[4]{(3x)^4} = |3x| = 3|x|$

81. $\sqrt[3]{8a^3} = \sqrt[3]{(2a)^3} = 2a$

83. $\sqrt[4]{\frac{1}{16}x^4} = \sqrt[4]{\left(\frac{1}{2}x\right)^4} = \left|\frac{1}{2}x\right| = \frac{1}{2}|x|$

85. $\sqrt[4]{x^{12}} = \sqrt[4]{(x^3)^4} = |x^3|$

87. $\sqrt[5]{-x^5} = \sqrt[5]{(-x)^5} = -x$

89. $\sqrt[3]{-27a^6} = \sqrt[3]{(-3a^2)^3} = -3a^2$

91. $f(x) = \sqrt{x+4}$; Shift $y = \sqrt{x}$ left 4.

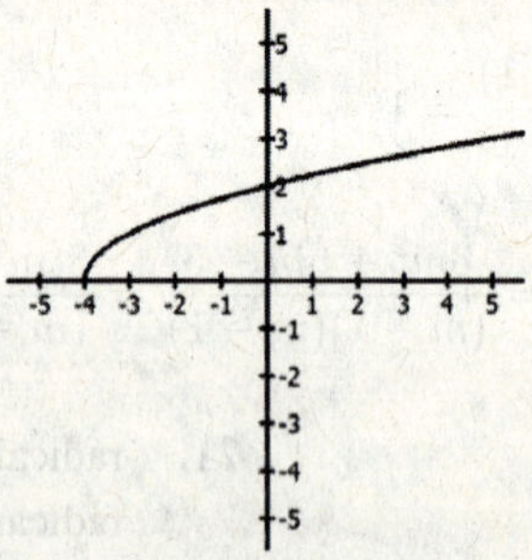

$D = [-4, \infty); R = [0, \infty)$

93. $f(x) = -\sqrt{x} - 3$; Reflect $y = \sqrt{x}$ about the x-axis and shift down 3.

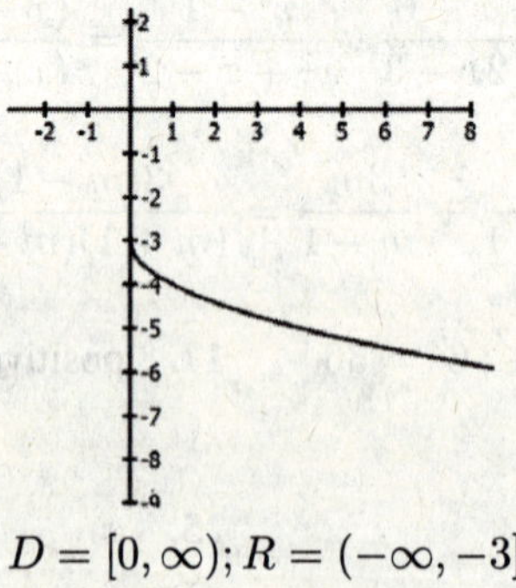

$D = [0, \infty); R = (-\infty, -3]$

© 2015 Cengage Learning. All Rights Reserved. May not be scanned, copied or duplicated, or posted to a publicly accessible website, in whole or in part.

SECTION 7.1

95. $f(x) = \sqrt[3]{x} - 1$; Shift $y = \sqrt[3]{x}$ down 1.

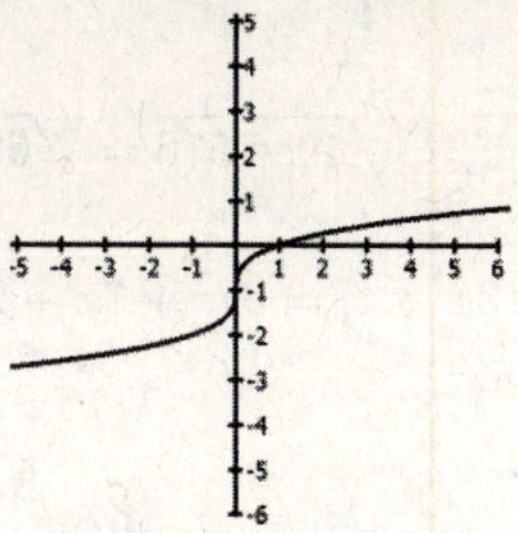

$D = (-\infty, \infty); R = (-\infty, \infty)$

97. $f(x) = -\sqrt[3]{x-1} - 2$; Reflect $y = \sqrt[3]{x}$ about the x-axis. Shift right 1 and down 2.

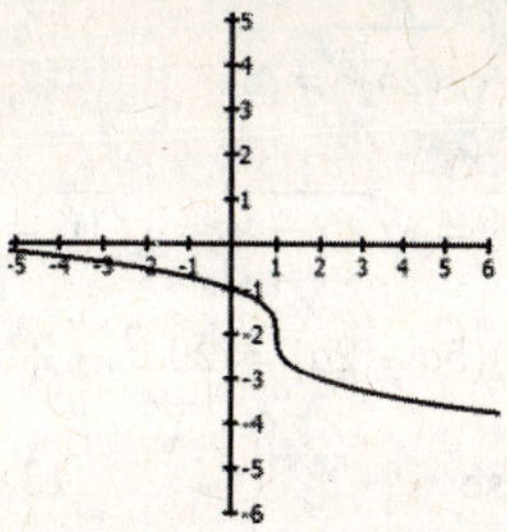

$D = (-\infty, \infty); R = (-\infty, \infty)$

99. $\sqrt{(-4)^2} = \sqrt{16} = 4$

101. $\sqrt{-36}$: not a real number

103. $\sqrt{(-5b)^2} = |-5b| = 5|b|$

105. $\sqrt{t^2 + 24t + 144} = \sqrt{(t+12)^2} = |t + 12|$

107. $\sqrt[3]{-\frac{1}{8}m^6n^3} = \sqrt[3]{\left(-\frac{1}{2}m^2n\right)^3} = -\frac{1}{2}m^2n$

109. $\sqrt[3]{0.008z^9} = \sqrt[3]{(0.2z^3)^3} = 0.2z^3$

111. $\sqrt[25]{(x+2)^{25}} = x + 2$

113. $\sqrt[8]{0.00000001x^{16}y^8} = \sqrt[8]{(0.1x^2y)^8}$
$= |0.1x^2y| = 0.1x^2|y|$

115. $\sqrt{12} \approx 3.4641$

117. $\sqrt{679.25} \approx 26.0624$

119. $t = 2\pi\sqrt{\frac{l}{32}} = 2\pi\sqrt{\frac{13}{32}} \approx 4.0048 \approx 4$ sec

121. mean $= \frac{2+5+5+6+7}{5} = \frac{25}{5} = 5$

Original term	Mean	Difference (term−mean)	Square of difference
2	5	−3	9
5	5	0	0
5	5	0	0
6	5	1	1
7	5	2	4

st. dev. $= \sqrt{\frac{9+0+0+1+4}{5}} \approx 1.67$

123. $s_{\overline{x}} = \frac{s}{\sqrt{N}} = \frac{65}{\sqrt{30}} \approx 11.8673$

125. $r = \sqrt{\frac{A}{\pi}} = \sqrt{\frac{9\pi}{\pi}} = \sqrt{9} = 3$ units

127. $t = \frac{\sqrt{s}}{4} = \frac{\sqrt{256}}{4} = \frac{16}{4} = 4$ seconds

129. $I = \sqrt{\frac{P}{18}} = \sqrt{\frac{980}{18}} \approx \sqrt{54.44} \approx 7.4$ amps

131. Answers may vary.

133. $\sqrt{x^2 - 4x + 4} = \sqrt{(x-2)^2} = |x - 2|$.
$|x - 2| = x - 2$ when $x - 2 \geq 0$, or $x \geq 2$.

© 2015 Cengage Learning. All Rights Reserved. May not be scanned, copied or duplicated, or posted to a publicly accessible website, in whole or in part.

Exercises 7.2 (page 471)

1. $\sqrt{625} = \sqrt{(25)^2} = 25$

3. $\sqrt{7^2 + 24^2} = \sqrt{49 + 576} = \sqrt{625} = 25$

5. $\sqrt{5^2 - 3^2} = \sqrt{25 - 9} = \sqrt{16} = 4$

7. $(2x + 5)(3x - 4) = 6x^2 + 7x - 20$

9. $(4a - 3b)(5a - 2b) = 20a^2 - 23ab + 6b^2$

11. hypotenuse

13. $a^2 + b^2 = c^2$

15. positive

17.
$$\begin{aligned} c^2 &= a^2 + b^2 \\ c^2 &= 6^2 + 8^2 \\ c^2 &= 36 + 64 \\ c^2 &= 100 \\ c &= \sqrt{100} \\ c &= 10 \text{ ft} \end{aligned}$$

19.
$$\begin{aligned} c^2 &= a^2 + b^2 \\ 82^2 &= a^2 + 18^2 \\ 6724 &= a^2 + 324 \\ 6400 &= a^2 \\ \sqrt{6400} &= a \\ 80 \text{ m} &= a \end{aligned}$$

21.
$$\begin{aligned} c^2 &= a^2 + b^2 \\ 50^2 &= 14^2 + b^2 \\ 2500 &= 196 + b^2 \\ 2304 &= b^2 \\ \sqrt{2304} &= b \\ 48 \text{ in.} &= b \end{aligned}$$

23.
$$\begin{aligned} c^2 &= a^2 + b^2 \\ c^2 &= \left(\sqrt{8}\right)^2 + \left(\sqrt{8}\right)^2 \\ c^2 &= 8 + 8 \\ c^2 &= 16 \\ c &= \sqrt{16} \\ c &= 4 \text{ mi} \end{aligned}$$

25.
$$\begin{aligned} d &= \sqrt{(x_2 - x_1)^2 + (y_2 - y_1)^2} \\ &= \sqrt{(3 - 0)^2 + (-4 - 0)^2} \\ &= \sqrt{3^2 + (-4)^2} \\ &= \sqrt{9 + 16} = \sqrt{25} = 5 \end{aligned}$$

27. $d = \sqrt{(x_2 - x_1)^2 + (y_2 - y_1)^2} = \sqrt{(-2 - 10)^2 + (3 - 8)^2} = \sqrt{(-12)^2 + (-5)^2} = \sqrt{144 + 25}$
$= \sqrt{169} = 13$

29. $d = \sqrt{(x_2 - x_1)^2 + (y_2 - y_1)^2} = \sqrt{[-1 - (-7)]^2 + (4 - 12)^2} = \sqrt{6^2 + (-8)^2} = \sqrt{36 + 64}$
$= \sqrt{100} = 10$

31. $d = \sqrt{(x_2 - x_1)^2 + (y_2 - y_1)^2} = \sqrt{(-3 - 12)^2 + [2 - (-6)]^2} = \sqrt{(-15)^2 + 8^2} = \sqrt{225 + 64}$
$= \sqrt{289} = 17$

33. $d = \sqrt{(x_2 - x_1)^2 + (y_2 - y_1)^2} = \sqrt{[-5 - (-3)]^2 + (-5 - 5)^2} = \sqrt{(-2)^2 + (-10)^2}$
$= \sqrt{4 + 100} = \sqrt{104} \approx 10.2$

35. $d = \sqrt{(x_2 - x_1)^2 + (y_2 - y_1)^2} = \sqrt{[4 - (-9)]^2 + (7 - 3)^2} = \sqrt{13^2 + 4^2} = \sqrt{169 + 16}$
$= \sqrt{185} \approx 13.6$

© 2015 Cengage Learning. All Rights Reserved. May not be scanned, copied or duplicated, or posted to a publicly accessible website, in whole or in part.

SECTION 7.2

37. Let $x =$ distance to 2nd

$x^2 = 90^2 + 90^2$

$x^2 = 8100 + 8100$

$x^2 = 16200$

$x \approx 127$ ft

39. Refer to the diagram provided. The 3rd baseman is at B, so $\overline{BC}$ has a length of 10 ft. Let $\overline{AC}$ and $\overline{AB}$ both have a length x.

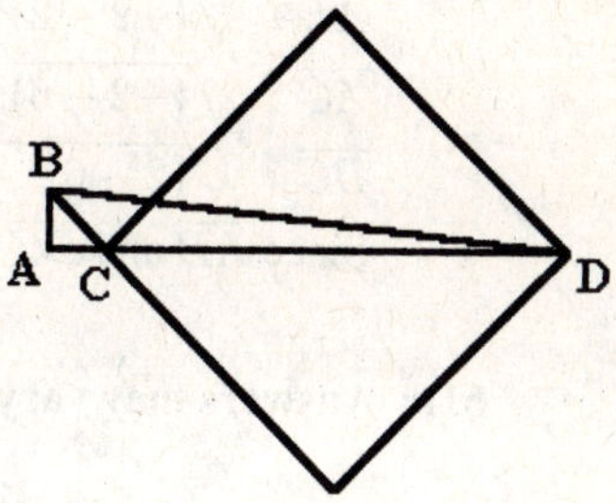

$x^2 + x^2 = 10^2$

$2x^2 = 100$

$x^2 = 50$

$x = \sqrt{50} \approx 7.1$ ft

From **#37**, the length of $\overline{CD} \approx 127.3$ ft, so $\overline{AD}$ has a length of about $127.3 + 7.1 = 134.4$ ft. Let y = the length of $\overline{BD}$.

$y^2 = (7.1)^2 + (134.4)^2$

$y^2 = 18113.77$

$y = \sqrt{18113.77} \approx 135$ ft

41. Let $x =$ the length of the diagonal.

$7^2 + 7^2 = x^2$

$49 + 49 = x^2$

$98 = x^2$

$\sqrt{98} = x$

$9.9 \text{ cm} \approx x$

43. Let $x =$ half the length of the stretched wire.

$x^2 = 20^2 + 1^2$

$x^2 = 401$

$x = 20.025$ ft

The stretched wire has a length of 40.05 ft.
It has been stretched by 0.05 ft.

45. $d^2 = 5^2 + 12^2$

$d^2 = 25 + 144$

$d^2 = 169$

$d = 13$ ft

47. $37^2 = 9^2 + h^2$

$1369 = 81 + h^2$

$1288 = h^2$

$35.9 = h$

The ladder will reach.

49. Let $x =$ direct distance from A to D.

$x^2 = 52^2 + (105 + 60)^2$

$x^2 = 52^2 + 165^2$

$x^2 = 29929$

$x = 173$ yd

51. $A = 6\sqrt[3]{V^2} = 6\sqrt[3]{8^2} = 6\sqrt[3]{64} = 6(4)$

$= 24 \text{ cm}^2$

53. $d = \sqrt{a^2 + b^2 + c^2} = \sqrt{12^2 + 24^2 + 17^2} = \sqrt{1009} = 31.76$ in. $\Rightarrow$ The racket will not fit.

55. $d = \sqrt{a^2 + b^2 + c^2} = \sqrt{21^2 + 21^2 + 3^2} = \sqrt{891} = 29.8$ in. $\Rightarrow$ The femur will not fit.

© 2015 Cengage Learning. All Rights Reserved. May not be scanned, copied or duplicated, or posted to a publicly accessible website, in whole or in part.

57. Let the points be represented by $A(5,1)$, $B(7,0)$ and $C(3,0)$. Find the length of $\overline{AB}$ and $\overline{AC}$:
$\overline{AB}: \sqrt{(5-7)^2+(1-0)^2} = \sqrt{(-2)^2+1^2} = \sqrt{5}$
$\overline{AC}: \sqrt{(5-3)^2+(1-0)^2} = \sqrt{(2)^2+1^2} = \sqrt{5}$
Since $\overline{AB}$ and $\overline{AC}$ have the same length, $(5,1)$ is equidistant from $(7,0)$ and $(3,0)$.

59. Let the points be represented by $A(-2,4)$, $B(2,8)$ and $C(6,4)$. Find the length of each side:
$\overline{AB}: \sqrt{(-2-2)^2+(4-8)^2} = \sqrt{(-4)^2+(-4)^2} = \sqrt{32}$
$\overline{AC}: \sqrt{(-2-6)^2+(4-4)^2} = \sqrt{(-8)^2+0^2} = \sqrt{64}$
$\overline{BC}: \sqrt{(2-6)^2+(8-4)^2} = \sqrt{(-4)^2+4^2} = \sqrt{32}$
Since $\overline{AB}$ and $\overline{BC}$ have the same length, the triangle is isosceles.

61. **Answers may vary.**

63. $I = \dfrac{703w}{h^2} = \dfrac{703(104)}{(54.1)^2} = \dfrac{73112}{2926.81} \approx 25$

65.
$$\begin{aligned} v^2 &= 64d \\ v^2 &= 64(4) \\ v^2 &= 256 \\ v &= \sqrt{256} = 16 \text{ ft/sec} \end{aligned}$$

Exercises 7.3 (page 480)

1. $-7^2 = -1 \cdot 7 \cdot 7 = -49$

3. $5^3 = 5 \cdot 5 \cdot 5 = 125$

5. $4^{-2} = \dfrac{1}{4^2} = \dfrac{1}{16}$

7. $\sqrt{4^3} = \sqrt{64} = 8$

9. $\sqrt[3]{8^2} = \sqrt[3]{64} = 4$

11.
$$\begin{aligned} 8x - 7 &\le 1 \\ 8x &\le 8 \\ x &\le 1 \end{aligned}$$

13.
$$\begin{aligned} \frac{4}{5}(r-3) &> \frac{2}{3}(r+2) \\ 15 \cdot \frac{4}{5}(r-3) &> 15 \cdot \frac{2}{3}(r+2) \\ 12(r-3) &> 10(r+2) \\ 2r &> 56 \\ r &> 28 \end{aligned}$$

15. Let x = pints of water added (0% alcohol).

Alcohol at start	+	Alcohol added	=	Alcohol at end

$$\begin{aligned} 0.20(5) + 0(x) &= 0.15(5+x) \\ 1 + 0 &= 0.75 + 0.15x \\ 0.25 &= 0.15x \\ x &= \frac{0.25}{0.15} = \frac{5}{3} \end{aligned}$$

$1\frac{2}{3}$ pints of water should be added.

17. $a \cdot a \cdot a \cdot a$

19. a^{mn}

21. $\dfrac{a^n}{b^n}$

23. $\dfrac{1}{a^n}$; 0

25. $\left(\dfrac{b}{a}\right)^n$

27. $|x|$

29. $5^{1/4} = \sqrt[4]{5}$

31. $8^{1/5} = \sqrt[5]{8}$

© 2015 Cengage Learning. All Rights Reserved. May not be scanned, copied or duplicated, or posted to a publicly accessible website, in whole or in part.

SECTION 7.3

33. $(13a)^{1/7} = \sqrt[7]{13a}$

35. $\left(\frac{1}{2}x^3y\right)^{1/4} = \sqrt[4]{\frac{1}{2}x^3y}$

37. $(6a^3b)^{1/4} = \sqrt[4]{6a^3b}$

39. $(x^2 + y^2)^{1/2} = \sqrt{x^2 + y^2}$

41. $49^{1/2} = \sqrt{49} = 7$

43. $27^{1/3} = \sqrt[3]{27} = 3$

45. $\left(\frac{1}{9}\right)^{1/2} = \sqrt{\frac{1}{9}} = \frac{1}{3}$

47. $\left(\frac{1}{8}\right)^{1/3} = \sqrt[3]{\frac{1}{8}} = \frac{1}{2}$

49. $-81^{1/4} = -\sqrt[4]{81} = -3$

51. $(-25)^{1/2} = \sqrt{-25}$; not a real number

53. $\sqrt{7} = 7^{1/2}$

55. $\sqrt[4]{3a} = (3a)^{1/4}$

57. $5\sqrt[7]{b} = 5b^{1/7}$

59. $\sqrt[6]{\frac{1}{7}abc} = \left(\frac{1}{7}abc\right)^{1/6}$

61. $\sqrt[5]{\frac{1}{2}mn} = \left(\frac{1}{2}mn\right)^{1/5}$

63. $\sqrt[3]{x^2 + y^2} = (x^2 + y^2)^{1/3}$

65. $(25y^2)^{1/2} = \left[(5y)^2\right]^{1/2} = |5y| = 5|y|$

67. $(243x^5)^{1/5} = \left[(3x)^5\right]^{1/5} = 3x$

69. $\left[(x+1)^4\right]^{1/4} = |x + 1|$

71. $(-81x^{12})^{1/4} \Rightarrow$ not a real number

73. $25^{3/2} = \left(25^{1/2}\right)^3 = 5^3 = 125$

75. $81^{3/4} = \left(81^{1/4}\right)^3 = 3^3 = 27$

77. $\left(-27x^3\right)^{4/3} = \left[\left(-27x^3\right)^{1/3}\right]^4 = [-3x]^4$
$= 81x^4$

79. $\left(\frac{1}{8}\right)^{2/3} = \left[\left(\frac{1}{8}\right)^{1/3}\right]^2 = \left(\frac{1}{2}\right)^2 = \frac{1}{4}$

81. $4^{-1/2} = \dfrac{1}{4^{1/2}} = \dfrac{1}{2}$

83. $4^{-3/2} = \dfrac{1}{4^{3/2}} = \dfrac{1}{\left(4^{1/2}\right)^3} = \dfrac{1}{2^3} = \dfrac{1}{8}$

85. $(16x^2)^{-3/2} = \dfrac{1}{(16x^2)^{3/2}} = \dfrac{1}{\left[(16x^2)^{1/2}\right]^3} = \dfrac{1}{(4x)^3} = \dfrac{1}{64x^3}$

87. $\left(-27y^3\right)^{-2/3} = \dfrac{1}{(-27y^3)^{2/3}} = \dfrac{1}{\left[(-27y^3)^{1/3}\right]^2} = \dfrac{1}{(-3y)^2} = \dfrac{1}{9y^2}$

89. $\left(\dfrac{1}{4}\right)^{-3/2} = \left(\dfrac{4}{1}\right)^{3/2} = 4^{3/2} = \left(4^{1/2}\right)^3 - 2^3 = 8$

91. $\left(\dfrac{27}{8}\right)^{-4/3} = \left(\dfrac{8}{27}\right)^{4/3} = \left[\left(\dfrac{8}{27}\right)^{1/3}\right]^4 = \left(\dfrac{2}{3}\right)^4 = \dfrac{16}{81}$

93. $5^{4/9}5^{4/9} = 5^{4/9+4/9} = 5^{8/9}$

95. $\left(4^{1/5}\right)^3 = 4^{(1/5)\cdot 3} = 4^{3/5}$

97. $6^{-2/3}6^{-4/3} = 6^{-6/3} = 6^{-2}$
$= \frac{1}{36}$

© 2015 Cengage Learning. All Rights Reserved. May not be scanned, copied or duplicated, or posted to a publicly accessible website, in whole or in part.

SECTION 7.3

99. $\dfrac{9^{4/5}}{9^{3/5}} = 9^{4/5-3/5} = 9^{1/5}$

101. $\dfrac{7^{1/2}}{7^0} = 7^{1/2-0} = 7^{1/2}$

103. $\dfrac{2^{5/6}2^{1/3}}{2^{1/2}} = \dfrac{2^{7/6}}{2^{1/2}} = 2^{4/6}$
$= 2^{2/3}$

105. $\left(x^{1/4} \cdot x^{3/2}\right)^8 = x^{8/4} \cdot x^{24/2} = x^{14}$

107. $\left(a^{2/3}\right)^{1/3} = a^{(2/3)(1/3)} = a^{2/9}$

109. $y^{1/3}\left(y^{2/3} + y^{5/3}\right) = y^{3/3} + y^{6/3} = y + y^2$

111. $x^{3/5}\left(x^{7/5} - x^{2/5} + 1\right) = x^{10/5} - x^{5/5} + x^{3/5} = x^2 - x + x^{3/5}$

113. $\left(x^{1/2} + y^{1/2}\right)\left(x^{1/2} - y^{1/2}\right) = x^{2/2} - x^{1/2}y^{1/2} + x^{1/2}y^{1/2} - y^{2/2} = x - y$

115. $\left(x^{2/3} + y^{2/3}\right)^2 = \left(x^{2/3} + y^{2/3}\right)\left(x^{2/3} + y^{2/3}\right) = x^{4/3} + x^{2/3}y^{2/3} + x^{2/3}y^{2/3} + y^{4/3}$
$= x^{4/3} + 2x^{2/3}y^{2/3} + y^{4/3}$

117. $\sqrt[6]{p^3} = \left(p^3\right)^{1/6} = p^{3/6} = p^{1/2} = \sqrt{p}$

119. $\sqrt[4]{25b^2} = \left(5^2b^2\right)^{1/4} = 5^{1/2}b^{1/2} = \sqrt{5b}$

121. $16^{1/4} = \sqrt[4]{16} = 2$

123. $32^{1/5} = \sqrt[5]{32} = 2$

125. $0^{1/3} = \sqrt[3]{0} = 0$

127. $(-27)^{1/3} = \sqrt[3]{-27} = -3$

129. $\left(25x^4\right)^{3/2} = \left[\left(25x^4\right)^{1/2}\right]^3 = \left(5x^2\right)^3$
$= 125x^6$

131. $\left(\dfrac{8x^3}{27}\right)^{2/3} = \left[\left(\dfrac{8x^3}{27}\right)^{1/3}\right]^2 = \left(\dfrac{2x}{3}\right)^2$
$= \dfrac{4x^2}{9}$

133. $\left(-32p^5\right)^{-2/5} = \dfrac{1}{\left(-32p^5\right)^{2/5}} = \dfrac{1}{\left[\left(-32p^5\right)^{1/5}\right]^2} = \dfrac{1}{(-2p)^2} = \dfrac{1}{4p^2}$

135. $\left(-\dfrac{8x^3}{27}\right)^{-1/3} = \left(-\dfrac{27}{8x^3}\right)^{1/3} = -\dfrac{3}{2x}$

137. $\left(a^{1/2}b^{1/3}\right)^{3/2} = a^{3/4}b^{1/2}$

139. $\left(mn^{-2/3}\right)^{-3/5} = m^{-3/5}n^{2/5} = \dfrac{n^{2/5}}{m^{3/5}}$

141. $\dfrac{\left(4x^3y\right)^{1/2}}{(9xy)^{1/2}} = \dfrac{2x^{3/2}y^{1/2}}{3x^{1/2}y^{1/2}} = \dfrac{2x}{3}$

143. $\left(27x^{-3}\right)^{-1/3} = (27)^{-1/3}x = \dfrac{1}{3}x$

145. $x^{4/3}\left(x^{2/3} + 3x^{5/3} - 4\right) = x^{6/3} + 3x^{9/3} - 4x^{4/3} = x^2 + 3x^3 - 4x^{4/3}$

© 2015 Cengage Learning. All Rights Reserved. May not be scanned, copied or duplicated, or posted to a publicly accessible website, in whole or in part.

SECTION 7.3

147. $\left(x^{-1/2}-x^{1/2}\right)^2=\left(x^{-1/2}-x^{1/2}\right)\left(x^{-1/2}-x^{1/2}\right)=x^{-2/2}-x^{1/2}x^{-1/2}-x^{1/2}x^{-1/2}+x^{2/2}$
$=x^{-1}-1-1+x=\frac{1}{x}-2+x$

Problems 149-155 are to be solved using a calculator. The keystrokes needed to solve each problem using a TI-83 graphing calculator appear in each solution. There may be other solutions. Keystrokes for other calculators may be slightly different.

149. [1] [5] [^] [(] [1] [÷] [3] [)] [ENTER] {2.4662...} ⇒ 2.47

151. [1] [.] [0] [4] [5] [^] [(] [1] [÷] [5] [)] [ENTER] {1.0088...} ⇒ 1.01

153. [1] [7] [^] [(] [(-)] [1] [÷] [2] [)] [ENTER] {0.2425...} ⇒ 0.24

155. [(] [(-)] [.] [2] [5] [)] [^] [(] [(-)] [1] [÷] [5] [)] [ENTER] {−1.3195...} ⇒ −1.32

157. Answers may vary.

159. $16^{2/4}=2^2=4;\ 16^{1/2}=4\Rightarrow$ Yes

Exercises 7.4 (page 489)

1. $28=4\cdot 7$

3. $18=9\cdot 2$

5. $16=8\cdot 2$

7. $54=27\cdot 2$

9. $12x^2+12x^2=24x^2$

11. $7x^3y(-3x^4y^{-5})=-21x^7y^{-4}=-\frac{21x^7}{y^4}$

13. $(3t+2)^2=(3t+2)(3t+2)$
$=9t^2+12t+4$

15.
$$\begin{array}{r}3p+\ \ 4+\frac{-5}{2p-5}\\ 2p-5\,\overline{)\,6p^2-\ 7p-\ 25}\\ \underline{6p^2-15p\qquad}\\ 8p-\ 25\\ \underline{8p-\ 20}\\ -5\end{array}$$

17. $\sqrt[n]{a}\sqrt[n]{b}$

19. like

21. $\sqrt{6}\sqrt{6}=\sqrt{36}=6$

23. $\sqrt{t}\sqrt{t}=\sqrt{t^2}=t$

25. $\sqrt[3]{5x^2}\sqrt[3]{25x}=\sqrt[3]{125x^3}$
$=5x$

27. $\frac{\sqrt{500}}{\sqrt{5}}=\sqrt{\frac{500}{5}}=\sqrt{100}$
$=10$

29. $\frac{\sqrt{98x^3}}{\sqrt{2x}}=\sqrt{\frac{98x^3}{2x}}=\sqrt{49x^2}=7x$

31. $\frac{\sqrt{180ab^4}}{\sqrt{5ab^2}}=\sqrt{\frac{180ab^4}{5ab^2}}=\sqrt{36b^2}=6b$

33. $\frac{\sqrt[3]{48}}{\sqrt[3]{6}}=\sqrt[3]{\frac{48}{6}}=\sqrt[3]{8}=2$

35. $\frac{\sqrt[3]{189a^4}}{\sqrt[3]{7a}}=\sqrt[3]{\frac{189a^4}{7a}}=\sqrt[3]{27a^3}=3a$

37. $\sqrt{20}=\sqrt{4\cdot 5}=\sqrt{4}\sqrt{5}=2\sqrt{5}$

39. $-\sqrt{200}=-\sqrt{100\cdot 2}=-\sqrt{100}\sqrt{2}$
$=-10\sqrt{2}$

© 2015 Cengage Learning. All Rights Reserved. May not be scanned, copied or duplicated, or posted to a publicly accessible website, in whole or in part.

SECTION 7.4

41. $\sqrt[3]{80} = \sqrt[3]{8 \cdot 10} = \sqrt[3]{8}\sqrt[3]{10} = 2\sqrt[3]{10}$

43. $\sqrt[3]{-81} = \sqrt[3]{-27 \cdot 3} = \sqrt[3]{-27}\sqrt[3]{3} = -3\sqrt[3]{3}$

45. $\sqrt[4]{32} = \sqrt[4]{16 \cdot 2} = \sqrt[4]{16}\sqrt[4]{2} = 2\sqrt[4]{2}$

47. $\sqrt[5]{96} = \sqrt[5]{32 \cdot 3} = \sqrt[5]{32}\sqrt[5]{3} = 2\sqrt[5]{3}$

49. $\sqrt{\dfrac{5}{49a^2}} = \dfrac{\sqrt{5}}{\sqrt{49a^2}} = \dfrac{\sqrt{5}}{7a}$

51. $\sqrt[3]{\dfrac{7a^3}{64}} = \dfrac{\sqrt[3]{7a^3}}{\sqrt[3]{64}} = \dfrac{\sqrt[3]{a^3}\sqrt[3]{7}}{4} = \dfrac{a\sqrt[3]{7}}{4}$

53. $$\sqrt[4]{\dfrac{3p^4}{10{,}000q^4}} = \dfrac{\sqrt[4]{3p^4}}{\sqrt[4]{10{,}000q^4}} = \dfrac{\sqrt[4]{p^4}\sqrt[4]{3}}{10q} = \dfrac{p\sqrt[4]{3}}{10q}$$

55. $$\sqrt[5]{\dfrac{3m^{15}}{32n^{10}}} = \dfrac{\sqrt[5]{3m^{15}}}{\sqrt[5]{32n^{10}}} = \dfrac{\sqrt[5]{m^{15}}\sqrt[5]{3}}{2n^2} = \dfrac{m^3\sqrt[5]{3}}{2n^2}$$

57. $\sqrt{63y^3} = \sqrt{9y^2 \cdot 7y} = \sqrt{9y^2}\sqrt{7y} = 3y\sqrt{7y}$

59. $\sqrt{48a^5} = \sqrt{16a^4 \cdot 3a} = \sqrt{16a^4}\sqrt{3a} = 4a^2\sqrt{3a}$

61. $-\sqrt{112a^3} = -\sqrt{16a^2 \cdot 7a} = -\sqrt{16a^2}\sqrt{7a} = -4a\sqrt{7a}$

63. $\sqrt{175a^2b^3} = \sqrt{25a^2b^2 \cdot 7b} = \sqrt{25a^2b^2}\sqrt{7b} = 5ab\sqrt{7b}$

65. $\dfrac{-\sqrt{150ab^4}}{\sqrt{2a}} = -\sqrt{\dfrac{150ab^4}{2a}} = -\sqrt{75b^4} = -\sqrt{25b^4}\sqrt{3} = -5b^2\sqrt{3}$

67. $\dfrac{\sqrt[3]{-108x^6y}}{\sqrt[3]{2y}} = \sqrt[3]{\dfrac{-108x^6y}{2y}} = \sqrt[3]{-54x^6} = \sqrt[3]{-27x^6}\sqrt[3]{2} = -3x^2\sqrt[3]{2}$

69. $\sqrt[3]{16x^{12}y^3} = \sqrt[3]{8x^{12}y^3 \cdot 2} = \sqrt[3]{8x^{12}y^3}\sqrt[3]{2} = 2x^4y\sqrt[3]{2}$

71. $\sqrt{\dfrac{z^2}{16x^2}} = \dfrac{\sqrt{z^2}}{\sqrt{16x^2}} = \dfrac{z}{4x}$

73. $\sqrt{3} + \sqrt{27} = \sqrt{3} + \sqrt{9}\sqrt{3} = \sqrt{3} + 3\sqrt{3} = 4\sqrt{3}$

75. $\sqrt{2} - \sqrt{8} = \sqrt{2} - \sqrt{4}\sqrt{2} = \sqrt{2} - 2\sqrt{2} = -\sqrt{2}$

77. $\sqrt{98} - \sqrt{50} = \sqrt{49}\sqrt{2} - \sqrt{25}\sqrt{2} = 7\sqrt{2} - 5\sqrt{2} = 2\sqrt{2}$

79. $3\sqrt{125} + 4\sqrt{45} = 3\sqrt{25}\sqrt{5} + 4\sqrt{9}\sqrt{5} = 3(5)\sqrt{5} + 4(3)\sqrt{5} = 15\sqrt{5} + 12\sqrt{5} = 27\sqrt{5}$

81. $$3\sqrt{54} + 2\sqrt{28} - 4\sqrt{96} = 3\sqrt{9}\sqrt{6} + 2\sqrt{4}\sqrt{7} - 4\sqrt{16}\sqrt{6} = 3(3)\sqrt{6} + 2(2)\sqrt{7} - 4(4)\sqrt{6} = 9\sqrt{6} + 4\sqrt{7} - 16\sqrt{6} = 4\sqrt{7} - 7\sqrt{6}$$

83. $\sqrt[3]{24} + \sqrt[3]{81} = \sqrt[3]{8}\sqrt[3]{3} + \sqrt[3]{27}\sqrt[3]{3} = 2\sqrt[3]{3} + 3\sqrt[3]{3} = 5\sqrt[3]{3}$

© 2015 Cengage Learning. All Rights Reserved. May not be scanned, copied or duplicated, or posted to a publicly accessible website, in whole or in part.

SECTION 7.4

85. $\sqrt[3]{32}-\sqrt[3]{108}=\sqrt[3]{8}\sqrt[3]{4}-\sqrt[3]{27}\sqrt[3]{4}=2\sqrt[3]{4}-3\sqrt[3]{4}=-\sqrt[3]{4}$

87. $2\sqrt[3]{125}-5\sqrt[3]{64}=2(5)-5(4)=10-20=-10$

89. $2\sqrt[3]{16}-\sqrt[3]{54}-3\sqrt[3]{128}=2\sqrt[3]{8}\sqrt[3]{2}-\sqrt[3]{27}\sqrt[3]{2}-3\sqrt[3]{64}\sqrt[3]{2}=2(2)\sqrt[3]{2}-3\sqrt[3]{2}-3(4)\sqrt[3]{2}$
$=4\sqrt[3]{2}-3\sqrt[3]{2}-12\sqrt[3]{2}=-11\sqrt[3]{2}$

91. $\sqrt[3]{81x^5}-\sqrt[3]{24x^5}=\sqrt[3]{27x^3}\sqrt[3]{3x^2}-\sqrt[3]{8x^3}\sqrt[3]{3x^2}=3x\sqrt[3]{3x^2}-2x\sqrt[3]{3x^2}=x\sqrt[3]{3x^2}$

93. $\sqrt{25yz^2}+\sqrt{9yz^2}=\sqrt{25z^2}\sqrt{y}+\sqrt{9z^2}\sqrt{y}=5z\sqrt{y}+3z\sqrt{y}=8z\sqrt{y}$

95. $\sqrt{y^5}-\sqrt{9y^5}-\sqrt{25y^5}=\sqrt{y^4}\sqrt{y}-\sqrt{9y^4}\sqrt{y}-\sqrt{25y^4}\sqrt{y}=y^2\sqrt{y}-3y^2\sqrt{y}-5y^2\sqrt{y}$
$=-7y^2\sqrt{y}$

97. $3\sqrt[3]{-2x^4}-\sqrt[3]{54x^4}=3\sqrt[3]{-x^3}\sqrt[3]{2x}-\sqrt[3]{27x^3}\sqrt[3]{2x}=-3x\sqrt[3]{2x}-3x\sqrt[3]{2x}=-6x\sqrt[3]{2x}$

99. $a=b=\frac{2}{3}$; $c=\frac{2}{3}\sqrt{2}$

101. $b=a=5\sqrt{2}$; $c=5\sqrt{2}\sqrt{2}=5(2)=10$

103. $a=b=\dfrac{5\sqrt{2}}{\sqrt{2}}=5$

105. $a=b=\dfrac{7\sqrt{2}}{\sqrt{2}}=7$

107. $b=5\sqrt{3}$; $c=2(5)=10$

109. $a=\dfrac{9\sqrt{3}}{\sqrt{3}}=9$; $c=2(9)=18$

111. $a=\dfrac{24}{2}=12$; $b=12\sqrt{3}$

113. $a=\dfrac{15}{2}$; $b=\dfrac{15}{2}\sqrt{3}$

115. $4\sqrt{2x}+6\sqrt{2x}=10\sqrt{2x}$

117. $\sqrt[4]{32x^{12}y^4}=\sqrt[4]{16x^{12}y^4\cdot 2}$
$=\sqrt[4]{16x^{12}y^4}\sqrt[4]{2}=2x^3y\sqrt[4]{2}$

119. $\sqrt[4]{\dfrac{5x}{16z^4}}=\dfrac{\sqrt[4]{5x}}{\sqrt[4]{16z^4}}=\dfrac{\sqrt[4]{5x}}{2z}$

121. $\sqrt{98}-\sqrt{50}-\sqrt{72}=\sqrt{49}\sqrt{2}-\sqrt{25}\sqrt{2}-\sqrt{36}\sqrt{2}=7\sqrt{2}-5\sqrt{2}-6\sqrt{2}=-4\sqrt{2}$

123. $3\sqrt[3]{27}+12\sqrt[3]{216}=3(3)+12(6)=9+72=81$

125. $23\sqrt[4]{768}+\sqrt[4]{48}=23\sqrt[4]{256}\sqrt[4]{3}+\sqrt[4]{16}\sqrt[4]{3}=23(4)\sqrt[4]{3}+2\sqrt[4]{3}=92\sqrt[4]{3}+2\sqrt[4]{3}=94\sqrt[4]{3}$

127. $4\sqrt[4]{243}-\sqrt[4]{48}=4\sqrt[4]{81}\sqrt[4]{3}-\sqrt[4]{16}\sqrt[4]{3}=4(3)\sqrt[4]{3}-2\sqrt[4]{3}=12\sqrt[4]{3}-2\sqrt[4]{3}=10\sqrt[4]{3}$

129. $6\sqrt[3]{5y}+3\sqrt[3]{5y}=9\sqrt[3]{5y}$

131. $10\sqrt[6]{12xyz}-\sqrt[6]{12xyz}=9\sqrt[6]{12xyz}$

© 2015 Cengage Learning. All Rights Reserved. May not be scanned, copied or duplicated, or posted to a publicly accessible website, in whole or in part.

SECTION 7.4

133. $\sqrt[5]{x^6y^2}+\sqrt[5]{32x^6y^2}+\sqrt[5]{x^6y^2}=\sqrt[5]{x^5}\sqrt[5]{xy^2}+\sqrt[5]{32x^5}\sqrt[5]{xy^2}+\sqrt[5]{x^5}\sqrt[5]{xy^2}$
$=x\sqrt[5]{xy^2}+2x\sqrt[5]{xy^2}+x\sqrt[5]{xy^2}=4x\sqrt[5]{xy^2}$

135. $\sqrt{x^2+2x+1}+\sqrt{x^2+2x+1}=\sqrt{(x+1)^2}+\sqrt{(x+1)^2}=x+1+x+1=2x+2$

137. $x=2.00$; $h=2\sqrt{2}\approx 2.83$

139. $h=2(5)=10.00$; $x=5\sqrt{3}\approx 8.66$

141. $x=\frac{9.37}{2}\approx 4.69$; $y\approx 4.69\sqrt{3}\approx 8.11$

143. $x=y=\frac{17.12}{\sqrt{2}}\approx 12.11$

145. $x=5\sqrt{3}\approx 8.66$ mm
$h=2x=2\left(5\sqrt{3}\right)=10\sqrt{3}\approx 17.32$ mm

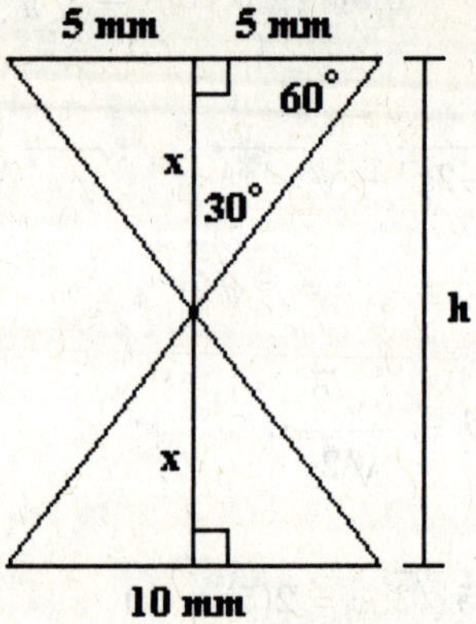

147. **Answers may vary.**

149. $\sqrt{3}+\sqrt{3^2}+\sqrt{3^3}+\sqrt{3^4}+\sqrt{3^5}=\sqrt{3}+3+\sqrt{3^2}\sqrt{3}+3^2+\sqrt{3^4}\sqrt{3}$
$=\sqrt{3}+3+3\sqrt{3}+9+9\sqrt{3}=12+13\sqrt{3}$

Exercises 7.5 (page 499)

1. $\sqrt{5}\sqrt{5}=\sqrt{25}=5$

3. $\sqrt{xy}\sqrt{xy}=\sqrt{x^2y^2}=xy$

5. $\sqrt[3]{5}\sqrt[3]{5}\sqrt[3]{5}=\sqrt[3]{125}=5$

7. $\sqrt[3]{a^2b}\sqrt[3]{ab^2}=\sqrt[3]{a^3b^3}=ab$

9.
$$\frac{2}{3-a}=1$$
$$2=3-a$$
$$a=1$$

© 2015 Cengage Learning. All Rights Reserved. May not be scanned, copied or duplicated, or posted to a publicly accessible website, in whole or in part.

SECTION 7.5

11. $$\begin{aligned}\frac{8}{b-2}+\frac{3}{2-b}&=-\frac{1}{b}\\ \frac{8}{b-2}+\frac{-3}{b-2}&=-\frac{1}{b}\\ \frac{5}{b-2}&=\frac{-1}{b}\\ 5b&=-b+2\\ 6b&=2\\ b&=\tfrac{2}{6}=\tfrac{1}{3}\end{aligned}$$

13. $2;\ \sqrt{7};\ \sqrt{5}$

15. $\sqrt{x}-1$

17. conjugate

19. $\sqrt{2}\sqrt{8}=\sqrt{16}=4$

21. $\sqrt{5}\sqrt{10}=\sqrt{50}=\sqrt{25}\sqrt{2}=5\sqrt{2}$

23. $$\begin{aligned}2\sqrt{3}\sqrt{6}=2\sqrt{18}=2\sqrt{9}\sqrt{2}&=2(3)\sqrt{2}\\ &=6\sqrt{2}\end{aligned}$$

25. $\sqrt[3]{5}\sqrt[3]{25}=\sqrt[3]{125}=5$

27. $$\begin{aligned}\sqrt[3]{5r^2s}\sqrt[3]{2r}=\sqrt[3]{10r^3s}&=\sqrt[3]{r^3}\sqrt[3]{10s}\\ &=r\sqrt[3]{10s}\end{aligned}$$

29. $\sqrt{ab^3}\sqrt{ab}=\sqrt{a^2b^4}=ab^2$

31. $\left(2\sqrt{3x^3y}\right)\left(5\sqrt{6xy^2}\right)=10\sqrt{18x^4y^3}=10\sqrt{9x^4y^2}\sqrt{2y}=10(3x^2y)\sqrt{2y}=30x^2y\sqrt{2y}$

33. $\sqrt{x(x+3)}\sqrt{x^3(x+3)}=\sqrt{x^4(x+3)^2}=x^2(x+3)$

35. $3\sqrt{5}\left(4-\sqrt{5}\right)=12\sqrt{5}-3\sqrt{25}=12\sqrt{5}-3(5)=12\sqrt{5}-15$

37. $3\sqrt{2}\left(4\sqrt{3}+2\sqrt{7}\right)=12\sqrt{6}+6\sqrt{14}$

39. $\left(\sqrt{2}+1\right)\left(\sqrt{2}-3\right)=\sqrt{4}-3\sqrt{2}+\sqrt{2}-3=2-2\sqrt{2}-3=-1-2\sqrt{2}$

41. $\left(3\sqrt{2}+2\right)\left(\sqrt{2}+1\right)=3\sqrt{4}+3\sqrt{2}+2\sqrt{2}+2=6+5\sqrt{2}+2=8+5\sqrt{2}$

43. $\left(4\sqrt{x}+3\right)\left(2\sqrt{x}-5\right)=8\sqrt{x^2}-20\sqrt{x}+6\sqrt{x}-15=8x-14\sqrt{x}-15$

45. $$\begin{aligned}\left(\sqrt{3a}+\sqrt{6}\right)\left(\sqrt{2a}-\sqrt{3}\right)&=\sqrt{6a^2}-\sqrt{9a}+\sqrt{12a}-\sqrt{18}\\ &=\sqrt{a^2}\sqrt{6}-\sqrt{9}\sqrt{a}+\sqrt{4}\sqrt{3a}-\sqrt{9}\sqrt{2}\\ &=a\sqrt{6}-3\sqrt{a}+2\sqrt{3a}-3\sqrt{2}\end{aligned}$$

47. $\sqrt{\frac{1}{7}}=\frac{\sqrt{1}}{\sqrt{7}}=\frac{1\sqrt{7}}{\sqrt{7}\sqrt{7}}=\frac{\sqrt{7}}{7}$

49. $\sqrt{\frac{2}{3}}=\frac{\sqrt{2}}{\sqrt{3}}=\frac{\sqrt{2}\sqrt{3}}{\sqrt{3}\sqrt{3}}=\frac{\sqrt{6}}{3}$

© 2015 Cengage Learning. All Rights Reserved. May not be scanned, copied or duplicated, or posted to a publicly accessible website, in whole or in part.

SECTION 7.5

51. $\frac{\sqrt{24}}{\sqrt{5}} = \frac{\sqrt{24}\sqrt{5}}{\sqrt{5}\sqrt{5}} = \frac{\sqrt{120}}{5} = \frac{\sqrt{4}\sqrt{30}}{5}$
$= \frac{2\sqrt{30}}{5}$

53. $\frac{\sqrt{7}}{\sqrt{12}} = \frac{\sqrt{7}\sqrt{3}}{\sqrt{12}\sqrt{3}} = \frac{\sqrt{21}}{\sqrt{36}} = \frac{\sqrt{21}}{6}$

55. $\frac{3}{\sqrt[3]{16}} = \frac{3\sqrt[3]{4}}{\sqrt[3]{16}\sqrt[3]{4}} = \frac{3\sqrt[3]{4}}{\sqrt[3]{64}} = \frac{3\sqrt[3]{4}}{4}$

57. $\frac{\sqrt[3]{7}}{\sqrt[3]{72}} = \frac{\sqrt[3]{7}\sqrt[3]{3}}{\sqrt[3]{72}\sqrt[3]{3}} = \frac{\sqrt[3]{21}}{\sqrt[3]{216}} = \frac{\sqrt[3]{21}}{6}$

59. $\frac{\sqrt{8x^2y}}{\sqrt{xy}} = \sqrt{\frac{8x^2y}{xy}} = \sqrt{8x} = 2\sqrt{2x}$

61. $\frac{\sqrt{10xy^2}}{\sqrt{2xy^3}} = \sqrt{\frac{10xy^2}{2xy^3}} = \sqrt{\frac{5}{y}} = \frac{\sqrt{5}\sqrt{y}}{\sqrt{y}\sqrt{y}}$
$= \frac{\sqrt{5y}}{y}$

63. $\frac{1}{\sqrt{2}-1} = \frac{1\left(\sqrt{2}+1\right)}{\left(\sqrt{2}-1\right)\left(\sqrt{2}+1\right)} = \frac{\sqrt{2}+1}{\sqrt{4}-1} = \frac{\sqrt{2}+1}{2-1} = \frac{\sqrt{2}+1}{1} = \sqrt{2}+1$

65. $\frac{\sqrt{2}}{\sqrt{5}+3} = \frac{\sqrt{2}\left(\sqrt{5}-3\right)}{\left(\sqrt{5}+3\right)\left(\sqrt{5}-3\right)} = \frac{\sqrt{2}\left(\sqrt{5}-3\right)}{\sqrt{25}-9} = \frac{\sqrt{2}\left(\sqrt{5}-3\right)}{5-9} = \frac{\sqrt{2}\left(\sqrt{5}-3\right)}{-4}$
$= \frac{\sqrt{10}-3\sqrt{2}}{-4}$
$= \frac{3\sqrt{2}-\sqrt{10}}{4}$

67. $\frac{\sqrt{3}+1}{\sqrt{3}-1} = \frac{\left(\sqrt{3}+1\right)\left(\sqrt{3}+1\right)}{\left(\sqrt{3}-1\right)\left(\sqrt{3}+1\right)} = \frac{\sqrt{9}+\sqrt{3}+\sqrt{3}+1}{\sqrt{9}-1} = \frac{4+2\sqrt{3}}{2} = \frac{2\left(2+\sqrt{3}\right)}{2}$
$= 2+\sqrt{3}$

69. $\frac{\sqrt{7}-\sqrt{2}}{\sqrt{2}+\sqrt{7}} = \frac{\left(\sqrt{7}-\sqrt{2}\right)\left(\sqrt{2}-\sqrt{7}\right)}{\left(\sqrt{2}+\sqrt{7}\right)\left(\sqrt{2}-\sqrt{7}\right)} = \frac{\sqrt{14}-7-2+\sqrt{14}}{\sqrt{4}-\sqrt{49}} = \frac{2\sqrt{14}-9}{-5} = \frac{9-2\sqrt{14}}{5}$

71. $\frac{2}{\sqrt{x}+1} = \frac{2(\sqrt{x}-1)}{(\sqrt{x}+1)(\sqrt{x}-1)} = \frac{2(\sqrt{x}-1)}{\sqrt{x^2}-1} = \frac{2(\sqrt{x}-1)}{x-1}$

73. $\frac{x}{\sqrt{x}-4} = \frac{x(\sqrt{x}+4)}{(\sqrt{x}-4)(\sqrt{x}+4)} = \frac{x(\sqrt{x}+4)}{\sqrt{x^2}-16} = \frac{x(\sqrt{x}+4)}{x-16}$

75. $\frac{\sqrt{x}-\sqrt{y}}{\sqrt{x}+\sqrt{y}} = \frac{(\sqrt{x}-\sqrt{y})(\sqrt{x}-\sqrt{y})}{(\sqrt{x}+\sqrt{y})(\sqrt{x}-\sqrt{y})} = \frac{\sqrt{x^2}-\sqrt{xy}-\sqrt{xy}+\sqrt{y^2}}{\sqrt{x^2}-\sqrt{y^2}} = \frac{x-2\sqrt{xy}+y}{x-y}$

© 2015 Cengage Learning. All Rights Reserved. May not be scanned, copied or duplicated, or posted to a publicly accessible website, in whole or in part.

SECTION 7.5

77. $\frac{\sqrt{3}+1}{2}=\frac{(\sqrt{3}+1)(\sqrt{3}-1)}{2(\sqrt{3}-1)}=\frac{\sqrt{9}-1}{2(\sqrt{3}-1)}=\frac{2}{2(\sqrt{3}-1)}=\frac{1}{\sqrt{3}-1}$

79. $\frac{\sqrt{x}+3}{x}=\frac{(\sqrt{x}+3)(\sqrt{x}-3)}{x(\sqrt{x}-3)}=\frac{\sqrt{x^2}-9}{x(\sqrt{x}-3)}=\frac{x-9}{x(\sqrt{x}-3)}$

81. $(3\sqrt[3]{9})(2\sqrt[3]{3})=6\sqrt[3]{27}=6(3)=18$

83. $\sqrt{5ab}\sqrt{5a}=\sqrt{25a^2b}=\sqrt{25a^2}\sqrt{b}$
$=5a\sqrt{b}$

85. $\sqrt[3]{a^5b}\sqrt[3]{16ab^5}=\sqrt[3]{16a^6b^6}=\sqrt[3]{8a^6b^6}\sqrt[3]{2}$
$=2a^2b^2\sqrt[3]{2}$

87. $\sqrt[3]{6x^2(y+z)^2}\sqrt[3]{18x(y+z)}=\sqrt[3]{108x^3(y+z)^3}=\sqrt[3]{27x^3(y+z)^3}\sqrt[3]{4}=3x(y+z)\sqrt[3]{4}$

89. $-2\sqrt{5x}(4\sqrt{2x}-3\sqrt{3})=-8\sqrt{10x^2}+6\sqrt{15x}=-8x\sqrt{10}+6\sqrt{15x}$

91. $(\sqrt{5z}+\sqrt{3})(\sqrt{5z}+\sqrt{3})=\sqrt{25z^2}+\sqrt{15z}+\sqrt{15z}+\sqrt{9}=5z+2\sqrt{15z}+3$

93. $(3\sqrt{2r}-2)^2=(3\sqrt{2r}-2)(3\sqrt{2r}-2)=9\sqrt{4r^2}-6\sqrt{2r}-6\sqrt{2r}+4=18r-12\sqrt{2r}+4$

95. $-2(\sqrt{3x}+\sqrt{3})^2=-2(\sqrt{3x}+\sqrt{3})(\sqrt{3x}+\sqrt{3})$
$=-2(\sqrt{9x^2}+\sqrt{9x}+\sqrt{9x}+\sqrt{9})$
$=-2(3x+3\sqrt{x}+3\sqrt{x}+3)=-2(3x+6\sqrt{x}+3)=-6x-12\sqrt{x}-6$

97. $\frac{3}{\sqrt[3]{9}}=\frac{3\sqrt[3]{3}}{\sqrt[3]{9}\sqrt[3]{3}}=\frac{3\sqrt[3]{3}}{\sqrt[3]{27}}=\frac{3\sqrt[3]{3}}{3}=\sqrt[3]{3}$

99. $\frac{1}{\sqrt[4]{4}}=\frac{1\sqrt[4]{4}}{\sqrt[4]{4}\sqrt[4]{4}}=\frac{\sqrt[4]{4}}{\sqrt[4]{16}}=\frac{\sqrt[4]{4}}{2}$

101. $\frac{1}{\sqrt[5]{16}}=\frac{1\sqrt[5]{2}}{\sqrt[5]{16}\sqrt[5]{2}}=\frac{\sqrt[5]{2}}{\sqrt[5]{32}}=\frac{\sqrt[5]{2}}{2}$

103. $\frac{\sqrt[3]{4a^2}}{\sqrt[3]{2ab}}=\sqrt[3]{\frac{4a^2}{2ab}}=\sqrt[3]{\frac{2a}{b}}=\frac{\sqrt[3]{2a}\sqrt[3]{b^2}}{\sqrt[3]{b}\sqrt[3]{b^2}}$
$=\frac{\sqrt[3]{2ab^2}}{b}$

105. $\frac{2z-1}{\sqrt{2z}-1}=\frac{(2z-1)(\sqrt{2z}+1)}{(\sqrt{2z}-1)(\sqrt{2z}+1)}=\frac{(2z-1)(\sqrt{2z}+1)}{\sqrt{4z^2}-1}=\frac{(2z-1)(\sqrt{2z}+1)}{2z-1}$
$=\sqrt{2z}+1$

107. $\frac{\sqrt{x}+\sqrt{y}}{\sqrt{x}}=\frac{(\sqrt{x}+\sqrt{y})(\sqrt{x}-\sqrt{y})}{\sqrt{x}(\sqrt{x}-\sqrt{y})}=\frac{\sqrt{x^2}-\sqrt{y^2}}{\sqrt{x^2}-\sqrt{xy}}=\frac{x-y}{x-\sqrt{xy}}$

© 2015 Cengage Learning. All Rights Reserved. May not be scanned, copied or duplicated, or posted to a publicly accessible website, in whole or in part.

109. If the area of the aperture is again cut in half, the area will equal $9\pi/4$ cm^2.

$$A = \pi r^2 \qquad f\text{-number} = \frac{f}{d}$$
$$\frac{9\pi}{4} = \pi\left(\frac{d}{2}\right)^2 \qquad = \frac{12}{3}$$
$$\frac{9\pi}{4} = \frac{\pi d^2}{4} \qquad = 4$$
$$\boxed{f/4}$$
$$36\pi = 4\pi d^2$$
$$\frac{36\pi}{4\pi} = d^2$$
$$9 = d^2$$
$$3 = d$$

111. $r = \sqrt{\frac{A}{\pi}} = \frac{\sqrt{A}}{\sqrt{\pi}} = \frac{\sqrt{A}\sqrt{\pi}}{\sqrt{\pi}\sqrt{\pi}} = \frac{\sqrt{A\pi}}{\pi}$

113. $\frac{8}{\sqrt{2}} = \frac{8\sqrt{2}}{\sqrt{2}\sqrt{2}} = \frac{8\sqrt{2}}{2} = 4\sqrt{2}$ cm

115. shorter leg: $\frac{6}{\sqrt{3}} = \frac{6\sqrt{3}}{\sqrt{3}\sqrt{3}} = \frac{6\sqrt{3}}{3} = 2\sqrt{3}$ ft; hypotenuse: $2\left(2\sqrt{3}\right) = 4\sqrt{3}$ ft

117. **Answers may vary.**

119.
$$\frac{\sqrt{x}-3}{4} = \frac{(\sqrt{x}-3)(\sqrt{x}+3)}{4(\sqrt{x}+3)}$$
$$= \frac{x-9}{4(\sqrt{x}+3)}$$

Exercises 7.6 (page 507)

1. $\left(x^{1/2}\right)^2 = x^{2/2} = x$

3. $\left[(x+3)^{1/2}\right]^2 = (x+3)^{2/2} = x+3$

5. $(x+7)^2 = (x+7)(x+7) = x^2 + 14x + 49$

7. $f(-2) = 3(-2)^2 - 4(-2) + 2 = 22$

9. $f(3) = 3(3)^2 - 4(3) + 2 = 17$

11. $x^n = y^n$; power rule

13. square

15. extraneous

17.
$$\sqrt{4x+5} = 3$$
$$\left(\sqrt{4x+5}\right)^2 = 3^2$$
$$4x+5 = 9$$
$$4x = 4$$
$$x = 1$$

The answer checks.

19.
$$\sqrt{4x-8}+4 = 10$$
$$\sqrt{4x-8} = 6$$
$$\left(\sqrt{4x-8}\right)^2 = 6^2$$
$$4x-8 = 36$$
$$4x = 44$$
$$x = 11$$

The answer checks.

© 2015 Cengage Learning. All Rights Reserved. May not be scanned, copied or duplicated, or posted to a publicly accessible website, in whole or in part.

SECTION 7.6

21.
$$\begin{aligned}\sqrt[3]{7n-1}&=3\\ \left(\sqrt[3]{7n-1}\right)^3&=3^3\\ 7n-1&=27\\ 7n&=28\\ n&=4\end{aligned}$$
The answer checks.

23.
$$\begin{aligned}x&=\frac{\sqrt{12x-5}}{2}\\ 2x&=\sqrt{12x-5}\\ (2x)^2&=\left(\sqrt{12x-5}\right)^2\\ 4x^2&=12x-5\\ 4x^2-12x+5&=0\\ (2x-1)(2x-5)&=0\end{aligned}$$
$2x-1=0$ **or** $2x-5=0$
$2x=1$ $\quad 2x=5$
$x=\frac{1}{2}$ $\quad x=\frac{5}{2}$
Both answers check.

25.
$$\begin{aligned}\sqrt{2y-5}+4&=y\\ \sqrt{2y-5}&=y-4\\ \left(\sqrt{2y-5}\right)^2&=(y-4)^2\\ 2y-5&=y^2-8y+16\\ 0&=y^2-10y+21\\ 0&=(y-7)(y-3)\end{aligned}$$
$y-7=0$ **or** $y-3=0$
$y=7$ $\quad y=3$
solution $\quad$ extraneous

27.
$$\begin{aligned}\sqrt{-5x+24}&=6-x\\ \left(\sqrt{-5x+24}\right)^2&=(6-x)^2\\ -5x+24&=36-12x+x^2\\ 0&=x^2-7x+12\\ 0&=(x-3)(x-4)\end{aligned}$$
$x-3=0$ **or** $x-4=0$
$x=3$ $\quad x=4$
solution $\quad$ solution

29.
$$\begin{aligned}\sqrt[3]{x^3-1}&=x-1\\ \left(\sqrt[3]{x^3-1}\right)^3&=(x-1)^3\\ x^3-1&=x^3-3x^2+3x-1\\ 3x^2-3x&=0\\ 3x(x-1)&=0\end{aligned}$$
$3x=0$ **or** $x-1=0$
$x=0$ $\quad x=1$
solution $\quad$ solution

31.
$$\begin{aligned}\sqrt[3]{x^3+56}-2&=x\\ \sqrt[3]{x^3+56}&=x+2\\ \left(\sqrt[3]{x^3+56}\right)^3&=(x+2)^3\\ x^3+56&=x^3+6x^2+12x+8\\ 0&=6x^2+12x-48\\ 0&=6(x-2)(x+4)\end{aligned}$$
$x-2=0$ **or** $x+4=0$
$x=2$ $\quad x=-4$
solution $\quad$ solution

33.
$$\begin{aligned}2\sqrt{4x+1}&=\sqrt{x+4}\\ \left(2\sqrt{4x+1}\right)^2&=\left(\sqrt{x+4}\right)^2\\ 4(4x+1)&=x+4\\ 16x+4&=x+4\\ 15x&=0\\ x&=0\end{aligned}$$
The answer checks.

35.
$$\begin{aligned}\sqrt{x+5}&=\sqrt{7-x}\\ \left(\sqrt{x+5}\right)^2&=\left(\sqrt{7-x}\right)^2\\ x+5&=7-x\\ 2x&=2\\ x&=1\end{aligned}$$
The answer checks.

© 2015 Cengage Learning. All Rights Reserved. May not be scanned, copied or duplicated, or posted to a publicly accessible website, in whole or in part.

SECTION 7.6

37.
$$1+\sqrt{z}=\sqrt{z+3}$$
$$\left(1+\sqrt{z}\right)^2=\left(\sqrt{z+3}\right)^2$$
$$1+2\sqrt{z}+z=z+3$$
$$2\sqrt{z}=2$$
$$\left(2\sqrt{z}\right)^2=2^2$$
$$4z=4$$
$$z=1$$
The answer checks.

39.
$$\sqrt{4s+1}-\sqrt{6s}=-1$$
$$\sqrt{4s+1}=\sqrt{6s}-1$$
$$\left(\sqrt{4s+1}\right)^2=\left(\sqrt{6s}-1\right)^2$$
$$4s+1=6s-2\sqrt{6s}+1$$
$$2\sqrt{6s}=2s$$
$$\left(2\sqrt{6s}\right)^2=(2s)^2$$
$$24s=4s^2$$
$$0=4s^2-24s$$
$$0=4s(s-6)$$
$4s=0$ **or** $s-6=0$

$s=0$ extraneous; $s=6$ solution

41.
$$\sqrt{y}+\sqrt{5}=\sqrt{y+15}$$
$$\left(\sqrt{y}+\sqrt{5}\right)^2=\left(\sqrt{y+15}\right)^2$$
$$y+2\sqrt{5y}+5=y+15$$
$$2\sqrt{5y}=10$$
$$2\sqrt{5y}=10$$
$$\left(2\sqrt{5y}\right)^2=10^2$$
$$20y=100$$
$$y=5$$
The answer checks.

43.
$$\sqrt{3x}-\sqrt{x+1}=\sqrt{x-2}$$
$$\left(\sqrt{3x}-\sqrt{x+1}\right)^2=\left(\sqrt{x-2}\right)^2$$
$$3x-2\sqrt{3x(x+1)}+x+1=x-2$$
$$3x+3=2\sqrt{3x^2+3x}$$
$$(3x+3)^2=\left(2\sqrt{3x^2+3x}\right)^2$$
$$(3x+3)^2=\left(2\sqrt{3x^2+3x}\right)^2$$
$$9x^2+18x+9=12x^2+12x$$
$$0=3x^2-6x-9$$
$$0=3(x-3)(x+1)$$
$x-3=0$ **or** $x+1=0$

$x=3$ solution; $x=-1$ extraneous

© 2015 Cengage Learning. All Rights Reserved. May not be scanned, copied or duplicated, or posted to a publicly accessible website, in whole or in part.

SECTION 7.6

45.
$$\begin{aligned}v &= \sqrt{2gh}\\ v^2 &= \left(\sqrt{2gh}\right)^2\\ v^2 &= 2gh\\ \frac{v^2}{2h} &= \frac{2gh}{2h}\\ \frac{v^2}{2h} &= g, \text{ or } g = \frac{v^2}{2h}\end{aligned}$$

47.
$$\begin{aligned}T &= 2\pi\sqrt{\frac{l}{32}}\\ T^2 &= \left(2\pi\sqrt{\frac{l}{32}}\right)^2\\ T^2 &= 4\pi^2\cdot\frac{l}{32}\\ 32\left(T^2\right) &= 32\left(4\pi^2\cdot\frac{l}{32}\right)\\ 32T^2 &= 4\pi^2 l\\ \frac{32T^2}{4\pi^2} &= \frac{4\pi^2 l}{4\pi^2}\\ \frac{8T^2}{\pi^2} &= l, \text{ or } l = \frac{8T^2}{\pi^2}\end{aligned}$$

49.
$$\begin{aligned}r &= \sqrt[3]{\frac{A}{P}} - 1\\ r+1 &= \sqrt[3]{\frac{A}{P}}\\ (r+1)^3 &= \left(\sqrt[3]{\frac{A}{P}}\right)^3\\ (r+1)^3 &= \frac{A}{P}\\ P(r+1)^3 &= P\left(\frac{A}{P}\right)\\ P(r+1)^3 &= A, \text{ or } A = P(r+1)^3\end{aligned}$$

51.
$$\begin{aligned}L_A &= L_B\sqrt{1-\frac{v^2}{c^2}}\\ L_A^2 &= \left(L_B\sqrt{1-\frac{v^2}{c^2}}\right)^2\\ L_A^2 &= L_B^2\left(1-\frac{v^2}{c^2}\right)\\ \frac{L_A^2}{L_B^2} &= 1-\frac{v^2}{c^2}\\ \frac{v^2}{c^2} &= 1-\frac{L_A^2}{L_B^2}\\ c^2\left(\frac{v^2}{c^2}\right) &= c^2\left(1-\frac{L_A^2}{L_B^2}\right)\\ v^2 &= c^2\left(1-\frac{L_A^2}{L_B^2}\right)\end{aligned}$$

53.
$$\begin{aligned}\sqrt{y+2} &= 4-y\\ \left(\sqrt{y+2}\right) &= (4-y)^2\\ y+2 &= 16-8y+y^2\\ 0 &= y^2-9y+14\\ 0 &= (y-2)(y-7)\end{aligned}$$
$y-2=0$ **or** $y-7=0$

$y=2$ $\qquad y=7$

solution $\qquad$ extraneous

55.
$$\begin{aligned}2\sqrt{x} &= \sqrt{5x-16}\\ \left(2\sqrt{x}\right)^2 &= \left(\sqrt{5x-16}\right)^2\\ 4x &= 5x-16\\ -x &= -16\\ x &= 16\end{aligned}$$
The answer checks.

© 2015 Cengage Learning. All Rights Reserved. May not be scanned, copied or duplicated, or posted to a publicly accessible website, in whole or in part.

SECTION 7.6

57.

$$\begin{aligned}\sqrt{2y+1} &= 1-2\sqrt{y}\\ \left(\sqrt{2y+1}\right)^2 &= \left(1-2\sqrt{y}\right)^2\\ 2y+1 &= 1-4\sqrt{y}+4y\\ 4\sqrt{y} &= 2y\\ \left(4\sqrt{y}\right)^2 &= (2y)^2\end{aligned}$$

$$\begin{aligned}16y &= 4y^2\\ 0 &= 4y^2-16y\\ 0 &= 4y(y-4)\end{aligned}$$

$4y = 0$ **or** $y - 4 = 0$

$y = 0$ $\quad$ $y = 4$

solution $\quad$ extraneous

59.

$$\begin{aligned}\sqrt{2x+5}+\sqrt{x+2} &= 5\\ \sqrt{2x+5} &= 5-\sqrt{x+2}\\ \left(\sqrt{2x+5}\right)^2 &= \left(5-\sqrt{x+2}\right)^2\\ 2x+5 &= 25-10\sqrt{x+2}+x+2\\ 10\sqrt{x+2} &= -x+22\\ \left(10\sqrt{x+2}\right)^2 &= (-x+22)^2\end{aligned}$$

$$\begin{aligned}\left(10\sqrt{x+2}\right)^2 &= (-x+22)^2\\ 100(x+2) &= x^2-44x+484\\ 0 &= x^2-144x+284\\ 0 &= (x-142)(x-2)\end{aligned}$$

$x - 142 = 0$ **or** $x - 2 = 0$

$x = 142$ $\quad$ $x = 2$

extraneous $\quad$ solution

61.

$$\begin{aligned}5r+4 &= \sqrt{5r+20}+4r\\ r+4 &= \sqrt{5r+20}\\ (r+4)^2 &= \left(\sqrt{5r+20}\right)^2\\ r^2+8r+16 &= 5r+20\\ r^2+3r-4 &= 0\\ (r+4)(r-1) &= 0\end{aligned}$$

$r + 4 = 0$ **or** $r - 1 = 0$

$r = -4$ $\quad$ $r = 1$

solution $\quad$ solution

63.

$$\begin{aligned}\sqrt{x}\sqrt{x+6} &= 4\\ \left(\sqrt{x}\sqrt{x+6}\right)^2 &= 4^2\\ x(x+6) &= 16\\ x^2+6x-16 &= 0\\ (x-2)(x+8) &= 0\end{aligned}$$

$x - 2 = 0$ **or** $x + 8 = 0$

$x = 2$ $\quad$ $x = -8$

solution $\quad$ extraneous

65.

$$\begin{aligned}\sqrt[4]{x^4+4x^2-4} &= -x\\ \left(\sqrt[4]{x^4+4x^2-4}\right)^4 &= (-x)^4\\ x^4+4x^2-4 &= x^4\\ 4x^2-4 &= 0\\ 4(x+1)(x-1) &= 0\end{aligned}$$

$x + 1 = 0$ **or** $x - 1 = 0$

$x = -1$ $\quad$ $x = 1$

solution $\quad$ extraneous

67.

$$\begin{aligned}2+\sqrt{u} &= \sqrt{2u+7}\\ \left(2+\sqrt{u}\right)^2 &= \left(\sqrt{2u+7}\right)^2\\ 4+4\sqrt{u}+u &= 2u+7\\ 4\sqrt{u} &= u+3\\ \left(4\sqrt{u}\right)^2 &= (u+3)^2\\ 16u &= u^2+6u+9\\ 0 &= u^2-10u+9\\ 0 &= (u-9)(u-1)\end{aligned}$$

$u - 9 = 0$ **or** $u - 1 = 0$

$u = 9$ $\quad$ $u = 1$

solution $\quad$ solution

© 2015 Cengage Learning. All Rights Reserved. May not be scanned, copied or duplicated, or posted to a publicly accessible website, in whole or in part.

SECTION 7.6

69.
$$\begin{aligned} u &= \sqrt[4]{u^4 - 6u^2 + 24} \\ u^4 &= \left(\sqrt[4]{u^4 - 6u^2 + 24}\right)^4 \\ u^4 &= u^4 - 6u^2 + 24 \\ 6u^2 - 24 &= 0 \\ 6(u+2)(u-2) &= 0 \end{aligned}$$
$u + 2 = 0$ **or** $u - 2 = 0$
$u = -2$ $\quad u = 2$
extraneous $\quad$ solution

71.
$$\begin{aligned} \sqrt{x-5} - \sqrt{x+3} &= 4 \\ \sqrt{x-5} &= \sqrt{x+3} + 4 \\ \left(\sqrt{x-5}\right)^2 &= \left(\sqrt{x+3}+4\right)^2 \\ x - 5 &= x + 3 + 8\sqrt{x+3} + 16 \\ -24 &= 8\sqrt{x+3} \\ -3 &= \sqrt{x+3} \\ (-3)^2 &= \left(\sqrt{x+3}\right)^2 \\ 9 &= x + 3 \\ 6 &= x \end{aligned}$$
extraneous $\Rightarrow \emptyset$

73.
$$\begin{aligned} \sqrt{x+8} - \sqrt{x-4} &= -2 \\ \sqrt{x+8} &= \sqrt{x-4} - 2 \\ \left(\sqrt{x+8}\right)^2 &= \left(\sqrt{x-4}-2\right)^2 \\ x + 8 &= x - 4 - 4\sqrt{x-4} + 4 \\ 4\sqrt{x-4} &= -8 \\ \sqrt{x-4} &= -2 \\ \left(\sqrt{x-4}\right)^2 &= (-2)^2 \\ x - 4 &= 4 \\ x &= 8 \end{aligned}$$
extraneous $\Rightarrow \emptyset$

75.
$$\begin{aligned} \sqrt{z-1} + \sqrt{z+2} &= 3 \\ \sqrt{z-1} &= 3 - \sqrt{z+2} \\ \left(\sqrt{z-1}\right)^2 &= \left(3 - \sqrt{z+2}\right)^2 \\ z - 1 &= 9 - 6\sqrt{z+2} + z + 2 \\ 6\sqrt{z+2} &= 12 \\ \sqrt{z+2} &= 2 \\ \left(\sqrt{z+2}\right)^2 &= 2^2 \\ z + 2 &= 4 \end{aligned}$$
$z = 2$: The answer checks.

77.
$$\begin{aligned} \sqrt{\sqrt{a} + \sqrt{a+8}} &= 2 \\ \left(\sqrt{\sqrt{a} + \sqrt{a+8}}\right)^2 &= 2^2 \\ \sqrt{a} + \sqrt{a+8} &= 4 \\ \left(\sqrt{a} + \sqrt{a+8}\right)^2 &= 4^2 \\ a + 2\sqrt{a(a+8)} + a + 8 &= 16 \\ 2\sqrt{a^2 + 8a} &= -2a + 8 \\ \left(2\sqrt{a^2+8a}\right)^2 &= (-2a+8)^2 \\ 4a^2 + 32a &= 4a^2 - 32a + 64 \\ 64a &= 64 \\ a &= 1 \end{aligned}$$
The answer checks.

79.
$$\begin{aligned} \frac{\sqrt{2x}}{\sqrt{x+2}} &= \sqrt{x-1} \\ \left(\frac{\sqrt{2x}}{\sqrt{x+2}}\right)^2 &= \left(\sqrt{x-1}\right)^2 \\ \frac{2x}{x+2} &= x - 1 \\ 2x &= x^2 + x - 2 \\ 0 &= x^2 - x - 2 \\ 0 &= (x-2)(x+1) \end{aligned}$$
$x - 2 = 0$ **or** $x + 1 = 0$
$x = 2$ $\quad x = -1$
solution $\quad$ extraneous

© 2015 Cengage Learning. All Rights Reserved. May not be scanned, copied or duplicated, or posted to a publicly accessible website, in whole or in part.

81.
$$\begin{aligned} s &= 1.45\sqrt{r} \\ 65 &= 1.45\sqrt{r} \\ 65^2 &= \left(1.45\sqrt{r}\right)^2 \\ 4225 &= 2.1025r \\ 2010 \text{ ft} &\approx r \end{aligned}$$

83.
$$\begin{aligned} v &= \sqrt[3]{\frac{P}{0.02}} \\ v &= \sqrt[3]{\frac{500}{0.02}} \\ v &= \sqrt[3]{25000} \\ v &\approx 29 \text{ mph} \end{aligned}$$

85. $r = 1 - \sqrt[n]{\frac{S}{C}} = 1 - \sqrt[15]{\frac{900}{22000}} \approx 1 - \sqrt[15]{0.040909} \approx 1 - 0.808 \approx 0.192 \approx 19\%$

87.
$$\begin{aligned} \sqrt{5x} &= \sqrt{100 - 3x^2} \\ \left(\sqrt{5x}\right)^2 &= \left(\sqrt{100 - 3x^2}\right)^2 \\ 5x &= 100 - 3x^2 \end{aligned}$$
$$3x^2 + 5x - 100 = 0$$
$$(x - 5)(3x + 20) = 0$$
$x - 5 = 0$ **or** $3x + 20 = 0$

$x = 5$ $\quad$ $3x = -20$

solution $\quad$ $x = -\frac{20}{3}$

does not make sense

The equilibrium price is \$5.

89.
$$\begin{aligned} r &= \sqrt[4]{\frac{8kl}{\pi R}} \\ r^4 &= \left(\sqrt[4]{\frac{8kl}{\pi R}}\right)^4 \\ r^4 &= \frac{8kl}{\pi R} \\ \pi R r^4 &= 8kl \\ R &= \frac{8kl}{\pi r^4} \end{aligned}$$

91. **Answers may vary.**

93.
$$\begin{aligned} \sqrt[3]{2x} &= \sqrt{x} \\ \left(\sqrt[3]{2x}\right)^2 &= \left(\sqrt{x}\right)^2 \\ \left[\left(\sqrt[3]{2x}\right)^2\right]^3 &= x^3 \\ 4x^2 &= x^3 \\ 0 &= x^3 - 4x^2 \\ 0 &= x^2(x - 4) \end{aligned}$$
$x = 0$ or $x = 4$

Exercises 7.7 (page 518)

1. $\sqrt{-9}$: not real

3. $\sqrt{a}$ if $a \geq 0$: real

5. $-\sqrt{-1}$: not real

7. $3 - \sqrt{5}$

9. $5 + 3x + 7 + 4x = 7x + 12$

11. $\frac{x^2 - x - 6}{9 - x^2} \cdot \frac{x^2 + x - 6}{x^2 - 4} = \frac{(x - 3)(x + 2)}{(3 + x)(3 - x)} \cdot \frac{(x + 3)(x - 2)}{(x + 2)(x - 2)} = -1$

© 2015 Cengage Learning. All Rights Reserved. May not be scanned, copied or duplicated, or posted to a publicly accessible website, in whole or in part.

SECTION 7.7

13. Let $w =$ the speed of the wind.

	Rate	Time	Dist.
Downwind	$200 + w$	$\frac{330}{200+w}$	330
Upwind	$200 - w$	$\frac{330}{200-w}$	330

$$\frac{330}{200+w} + \frac{330}{200-w} = \frac{10}{3}$$
$$\left(\frac{330}{200+w} + \frac{330}{200-w}\right)3(200+w)(200-w) = \frac{10}{3} \cdot 3(200+w)(200-w)$$
$$330(3)(200-w) + 330(3)(200+w) = 10(40{,}000 - w^2)$$
$$198{,}000 - 990w + 198{,}000 + 990w = 400{,}000 - 10w^2$$
$$10w^2 = 4000$$
$$w^2 = 400$$
$$w = 20 \text{ mph}$$

15. i

17. $-i$

19. imaginary

21. $\frac{\sqrt{a}}{\sqrt{b}}$

23. 5; 7

25. conjugates

27. $\sqrt{-25} = \sqrt{25(-1)} = \sqrt{25}\sqrt{-1} = 5i$

29. $\sqrt{-121} = \sqrt{121(-1)} = \sqrt{121}\sqrt{-1} = 11i$

31. $\sqrt{-7} = \sqrt{7(-1)} = \sqrt{7}\sqrt{-1} = i\sqrt{7}$

33. $\sqrt{-8} = \sqrt{8(-1)} = \sqrt{8}\sqrt{-1} = \sqrt{4}\sqrt{2} \cdot i$
$= 2i\sqrt{2}$

35. $9 = 9 + 0i$

37. $9 + \sqrt{-9} = 9 + 3i$

39. $3 + 7i \stackrel{?}{=} \sqrt{9} + (5+2)i$
$3 + 7i \stackrel{?}{=} 3 + 7i$
They are equal.

41. $\sqrt{4} + \sqrt{-4} \stackrel{?}{=} 2 - 2i$
$2 + \sqrt{-1 \cdot 4} \stackrel{?}{=} 2 - 2i$
$2 + 2i = 2 - 2i$
They are not equal.

43. $(4 + 7i) + (3 - 5i) = 4 + 7i + 3 - 5i$
$= 7 + 2i$

45. $(7 - 3i) - (4 + 2i) = 7 - 3i - 4 - 2i$
$= 3 - 5i$

47. $(8 + 5i) + (7 + 2i) = 8 + 5i + 7 + 2i$
$= 15 + 7i$

49. $(1 + i) - 2i + (5 - 7i) = 1 + i - 2i + 5 - 7i$
$= 6 - 8i$

51. $6i(3 - 4i) = 18i - 24i^2 = 18i - 24(-1) = 18i + 24 = 24 + 18i$

53. $-5i(5 - 5i) = -25i + 25i^2 = -25i + 25(-1) = -25i - 25 = -25 - 25i$

55. $(2 + i)(3 - i) = 6 - 2i + 3i - i^2 = 6 + i - (-1) = 6 + i + 1 = 7 + i$

© 2015 Cengage Learning. All Rights Reserved. May not be scanned, copied or duplicated, or posted to a publicly accessible website, in whole or in part.

SECTION 7.7

57. $(5+2i)(4-6i)=20-30i+8i-12i^2=20-22i-12(-1)=20-22i+12=32-22i$

59. $(4+\sqrt{3}i)(3-\sqrt{3}i)=12-4\sqrt{3}i+3\sqrt{3}i-3i^2=12-\sqrt{3}i-3(-1)=15-\sqrt{3}i$

61. $(4+3i)^2=(4+3i)(4+3i)=16+12i+12i+9i^2=16+24i-9=7+24i$

63. $(6-5i)(6+5i)=36+30i-30i-25i^2=36-25(-1)=36+25=61$

65. $(3+4i)(3-4i)=9-12i+12i-16i^2=9-16(-1)=9+16=25$

67. $\dfrac{5}{2-i}=\dfrac{5(2+i)}{(2-i)(2+i)}=\dfrac{5(2+i)}{4-i^2}=\dfrac{5(2+i)}{5}=2+i$

69. $\dfrac{10}{5-i}=\dfrac{10(5+i)}{(5-i)(5+i)}=\dfrac{10(5+i)}{25-i^2}=\dfrac{10(5+i)}{26}=\dfrac{5(5+i)}{13}=\dfrac{25+5i}{13}=\dfrac{25}{13}+\dfrac{5}{13}i$

71. $\dfrac{3-2i}{3+2i}=\dfrac{(3-2i)(3-2i)}{(3+2i)(3-2i)}=\dfrac{9-6i-6i+4i^2}{9-4i^2}=\dfrac{5-12i}{13}=\dfrac{5}{13}-\dfrac{12}{13}i$

73. $\dfrac{3+2i}{3+i}=\dfrac{(3+2i)(3-i)}{(3+i)(3-i)}=\dfrac{9-3i+6i-2i^2}{9-i^2}=\dfrac{11+3i}{10}=\dfrac{11}{10}+\dfrac{3}{10}i$

75. $\dfrac{-12}{7-\sqrt{-1}}=\dfrac{-12}{7-i}=\dfrac{-12(7+i)}{(7-i)(7+i)}=\dfrac{-84-12i}{49-i^2}=\dfrac{-84-12i}{50}=\dfrac{-84}{50}-\dfrac{12}{50}i=-\dfrac{42}{25}-\dfrac{6}{25}i$

77. $\dfrac{7}{5-\sqrt{-9}}=\dfrac{7}{5-3i}=\dfrac{7(5+3i)}{(5-3i)(5+3i)}=\dfrac{7(5+3i)}{25-9i^2}=\dfrac{7(5+3i)}{34}=\dfrac{35+21i}{34}=\dfrac{35}{34}+\dfrac{21}{34}i$

79. $i^5=i^4i^1=1i=i$

81. $i^{10}=i^8i^2=\left(i^4\right)^2i^2=1^2(-1)=-1$

83. $i^{37}=i^{36}i^1=\left(i^4\right)^9i=1^9i=i$

85. $i^{27}=i^{24}i^3=\left(i^4\right)^6i^3=1^6i^3=i^3=-i$

87. $i^{64}=\left(i^4\right)^{16}=1^{16}=1$

89. $i^{97}=i^{96}i^1=\left(i^4\right)^{24}i=1^{24}i=i$

91. $5i^3+2i^2=5(-i)+2(-1)=-2-5i$

93. $\dfrac{1}{i}=\dfrac{1i^3}{ii^3}=\dfrac{i^3}{i^4}=\dfrac{i^3}{1}=i^3=-i=0-i$

95. $\dfrac{4}{5i^3}=\dfrac{4i}{5i^3i}=\dfrac{4i}{5i^4}=\dfrac{4i}{5(1)}=\dfrac{4}{5}i=0+\dfrac{4}{5}i$

97. $\dfrac{3i}{8\sqrt{-9}}=\dfrac{3i}{8(3i)}=\dfrac{1}{8}=\dfrac{1}{8}+0i$

99. $\dfrac{-3}{5i^5}=\dfrac{-3i^3}{5i^5i^3}=\dfrac{-3(-i)}{5i^8}=\dfrac{3i}{5}=0+\dfrac{3}{5}i$

101. $|6+8i|=\sqrt{6^2+8^2}=\sqrt{36+64}$
$=\sqrt{100}=10$

© 2015 Cengage Learning. All Rights Reserved. May not be scanned, copied or duplicated, or posted to a publicly accessible website, in whole or in part.

SECTION 7.7

103. $|12-5i| = \sqrt{12^2+(-5)^2} = \sqrt{144+25}$
$= \sqrt{169} = 13$

105. $|5+7i| = \sqrt{5^2+7^2} = \sqrt{25+49} = \sqrt{74}$

107. $\left|\frac{3}{5}-\frac{4}{5}i\right| = \sqrt{\left(\frac{3}{5}\right)^2+\left(\frac{4}{5}\right)^2}$
$= \sqrt{\frac{9}{25}+\frac{16}{25}}$
$= \sqrt{\frac{25}{25}} = \sqrt{1} = 1$

109. $8+5i \stackrel{?}{=} 2^3+\sqrt{25}i^3$
$8+5i \stackrel{?}{=} 8+5(-i)$
$8+5i \stackrel{?}{=} 8-5i$; not equal

111. $(8-\sqrt{-1})(-2-\sqrt{-16}) = (8-i)(-2-4i) = -16-32i+2i+4i^2 = -16-30i-4$
$= -20-30i$

113. $(2+3i)^2 = (2+3i)(2+3i) = 4+6i+6i+9i^2 = 4+12i-9 = -5+12i$

115. $\frac{5i}{6+2i} = \frac{5i(6-2i)}{(6+2i)(6-2i)} = \frac{30i-10i^2}{36-4i^2} = \frac{10+30i}{40} = \frac{10}{40}+\frac{30}{40}i = \frac{1}{4}+\frac{3}{4}i$

117. $\frac{\sqrt{5}-\sqrt{3}i}{\sqrt{5}+\sqrt{3}i} = \frac{\left(\sqrt{5}-\sqrt{3}i\right)\left(\sqrt{5}-\sqrt{3}i\right)}{\left(\sqrt{5}+\sqrt{3}i\right)\left(\sqrt{5}-\sqrt{3}i\right)} = \frac{5-\sqrt{15}i-\sqrt{15}i+3i^2}{5-3i^2} = \frac{2-2\sqrt{15}i}{8}$
$= \frac{2}{8}-\frac{2\sqrt{15}}{8}i$
$= \frac{1}{4}-\frac{\sqrt{15}}{4}i$

119. $\left(\frac{i}{3+2i}\right)^2 = \frac{i^2}{(3+2i)^2} = \frac{-1}{(3+2i)(3+2i)} = \frac{-1}{9+12i+4i^2} = \frac{-1}{5+12i} = \frac{-1(5-12i)}{(5+12i)(5-12i)}$
$= \frac{-5+12i}{25-144i^2}$
$= -\frac{5}{169}+\frac{12}{169}i$

121. $(5+3i)-(3-5i)+\sqrt{-1} = 5+3i-3+5i+i = 2+9i$

123. $(-8-\sqrt{3}i)-(7-3\sqrt{3}i) = -8-\sqrt{3}i-7+3\sqrt{3}i = \ 15+2\sqrt{3}i$

125. $(2+i)(2-i)(1+i) = (4-2i+2i-i^2)(1+i) = 5(1+i) = 5+5i$

127. $(3+i)[(3-2i)+(2+i)] = (3+i)(5-i) = 15-3i+5i-i^2 = 16+2i$

129. $V = IR = (2-3i)(2+i) = 4+2i-6i-3i^2 = 4-4i-3(-1) = 4-4i+3 = 7-4i$ volts

© 2015 Cengage Learning. All Rights Reserved. May not be scanned, copied or duplicated, or posted to a publicly accessible website, in whole or in part.

131. $Z = \dfrac{V}{I} = \dfrac{1.7 + 0.5i}{0.5i} = \dfrac{(1.7 + 0.5i)i}{(0.5i)i} = \dfrac{1.7i + 0.5i^2}{0.5i^2} = \dfrac{-0.5 + 1.7i}{-0.5} = 1 - 3.4i$

133. Answers may vary.

135.
$$\begin{aligned} x^2 - 2x + 26 &= 0 \\ (1-5i)^2 - 2(1-5i) + 26 &= 0 \\ (1-5i)(1-5i) - 2 + 10i + 26 &= 0 \\ 1 - 10i + 25i^2 + 24 + 10i &= 0 \\ 1 - 10i - 25 + 24 + 10i &= 0 \\ 0 &= 0 \end{aligned}$$

137.
$$\begin{aligned} x^4 - 3x^2 - 4 &= 0 \\ i^4 - 3i^2 - 4 &= 0 \\ 1 - 3(-1) - 4 &= 0 \\ 1 + 3 - 4 &= 0 \\ 0 &= 0 \end{aligned}$$

139. $\dfrac{3-i}{2} = \dfrac{(3-i)(3+i)}{2(3+i)} = \dfrac{9 - i^2}{2(3+i)} = \dfrac{10}{2(3+i)} = \dfrac{5}{3+i}$

Chapter 7 Review (page 524)

1. $\sqrt{81} = \sqrt{9^2} = 9$

2. $-\sqrt{169} = -\sqrt{13^2} = -13$

3. $-\sqrt{36} = -\sqrt{6^2} = -6$

4. $\sqrt{225} = \sqrt{15^2} = 15$

5. $\sqrt[3]{-27} = \sqrt[3]{(-3)^3} = -3$

6. $-\sqrt[3]{216} = -\sqrt[3]{6^3} = -6$

7. $\sqrt[4]{625} = \sqrt[4]{5^4} = 5$

8. $\sqrt[5]{-32} = \sqrt[5]{(-2)^5} = -2$

9. $\sqrt{25x^2} = \sqrt{5^2x^2} = |5x| = 5|x|$

10. $\sqrt{x^2 + 6x + 9} = \sqrt{(x+3)^2} = |x+3|$

11. $\sqrt[3]{27a^6b^3} = 3a^2b$

12. $\sqrt[4]{256x^8y^4} = |4x^2y| = 4x^2|y|$

13. $y = f(x) = \sqrt{x+2}$; Shift $y = \sqrt{x}$ left 2.
D: $[-2, \infty)$, R: $[0, \infty)$

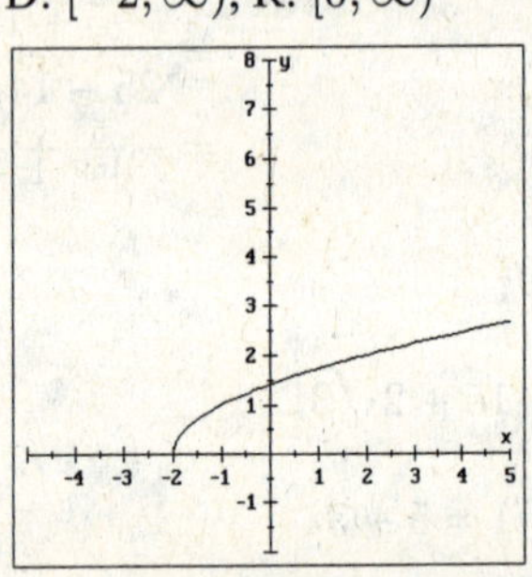

14. $y = f(x) = -\sqrt{x-1}$; Reflect $y = \sqrt{x}$ about the x-axis and shift right 1.
D: $[1, \infty)$, R: $(-\infty, 0]$

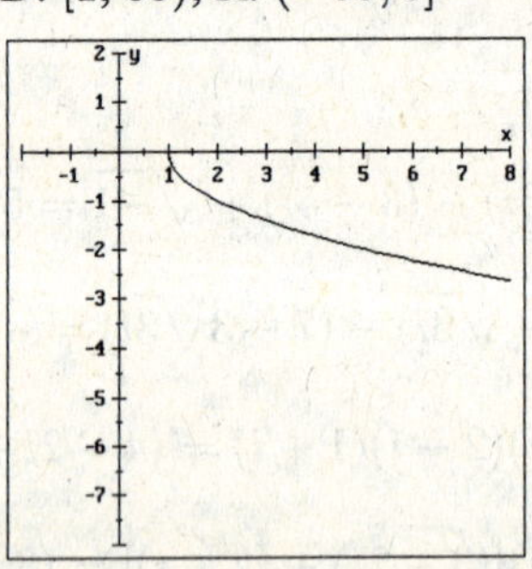

© 2015 Cengage Learning. All Rights Reserved. May not be scanned, copied or duplicated, or posted to a publicly accessible website, in whole or in part.

CHAPTER 7 REVIEW

15. $y = f(x) = -\sqrt{x} + 2$; Reflect $y = \sqrt{x}$ about the x-axis and shift up 2.
D: $[0, \infty)$, R: $(-\infty, 2]$

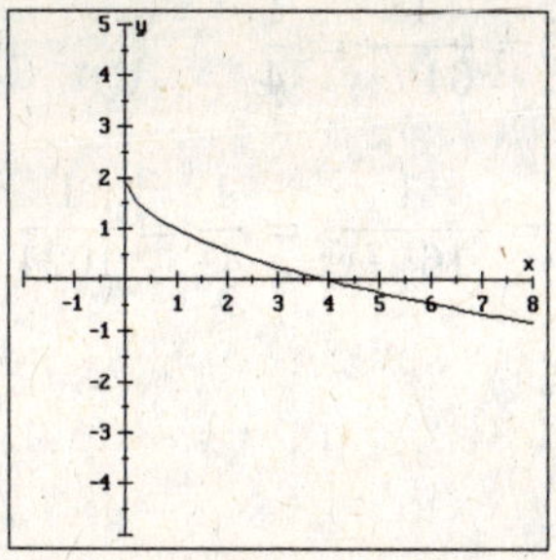

16. $y = f(x) = -\sqrt[3]{x} + 3$; Reflect $y = \sqrt[3]{x}$ about the x-axis and shift up 3.
D: $(-\infty, \infty)$, R: $(-\infty, \infty)$

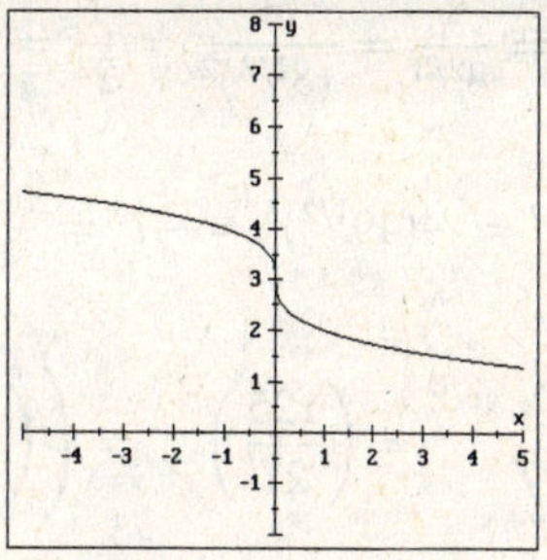

17. mean $= \frac{4+8+12+16+20}{5} = \frac{60}{5} = 12$

18.

Original term	Mean	Difference (term−mean)	Square of difference
4	12	−8	64
8	12	−4	16
12	12	0	0
16	12	4	16
20	12	8	64

$$s = \sqrt{\frac{64+16+0+16+64}{5}} = \sqrt{\frac{160}{5}} = \sqrt{32} \approx 5.7$$

19. $d = 1.4\sqrt{h}$
$d = 1.4\sqrt{4.7} \approx 3$ miles

20. $d = 1.4\sqrt{h}$
$4 = 1.4\sqrt{h}$
$16 = 1.96h$
$8.2 \text{ ft} = h$

21. Let $x =$ one-half of d.
$125^2 = x^2 + 117^2$
$15625 = x^2 + 13689$
$1936 = x^2$
$44 = x$
$d = 2x = 2(44) = 88$ yd

22. Let $x =$ one-half of d.
$8900^2 = x^2 + 3900^2$
$79{,}210{,}000 = x^2 + 15{,}210{,}000$
$64{,}000{,}000 = x^2$
$8000 = x$
$d = 2x = 2(8000) = 16{,}000$ yd (≈ 9 mi)

23. $d = \sqrt{(x_2 - x_1)^2 + (y_2 - y_1)^2} = \sqrt{(9-1)^2 + (-1-5)^2} = \sqrt{8^2 + (-6)^2} = \sqrt{100} = 10$

24. $d = \sqrt{(x_2 - x_1)^2 + (y_2 - y_1)^2} = \sqrt{(-2-(-4))^2 + (8-6)^2} = \sqrt{(2)^2 + (2)^2} = \sqrt{8} \approx 2.83$

25. $81^{1/2} = 9$

26. $-49^{1/2} = -7$

27. $9^{3/2} = \left(9^{1/2}\right)^3 = 3^3 = 27$

© 2015 Cengage Learning. All Rights Reserved. May not be scanned, copied or duplicated, or posted to a publicly accessible website, in whole or in part.

CHAPTER 7 REVIEW

28. $16^{3/2} = \left(16^{1/2}\right)^3 = 4^3 = 64$

29. $(-27)^{1/3} = -3$

30. $-8^{2/3} = -(8^{1/3})^2 = -2^2 = -4$

31. $8^{-2/3} = \dfrac{1}{8^{2/3}} = \dfrac{1}{(8^{1/3})^2} = \dfrac{1}{2^2} = \dfrac{1}{4}$

32. $64^{-1/3} = \dfrac{1}{64^{1/3}} = \dfrac{1}{4}$

33. $-49^{5/2} = -(49^{1/2})^5 = -7^5 = -16{,}807$

34. $\dfrac{1}{16^{5/2}} = \dfrac{1}{(16^{1/2})^5} = \dfrac{1}{4^5} = \dfrac{1}{1024}$

35. $\left(\dfrac{27}{125}\right)^{-2/3} = \left(\dfrac{125}{27}\right)^{2/3} = \left(\left(\dfrac{125}{27}\right)^{1/3}\right)^2 = \left(\dfrac{5}{3}\right)^2 = \dfrac{25}{9}$

36. $\left(\dfrac{4}{9}\right)^{-3/2} = \left(\dfrac{9}{4}\right)^{3/2} = \left[\left(\dfrac{9}{4}\right)^{1/2}\right]^3 = \left(\dfrac{3}{2}\right)^3 = \dfrac{27}{8}$

37. $(27x^3y)^{1/3} = 3xy^{1/3}$

38. $(16x^2y^4)^{1/4} = 2x^{1/2}y$

39. $(25x^3y^4)^{3/2} = 125x^{9/2}y^6$

40. $(8u^2v^3)^{-2/3} = \dfrac{1}{4u^{4/3}v^2}$

41. $5^{1/4}5^{1/2} = 5^{1/4+1/2} = 5^{3/4}$

42. $a^{5/9}a^{2/9} = a^{5/9+2/9} = a^{7/9}$

43. $u^{1/2}\left(u^{1/2} - u^{-1/2}\right) = u^{2/2} - u^0 = u - 1$

44. $v^{2/3}\left(v^{1/3} + v^{4/3}\right) = v^{3/3} + v^{6/3} = v + v^2$

45. $\left(x^{1/2} + y^{1/2}\right)^2 = \left(x^{1/2} + y^{1/2}\right)\left(x^{1/2} + y^{1/2}\right) = x^{2/2} + x^{1/2}y^{1/2} + x^{1/2}y^{1/2} + y^{2/2} = x + 2x^{1/2}y^{1/2} + y$

46. $\left(a^{2/3} + b^{2/3}\right)\left(a^{2/3} - b^{2/3}\right) = a^{4/3} - a^{2/3}b^{2/3} + a^{2/3}b^{2/3} - b^{4/3} = a^{4/3} - b^{4/3}$

47. $\sqrt[6]{5^2} = 5^{2/6} = 5^{1/3} = \sqrt[3]{5}$

48. $\sqrt[8]{x^4} = x^{4/8} = x^{1/2} = \sqrt{x}$

49. $\sqrt[9]{27a^3b^6} = \left(3^3a^3b^6\right)^{1/9} = 3^{3/9}a^{3/9}b^{6/9} = 3^{1/3}a^{1/3}b^{2/3} = \sqrt[3]{3ab^2}$

50. $\sqrt[4]{25a^2b^2} = 5^{2/4}a^{2/4}b^{2/4} = 5^{1/2}a^{1/2}b^{1/2} = \sqrt{5ab}$

51. $\sqrt{176} = \sqrt{16 \cdot 11} = \sqrt{16}\sqrt{11} = 4\sqrt{11}$

52. $\sqrt[3]{250} = \sqrt[3]{125 \cdot 2} = \sqrt[3]{125}\sqrt[3]{2} = 5\sqrt[3]{2}$

53. $\sqrt[4]{32} = \sqrt[4]{16 \cdot 2} = \sqrt[4]{16}\sqrt[4]{2} = 2\sqrt[4]{2}$

54. $\sqrt[5]{96} = \sqrt[5]{32 \cdot 3} = \sqrt[5]{32}\sqrt[5]{3} = 2\sqrt[5]{3}$

55. $\sqrt{8x^3} = \sqrt{4x^2}\sqrt{2x} = 2x\sqrt{2x}$

56. $\sqrt{24x^6y^5} = \sqrt{4x^6y^4}\sqrt{6y} = 2x^3y^2\sqrt{6y}$

57. $\sqrt[3]{16x^5y^4} = \sqrt[3]{8x^3y^3}\sqrt[3]{2x^2y} = 2xy\sqrt[3]{2x^2y}$

58. $\sqrt[3]{54x^7y^3} = \sqrt[3]{27x^6y^3}\sqrt[3]{2x} = 3x^2y\sqrt[3]{2x}$

59. $\dfrac{\sqrt{32x^3}}{\sqrt{2x}} = \sqrt{\dfrac{32x^3}{2x}} = \sqrt{16x^2} = 4x$

60. $\dfrac{\sqrt[3]{16x^5}}{\sqrt[3]{2x^2}} = \sqrt[3]{\dfrac{16x^5}{2x^2}} = \sqrt[3]{8x^3} = 2x$

© 2015 Cengage Learning. All Rights Reserved. May not be scanned, copied or duplicated, or posted to a publicly accessible website, in whole or in part.

CHAPTER 7 REVIEW

61. $\sqrt[3]{\dfrac{2a^2b}{27x^3}} = \dfrac{\sqrt[3]{2a^2b}}{3x}$

62. $\sqrt{\dfrac{17xy}{64a^4}} = \dfrac{\sqrt{17xy}}{8a^2}$

63. $\sqrt{12} + \sqrt{27} = \sqrt{4}\sqrt{3} + \sqrt{9}\sqrt{3}$
$= 2\sqrt{3} + 3\sqrt{3} = 5\sqrt{3}$

64. $\sqrt{18} - \sqrt{32} = \sqrt{9}\sqrt{2} - \sqrt{16}\sqrt{2}$
$= 3\sqrt{2} - 4\sqrt{2} = -\sqrt{2}$

65. $2\sqrt[3]{3} - \sqrt[3]{24} = 2\sqrt[3]{3} - \sqrt[3]{8}\sqrt[3]{3} = 2\sqrt[3]{3} - 2\sqrt[3]{3} = 0$

66. $\sqrt[4]{32} + 2\sqrt[4]{162} = \sqrt[4]{16}\sqrt[4]{2} + 2\sqrt[4]{81}\sqrt[4]{2} = 2\sqrt[4]{2} + 2(3)\sqrt[4]{2} = 2\sqrt[4]{2} + 6\sqrt[4]{2} = 8\sqrt[4]{2}$

67. $2x\sqrt{8} + 2\sqrt{200x^2} + \sqrt{50x^2} = 2x\sqrt{4}\sqrt{2} + 2\sqrt{100x^2}\sqrt{2} + \sqrt{25x^2}\sqrt{2}$
$= 2x(2)\sqrt{2} + 2(10x)\sqrt{2} + 5x\sqrt{2}$
$= 4x\sqrt{2} + 20x\sqrt{2} + 5x\sqrt{2} = 29x\sqrt{2}$

68. $3\sqrt{27a^3} - 2a\sqrt{3a} + 5\sqrt{75a^3} = 3\sqrt{9a^2}\sqrt{3a} - 2a\sqrt{3a} + 5\sqrt{25a^2}\sqrt{3a}$
$= 3(3a)\sqrt{3a} - 2a\sqrt{3a} + 5(5a)\sqrt{3a}$
$= 9a\sqrt{3a} - 2a\sqrt{3a} + 25a\sqrt{3a} = 32a\sqrt{3a}$

69. $\sqrt[3]{54} - 3\sqrt[3]{16} + 4\sqrt[3]{128} = \sqrt[3]{27}\sqrt[3]{2} - 3\sqrt[3]{8}\sqrt[3]{2} + 4\sqrt[3]{64}\sqrt[3]{2}$
$= 3\sqrt[3]{2} - 3(2)\sqrt[3]{2} + 4(4)\sqrt[3]{2} = 3\sqrt[3]{2} - 6\sqrt[3]{2} + 16\sqrt[3]{2} = 13\sqrt[3]{2}$

70. $2\sqrt[4]{32x^5} + 4\sqrt[4]{162x^5} - 5x\sqrt[4]{512x} = 2\sqrt[4]{16x^4}\sqrt[4]{2x} + 4\sqrt[4]{81x^4}\sqrt[4]{2x} - 5x\sqrt[4]{256}\sqrt[4]{2x}$
$= 2(2x)\sqrt[4]{2x} + 4(3x)\sqrt[4]{2x} - 5x(4)\sqrt[4]{2x}$
$= 4x\sqrt[4]{2x} + 12x\sqrt[4]{2x} - 20x\sqrt[4]{2x} = -4x\sqrt[4]{2x}$

71. hypotenuse $= 7\sqrt{2}$ m

72. shorter leg $= \frac{1}{2}\left(12\sqrt{3}\right) = 6\sqrt{3}$ cm
longer leg $= \sqrt{3}\left(6\sqrt{3}\right) = 6(3) = 18$ cm

73. $x = 5\sqrt{2} \approx 7.07$ in.

74. $x = \sqrt{3}(\frac{1}{2} \cdot 10) = 5\sqrt{3} \approx 8.66$ cm

75. $\left(5\sqrt{3}\right)\left(2\sqrt{5}\right) = 10\sqrt{15}$

76. $2\sqrt{6}\sqrt{216} = 2\sqrt{6}\sqrt{36}\sqrt{6} = 2(6)(6) = 72$

77. $\sqrt{5x}\sqrt{4x} = \sqrt{20x^2} = \sqrt{4x^2}\sqrt{5} = 2x\sqrt{5}$

78. $\sqrt[3]{3x^2}\sqrt[3]{9x^4} = \sqrt[3]{27x^6} = 3x^2$

79. $-\sqrt[3]{2x^2}\sqrt[3]{4x} = -\sqrt[3]{8x^3} = -2x$

80. $-\sqrt[4]{256x^5y^{11}}\sqrt[4]{625x^9y^3} = -\sqrt[4]{256x^4y^8}\sqrt[4]{xy^3}\sqrt[4]{625x^8}\sqrt[4]{xy^3} = -4xy^2(5x^2)\sqrt[4]{x^2y^6}$
$= -20x^3y^2\sqrt[4]{y^4}\sqrt[4]{x^2y^2}$
$= -20x^3y^3\sqrt[4]{x^2y^2} = -20x^3y^3\sqrt{xy}$

© 2015 Cengage Learning. All Rights Reserved. May not be scanned, copied or duplicated, or posted to a publicly accessible website, in whole or in part.

81. $\sqrt{3}\left(\sqrt{27}-4\right)=\sqrt{81}-4\sqrt{3}=9-4\sqrt{3}$

82. $\sqrt{5}\left(\sqrt{5}-6\right)=\sqrt{25}-6\sqrt{5}=5-6\sqrt{5}$

83. $\sqrt{5}\left(\sqrt{2}-1\right)=\sqrt{10}-\sqrt{5}$

84. $\sqrt{3}\left(\sqrt{3}+\sqrt{2}\right)=\sqrt{9}+\sqrt{6}=3+\sqrt{6}$

85. $\left(\sqrt{5}+1\right)\left(\sqrt{5}-1\right)=\sqrt{25}-\sqrt{5}+\sqrt{5}-1=4$

86. $\left(\sqrt{3}+\sqrt{2}\right)\left(\sqrt{3}+\sqrt{2}\right)=\sqrt{9}+\sqrt{6}+\sqrt{6}+\sqrt{4}=5+2\sqrt{6}$

87. $(\sqrt{x}+\sqrt{y})(\sqrt{x}-\sqrt{y})=\sqrt{x^2}-\sqrt{xy}+\sqrt{xy}-\sqrt{y^2}=x-y$

88. $(2\sqrt{u}+3)(3\sqrt{u}-4)=6\sqrt{u^2}-8\sqrt{u}+9\sqrt{u}-12=6u+\sqrt{u}-12$

89. $\dfrac{1}{\sqrt{3}}=\dfrac{1\sqrt{3}}{\sqrt{3}\sqrt{3}}=\dfrac{\sqrt{3}}{3}$

90. $\dfrac{\sqrt{3}}{\sqrt{5}}=\dfrac{\sqrt{3}\sqrt{5}}{\sqrt{5}\sqrt{5}}=\dfrac{\sqrt{15}}{5}$

91. $\dfrac{x}{\sqrt{xy}}=\dfrac{x\sqrt{xy}}{\sqrt{xy}\sqrt{xy}}=\dfrac{x\sqrt{xy}}{xy}=\dfrac{\sqrt{xy}}{y}$

92. $\dfrac{\sqrt[3]{uv}}{\sqrt[3]{u^5v^7}}=\dfrac{\sqrt[3]{uv}\sqrt[3]{uv^2}}{\sqrt[3]{u^5v^7}\sqrt[3]{uv^2}}=\dfrac{\sqrt[3]{u^2v^3}}{\sqrt[3]{u^6v^9}}=\dfrac{v\sqrt[3]{u^2}}{u^2v^3}=\dfrac{\sqrt[3]{u^2}}{u^2v^2}$

93. $\dfrac{2}{\sqrt{2}-1}=\dfrac{2\left(\sqrt{2}+1\right)}{\left(\sqrt{2}-1\right)\left(\sqrt{2}+1\right)}=\dfrac{2\left(\sqrt{2}+1\right)}{\sqrt{4}-1}=\dfrac{2\left(\sqrt{2}+1\right)}{2-1}=\dfrac{2\left(\sqrt{2}+1\right)}{1}=2\left(\sqrt{2}+1\right)$

94. $\dfrac{\sqrt{2}}{\sqrt{3}-1}=\dfrac{\sqrt{2}\left(\sqrt{3}+1\right)}{\left(\sqrt{3}-1\right)\left(\sqrt{3}+1\right)}=\dfrac{\sqrt{2}\left(\sqrt{3}+1\right)}{\sqrt{9}-1}=\dfrac{\sqrt{2}\left(\sqrt{3}+1\right)}{3-1}=\dfrac{\sqrt{2}\left(\sqrt{3}+1\right)}{2}$

$=\dfrac{\sqrt{6}+\sqrt{2}}{2}$

95. $\dfrac{2x-32}{\sqrt{x}+4}=\dfrac{(2x-32)(\sqrt{x}-4)}{(\sqrt{x}+4)(\sqrt{x}-4)}=\dfrac{2(x-16)(\sqrt{x}-4)}{\sqrt{x^2}-16}=\dfrac{2(x-16)(\sqrt{x}-4)}{x-16}=2(\sqrt{x}-4)$

96. $\dfrac{\sqrt{a}+1}{\sqrt{a}-1}=\dfrac{(\sqrt{a}+1)(\sqrt{a}+1)}{(\sqrt{a}-1)(\sqrt{a}+1)}=\dfrac{\sqrt{a^2}+\sqrt{a}+\sqrt{a}+1}{\sqrt{a^2}-1}=\dfrac{a+2\sqrt{a}+1}{a-1}$

97. $\dfrac{\sqrt{3}}{5}=\dfrac{\sqrt{3}\sqrt{3}}{5\sqrt{3}}=\dfrac{3}{5\sqrt{3}}$

98. $\dfrac{\sqrt[3]{9}}{3}=\dfrac{\sqrt[3]{9}\sqrt[3]{3}}{3\sqrt[3]{3}}=\dfrac{\sqrt[3]{27}}{3\sqrt[3]{3}}=\dfrac{3}{3\sqrt[3]{3}}=\dfrac{1}{\sqrt[3]{3}}$

© 2015 Cengage Learning. All Rights Reserved. May not be scanned, copied or duplicated, or posted to a publicly accessible website, in whole or in part.

CHAPTER 7 REVIEW

99. $\dfrac{3-\sqrt{x}}{2} = \dfrac{(3-\sqrt{x})(3+\sqrt{x})}{2(3+\sqrt{x})} = \dfrac{9-\sqrt{x^2}}{2(3+\sqrt{x})} = \dfrac{9-x}{2(3+\sqrt{x})}$

100. $\dfrac{\sqrt{a}-\sqrt{b}}{\sqrt{a}} = \dfrac{\left(\sqrt{a}-\sqrt{b}\right)\left(\sqrt{a}+\sqrt{b}\right)}{\sqrt{a}\left(\sqrt{a}+\sqrt{b}\right)} = \dfrac{\sqrt{a^2}-\sqrt{b^2}}{\sqrt{a^2}+\sqrt{ab}} = \dfrac{a-b}{a+\sqrt{ab}}$

101.
$$\begin{aligned} \sqrt{3y-11} &= \sqrt{y+7} \\ \left(\sqrt{3y-11}\right)^2 &= \left(\sqrt{y+7}\right)^2 \\ 3y-11 &= y+7 \\ 2y &= 18 \\ y &= 9 \end{aligned}$$
The answer checks.

102.
$$\begin{aligned} u &= \sqrt{25u-144} \\ u^2 &= \left(\sqrt{25u-144}\right)^2 \\ u^2 &= 25u-144 \\ u^2-25u+144 &= 0 \\ (u-9)(u-16) &= 0 \end{aligned}$$
$u=9$ or $u=16$; Both answers check.

103.
$$\begin{aligned} \sqrt{2x+6}+1 &= x \\ \sqrt{2x+6} &= x-1 \\ \left(\sqrt{2x+6}\right)^2 &= (x-1)^2 \\ 2x+6 &= x^2-2x+1 \\ 0 &= x^2-4x-5 \\ 0 &= (x-5)(x+1) \end{aligned}$$
$x-5=0$ **or** $x+1=0$

$x=5$ $\quad x=-1$

solution $\quad$ extraneous

104.
$$\begin{aligned} \sqrt{z+1}+\sqrt{z} &= 2 \\ \sqrt{z+1} &= 2-\sqrt{z} \\ \left(\sqrt{z+1}\right)^2 &= \left(2-\sqrt{z}\right)^2 \\ z+1 &= 4-4\sqrt{z}+z \\ 4\sqrt{z} &= 3 \\ \left(4\sqrt{z}\right)^2 &= 3^2 \\ 16z &= 9 \end{aligned}$$
$z=\frac{9}{16}$; The answer checks.

105.
$$\begin{aligned} \sqrt{2x+5}-\sqrt{2x} &= 1 \\ \sqrt{2x+5} &= 1+\sqrt{2x} \\ \left(\sqrt{2x+5}\right)^2 &= \left(1+\sqrt{2x}\right)^2 \\ 2x+5 &= 1+2\sqrt{2x}+2x \\ 4 &= 2\sqrt{2x} \\ 4^2 &= \left(2\sqrt{2x}\right)^2 \\ 16 &= 8x \end{aligned}$$
$2=x$; The answer checks.

106.
$$\begin{aligned} \sqrt[3]{x^3+8} &= x+2 \\ \left(\sqrt[3]{x^3+8}\right)^3 &= (x+2)^3 \\ x^3+8 &= x^3+6x^2+12x+8 \\ 0 &= 6x^2+12x \\ 0 &= 6x(x+2) \end{aligned}$$
$x=0$ or $x=-2$; Both answers check.

107. $(6+3i)+(4-15i) = 6+3i+4-15i = 10-12i$

108. $(-5-22i)-(-7+28i) = -5-22i+7-28i = 2-50i$

109. $(-32+\sqrt{-144})-(64+\sqrt{-81}) = -32+\sqrt{144i^2}-64-\sqrt{81i^2} = -32+12i-64-9i$
$= -96+3i$

© 2015 Cengage Learning. All Rights Reserved. May not be scanned, copied or duplicated, or posted to a publicly accessible website, in whole or in part.

110. $(-8+\sqrt{-8})+(6-\sqrt{-32}) = -8+\sqrt{4i^2}\sqrt{2}+6-\sqrt{16i^2}\sqrt{2}$
$= -8+2i\sqrt{2}+6-4i\sqrt{2} = -2-2\sqrt{2}i$

111. $(2-7i)(-3+4i) = -6+8i+21i-28i^2 = -6+29i+28 = 22+29i$

112. $(-4+5i)(3+2i) = -12-8i+15i+10i^2 = -12+7i-10 = -22+7i$

113. $(5-\sqrt{-27})(-6+\sqrt{-12}) = (5-3i\sqrt{3})(-6+2i\sqrt{3}) = -30+10i\sqrt{3}+18i\sqrt{3}-6i^2(3)$
$= -30+28i\sqrt{3}+18 = -12+28\sqrt{3}i$

114. $(2+\sqrt{-128})(3-\sqrt{-98}) = (2+8i\sqrt{2})(3-7i\sqrt{2}) = 6-14i\sqrt{2}+24i\sqrt{2}-56i^2(2)$
$= 6+10i\sqrt{2}+112 = 118+10\sqrt{2}i$

115. $\frac{3}{4i} = \frac{3}{4i}\cdot\frac{i}{i} = \frac{3i}{4i^2} = \frac{3i}{-4} = -\frac{3}{4}i = 0-\frac{3}{4}i$

116. $\frac{-2}{5i^3} = \frac{-2}{5i^3}\cdot\frac{i}{i} = \frac{-2i}{5i^4} = \frac{-2i}{5} = 0-\frac{2}{5}i$

117. $\frac{6}{2+i} = \frac{6}{2+i}\cdot\frac{2-i}{2-i} = \frac{6(2-i)}{4-i^2} = \frac{6(2-i)}{5} = \frac{12-6i}{5} = \frac{12}{5}-\frac{6}{5}i$

118. $\frac{7}{3-i} = \frac{7}{3-i}\cdot\frac{3+i}{3+i} = \frac{7(3+i)}{9-i^2} = \frac{7(3+i)}{10} = \frac{21+7i}{10} = \frac{21}{10}+\frac{7}{10}i$

119. $\frac{4+i}{4-i} = \frac{4+i}{4-i}\cdot\frac{4+i}{4+i} = \frac{16+8i+i^2}{16-i^2} = \frac{15+8i}{17} = \frac{15}{17}+\frac{8}{17}i$

120. $\frac{3-i}{3+i} = \frac{3-i}{3+i}\cdot\frac{3-i}{3-i} = \frac{9-6i+i^2}{9-i^2} = \frac{8-6i}{10} = \frac{8}{10}-\frac{6}{10}i = \frac{4}{5}-\frac{3}{5}i$

121. $\frac{3}{5+\sqrt{-4}} = \frac{3}{5+2i} = \frac{3}{5+2i}\cdot\frac{5-2i}{5-2i} = \frac{3(5-2i)}{25-4i^2} = \frac{3(5-2i)}{29} = \frac{15-6i}{29} = \frac{15}{29}-\frac{6}{29}i$

122. $\frac{2}{3-\sqrt{-9}} = \frac{2}{3-3i} = \frac{2}{3-3i}\cdot\frac{3+3i}{3+3i} = \frac{2(3+3i)}{9-9i^2} = \frac{2(3+3i)}{18} = \frac{3+3i}{9} = \frac{3}{9}+\frac{3}{9}i = \frac{1}{3}+\frac{1}{3}i$

123. $|9+12i| = \sqrt{9^2+12^2} = \sqrt{81+144} = \sqrt{225} = 15$

124. $|24-10i| = \sqrt{24^2+(-10)^2} = \sqrt{576+100} = \sqrt{676} = 26$

125. $i^{39} = i^{36}i^3 = (i^4)^9 i^3 = 1^9(-i) = -i$

126. $i^{52} = (i^4)^{13} = 1^{13} = 1$

Chapter 7 Test (page 530)

1. $\sqrt{36} = \sqrt{6^2} = 6$

2. $\sqrt{125} = \sqrt{25}\sqrt{5} = 5\sqrt{5}$

© 2015 Cengage Learning. All Rights Reserved. May not be scanned, copied or duplicated, or posted to a publicly accessible website, in whole or in part.

CHAPTER 7 TEST

3. $\sqrt{4x^2} = \sqrt{(2x)^2} = |2x| = 2|x|$

4. $\sqrt[3]{8x^3} = \sqrt[3]{(2x)^3} = 2x$

5. $f(x) = \sqrt{x-2}$; Shift $y = \sqrt{x}$ right 2.

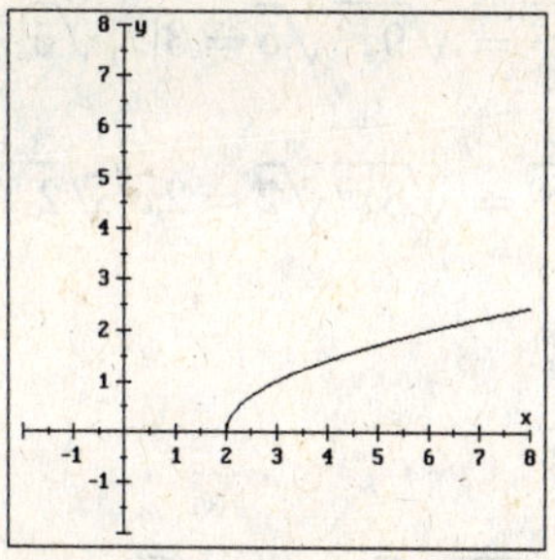

domain $= [2, \infty)$, range $= [0, \infty)$

6. $f(x) = \sqrt[3]{x} + 3$; Shift $y = \sqrt[3]{x}$ up 3.

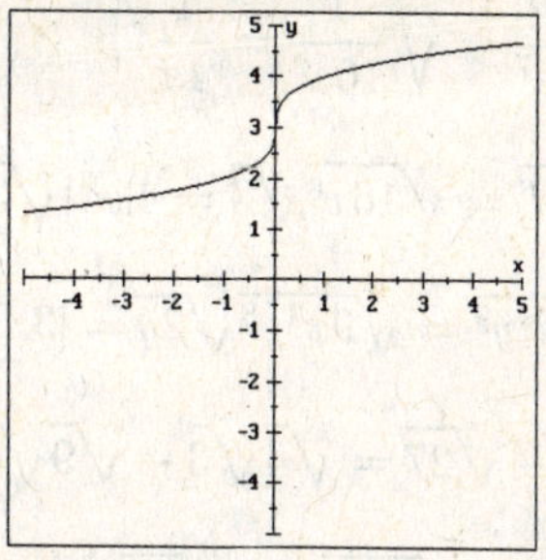

domain $= (-\infty, \infty)$, range $= (-\infty, \infty)$

7.
$$\begin{aligned} 53^2 &= 45^2 + h^2 \\ 2809 &= 2025 + h^2 \\ 784 &= h^2 \\ 28 \text{ in.} &= h \end{aligned}$$

8.
$$\begin{aligned} 2^2 &= \left(\frac{w}{2}\right)^2 + (1.9)^2 \\ 4 &= \frac{w^2}{4} + 3.61 \\ 0.39 &= \frac{w^2}{4} \\ 1.56 &= w^2 \\ 1.25 \text{ meters} &= w \end{aligned}$$

9. $d = \sqrt{(x_2 - x_1)^2 + (y_2 - y_1)^2} = \sqrt{(0-6)^2 + (0-8)^2} = \sqrt{(-6)^2 + (-8)^2} = \sqrt{100} = 10$

10. $d = \sqrt{(x_2 - x_1)^2 + (y_2 - y_1)^2} = \sqrt{(22-(-2))^2 + (12-5)^2} = \sqrt{24^2 + 7^2} = \sqrt{625} = 25$

11. $81^{1/4} = 3$

12. $-64^{2/3} = -\left(64^{1/3}\right)^2 = -4^2 = -16$

13. $36^{-3/2} = \dfrac{1}{36^{3/2}} = \dfrac{1}{\left(36^{1/2}\right)^3} = \dfrac{1}{6^3} = \dfrac{1}{216}$

14. $\left(-\dfrac{8}{27}\right)^{-2/3} = \left(-\dfrac{27}{8}\right)^{2/3} = \left[\left(-\dfrac{27}{8}\right)^{1/3}\right]^2 = \left(-\dfrac{3}{2}\right)^2 = \dfrac{9}{4}$

15. $\dfrac{2^{5/3}2^{1/6}}{2^{1/2}} = \dfrac{2^{10/6}2^{1/6}}{2^{3/6}} = 2^{10/6+1/6-3/6} = 2^{8/6} = 2^{4/3}$

16. $\dfrac{\left(8x^3y\right)^{1/2}\left(8xy^5\right)^{1/2}}{\left(x^3y^6\right)^{1/3}} = \dfrac{8^{1/2}x^{3/2}y^{1/2}8^{1/2}x^{1/2}y^{5/2}}{x^{3/3}y^{6/3}} = \dfrac{8^{2/2}x^{4/2}y^{6/2}}{xy^2} = \dfrac{8x^2y^3}{xy^2} = 8xy$

17. $\sqrt{108} = \sqrt{36}\sqrt{3} = 6\sqrt{3}$

18. $\sqrt{250x^3y^5} = \sqrt{25x^2y^4}\sqrt{10xy} = 5xy^2\sqrt{10xy}$

© 2015 Cengage Learning. All Rights Reserved. May not be scanned, copied or duplicated, or posted to a publicly accessible website, in whole or in part.

CHAPTER 7 TEST

19. $\dfrac{\sqrt[3]{240x^{10}y^4}}{\sqrt[3]{6xy}} = \sqrt[3]{\dfrac{240x^{10}y^4}{6xy}} = \sqrt[3]{40x^9y^3} = \sqrt[3]{8x^9y^3}\sqrt[3]{5} = 2x^3y\sqrt[3]{5}$

20. $\sqrt{\dfrac{3a^5}{48a^7}} = \sqrt{\dfrac{1}{16a^2}} = \dfrac{1}{4a}$

21. $\sqrt{45x^2} = \sqrt{9x^2}\sqrt{5} = 3|x|\sqrt{5}$

22. $\sqrt{48x^6} = \sqrt{16x^6}\sqrt{3} = 4|x^3|\sqrt{3}$

23. $\sqrt[3]{16x^9} = \sqrt[3]{8x^9}\sqrt[3]{2} = 2x^3\sqrt[3]{2}$

24. $\sqrt{18x^4y^9} = \sqrt{9x^4y^8}\sqrt{2y} = |3x^2y^4|\sqrt{2y} = \quad 3x^2y^4\sqrt{2y}$

25. $\sqrt{12} - \sqrt{27} = \sqrt{4}\sqrt{3} - \sqrt{9}\sqrt{3} = 2\sqrt{3} - 3\sqrt{3} = -\sqrt{3}$

26. $$\begin{aligned} 2\sqrt[3]{40} - \sqrt[3]{5000} + 4\sqrt[3]{625} = 2\sqrt[3]{8}\sqrt[3]{5} - \sqrt[3]{1000}\sqrt[3]{5} + 4\sqrt[3]{125}\sqrt[3]{5} &= 2(2)\sqrt[3]{5} - 10\sqrt[3]{5} + 4(5)\sqrt[3]{5} \\ &= 4\sqrt[3]{5} - 10\sqrt[3]{5} + 20\sqrt[3]{5} \\ &= 14\sqrt[3]{5} \end{aligned}$$

27. $$\begin{aligned} 2\sqrt{48y^5} - 3y\sqrt{12y^3} = 2\sqrt{16y^4}\sqrt{3y} - 3y\sqrt{4y^2}\sqrt{3y} &= 2(4y^2)\sqrt{3y} - 3y(2y)\sqrt{3y} \\ &= 8y^2\sqrt{3y} - 6y^2\sqrt{3y} = 2y^2\sqrt{3y} \end{aligned}$$

28. $\sqrt[4]{768z^5} + z\sqrt[4]{48z} = \sqrt[4]{256z^4}\sqrt[4]{3z} + z\sqrt[4]{16}\sqrt[4]{3z} = 4z\sqrt[4]{3z} + 2z\sqrt[4]{3z} = 6z\sqrt[4]{3z}$

29. $-2\sqrt{xy}(3\sqrt{x} + \sqrt{xy^3}) = -6\sqrt{x^2y} - 2\sqrt{x^2y^4} = -6x\sqrt{y} - 2xy^2$

30. $\left(3\sqrt{2} + \sqrt{3}\right)\left(2\sqrt{2} - 3\sqrt{3}\right) = 6\sqrt{4} - 9\sqrt{6} + 2\sqrt{6} - 3\sqrt{9} = 12 - 7\sqrt{6} - 9 = 3 - 7\sqrt{6}$

31. $\dfrac{1}{\sqrt{5}} = \dfrac{1\sqrt{5}}{\sqrt{5}\sqrt{5}} = \dfrac{\sqrt{5}}{5}$

32. $\dfrac{3t-1}{\sqrt{3t}-1} = \dfrac{(3t-1)\left(\sqrt{3t}+1\right)}{\left(\sqrt{3t}-1\right)\left(\sqrt{3t}+1\right)} = \dfrac{(3t-1)\left(\sqrt{3t}+1\right)}{\sqrt{9t^2}-1} = \dfrac{(3t-1)\left(\sqrt{3t}+1\right)}{3t-1} = \sqrt{3t}+1$

33. $\dfrac{\sqrt{3}}{\sqrt{7}} = \dfrac{\sqrt{3}\sqrt{3}}{\sqrt{7}\sqrt{3}} = \dfrac{3}{\sqrt{21}}$

34. $\dfrac{\sqrt{a}+\sqrt{b}}{\sqrt{a}-\sqrt{b}} = \dfrac{\left(\sqrt{a}+\sqrt{b}\right)\left(\sqrt{a}-\sqrt{b}\right)}{\left(\sqrt{a}-\sqrt{b}\right)\left(\sqrt{a}-\sqrt{b}\right)} = \dfrac{\sqrt{a^2}-\sqrt{b^2}}{\sqrt{a^2}-\sqrt{ab}-\sqrt{ab}+\sqrt{b^2}} = \dfrac{a-b}{a-2\sqrt{ab}+b}$

© 2015 Cengage Learning. All Rights Reserved. May not be scanned, copied or duplicated, or posted to a publicly accessible website, in whole or in part.

35.
$$\sqrt[3]{6n+4}-4=0$$
$$\sqrt[3]{6n+4}=4$$
$$\left(\sqrt[3]{6n+4}\right)^3=4^3$$
$$6n+4=64$$
$$6n=60$$
$$n=10$$
The answer checks.

36.
$$1-\sqrt{u}=\sqrt{u-3}$$
$$\left(1-\sqrt{u}\right)^2=\left(\sqrt{u-3}\right)^2$$
$$1-2\sqrt{u}+u=u-3$$
$$4=2\sqrt{u}$$
$$4^2=\left(2\sqrt{u}\right)^2$$
$$16=4u$$
$$4=u$$
extraneous $\Rightarrow \emptyset$

37. $(2+4i)+(-3+7i)=2+4i-3+7i=-1+11i$

38. $(3-\sqrt{-9})-(-1+\sqrt{-16})=3-3i+1-4i=4-7i$

39. $2i(3-4i)=6i-8i^2=6i+8=8+6i$

40. $(3+2i)(-4-i)=-12-3i-8i-2i^2=-12-11i+2=-10-11i$

41. $\dfrac{1}{\sqrt{2}i}=\dfrac{1}{\sqrt{2}i}\cdot\dfrac{i\sqrt{2}}{i\sqrt{2}}=\dfrac{i\sqrt{2}}{2i^2}=-\dfrac{\sqrt{2}i}{2}=0-\dfrac{\sqrt{2}}{2}i$

42. $\dfrac{2+i}{3-i}=\dfrac{2+i}{3-i}\cdot\dfrac{3+i}{3+i}=\dfrac{6+5i+i^2}{9-i^2}=\dfrac{5+5i}{10}=\dfrac{1}{2}+\dfrac{1}{2}i$

Exercises 8.1 (page 544)

1. $(x+5)(x-5)=x^2-5x+5x-25=x^2-25$

3. $(x+2)^2=(x+2)(x+2)=x^2+2x+2x+4=x^2+4x+2$

5. $(x-7)^2=(x-7)(x-7)=x^2-7x-7x+49=x^2-14x+49$

7.
$$\frac{t+9}{2}+\frac{t+2}{5}=\frac{8}{5}+4t$$
$$5(t+9)+2(t+2)=2(8)+40t$$
$$5t+45+2t+4=16+40t$$
$$33=33t$$
$$1=t$$

9.
$$3(t-3)+3t\le 2(t+1)+t+1$$
$$3t-9+3t\le 2t+2+t+1$$
$$3t\le 12$$
$$t\le 4$$

11. $x=\sqrt{c};\ x=-\sqrt{c}$

13. positive or negative

© 2015 Cengage Learning. All Rights Reserved. May not be scanned, copied or duplicated, or posted to a publicly accessible website, in whole or in part.

SECTION 8.1

15.
$$\begin{aligned} x^2 - 81 &= 0 \\ (x+9)(x-9) &= 0 \\ x+9 = 0 \quad &\textbf{or} \quad x-9=0 \\ x = -9 \quad & \quad\quad x = 9 \end{aligned}$$

17.
$$\begin{aligned} 2y^2 - 72 &= 0 \\ 2(y^2-36) &= 0 \\ 2(y+6)(y-6) &= 0 \\ y+6=0 \quad &\textbf{or} \quad y-6=0 \\ y=-6 \quad & \quad\quad y=6 \end{aligned}$$

19.
$$\begin{aligned} y^2+7y+12 &= 0 \\ (y+3)(y+4) &= 0 \\ y+3=0 \quad &\textbf{or} \quad y+4=0 \\ y=-3 \quad & \quad\quad y=-4 \end{aligned}$$

21.
$$\begin{aligned} 6s^2+11s-10 &= 0 \\ (2s+5)(3s-2) &= 0 \\ 2s+5=0 \quad &\textbf{or} \quad 3s-2=0 \\ s=-\tfrac{5}{2} \quad & \quad\quad s=\tfrac{2}{3} \end{aligned}$$

23.
$$\begin{aligned} x^2 &= 36 \\ x &= \pm\sqrt{36} = \pm 6 \end{aligned}$$

25.
$$\begin{aligned} z^2 &= 5 \\ z &= \pm\sqrt{5} \end{aligned}$$

27.
$$\begin{aligned} (y+3)^2 &= 9 \\ y+3 &= \pm\sqrt{9} \\ y+3 &= \pm 3 \\ y &= -3 \pm 3 \\ y = 0 &\text{ or } y = -6 \end{aligned}$$

29.
$$\begin{aligned} (x-2)^2 - 5 &= 0 \\ (x-2)^2 &= 5 \\ x-2 &= \pm\sqrt{5} \\ x &= 2 \pm \sqrt{5} \end{aligned}$$

31.
$$\begin{aligned} 16p^2 + 49 &= 0 \\ 16p^2 &= -49 \\ p^2 &= -\frac{49}{16} \\ p &= \pm\sqrt{-\frac{49}{16}} \\ p &= \pm\frac{7}{4}i \end{aligned}$$

33.
$$\begin{aligned} 4m^2 + 81 &= 0 \\ 4m^2 &= -81 \\ m^2 &= -\frac{81}{4} \\ m &= \pm\sqrt{-\frac{81}{4}} \\ m &= \pm\frac{9}{2}i \end{aligned}$$

35. $x^2 + 4x;\ \frac{1}{2}\cdot 4 = 2;\ 2^2 = \boxed{4}$

37. $x^2 - 3x;\ \frac{1}{2}\cdot 3 = \frac{3}{2};\ \left(\frac{3}{2}\right)^2 = \boxed{\frac{9}{4}}$

39.
$$\begin{aligned} x^2+4x-12 &= 0 \\ x^2+4x &= 12 \\ x^2+4x+4 &= 12+4 \\ (x+2)^2 &= 16 \\ x+2 &= \pm 4 \\ x &= -2 \pm 4 \\ x = 2 &\text{ or } x = -6 \end{aligned}$$

41.
$$\begin{aligned} x^2-6x+8 &= 0 \\ x^2-6x &= -8 \\ x^2-6x+9 &= -8+9 \\ (x-3)^2 &= 1 \\ x-3 &= \pm 1 \\ x &= 3 \pm 1 \\ x = 4 &\text{ or } x = 2 \end{aligned}$$

© 2015 Cengage Learning. All Rights Reserved. May not be scanned, copied or duplicated, or posted to a publicly accessible website, in whole or in part.

43.
$$\begin{aligned}6x^2+11x+3&=0\\6x^2+11x&=-3\\x^2+\frac{11}{6}x&=-\frac{1}{2}\\x^2+\frac{11}{6}x+\frac{121}{144}&=-\frac{1}{2}+\frac{121}{144}\\\left(x+\frac{11}{12}\right)^2&=\frac{49}{144}\\x+\frac{11}{12}&=\pm\frac{7}{12}\\x&=-\frac{11}{12}\pm\frac{7}{12}\end{aligned}$$
$x=-\frac{4}{12}=-\frac{1}{3}$ or $x=-\frac{18}{12}=-\frac{3}{2}$

45.
$$\begin{aligned}6x^2-7x-5&=0\\6x^2-7x&=5\\x^2-\frac{7}{6}x&=\frac{5}{6}\\x^2-\frac{7}{6}x+\frac{49}{144}&=\frac{5}{6}+\frac{49}{144}\\\left(x-\frac{7}{12}\right)^2&=\frac{169}{144}\\x-\frac{7}{12}&=\pm\frac{13}{12}\\x&=\frac{7}{12}\pm\frac{13}{12}\end{aligned}$$
$x=\frac{20}{12}=\frac{5}{3}$ or $x=-\frac{6}{12}=-\frac{1}{2}$

47.
$$\begin{aligned}9-6r&=8r^2\\-8r^2-6r&=-9\\r^2+\frac{3}{4}r&=\frac{9}{8}\\r^2+\frac{3}{4}r+\frac{9}{64}&=\frac{9}{8}+\frac{9}{64}\\\left(r+\frac{3}{8}\right)^2&=\frac{81}{64}\\r+\frac{3}{8}&=\pm\frac{9}{8}\\r&=-\frac{3}{8}\pm\frac{9}{8}\end{aligned}$$
$r=\frac{6}{8}=\frac{3}{4}$ or $r=-\frac{12}{8}=-\frac{3}{2}$

49.
$$\begin{aligned}x+1&=2x^2\\-2x^2+x&=-1\\x^2-\frac{1}{2}x&=\frac{1}{2}\\x^2-\frac{1}{2}x+\frac{1}{16}&=\frac{1}{2}+\frac{1}{16}\\\left(x-\frac{1}{4}\right)^2&=\frac{9}{16}\\x-\frac{1}{4}&=\pm\frac{3}{4}\\x&=\frac{1}{4}\pm\frac{3}{4}\end{aligned}$$
$x=\frac{4}{4}=1$ or $x=-\frac{2}{4}=-\frac{1}{2}$

51.
$$\begin{aligned}\frac{7x+1}{5}&=-x^2\\7x+1&=-5x^2\\5x^2+7x&=-1\\x^2+\frac{7}{5}x&=-\frac{1}{5}\\x^2+\frac{7}{5}x+\frac{49}{100}&=-\frac{1}{5}+\frac{49}{100}\\\left(x+\frac{7}{10}\right)^2&=\frac{29}{100}\\x+\frac{7}{10}&=\pm\frac{\sqrt{29}}{10}\\x&=-\frac{7}{10}\pm\frac{\sqrt{29}}{10}\end{aligned}$$

53.
$$\begin{aligned}3x^2-6x+1&=0\\3x^2-6x&=-1\\x^2-2x&=-\frac{1}{3}\\x^2-2x+1&=-\frac{1}{3}+1\\(x-1)^2&=\frac{2}{3}\\x-1&=\pm\sqrt{\frac{2}{3}}\\x-1&=\pm\frac{\sqrt{2}}{\sqrt{3}}\cdot\frac{\sqrt{3}}{\sqrt{3}}\\x-1&=\pm\frac{\sqrt{6}}{3}\\x&=1\pm\frac{\sqrt{6}}{3}\end{aligned}$$

© 2015 Cengage Learning. All Rights Reserved. May not be scanned, copied or duplicated, or posted to a publicly accessible website, in whole or in part.

55.
$$\begin{aligned} p^2+2p+2&=0\\ p^2+2p&=-2\\ p^2+2p+1&=-2+1\\ (p+1)^2&=-1\\ p+1&=\pm\sqrt{-1}\\ p+1&=\pm i\\ p&=-1\pm i \end{aligned}$$

57.
$$\begin{aligned} y^2+8y+18&=0\\ y^2+8y&=-18\\ y^2+8y+16&=-18+16\\ (y+4)^2&=-2\\ y+4&=\pm\sqrt{-2}\\ y&=-4\pm i\sqrt{2} \end{aligned}$$

59.
$$\begin{aligned} 9x-8&=x^2\\ 0&=x^2-9x+8\\ 0&=(x-8)(x-1) \end{aligned}$$
$$x-8=0 \quad \textbf{or} \quad x-1=0$$
$$x=8 \qquad\qquad x=1$$

61.
$$\begin{aligned} 3x^2-16&=0\\ 3x^2&=16\\ x^2&=\frac{16}{3}\\ x&=\pm\sqrt{\frac{16}{3}}=\pm\frac{4}{\sqrt{3}}=\pm\frac{4\sqrt{3}}{3} \end{aligned}$$

63.
$$\begin{aligned} (s-7)^2-9&=0\\ (s-7)^2&=9\\ s-7&=\pm\sqrt{9}\\ s-7&=\pm 3\\ s&=7\pm 3 \end{aligned}$$
$s=10$ or $s=4$

65.
$$\begin{aligned} (x+5)^2-3&=0\\ (x+5)^2&=3\\ x+5&=\pm\sqrt{3}\\ x&=-5\pm\sqrt{3} \end{aligned}$$

67.
$$\begin{aligned} 2z^2-5z+2&=0\\ (2z-1)(z-2)&=0 \end{aligned}$$
$$2z-1=0 \quad \textbf{or} \quad z-2=0$$
$$z=\tfrac{1}{2} \qquad\qquad z=2$$

69.
$$\begin{aligned} 3m^2-2m-3&=0\\ m^2-\frac{2}{3}m&=1\\ m^2-\frac{2}{3}m+\frac{1}{9}&=1+\frac{1}{9}\\ \left(m-\frac{1}{3}\right)^2&=\frac{10}{9}\\ m-\frac{1}{3}&=\pm\sqrt{\frac{10}{9}}\\ m&=\frac{1}{3}\pm\frac{\sqrt{10}}{3} \end{aligned}$$

71.
$$\begin{aligned} 5x^2+15x&=0\\ 5x(x+3)&=0 \end{aligned}$$
$$5x=0 \quad \textbf{or} \quad x+3=0$$
$$x=0 \qquad\qquad x=-3$$

73.
$$\begin{aligned} x^2+5x+4&=0\\ (x+4)(x+1)&=0 \end{aligned}$$
$$x+4=0 \quad \textbf{or} \quad x+1=0$$
$$x=-4 \qquad\qquad x=-1$$

© 2015 Cengage Learning. All Rights Reserved. May not be scanned, copied or duplicated, or posted to a publicly accessible website, in whole or in part.

SECTION 8.1

75.
$$\begin{aligned} x^2 - 9x - 10 &= 0 \\ (x+1)(x-10) &= 0 \\ x+1 = 0 \quad &\textbf{or} \quad x - 10 = 0 \\ x = -1 \quad &\quad\quad\quad x = 10 \end{aligned}$$

77.
$$\begin{aligned} 2x^2 - x + 8 &= 0 \\ 2x^2 - x &= -8 \\ x^2 - \frac{1}{2}x &= -4 \\ x^2 - \frac{1}{2}x + \frac{1}{16} &= -4 + \frac{1}{16} \\ \left(x - \frac{1}{4}\right)^2 &= \frac{-63}{16} \\ x - \frac{1}{4} &= \pm\frac{\sqrt{-63}}{4} \\ x &= \frac{1}{4} \pm \frac{i\sqrt{9}\sqrt{7}}{4} \\ x &= \frac{1}{4} \pm \frac{3\sqrt{7}}{4}i \end{aligned}$$

79.
$$\begin{aligned} 2d^2 &= 3h \\ d^2 &= \frac{3h}{2} \\ d &= \sqrt{\frac{3h}{2}} \\ d &= \frac{\sqrt{3h}}{\sqrt{2}} \cdot \frac{\sqrt{2}}{\sqrt{2}} \\ d &= \frac{\sqrt{6h}}{2} \end{aligned}$$

81.
$$\begin{aligned} E &= mc^2 \\ \frac{E}{m} &= c^2 \\ \sqrt{\frac{E}{m}} &= c \\ \frac{\sqrt{E}}{\sqrt{m}} \cdot \frac{\sqrt{m}}{\sqrt{m}} &= c \\ \frac{\sqrt{Em}}{m} &= c \end{aligned}$$

83.
$$\begin{aligned} f(x) &= 0 \\ 2x^2 + x - 5 &= 0 \\ x^2 + \frac{1}{2}x &= \frac{5}{2} \\ x^2 + \frac{1}{2}x + \frac{1}{16} &= \frac{5}{2} + \frac{1}{16} \\ \left(x + \frac{1}{4}\right)^2 &= \frac{41}{16} \\ x + \frac{1}{4} &= \pm\sqrt{\frac{41}{16}} \\ x + \frac{1}{4} &- \pm\frac{\sqrt{41}}{4} \\ x &= -\frac{1}{4} \pm \frac{\sqrt{41}}{4} \end{aligned}$$

85.
$$\begin{aligned} f(x) &= 0 \\ x^2 + x - 3 &= 0 \\ x^2 + x &= 3 \\ x^2 + x + \frac{1}{4} &= 3 + \frac{1}{4} \\ \left(x + \frac{1}{2}\right)^2 &= \frac{13}{4} \\ x + \frac{1}{2} &= \pm\sqrt{\frac{13}{4}} \\ x &= -\frac{1}{2} \pm \frac{\sqrt{13}}{2} \end{aligned}$$

© 2015 Cengage Learning. All Rights Reserved. May not be scanned, copied or duplicated, or posted to a publicly accessible website, in whole or in part.

87.
$$\begin{aligned} s &= 16t^2 \\ 256 &= 16t^2 \\ 16 &= t^2 \\ \pm\sqrt{16} &= t \\ \pm 4 &= t \end{aligned}$$
$t = 4$ is the only answer that makes sense, so it will take 4 seconds.

89.
$$\begin{aligned} s^2 &= 10.5l \\ s^2 &= 10.5(495) \\ s^2 &= 5197.5 \\ s &= \pm\sqrt{5197.5} \\ s &\approx \pm 72.09 \end{aligned}$$
s must be positive, so the speed was about 72 mph.

91.
$$\begin{aligned} A &= P(1+r)^t \\ 9193.60 &= 8500(1+r)^2 \\ \frac{9193.60}{8500} &= (1+r)^2 \\ 1.0816 &= (1+r)^2 \\ \pm\sqrt{1.0816} &= \sqrt{(1+r)^2} \\ \pm 1.04 &= 1+r \\ -1 \pm 1.04 &= r \end{aligned}$$
$r = 0.04$ or $r = -2.04$; r must be positive, so $r = 0.04$, or 4%.

93.
$$\begin{aligned} A &= 100 \\ lw &= 100 \\ (1.9x)(x) &= 100 \\ 1.9x^2 &= 100 \\ x^2 &= \frac{100}{1.9} \\ x &= \pm\sqrt{\frac{100}{1.9}} \approx \pm 7.25 \end{aligned}$$
x must be positive, so $x = 7\frac{1}{4}$ ft.
$1.9x \approx 1.9(7.25) \approx 13.775$
$13.775 \approx 13.75$, which is $13\frac{3}{4}$ ft

95. **Answers may vary.**

97. $\left(\frac{1}{2}\sqrt{3}\right)^2 = \left(\frac{\sqrt{3}}{2}\right)^2 = \frac{3}{4}$

Exercises 8.2 (page 551)

1. $5x^2 - 3x = 4 \Rightarrow 5x^2 - 3x - 4 = 0$
$a = 5, b = -3, c = -4$

3.
$$\begin{aligned} Ax + By &= C \\ By &= -Ax + C \\ y &= \frac{-Ax + C}{B} \end{aligned}$$

5. $\sqrt{45} = \sqrt{9}\sqrt{5} = 3\sqrt{5}$

7. $\frac{5}{\sqrt{5}} = \frac{5}{\sqrt{5}} \cdot \frac{\sqrt{5}}{\sqrt{5}} = \frac{5\sqrt{5}}{5} = \sqrt{5}$

9. 7; −4; −9

© 2015 Cengage Learning. All Rights Reserved. May not be scanned, copied or duplicated, or posted to a publicly accessible website, in whole or in part.

SECTION 8.2

11. $x^2 + 6x + 5 = 0$

$a = 1, b = 6, c = 5$

$$x = \frac{-b \pm \sqrt{b^2 - 4ac}}{2a} = \frac{-6 \pm \sqrt{6^2 - 4(1)(5)}}{2(1)} = \frac{-6 \pm \sqrt{36 - 20}}{2} = \frac{-6 \pm \sqrt{16}}{2} = \frac{-6 \pm 4}{2}$$

$x = -\frac{10}{2} = -5$ or $x = -\frac{2}{2} = -1$

13. $x^2 - 5x - 14 = 0$

$a = 1, b = -5, c = -14$

$$x = \frac{-b \pm \sqrt{b^2 - 4ac}}{2a} = \frac{5 \pm \sqrt{(-5)^2 - 4(1)(-14)}}{2(1)} = \frac{5 \pm \sqrt{25 + 56}}{2} = \frac{5 \pm \sqrt{81}}{2} = \frac{5 \pm 9}{2}$$

$x = \frac{14}{2} = 7$ or $x = -\frac{4}{2} = -2$

15. $x^2 + 9x = -20$

$x^2 + 9x + 20 = 0$

$a = 1, b = 9, c = 20$

$$x = \frac{-b \pm \sqrt{b^2 - 4ac}}{2a} = \frac{-9 \pm \sqrt{9^2 - 4(1)(20)}}{2(1)} = \frac{-9 \pm \sqrt{81 - 80}}{2} = \frac{-9 \pm \sqrt{1}}{2} = \frac{-9 \pm 1}{2}$$

$x = -\frac{8}{2} = -4$ or $x = -\frac{10}{2} = -5$

17. $2x^2 - x - 3 = 0$

$a = 2, b = -1, c = -3$

$$x = \frac{-b \pm \sqrt{b^2 - 4ac}}{2a} = \frac{1 \pm \sqrt{(-1)^2 - 4(2)(-3)}}{2(2)} = \frac{1 \pm \sqrt{1 + 24}}{4} = \frac{1 \pm \sqrt{25}}{4} = \frac{1 \pm 5}{4}$$

$x = \frac{6}{4} = \frac{3}{2}$ or $x = -\frac{4}{4} = -1$

19. $15x^2 - 14x = 8$

$15x^2 - 14x - 8 = 0$

$a = 15, b = -14, c = -8$

$$x = \frac{-b \pm \sqrt{b^2 - 4ac}}{2a} = \frac{14 \pm \sqrt{(-14)^2 - 4(15)(-8)}}{2(15)} = \frac{14 \pm \sqrt{196 + 480}}{30} = \frac{14 \pm \sqrt{676}}{30} = \frac{14 \pm 26}{30}$$

$x = \frac{40}{30} = \frac{4}{3}$ or $x = -\frac{12}{30} = -\frac{2}{5}$

21. $8u = -4u^2 - 3$

$4u^2 + 8u + 3 = 0$

$a = 4, b = 8, c = 3$

$$u = \frac{-b \pm \sqrt{b^2 - 4ac}}{2a} = \frac{-8 \pm \sqrt{8^2 - 4(4)(3)}}{2(4)} = \frac{-8 \pm \sqrt{64 - 48}}{8} = \frac{-8 \pm \sqrt{16}}{8} = \frac{-8 \pm 4}{8}$$

$u = \frac{-4}{8} = -\frac{1}{2}$ or $u = \frac{-12}{8} = -\frac{3}{2}$

© 2015 Cengage Learning. All Rights Reserved. May not be scanned, copied or duplicated, or posted to a publicly accessible website, in whole or in part.

SECTION 8.2

23. $5x^2 + 5x + 1 = 0$

$a = 5, b = 5, c = 1$

$$\begin{aligned} x &= \frac{-b \pm \sqrt{b^2 - 4ac}}{2a} \\ &= \frac{-5 \pm \sqrt{5^2 - 4(5)(1)}}{2(5)} \\ &= \frac{-5 \pm \sqrt{25 - 20}}{10} \\ &= \frac{-5 \pm \sqrt{5}}{10} \\ &= \frac{-5}{10} \pm \frac{\sqrt{5}}{10} = -\frac{1}{2} \pm \frac{\sqrt{5}}{10} \end{aligned}$$

25. $5x^2 + 2x - 1 = 0$

$a = 5, b = 2, c = -1$

$$\begin{aligned} x &= \frac{-b \pm \sqrt{b^2 - 4ac}}{2a} \\ &= \frac{-2 \pm \sqrt{2^2 - 4(5)(-1)}}{2(5)} \\ &= \frac{-2 \pm \sqrt{4 + 20}}{10} \\ &= \frac{-2 \pm \sqrt{24}}{10} \\ &= \frac{-2}{10} \pm \frac{\sqrt{4}\sqrt{6}}{10} \\ &= -\frac{1}{5} \pm \frac{2\sqrt{6}}{10} = -\frac{1}{5} \pm \frac{\sqrt{6}}{5} \end{aligned}$$

27. $x^2 + 2x + 2 = 0$

$a = 1, b = 2, c = 2$

$$\begin{aligned} x &= \frac{-b \pm \sqrt{b^2 - 4ac}}{2a} \\ &= \frac{-2 \pm \sqrt{2^2 - 4(1)(2)}}{2(1)} \\ &= \frac{-2 \pm \sqrt{4 - 8}}{2} \\ &= \frac{-2 \pm \sqrt{-4}}{2} \\ &= \frac{-2 \pm \sqrt{-1 \cdot 4}}{2} \\ &= \frac{-2 \pm 2i}{2} = \frac{-2}{2} \pm \frac{2i}{2} = -1 \pm i \end{aligned}$$

29. $x^2 + 5x + 7 = 0$

$a = 1, b = 5, c = 7$

$$\begin{aligned} x &= \frac{-b \pm \sqrt{b^2 - 4ac}}{2a} \\ &= \frac{-5 \pm \sqrt{5^2 - 4(1)(7)}}{2(1)} \\ &= \frac{-5 \pm \sqrt{25 - 28}}{2} \\ &= \frac{-5 \pm \sqrt{-3}}{2} \\ &= \frac{-5 \pm \sqrt{-1 \cdot 3}}{2} \\ &= \frac{-5 \pm i\sqrt{3}}{2} = -\frac{5}{2} \pm \frac{\sqrt{3}}{2}i \end{aligned}$$

© 2015 Cengage Learning. All Rights Reserved. May not be scanned, copied or duplicated, or posted to a publicly accessible website, in whole or in part.

SECTION 8.2

31.

$$
\begin{aligned}
3x^2 - 4x &= -2 \\
3x^2 - 4x + 2 &= 0 \\
a = 3, b &= -4, c = 2 \\
x &= \frac{-b \pm \sqrt{b^2 - 4ac}}{2a} \\
&= \frac{4 \pm \sqrt{(-4)^2 - 4(3)(2)}}{2(3)} \\
&= \frac{4 \pm \sqrt{16 - 24}}{6} \\
&= \frac{4 \pm \sqrt{-8}}{6} \\
&= \frac{4 \pm \sqrt{-1 \cdot 4 \cdot 2}}{6} \\
&= \frac{4 \pm 2i\sqrt{2}}{6} \\
&= \frac{4}{6} \pm \frac{2\sqrt{2}}{6}i = \frac{2}{3} \pm \frac{\sqrt{2}}{3}i
\end{aligned}
$$

33.

$$
\begin{aligned}
3x^2 - 2x &= -3 \\
3x^2 - 2x + 3 &= 0 \\
a = 3, b &= -2, c = 3 \\
x &= \frac{-b \pm \sqrt{b^2 - 4ac}}{2a} \\
&= \frac{2 \pm \sqrt{(-2)^2 - 4(3)(3)}}{2(3)} \\
&= \frac{2 \pm \sqrt{4 - 36}}{6} \\
&= \frac{2 \pm \sqrt{-32}}{6} \\
&= \frac{2 \pm \sqrt{-1 \cdot 16 \cdot 2}}{6} \\
&= \frac{2 \pm 4i\sqrt{2}}{6} \\
&= \frac{2}{6} \pm \frac{4\sqrt{2}}{6}i = \frac{1}{3} \pm \frac{2\sqrt{2}}{3}i
\end{aligned}
$$

35.

$$
\begin{aligned}
C &= \frac{N^2 - N}{2} \\
2C &= N^2 - N \\
N^2 - N - 2C &= 0 \\
a = 1, b &= -1, c = -2C \\
N &= \frac{-b \pm \sqrt{b^2 - 4ac}}{2a} \\
&= \frac{1 \pm \sqrt{(-1)^2 - 4(1)(-2C)}}{2(1)} \\
&= \frac{1 \pm \sqrt{1 + 8C}}{2} \\
&= \frac{1}{2} \pm \frac{\sqrt{1 + 8C}}{2}
\end{aligned}
$$

37.

$$
\begin{aligned}
x^2 - kx &= -ay \\
x^2 - kx + ay &= 0 \\
a = 1, b &= -k, c = ay \\
x &= \frac{-b \pm \sqrt{b^2 - 4ac}}{2a} \\
&= \frac{k \pm \sqrt{(-k)^2 - 4(1)(ay)}}{2(1)} \\
&= \frac{k \pm \sqrt{k^2 - 4ay}}{2} = \frac{k}{2} \pm \frac{\sqrt{k^2 - 4ay}}{2}
\end{aligned}
$$

© 2015 Cengage Learning. All Rights Reserved. May not be scanned, copied or duplicated, or posted to a publicly accessible website, in whole or in part.

SECTION 8.2

39. $6x^2 - 5x - 4 = 0$

$a = 6, b = -5, c = -4$

$$x = \frac{-b \pm \sqrt{b^2 - 4ac}}{2a} = \frac{5 \pm \sqrt{(-5)^2 - 4(6)(-4)}}{2(6)} = \frac{5 \pm \sqrt{25 + 96}}{12} = \frac{5 \pm \sqrt{121}}{12} = \frac{5 \pm 11}{12}$$

$x = \frac{16}{12} = \frac{4}{3}$ or $x = -\frac{6}{12} = -\frac{1}{2}$

41. $\frac{x^2}{2} + \frac{5}{2}x = -1$

$x^2 + 5x = -2$

$x^2 + 5x + 2 = 0$

$a = 1, b = 5, c = 2$

$$x = \frac{-b \pm \sqrt{b^2 - 4ac}}{2a} = \frac{-5 \pm \sqrt{5^2 - 4(1)(2)}}{2(1)} = \frac{-5 \pm \sqrt{25 - 8}}{2} = \frac{-5 \pm \sqrt{17}}{2} = -\frac{5}{2} \pm \frac{\sqrt{17}}{2}$$

43. $3x^2 - 2 = 2x$

$3x^2 - 2x - 2 = 0$

$a = 3, b = -2, c = -2$

$$x = \frac{-b \pm \sqrt{b^2 - 4ac}}{2a} = \frac{2 \pm \sqrt{(-2)^2 - 4(3)(-2)}}{2(3)} = \frac{2 \pm \sqrt{4 + 24}}{6} = \frac{2 \pm \sqrt{28}}{6} = \frac{2 \pm \sqrt{4}\sqrt{7}}{6} = \frac{2}{6} \pm \frac{2\sqrt{7}}{6} = \frac{1}{3} \pm \frac{\sqrt{7}}{3}$$

45. $16y^2 + 8y - 3 = 0$

$a = 16, b = 8, c = -3$

$$y = \frac{-b \pm \sqrt{b^2 - 4ac}}{2a} = \frac{-8 \pm \sqrt{8^2 - 4(16)(-3)}}{2(16)} = \frac{-8 \pm \sqrt{64 + 192}}{32} = \frac{-8 \pm \sqrt{256}}{32} = \frac{-8 \pm 16}{32}$$

$y = \frac{8}{32} = \frac{1}{4}$ or $y = \frac{-24}{32} = -\frac{3}{4}$

© 2015 Cengage Learning. All Rights Reserved. May not be scanned, copied or duplicated, or posted to a publicly accessible website, in whole or in part.

47.
$$\begin{aligned} f(x) &= 0 \\ 4x^2 + 4x - 19 &= 0 \\ a = 4, b = 4, c &= -19 \\ x &= \frac{-b \pm \sqrt{b^2 - 4ac}}{2a} \\ &= \frac{-4 \pm \sqrt{4^2 - 4(4)(-19)}}{2(4)} \\ &= \frac{-4 \pm \sqrt{16 + 304}}{8} \\ &= \frac{-4 \pm \sqrt{320}}{8} \\ &= -\frac{4}{8} \pm \frac{8\sqrt{5}}{8} = -\frac{1}{2} \pm \sqrt{5} \end{aligned}$$

49.
$$\begin{aligned} f(x) &= 0 \\ 3x^2 + 2x + 2 &= 0 \\ a = 3, b = 2, c &= 2 \\ x &= \frac{-b \pm \sqrt{b^2 - 4ac}}{2a} \\ &= \frac{-2 \pm \sqrt{2^2 - 4(3)(2)}}{2(3)} \\ &= \frac{-2 \pm \sqrt{4 - 24}}{6} \\ &= \frac{-2 \pm \sqrt{-20}}{6} \\ &= -\frac{2}{6} \pm \frac{2i\sqrt{5}}{6} = -\frac{1}{3} \pm \frac{\sqrt{5}}{3}i \end{aligned}$$

51.
$$\begin{aligned} 0.7x^2 - 3.5x - 25 &= 0 \\ a = 0.7, b = -3.5, c &= -25 \\ x &= \frac{-b \pm \sqrt{b^2 - 4ac}}{2a} \\ &= \frac{3.5 \pm \sqrt{(-3.5)^2 - 4(0.7)(-25)}}{2(0.7)} \\ &= \frac{3.5 \pm \sqrt{12.25 + 70}}{1.4} \\ &= \frac{3.5 \pm \sqrt{82.25}}{1.4} \\ &= \frac{3.5 \pm 9.069}{1.4} \end{aligned}$$
$x = 8.98$ or $x = -3.98$

53.
$$\begin{aligned} (x - 3)(x - 5) &= 0 \\ x^2 - 8x + 15 &= 0 \end{aligned}$$

55.
$$\begin{aligned} (x - 2)(x - 3)(x + 4) &= 0 \\ (x^2 - 5x + 6)(x + 4) &= 0 \\ x^3 - x^2 - 14x + 24 &= 0 \end{aligned}$$

57.
length $\cdot$ width $=$ Area
$$\begin{aligned} (x + 4)x &= 96 \\ x^2 + 4x - 96 &= 0 \\ (x + 12)(x - 8) &= 0 \end{aligned}$$
$x = -12$ or $x = 8$
Since the width is positive, the dimensions are 8 ft by 12 ft.

59. Let $s =$ the length of a side.
Area $=$ perimeter
$$\begin{aligned} s^2 &= 4s \\ s^2 - 4s &= 0 \\ s(s - 4) &= 0 \end{aligned}$$
$s = 0$ or $s = 4$
Since the length cannot be 0, the length of a side is 4 units.

© 2015 Cengage Learning. All Rights Reserved. May not be scanned, copied or duplicated, or posted to a publicly accessible website, in whole or in part.

61. Let b represent the base. Then $3b+5$ represents the height.

$$\tfrac{1}{2}\text{base}\cdot\text{height} = \text{Area}$$
$$\tfrac{1}{2}b(3b+5) = 6$$
$$b(3b+5) = 12$$
$$3b^2+5b-12 = 0$$
$$(3b-4)(b+3) = 0$$
$b=\frac{4}{3}$ or $b=-3 \Rightarrow$ Since the base is positive, it must be $\frac{4}{3}$ cm.

63. Let x and $x+2$ represent the integers.

$$x(x+2) = 168$$
$$x^2+2x-168 = 0$$
$$(x+14)(x-12) = 0$$
$x=-14$ or $x=12$

The integers are 12 & 14, or -14 & -12.

65. Let x and $x+1$ represent the integers.

$$x^2+(x+1)^2 = 85$$
$$x^2+x^2+2x+1 = 85$$
$$2x^2+2x-84 = 0$$
$$2(x+7)(x-6) = 0$$
$x=-7$ or $x=6$

The integers are 6.& 7, or -7 & -6.

67. Let $r =$ the slower rate. Then $r+20 =$ the faster rate.

	Rate	Time	Dist.
Slower	r	$\frac{150}{r}$	150
Faster	$r+20$	$\frac{150}{r+20}$	150

$$\boxed{\text{Faster time}}+2=\boxed{\text{Slower time}}$$
$$\frac{150}{r+20}+2=\frac{150}{r}$$
$$\left(\frac{150}{r+20}+2\right)(r)(r+20)=\frac{150}{r}\cdot r(r+20)$$
$$150r+2r(r+20)=150(r+20)$$
$$2r^2+40r-3000=0$$
$$2(r+50)(r-30)=0$$
$r=-50$ or $r=30$ $\quad$ $r=30$ is the only answer that makes sense.

Her original speed was 30 mph.

69. Let $x =$ the number of 10¢ increases. Then the ticket price will be $4+0.10x$, while the projected attendance will be $300-5x$, for total receipts of $(4+0.10x)(300-5x)$.

$$\text{Total} = 1248$$
$$(4+0.10x)(300-5x) = 1248$$
$$1200+10x-0.5x^2 = 1248$$
$$-0.5x^2+10x-48 = 0$$
$$x^2-20x+96 = 0$$
$$(x-12)(x-8) = 0$$
$x=12$ or $x=8 \Rightarrow 4+0.10(12)=5.20;\ 4+0.10(8)=4.80$

The ticket price would be either \$5.20 or \$4.80.

© 2015 Cengage Learning. All Rights Reserved. May not be scanned, copied or duplicated, or posted to a publicly accessible website, in whole or in part.

71. Let $x =$ the number of additional subscribers. Then the profit per subscriber will be $20 + 0.01x$, for a total profit of $(20 + 0.01x)(3000 + x)$.

$$\begin{aligned} \text{Total profit} &= 120000 \\ (20 + 0.01x)(3000 + x) &= 120000 \\ 60{,}000 + 50x + 0.01x^2 &= 120000 \\ 0.01x^2 + 50x - 60{,}000 &= 0 \\ x^2 + 5000x - 6{,}000{,}000 &= 0 \\ (x + 6000)(x - 1000) &= 0 \end{aligned}$$

$x = -6000$ (impossible) or $x = 1000 \Rightarrow$ The total number of subscribers would be 4000.

73. Let $w =$ the constant width.

$$\begin{aligned} \text{Frame area} &= \text{Picture area} \\ (12 + 2w)(10 + 2w) - 12(10) &= 12(10) \\ 120 + 44w + 4w^2 - 240 &= 0 \\ 4w^2 + 44w - 120 &= 0 \\ 4(w^2 + 11w - 30) &= 0 \\ w^2 + 11w - 30 &= 0 \Rightarrow a = 1, b = 11, c = -30 \end{aligned}$$

$$w = \frac{-11 \pm \sqrt{11^2 - 4(1)(-30)}}{2(1)}$$

$$= \frac{-11 \pm \sqrt{121 + 120}}{2} = \frac{-11 \pm \sqrt{241}}{2} = \frac{-11 \pm 15.52}{2} = \frac{4.52}{2} \text{ or } \frac{-26.52}{2} \text{ (impossible)}$$

$w = 2.26$ in.

75.

$$\begin{aligned} P &= -0.0072x^2 + 0.4904x + 58.2714 \\ 65 &= -0.0072x^2 + 0.4904x + 58.2714 \\ 0 &= -0.0072x^2 + 0.4904x - 6.7286 \end{aligned}$$

$a = -0.0072$, $b = 0.4904$, $c = -6.7286$

$$x = \frac{-b \pm \sqrt{b^2 - 4ac}}{2a}$$

$$= \frac{-0.4904 \pm \sqrt{(0.4904)^2 - 4(-0.0072)(-6.7286)}}{2(-0.0072)} \approx \frac{-0.4904 \pm \sqrt{0.0467}}{-0.0144} \approx 19.05 \text{ or } 49.06$$

Since $x \leq 42$, the only answer that works is 19.05. The model indicates the desired result happened in 1985.

© 2015 Cengage Learning. All Rights Reserved. May not be scanned, copied or duplicated, or posted to a publicly accessible website, in whole or in part.

77. Let [H⁺] (and then [A⁻]) $= x$ and [HA] $= 0.1 - x$.

$$\frac{[H^+][A^-]}{[HA]} = 4 \times 10^{-4}$$
$$\frac{x^2}{0.1 - x} = 4 \times 10^{-4}$$
$$x^2 = 4 \times 10^{-5} - (4 \times 10^{-4})x$$
$$x^2 + (4 \times 10^{-4})x - 4 \times 10^{-5} = 0$$
$$x = \frac{-b \pm \sqrt{b^2 - 4ac}}{2a} = \frac{-4 \times 10^{-4} \pm \sqrt{(4 \times 10^{-4})^2 - 4(1)(-4 \times 10^{-5})}}{2(1)}$$
$$\approx \frac{-4 \times 10^{-4} \pm 0.012655}{2} \approx \frac{0.012255}{2} \text{ or } -\frac{0.013055}{2} \text{ (impossible)}$$

The concentration is about 0.00613 M $= 6.13 \times 10^{-3}$ M.

79. Answers may vary.

81. $x^2 + 2\sqrt{2}x - 6 = 0$

$a = 1, b = 2\sqrt{2}, c = -6$

$$x = \frac{-b \pm \sqrt{b^2 - 4ac}}{2a}$$
$$= \frac{-2\sqrt{2} \pm \sqrt{\left(2\sqrt{2}\right)^2 - 4(1)(-6)}}{2(1)}$$
$$= \frac{-2\sqrt{2} \pm \sqrt{8 + 24}}{2}$$
$$= \frac{-2\sqrt{2} \pm \sqrt{32}}{2} = \frac{-2\sqrt{2} \pm 4\sqrt{2}}{2}$$
$$x = \frac{2\sqrt{2}}{2} = \sqrt{2} \text{ or } x = \frac{-6\sqrt{2}}{2} = -3\sqrt{2}$$

83. $x^2 - 3ix - 2 = 0$

$a = 1, b = -3i, c = -2$

$$x = \frac{-b \pm \sqrt{b^2 - 4ac}}{2a}$$
$$= \frac{3i \pm \sqrt{(-3i)^2 - 4(1)(-2)}}{2(1)}$$
$$= \frac{3i \pm \sqrt{9i^2 + 8}}{2}$$
$$= \frac{3i \pm \sqrt{-1}}{2} = \frac{3i \pm i}{2}$$
$$x = \frac{4i}{2} = 2i \text{ or } x = \frac{2i}{2} = i$$

Exercises 8.3 (page 560)

1. $b^2 - 4ac = (1)^2 - 4(1)(1)$

$= 1 - 4 = -3$

3. If $1 - i$ is a solution, then so is $1 + i$.

Sum: $(1 - i) + (1 + i) = 2 = -b$

Product: $(1 + i)(1 - i) = 1 - i^2 = 2 = c$

Yes, $1 - i$ is a solution.

5. If $-1 + i$ is a solution, then so is $-1 - i$.

Sum: $(-1 + i) + (-1 - i) = -2 \neq -b$

No, $-1 + i$ is not a solution.

7.
$$\frac{1}{4} + \frac{1}{t} = \frac{1}{2t}$$
$$\left(\frac{1}{4} + \frac{1}{t}\right)4t = \frac{1}{2t} \cdot 4t$$
$$t + 4 = 2$$
$$t = -2$$

© 2015 Cengage Learning. All Rights Reserved. May not be scanned, copied or duplicated, or posted to a publicly accessible website, in whole or in part.

SECTION 8.3

9. $m = \dfrac{\Delta y}{\Delta x} = \dfrac{1-(-2)}{4-(-3)} = \dfrac{3}{7}$

11. $b^2 - 4ac$

13. rational; unequal

15. $9x^2 + 6x + 1 = 0; a = 9, b = 6, c = 1$
$$\begin{aligned} b^2 - 4ac &= (6)^2 - 4(9)(1) \\ &= 36 - 36 = 0 \end{aligned}$$
The solutions are rational and equal.

17. $10x^2 + 2x + 3 = 0; a = 10, b = 2, c = 3$
$$\begin{aligned} b^2 - 4ac &= 2^2 - 4(10)(3) \\ &= 4 - 120 = -116 \end{aligned}$$
The solutions are complex conjugates.

19.
$$\begin{aligned} 6x^2 &= 8x - 1 \\ 6x^2 - 8x + 1 &= 0; a = 6, b = -8, c = 1 \\ b^2 - 4ac &= (-8)^2 - 4(6)(1) \\ &= 64 - 24 = 40 \end{aligned}$$
The solutions are irrational and unequal.

21.
$$\begin{aligned} x(2x-3) &= 20 \\ 2x^2 - 3x - 20 &= 0; a = 2, b = -3, c = -20 \\ b^2 - 4ac &= (-3)^2 - 4(2)(-20) \\ &= 9 + 160 = 169 \end{aligned}$$
The solutions are rational and unequal.

23. $x^2 + kx + 16 = 0; a = 1, b = k, c = 16$

Set the discriminant equal to 0:
$$\begin{aligned} b^2 - 4ac &= 0 \\ k^2 - 4(1)(16) &= 0 \\ k^2 - 64 &= 0 \\ k^2 &= 64 \\ k &= \pm 8 \end{aligned}$$

25.
$$\begin{aligned} 4x^2 + 9 &= kx \\ 4x^2 - kx + 9 &= 0; a = 4, b = -k, c = 9 \end{aligned}$$
Set the discriminant equal to 0:
$$\begin{aligned} b^2 - 4ac &= 0 \\ (-k)^2 - 4(4)(9) &= 0 \\ k^2 - 144 &= 0 \\ k^2 &= 144 \\ k &= \pm 12 \end{aligned}$$

27.
$$\begin{aligned} x^4 - 17x^2 + 16 &= 0 \\ (x^2 - 16)(x^2 - 1) &= 0 \end{aligned}$$
$$\begin{aligned} x^2 - 16 &= 0 \quad \textbf{or} \quad & x^2 - 1 &= 0 \\ x^2 &= 16 & x^2 &= 1 \\ x &= \pm 4 & x &= \pm 1 \end{aligned}$$

29.
$$\begin{aligned} x^4 - 3x^2 &= -2 \\ x^4 - 3x^2 + 2 &= 0 \\ (x^2 - 2)(x^2 - 1) &= 0 \end{aligned}$$
$$\begin{aligned} x^2 - 2 &= 0 \quad \textbf{or} \quad & x^2 - 1 &= 0 \\ x^2 &= 2 & x^2 &= 1 \\ x &= \pm\sqrt{2} & x &= \pm 1 \end{aligned}$$

31.
$$\begin{aligned} x^4 &= 6x^2 - 5 \\ x^4 - 6x^2 + 5 &= 0 \\ (x^2 - 5)(x^2 - 1) &= 0 \end{aligned}$$
$$\begin{aligned} x^2 - 5 &= 0 \quad \textbf{or} \quad & x^2 - 1 &= 0 \\ x^2 &= 5 & x^2 &= 1 \\ x &= \pm\sqrt{5} & x &= \pm 1 \end{aligned}$$

33.
$$\begin{aligned} 2x^4 - 10x^2 &= -8 \\ 2x^4 - 10x^2 + 8 &= 0 \\ 2(x^2 - 4)(x^2 - 1) &= 0 \end{aligned}$$
$$\begin{aligned} x^2 - 4 &= 0 \quad \textbf{or} \quad & x^2 - 1 &= 0 \\ x^2 &= 4 & x^2 &= 1 \\ x &= \pm 2 & x &= \pm 1 \end{aligned}$$

© 2015 Cengage Learning. All Rights Reserved. May not be scanned, copied or duplicated, or posted to a publicly accessible website, in whole or in part.

SECTION 8.3

35.
$$x - 7\sqrt{x} + 10 = 0$$
$$(\sqrt{x} - 2)(\sqrt{x} - 5) = 0$$
$$\sqrt{x} - 2 = 0 \quad \textbf{or} \quad \sqrt{x} - 5 = 0$$
$$\sqrt{x} = 2 \qquad \sqrt{x} = 5$$
$$\left(\sqrt{x}\right)^2 = 2^2 \qquad \left(\sqrt{x}\right)^2 = 5^2$$
$$x = 4 \qquad x = 25$$
Solution   Solution

37.
$$2x - \sqrt{x} = 3$$
$$2x - \sqrt{x} - 3 = 0$$
$$(2\sqrt{x} - 3)(\sqrt{x} + 1) = 0$$
$$2\sqrt{x} - 3 = 0 \quad \textbf{or} \quad \sqrt{x} + 1 = 0$$
$$2\sqrt{x} = 3 \qquad \sqrt{x} = -1$$
$$\left(2\sqrt{x}\right)^2 = 3^2 \qquad \left(\sqrt{x}\right)^2 = (-1)^2$$
$$4x = 9 \qquad x = 1$$
$$x = \tfrac{9}{4} \qquad \text{Extraneous}$$
Solution

39.
$$2x + x^{1/2} - 3 = 0$$
$$\left(2x^{1/2} + 3\right)\left(x^{1/2} - 1\right) = 0$$
$$2x^{1/2} + 3 = 0 \quad \textbf{or} \quad x^{1/2} - 1 = 0$$
$$2x^{1/2} = -3 \qquad x^{1/2} = 1$$
$$x^{1/2} = -\tfrac{3}{2} \qquad \left(x^{1/2}\right)^2 = 1^2$$
$$\left(x^{1/2}\right)^2 = \left(-\tfrac{3}{2}\right)^2 \qquad x = 1$$
$$x = \tfrac{9}{4} \qquad \text{Solution}$$
Extraneous

41.
$$3x + 5x^{1/2} + 2 = 0$$
$$\left(3x^{1/2} + 2\right)\left(x^{1/2} + 1\right) = 0$$
$$3x^{1/2} + 2 = 0 \quad \textbf{or} \quad x^{1/2} + 1 = 0$$
$$3x^{1/2} = -2 \qquad x^{1/2} = -1$$
$$x^{1/2} = -\tfrac{2}{3} \qquad \left(x^{1/2}\right)^2 = (-1)^2$$
$$\left(x^{1/2}\right)^2 = \left(-\tfrac{2}{3}\right)^2 \qquad x = 1$$
$$x = \tfrac{4}{9} \qquad \text{Extraneous}$$
Extraneous

solution set: $\emptyset$

43.
$$x^{2/3} + 5x^{1/3} + 6 = 0$$
$$\left(x^{1/3} + 2\right)\left(x^{1/3} + 3\right) = 0$$
$$x^{1/3} + 2 = 0 \quad \textbf{or} \quad x^{1/3} + 3 = 0$$
$$x^{1/3} = -2 \qquad x^{1/3} = -3$$
$$\left(x^{1/3}\right)^3 = (-2)^3 \qquad \left(x^{1/3}\right)^3 = (-3)^3$$
$$x = -8 \qquad x = -27$$
Solution   Solution

45.
$$3m^{2/3} - m^{1/3} - 2 = 0$$
$$\left(3m^{1/3} + 2\right)\left(m^{1/3} - 1\right) = 0$$
$$3m^{1/3} + 2 = 0 \quad \textbf{or} \quad m^{1/3} - 1 = 0$$
$$3m^{1/3} = -2 \qquad m^{1/3} = 1$$
$$m^{1/3} = -\tfrac{2}{3} \qquad \left(m^{1/3}\right)^3 = 1^3$$
$$\left(m^{1/3}\right)^3 = \left(-\tfrac{2}{3}\right)^3 \qquad m = 1$$
$$m = -\tfrac{8}{27} \qquad \text{Solution}$$
Solution

47.
$$x + 10 + \frac{9}{x} = 0$$
$$x\left(x + 10 + \frac{9}{x}\right) = x(0)$$
$$x^2 + 10x + 9 = 0$$
$$(x + 9)(x + 1) = 0$$
$$x + 9 = 0 \quad \textbf{or} \quad x + 1 = 0$$
$$x = -9 \qquad x = -1$$

49.
$$x + 3 = \frac{28}{x}$$
$$x + 3 - \frac{28}{x} = 0$$
$$x\left(x + 3 - \frac{28}{x}\right) = x(0)$$
$$x^2 + 3x - 28 = 0$$
$$(x + 7)(x - 4) = 0$$
$$x + 7 = 0 \quad \textbf{or} \quad x - 4 = 0$$
$$x = -7 \qquad x = 4$$

© 2015 Cengage Learning. All Rights Reserved. May not be scanned, copied or duplicated, or posted to a publicly accessible website, in whole or in part.

SECTION 8.3

51.
$$\frac{1}{x-1}+\frac{3}{x+1}=2$$
$$\left(\frac{1}{x-1}+\frac{3}{x+1}\right)(x-1)(x+1)=2(x+1)(x-1)$$
$$1(x+1)+3(x-1)=2(x^2-1)$$
$$x+1+3x-3=2x^2-2$$
$$0=2x^2-4x$$
$$0=2x(x-2) \qquad 2x=0 \ \textbf{or} \ x-2=0$$
$$x=0 \qquad x=2$$

53.
$$\frac{1}{x+2}+\frac{24}{x+3}=13$$
$$\left(\frac{1}{x+2}+\frac{24}{x+3}\right)(x+2)(x+3)=13(x+2)(x+3)$$
$$1(x+3)+24(x+2)=13(x^2+5x+6)$$
$$x+3+24x+48=13x^2+65x+78$$
$$0=13x^2+40x+27$$
$$0=(13x+27)(x+1) \qquad 13x+27=0 \ \textbf{or} \ x+1=0$$
$$13x=-27 \qquad x=-1$$
$$x=-\tfrac{27}{13}$$

55.
$$x^{-4}-2x^{-2}+1=0$$
$$(x^{-2}-1)(x^{-2}-1)=0$$
$$x^{-2}-1=0 \ \textbf{or} \ x^{-2}-1=0$$
$$x^{-2}=1 \qquad x^{-2}=1$$
$$\frac{1}{x^2}=1 \qquad \frac{1}{x^2}=1$$
$$1=x^2 \qquad 1=x^2$$
$$\pm 1=x \qquad \pm 1=x$$
(double roots)

57.
$$8a^{-2}-10a^{-1}-3=0$$
$$(2a^{-1}-3)(4a^{-1}+1)=0$$
$$2a^{-1}-3=0 \ \textbf{or} \ 4a^{-1}+1=0$$
$$2a^{-1}=3 \qquad 4a^{-1}=-1$$
$$\frac{2}{a}=3 \qquad \frac{4}{a}=-1$$
$$2=3a \qquad 4=-a$$
$$\tfrac{2}{3}=a \qquad -4=a$$

59.
$$x^2+y^2=r^2$$
$$x^2=r^2-y^2$$
$$x=\pm\sqrt{r^2-y^2}$$

61. $xy^2+5xy+3=0;\ a=x, b=5x, c=3$
$$y=\frac{-b\pm\sqrt{b^2-4ac}}{2a}$$
$$=\frac{-5x\pm\sqrt{(5x)^2-4(x)(3)}}{2x}$$
$$=\frac{-5x\pm\sqrt{25x^2-12x}}{2x}$$

© 2015 Cengage Learning. All Rights Reserved. May not be scanned, copied or duplicated, or posted to a publicly accessible website, in whole or in part.

SECTION 8.3

63. $12x^2 - 5x - 2 = 0; a = 12, b = -5, c = -2$

$(4x+1)(3x-2) = 0$

$4x + 1 = 0$ **or** $3x - 2 = 0$

$4x = -1 \qquad 3x = 2$

$x = -\frac{1}{4} \qquad x = \frac{2}{3}$

$$-\frac{b}{a} = -\frac{-5}{12} = \frac{5}{12} \qquad \frac{c}{a} = \frac{-2}{12} = -\frac{1}{6}$$

$$-\frac{1}{4} + \frac{2}{3} = -\frac{3}{12} + \frac{8}{12} = \frac{5}{12} \qquad \left(-\frac{1}{4}\right)\left(\frac{2}{3}\right) = -\frac{1}{6}$$

65. $2x^2 + 5x + 1 = 0; a = 2, b = 5, c = 1; -\frac{b}{a} = -\frac{5}{2}; \frac{c}{a} = \frac{1}{2}$

$$x = \frac{-b \pm \sqrt{b^2 - 4ac}}{2a} = \frac{-5 \pm \sqrt{5^2 - 4(2)(1)}}{2(2)} = \frac{-5 \pm \sqrt{17}}{4} = -\frac{5}{4} \pm \frac{\sqrt{17}}{4}$$

$$\frac{-5+\sqrt{17}}{4} + \frac{-5-\sqrt{17}}{4} = \frac{-10}{4} = -\frac{5}{2}$$

$$\left(\frac{-5+\sqrt{17}}{4}\right)\left(\frac{-5-\sqrt{17}}{4}\right) = \frac{25 + 5\sqrt{17} - 5\sqrt{17} - 17}{16} = \frac{8}{16} = \frac{1}{2}$$

67. $3x^2 - 2x + 4 = 0; a = 3, b = -2, c = 4; -\frac{b}{a} = -\frac{-2}{3} = \frac{2}{3}; \frac{c}{a} = \frac{4}{3}$

$$x = \frac{-b \pm \sqrt{b^2 - 4ac}}{2a} = \frac{-(-2) \pm \sqrt{(-2)^2 - 4(3)(4)}}{2(3)} = \frac{2 \pm \sqrt{-44}}{6} = \frac{1}{3} \pm \frac{i\sqrt{11}}{3}$$

$$\frac{1+i\sqrt{11}}{3} + \frac{1-i\sqrt{11}}{3} = \frac{2}{3}$$

$$\left(\frac{1+i\sqrt{11}}{3}\right)\left(\frac{1-i\sqrt{11}}{3}\right) = \frac{1 - i\sqrt{11} + i\sqrt{11} - 11i^2}{9} = \frac{1+11}{9} = \frac{12}{9} = \frac{4}{3}$$

69. $x^2 + 2x + 5 = 0; a = 1, b = 2, c = 5; -\frac{b}{a} = -\frac{2}{1} = -2; \frac{c}{a} = \frac{5}{1} = 5$

$$x = \frac{-b \pm \sqrt{b^2 - 4ac}}{2a} = \frac{-2 \pm \sqrt{2^2 - 4(1)(5)}}{2(1)} = \frac{-2 \pm \sqrt{-16}}{2} = \frac{-2 \pm 4i}{2} = -1 \pm 2i$$

$(-1+2i) + (-1-2i) = -2;\ (-1+2i)(-1-2i) = 1 + 2i - 2i - 4i^2 = 1 + 4 = 5$

71. $1492x^2 + 1776x - 1984 = 0$

$a = 1492, b = 1776, c = -1984$

$b^2 - 4ac = (1776)^2 - 4(1492)(-1984)$

$= 3{,}154{,}176 + 11{,}840{,}512$

$= 14{,}994{,}688$

The solutions are real numbers.

73. $4x - 5\sqrt{x} - 9 = 0$

$(4\sqrt{x} - 9)(\sqrt{x} + 1) = 0$

$4\sqrt{x} - 9 = 0$ **or** $\sqrt{x} + 1 = 0$

$4\sqrt{x} = 9 \qquad \sqrt{x} = -1$

$(4\sqrt{x})^2 = 9^2 \qquad (\sqrt{x})^2 = (-1)^2$

$16x = 81 \qquad x = 1$

$x = \frac{81}{16}$ Solution; $x = 1$ Extraneous

© 2015 Cengage Learning. All Rights Reserved. May not be scanned, copied or duplicated, or posted to a publicly accessible website, in whole or in part.

SECTION 8.3

75.

$$3x^{2/3} - x^{1/3} - 2 = 0$$
$$\left(x^{1/3} - 1\right)\left(3x^{1/3} + 2\right) = 0$$
$$x^{1/3} - 1 = 0 \quad \textbf{or} \quad 3x^{1/3} + 2 = 0$$
$$x^{1/3} = 1 \qquad 3x^{1/3} = -2$$
$$\left(x^{1/3}\right)^3 = (1)^3 \qquad \left(3x^{1/3}\right)^3 = (-2)^3$$
$$x = 1 \qquad 27x = -8$$
$$\text{Solution} \qquad x = -\tfrac{8}{27}$$
$$\text{Solution}$$

77.

$$2x^4 + 24 = 26x^2$$
$$2x^4 - 26x^2 + 24 = 0$$
$$2(x^2 - 12)(x^2 - 1) = 0$$
$$x^2 - 12 = 0 \quad \textbf{or} \quad x^2 - 1 = 0$$
$$x^2 = 12 \qquad x^2 = 1$$
$$x = \pm\sqrt{12} \qquad x = \pm 1$$
$$= \pm 2\sqrt{3}$$

79.

$$x^4 - 24 = -2x^2$$
$$x^4 + 2x^2 - 24 = 0$$
$$(x^2 - 4)(x^2 + 6) = 0$$
$$x^2 - 4 = 0 \quad \textbf{or} \quad x^2 + 6 = 0$$
$$x^2 = 4 \qquad x^2 = -6$$
$$x = \pm\sqrt{4} \qquad x = \pm\sqrt{-6}$$
$$x = \pm 2 \qquad x = \pm i\sqrt{6}$$

81. Let $y = 2x - 1$.

$$4(2x - 1)^2 - 3(2x - 1) - 1 = 0$$
$$4y^2 - 3y - 1 = 0$$
$$(4y + 1)(y - 1) = 0$$
$$4y + 1 = 0 \quad \textbf{or} \quad y - 1 = 0$$
$$4y = -1 \qquad y = 1$$
$$y = -\tfrac{1}{4} \qquad y = 1$$
$$2x - 1 = -\tfrac{1}{4} \qquad 2x - 1 = 1$$
$$2x = \tfrac{3}{4} \qquad 2x = 2$$
$$x = \tfrac{3}{8} \qquad x = 1$$

83.

$$x^{-2/3} - 2x^{-1/3} - 3 = 0$$
$$(x^{-1/3} - 3)(x^{-1/3} + 1) = 0$$
$$x^{-1/3} - 3 = 0 \quad \textbf{or} \quad x^{-1/3} + 1 = 0$$
$$x^{-1/3} = 3 \qquad x^{-1/3} = -1$$
$$\left(x^{-1/3}\right)^{-3} = 3^{-3} \qquad \left(x^{-1/3}\right)^{-3} = (-1)^{-3}$$
$$x = \tfrac{1}{3^3} \qquad x = \tfrac{1}{(-1)^3}$$
$$x = \tfrac{1}{27} \qquad x = -1$$

85.

$$x + \frac{2}{x - 2} = 0$$
$$\left(x + \frac{2}{x - 2}\right)(x - 2) = 0(x - 2)$$
$$x(x - 2) + 2 = 0$$
$$x^2 - 2x + 2 = 0$$
$$x^2 - 2x = -2$$
$$x^2 - 2x + 1 = -2 + 1$$
$$(x - 1)^2 = -1$$
$$x - 1 = \pm\sqrt{-1}$$
$$x = 1 \pm i$$

© 2015 Cengage Learning. All Rights Reserved. May not be scanned, copied or duplicated, or posted to a publicly accessible website, in whole or in part.

SECTION 8.3

87. Let $y = m + 1$.

$$\begin{aligned} 8(m+1)^{-2} - 30(m+1)^{-1} + 7 &= 0 \\ 8y^{-2} - 30y^{-1} + 7 &= 0 \\ (4y^{-1} - 1)(2y^{-1} - 7) &= 0 \end{aligned}$$

$$\begin{array}{lcl} 4y^{-1} - 1 = 0 & \textbf{or} & 2y^{-1} - 7 = 0 \\ 4y^{-1} = 1 & & 2y^{-1} = 7 \\ \frac{4}{y} = 1 & & \frac{2}{y} = 7 \\ 4 = y & & 2 = 7y \\ & & \frac{2}{7} = y \\ 4 = m + 1 & & \frac{2}{7} = m + 1 \\ 3 = m & & -\frac{5}{7} = m \end{array}$$

89.

$$\begin{aligned} I &= \frac{k}{d^2} \\ Id^2 &= k \\ d^2 &= \frac{k}{I} \\ d &= \pm\sqrt{\frac{k}{I}} = \pm\frac{\sqrt{kI}}{I} \end{aligned}$$

91.

$$\begin{aligned} \sigma &= \sqrt{\frac{\Sigma x^2}{N} - \mu^2} \\ \sigma^2 &= \frac{\Sigma x^2}{N} - \mu^2 \\ \mu^2 &= \frac{\Sigma x^2}{N} - \sigma^2 \end{aligned}$$

93. $(k-1)x^2 + (k-1)x + 1 = 0$

$a = k - 1, b = k - 1, c = 1$

Set the discriminant equal to 0:

$$\begin{aligned} b^2 - 4ac &= 0 \\ (k-1)^2 - 4(k-1)(1) &= 0 \\ k^2 - 2k + 1 - 4k + 4 &= 0 \\ k^2 - 6k + 5 &= 0 \\ (k-5)(k-1) &= 0 \end{aligned}$$

$$\begin{array}{lcl} k - 5 = 0 & \textbf{or} & k - 1 = 0 \\ k = 5 & & k = 1 \\ & & \text{doesn't work} \end{array}$$

95. $(k+4)x^2 + 2kx + 9 = 0$

$a = k + 4, b = 2k, c = 9$

Set the discriminant equal to 0:

$$\begin{aligned} b^2 - 4ac &= 0 \\ (2k)^2 - 4(k+4)(9) &= 0 \\ 4k^2 - 36(k+4) &= 0 \\ 4k^2 - 36k - 144 &= 0 \\ 4(k^2 - 9k - 36) &= 0 \\ 4(k-12)(k+3) &= 0 \end{aligned}$$

$$\begin{array}{lcl} k - 12 = 0 & \textbf{or} & k + 3 = 0 \\ k = 12 & & k = -3 \end{array}$$

97.

$$\begin{aligned} 3x^2 + 4x &= k \\ 3x^2 + 4x - k &= 0 \end{aligned}$$

$a = 3, b = 4, c = -k$

Set the discriminant less than 0:

$$\begin{aligned} b^2 - 4ac &< 0 \\ 4^2 - 4(3)(-k) &< 0 \\ 16 + 12k &< 0 \\ 12k &< -16 \\ k &< -\frac{16}{12}, \text{ or } k < -\frac{4}{3} \end{aligned}$$

99. **Answers may vary.**

101. No

© 2015 Cengage Learning. All Rights Reserved. May not be scanned, copied or duplicated, or posted to a publicly accessible website, in whole or in part.

Exercises 8.4 (page 574)

1. $x = 3$

3. $x = -1$

5.
$$f(x) = 0$$
$$x^2 - 9 = 0$$
$$x^2 = 9$$
$$x = \pm 3$$

7.
$$3x + 5 = 5x - 15$$
$$-2x = -20$$
$$x = 10$$

9. Let t = the time of the second train.
Then $t + 3$ = the time of the first train.

	Rate	Time	Dist.
First	30	$t+3$	$30(t+3)$
Second	55	t	$55t$

1st distance = 2nd distance
$$30(t+3) = 55t$$
$$30t + 90 = 55t$$
$$-25t = -90$$
$$t = \frac{-90}{-25} = \frac{18}{5} = 3\tfrac{3}{5} \text{ hours}$$

11. $f(x) = ax^2 + bx + c$; $a \neq 0$

13. maximum; minimum; vertex

15. upward

17. to the right

19. upward

21. $f(x) = x^2$
vertex: $(0, 0)$; opens U

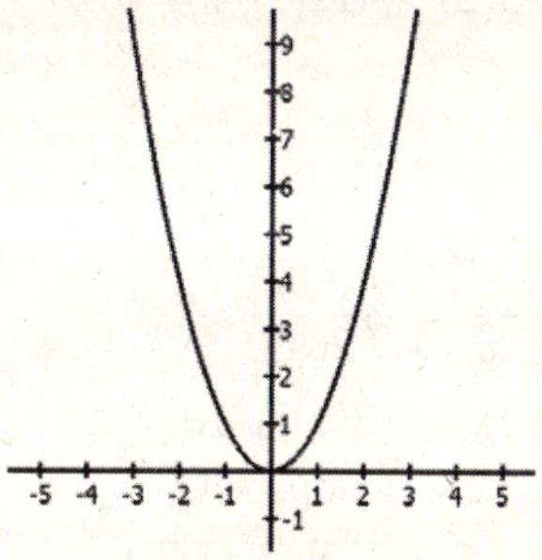

23. $f(x) = -2x^2$
vertex: $(0, 0)$; opens D

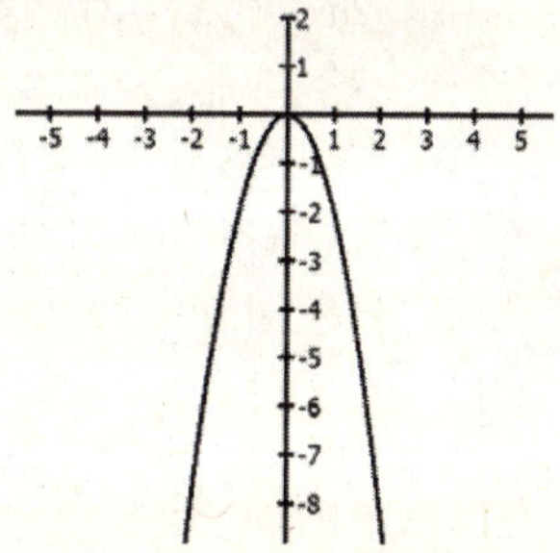

25. $f(x) = x^2 + 2$
Shift $f(x) = x^2$ U2.

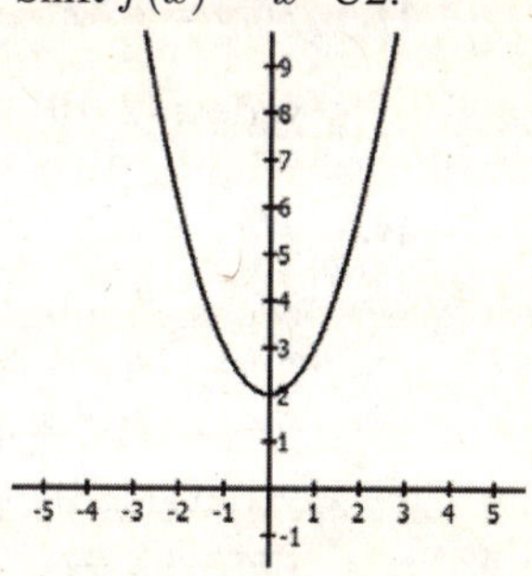

27. $f(x) = x^2 - 1$
Shift $f(x) = x^2$ D1.

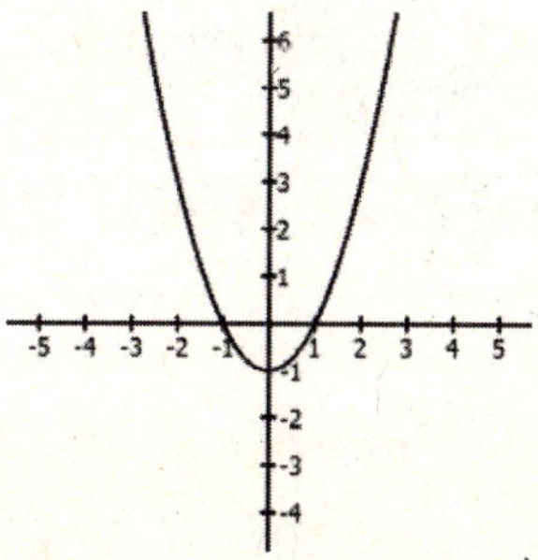

29. $f(x) = -(x-2)^2$; opens D
Shift $f(x) = x^2$ R2.

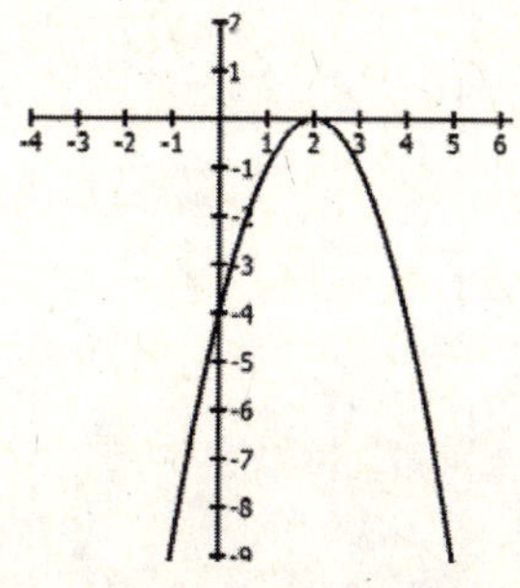

31. $f(x) = -(x-3)^2$; opens D
Shift $f(x) = x^2$ R3.

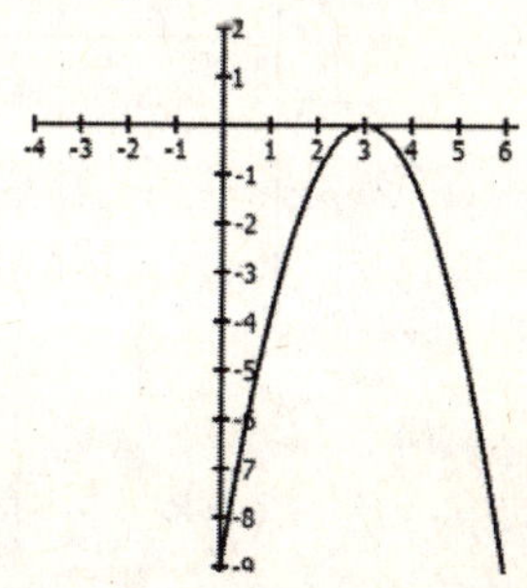

© 2015 Cengage Learning. All Rights Reserved. May not be scanned, copied or duplicated, or posted to a publicly accessible website, in whole or in part.

SECTION 8.4

33. $f(x) = (x-3)^2 + 2$

Shift $f(x) = x^2$ U2 and R3.

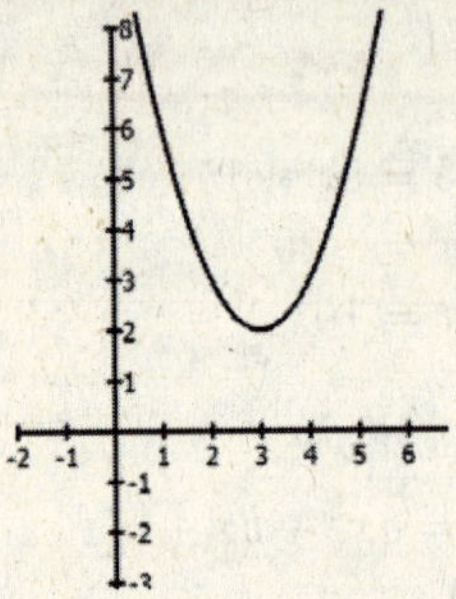

35. $y = (x-4)^2 + 5$

Shift $y = x^2$ U5 and R4.

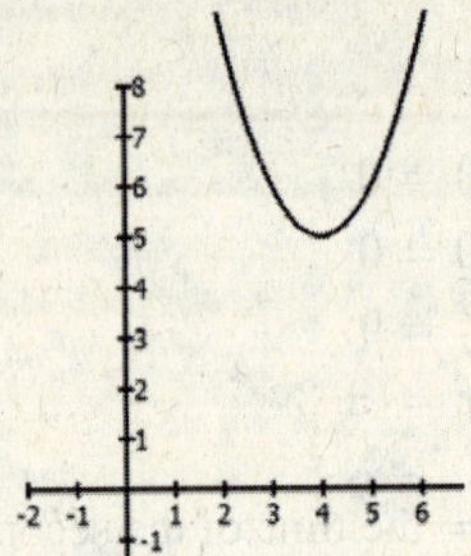

37. $y = (x-1)^2 + 2; V(1, 2)$; axis: $x = 1$

39. $y = 2(x+3)^2 - 4; V(-3, -4)$; axis: $x = -3$

41. $y = -3x^2 \Rightarrow y = -3(x-0)^2 + 0; V(0, 0)$; axis: $x = 0$

43. $y = 2x^2 - 4x$

$y = 2(x^2 - 2x)$

$y = 2(x^2 - 2x + 1) - 2$

$y = 2(x-1)^2 - 2$

$V(1, -2)$; axis: $x = 1$

45. $f(x) = 3x^2 + 6x + 1$

$x = -\dfrac{b}{2a} = -\dfrac{6}{2(3)} = -1$

$y = 3(-1)^2 + 6(-1) + 1 = -2$

vertex: $(-1, -2)$

47. $f(x) = -3x^2 + 12x + 4; x = -\dfrac{b}{2a} = -\dfrac{12}{2(-3)} = 2; y = -3(2)^2 + 12(2) + 4 = 16$

vertex: $(2, 16)$

49. $f(x) = -2x^2 + 4x + 1$

$x = -\dfrac{b}{2a} = -\dfrac{4}{2(-2)} = 1$

$y = -2(1)^2 + 4(1) + 1 = 3$

vertex: $(1, 3)$; opens D

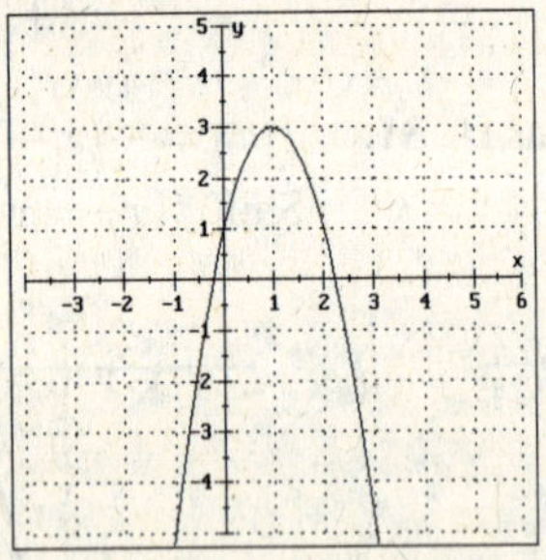

51. $f(x) = 3x^2 - 12x + 10$

$x = -\dfrac{b}{2a} = -\dfrac{-12}{2(3)} = 2$

$y = 3(2)^2 - 12(2) + 10 = -2$

vertex: $(2, -2)$; opens U

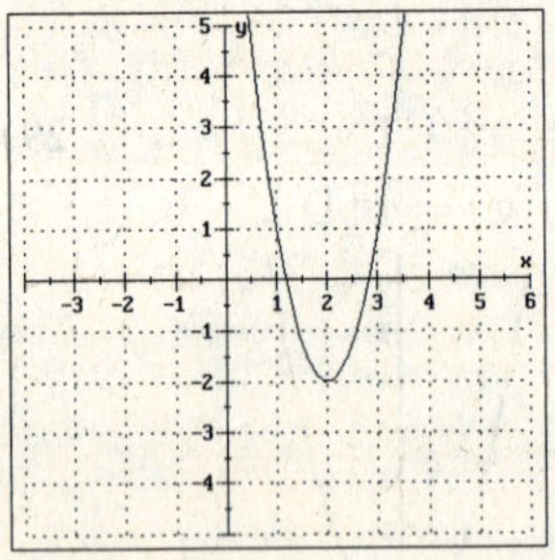

© 2015 Cengage Learning. All Rights Reserved. May not be scanned, copied or duplicated, or posted to a publicly accessible website, in whole or in part.

SECTION 8.4

53. $y = -4x^2 + 16x + 5$
$y = -4(x^2 - 4x) + 5$
$y = -4(x^2 - 4x + 4) + 5 + 16$
$y = -4(x - 2)^2 + 21$
$V(2, 21)$; axis: $x = 2$

55. $y - 7 = 6x^2 - 5x$
$y = 6x^2 - 5x + 7$
$$x = -\frac{b}{2a} = -\frac{-5}{2(6)} = \frac{5}{12}$$
$y = 6x^2 - 5x + 7$
$$= 6\left(\frac{5}{12}\right)^2 - 5\left(\frac{5}{12}\right) + 7$$
$$= \frac{25}{24} - \frac{25}{12} + 7 = \frac{143}{24}$$
$V\left(\frac{5}{12}, \frac{143}{24}\right)$; axis: $x = \frac{5}{12}$

57. $f(x) = x^2 + x - 6$
$f(x) = \left(x + \frac{1}{2}\right)^2 - \frac{25}{4}$
vertex: $\left(-\frac{1}{2}, -\frac{25}{4}\right)$; opens U

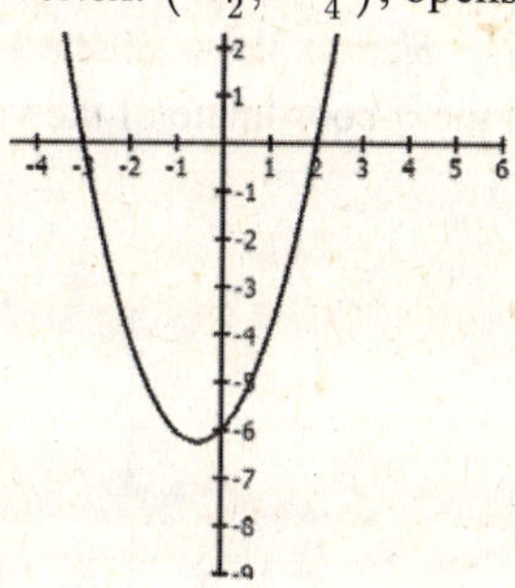

59. $y = 2x^2 - x + 1$

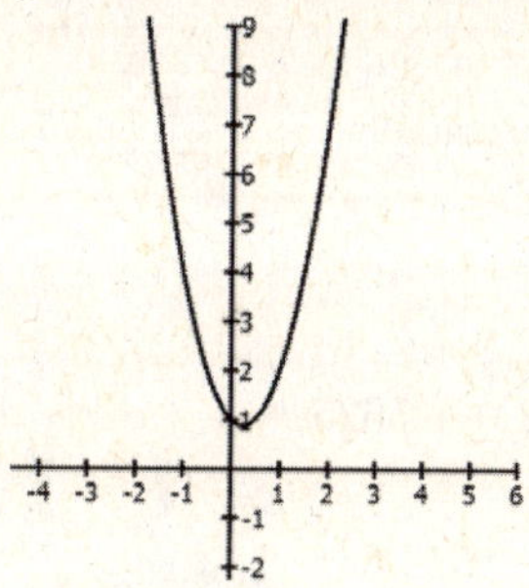

$V(0.25, 0.88)$

61. $y = 7 + x - x^2$

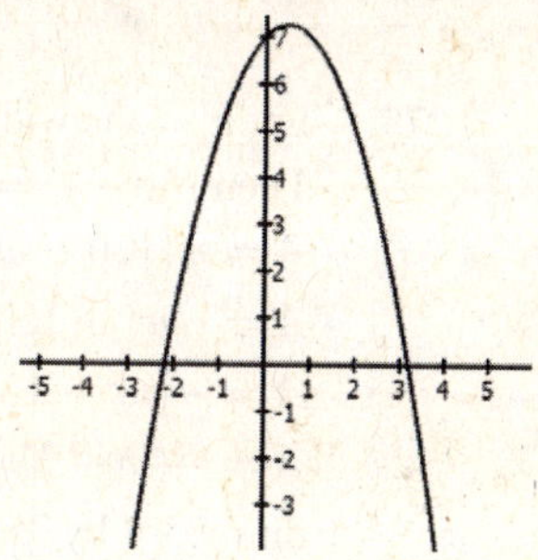

$V(0.5, 7.25)$

63. $y = x^2 + 9x - 10$

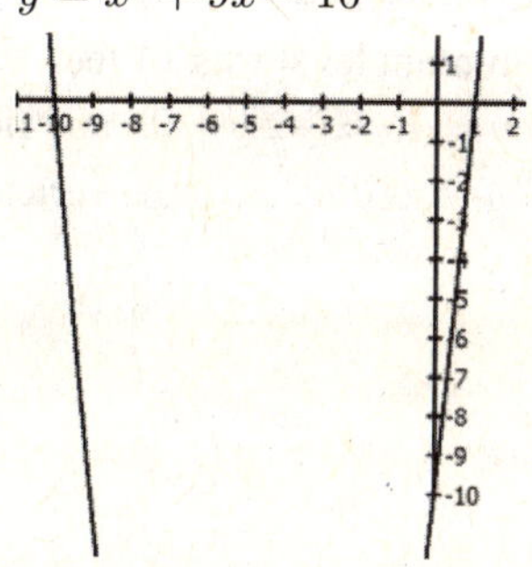

solution set: $\{-10, 1\}$

65. $y = 0.5x^2 - 0.7x - 3$

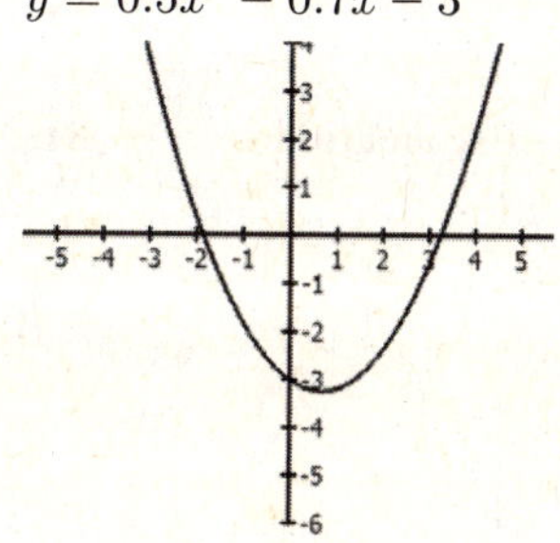

solution set: $\{-1.85, 3.25\}$

67. $y - 3 = (x + 7)^2$
$y = (x + 7)^2 + 3$
$V(-7, 3)$

69. Since the graph of the height equation is a parabola, the max. height occurs at the vertex.
$s = 48t - 16t^2$
$s = -16(t^2 - 3t)$
$s = -16\left(t^2 - 3t + \frac{9}{4}\right) + 36$
$s = -16\left(t - \frac{3}{2}\right)^2 + 36$

$V\left(\frac{3}{2}, 36\right) \Rightarrow$ max. height $= 36$ ft
The maximum height is 36 feet, which occurs after $\frac{3}{2}$ seconds (1.5 seconds).

© 2015 Cengage Learning. All Rights Reserved. May not be scanned, copied or duplicated, or posted to a publicly accessible website, in whole or in part.

71. Let $w =$ the width of the rectangle.
Then $200 - w =$ the length.
$A = w(200 - w)$
$A = -w^2 + 200w$
$A = -(w^2 - 200w + 10000) + 10000$
$A = -(w - 100)^2 + 10000$
dim: 100 ft by 100 ft; area $= 10{,}000$ ft^2

73. Replace p with x.
Graph $y = 50x(1 - x)$ and find the x-coordinate(s) when $y = 9.375$.

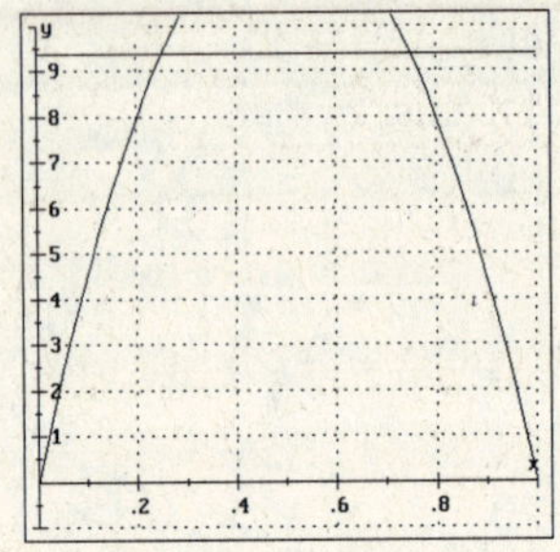

$p = 0.25$ or $p = 0.75$

75. Let $w =$ the width of the rectangle.
Then $150 - w =$ the length.
$A = w(150 - w)$
$A = -w^2 + 150w$
$A = -(w^2 - 150w + 5625) + 5625$
$A = -(w - 75)^2 + 5625$
dim: 75 ft by 75 ft; area $= 5625$ ft^2

77. Replace t with x.
Graph $y = H = 3.3x^2 - 59.4x + 281.3$ and find the y-coordinate of the vertex:

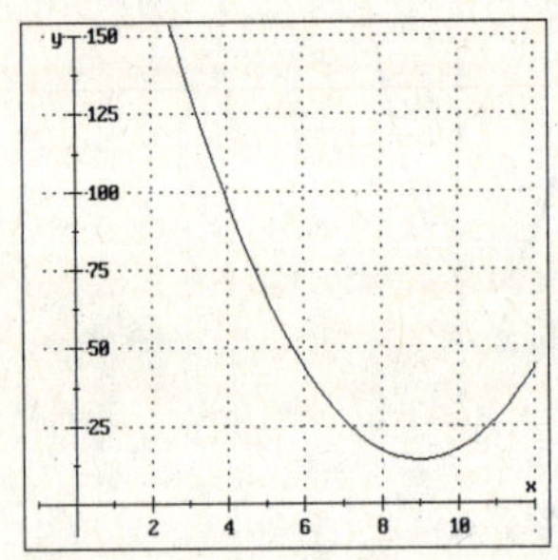

The minimum level was 14 feet.

79. Graph $y = R = -\frac{x^2}{1000} + 10x$ and find the x-coordinate of the vertex:

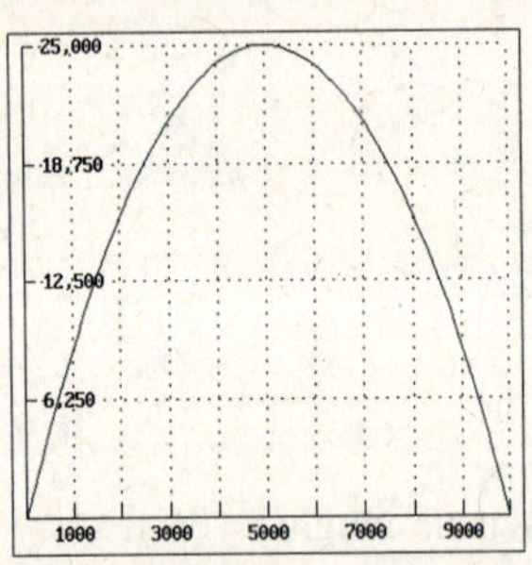

5000 stereos should be sold.

81. Graph $y = R = -\frac{x^2}{728} + 9x$ and find the x- and y- coordinates of the vertex:

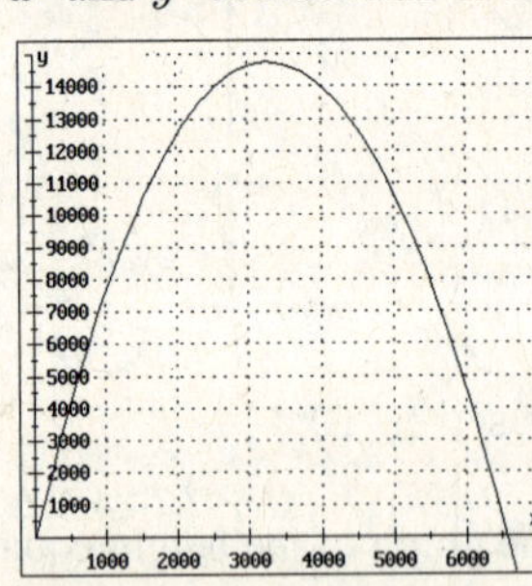

Max. revenue $= \$14{,}742$; # radios $= 3276$

© 2015 Cengage Learning. All Rights Reserved. May not be scanned, copied or duplicated, or posted to a publicly accessible website, in whole or in part.

83. Let $x =$ the number of \$1 increases to the price. Then the sales will be $4000 - 100x$, and the revenue will be $(30 + x)(4000 - 100x)$. Find the vertex of the parabola $y = (30 + x)(4000 - 100x)$.

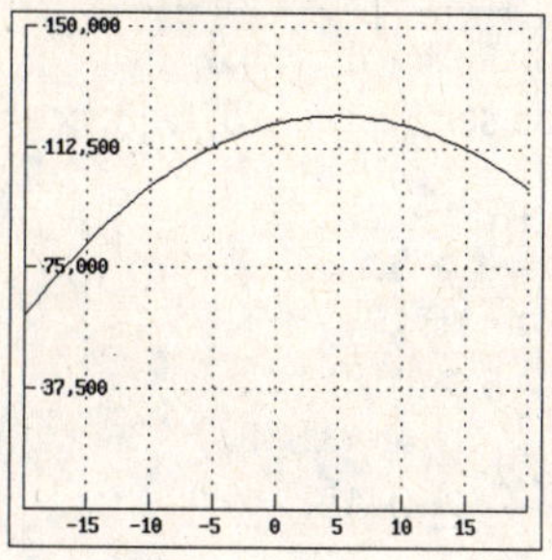

The price should increase \$5, to a total of \$35.

85. **Answers may vary.**

87. Graph $y = x^2 + x + 1$ to find x-intercept(s):

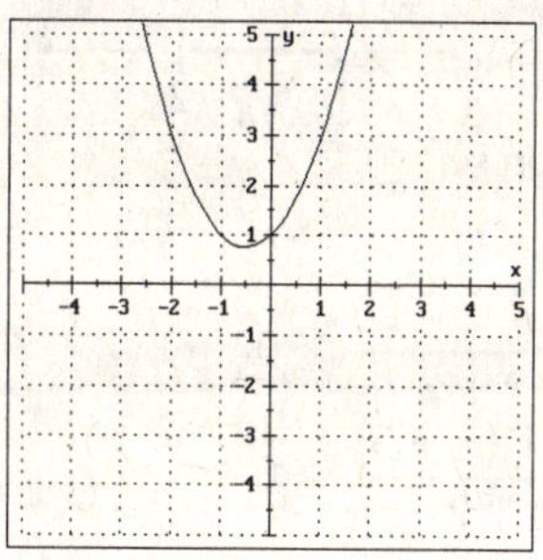

There are no x-intercepts, which means there is no solution to the equation.

Exercises 8.5 (page 586)

1. $x - 2 = 0$
$x = 2$

3. $x - 2 < 0$
$x < 2$

5. $x + 3 > 0$
$x > -3$

7. $x + 3 = 0$
$x = -3$

9. $y = kx^2$

11. $y = kab$

13. $y = -7x + 2$
$m = -7$

15. greater

17. quadratic

19. undefined

21. sign

23. $x^2 - 5x + 4 < 0$
$(x - 4)(x - 1) < 0$

$x - 4$ – – – – – – – 0 ++++
$x - 1$ – – – 0++++++++++

←—(——)—→
1 4

solution set: $(1, 4)$

25. $x^2 - 8x + 15 > 0$
$(x - 5)(x - 3) > 0$

$x - 5$ – – – – – – – – –0++++
$x - 3$ – – – 0++++++++++

←—)——(—→
3 5

solution set: $(-\infty, 3) \cup (5, \infty)$

© 2015 Cengage Learning. All Rights Reserved. May not be scanned, copied or duplicated, or posted to a publicly accessible website, in whole or in part.

SECTION 8.5

27. $x^2 + x - 12 \le 0$

$(x-3)(x+4) \le 0$

$x - 3$ − − − − − − − − 0++++

$x + 4$ − − − 0++++++++++

⟵ [——] ⟶

-4 3

solution set: $[-4, 3]$

29. $x^2 + 2x \ge 15$

$x^2 + 2x - 15 \ge 0$

$(x-3)(x+5) \ge 0$

$x - 3$ − − − − − − − − 0++++

$x + 5$ − − − 0++++++++++

⟵]——[⟶

-5 3

solution set: $(-\infty, -5] \cup [3, \infty)$

31. $\dfrac{3}{x} < 6$

$\dfrac{3}{x} - 6 < 0$

$\dfrac{3}{x} - \dfrac{6x}{x} < 0$

$\dfrac{3 - 6x}{x} < 0$

$3 - 6x$ +++++++++ 0 − − −

x − − − 0++++++++++

⟵)——(⟶

0 $\frac{1}{2}$

solution set: $(-\infty, 0) \cup \left(\frac{1}{2}, \infty\right)$

33. $\dfrac{10}{x} \ge 5$

$\dfrac{10}{x} - 5 \ge 0$

$\dfrac{10}{x} - \dfrac{5x}{x} \ge 0$

$\dfrac{10 - 5x}{x} \ge 0$

$10 - 5x$ ++++++++++0 − − −

x − − − 0+++++++++++

⟵ (——] ⟶

0 2

solution set: $(0, 2]$

35. $\dfrac{x^2 - 5x + 4}{x + 3} < 0$

$\dfrac{(x-4)(x-1)}{x+3} < 0$

$x - 4$ − − − − − − − − − − − 0++++

$x - 1$ − − − − − − − − 0+++ +++++

$x + 3$ − − − 0++++++ ++++ +++++

⟵)——(——) ⟶

-3 1 4

solution set: $(-\infty, -3) \cup (1, 4)$

37. $\dfrac{x^2 + x - 20}{x + 2} \ge 0$

$\dfrac{(x-4)(x+5)}{x+2} \ge 0$

$x - 4$ − − − − − − − − − − − − 0++++

$x + 2$ − − − − − − − − 0++++ +++++

$x + 5$ − − − 0+++++++++++ +++++

⟵ [——)——[⟶

-5 -2 4

Include when the numerator $= 0$.

solution set: $[-5, -2) \cup [4, \infty)$

© 2015 Cengage Learning. All Rights Reserved. May not be scanned, copied or duplicated, or posted to a publicly accessible website, in whole or in part.

SECTION 8.5

39.
$$\frac{3}{x-2} < \frac{4}{x}$$
$$\frac{3}{x-2} - \frac{4}{x} < 0$$
$$\frac{3x}{x(x-2)} - \frac{4(x-2)}{x(x-2)} < 0$$
$$\frac{-x+8}{x(x-2)} < 0$$

$-x+8$ +++++++++++++++0− − −
$x-2$ − − − − − − − 0++++++++++
x − − − 0++++ ++++++++++

← (——) — (→
0 2 8

solution set: $(0, 2) \cup (8, \infty)$

41.
$$\frac{-5}{x+2} \geq \frac{4}{2-x}$$
$$\frac{-5}{x+2} - \frac{4}{2-x} \geq 0$$
$$\frac{-5(2-x)}{(x+2)(2-x)} - \frac{4(x+2)}{(x+2)(2-x)} \geq 0$$
$$\frac{x-18}{(x+2)(2-x)} \geq 0$$

$x-18$ − − − − − − − − − − − − − 0++++
$2-x$ +++++++++++++ 0− − − − − −
$x+2$ − − − 0++++++++++++++++++

←) —— (——] →
−2 2 18

Include when the numerator = 0.
solution set: $(-\infty, -2) \cup (2, 18]$

43. $y < x^2 + 1$

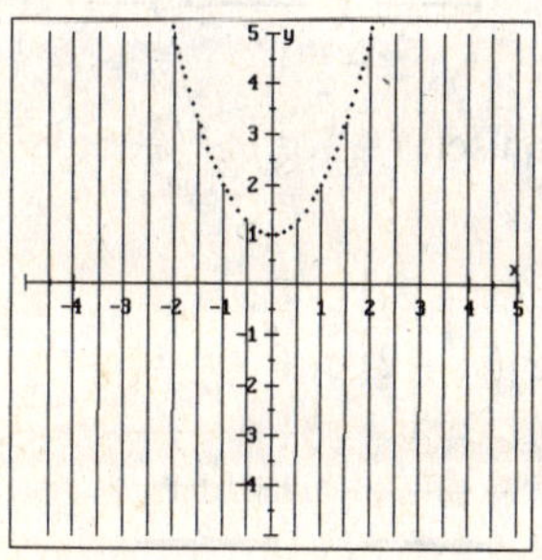

45. $y \leq x^2 + 5x + 6$

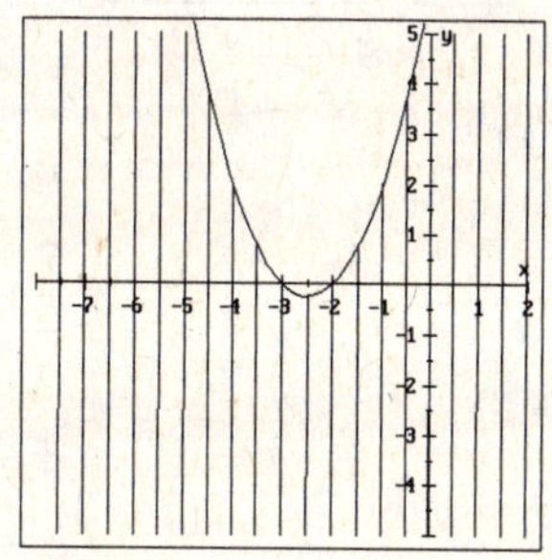

47. $y < |x + 4|$

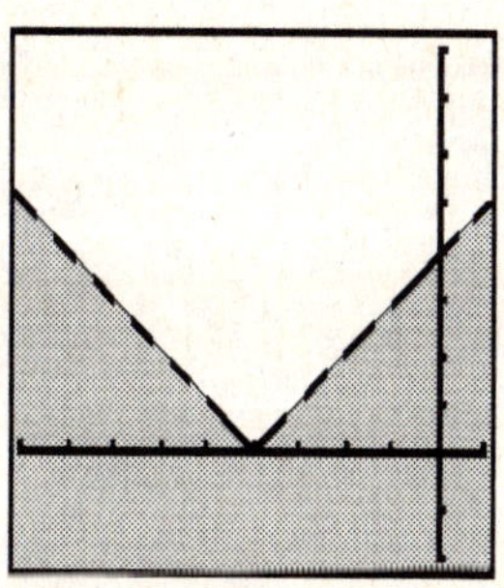

49. $y \leq -|x| + 2$

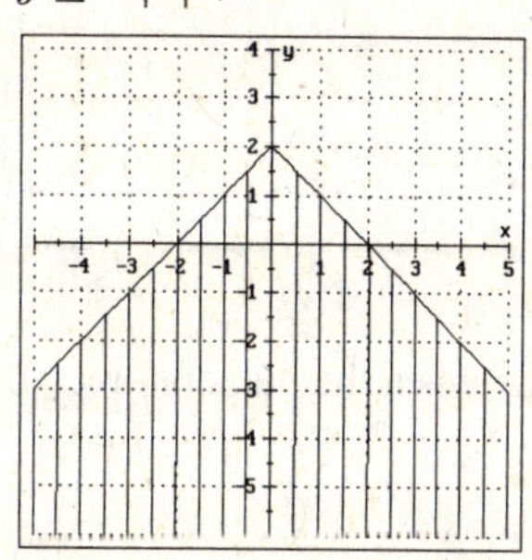

© 2015 Cengage Learning. All Rights Reserved. May not be scanned, copied or duplicated, or posted to a publicly accessible website, in whole or in part.

SECTION 8.5

51. $x^2 - 10x < -25$

$x^2 - 10x + 25 < 0$

$(x-5)(x-5) < 0$

$x - 5$ ---- 0 +++++++

$x - 5$ ---- 0 +++++++

←———— 5 ————→

Since the product is never negative, there is no solution.

53. $x^2 \geq 9$

$x^2 - 9 \geq 0$

$(x-3)(x+3) \geq 0$

$x - 3$ --------0++++

$x + 3$ --- 0++++++++++

←——] —— [——→ (at -3 and 3)

solution set: $(-\infty, -3] \cup [3, \infty)$

55. $\dfrac{x^2 - 6x + 9}{x+4} < 0$

$\dfrac{(x-3)(x-3)}{x+4} < 0$

$x - 3$ -------- 0++++

$x - 3$ -------- 0++++

$x + 4$ --- 0++++++ +++++

←——) ———— 3 ——→ (at -4)

solution set: $(-\infty, -4)$

57. $\dfrac{6x^2 - 5x + 1}{2x+1} > 0$

$\dfrac{(2x-1)(3x-1)}{2x+1} > 0$

$2x - 1$ -------- ---- 0++++

$3x - 1$ -------- 0++++++++

$2x + 1$ --- 0+++++++++++++++

←—— (——) —— (——→ (at $-\frac{1}{2}$, $\frac{1}{3}$, $\frac{1}{2}$)

solution set: $\left(-\frac{1}{2}, \frac{1}{3}\right) \cup \left(\frac{1}{2}, \infty\right)$

59. $\dfrac{7}{x-3} \geq \dfrac{2}{x+4}$

$\dfrac{7}{x-3} - \dfrac{2}{x+4} \geq 0$

$\dfrac{7(x+4)}{(x-3)(x+4)} - \dfrac{2(x-3)}{(x-3)(x+4)} \geq 0$

$\dfrac{5x+34}{(x-3)(x+4)} \geq 0$

$x - 3$ -------- ---- 0++++

$x + 4$ -------- 0++++ +++++

$5x + 34$ --- 0++++++ +++++ +++++

←—— [——) —— (——→ (at $-\frac{34}{5}$, -4, 3)

Include when the numerator $= 0$.

solution set: $\left[-\frac{34}{5}, -4\right) \cup (3, \infty)$

61. $(x+2)^2 > 0$

$(x+2)(x+2) > 0$

$x + 2$ -----0+++++++

$x + 2$ -----0+++++++

←——) (——→ (at -2)

solution set: $(-\infty, -2) \cup (-2, \infty)$

63. $3x^2 - 75 < 0$

$3(x-5)(x+5) < 0$

$x - 5$ -------- 0++++

$x + 5$ --- 0++++++ ++++

←—— (——) ——→ (at -5 and 5)

solution set: $(-5, 5)$

© 2015 Cengage Learning. All Rights Reserved. May not be scanned, copied or duplicated, or posted to a publicly accessible website, in whole or in part.

SECTION 8.5

65.

$$-\frac{5}{x} < 3$$
$$-\frac{5}{x} - 3 < 0$$
$$\frac{-5}{x} - \frac{3x}{x} < 0$$
$$\frac{-5-3x}{x} < 0$$

$-5-3x$ +++ 0 − − − − − − − − −

x − − − − − − − − − 0++++

⟵) —— (⟶ at $-\frac{5}{3}$ and 0

solution set: $(-\infty, -\frac{5}{3}) \cup (0, \infty)$

67.

$$\frac{x}{x+4} \le \frac{1}{x+1}$$
$$\frac{x}{x+4} - \frac{1}{x+1} \le 0$$
$$\frac{x(x+1)}{(x+4)(x+1)} - \frac{1(x+4)}{(x+4)(x+1)} \le 0$$
$$\frac{x^2-4}{(x+4)(x+1)} \le 0$$
$$\frac{(x+2)(x-2)}{(x+4)(x+1)} \le 0$$

$x-2$ − − − − − − − − − − − − − − 0+++

$x+1$ − − − − − − − − − − 0++++ ++++

$x+2$ − − − − − − 0++++ +++++ ++++

$x+4$ − − 0+++++++++ +++++ ++++

⟵ (——] —— (——] ⟶ at -4, -2, -1, 2

Include when the numerator $= 0$.

solution set: $(-4, -2] \cup (-1, 2]$

69.

$$\frac{x}{x+16} > \frac{1}{x+1}$$
$$\frac{x}{x+16} - \frac{1}{x+1} > 0$$
$$\frac{x(x+1)}{(x+16)(x+1)} - \frac{1(x+16)}{(x+16)(x+1)} > 0$$
$$\frac{x^2-16}{(x+16)(x+1)} > 0$$
$$\frac{(x+4)(x-4)}{(x+16)(x+1)} > 0$$

$x-4$ − − − − − − − − − − − − − − − − 0+++

$x+1$ − − − − − − − − − − − 0++++ ++++

$x+4$ − − − − − − − 0++++ +++++ ++++

$x+16$ − − − 0++++ +++++ +++++ ++++

⟵) —— (——) —— (⟶ at -16, -4, -1, 4

solution set: $(-\infty, -16) \cup (-4, -1) \cup (4, \infty)$

71. $y \ge (x-1)^2$

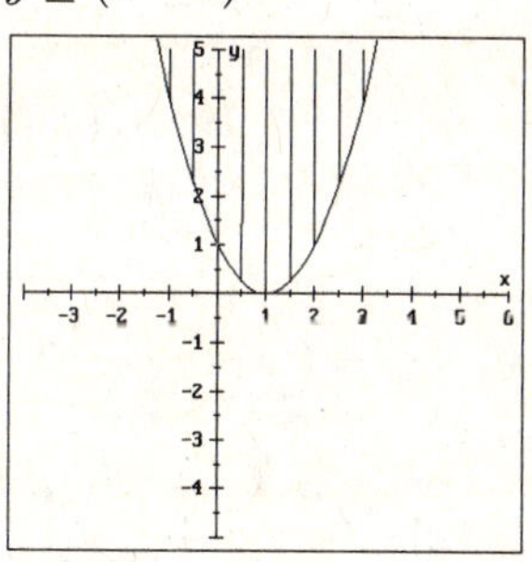

73. $-x^2 - y + 6 > -x$

$-x^2 + x + 6 > y$

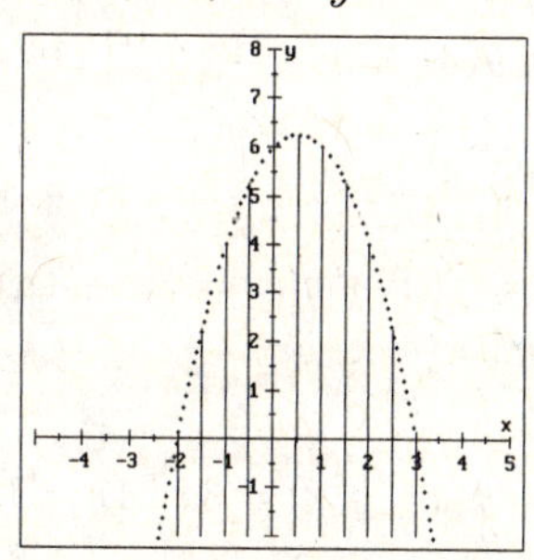

© 2015 Cengage Learning. All Rights Reserved. May not be scanned, copied or duplicated, or posted to a publicly accessible website, in whole or in part.

75. $x^2 - 2x - 3 < 0$
Graph $y = x^2 - 2x - 3$ and find the x-coordinates of points below the x-axis.

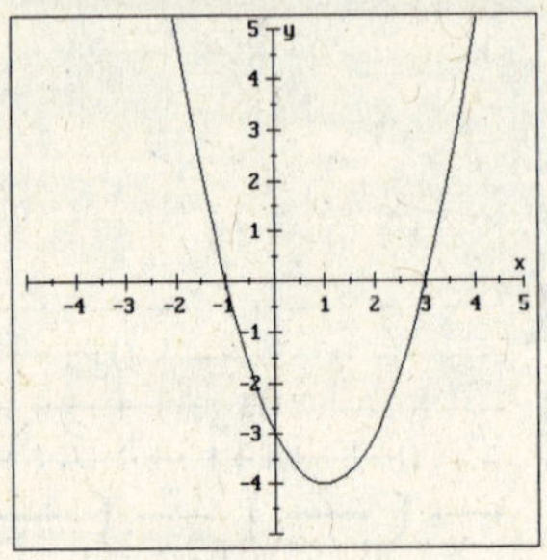

$(-1, 3)$

77. $\frac{x+3}{x-2} > 0$
Graph $y = (x+3)/(x-2)$ and find the x-coordinates of points above the x-axis.

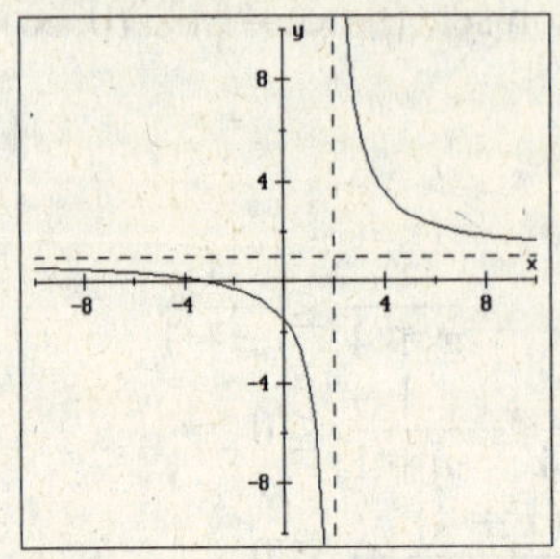

$(-\infty, -3) \cup (2, \infty)$

79. **Answers may vary.**

81. It will be positive if 4, 2 or 0 factors are negative.

Exercises 8.6 (page 594)

1. $f(-3) = 2(-3) + 3 = -3$

3. $g(x+1) = 4(x+1)^2 - 2 = 4(x^2 + 2x + 1) - 2 = 4x^2 + 8x + 4 - 2 = 4x^2 + 8x + 2$

5. $$\begin{aligned}(2x+3) + (4x^2 - 2) &= 2x + 3 + 4x^2 - 2 \\ &= 4x^2 + 2x + 1\end{aligned}$$

7. $$\begin{aligned}(2x+3) - (4x^2 - 2) &= 2x + 3 - 4x^2 + 2 \\ &= -4x^2 + 2x + 5\end{aligned}$$

9. $$\frac{5x^2 - 13x - 6}{9 - x^2} = \frac{(5x+2)(x-3)}{(3+x)(3-x)} = -\frac{5x+2}{x+3}$$

11. $$\begin{aligned}\frac{8 + 2x - x^2}{12 + x - 3x^2} \div \frac{3x^2 + 5x - 2}{3x - 1} &= \frac{x^2 - 2x - 8}{3x^2 - x - 12} \cdot \frac{3x - 1}{3x^2 + 5x - 2} \\ &= \frac{(x-4)(x+2)}{3x^2 - x - 12} \cdot \frac{3x - 1}{(3x-1)(x+2)} = \frac{x - 4}{3x^2 - x - 12}\end{aligned}$$

13. $f(x) + g(x)$

15. $f(x)g(x)$

17. domain

19. $f(g(x))$

21. given point

23. $f + g = f(x) + g(x) = 3x + 4x = 7x$
domain $= (-\infty, \infty)$

25. $g - f = g(x) - f(x) = 4x - 3x = x$
domain $= (-\infty, \infty)$

27. $f \cdot g = f(x) \cdot g(x) = 3x \cdot 4x = 12x^2$
domain $= (-\infty, \infty)$

29. $g/f = \dfrac{g(x)}{f(x)} = \dfrac{4x}{3x} = \dfrac{4}{3}$ (for $x \neq 0$)
domain $= (-\infty, 0) \cup (0, \infty)$

31. $f + g = f(x) + g(x) = 2x + 1 + x - 3 = 3x - 2$; domain $= (-\infty, \infty)$

© 2015 Cengage Learning. All Rights Reserved. May not be scanned, copied or duplicated, or posted to a publicly accessible website, in whole or in part.

SECTION 8.6

33. $g - f = g(x) - f(x) = (x-3) - (2x+1) = x - 3 - 2x - 1 = -x - 4$; domain $= (-\infty, \infty)$

35. $f \cdot g = f(x) \cdot g(x) = (2x+1)(x-3) = 2x^2 - 5x - 3$; domain $= (-\infty, \infty)$

37. $g/f = \dfrac{g(x)}{f(x)} = \dfrac{x-3}{2x+1}$; domain $= \left(-\infty, -\dfrac{1}{2}\right) \cup \left(-\dfrac{1}{2}, \infty\right)$

39. $(f \circ g)(4) = f(g(4)) = f(4^2 - 1) = f(15) = 2(15) + 1 = 31$

41. $(g \circ f)(-3) = g(f(-3)) = g(2(-3)+1) = g(-5) = (-5)^2 - 1 = 24$

43. $(f \circ g)(0) = f(g(0)) = f(0^2 - 1) = f(-1) = 2(-1) + 1 = -1$

45. $(f \circ g)\left(\dfrac{1}{4}\right) = f\left(g\left(\dfrac{1}{4}\right)\right) = f\left(\left(\dfrac{1}{4}\right)^2 - 1\right) = f\left(-\dfrac{15}{16}\right) = 2\left(-\dfrac{15}{16}\right) + 1 = -\dfrac{15}{8} + 1 = -\dfrac{7}{8}$

47. $(f \circ g)(x) = f(g(x)) = f(x^2 - 1) = 2(x^2 - 1) + 1 = 2x^2 - 2 + 1 = 2x^2 - 1$

49. $(g \circ f)(3x) = g(f(3x)) = g(2(3x)+1) = g(6x+1) = (6x+1)^2 - 1 = 36x^2 + 12x + 1 - 1$
$= 36x^2 + 12x$

51. $\dfrac{f(x+h) - f(x)}{h} = \dfrac{4(x+h) + 5 - (4x+5)}{h} = \dfrac{4x + 4h + 5 - 4x - 5}{h} = \dfrac{4h}{h} = 4$

53. $\dfrac{f(x+h) - f(x)}{h} = \dfrac{(x+h)^2 - x^2}{h} = \dfrac{x^2 + 2xh + h^2 - x^2}{h} = \dfrac{2xh + h^2}{h} = 2x + h$

55. $\dfrac{f(x+h) - f(x)}{h} = \dfrac{2(x+h)^2 - 1 - (2x^2 - 1)}{h} = \dfrac{2x^2 + 4xh + 2h^2 - 1 - 2x^2 + 1}{h}$
$= \dfrac{4xh + 2h^2}{h} = 4x + 2h$

57. $\dfrac{f(x+h) - f(x)}{h} = \dfrac{(x+h)^2 + (x+h) - (x^2 + x)}{h} = \dfrac{x^2 + 2xh + h^2 + x + h - x^2 - x}{h}$
$= \dfrac{2xh + h^2 + h}{h} = 2x + h + 1$

59. $\dfrac{f(x+h) - f(x)}{h} = \dfrac{(x+h)^2 + 3(x+h) - 4 - (x^2 + 3x - 4)}{h}$
$= \dfrac{x^2 + 2xh + h^2 + 3x + 3h - 4 - x^2 - 3x + 4}{h}$
$= \dfrac{2xh + h^2 + 3h}{h} = 2x + h + 3$

© 2015 Cengage Learning. All Rights Reserved. May not be scanned, copied or duplicated, or posted to a publicly accessible website, in whole or in part.

61. $$\begin{aligned}\frac{f(x+h)-f(x)}{h} &= \frac{2(x+h)^2+3(x+h)-7-(2x^2+3x-7)}{h} \\ &= \frac{2x^2+4xh+2h^2+3x+3h-7-2x^2-3x+7}{h} \\ &= \frac{4xh+2h^2+3h}{h} = 4x+2h+3\end{aligned}$$

63. $f-g=f(x)-g(x)=(3x-2)-(2x^2+1)=3x-2-2x^2-1=-2x^2+3x-3$
domain $=(-\infty,\infty)$

65. $f/g=\dfrac{f(x)}{g(x)}=\dfrac{3x-2}{2x^2+1}$; domain $=(-\infty,\infty)$

67. $f-g=f(x)-g(x)=(x^2-1)-(x^2-4)=x^2-1-x^2+4=3$; domain $=(-\infty,\infty)$

69. $g/f=\dfrac{g(x)}{f(x)}=\dfrac{x^2-4}{x^2-1}=\dfrac{(x+2)(x-2)}{(x+1)(x-1)}$; domain $=(-\infty,-1)\cup(-1,1)\cup(1,\infty)$

71. $(f\circ g)(4)=f(g(4))=f(4^2+4)=f(20)=3(20)-2=58$

73. $(g\circ f)(-2)=g(f(-2))=g(3(-2)-2)=g(-8)=(-8)^2+(-8)=56$

75. $(g\circ f)(0)=g(f(0))=g(3(0)-2)=g(-2)=(-2)^2+(-2)=2$

77. $$\begin{aligned}(g\circ f)(x)=g(f(x))=g(3x-2)=(3x-2)^2+3x-2 &= 9x^2-12x+4+3x-2 \\ &= 9x^2-9x+2\end{aligned}$$

79. $$\frac{f(x)-f(a)}{x-a}=\frac{(4x-7)-(4a-7)}{x-a}=\frac{4x-7-4a+7}{x-a}=\frac{4x-4a}{x-a}=\frac{4(x-a)}{x-a}=4$$

81. $$\frac{f(x)-f(a)}{x-a}=\frac{x^2-a^2}{x-a}=\frac{(x+a)(x-a)}{x-a}=x+a$$

83. $$\begin{aligned}\frac{f(x)-f(a)}{x-a}=\frac{(2x^2-1)-(2a^2-1)}{x-a}=\frac{2x^2-1-2a^2+1}{x-a} &= \frac{2x^2-2a^2}{x-a} \\ &= \frac{2(x+a)(x-a)}{x-a} \\ &= 2(x+a)=2x+2a\end{aligned}$$

85. $$\begin{aligned}\frac{f(x)-f(a)}{x-a}=\frac{(x^2+x)-(a^2+a)}{x-a}=\frac{x^2+x-a^2-a}{x-a} &= \frac{x^2-a^2+x-a}{x-a} \\ &= \frac{(x+a)(x-a)+1(x-a)}{x-a} \\ &= \frac{(x-a)(x+a+1)}{x-a}=x+a+1\end{aligned}$$

© 2015 Cengage Learning. All Rights Reserved. May not be scanned, copied or duplicated, or posted to a publicly accessible website, in whole or in part.

SECTION 8.6

87. $$\frac{f(x)-f(a)}{x-a}=\frac{(x^2+3x-4)-(a^2+3a-4)}{x-a}=\frac{x^2+3x-4-a^2-3a+4}{x-a}$$
$$=\frac{x^2-a^2+3x-3a}{x-a}$$
$$=\frac{(x+a)(x-a)+3(x-a)}{x-a}$$
$$=\frac{(x-a)(x+a+3)}{x-a}=x+a+3$$

89. $$\frac{f(x)-f(a)}{x-a}=\frac{(2x^2+3x-7)-(2a^2+3a-7)}{x-a}=\frac{2x^2+3x-7-2a^2-3a+7}{x-a}$$
$$=\frac{2x^2-2a^2+3x-3a}{x-a}$$
$$=\frac{2(x+a)(x-a)+3(x-a)}{x-a}$$
$$=\frac{(x-a)(2(x+a)+3)}{x-a}$$
$$=2(x+a)+3=2x+2a+3$$

91. $(f\circ g)(x)=f(g(x))=f(2x-5)=(2x-5)+1=2x-4$
$(g\circ f)(x)=g(f(x))=g(x+1)=2(x+1)-5=2x+2-5=2x-3$

93. $f(a)=a^2+2a-3;\ f(h)=h^2+2h-3\Rightarrow f(a)+f(h)=a^2+h^2+2a+2h-6$
$f(a+h)=(a+h)^2+2(a+h)-3=a^2+2ah+h^2+2a+2h-3$
$=a^2+h^2+2ah+2a+2h-3$

95. $$\frac{f(x+h)-f(x)}{h}=\frac{(x+h)^3-1-(x^3-1)}{h}=\frac{x^3+3x^2h+3xh^2+h^3-1-x^3+1}{h}$$
$$=\frac{3x^2h+3xh^2+h^3}{h}$$
$$=\frac{h(3x^2+3xh+h^2)}{h}=3x^2+3xh+h^2$$

97. $F(t)=2700-200t;\ C(F)=\frac{5}{9}(F-32)$
$C(F(t))=C(2700-200t)=\frac{5}{9}(2700-200t-32)=\frac{5}{9}(2668-200t)$

99. **Answers may vary.**

101. It is associative. Examples will vary.

Exercises 8.7 (page 603)

1. $f(0)=2(0)+1=1;\ (0,1)$

3. $f(3)=2(3)+1=7;\ (3,7)$

5. $f(-1)=2(-1)+1=-1;\ (-1,-1)$

7. $7-\sqrt{-49}=7-\sqrt{49i^2}=7-7i$

9. $(3+4i)(2-3i)=6-9i+8i-12i^2=6-i-12(-1)=6-i+12=18-i$

© 2015 Cengage Learning. All Rights Reserved. May not be scanned, copied or duplicated, or posted to a publicly accessible website, in whole or in part.

SECTION 8.7

11. $|6-8i| = \sqrt{6^2+(-8)^2} = \sqrt{36+64} = \sqrt{100} = 10$

13. one-to-one

15. 2

17. x

19. Each input has a different output. one-to-one

21. The inputs $x = 2$ and $x = -2$ have the same output. not one-to-one

23. one-to-one

25. one-to-one

27. not one-to-one

29. one-to-one

31. inverse $= \{(3,4),(2,3),(1,2)\}$. Since each x-coordinate is paired with only one y-coordinate, the inverse relation **is a function**.

33. inverse $= \{(2,1),(3,2),(3,1),(5,1)\}$. Since $x = 3$ is paired with more than one y-coordinate, the inverse relation **is not a function**.

35.

$$\begin{aligned} f(x) &= 4x+1 \\ y &= 4x+1 \\ x &= 4y+1 \\ x-1 &= 4y \\ \frac{x-1}{4} &= y \\ \frac{1}{4}x - \frac{1}{4} &= y \\ f^{-1}(x) &= \tfrac{1}{4}x - \tfrac{1}{4} \end{aligned}$$

$f \circ f^{-1}$

$$\begin{aligned} f[f^{-1}(x)] &= f\left(\tfrac{1}{4}x - \tfrac{1}{4}\right) \\ &= 4 \cdot \left(\tfrac{1}{4}x - \tfrac{1}{4}\right) + 1 \\ &= x - 1 + 1 \\ &= x \end{aligned}$$

$f^{-1} \circ f$

$$\begin{aligned} f^{-1}[f(x)] &= f^{-1}(4x+1) \\ &= \tfrac{1}{4}(4x+1) - \tfrac{1}{4} \\ &= x + \tfrac{1}{4} - \tfrac{1}{4} \\ &= x \end{aligned}$$

37.

$$\begin{aligned} x+4 &= 5y \\ y = f(x) &= \frac{x+4}{5} \\ x &= \frac{y+4}{5} \\ 5x &= y+4 \\ 5x-4 &= y \\ f^{-1}(x) &= 5x-4 \end{aligned}$$

$f \circ f^{-1}$

$$\begin{aligned} f[f^{-1}(x)] &= f(5x-4) \\ &= \frac{(5x-4)+4}{5} \\ &= \frac{5x}{5} \\ &= x \end{aligned}$$

$f^{-1} \circ f$

$$\begin{aligned} f^{-1}[f(x)] &= f^{-1}\left(\frac{x+4}{5}\right) \\ &= 5 \cdot \left(\frac{x+4}{5}\right) - 4 \\ &= x + 4 - 4 \\ &= x \end{aligned}$$

39.

$$\begin{aligned} y &= 4x+3 \\ x &= 4y+3 \\ x-3 &= 4y \\ \frac{x-3}{4} &= y;\ f^{-1}(x) = \frac{x-3}{4} \end{aligned}$$

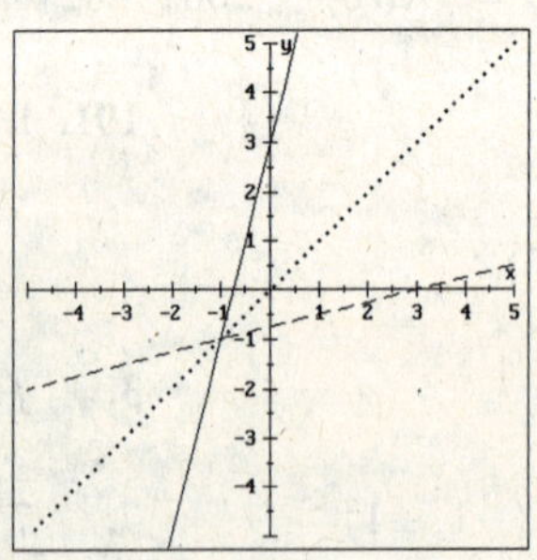

© 2015 Cengage Learning. All Rights Reserved. May not be scanned, copied or duplicated, or posted to a publicly accessible website, in whole or in part.

SECTION 8.7

41.
$$x = \frac{y-2}{3}$$
$$y = \frac{x-2}{3};\ f^{-1}(x) = \frac{x-2}{3}$$

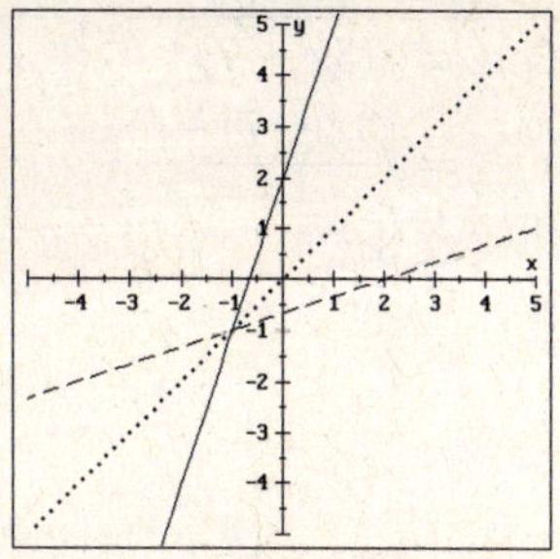

43.
$$y = x^2 + 9$$
$$x = y^2 + 9$$
$$x - 9 = y^2$$
$$\pm\sqrt{x-9} = y$$
The relation **is not** a function.

45.
$$y = x^3$$
$$x = y^3$$
$$\sqrt[3]{x} = y;\ f^{-1}(x) = \sqrt[3]{x}$$
The relation **is** a function.

47. $y = x^2 + 1;\ x \geq 0$

inverse: $x = y^2 + 1;\ y \geq 0$
$$x - 1 = y^2;\ y \geq 0$$
$$\pm\sqrt{x-1} = y;\ y \geq 0$$
$$\sqrt{x-1} = y$$

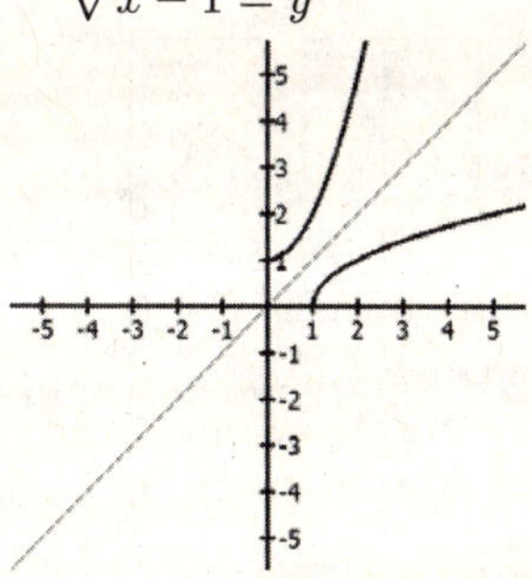

49. $y = \sqrt{x}, x \geq 0$

inverse: $x = \sqrt{y}, y \geq 0$
$$x^2 = y, y \geq 0$$

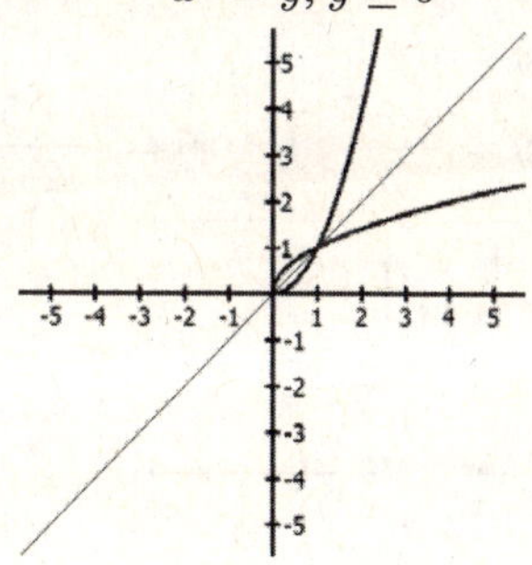

51. inverse $= \{(1,1), (4,2), (9,3), (16,4)\}$. Since each x-coordinate is paired with only one y-coordinate, the inverse relation **is a function**.

53.
$$y = -\sqrt[3]{x}$$
$$x = -\sqrt[3]{y}$$
$$-x = \sqrt[3]{y}$$
$$(-x)^3 = \left(\sqrt[3]{y}\right)^3$$
$$-x^3 = y, \text{ or } f^{-1}(x) = -x^3$$

55.
$$y = 2x^3 - 3$$
$$x = 2y^3 - 3$$
$$x + 3 = 2y^3$$
$$\frac{x+3}{2} = y^3$$
$$\sqrt[3]{\frac{x+3}{2}} = \sqrt[3]{y^3}, \text{ or } y = f^{-1}(x) = \sqrt[3]{\frac{x+3}{2}}$$

© 2015 Cengage Learning. All Rights Reserved. May not be scanned, copied or duplicated, or posted to a publicly accessible website, in whole or in part.

SECTION 8.7

57.

$$\begin{aligned} f(x) &= \frac{x-7}{3} \\ y &= \frac{x-7}{3} \\ x &= \frac{y-7}{3} \\ 3x &= y-7 \\ 3x+7 &= y \\ f^{-1}(x) &= 3x+7 \end{aligned}$$

$f \circ f^{-1}$

$$\begin{aligned} f\left[f^{-1}(x)\right] &= f(3x+7) \\ &= \frac{(3x+7)-7}{3} \\ &= \frac{3x}{3} \\ &= x \end{aligned}$$

$f^{-1} \circ f$

$$\begin{aligned} f^{-1}[f(x)] &= f^{-1}\left(\frac{x-7}{3}\right) \\ &= 3 \cdot \left(\frac{x-7}{3}\right) + 7 \\ &= x - 7 + 7 \\ &= x \end{aligned}$$

59.

$$\begin{aligned} 4x - 5y &= 20 \\ 5y &= 4x - 20 \\ y = f(x) &= \tfrac{4}{5}x - 4 \\ x &= \tfrac{4}{5}y - 4 \\ x + 4 &= \tfrac{4}{5}y \\ \tfrac{5}{4}(x+4) &= y \\ f^{-1}(x) &= \tfrac{5}{4}x + 5 \end{aligned}$$

$f \circ f^{-1}$

$$\begin{aligned} f\left[f^{-1}(x)\right] &= f\left(\tfrac{5}{4}x + 5\right) \\ &= \tfrac{4}{5}\left(\tfrac{5}{4}x + 5\right) - 4 \\ &= x + 4 - 4 \\ &= x \end{aligned}$$

$f^{-1} \circ f$

$$\begin{aligned} f^{-1}[f(x)] &= f^{-1}\left(\tfrac{4}{5}x - 4\right) \\ &= \tfrac{5}{4}\left(\tfrac{4}{5}x - 4\right) + 5 \\ &= x - 5 + 5 \\ &= x \end{aligned}$$

61.

$$\begin{aligned} 3x - y &= 5 \\ 3y - x &= 5 \\ 3y &= x + 5 \\ y &= \frac{x+5}{3};\ f^{-1}(x) = \frac{x+5}{2} \end{aligned}$$

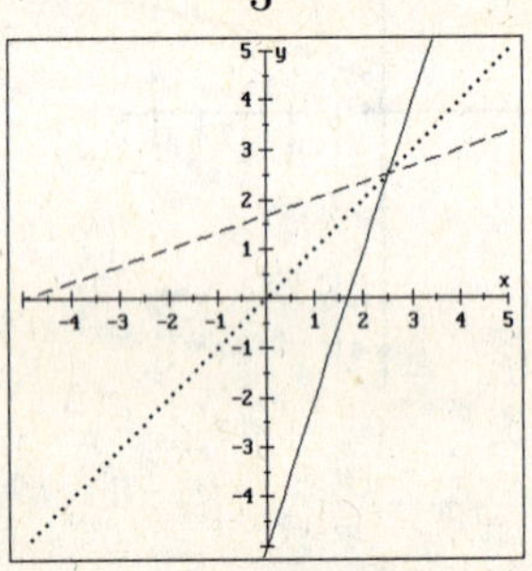

63.

$$\begin{aligned} 3(x+y) &= 2x+4 \\ 3(y+x) &= 2y+4 \\ 3y+3x &= 2y+4 \\ y &= 4-3x;\ f^{-1}(x) = -3x+4 \end{aligned}$$

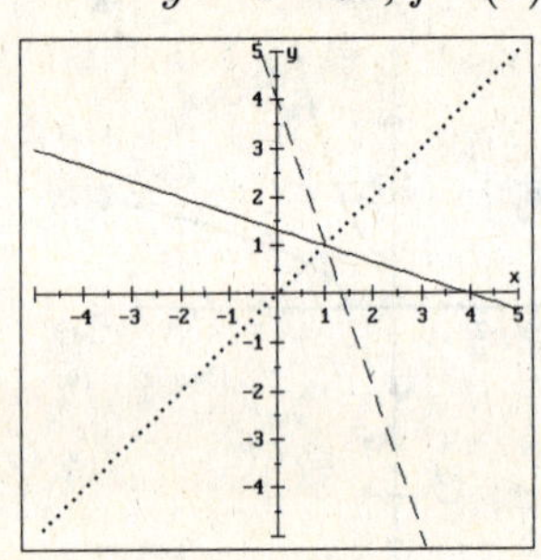

65. **Answers may vary.**

67.

$$\begin{aligned} y &= \frac{x+1}{x-1} \\ x &= \frac{y+1}{y-1} \\ x(y-1) &= y+1 \\ xy - x &= y+1 \\ xy - y &= x+1 \\ y(x-1) &= x+1 \\ y &= \frac{x+1}{x-1}, \text{ or } f^{-1}(x) = \frac{x+1}{x-1} \end{aligned}$$

© 2015 Cengage Learning. All Rights Reserved. May not be scanned, copied or duplicated, or posted to a publicly accessible website, in whole or in part.

Chapter 8 Review (page 608)

1.
$$\begin{aligned} 12x^2+x-6&=0\\ (4x+3)(3x-2)&=0\\ 4x+3=0 \quad &\textbf{or} \quad 3x-2=0\\ x=-\tfrac{3}{4} \quad &\quad\quad x=\tfrac{2}{3} \end{aligned}$$

2.
$$\begin{aligned} 6x^2+17x+5&=0\\ (2x+5)(3x+1)&=0\\ 2x+5=0 \quad &\textbf{or} \quad 3x+1=0\\ x=-\tfrac{5}{2} \quad &\quad\quad x=-\tfrac{1}{3} \end{aligned}$$

3.
$$\begin{aligned} 15x^2+2x-8&=0\\ (3x-2)(5x+4)&=0\\ 3x-2=0 \quad &\textbf{or} \quad 5x+4=0\\ x=\tfrac{2}{3} \quad &\quad\quad x=-\tfrac{4}{5} \end{aligned}$$

4.
$$\begin{aligned} (x+3)^2&=16\\ x+3&=\pm\sqrt{16}\\ x+3&=\pm 4\\ x&=-3\pm 4\\ x=1 \text{ or } &x=-7 \end{aligned}$$

5.
$$\begin{aligned} x^2+8x+12&=0\\ x^2+8x&=-12\\ x^2+8x+16&=-12+16\\ (x+4)^2&=4\\ x+4&=\pm 2\\ x&=-4\pm 2\\ x=-2 \text{ or } &x=-6 \end{aligned}$$

6.
$$\begin{aligned} 2x^2-9x+7&=0\\ x^2-\frac{9}{2}x+\frac{7}{2}&=0\\ x^2-\frac{9}{2}x&=-\frac{7}{2}\\ x^2-\frac{9}{2}x+\frac{81}{16}&=-\frac{56}{16}+\frac{81}{16}\\ \left(x-\frac{9}{4}\right)^2&=\frac{25}{16}\\ x-\frac{9}{4}&=\pm\frac{5}{4}\\ x&=\frac{9}{4}\pm\frac{5}{4}\\ x=\frac{7}{2} \text{ or } &x=1 \end{aligned}$$

7.
$$\begin{aligned} 2x^2-x-5&=0\\ x^2-\frac{1}{2}x-\frac{5}{2}&=0\\ x^2-\frac{1}{2}x&=\frac{5}{2}\\ x^2-\frac{1}{2}x+\frac{1}{16}&=\frac{5}{2}+\frac{1}{16}\\ \left(x-\frac{1}{4}\right)^2&=\frac{41}{16}\\ x-\frac{1}{4}&=\pm\frac{\sqrt{41}}{4}\\ x&=\frac{1}{4}\pm\frac{\sqrt{41}}{4} \end{aligned}$$

8.
$$\begin{aligned} &x^2-5x-6=0\\ &a=1, b=-5, c=-6\\ x&=\frac{-b\pm\sqrt{b^2-4ac}}{2a}\\ &=\frac{-(-5)\pm\sqrt{(-5)^2-4(1)(-6)}}{2(1)}\\ &=\frac{5\pm\sqrt{25+24}}{2}\\ &=\frac{5\pm\sqrt{49}}{2}=\frac{5\pm 7}{2}\\ x&=\frac{12}{2}=6 \text{ or } x=\frac{-2}{2}=-1 \end{aligned}$$

© 2015 Cengage Learning. All Rights Reserved. May not be scanned, copied or duplicated, or posted to a publicly accessible website, in whole or in part.

CHAPTER 8 REVIEW

9.
$$\begin{aligned} x^2 &= 7x \\ x^2 - 7x &= 0 \\ a = 1, b &= -7, c = 0 \\ x &= \frac{-b \pm \sqrt{b^2 - 4ac}}{2a} \\ &= \frac{-(-7) \pm \sqrt{(-7)^2 - 4(1)(0)}}{2(1)} \\ &= \frac{7 \pm \sqrt{49 + 0}}{2} \\ &= \frac{7 \pm \sqrt{49}}{2} = \frac{7 \pm 7}{2} \\ x &= \frac{14}{2} = 7 \text{ or } x = \frac{0}{2} = 0 \end{aligned}$$

10.
$$\begin{aligned} 2x^2 + 13x - 7 &= 0 \\ a = 2, b &= 13, c = -7 \\ x &= \frac{-b \pm \sqrt{b^2 - 4ac}}{2a} \\ &= \frac{-(13) \pm \sqrt{13^2 - 4(2)(-7)}}{2(2)} \\ &= \frac{-13 \pm \sqrt{169 + 56}}{4} \\ &= \frac{-13 \pm \sqrt{225}}{4} = \frac{-13 \pm 15}{4} \\ x &= \frac{2}{4} = \frac{1}{2} \text{ or } x = \frac{-28}{4} = -7 \end{aligned}$$

11.
$$\begin{aligned} 5x^2 - 2x - 3 &= 0 \\ a = 5, b &= -2, c = -3 \\ x &= \frac{-b \pm \sqrt{b^2 - 4ac}}{2a} \\ &= \frac{-(-2) \pm \sqrt{(-2)^2 - 4(5)(-3)}}{2(5)} \\ &= \frac{2 \pm \sqrt{4 + 60}}{10} \\ &= \frac{2 \pm \sqrt{64}}{10} = \frac{2 \pm 8}{10} \\ x &= \frac{10}{10} = 1 \text{ or } x = \frac{-6}{10} = -\frac{3}{5} \end{aligned}$$

12.
$$\begin{aligned} 2x^2 &= x + 2 \\ 2x^2 - x - 2 &= 0 \\ a = 2, b &= -1, c = -2 \\ x &= \frac{-b \pm \sqrt{b^2 - 4ac}}{2a} \\ &= \frac{-(-1) \pm \sqrt{(-1)^2 - 4(2)(-2)}}{2(2)} \\ &= \frac{1 \pm \sqrt{1 + 16}}{4} \\ &= \frac{1 \pm \sqrt{17}}{4} = \frac{1}{4} \pm \frac{\sqrt{17}}{4} \end{aligned}$$

13.
$$\begin{aligned} x^2 + x + 2 &= 0 \\ a = 1, b &= 1, c = 2 \\ x &= \frac{-b \pm \sqrt{b^2 - 4ac}}{2a} \\ &= \frac{-1 \pm \sqrt{1^2 - 4(1)(2)}}{2(1)} \\ &= \frac{-1 \pm \sqrt{1 - 8}}{2} \\ &= \frac{-1 \pm \sqrt{-7}}{2} = -\frac{1}{2} \pm \frac{\sqrt{7}}{2}i \end{aligned}$$

14. Let w represent the original width.
Then $w + 2$ represents the original length.
The new dimensions are then $2w$ and
$2(w + 2) = 2w + 4$.

Old Area + 72 = New Area

$$\begin{aligned} w(w + 2) + 72 &= 2w(2w + 4) \\ w^2 + 2w + 72 &= 4w^2 + 8w \\ 0 &= 3w^2 + 6w - 72 \\ 0 &= 3(w + 6)(w - 4) \end{aligned}$$

$w = -6$ or $w = 4$

Since the width is positive, the dimensions are 4 cm by 6 cm.

© 2015 Cengage Learning. All Rights Reserved. May not be scanned, copied or duplicated, or posted to a publicly accessible website, in whole or in part.

15. Let w represent the original width.
Then $w+1$ represents the original length.
The new dimensions are then $2w$ and $3(w+1)=3w+3$.

Old Area + 30 = New Area

$$w(w+1)+30=2w(3w+3)$$
$$w^2+w+30=6w^2+6w$$
$$0=5w^2+5w-30$$
$$0=5(w+3)(w-2)$$

$w=-3$ or $w=2$

Since the width is positive, the dimensions are 2 ft by 3 ft.

16. When the rocket hits the ground, $h=0$:

$$h=112t-16t^2$$
$$0=112t-16t^2$$
$$0=16t(7-t)$$
$$t=0 \text{ or } t=7$$

It hits the ground after 7 seconds.

17. The maximum height occurs at the vertex:

$$h=112t-16t^2$$
$$h=-16t^2+112t$$
$$h=-16(t^2-7t)$$
$$h=-16\left(t^2-7t+\frac{49}{4}\right)+196$$
$$h=-16\left(t^2-7t+\frac{49}{4}\right)+196$$
$$h=-16\left(t-\frac{7}{2}\right)^2+196$$

Vertex: $\left(\frac{7}{2},196\right) \Rightarrow$ max. height = 196 ft

18. $2x^2+5x-4=0$

$a=2, b=5, c=-4$

$$b^2-4ac=5^2-4(2)(-4)$$
$$=25+32=57$$

two irrational unequal solutions

19. $4x^2-5x+7=0$

$a=4, b=-5, c=7$

$$b^2-4ac=(-5)^2-4(4)(7)$$
$$=25-112=-87$$

two complex conjugate solutions

20. $(k-8)x^2+(k+16)x=-49$

$(k-8)x^2+(k+16)x+49=0$

$a=k-8, b=k+16, c=49$

Set the discriminant equal to 0:

$$b^2-4ac=0$$
$$(k+16)^2-4(k-8)(49)=0$$
$$k^2+32k+256-196k+1568=0$$
$$k^2-164k+1824=0$$
$$(k-12)(k-152)=0$$

$k-12=0$ **or** $k-152=0$

$k=12$ $\quad$ $k=152$

21. $3x^2+4x=k+1$

$3x^2+4x-k-1=0$

$a=3, b=4, c=-k-1$

Set the discriminant ≥ 0:

$$b^2-4ac\geq 0$$
$$4^2-4(3)(-k-1)\geq 0$$
$$16+12k+12\geq 0$$
$$12k\geq -28$$
$$k\geq -\frac{28}{12}$$
$$k\geq -\frac{7}{3}$$

© 2015 Cengage Learning. All Rights Reserved. May not be scanned, copied or duplicated, or posted to a publicly accessible website, in whole or in part.

22.
$$x - 8x^{1/2} + 7 = 0$$
$$\left(x^{1/2} - 7\right)\left(x^{1/2} - 1\right) = 0$$
$$x^{1/2} - 7 = 0 \quad \textbf{or} \quad x^{1/2} - 1 = 0$$
$$x^{1/2} = 7 \qquad x^{1/2} = 1$$
$$\left(x^{1/2}\right)^2 = (7)^2 \qquad \left(x^{1/2}\right)^2 = 1^2$$
$$x = 49 \qquad x = 1$$
Solution. Solution.

23.
$$a^{2/3} + a^{1/3} - 6 = 0$$
$$\left(a^{1/3} - 2\right)\left(a^{1/3} + 3\right) = 0$$
$$a^{1/3} - 2 = 0 \quad \textbf{or} \quad a^{1/3} + 3 = 0$$
$$a^{1/3} = 2 \qquad a^{1/3} = -3$$
$$\left(a^{1/3}\right)^3 = (2)^3 \qquad \left(a^{1/3}\right)^3 = (-3)^3$$
$$a = 8 \qquad a = -27$$
Solution. Solution.

24.
$$\frac{1}{x+1} - \frac{1}{x} = -\frac{1}{x+1}$$
$$\left(\frac{1}{x+1} - \frac{1}{x}\right)(x)(x+1) = -\frac{1}{x+1}(x)(x+1)$$
$$1(x) - 1(x+1) = -x$$
$$x - x - 1 = -x$$
$$-1 = -x$$
$$1 = x$$

25.
$$\frac{6}{x+2} + \frac{6}{x+1} = 5$$
$$\left(\frac{6}{x+2} + \frac{6}{x+1}\right)(x+2)(x+1) = 5(x+2)(x+1)$$
$$6(x+1) + 6(x+2) = 5(x^2 + 3x + 2)$$
$$6x + 6 + 6x + 12 = 5x^2 + 15x + 10$$
$$0 = 5x^2 + 3x - 8$$
$$0 = (5x+8)(x-1) \qquad 5x + 8 = 0 \quad \textbf{or} \quad x - 1 = 0$$
$$x = -\tfrac{8}{5} \qquad x = 1$$

26. $3x^2 - 14x + 3 = 0$

$\text{sum} = -\frac{b}{a} = -\frac{-14}{3} = \frac{14}{3}$

27. $3x^2 - 14x + 3 = 0$

$\text{product} = \frac{c}{a} = \frac{3}{3} = 1$

28. $y = 2x^2 - 3$

$y = 2(x - 0)^2 - 3$

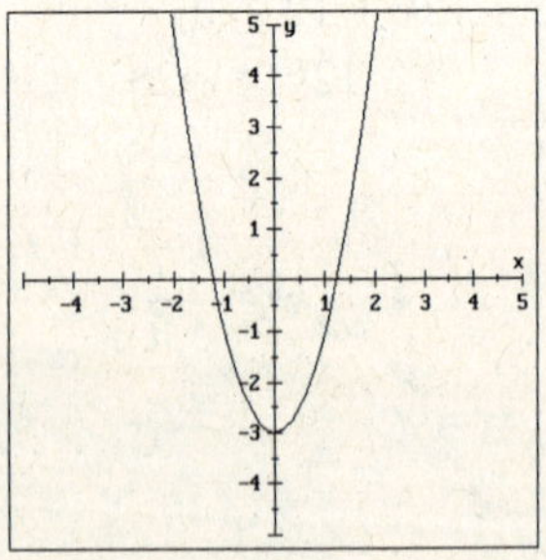

vertex: (0, −3)

29. $y = -2x^2 - 1$

$y = -2(x - 0)^2 - 1$

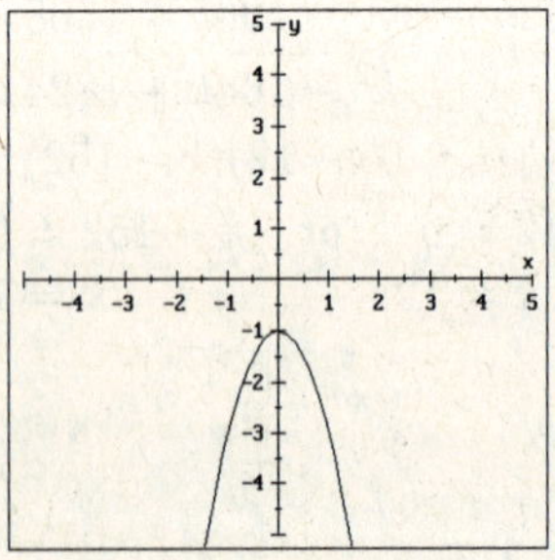

vertex: (0, −1)

© 2015 Cengage Learning. All Rights Reserved. May not be scanned, copied or duplicated, or posted to a publicly accessible website, in whole or in part.

CHAPTER 8 REVIEW

30. $y = -4(x-2)^2 + 1$

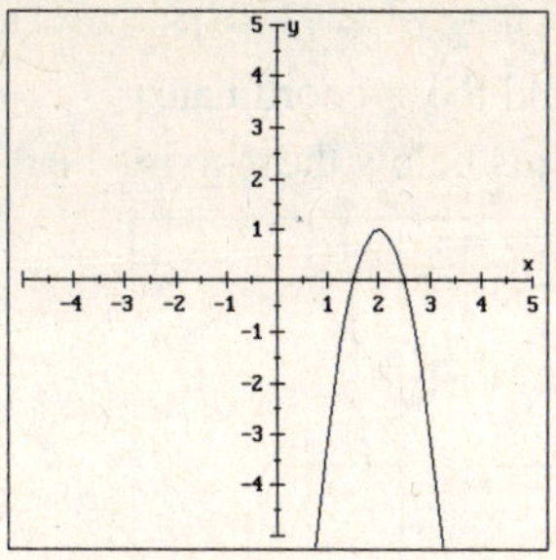

vertex: $(2, 1)$

31. $y = 5x^2 + 10x - 1$
$= 5(x^2 + 2x) - 1$
$= 5(x^2 + 2x + 1) - 1 - 5$
$= 5(x+1)^2 - 6$

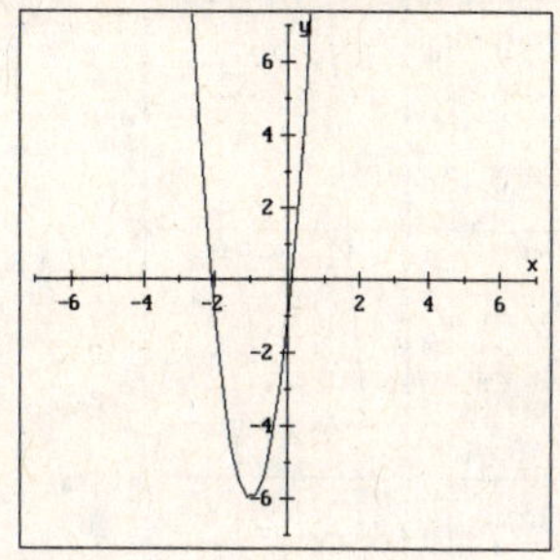

vertex: $(-1, -6)$

32. $y = 4x^2 - 16x - 3 = 4(x^2 - 4x) - 3 = 4(x^2 - 4x + 4) - 3 - 16 = 3(x-2)^2 - 19$
vertex: $(2, -19)$; axis: $x = 2$

33. $x^2 + 2x - 35 > 0$
$(x+7)(x-5) > 0$

$x - 5$ − − − − − − − − 0 + + + +
$x + 7$ − − − 0 + + + + + + + + + +
⟵) —— (⟶ at -7 and 5

solution set: $(-\infty, -7) \cup (5, \infty)$

34. $x^2 + 7x - 18 < 0$
$(x-2)(x+9) < 0$

$x - 2$ − − − − − − − − 0 + + + +
$x + 9$ − − − 0 + + + + + + + + + +
⟵ (——) ⟶ at -9 and 2

solution set: $(-9, 2)$

35. $\frac{3}{x} \le 5$
$\frac{3}{x} - 5 \le 0$
$\frac{3}{x} - \frac{5x}{x} \le 0$
$\frac{3-5x}{x} \le 0$

$3 - 5x$ + + + + + + + + + 0 − − −
x − − − 0 + + + + + + + + + +
⟵) —— [⟶ at 0 and $\frac{3}{5}$

solution set: $(-\infty, 0) \cup \left[\frac{3}{5}, \infty\right)$

36. $\frac{2x^2 + 5x - 7}{x-4} > 0$
$\frac{(2x+7)(x-1)}{x-4} > 0$

$x - 4$ − − − − − − − − − − − − 0 + + + +
$x - 1$ − − − − − − − − 0 + + + + + + + + +
$2x + 7$ − − − 0 + + + + + + + + + + + + + + + +
⟵ (——) — (⟶ at $-\frac{7}{2}$, 1 and 4

solution set: $\left(-\frac{7}{2}, 1\right) \cup (4, \infty)$

© 2015 Cengage Learning. All Rights Reserved. May not be scanned, copied or duplicated, or posted to a publicly accessible website, in whole or in part.

37. $x^2 + 2x - 35 > 0$
Graph $y = x^2 + 2x - 35$
and find the x-coordinates
of points above the x-axis.

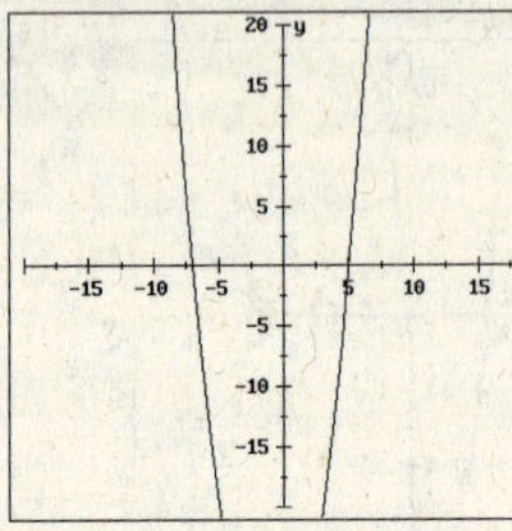

$(-\infty, -7) \cup (5, \infty)$

38. $x^2 + 7x - 18 < 0$
Graph $y = x^2 + 7x - 18$
and find the x-coordinates
of points below the x-axis.

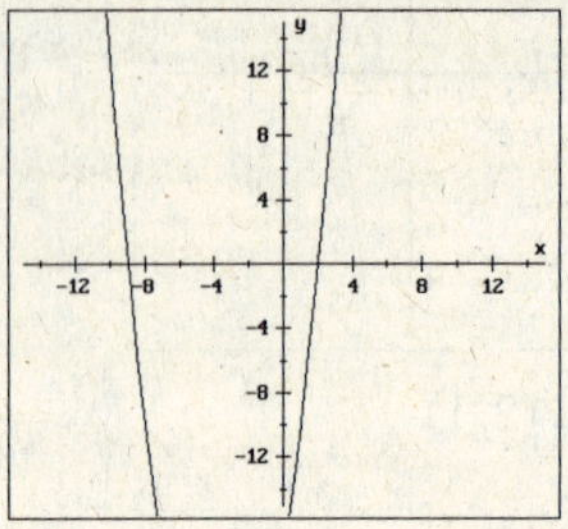

$(-9, 2)$

39. $\frac{3}{x} \le 5 \Rightarrow \frac{3}{x} - 5 \le 0$
Graph $y = (3/x) - 5$
and find the x-coordinates
of points below or on the x-axis.

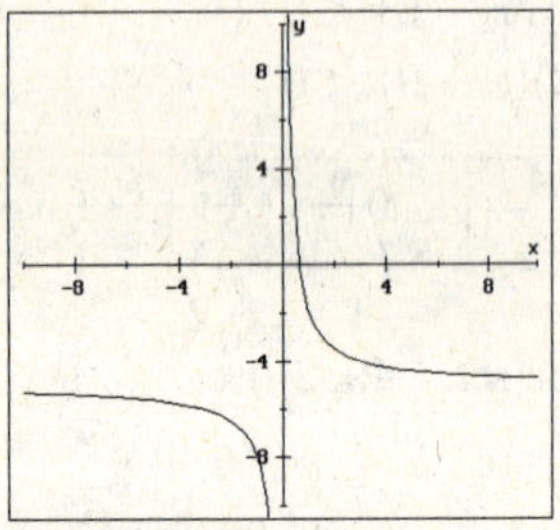

$(-\infty, 0) \cup \left[\frac{3}{5}, \infty\right)$

40. $\frac{2x^2-x-28}{x-1} > 0$
Graph $y = \left(2x^2 - x - 28\right)/(x-1)$
and find the x-coordinates
of points above the x-axis.

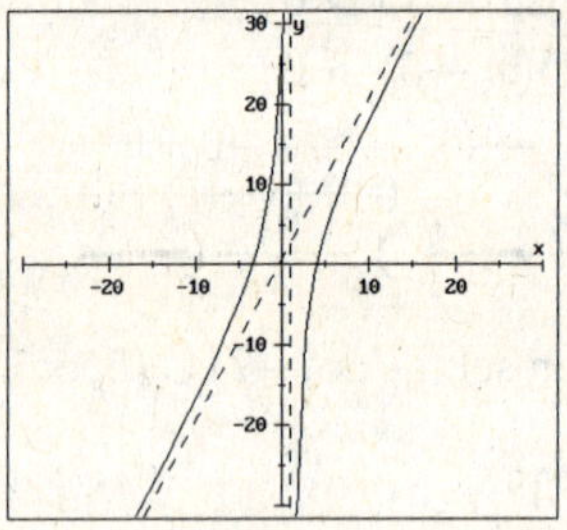

$\left(-\frac{7}{2}, 1\right) \cup (4, \infty)$

41. $y < \dfrac{1}{2}x^2 - 1$

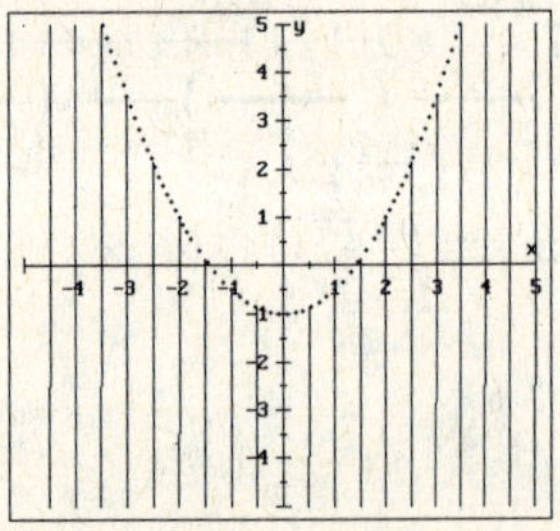

42. $y \ge -|x|$

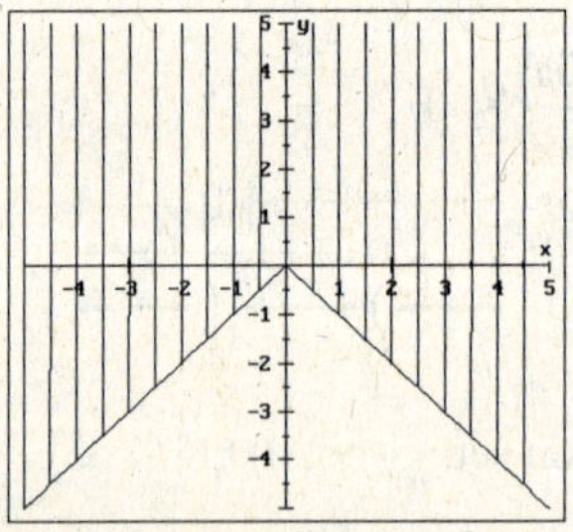

43. $f + g = f(x) + g(x) = 2x + x + 1 = 3x + 1$; domain $= (-\infty, \infty)$

44. $f - g = f(x) - g(x) = 2x - (x+1) = x - 1$; domain $= (-\infty, \infty)$

© 2015 Cengage Learning. All Rights Reserved. May not be scanned, copied or duplicated, or posted to a publicly accessible website, in whole or in part.

45. $f \cdot g = f(x)g(x) = 2x(x+1) = 2x^2 + 2x$; domain $= (-\infty, \infty)$

46. $f/g = \dfrac{f(x)}{g(x)} = \dfrac{2x}{x+1}$ $(x \neq -1)$; domain $= (-\infty, -1) \cup (-1, \infty)$

47. $(f \circ g)(3) = f(g(3)) = f(3+1) = f(4) = 2(4) = 8$

48. $(g \circ f)(-2) = g(f(-2)) = g(2(-2)) = g(-4) = -4 + 1 = -3$

49. $(f \circ g)(x) = f(g(x)) = f(x+1) = 2(x+1)$

50. $(g \circ f)(x) = g(f(x)) = g(2x) = 2x + 1$

51. $f(x) = 2(x-3)$

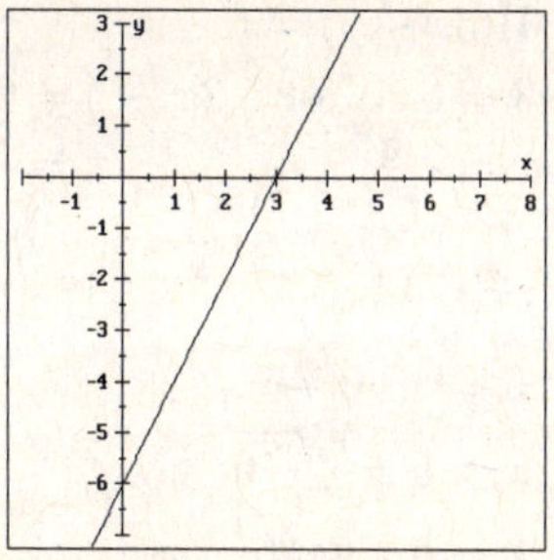

one-to-one

52. $f(x) = x(2x-3)$

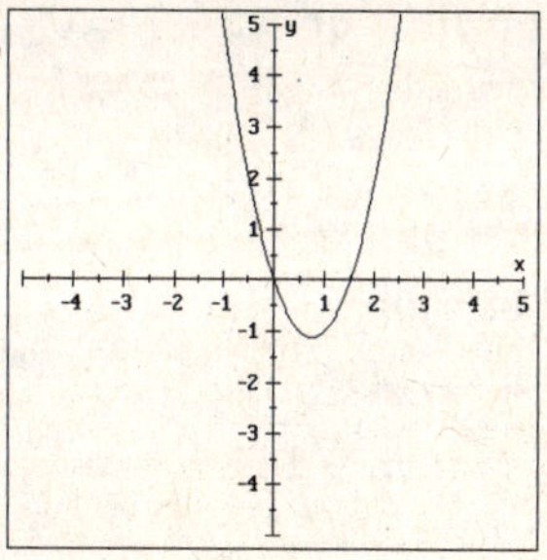

not one-to-one

53. $f(x) = -3(x-2)^2 + 5$

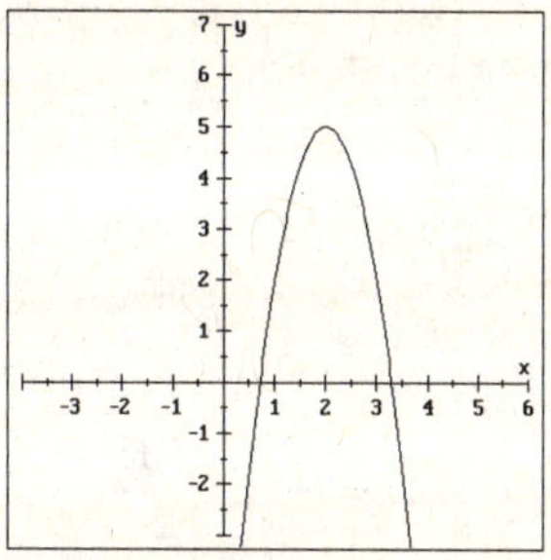

not one-to-one

54. $f(x) = |x|$

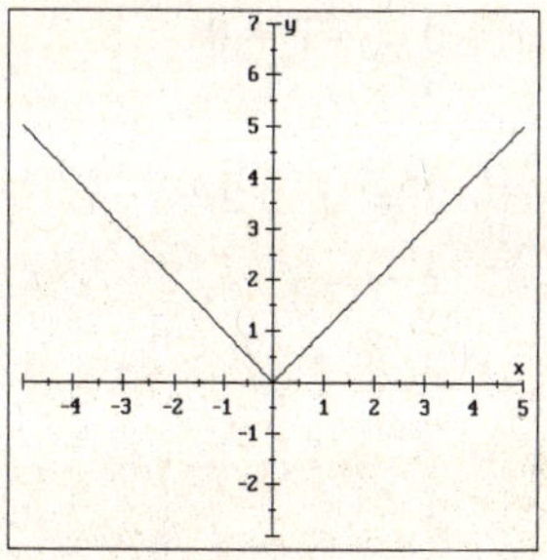

not one-to-one

55.
$$\begin{aligned} y &= 7x - 2 \\ x &= 7y - 2 \\ x + 2 &= 7y \\ \tfrac{x+2}{7} &= y, \text{ or } y = f^{-1}(x) = \tfrac{x+2}{7} \end{aligned}$$

56.
$$\begin{aligned} y &= 4x + 5 \\ x &= 4y + 5 \\ x - 5 &= 4y \\ \tfrac{x-5}{4} &= y, \text{ or } y = f^{-1}(x) = \tfrac{x-5}{4} \end{aligned}$$

© 2015 Cengage Learning. All Rights Reserved. May not be scanned, copied or duplicated, or posted to a publicly accessible website, in whole or in part.

57.
$$\begin{aligned} y &= 2x^2 - 1 \\ x &= 2y^2 - 1 \\ x + 1 &= 2y^2 \\ \tfrac{x+1}{2} &= y^2 \\ \sqrt{\tfrac{x+1}{2}} &= y, \text{ or } y = f^{-1}(x) = \sqrt{\tfrac{x+1}{2}} \end{aligned}$$

Chapter 8 Test (page 615)

1.
$$\begin{aligned} x^2 - 3x - 28 &= 0 \\ (x+4)(x-7) &= 0 \\ x + 4 = 0 \quad &\textbf{or} \quad x - 7 = 0 \\ x = -4 \quad & \quad x = 7 \end{aligned}$$

2.
$$\begin{aligned} x(6x+19) &= -15 \\ 6x^2 + 19x + 15 &= 0 \\ (2x+3)(3x+5) &= 0 \\ 2x + 3 = 0 \quad &\textbf{or} \quad 3x + 5 = 0 \\ x = -\tfrac{3}{2} \quad & \quad x = -\tfrac{5}{3} \end{aligned}$$

3.
$$\begin{aligned} x^2 - 144 &= 0 \\ x^2 &= 144 \\ x &= \pm\sqrt{144} \\ x &= \pm 12 \end{aligned}$$

4.
$$\begin{aligned} (x-5)^2 &= -6 \\ x - 5 &= \pm\sqrt{-6} \\ x &= 5 \pm \sqrt{-6} \\ x &= 5 \pm i\sqrt{6} \end{aligned}$$

5.
$$\begin{aligned} x^2 + 6x + 7 &= 0 \\ x^2 + 6x &= -7 \\ x^2 + 6x + 9 &= -7 + 9 \\ (x+3)^2 &= 2 \\ x + 3 &= \pm\sqrt{2} \\ x &= -3 \pm \sqrt{2} \end{aligned}$$

6.
$$\begin{aligned} x^2 - 5x - 3 &= 0 \\ x^2 - 5x &= 3 \\ x^2 - 5x + \frac{25}{4} &= 3 + \frac{25}{4} \\ \left(x - \frac{5}{2}\right)^2 &= \frac{37}{4} \\ x - \frac{5}{2} &= \pm\sqrt{\frac{37}{4}} \\ x &= \frac{5}{2} \pm \frac{\sqrt{37}}{2} \end{aligned}$$

7.
$$\begin{aligned} 2x^2 + 5x + 1 &= 0 \\ a = 2, b &= 5, c = 1 \\ x &= \frac{-b \pm \sqrt{b^2 - 4ac}}{2a} \\ &= \frac{-5 \pm \sqrt{5^2 - 4(2)(1)}}{2(2)} \\ &= \frac{-5 \pm \sqrt{25 - 8}}{4} \\ &= \frac{-5 \pm \sqrt{17}}{4} = -\frac{5}{4} \pm \frac{\sqrt{17}}{4} \end{aligned}$$

8.
$$\begin{aligned} x^2 - 2x + 6 &= 0 \\ a = 1, b &= -2, c = 6 \\ x &= \frac{-b \pm \sqrt{b^2 - 4ac}}{2a} \\ &= \frac{-(-2) \pm \sqrt{(-2)^2 - 4(1)(6)}}{2(1)} \\ &= \frac{2 \pm \sqrt{4 - 24}}{2} \\ &= \frac{2 \pm \sqrt{-20}}{2} = \frac{2 \pm 2i\sqrt{5}}{2} = 1 \pm i\sqrt{5} \end{aligned}$$

© 2015 Cengage Learning. All Rights Reserved. May not be scanned, copied or duplicated, or posted to a publicly accessible website, in whole or in part.

9. $4x^2 + 3x + 10 = 0$

$a = 4, b = 3, c = 10$

$$\begin{aligned} b^2 - 4ac &= 3^2 - 4(4)(10) \\ &= 9 - 160 = -151 \end{aligned}$$

nonreal solutions

10. $4x^2 - 2kx + k - 1 = 0$

$a = 4, b = -2k, c = k - 1$

Set the discriminant equal to 0:

$$\begin{aligned} b^2 - 4ac &= 0 \\ (-2k)^2 - 4(4)(k-1) &= 0 \\ 4k^2 - 16k + 16 &= 0 \\ 4(k-2)(k-2) &= 0 \end{aligned}$$

$k - 2 = 0$ **or** $k - 2 = 0$

$k = 2 \qquad\qquad k = 2$

11. Let $x =$ the length of the shorter leg.

Then $x + 14 =$ the other length.

$$\begin{aligned} x^2 + (x+14)^2 &= 26^2 \\ x^2 + x^2 + 28x + 196 &= 676 \\ 2x^2 + 28x - 480 &= 0 \\ 2(x+24)(x-10) &= 0 \end{aligned}$$

$x = -24$ or $x = 10$

The shorter leg is 10 inches long.

12.

$$\begin{aligned} 2y - 3y^{1/2} + 1 &= 0 \\ \left(2y^{1/2} - 1\right)\left(y^{1/2} - 1\right) &= 0 \end{aligned}$$

$2y^{1/2} - 1 = 0$ **or** $y^{1/2} - 1 = 0$

$2y^{1/2} = 1 \qquad y^{1/2} = 1$

$\left(y^{1/2}\right)^2 = \left(\frac{1}{2}\right)^2 \qquad \left(y^{1/2}\right)^2 = 1^2$

$y = \frac{1}{4} \qquad y = 1$

Solution $\qquad$ Solution

13. $y = \dfrac{1}{2}x^2 - 4 = \dfrac{1}{2}(x - 0)^2 - 4$

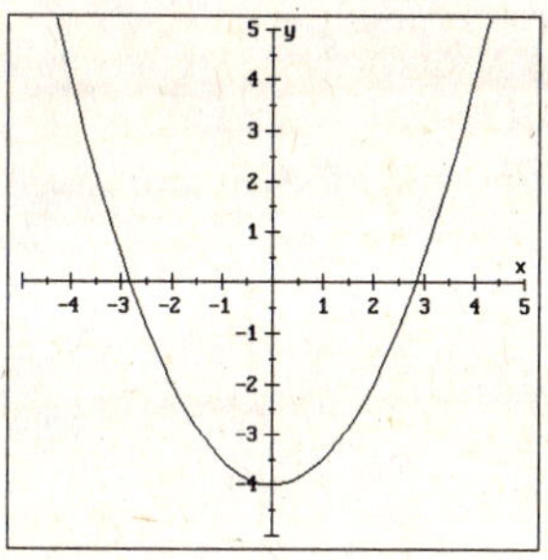

vertex: $(0, -4)$

14.

$$\begin{aligned} y &= -3x^2 + 12x - 5 \\ &= -3\left(x^2 - 4x\right) - 5 \\ &= -3\left(x^2 - 4x + 4\right) - 5 + 12 \\ &= -3(x-2)^2 + 7 \end{aligned}$$

Vertex: (2, 7); axis: $x = 2$

15. $y \le -x^2 + 3$

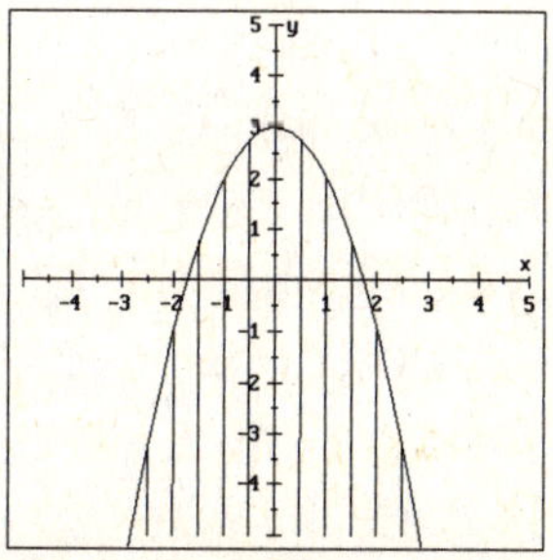

16.

$$\begin{aligned} x^2 - 2x - 8 &> 0 \\ (x+2)(x-4) &> 0 \end{aligned}$$

$x \quad 4$ ---------0 ++++

$x + 2$ --- 0++++++ ++++

⟵) —— (⟶

$-2 \qquad 4$

solution set: $(-\infty, -2) \cup (4, \infty)$

© 2015 Cengage Learning. All Rights Reserved. May not be scanned, copied or duplicated, or posted to a publicly accessible website, in whole or in part.

17. $\frac{x-2}{x+3} \le 0$

$x-2$ − − − − − − − 0 ++++
$x+3$ − − − 0++++++ ++++

⟵ (——] ⟶
−3 2

solution set: $(-3, 2]$

18. $g+f = g(x)+f(x) = x-1+4x = 5x-1$; domain $= (-\infty, \infty)$

19. $f-g = f(x)-g(x) = 4x-(x-1) = 3x+1$; domain $= (-\infty, \infty)$

20. $g \cdot f = g(x)f(x) = (x-1)4x = 4x^2-4x$; domain $= (-\infty, \infty)$

21. $g/f = \frac{g(x)}{f(x)} = \frac{x-1}{4x}$; domain $= (-\infty, 0) \cup (0, \infty)$

22. $(g \circ f)(1) = g(f(1)) = g(4(1))$
$= g(4) = 4-1 = 3$

23. $(f \circ g)(0) = f(g(0)) = f(0-1)$
$= f(-1) = 4(-1) = -4$

24. $(f \circ g)(-1) = f(g(-1)) = f(-1-1) = f(-2) = 4(-2) = -8$

25. $(g \circ f)(-2) = g(f(-2)) = g(4(-2)) = g(-8) = -8-1 = -9$

26. $(f \circ g)(x) = f(g(x)) = f(x-1) = 4(x-1)$

27. $(g \circ f)(x) = g(f(x)) = g(4x) = 4x-1$

28. $3x+2y = 12$
$3y+2x = 12$
$3y = -2x+12$
$y = \frac{-2x+12}{3}$; $f^{-1}(x) = \frac{-2x+12}{3}$

Cumulative Review Exercises (page 616)

1. $y = f(x) = 5x^2-2$
domain $= (-\infty, \infty)$
range $= [-2, \infty)$

2. $y = f(x) = -|x-4|$
domain $= (-\infty, \infty)$
range $= (-\infty, 0]$

3. $y-y_1 = m(x-x_1)$
$y+5 = 4(x+3)$
$y = 4x+7$

© 2015 Cengage Learning. All Rights Reserved. May not be scanned, copied or duplicated, or posted to a publicly accessible website, in whole or in part.

CUMULATIVE REVIEW EXERCISES

4. $2x + 3y = 6$ $\qquad$ $y - y_1 = m(x - x_1)$

$$3y = -2x + 6 \qquad y + 2 = -\frac{2}{3}(x - 0)$$

$$y = -\frac{2}{3}x + 2 \qquad y = -\frac{2}{3}x - 2$$

5. $(2a^2 + 4a - 7) - 2(3a^2 - 4a) = 2a^2 + 4a - 7 - 6a^2 + 8a = -4a^2 + 12a - 7$

6. $(4x - 3)(5x + 2) = 20x^2 + 8x - 15x - 6 = 20x^2 - 7x - 6$

7. $x^4 - 16y^4 = (x^2 + 4y^2)(x^2 - 4y^2) = (x^2 + 4y^2)(x + 2y)(x - 2y)$

8. $15x^2 - 2x - 8 = (5x - 4)(3x + 2)$

9.
$$x^2 - 8x - 9 = 0$$
$$(x - 9)(x + 1) = 0$$
$$x - 9 = 0 \quad \textbf{or} \quad x + 1 = 0$$
$$x = 9 \qquad\qquad x = -1$$

10.
$$6a^3 - 2a = a^2$$
$$6a^3 - a^2 - 2a = 0$$
$$a(6a^2 - a - 2) = 0$$
$$a(3a - 2)(2a + 1) = 0$$
$$a = 0 \quad \textbf{or} \quad 3a - 2 = 0 \quad \textbf{or} \quad 2a + 1 = 0$$
$$a = \tfrac{2}{3} \qquad\qquad a = -\tfrac{1}{2}$$

11. $\sqrt{36a^2b^4} = 6ab^2$

12. $\sqrt{48t^3} = \sqrt{16t^2}\sqrt{3t} = 4t\sqrt{3t}$

13. $\sqrt[3]{-64y^6} = -4y^2$

14. $\sqrt[3]{\dfrac{128x^4}{2x}} = \sqrt[3]{64x^3} = 4x$

15. $8^{-1/3} = \dfrac{1}{8^{1/3}} = \dfrac{1}{2}$

16. $27^{2/3} = \left(27^{1/3}\right)^2 = 3^2 = 9$

17. $\dfrac{y^{2/3}y^{5/3}}{y^{1/3}} = \dfrac{y^{7/3}}{y^{1/3}} = y^{6/3} = y^2$

18. $\dfrac{x^{5/3}x^{1/2}}{x^{3/4}} = \dfrac{x^{13/6}}{x^{3/4}} = x^{17/12}$

19. $f(x) = \sqrt{x - 2}$; Shift $y = \sqrt{x}$ right 2.

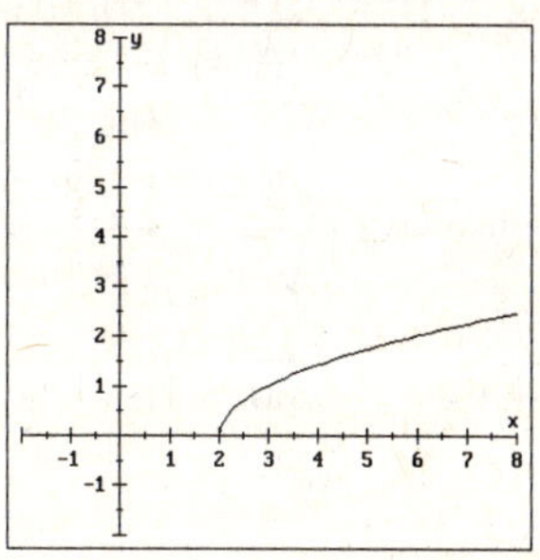

$D = [2, \infty)$, $R = [0, \infty)$

20. $f(x) = -\sqrt{x + 2}$; Reflect $y = \sqrt{x}$ about the x-axis and shift left 2.

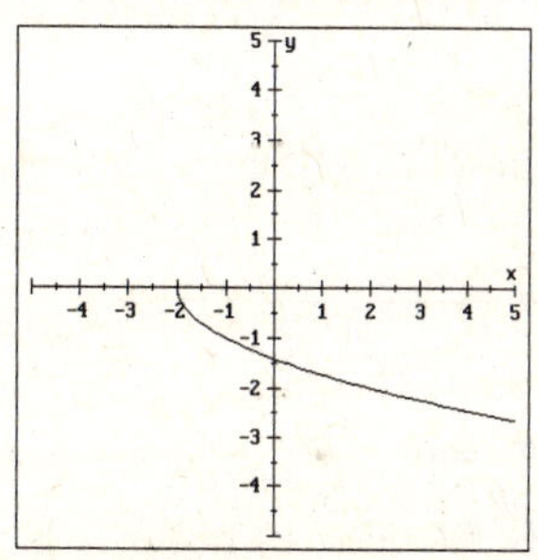

$D = [-2, \infty)$, $R = (-\infty, 0]$

© 2015 Cengage Learning. All Rights Reserved. May not be scanned, copied or duplicated, or posted to a publicly accessible website, in whole or in part.

CUMULATIVE REVIEW EXERCISES

21. $\left(x^{2/3}-x^{1/3}\right)\left(x^{2/3}+x^{1/3}\right)=x^{4/3}+x^{3/3}-x^{3/3}-x^{2/3}=x^{4/3}-x^{2/3}$

22. $\left(x^{-1/2}+x^{1/2}\right)^2=\left(x^{-1/2}+x^{1/2}\right)\left(x^{-1/2}+x^{1/2}\right)=x^{-2/2}+x^0+x^0+x^{2/2}=x+2+\frac{1}{x}$

23. $\sqrt{50}-\sqrt{8}+\sqrt{32}=\sqrt{25}\sqrt{2}-\sqrt{4}\sqrt{2}+\sqrt{16}\sqrt{2}=5\sqrt{2}-2\sqrt{2}+4\sqrt{2}=7\sqrt{2}$

24.
$$\begin{aligned}-3\sqrt[4]{32}-2\sqrt[4]{162}+5\sqrt[4]{48}&=-3\sqrt[4]{16}\sqrt[4]{2}-2\sqrt[4]{81}\sqrt[4]{2}+5\sqrt[4]{16}\sqrt[4]{3}\\&=-3(2)\sqrt[4]{2}-2(3)\sqrt[4]{2}+5(2)\sqrt[4]{3}\\&=-6\sqrt[4]{2}-6\sqrt[4]{2}+10\sqrt[4]{3}=-12\sqrt[4]{2}+10\sqrt[4]{3}\end{aligned}$$

25. $3\sqrt{2}(2\sqrt{3}-4\sqrt{12})=6\sqrt{6}-12\sqrt{24}=6\sqrt{6}-12\sqrt{4}\sqrt{6}=6\sqrt{6}-24\sqrt{6}=-18\sqrt{6}$

26. $\dfrac{5}{\sqrt[3]{x}}=\dfrac{5}{\sqrt[3]{x}}\cdot\dfrac{\sqrt[3]{x^2}}{\sqrt[3]{x^2}}=\dfrac{5\sqrt[3]{x^2}}{\sqrt[3]{x^3}}=\dfrac{5\sqrt[3]{x^2}}{x}$

27. $\dfrac{\sqrt{x}+2}{\sqrt{x}-1}=\dfrac{\sqrt{x}+2}{\sqrt{x}-1}\cdot\dfrac{\sqrt{x}+1}{\sqrt{x}+1}=\dfrac{x+3\sqrt{x}+2}{x-1}$

28. $\sqrt[6]{x^3y^3}=(x^3y^3)^{1/6}=x^{3/6}y^{3/6}=x^{1/2}y^{1/2}=\sqrt{xy}$

29.
$$\begin{aligned}5\sqrt{x+2}&=x+8\\\left(5\sqrt{x+2}\right)^2&=(x+8)^2\\25(x+2)&=x^2+16x+64\\25x+50&=x^2+16x+64\\0&=x^2-9x+14\\0&=(x-7)(x-2)\end{aligned}$$
$x=7$ or $x=2$ (Both check.)

30.
$$\begin{aligned}\sqrt{x}+\sqrt{x+2}&=2\\\sqrt{x}&=2-\sqrt{x+2}\\\left(\sqrt{x}\right)^2&=\left(2-\sqrt{x+2}\right)^2\\x&=4-4\sqrt{x+2}+x+2\\4\sqrt{x+2}&=6\\\left(4\sqrt{x+2}\right)^2&=6^2\\16(x+2)&=36\\16x+32&=36\\16x&=4\\x&=\tfrac{4}{16}=\tfrac{1}{4}\end{aligned}$$

31. hypotenuse $=3\sqrt{2}$ in.

32. hypotenuse $=2\cdot\dfrac{3}{\sqrt{3}}=\dfrac{6\sqrt{3}}{3}=2\sqrt{3}$ in.

33. $d=\sqrt{(-2-4)^2+(6-14)^2}=\sqrt{(-6)^2+(-8)^2}=\sqrt{36+64}=\sqrt{100}=10$

34.
$$\begin{aligned}x^2+25&=0\\x^2&=-25\\x&=\pm\sqrt{-25}=\pm 5i\end{aligned}$$

© 2015 Cengage Learning. All Rights Reserved. May not be scanned, copied or duplicated, or posted to a publicly accessible website, in whole or in part.

CUMULATIVE REVIEW EXERCISES

35.
$$\begin{aligned} 2x^2 + x - 3 &= 0 \\ x^2 + \frac{1}{2}x - \frac{3}{2} &= 0 \\ x^2 + \frac{1}{2}x &= \frac{3}{2} \\ x^2 + \frac{1}{2}x + \frac{1}{16} &= \frac{3}{2} + \frac{1}{16} \\ \left(x + \frac{1}{4}\right)^2 &= \frac{25}{16} \\ x + \frac{1}{4} &= \pm\frac{5}{4} \\ x &= -\frac{1}{4} \pm \frac{5}{4} \end{aligned}$$
$$x = \frac{4}{4} = 1 \quad \text{or} \quad x = -\frac{6}{4} = -\frac{3}{2}$$

36.
$$\begin{aligned} &3x^2 + 4x - 1 = 0 \\ &a = 3, b = 4, c = -1 \\ x &= \frac{-b \pm \sqrt{b^2 - 4ac}}{2a} \\ &= \frac{-4 \pm \sqrt{4^2 - 4(3)(-1)}}{2(3)} \\ &= \frac{-4 \pm \sqrt{16+12}}{6} \\ &= \frac{-4 \pm \sqrt{28}}{6} \\ &= \frac{-4 \pm 2\sqrt{7}}{6} = -\frac{2}{3} \pm \frac{\sqrt{7}}{3} \end{aligned}$$

37. $y = \frac{1}{2}x^2 + 5 = \frac{1}{2}(x-0)^2 + 5$

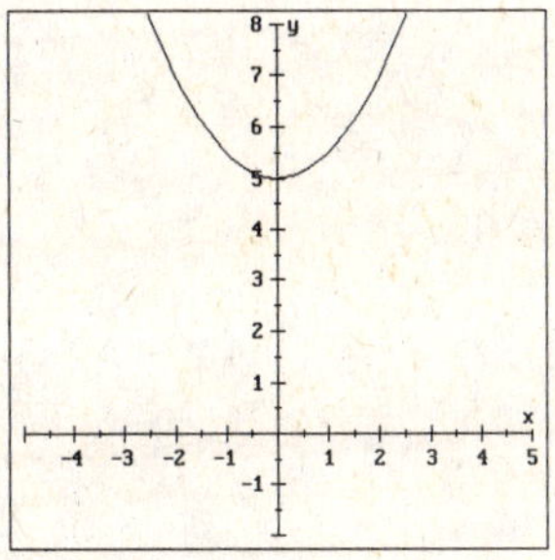

vertex: $(0, 5)$; axis: $x = 0$

38. $y \le -x^2 + 3$

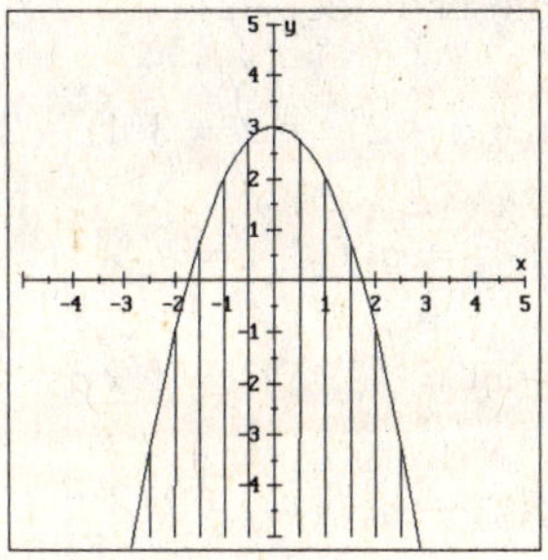

vertex: $(0, 3)$

39. $(5 + 7i) + (8 - 10i) = 5 + 7i + 8 - 10i = 13 - 3i$

40. $(7 - 4i) - (12 + 3i) = 7 - 4i - 12 - 3i = -5 - 7i$

41. $(6 + 5i)(6 - 5i) = 36 - 30i + 30i - 25i^2 = 36 + 25 = 61$

42. $(3 + i)(3 - 3i) = 9 - 9i + 3i - 3i^2 = 9 - 6i + 3 = 12 - 6i$

43.
$$\begin{aligned} (3 - 2i) - (4 + i)^2 = 3 - 2i - (16 + 8i + i^2) = 3 - 2i \quad (15 + 8i) &= 3 - 2i - 15 - 8i \\ &= -12 - 10i \end{aligned}$$

44. $\frac{5}{3-i} = \frac{5}{3-i} \cdot \frac{3+i}{3+i} = \frac{5(3+i)}{9-i^2} = \frac{5(3+i)}{10} = \frac{3+i}{2} = \frac{3}{2} + \frac{1}{2}i$

45. $|3 + 2i| = \sqrt{3^2 + 2^2} = \sqrt{9+4} = \sqrt{13}$

46. $|5 - 6i| = \sqrt{5^2 + (-6)^2} = \sqrt{25+36} = \sqrt{61}$

© 2015 Cengage Learning. All Rights Reserved. May not be scanned, copied or duplicated, or posted to a publicly accessible website, in whole or in part.

47.
$$2x^2 + 4x = k$$
$$2x^2 + 4x - k = 0$$
$$a = 2, b = 4, c = -k$$
Set the discriminant equal to 0:
$$b^2 - 4ac = 0$$
$$4^2 - 4(2)(-k) = 0$$
$$16 + 8k = 0$$
$$8k = -16$$
$$k = -2$$

48.
$$a - 7a^{1/2} + 12 = 0$$
$$\left(a^{1/2} - 3\right)\left(a^{1/2} - 4\right) = 0$$
$$a^{1/2} - 3 = 0 \quad \textbf{or} \quad a^{1/2} - 4 = 0$$
$$a^{1/2} = 3 \qquad a^{1/2} = 4$$
$$\left(a^{1/2}\right)^2 = 3^2 \qquad \left(a^{1/2}\right)^2 = 4^2$$
$$a = 9 \qquad a = 16$$
Solution   Solution

49.
$$x^2 - x - 6 > 0$$
$$(x + 2)(x - 3) > 0$$
$x - 3$ −−−−−−−−0 ++++
$x + 2$ −−− 0++++++ ++++
⟵) ——— (⟶
−2   3
solution set: $(-\infty, -2) \cup (3, \infty)$

50.
$$\frac{x - 3}{x + 2} \le 0$$
$x - 3$ −−−−−−−0 ++++
$x + 2$ −−−0++++++ ++++
⟵ (———] ⟶
−2   3
solution set: $(-2, 3]$

51. $f(-2) = 3(-2)^2 + 2 = 3(4) + 2 = 14$

52. $(g \circ f)(4) = g(f(4)) = g(3(4)^2 + 2) = g(50) = 2(50) - 1 = 99$

53. $(f \circ g)(x) = f(g(x)) = f(2x - 1) = 3(2x - 1)^2 + 2 = 3(4x^2 - 4x + 1) + 2 = 12x^2 - 12x + 5$

54. $(g \circ f)(x) = g(f(x)) = g(3x^2 + 2) = 2(3x^2 + 2) - 1 = 6x^2 + 3$

55.
$$y = 3x + 2$$
$$x = 3y + 2$$
$$x - 2 = 3y$$
$$\frac{x - 2}{3} = y, \text{ or } y = f^{-1}(x) = \frac{x - 2}{3}$$

56.
$$y = x^3 + 4$$
$$x = y^3 + 4$$
$$x - 4 = y^3$$
$$\sqrt[3]{x - 4} = y, \text{ or } y = f^{-1}(x) = \sqrt[3]{x - 4}$$

Exercises 9.1 (page 629)

1. $3^x = 3^2 = 9$

3. $3(6^x) = 3(6^2) = 3(36) = 108$

5. $3^x = 3^{-2} = \frac{1}{3^2} = \frac{1}{9}$

7. $3(6^x) = 3(6^{-2}) = 3\left(\frac{1}{6^2}\right) = 3\left(\frac{1}{36}\right) = \frac{1}{12}$

9.
$$3x + 2x - 20 = 180$$
$$5x = 200$$
$$x = 40$$

11. $m(\angle 2) = 3x = 3(40) = 120°$

13. exponential

15. $(0, \infty)$

17. increasing

© 2015 Cengage Learning. All Rights Reserved. May not be scanned, copied or duplicated, or posted to a publicly accessible website, in whole or in part.

SECTION 9.1

19. $P\left(1+\frac{r}{k}\right)^{kt}$

21. $(1, b)$

23. left

25. $y = f(x) = 3^x$
through $(0, 1)$ and $(1, 3)$

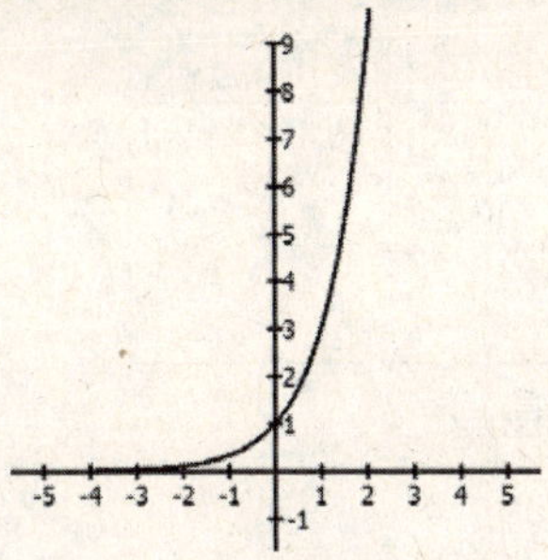

27. $y = f(x) = \left(\frac{5}{2}\right)^x$
through $(0, 1)$ and $\left(1, \frac{5}{2}\right)$

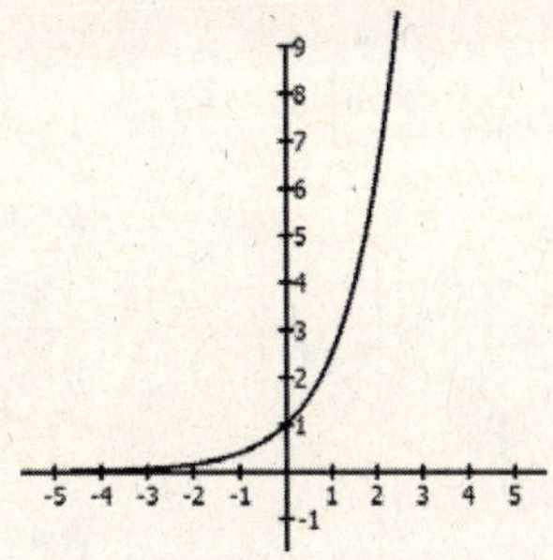

29. $y = f(x) = \left(\frac{1}{3}\right)^x$
through $(0, 1)$ and $\left(1, \frac{1}{3}\right)$

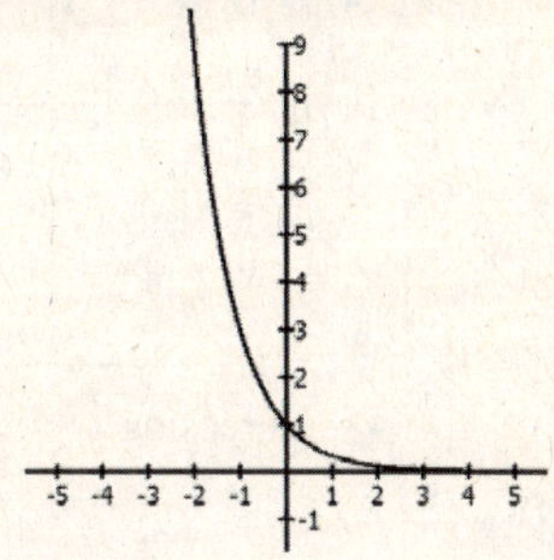

31. $y = f(x) = \left(\frac{2}{5}\right)^x$
through $(0, 1)$ and $\left(1, \frac{2}{5}\right)$

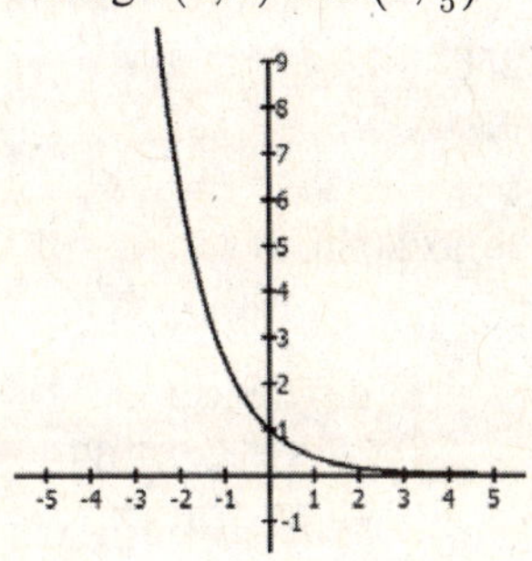

33.
$$y = b^x$$
$$\frac{1}{2} = b^1$$
$$\frac{1}{2} = b$$

35.
$$y = b^x$$
$$3 = b^1$$
$$3 = b$$

37. $f(x) = 3^x - 2$
Shift $y = 3^x$ down 2.

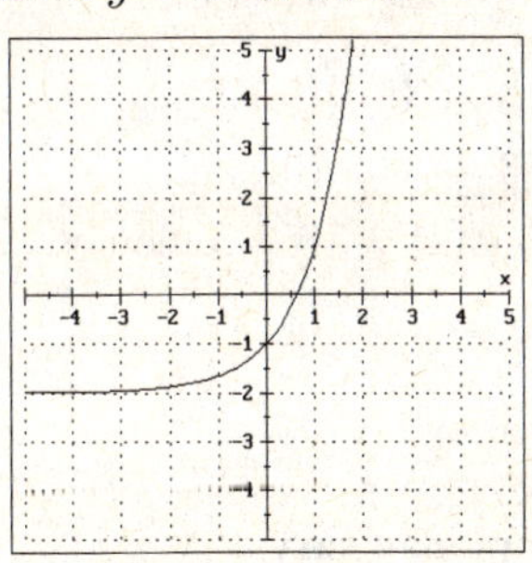

39. $f(x) = 3^{(x-1)}$
Shift $y = 3^x$ right 1.

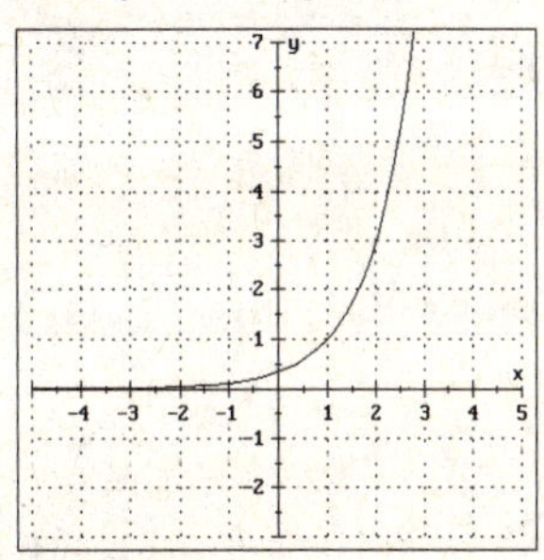

41.
$$y = b^x$$
$$2 = b^1$$
$$2 = b$$

© 2015 Cengage Learning. All Rights Reserved. May not be scanned, copied or duplicated, or posted to a publicly accessible website, in whole or in part.

43. $y = f(x) = \frac{1}{2}(3^{x/2})$

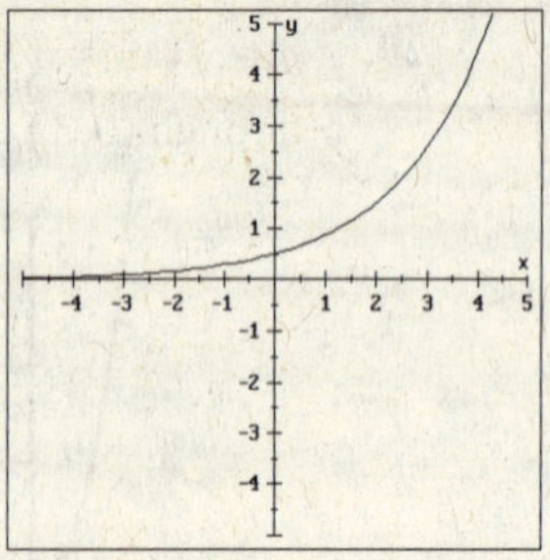

increasing

45. $y = f(x) = 2(3^{-x/2})$

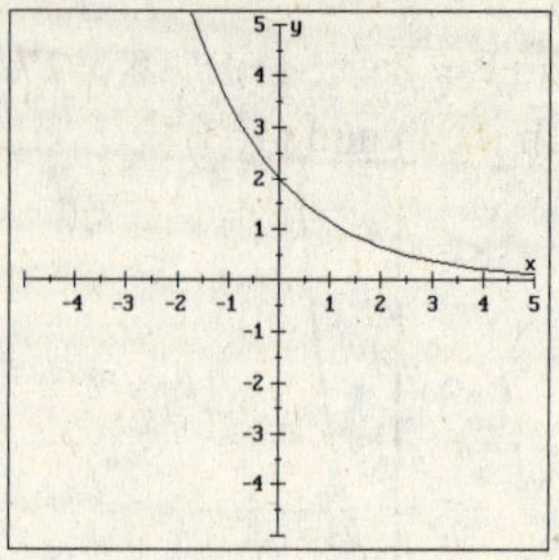

decreasing

47. $S(n) = 1.6(1.173)^n = 1.6(1.173)^1$
$= 1.8768$

There were about 1.9 billion users in 2005.

49. $A = A_0\left(\frac{2}{3}\right)^t$

$A = A_0\left(\frac{2}{3}\right)^8$

$A = \frac{256}{6561}A_0$

51. $A = P(1 + \frac{r}{k})^{kt} = 10{,}000(1 + \frac{0.08}{4})^{4(10)} = 10{,}000(1.02)^{40} \approx \$22{,}080.40$

53. $A = P\left(1 + \frac{r}{k}\right)^{kt}$
$= 100000\left(1 + \frac{0.04}{4}\right)^{4(10)}$
$= 100000(1.01)^{40}$
$\approx \$148{,}886.37$

$A = P\left(1 + \frac{r}{k}\right)^{kt}$
$= 100000\left(1 + \frac{0.045}{4}\right)^{4(10)}$
$= 100000(1.01125)^{40}$
$\approx \$156{,}437.69$

difference:
$= \$156{,}437.69 - \$148{,}886.37$
$= \$7{,}551.32$

55. $A = P\left(1 + \frac{r}{k}\right)^{kt}$
$= 1\left(1 + \frac{0.05}{1}\right)^{1(300)}$
$= 1(1.05)^{300}$
$\approx \$2{,}273{,}996.13$

57. $C = (3 \times 10^{-4})(0.7)^t$
$= (3 \times 10^{-4})(0.7)^5$
$\approx 5.0421 \times 10^{-5}$ coulombs

59. $A = P\left(1 + \frac{r}{k}\right)^{kt}$
$= 4700\left(1 + \frac{-0.25}{1}\right)^{1(5)}$
$= 4700(0.75)^5$
$\approx \$1115.33$

61. **Answers may vary.**

63. If the base were 0, then the function would not be defined for $x = 0 \Rightarrow y = 0^0$.

© 2015 Cengage Learning. All Rights Reserved. May not be scanned, copied or duplicated, or posted to a publicly accessible website, in whole or in part.

Exercises 9.2 (page 637)

Problems 1-6 are to be solved using a calculator. The keystrokes needed to solve each problem using a TI-84 graphing calculator appear in each solution. There may be other solutions. Keystrokes for other calculators may be slightly different.

1. $e^0 \Rightarrow$ [2nd] [LN] [0] [ENTER] $\{1\}$

3. $2e^3 \Rightarrow$ [2] [×] [2nd] [LN] [3] [ENTER] $\{40.17\}$

5. $e^{-1.5} \Rightarrow$ [2nd] [LN] [(–)] [1] [.] [5] [ENTER] $\{0.22\}$

7. $\sqrt{320x^7} = \sqrt{64x^6}\sqrt{5x} = 8x^3\sqrt{5x}$

9. $5\sqrt{98y^3} - 4y\sqrt{18y} = 5\sqrt{49y^2}\sqrt{2y} - 4y\sqrt{9}\sqrt{2y} = 5(7y)\sqrt{2y} - 4y(3)\sqrt{2y}$
$= 35y\sqrt{2y} - 12y\sqrt{2y} = 23y\sqrt{2y}$

11. 2.72

13. increasing

15. $A = Pe^{rt}$

17. $y = f(x) = e^x + 1$
Shift $y = e^x$ up 1.

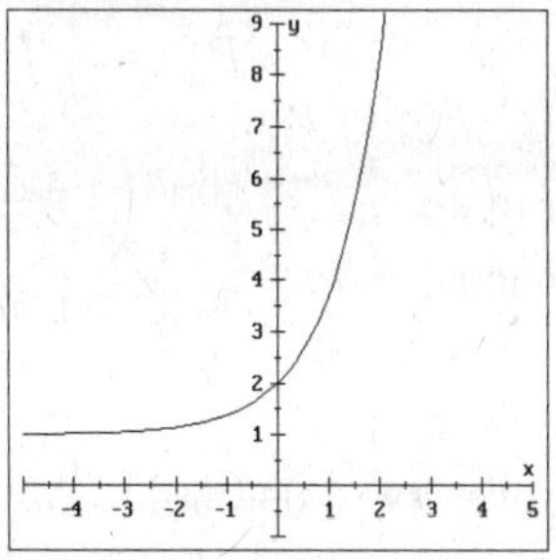

19. $y = f(x) = e^{x+3}$
Shift $y = e^x$ left 3.

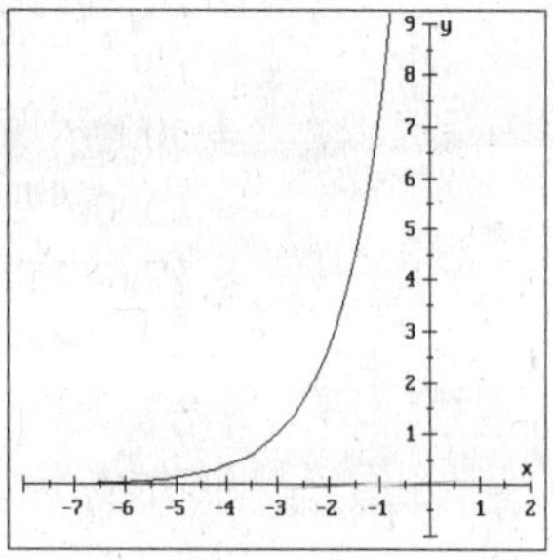

21. The graph should be increasing. The graph could not look like this.

23. The graph should go through the point $(0, 1)$. The graph could not look like this.

25. $y = f(x) = -e^x$; Reflect $y = e^x$ about the x-axis.

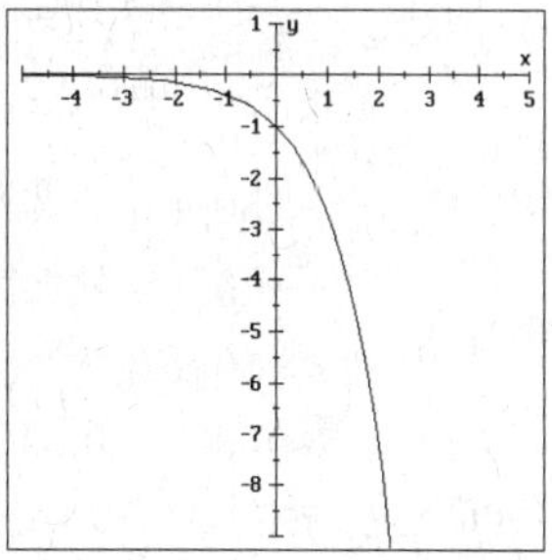

27. $y = f(x) = 2e^x$; Stretch $y = e^x$ vertically by a factor of 2.

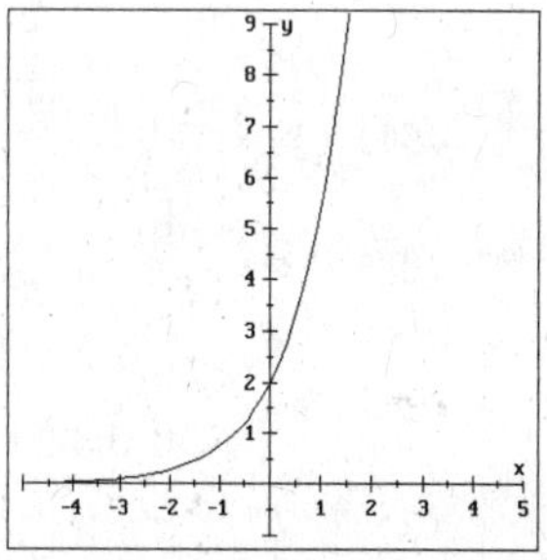

© 2015 Cengage Learning. All Rights Reserved. May not be scanned, copied or duplicated, or posted to a publicly accessible website, in whole or in part.

SECTION 9.2

29.
$$A = Pe^{rt} = 9000e^{0.04(20)} = 9000e^{0.8} \approx \$20{,}029.87$$

31.
$$A = Pe^{rt}$$
$$25000 = Pe^{0.05(15)}$$
$$25000 = Pe^{0.75}$$
$$25000 \approx P(2.117)$$
$$P \approx \frac{25000}{2.117}$$
$$P \approx \$11{,}809.16$$

33.
$$A = Pe^{rt} = 6e^{0.019(30)} = 6e^{0.57} \approx 10.6 \text{ billion people}$$

35.
$$A = Pe^{rt} = 6e^{0.019(50)} = 6e^{0.95} \approx 6(2.6)$$
It will increase by a factor of about 2.6.

37.
$$y = 1000e^{0.02x}$$
$$y = 31x + 2000$$

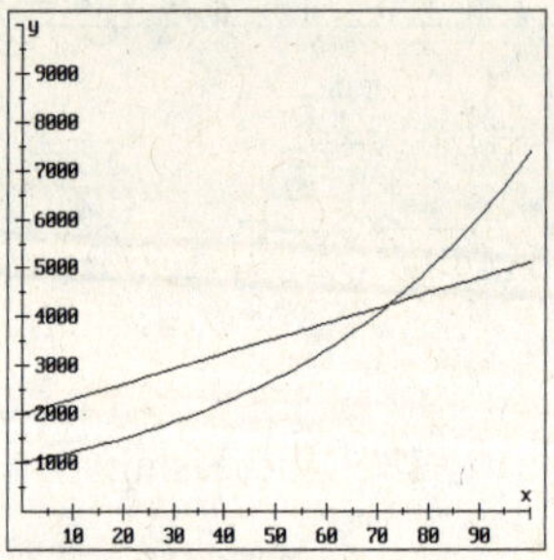

about 72 years

39.
$$A = A_0e^{-0.087t} = 75e^{-0.087(40)} = 75e^{-3.48} \approx 2.31 \text{ grams}$$

41.
$$A = A_0e^{-0.00000693t} = 2500e^{-0.00000693(100)} = 2500e^{-0.000693} \approx 2498.27 \text{ grams}$$

43.
$$A = Pe^{rt} = 5000e^{0.085(5)} = 5000e^{0.425} \approx \$7{,}647.95 \quad \text{(continuous)}$$

$$A = P\left(1 + \frac{r}{k}\right)^{kt} = 5000\left(1 + \frac{0.085}{1}\right)^{1(5)} = 5000(1.085)^5 \approx \$7{,}518.28 \quad \text{(annual)}$$

45.
$$P = 8000e^{-0.008t} = 8000e^{-0.008(50)} = 8000e^{-0.4} \approx 5363$$

47.
$$x = 0.08\left(1 - e^{-0.1t}\right) = 0.08\left(1 - e^{-0.1(30)}\right) = 0.08\left(1 - e^{-3}\right) \approx 0.08(1 - 0.049787) \approx 0.08(0.950213) \approx 0.076$$

49.
$$v = 50\left(1 - e^{-0.2t}\right) = 50\left(1 - e^{-0.2(0)}\right) = 50\left(1 - e^{0}\right) = 50(1 - 1) = 50(0) = 0 \text{ mps}$$

51.
$$v = 50\left(1 - e^{-0.2t}\right) = 50\left(1 - e^{-0.2(2)}\right) = 50\left(1 - e^{-0.4}\right) \approx 50(1 - 0.67032) \approx 50(0.32968) \approx 16.5 \text{ mps}$$

$$v = 50\left(1 - e^{-0.3t}\right) = 50\left(1 - e^{-0.3(2)}\right) = 50\left(1 - e^{-0.6}\right) \approx 50(1 - 0.54881) \approx 50(0.45119) \approx 22.6 \text{ mps} \Rightarrow \text{faster}$$

© 2015 Cengage Learning. All Rights Reserved. May not be scanned, copied or duplicated, or posted to a publicly accessible website, in whole or in part.

53. $A = Pe^{rt}$
$= 4570e^{-0.06(6.5)}$
$\approx \$3094.15$

55. **Answers may vary.**

57. $e \approx 2.7182;$
$1+1+\frac{1}{2}+\frac{1}{2\cdot 3}+\frac{1}{2\cdot 3\cdot 4}+\frac{1}{2\cdot 3\cdot 4\cdot 5}$
≈ 2.7167

59.
$$\begin{aligned} e^{t+10} &= ke^t \\ e^t \cdot e^{10} &= ke^t \\ e^{10}e^t &= ke^t \\ k &= e^{10} \end{aligned}$$

Exercises 9.3 (page 647)

1. $2^{\boxed{3}} = 8$

3. $\boxed{5}^3 = 125$

5. $5^{\boxed{2}} = 25$

7. $2^{\boxed{5}} = 32$

9. $\boxed{2}^{-2} = \frac{1}{4}$

11.
$$\begin{aligned} \sqrt{3x-4} &= \sqrt{-7x+2} \\ 3x-4 &= -7x+2 \\ 10x &= 6 \\ x &= \tfrac{6}{10} = \tfrac{3}{5} \end{aligned}$$
$\frac{3}{5}$ is extraneous $\Rightarrow \emptyset$

13.
$$\begin{aligned} 3-\sqrt{t-3} &= \sqrt{t} \\ \left(3-\sqrt{t-3}\right)^2 &= t \\ 9-6\sqrt{t-3}+t-3 &= t \\ 6 &= 6\sqrt{t-3} \\ 1 &= \sqrt{t-3} \\ 1 &= t-3 \\ 4 &= t \end{aligned}$$

15. $(0, \infty)$

17. x

19. exponent

21. $(b, 1); (1, 0)$

23. $20 \log \frac{E_O}{E_I}$

25. $\log_4 64 = 3 \Rightarrow 4^3 = 64$

27. $\log_{1/2} \frac{1}{8} = 3 \Rightarrow \left(\frac{1}{2}\right)^3 = \frac{1}{8}$

29. $\log_4 \frac{1}{64} = -3 \Rightarrow 4^{-3} = \frac{1}{64}$

31. $\log_{1/2} \frac{1}{8} = 3 \Rightarrow \left(\frac{1}{2}\right)^3 = \frac{1}{8}$

33. $7^2 = 49 \Rightarrow \log_7 49 = 2$

35. $6^{-2} = \frac{1}{36} \Rightarrow \log_6 \frac{1}{36} = -2$

37. $\left(\frac{1}{2}\right)^{-5} = 32 \Rightarrow \log_{1/2} 32 = -5$

39. $x^y = z \Rightarrow \log_x z = y$

41. $\log_2 32 = x \Rightarrow 2^x = 32 \Rightarrow x = 5$

43. $\log_5 125 = x \Rightarrow 5^x = 125 \Rightarrow x = 3$

45. $\log_7 x = 2 \Rightarrow 7^2 = x \Rightarrow x = 49$

© 2015 Cengage Learning. All Rights Reserved. May not be scanned, copied or duplicated, or posted to a publicly accessible website, in whole or in part.

SECTION 9.3

47. $\log_6 x = 1 \Rightarrow 6^1 = x \Rightarrow x = 6$

49. $\log_{25} x = \dfrac{1}{2} \Rightarrow 25^{1/2} = x \Rightarrow x = 5$

51. $\log_5 x = -2 \Rightarrow 5^{-2} = x \Rightarrow x = \dfrac{1}{25}$

53. $\log_x 5^3 = 3 \Rightarrow x^3 = 5^3 \Rightarrow x = 5$

55. $\log_x \dfrac{9}{4} = 2 \Rightarrow x^2 = \dfrac{9}{4} \Rightarrow x = \dfrac{3}{2}$

57. $\log_{1/2} \dfrac{1}{8} = x \Rightarrow \left(\dfrac{1}{2}\right)^x = \dfrac{1}{8} \Rightarrow x = 3$

59. $\log_{1/4} x = 3 \Rightarrow \left(\dfrac{1}{4}\right)^3 = x \Rightarrow x = \dfrac{1}{64}$

61. $y = f(x) = \log_3 x$
through $(1, 0)$ and $(3, 1)$

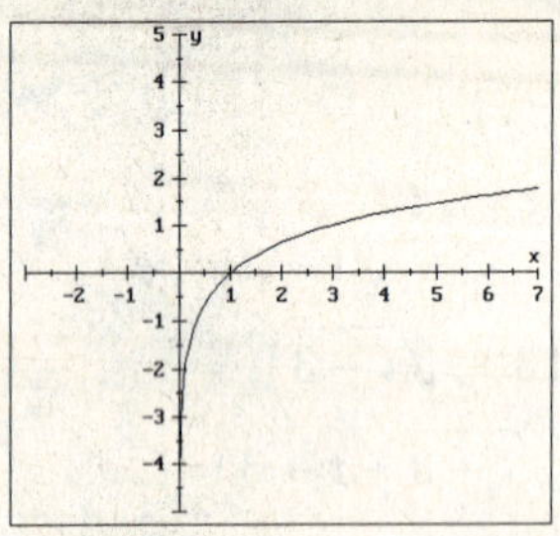

increasing

63. $y = f(x) = \log_{1/2} x$
through $(1, 0)$ and $(\frac{1}{2}, 1)$

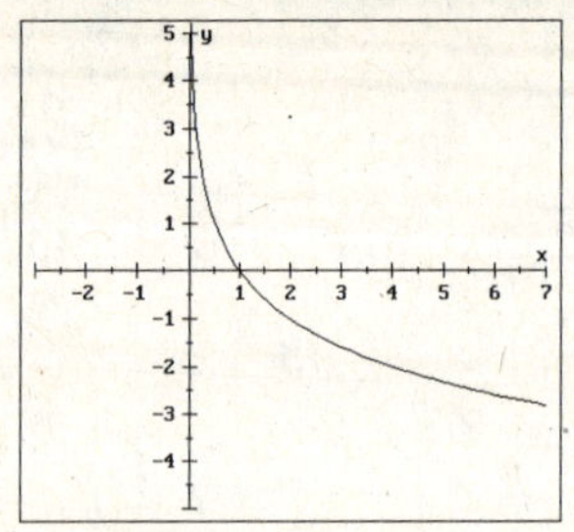

decreasing

65. $y = f(x) = 2^x$
$y = g(x) = \log_2 x$

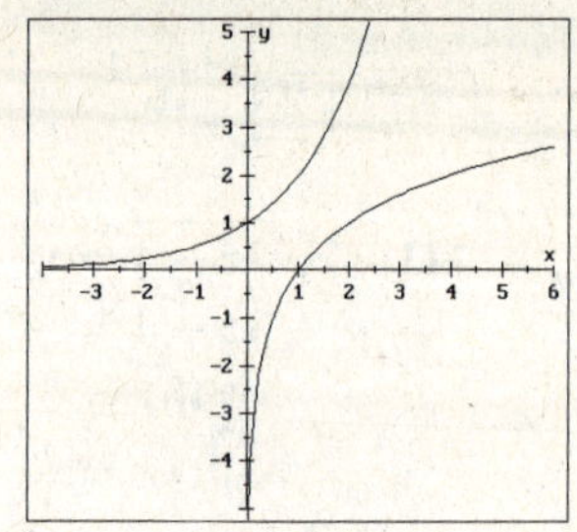

67. $y = f(x) = \left(\dfrac{1}{4}\right)^x$
$y = g(x) = \log_{1/4} x$

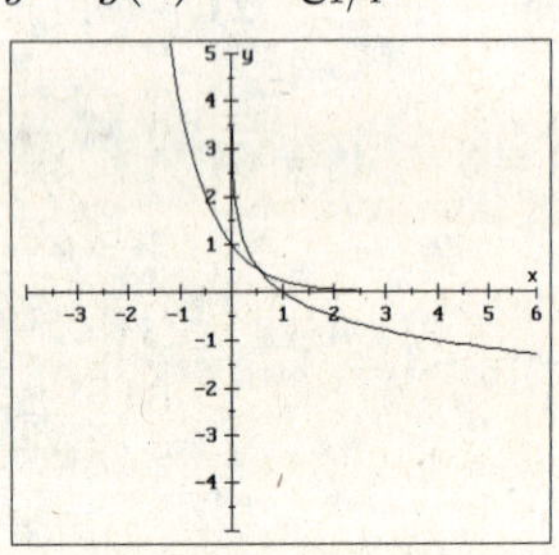

69. $y = f(x) = 3 + \log_3 x$
Shift $y = \log_3 x$ up 3.

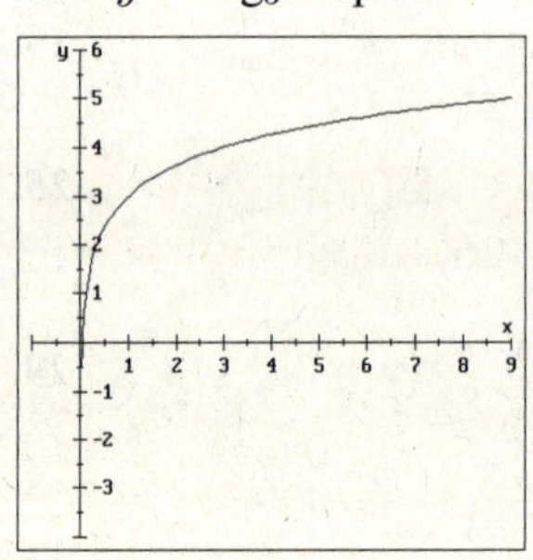

71. $y = f(x) = \log_{1/2}(x - 2)$
Shift $y = \log_{1/2} x$ right 2.

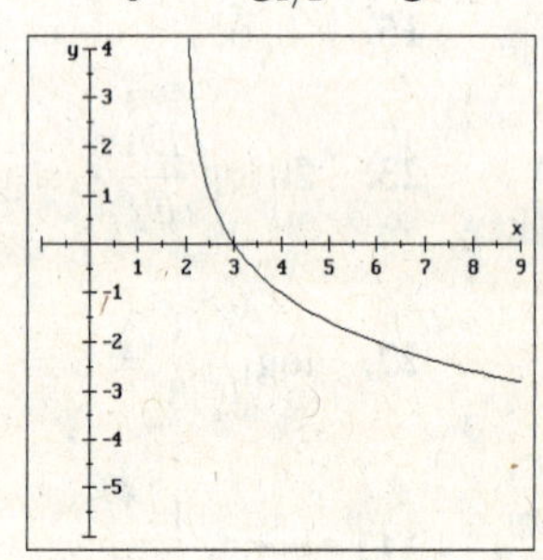

73. $\log 8.25 \approx 0.9165$

75. $\log 0.00867 \approx -2.0620$

77. $\log y = 4.24 \Rightarrow y = 17{,}378.01$

79. $\log y = -3.71 \Rightarrow y = 0.00$

81. $\log_{36} x = -\dfrac{1}{2} \Rightarrow 36^{-1/2} = x \Rightarrow x = \dfrac{1}{6}$

83. $\log_{1/2} 8 = x \Rightarrow \left(\dfrac{1}{2}\right)^x = 8 \Rightarrow x = -3$

© 2015 Cengage Learning. All Rights Reserved. May not be scanned, copied or duplicated, or posted to a publicly accessible website, in whole or in part.

85. $\log_{1/2} x = -3 \Rightarrow \left(\frac{1}{2}\right)^{-3} = x \Rightarrow x = 8$

87. $\log_4 2 = x \Rightarrow 4^x = 2 \Rightarrow x = \frac{1}{2}$

89. $\log_{100} \frac{1}{1000} = x \Rightarrow 100^x = \frac{1}{1000} \Rightarrow$
$x = -\frac{3}{2}$

91. $\log_{27} 9 = x \Rightarrow 27^x = 9 \Rightarrow x = \frac{2}{3}$

93. $\log_{2\sqrt{2}} x = 2 \Rightarrow (2\sqrt{2})^2 = x \Rightarrow x = 8$

95. $\log_x \frac{1}{64} = -3 \Rightarrow x^{-3} = \frac{1}{64} \Rightarrow x = 4$

97. $2^{\log_2 4} = x \Rightarrow x = 4$

99. $x^{\log_4 6} = 6 \Rightarrow x = 4$

101. $\log 10^3 = x \Rightarrow 10^x = 10^3 \Rightarrow x = 3$

103. $10^{\log x} = 100 \Rightarrow \log x = 2 \Rightarrow x = 100$

105. $\log y = 1.4023 \Rightarrow y = 25.25$

107. $\log y = \log 8 \Rightarrow \log y = 0.9030 \Rightarrow y = 8$

109. $\log_b 9 = 2 \Rightarrow b^2 = 9 \Rightarrow b = 3$

111. $\log_b 2 = 0 \Rightarrow b^0 = 2$
No such value exists.

113. $R = \log \frac{A}{P} = \log \frac{5000}{0.2} = \log 25{,}000$
≈ 4.4

115. $R = \log \frac{A}{P} = \log \frac{2500}{0.25} = \log 10000$
$= 4$

117. $n = \frac{\log V - \log C}{\log\left(1 - \frac{2}{N}\right)} = \frac{\log 2000 - \log 17000}{\log\left(1 - \frac{2}{5}\right)} \approx \frac{-0.929419}{-0.221849} \approx 4.2$ years old

119. $n = \frac{\log\left[\frac{Ar}{P} + 1\right]}{\log(1 + r)} = \frac{\log\left[\frac{20{,}000(0.12)}{1000} + 1\right]}{\log(1 + 0.12)} = \frac{\log 3.4}{\log 1.12} \approx 10.8$ years

121. dB gain $= 20 \log \frac{E_O}{E_I} = 20 \log \frac{20}{0.71} = 20 \log 28.169 \approx 29.0$ dB

123. dB gain $= 20 \log \frac{E_O}{E_I} = 20 \log \frac{30}{0.1} = 20 \log 300 \approx 49.5$ dB

125-129. Answers may vary.

Exercises 9.4 (page 654)

1. $\log_3 9 = 2 \Rightarrow 3^2 = 9$

3. $2^3 = 8 \Rightarrow \log_2 8 = 3$

5. $y = mx + b$
$y = -7x + 2$

© 2015 Cengage Learning. All Rights Reserved. May not be scanned, copied or duplicated, or posted to a publicly accessible website, in whole or in part.

SECTION 9.4

7. $3x + 2y = 9$

$$2y = -3x + 9$$

$$y = -\frac{3}{2}x + \frac{9}{2} \Rightarrow m = -\frac{3}{2}$$

Use the parallel slope:

$$y - y_1 = m(x - x_1)$$

$$y - 5 = -\frac{3}{2}(x - (-3))$$

$$y - 5 = -\frac{3}{2}x - \frac{9}{2}$$

$$y = -\frac{3}{2}x + \frac{1}{2}$$

9. $y = 6$

11. $\dfrac{x+1}{x} + \dfrac{x-1}{x+1} = \dfrac{(x+1)(x+1)}{x(x+1)} + \dfrac{(x-1)x}{(x+1)x} = \dfrac{x^2+2x+1}{x(x+1)} + \dfrac{x^2-x}{x(x+1)} = \dfrac{2x^2+x+1}{x(x+1)}$

13. $\dfrac{1+\frac{y}{x}}{\frac{y}{x}-1} = \dfrac{(1+\frac{y}{x})x}{(\frac{y}{x}-1)x} = \dfrac{x+y}{y-x}$

15. $(0, \infty); (-\infty, \infty)$ **17.** 10

19. $\dfrac{\ln 2}{r}$

21. $\ln 25.25 \approx 3.2288$

23. $\ln 9.89 \approx 2.2915$

25. $\ln y = 2.3015 \Rightarrow y = 9.9892$

27. $\ln y = 3.17 \Rightarrow y = 23.8075$

29. $y = -\ln x$

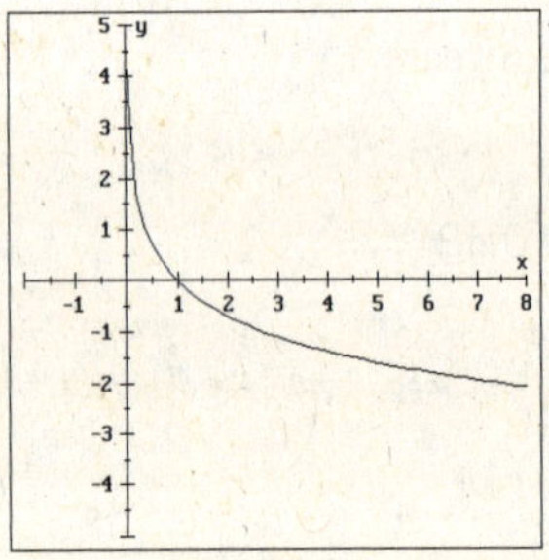

31. $y = \ln(-x)$

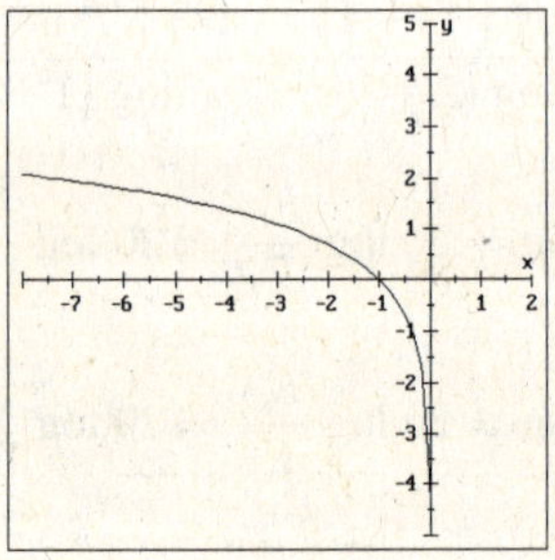

33. The graph must be increasing. The graph could not look like this.

35. The graph must go through $(1, 0)$. The graph could not look like this.

37. $\log(\ln 3) \approx \log(1.0986) \approx 0.0408$

39. $\ln(\log 0.7) = \ln(-0.1549) \Rightarrow$ no real value

41. $\ln y = -4.72 \Rightarrow y = 0.0089$

43. $\log y = \ln 4 \Rightarrow \log y \approx 1.3863 \Rightarrow y \approx 24.3385$
(The answer will vary if rounding is used on the calculator.)

© 2015 Cengage Learning. All Rights Reserved. May not be scanned, copied or duplicated, or posted to a publicly accessible website, in whole or in part.

45. $t = \frac{\ln 2}{r} = \frac{\ln 2}{0.12} \approx 5.8$ years

47. $t = \frac{\ln 2}{r} = \frac{\ln 2}{0.05} \approx 13.9$ years

49. $t = -\frac{1}{0.9}\ln\frac{50 - T_r}{212 - T_r} = -\frac{1}{0.9}\ln\frac{50-38}{212-38} = -\frac{1}{0.9}\ln\frac{12}{174} \approx -\frac{1}{0.9}(-2.6741) \approx 3.0$ hours

51. **Answers may vary.**

53. $P = P_0 e^{rt} = P_0 e^{r\frac{\ln 3}{r}} = P_0 e^{\ln 3} = 3P_0$

55. Let $t = \frac{\ln 5}{r} \Rightarrow P = P_0 e^{rt} = P_0 e^{r\frac{\ln 5}{r}} = P_0 e^{\ln 5} = 5P_0$

Exercises 9.5 (page 664)

1. $\log_5 25 = x \Rightarrow 5^x = 25 \Rightarrow x = 2$

3. $\log_7 x = 3 \Rightarrow 7^3 = x \Rightarrow x = 343$

5. $\log_9 x = \frac{1}{2} \Rightarrow 9^{1/2} = x \Rightarrow x = 3$

7. $\log_{1/3} x = 2 \Rightarrow \left(\frac{1}{3}\right)^2 = x \Rightarrow x = \frac{1}{9}$

9. $\log_x \frac{1}{4} = -2 \Rightarrow x^{-2} = \frac{1}{4} \Rightarrow x = 2$

11. $d = \sqrt{(-2-4)^2 + (3-(-4))^2} = \sqrt{(-6)^2 + 7^2} = \sqrt{36+49} = \sqrt{85}$

13. Use the slope $m = -\frac{7}{6}$ from **#10.**

$$y - y_1 = m(x - x_1)$$
$$y - 3 = -\tfrac{7}{6}(x+2)$$
$$y - 3 = -\tfrac{7}{6}x - \tfrac{7}{3}$$
$$y = -\tfrac{7}{6}x + \tfrac{2}{3}$$

15. 1

17. x

19. –

21. x

23. =

25. 0

27. 2

29. 10

31. 1

Problems 33-35 are to be solved using a calculator. The keystrokes needed to solve each problem using a TI-84 graphing calculator appear in each solution. There may be other solutions. Keystrokes for other calculators may be slightly different.

33. [log] [2] [.] [5] [×] [3] [.] [7] [ENTER] {0.96614}
[log] [2] [.] [5] [)] [+] [log] [3] [.] [7] [)] [ENTER] {0.96614}

35. [ln] [2] [.] [2] [5] [^] [4] [ENTER] {3.24372}
[4] [ln] [2] [.] [2] [5] [ENTER] {3.24372}

37. $\log_b 7xy = \log_b 7 + \log_b x + \log_b y$

39. $\log_b \frac{5x}{y} = \log_b 5x - \log_b y = \log_b 5 + \log_b x - \log_b y$

© 2015 Cengage Learning. All Rights Reserved. May not be scanned, copied or duplicated, or posted to a publicly accessible website, in whole or in part.

SECTION 9.5

41. $\log_b x^3y^2 = \log_b x^3 + \log_b y^2 = 3\log_b x + 2\log_b y$

43. $$\log_b \frac{x^3\sqrt{z}}{y^5} = \log_b \frac{x^3z^{1/2}}{y^5} = \log_b\left(x^3z^{1/2}\right) - \log_b y^5 = \log_b x^3 + \log_b z^{1/2} - \log_b y^5$$
$$= 3\log_b x + \frac{1}{2}\log_b z - 5\log_b y$$

45. $\log_b(x-3) - 5\log_b x = \log_b(x-3) - \log_b x^5 = \log_b \dfrac{x-3}{x^5}$

47. $7\log_b x + \dfrac{1}{2}\log_b z = \log_b x^7 + \log_b z^{1/2} = \log_b x^7z^{1/2} = \log_b\left(x^7\sqrt{z}\right)$

49. $\log 28 = \log 4\cdot 7 = \log 4 + \log 7 = 0.6021 + 0.8451 = 1.4472$

51. $\log 2.25 = \log \dfrac{9}{4} = \log 9 - \log 4 = 0.9542 - 0.6021 = 0.3521$

53. $$\log \frac{49}{36} = \log \frac{7^2}{4\cdot 9} = \log 7^2 - \log 4\cdot 9 = 2\log 7 - (\log 4 + \log 9)$$
$$= 2(0.8451) - 0.6021 - 0.9542 = 0.1339$$

55. $\log 252 = \log 4\cdot 7\cdot 9 = \log 4 + \log 7 + \log 9 = 0.6021 + 0.8451 + 0.9542 = 2.4014$

57. $\log_3 7 = \dfrac{\log 7}{\log 3} \approx 1.7712$

59. $\log_{1/3} 3 = \dfrac{\log 3}{\log \frac{1}{3}} \approx -1.0000$

61. $\log_3 8 = \dfrac{\log 8}{\log 3} \approx 1.8928$

63. $\log_{\sqrt{2}} \sqrt{5} = \dfrac{\log\sqrt{5}}{\log\sqrt{2}} \approx 2.3219$

65. 0

67. 7

69. 10

71. 1

73. $\log_b\left(\dfrac{xy}{z}\right)^{1/3} = \frac{1}{3}\log_b\left(\dfrac{xy}{z}\right) = \frac{1}{3}\left(\log_b x + \log_b y - \log_b z\right)$

75. $$\log_b \frac{\sqrt[3]{x}}{\sqrt[4]{yz}} = \log_b \frac{x^{1/3}}{(yz)^{1/4}} = \log_b x^{1/3} - \log_b (yz)^{1/4} = \frac{1}{3}\log_b x - \frac{1}{4}\log_b yz$$
$$= \frac{1}{3}\log_b x - \frac{1}{4}(\log_b y + \log_b z)$$
$$= \frac{1}{3}\log_b x - \frac{1}{4}\log_b y - \frac{1}{4}\log_b z$$

77. $$3\log_b(x+2) - 4\log_b y + \frac{1}{2}\log_b(x+1) = \log_b(x+2)^3 + \log_b y^{-4} + \log_b(x+1)^{1/2}$$
$$= \log_b(x+2)^3y^{-4}(x+1)^{1/2} = \log_b \frac{(x+2)^3\sqrt{x+1}}{y^4}$$

© 2015 Cengage Learning. All Rights Reserved. May not be scanned, copied or duplicated, or posted to a publicly accessible website, in whole or in part.

SECTION 9.5

79. $\log_b\left(\frac{x}{z}+x\right)-\log_b\left(\frac{y}{z}+y\right)=\log_b\frac{\frac{x}{z}+x}{\frac{y}{z}+y}=\log_b\frac{x+xz}{y+yz}=\log_b\frac{x(1+z)}{y(1+z)}=\log_b\frac{x}{y}$

81. $\log 112=\log 4^2\cdot 7=\log 4^2+\log 7=2\log 4+\log 7=2(0.6021)+0.8451=2.0493$

83. $\log\frac{144}{49}=\log\frac{16\cdot 9}{7^2}=\log\frac{4^2\cdot 9}{7^2}=\log 4^2+\log 9-\log 7^2=2\log 4+\log 9-2\log 7$
$=2(0.6021)+0.9542-2(0.8451)$
$=0.4682$

85. [log] [√] [2] [4] [.] [3] [)] [ENTER] {0.69280}
[.] [5] [log] [2] [4] [.] [3] [)] [ENTER] {0.69280}

87. $\log_b 0=1\Rightarrow b^1=0\Rightarrow b=0\Rightarrow$ FALSE $(b^1=b)$

89. $\log_b xy=(\log_b x)(\log_b y)\Rightarrow$ FALSE $(\log_b xy=\log_b x+\log_b y)$

91. $\log_7 7^7=7\Rightarrow 7^7=7^7\Rightarrow$ TRUE

93. $\frac{\log_b A}{\log_b B}=\log_b A-\log_b B\Rightarrow$ FALSE $\left(\log_b\frac{A}{B}=\log_b A-\log_b B\right)$

95. $3\log_b\sqrt[3]{a}=3\log_b a^{1/3}=\frac{1}{3}\cdot 3\log_b a=\log_b a\Rightarrow$ TRUE

97. $\log_b\frac{1}{a}=\log_b 1-\log_b a=0-\log_b a=-\log_b a\Rightarrow$ TRUE

99. $\text{pH}=-\log[\text{H}^+]=-\log(1.7\times 10^{-5})\approx 4.77$

101. low pH:
$\text{pH}=-\log[\text{H}^+]$
$6.8=-\log[\text{H}^+]$
$-6.8=\log[\text{H}^+]$
$[\text{H}^+]=1.5849\times 10^{-7}$

high pH:
$\text{pH}=-\log[\text{H}^+]$
$7.6=-\log[\text{H}^+]$
$-7.6=\log[\text{H}^+]$
$[\text{H}^+]=2.5119\times 10^{-8}$

103. $k\ln 2I=k(\ln 2+\ln I)$
$=k\ln 2+k\ln I$
$=k\ln 2+L$
The loudness increases by $k\ln 2$.

105. $L=3k\ln I=k\cdot 3\ln I$
$=k\ln I^3$
The intensity must be cubed.

107. Answers may vary.

109. $\ln(e^x)=\log_e(e^x)=x$

© 2015 Cengage Learning. All Rights Reserved. May not be scanned, copied or duplicated, or posted to a publicly accessible website, in whole or in part.

SECTION 9.5

111. Let $\log_{b^2} x = y$. Then

$$\begin{aligned} (b^2)^y &= x \\ b^{2y} &= x \\ (b^{2y})^{1/2} &= x^{1/2} \\ b^y &= x^{1/2} \\ \log_b x^{1/2} &= y \\ \frac{1}{2}\log_b x &= y \end{aligned}$$

Exercises 9.6 (page 675)

1. $$\begin{aligned} \log x &= 2 \\ x &= 10^2 \end{aligned}$$

3. $$\begin{aligned} \log(x+1) &= 1 \\ x+1 &= 10^1 \end{aligned}$$

5. [log] [4] [)] [÷] [log] [3] [)] [ENTER] {1.2619}

7. [(–)] [log] [5] [÷] [7] [)] [ENTER] {0.1461}

9. $$\begin{aligned} 4x^2 - 64x &= 0 \\ 4x(x-16) &= 0 \\ 4x = 0 \quad \textbf{or} \quad x - 16 &= 0 \\ x = 0 \qquad\qquad x &= 16 \end{aligned}$$

11. $$\begin{aligned} 3p^2 + 10p &= 8 \\ 3p^2 + 10p - 8 &= 0 \\ (3p-2)(p+4) &= 0 \\ 3p - 2 = 0 \quad \textbf{or} \quad p + 4 &= 0 \\ p = \tfrac{2}{3} \qquad\qquad p &= -4 \end{aligned}$$

13. exponential

15. A_0e^{-kt}

17. $$\begin{aligned} 3^x &= 8 \\ \log 3^x &= \log 8 \\ x\log 3 &= \log 8 \\ x &= \frac{\log 8}{\log 3} \\ x &\approx 1.8928 \end{aligned}$$

19. $$\begin{aligned} e^t &= 50 \\ \ln e^t &= \ln 50 \\ t\ln e &= \ln 50 \\ t &= \ln 50 \\ t &\approx 3.9120 \end{aligned}$$

21. $$\begin{aligned} 7 &= 4.3(1.01)^t \\ \frac{7}{4.3} &= (1.01)^t \\ \log\frac{7}{4.3} &= \log(1.01)^t \\ \log\frac{7}{4.3} &= t\log 1.01 \\ \frac{\log\frac{7}{4.3}}{\log 1.01} &= t \\ 48.9728 &\approx t \end{aligned}$$

23. $$\begin{aligned} 5 &= 2.1(1.04)^t \\ \frac{5}{2.1} &= (1.04)^t \\ \log\frac{5}{2.1} &= \log(1.04)^t \\ \log\frac{5}{2.1} &= t\log 1.04 \\ \frac{\log\frac{5}{2.1}}{\log 1.04} &= t \\ 22.1184 &\approx t \end{aligned}$$

© 2015 Cengage Learning. All Rights Reserved. May not be scanned, copied or duplicated, or posted to a publicly accessible website, in whole or in part.

SECTION 9.6

25.
$$\begin{aligned} 6^{x-2} &= 4 \\ \log 6^{x-2} &= \log 4 \\ (x-2)\log 6 &= \log 4 \\ x-2 &= \frac{\log 4}{\log 6} \\ x &= \frac{\log 4}{\log 6}+2 \\ x &\approx 2.7737 \end{aligned}$$

27.
$$\begin{aligned} 4^{x+1} &= 7^x \\ \log 4^{x+1} &= \log 7^x \\ (x+1)\log 4 &= x\log 7 \\ x\log 4+\log 4 &= x\log 7 \\ \log 4 &= x\log 7 - x\log 4 \\ \log 4 &= x(\log 7-\log 4) \\ \frac{\log 4}{\log 7-\log 4} &= x \\ 2.4772 &\approx x \end{aligned}$$

29.
$$\begin{aligned} 5^{x^2+2x} &= 125 \\ 5^{x^2+2x} &= 5^3 \\ x^2+2x &= 3 \\ x^2+2x-3 &= 0 \\ (x+3)(x-1) &= 0 \\ x+3=0 \quad &\textbf{or} \quad x-1=0 \\ x=-3 \quad & \quad\quad x=1 \end{aligned}$$

31.
$$\begin{aligned} 3^{x^2+4x} &= \frac{1}{81} \\ 3^{x^2+4x} &= 3^{-4} \\ x^2+4x &= -4 \\ x^2+4x+4 &= 0 \\ (x+2)(x+2) &= 0 \\ x+2=0 \quad &\textbf{or} \quad x+2=0 \\ x=-2 \quad & \quad\quad x=-2 \end{aligned}$$

33. $2^{x+1}=7 \Rightarrow$ Graph $y=2^{x+1}$ and $y=7$.

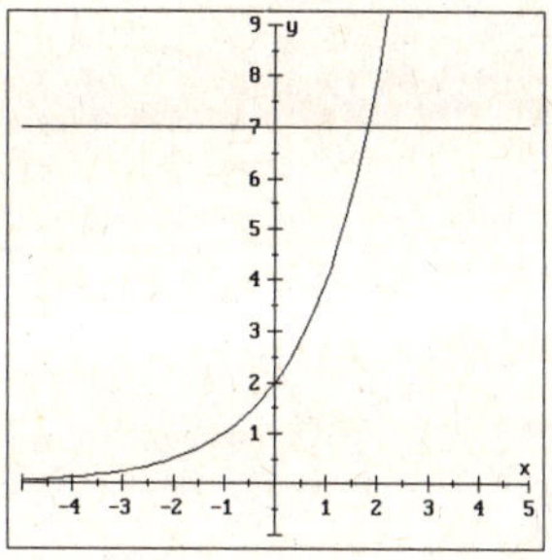

$x \approx 1.8$

35. $2^{x^2-2x}-8=0 \Rightarrow$ Graph $y=2^{x^2-2x}-8$ and find any x-intercept(s).

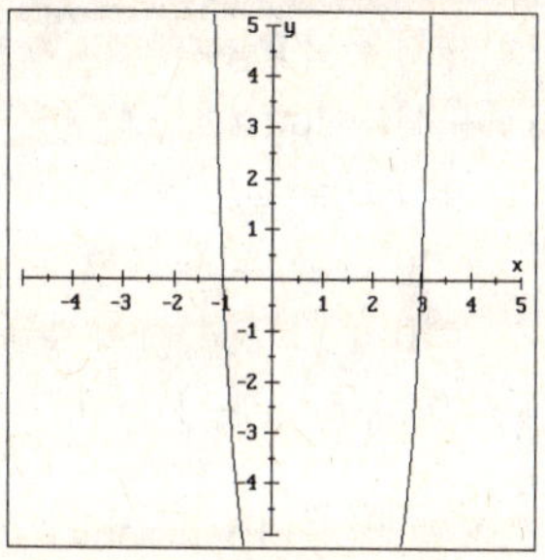

$x=3$ or $x=-1$

37.
$$\begin{aligned} \log 9x &= \log 27 \\ 9x &= 27 \\ x &= 3 \end{aligned}$$

39.
$$\begin{aligned} \log(4x+3) &= \log(x+9) \\ 4x+3 &= x+9 \\ 3x &= 6 \\ x &= 2 \end{aligned}$$

41.
$$\begin{aligned} \log x^2 &= 2 \\ 10^2 &= x^2 \\ 100 &= x^2 \\ \pm 10 &= x \end{aligned}$$

© 2015 Cengage Learning. All Rights Reserved. May not be scanned, copied or duplicated, or posted to a publicly accessible website, in whole or in part.

43.
$$\begin{aligned}\log x + \log(x-3) &= 1\\ \log x(x-3) &= 1\\ 10^1 &= x(x-3)\\ 0 &= x^2 - 3x - 10\\ 0 &= (x-5)(x+2)\end{aligned}$$
$x - 5 = 0$ **or** $x + 2 = 0$
$x = 5$ $\quad x = -2$
Extraneous

45.
$$\begin{aligned}\log x + \log(x-15) &= 2\\ \log x(x-15) &= 2\\ 10^2 &= x(x-15)\\ 0 &= x^2 - 15x - 100\\ 0 &= (x-20)(x+5)\end{aligned}$$
$x - 20 = 0$ **or** $x + 5 = 0$
$x = 20$ $\quad x = -5$
Extraneous

47.
$$\begin{aligned}\log(x+90) &= 3 - \log x\\ \log x + \log(x+90) &= 3\\ \log x(x+90) &= 3\\ 10^3 &= x(x+90)\\ 0 &= x^2 + 90x - 1000\\ 0 &= (x-10)(x+100)\end{aligned}$$
$x - 10 = 0$ **or** $x + 100 = 0$
$x = 10$ $\quad x = -100$
Extraneous

49.
$$\begin{aligned}\frac{\log(2x+1)}{\log(x-1)} &= 2\\ \log(2x+1) &= 2\log(x-1)\\ \log(2x+1) &= \log(x-1)^2\\ 2x+1 &= (x-1)^2\\ 2x+1 &= x^2 - 2x + 1\\ 0 &= x^2 - 4x\\ 0 &= x(x-4)\end{aligned}$$
$x = 0$ **or** $x - 4 = 0$
Extraneous $\quad x = 4$

51.
$$\begin{aligned}\frac{\log(3x+4)}{\log x} &= 2\\ \log(3x+4) &= 2\log x\\ \log(3x+4) &= \log x^2\\ 3x+4 &= x^2\\ 0 &= x^2 - 3x - 4\\ 0 &= (x-4)(x+1)\end{aligned}$$
$x - 4 = 0$ **or** $x + 1 = 0$
$x = 4$ $\quad x = -1$
Extraneous

53. $\log x + \log(x-15) = 2$
Graph $y = \log x + \log(x-15)$ and $y = 2$.

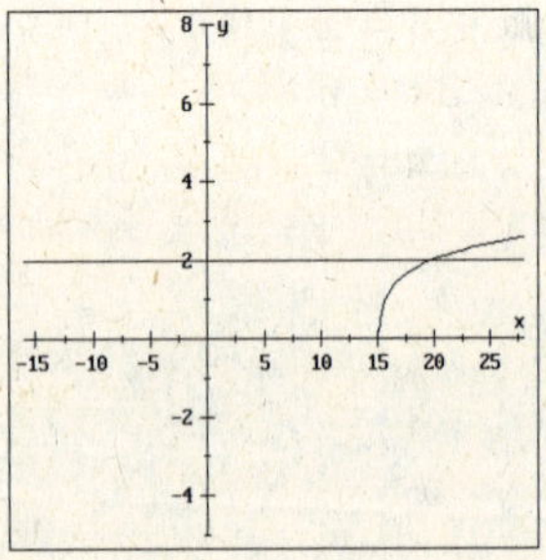

$x = 20$

55. $\ln(2x+5) - \ln 3 = \ln(x-1) \Rightarrow$ Graph
$y = \ln(2x+5) - \ln 3$ and $y = \ln(x-1)$.

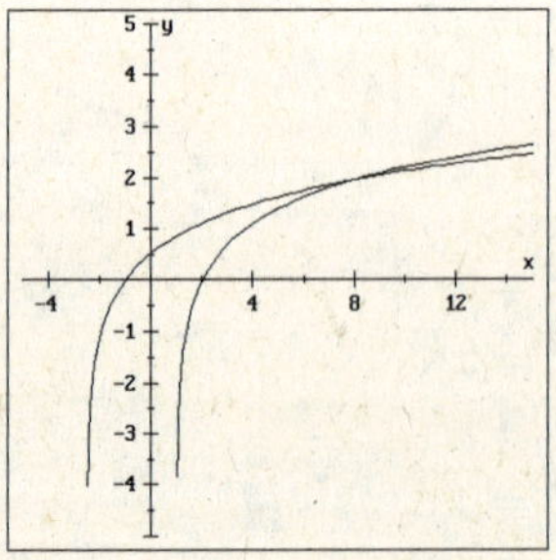

$x = 8$

© 2015 Cengage Learning. All Rights Reserved. May not be scanned, copied or duplicated, or posted to a publicly accessible website, in whole or in part.

57.
$$\begin{aligned} 5^x &= 4^x \\ \log 5^x &= \log 4^x \\ x \log 5 &= x \log 4 \\ 0 &= x \log 4 - x \log 5 \\ 0 &= x(\log 4 - \log 5) \\ \frac{0}{\log 4 - \log 5} &= x \\ 0 &= x \end{aligned}$$

59.
$$\begin{aligned} 7^{x^2} &= 10 \\ \log 7^{x^2} &= \log 10 \\ x^2 \log 7 &= \log 10 \\ x^2 &= \frac{\log 10}{\log 7} \\ x^2 &\approx 1.1833 \\ x &\approx \pm 1.0878 \end{aligned}$$

61.
$$\begin{aligned} 8^{x^2} &= 9^x \\ \log 8^{x^2} &= \log 9^x \\ x^2 \log 8 &= x \log 9 \\ x^2 \log 8 - x \log 9 &= 0 \\ x(x \log 8 - \log 9) &= 0 \end{aligned}$$
$$x = 0 \quad \textbf{or} \quad x \log 8 - \log 9 = 0$$
$$\begin{aligned} x \log 8 &= \log 9 \\ x &= \frac{\log 9}{\log 8} \\ x &\approx 1.0566 \end{aligned}$$

63.
$$\begin{aligned} \log(5 - 4x) - \log(x + 25) &= 0 \\ \log(5 - 4x) &= \log(x + 25) \\ 5 - 4x &= x + 25 \\ -20 &= 5x \\ -4 &= x \end{aligned}$$

65.
$$\begin{aligned} \log(x - 6) - \log(x - 2) &= \log \frac{5}{x} \\ \log \frac{x-6}{x-2} &= \log \frac{5}{x} \\ \frac{x-6}{x-2} &= \frac{5}{x} \\ x(x - 6) &= 5(x - 2) \\ x^2 - 6x &= 5x - 10 \\ x^2 - 11x + 10 &= 0 \\ (x - 10)(x - 1) &= 0 \end{aligned}$$
$$x - 10 = 0 \quad \textbf{or} \quad x - 1 = 0$$
$$x = 10 \qquad\qquad x = 1$$
Extraneous

67.
$$\begin{aligned} 4^{x+2} - 4^x &= 15 \\ 4^x 4^2 - 4^x &= 15 \\ 16 \cdot 4^x - 4^x &= 15 \\ 15 \cdot 4^x &= 15 \\ 4^x &= 1 \\ x &= 0 \end{aligned}$$

© 2015 Cengage Learning. All Rights Reserved. May not be scanned, copied or duplicated, or posted to a publicly accessible website, in whole or in part.

69.
$$\begin{aligned}\frac{\log(5x+6)}{2} &= \log x\\ \log(5x+6) &= 2\log x\\ \log(5x+6) &= \log x^2\\ 5x+6 &= x^2\\ 0 &= x^2-5x-6\\ 0 &= (x-6)(x+1)\end{aligned}$$
$$x-6=0 \quad \textbf{or} \quad x+1=0$$
$$x=6 \qquad\qquad x=-1$$
Extraneous

71.
$$\begin{aligned}\log_3 x &= \log_3\left(\frac{1}{x}\right)+4\\ \log_3 x &= \log_3\left(\frac{1}{x}\right)+\log_3 81\\ \log_3 x &= \log_3\left(\frac{81}{x}\right)\\ x &= \frac{81}{x}\\ x^2 &= 81\\ x &= 9 \text{ } (-9 \text{ is extraneous.})\end{aligned}$$

73.
$$\begin{aligned}2(3^x) &= 6^{2x}\\ \log 2(3^x) &= \log 6^{2x}\\ \log 2+\log 3^x &= 2x\log 6\\ \log 2+x\log 3 &= 2x\log 6\\ \log 2 &= 2x\log 6-x\log 3\\ \log 2 &= x(2\log 6-\log 3)\\ \frac{\log 2}{2\log 6-\log 3} &= x\\ 0.2789 &\approx x\end{aligned}$$

75.
$$\begin{aligned}\log x^2 &= (\log x)^2\\ 2\log x &= (\log x)^2\\ 0 &= (\log x)^2-2\log x\\ 0 &= \log x(\log x-2)\end{aligned}$$
$$\log x=0 \quad \textbf{or} \quad \log x-2=0$$
$$x=1 \qquad\qquad \log x=2$$
$$x=100$$

77.
$$\begin{aligned}2\log_2 x &= 3+\log_2(x-2)\\ \log_2 x^2-\log_2(x-2) &= 3\\ \log_2\frac{x^2}{x-2} &= 3\\ \frac{x^2}{x-2} &= 8\\ x^2 &= 8(x-2)\\ x^2 &= 8x-16\\ x^2-8x+16 &= 0\\ (x-4)(x-4) &= 0\end{aligned}$$
$$x-4=0 \quad \textbf{or} \quad x-4=0$$
$$x=4 \qquad\qquad x=4$$

79.
$$\begin{aligned}\log(7y+1) &= 2\log(y+3)-\log 2\\ \log(7y+1) &= \log(y+3)^2-\log 2\\ \log(7y+1) &= \log\frac{y^2+6y+9}{2}\\ 7y+1 &= \frac{y^2+6y+9}{2}\\ 2(7y+1) &= y^2+6y+9\\ 14y+2 &= y^2+6y+9\\ 0 &= y^2-8y+7\\ 0 &= (y-7)(y-1)\end{aligned}$$
$$y-7=0 \quad \textbf{or} \quad y-1=0$$
$$y=7 \qquad\qquad y=1$$

81.
$$\begin{aligned}\log\frac{4x+1}{2x+9} &= 0\\ 10^0 &= \frac{4x+1}{2x+9}\\ 1 &= \frac{4x+1}{2x+9}\\ 2x+9 &= 4x+1\\ 8 &= 2x\\ 4 &= x\end{aligned}$$

83.
$$\begin{aligned}A &= A_0e^{-0.013t}\\ 0.5A_0 &= A_0e^{-0.013t}\\ 0.5 &= e^{-0.013t}\\ \ln 0.5 &= \ln e^{-0.013t}\\ \ln 0.5 &= -0.013t\\ \frac{\ln 0.5}{-0.013} &= t\\ 53 \text{ days} &\approx t\end{aligned}$$

© 2015 Cengage Learning. All Rights Reserved. May not be scanned, copied or duplicated, or posted to a publicly accessible website, in whole or in part.

SECTION 9.6

85.
$$\begin{aligned} A &= A_0 2^{-t/h} \\ 0.80A_0 &= A_0 2^{-2/h} \\ 0.80 &= 2^{-2/h} \\ \log 0.80 &= \log 2^{-2/h} \\ \log 0.80 &= -\frac{2}{h}\log 2 \\ h \log 0.80 &= -2 \log 2 \\ h &= \frac{-2 \log 2}{\log 0.80} \\ h &\approx 6.2 \text{ years} \end{aligned}$$

87.
$$\begin{aligned} A &= A_0 2^{-t/h} \\ 0.60A_0 &= A_0 2^{-t/5700} \\ 0.60 &= 2^{-t/5700} \\ \log 0.60 &= \log 2^{-t/5700} \\ \log 0.60 &= -\frac{t}{5700}\log 2 \\ \frac{\log 0.60}{-\log 2} &= \frac{t}{5700} \\ -5700\frac{\log 0.60}{\log 2} &= t \\ 4200 \text{ years} &\approx t \end{aligned}$$

89.
$$\begin{aligned} P &= P_0 e^{kt} \\ 2P_0 &= P_0 e^{5k} \\ 2 &= e^{5k} \\ \ln 2 &= \ln e^{5k} \\ \ln 2 &= 5k \\ \frac{\ln 2}{5} &= k \end{aligned}$$

$$\begin{aligned} P &= P_0 e^{kt} \\ 1{,}000{,}000 &= 30{,}000 e^{\frac{\ln 2}{5}t} \\ 33.333 &\approx e^{\frac{\ln 2}{5}t} \\ \ln 33.333 &\approx \ln e^{\frac{\ln 2}{5}t} \\ \ln 33.333 &\approx \frac{\ln 2}{5}t \\ \frac{5 \ln 33.333}{\ln 2} &\approx t \\ 25.3 \text{ years} &\approx t \end{aligned}$$

91.
$$\begin{aligned} P &= P_0 e^{kt} \\ 2P_0 &= P_0 e^{24k} \\ 2 &= e^{24k} \\ \ln 2 &= \ln e^{24k} \\ \ln 2 &= 24k \\ \frac{\ln 2}{24} &= k \end{aligned}$$

$$\begin{aligned} P &= P_0 e^{kt} \\ P &= P_0 e^{\frac{\ln 2}{24}(36)} \\ P &= P_0 e^{\frac{3\ln 2}{2}} \\ P &= P_0(2.828) \end{aligned}$$

It will be about 2.828 times larger.

93. $n = \frac{1}{\log 2}\left(\log \frac{B}{b}\right) = \frac{1}{\log 2}\log \frac{5 \times 10^6}{500} = \frac{1}{\log 2}\log 10{,}000 \approx 13.3$ generations

95.
$$\begin{aligned} A &= A_0 2^{-t/h} \\ 0.20A_0 &= A_0 2^{-t/18.4} \\ 0.20 &= 2^{-t/18.4} \\ \log 0.20 &= \log 2^{-t/18.4} \\ \log 0.20 &= -\frac{t}{18.4}\log 2 \\ \frac{\log 0.20}{-\log 2} &= \frac{t}{18.4} \\ -18.4\frac{\log 0.20}{\log 2} &= t \\ 42.7 \text{ days} &\approx t \end{aligned}$$

97.
$$\begin{aligned} A &= P\left(1 + \frac{r}{k}\right)^{kt} \\ 800 &= 500\left(1 + \frac{0.085}{2}\right)^{2t} \\ 1.6 &= (1.0425)^{2t} \\ \log 1.6 &= \log (1.0425)^{2t} \\ \log 1.6 &= 2t \log (1.0425) \\ \frac{\log 1.6}{2 \log 1.0425} &= t \\ 5.6 \text{ years} &\approx t \end{aligned}$$

© 2015 Cengage Learning. All Rights Reserved. May not be scanned, copied or duplicated, or posted to a publicly accessible website, in whole or in part.

99.
$$\begin{aligned} A &= P\left(1+\frac{r}{k}\right)^{kt} \\ 2100 &= 1300\left(1+\frac{0.09}{4}\right)^{4t} \\ \frac{2100}{1300} &= (1.0225)^{4t} \\ \log\frac{2100}{1300} &= \log\,(1.0225)^{4t} \\ \log\frac{21}{13} &= 4t\log\,(1.0225) \\ \frac{\log\frac{21}{13}}{4\log 1.0225} &= t \\ 5.4 \text{ years} &\approx t \end{aligned}$$

101. doubling time
$$\begin{aligned} = t &= \frac{\ln 2}{r} \\ &= \frac{100\ln 2}{100r} \\ &\approx \frac{70}{100r} \\ &= \frac{70}{r\text{, written as a }\%} \end{aligned}$$

103. Answers may vary.

105. Since the logarithm of a negative number is not defined (as a real number), the values $x-3$ and x^2+2 must be nonnegative. Since x^2+2 is always greater than 0, the only restriction is that $x-3>0$, or $x>3$. Thus, x cannot be a solution if $x\le 3$.

Chapter 9 Review (page 681)

1.

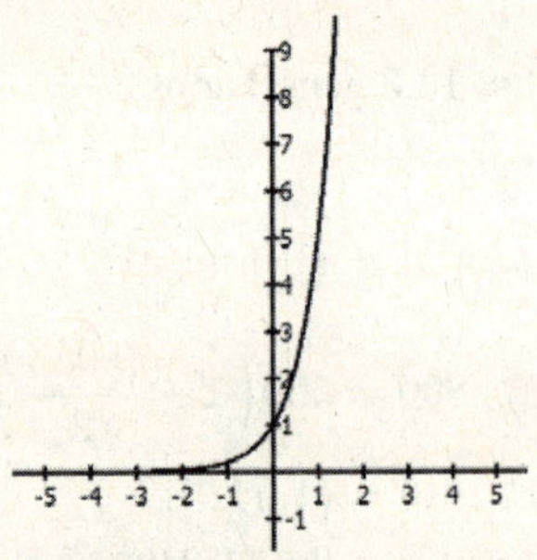

2.

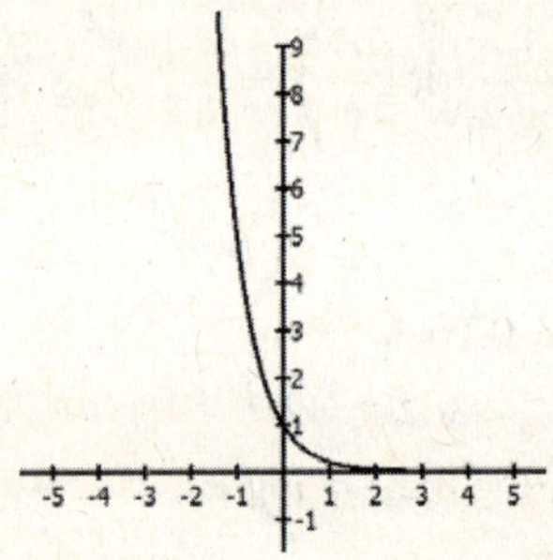

3. $y=3^x$; through $(0,1)$ and $(1,3)$

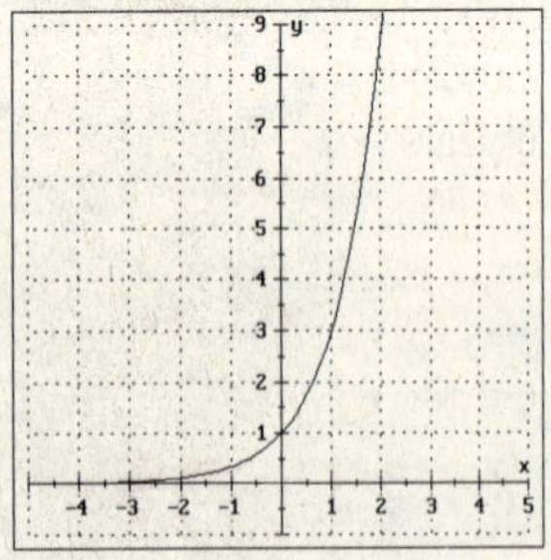

4. $y=\left(\frac{1}{3}\right)^x$; through $(0,1)$ and $\left(1,\frac{1}{3}\right)$

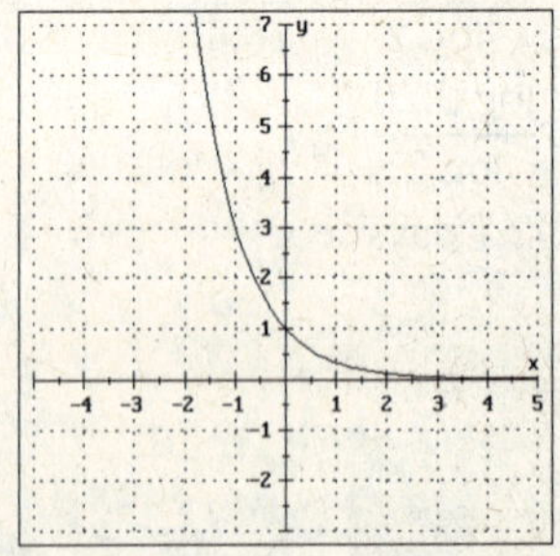

© 2015 Cengage Learning. All Rights Reserved. May not be scanned, copied or duplicated, or posted to a publicly accessible website, in whole or in part.

CHAPTER 9 REVIEW

5. The graph will go through $(0, 1)$ and $(1, 6)$, so $x = 1$ and $y = 6$.

6. domain $= (-\infty, \infty)$; range $= (0, \infty)$

7. $y = f(x) = \left(\frac{1}{2}\right)^x - 2$
Shift $y = \left(\frac{1}{2}\right)^x$ down 2.

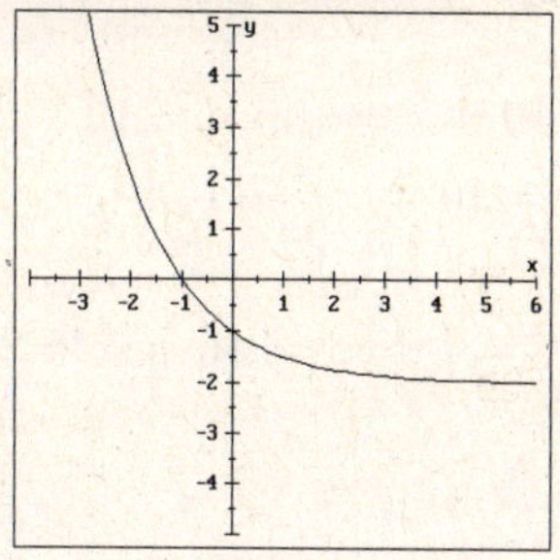

8. $y = f(x) = \left(\frac{1}{2}\right)^{x+2}$
Shift $y = \left(\frac{1}{2}\right)^x$ left 2.

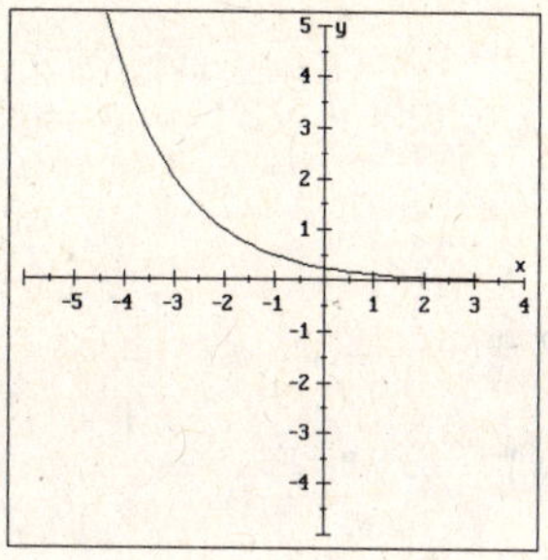

9. $A = P\left(1 + \frac{r}{k}\right)^{kt} = 25000\left(1 + \frac{0.05}{4}\right)^{4 \cdot 35} = 25000(1.0125)^{140} \approx \$142{,}312.97$

10. $A = Pe^{rt} = 25000e^{0.05(35)} = 25000e^{1.75} \approx \$143{,}865.07$

11. $y = f(x) = e^x + 1$; Shift $y = e^x$ up 1.

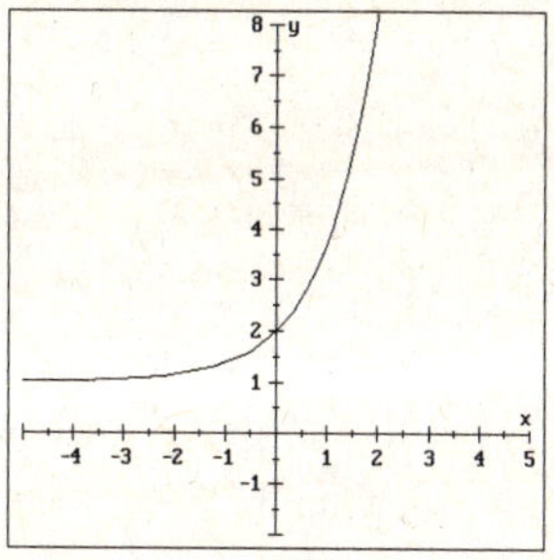

12. $y = f(x) = e^{x-3}$; Shift $y = e^x$ right 3.

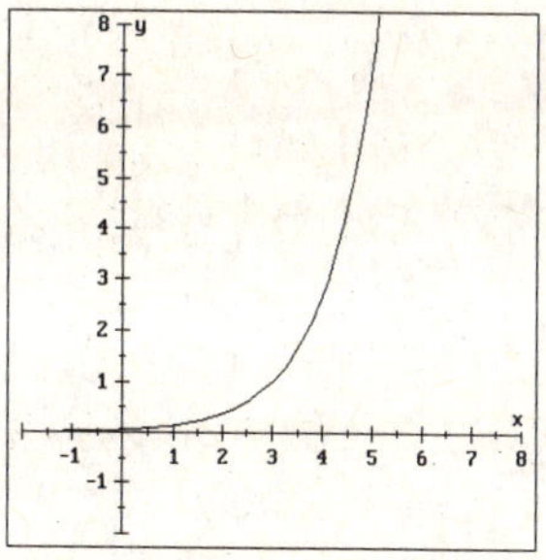

13. $P = P_0e^{kt} = 275{,}000{,}000e^{0.015(50)}$
$\approx 582{,}000{,}000$

14. $A = A_0e^{-0.0244t} = 50e^{-0.0244(20)} = 50e^{-0.488}$
≈ 30.69 g

15. domain $= (0, \infty)$; range $= (-\infty, \infty)$

16. **Answers will vary.**

17. $\log_4 16 = x \Rightarrow 4^x = 16 \Rightarrow \log_4 16 = 2$

18. $\log_{16} \frac{1}{4} = x \Rightarrow 16^x = \frac{1}{4} \Rightarrow \log_{16} \frac{1}{4} = -\frac{1}{2}$

19. $\log_\pi 1 = x \Rightarrow \pi^x = 1 \Rightarrow \log_\pi 1 = 0$

20. $\log_5 0.04 = x \Rightarrow \log_5 \frac{1}{25} = x \Rightarrow 5^x = \frac{1}{25}$
$\log_5 0.04 = -2$

© 2015 Cengage Learning. All Rights Reserved. May not be scanned, copied or duplicated, or posted to a publicly accessible website, in whole or in part.

21. $\log_b \sqrt{b} = \log_b b^{1/2} = x \Rightarrow b^x = b^{1/2}$
$\log_b \sqrt{b} = \frac{1}{2}$

22. $\log_a \sqrt[3]{x} = \log_a a^{1/3} = x \Rightarrow a^x = a^{1/3}$
$\log_a \sqrt[3]{a} = \frac{1}{3}$

23. $\log_3 x = 4 \Rightarrow 3^4 = x \Rightarrow x = 81$

24. $\log_{\sqrt{3}} x = 4 \Rightarrow \left(\sqrt{3}\right)^4 = x \Rightarrow x = 9$

25. $\log_{\sqrt{3}} x = 6 \Rightarrow \left(\sqrt{3}\right)^6 = x \Rightarrow x = 27$

26. $\log_{0.1} 10 = x \Rightarrow (0.1)^x = 10$
$\left(\frac{1}{10}\right)^x = 10 \Rightarrow x = -1$

27. $\log_x 2 = -\frac{1}{3} \Rightarrow x^{-1/3} = 2$
$\left(x^{-1/3}\right)^{-3} = 2^{-3} \Rightarrow x = \frac{1}{8}$

28. $\log_x 16 = 4 \Rightarrow x^4 = 16 \Rightarrow x = 2$

29. $\log_{0.25} x = -1 \Rightarrow (0.25)^{-1} = x$
$\left(\frac{1}{4}\right)^{-1} = x \Rightarrow x = 4$

30. $\log_{0.125} x = -\frac{1}{3} \Rightarrow (0.125)^{-1/3} = x$
$\left(\frac{1}{8}\right)^{-1/3} = x \Rightarrow x = 2$

31. $\log_{\sqrt{2}} 32 = x \Rightarrow \left(\sqrt{2}\right)^x = 32$
$\left(2^{1/2}\right)^x = 2^5 \Rightarrow \frac{1}{2}x = 5 \Rightarrow x = 10$

32. $\log_{\sqrt{5}} x = -4 \Rightarrow \left(\sqrt{5}\right)^{-4} = x$
$\left(5^{1/2}\right)^{-4} = x \Rightarrow 5^{-2} = x \Rightarrow x = \frac{1}{25}$

33. $\log_{\sqrt{3}} 9\sqrt{3} = x \Rightarrow \left(\sqrt{3}\right)^x = 9\sqrt{3}$
$\left(3^{1/2}\right)^x = 3^{5/2} \Rightarrow x = 5$

34. $\log_{\sqrt{5}} 5\sqrt{5} = x \Rightarrow \left(\sqrt{5}\right)^x = 5\sqrt{5}$
$\left(5^{1/2}\right)^x = 5^{3/2} \Rightarrow x = 3$

35. $y = f(x) = \log(x-2)$
Shift $y = \log x$ right 2.

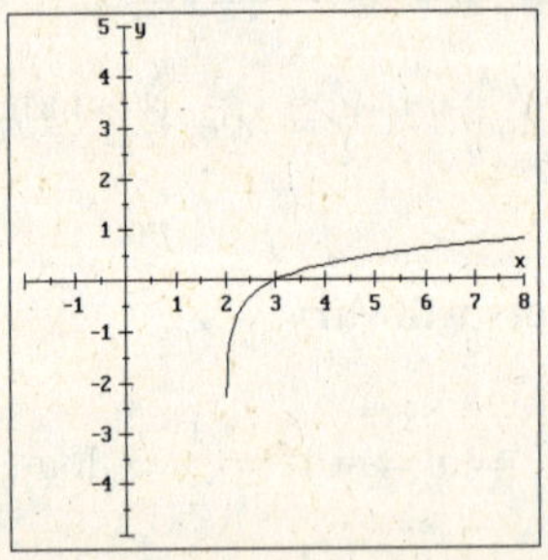

36. $y = f(x) = 3 + \log x$
Shift $y = \log x$ up 3.

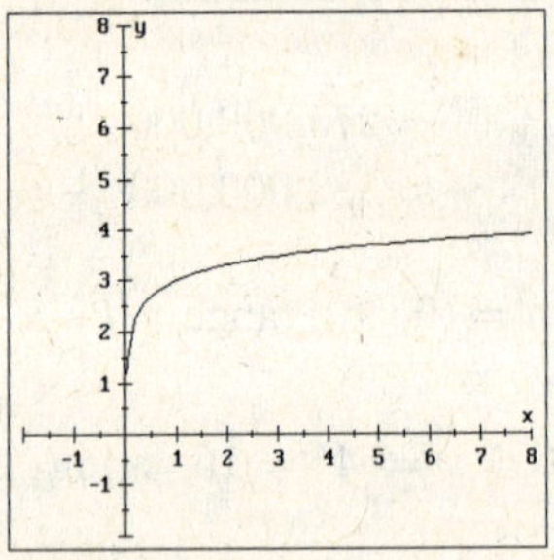

© 2015 Cengage Learning. All Rights Reserved. May not be scanned, copied or duplicated, or posted to a publicly accessible website, in whole or in part.

37. $y = 4^x$
$y = \log_4 x$

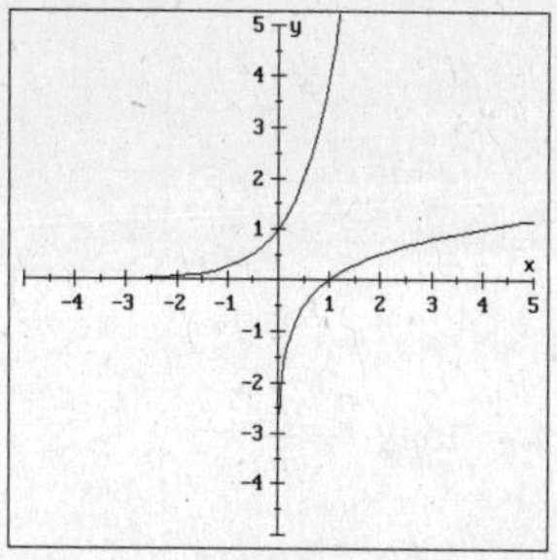

38. $y = \left(\frac{1}{3}\right)^x$
$y = \log_{1/3} x$

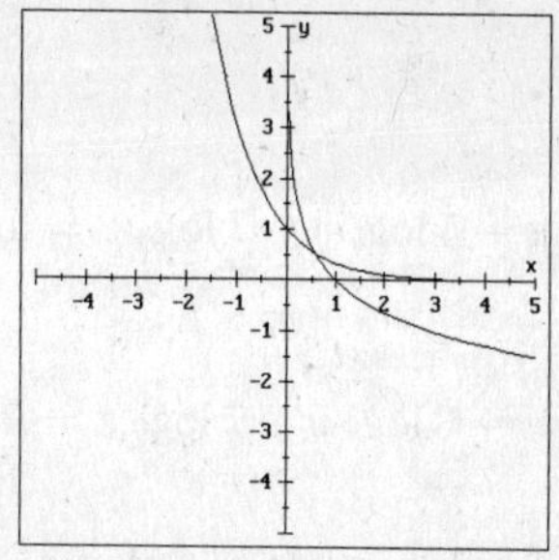

39. dB gain $= 20 \log \frac{E_O}{E_I} = 20 \log \frac{18}{0.04} = 20 \log 450 \approx 53$ dB

40. $R = \log \frac{A}{P} = \log \frac{7500}{0.3} = \log 25{,}000 \approx 4.4$

41. $\ln 362 \approx 5.8916$

42. $\ln(\log 7.85) \approx \ln 0.8949 \approx -0.1111$

43. $\ln x = 2.336 \Rightarrow x = 10.3398$

44. $\ln x = \log 8.8 \Rightarrow x = 2.5715$

45. $y = f(x) = 1 + \ln x$
Shift $y = \ln x$ up 1.

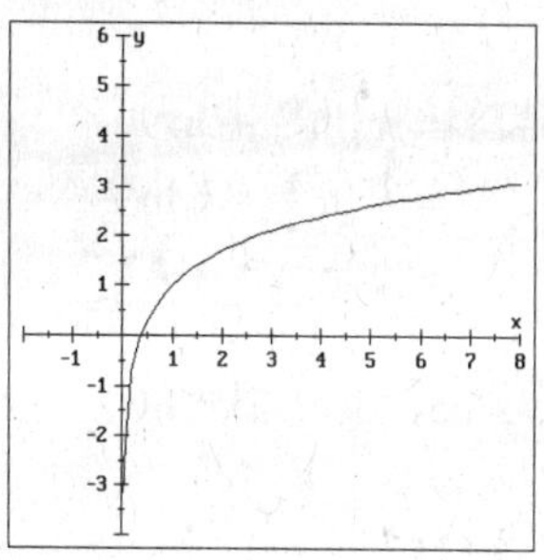

46. $y = f(x) = \ln(x + 1)$
Shift $y = \ln x$ left 1.

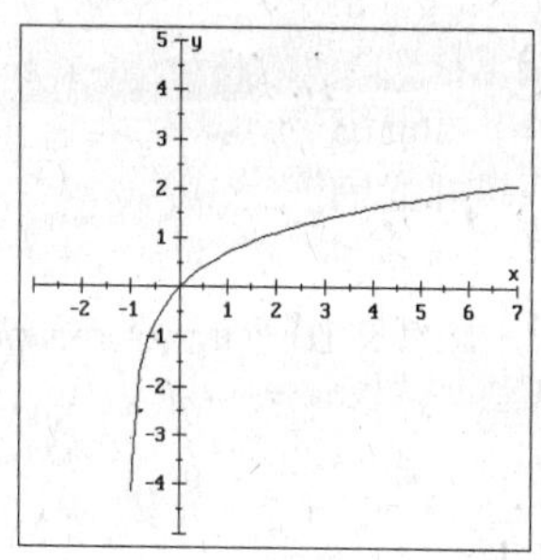

47. $t = \dfrac{\ln 2}{r} = \dfrac{\ln 2}{0.03} \approx 23$ years

48. $\log_5 1 = 0$

49. $\log_9 9 = 1$

50. $\log_7 7^3 = 3$

51. $8^{\log_8 5} = 5$

52. $\log 10^5 = 5$

53. $\ln 1 = 0$

54. $10^{\log_{10} 7} = 7$

55. $e^{\ln 2} = 2$

56. $\log_a a^6 = 6$

57. $\ln e^3 = 3$

58. $\log_b \dfrac{x^5 y^4}{z^2} = \log_b x^5 + \log_b y^4 - \log_b z^2 = 5 \log_b x + 4 \log_b y - 2 \log_b z$

© 2015 Cengage Learning. All Rights Reserved. May not be scanned, copied or duplicated, or posted to a publicly accessible website, in whole or in part.

59. $\log_b \sqrt{\dfrac{x}{yz^2}} = \log_b \left(\dfrac{x}{yz^2}\right)^{1/2} = \dfrac{1}{2}\log_b \dfrac{x}{yz^2} = \dfrac{1}{2}(\log_b x - \log_b y - \log_b z^2)$
$= \dfrac{1}{2}(\log_b x - \log_b y - 2\log_b z)$

60. $3\log_b x - 5\log_b y + 7\log_b z = \log_b x^3 - \log_b y^5 + \log_b z^7 = \log_b \dfrac{x^3 z^7}{y^5}$

61. $\dfrac{1}{2}\log_b x + 3\log_b y - 7\log_b z = \log_b x^{1/2} + \log_b y^3 - \log_b z^7 = \log_b \dfrac{y^3\sqrt{x}}{z^7}$

62. $\log abc = \log a + \log b + \log c = 0.6 + 0.36 + 2.4 = 3.36$

63. $\log a^2 b = \log a^2 + \log b = 2\log a + \log b = 2(0.6) + 0.36 = 1.56$

64. $\log \dfrac{ac}{b} = \log a + \log c - \log b = 0.6 + 2.4 - 0.36 = 2.64$

65. $\log \dfrac{a^2}{c^3 b^2} = \log a^2 - \log c^3 - \log b^2 = 2\log a - 3\log c - 2\log b$
$= 2(0.6) - 3(2.4) - 2(0.36) = -6.72$

66. $\log_5 17 = \dfrac{\log 17}{\log 5} \approx 1.7604$

67.
$$\begin{aligned} \text{pH} &= -\log[\text{H}^+] \\ 3.1 &= -\log[\text{H}^+] \\ -3.1 &= \log[\text{H}^+] \\ [\text{H}^+] &= 7.94 \times 10^{-4} \text{ gram-ions per liter} \end{aligned}$$

68.
$$\begin{aligned} k\ln\left(\tfrac{1}{2}I\right) &= k\left(\ln\tfrac{1}{2} + \ln I\right) \\ &= k\ln\tfrac{1}{2} + k\ln I \\ &= k\ln 2^{-1} + L \\ &= -k\ln 2 + L \end{aligned}$$

The loudness decreases by $k\ln 2$.

69.
$$\begin{aligned} 3^x &= 7 \\ \log 3^x &= \log 7 \\ x\log 3 &= \log 7 \\ x &= \frac{\log 7}{\log 3} \approx 1.7712 \end{aligned}$$

70.
$$\begin{aligned} 5^{x+2} &= 625 \\ 5^{x+2} &= 5^4 \\ x+2 &= 4 \\ x &= 2 \end{aligned}$$

© 2015 Cengage Learning. All Rights Reserved. May not be scanned, copied or duplicated, or posted to a publicly accessible website, in whole or in part.

71.
$$\begin{aligned}25 &= 5.5(1.05)^t\\ \frac{25}{5.5} &= (1.05)^t\\ \log\frac{25}{5.5} &= \log(1.05)^t\\ \log\frac{25}{5.5} &= t\log 1.05\\ \frac{\log\frac{25}{5.5}}{\log 1.05} &= t\\ 31.0335 &\approx t\end{aligned}$$

72.
$$\begin{aligned}4^{2t-1} &= 64\\ 4^{2t-1} &= 4^3\\ 2t-1 &= 3\\ 2t &= 4\\ t &= 2\end{aligned}$$

73.
$$\begin{aligned}2^x &= 3^{x-1}\\ \log 2^x &= \log 3^{x-1}\\ x\log 2 &= (x-1)\log 3\\ x\log 2 &= x\log 3 - \log 3\\ \log 3 &= x\log 3 - x\log 2\\ \log 3 &= x(\log 3 - \log 2)\\ \frac{\log 3}{\log 3 - \log 2} &= x\\ 2.7095 &\approx x\end{aligned}$$

74.
$$\begin{aligned}3^{x^2+4x} &= \frac{1}{27}\\ 3^{x^2+4x} &= 3^{-3}\\ x^2+4x &= -3\\ x^2+4x+3 &= 0\\ (x+3)(x+1) &= 0\end{aligned}$$
$x+3=0$ **or** $x+1=0$
$x=-3$ $\quad x=-1$

75.
$$\begin{aligned}\log x + \log(29-x) &= 2\\ \log x(29-x) &= 2\\ 10^2 &= x(29-x)\\ 100 &= 29x - x^2\\ x^2 - 29x + 100 &= 0\\ (x-25)(x-4) &= 0\end{aligned}$$
$x-25=0$ **or** $x-4=0$
$x=25$ $\quad x=4$

76.
$$\begin{aligned}\log_2 x + \log_2(x-2) &= 3\\ \log_2 x(x-2) &= 3\\ x(x-2) &= 2^3\\ x^2-2x &= 8\\ x^2-2x-8 &= 0\\ (x-4)(x+2) &= 0\end{aligned}$$
$x-4=0$ **or** $x+2=0$
$x=4$ $\quad x=-2$
Extraneous

77.
$$\begin{aligned}\log_2(x+2)+\log_2(x-1) &= 2\\ \log_2(x+2)(x-1) &= 2\\ (x+2)(x-1) &= 2^2\\ x^2+x-2 &= 4\\ x^2+x-6 &= 0\\ (x-2)(x+3) &= 0\end{aligned}$$
$x-2=0$ **or** $x+3=0$
$x=2$ $\quad x=-3$
Extraneous

78.
$$\begin{aligned}\frac{\log(7x-12)}{\log x} &= 2\\ \log(7x-12) &= 2\log x\\ \log(7x-12) &= \log x^2\\ 7x-12 &= x^2\\ 0 &= x^2-7x+12\\ 0 &= (x-4)(x-3)\end{aligned}$$
$x-4=0$ **or** $x-3=0$
$x=4$ $\quad x=3$

© 2015 Cengage Learning. All Rights Reserved. May not be scanned, copied or duplicated, or posted to a publicly accessible website, in whole or in part.

79.
$$\begin{aligned}\log x + \log(x-5) &= \log 6\\ \log x(x-5) &= \log 6\\ x(x-5) &= 6\\ x^2 - 5x - 6 &= 0\\ (x-6)(x+1) &= 0\\ x-6=0 \quad &\textbf{or} \quad x+1=0\\ x=6 \quad & \quad x=-1\\ & \quad \text{Extraneous}\end{aligned}$$

80.
$$\begin{aligned}\log 3 - \log(x-1) &= -1\\ \log\frac{3}{x-1} &= -1\\ \frac{3}{x-1} &= 10^{-1}\\ \frac{3}{x-1} &= \frac{1}{10}\\ 30 &= x-1\\ 31 &= x\end{aligned}$$

81.
$$\begin{aligned}e^{x\ln 2} &= 9\\ e^{\ln 2^x} &= 9\\ 2^x &= 9\\ \ln 2^x &= \ln 9\\ x\ln 2 &= \ln 9\\ x &= \frac{\ln 9}{\ln 2} \approx 3.1699\end{aligned}$$

82.
$$\begin{aligned}\ln x &= \ln(x-1)\\ x &= x-1\\ 0 &= -1\end{aligned}$$
There is no solution. $\emptyset$

83.
$$\begin{aligned}\ln x &= \ln(x-1)+1\\ \ln x - \ln(x-1) &= 1\\ \ln\frac{x}{x-1} &= 1\\ \frac{x}{x-1} &= e^1\\ x &= e(x-1)\\ x &= ex - e\\ e &= ex - x\\ e &= x(e-1)\\ \frac{e}{e-1} &= x\\ 1.5820 &\approx x\end{aligned}$$

84.
$$\begin{aligned}\ln x &= \log_{10} x\\ \ln x &= \frac{\ln x}{\ln 10}\\ \ln x \ln 10 &= \ln x\\ \ln x \ln 10 - \ln x &= 0\\ \ln x(\ln 10 - 1) &= 0\\ \ln x &= 0\\ x &= 1\end{aligned}$$

85.
$$\begin{aligned}A &= A_0 2^{-t/h}\\ \frac{2}{3}A_0 &= A_0 2^{-t/5730}\\ \frac{2}{3} &= 2^{-t/5730}\\ \log\frac{2}{3} &= \log 2^{-t/5730}\\ \log\frac{2}{3} &= -\frac{t}{5730}\log 2\\ \frac{\log\frac{2}{3}}{-\log 2} &= \frac{t}{5730}\\ -5730\frac{\log\frac{2}{3}}{\log 2} &= t\end{aligned}$$
3352 years $\approx t$ [or about 3400 years]

© 2015 Cengage Learning. All Rights Reserved. May not be scanned, copied or duplicated, or posted to a publicly accessible website, in whole or in part.

Chapter 9 Test (page 686)

1. $f(x) = 2^x + 1$; Shift $y = 2^x$ up 1.

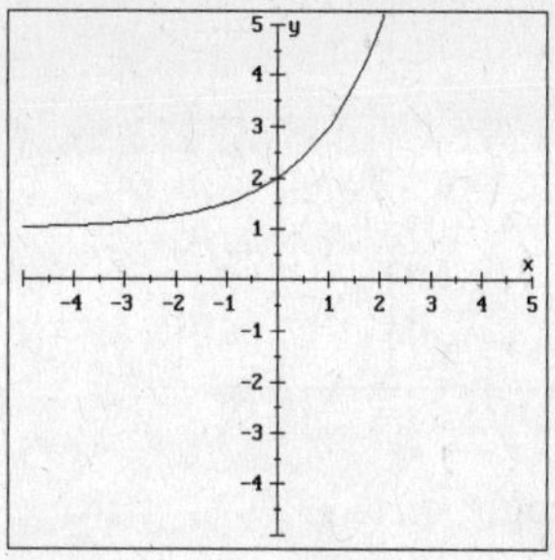

2. $f(x) = 2^{-x}$; Reflect $y = 2^x$ about y-axis.

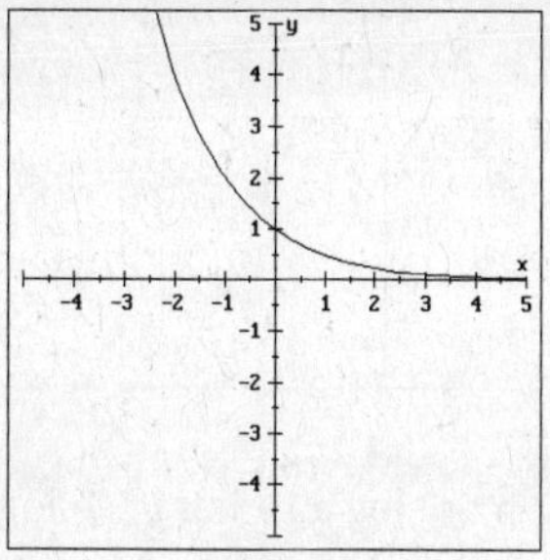

3. $A = A_0(2)^{-t} = 3(2)^{-6} = \frac{3}{2^6} = \frac{3}{64}$ gram

4. $A = A_0\left(1 + \frac{r}{k}\right)^{kt} = 1000\left(1 + \frac{0.06}{2}\right)^{2(1)} = 1000(1.03)^2 \approx \$1060.90$

5. $f(x) = e^x$

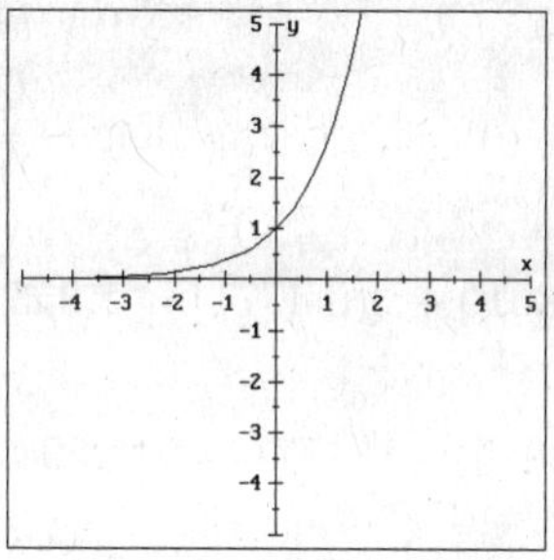

6. $A = A_0e^{rt} = 2000e^{(0.08)10}$
$= 2000e^{0.8}$
$\approx \$4451.08$

7. $\log_9 81 = x \Rightarrow 9^x = 81 \Rightarrow x = 2$

8. $\log_x 81 = 4 \Rightarrow x^4 = 81 \Rightarrow x = 3$

9. $\log_3 x = -3 \Rightarrow 3^{-3} = x \Rightarrow x = \frac{1}{27}$

10. $\log_x 49 = 2 \Rightarrow x^2 = 49 \Rightarrow x = 7$

11. $\log_{3/2} \frac{9}{4} = x \Rightarrow \left(\frac{3}{2}\right)^x = \frac{9}{4} \Rightarrow x = 2$

12. $\log_{2/3} x = -3 \Rightarrow \left(\frac{2}{3}\right)^{-3} = x \Rightarrow x = \frac{27}{8}$

© 2015 Cengage Learning. All Rights Reserved. May not be scanned, copied or duplicated, or posted to a publicly accessible website, in whole or in part.

CHAPTER 9 TEST

13. $f(x) = -\log_3 x$

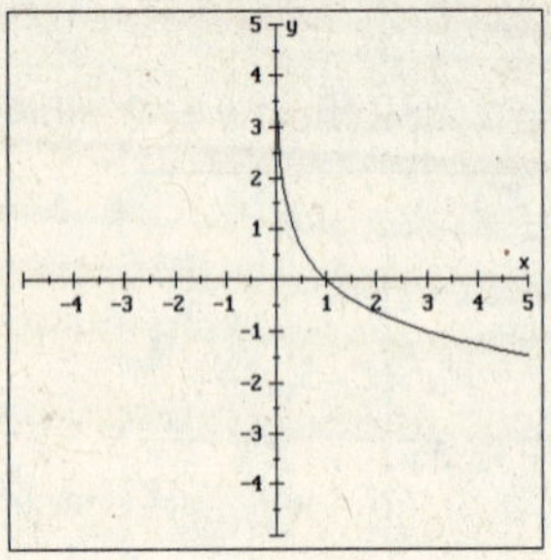

14. $f(x) = \ln x$

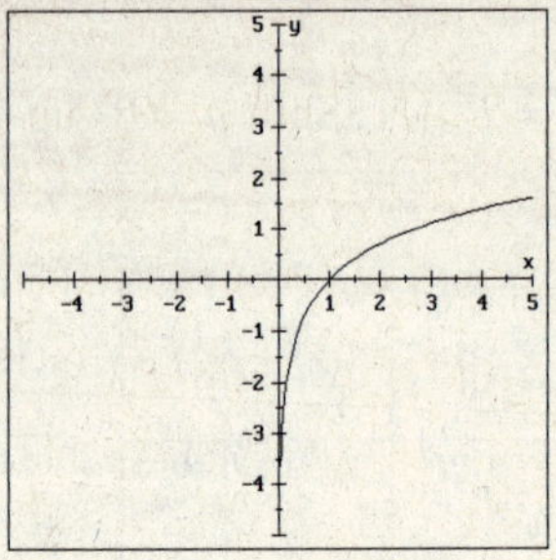

15. $\log xy^3z^4 = \log x + \log y^3 + \log z^4 = \log x + 3\log y + 4\log z$

16. $\ln\sqrt{\dfrac{a}{b^2c}} = \ln\left(\dfrac{a}{b^2c}\right)^{1/2} = \dfrac{1}{2}\ln\dfrac{a}{b^2c} = \dfrac{1}{2}(\ln a - \ln b^2 - \ln c) = \dfrac{1}{2}(\ln a - 2\ln b - \ln c)$

17. $\dfrac{1}{2}\log(a+2) + \log b - 3\log c = \log(a+2)^{1/2} + \log b - \log c^3 = \log\dfrac{b\sqrt{a+2}}{c^3}$

18. $\dfrac{1}{3}(\log a - 2\log b) - \log c = \dfrac{1}{3}(\log a - \log b^2) - \log c = \dfrac{1}{3}\log\dfrac{a}{b^2} - \log c = \log\sqrt[3]{\dfrac{a}{b^2}} - \log c$

$= \log\dfrac{\sqrt[3]{a}}{c\sqrt[3]{b^2}}$

19. $\log 18 = \log 2\cdot 3^2 = \log 2 + \log 3^2 = \log 2 + 2\log 3 = 0.3010 + 2(0.4771) = 1.2552$

20. $\log\dfrac{8}{3} = \log\dfrac{2^3}{3} = \log 2^3 - \log 3 = 3\log 2 - \log 3 = 3(0.3010) - 0.4771 = 0.4259$

21. $\log_3 2 = \dfrac{\log 2}{\log 3}$ or $\dfrac{\ln 2}{\ln 3}$

22. $\log_\pi e = \dfrac{\log e}{\log \pi}$ or $\dfrac{\ln e}{\ln \pi} = \dfrac{1}{\ln \pi}$

23. $\log_a ab = \log_a a + \log_a b = 1 + \log_a b \Rightarrow$ TRUE

24. $\dfrac{\log a}{\log b} = \log a - \log b \Rightarrow$ FALSE $\left(\log\dfrac{a}{b} = \log a - \log b\right)$

25. $\log a^{-3} = -3\log a \neq \dfrac{1}{3\log a} \Rightarrow$ FALSE

26. $\ln(-x) = -\ln x \Rightarrow$ FALSE (You cannot take the logarithm of a negative number.)

27. $\text{pH} = -\log[\text{H}^+] = -\log(3.7\times 10^{-7}) \approx 6.4$

28. $\text{dB gain} = 20\log\dfrac{E_O}{E_I} = 20\log\dfrac{60}{0.3} = 20\log 200 \approx 46$

© 2015 Cengage Learning. All Rights Reserved. May not be scanned, copied or duplicated, or posted to a publicly accessible website, in whole or in part.

29.
$$\begin{aligned} 7^x &= 4 \\ \log 7^x &= \log 4 \\ x \log 7 &= \log 4 \\ x &= \frac{\log 4}{\log 7} \\ x &\approx 0.7124 \end{aligned}$$

30.
$$\begin{aligned} 3^{x-1} &= 100^x \\ \log 3^{x-1} &= \log 100^x \\ (x-1)\log 3 &= x \log 100 \\ x \log 3 - \log 3 &= 2x \\ x \log 3 - 2x &= \log 3 \\ (\log 3 - 2) &= \log 3 \\ x &= \frac{\log 3}{\log 3 - 2} \\ x &\approx -0.3133 \end{aligned}$$

31.
$$\begin{aligned} \log(3x+1) &= \log(5x-7) \\ 3x+1 &= 5x-7 \\ -2x &= -8 \\ x &= 4 \end{aligned}$$

32.
$$\begin{aligned} \log x + \log(x-9) &= 1 \\ \log x(x-9) &= 1 \\ x(x-9) &= 10 \\ x^2 - 9x - 10 &= 0 \\ (x-10)(x+1) &= 0 \end{aligned}$$
$$x - 10 = 0 \quad \textbf{or} \quad x + 1 = 0$$
$$x = 10 \qquad\qquad x = -1$$
Extraneous

Exercises 10.1 (page 698)

1. $a = -5$: downward

3. $a = 1$: upward

5. $x^2 + 14x + 49 = (x+7)^2$

7. $x^2 - 20x + 100 = (x - 10)$

9.
$$|2x - 5| = 7$$
$$2x - 5 = 7 \quad \textbf{or} \quad 2x - 5 = -7$$
$$2x = 12 \qquad\qquad 2x = -2$$
$$x = 6 \qquad\qquad x = -1$$

11.
$$|3x + 4| = |5x - 2|$$
$$3x + 4 = 5x - 2 \quad \textbf{or} \quad 3x + 4 = -(5x - 2)$$
$$-2x = -6 \qquad\qquad 3x + 4 = -5x + 2$$
$$x = 3 \qquad\qquad 8x = -2$$
$$x = -\tfrac{1}{4}$$

13. conic

15. standard; $(0, 3)$; 4

17. circle; general

19. parabola; $(3, 2)$; right

© 2015 Cengage Learning. All Rights Reserved. May not be scanned, copied or duplicated, or posted to a publicly accessible website, in whole or in part.

21. $x^2 + y^2 = 9$

$C\ (0, 0);\ r = \sqrt{9} = 3$

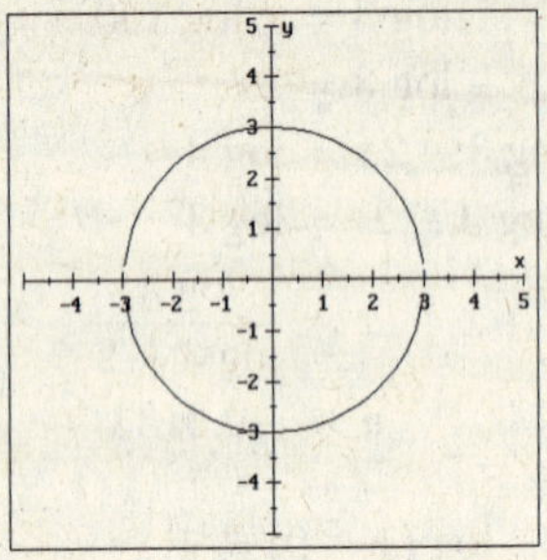

23. $(x - 2)^2 + y^2 = 9$

$C\ (2, 0);\ r = \sqrt{9} = 3$

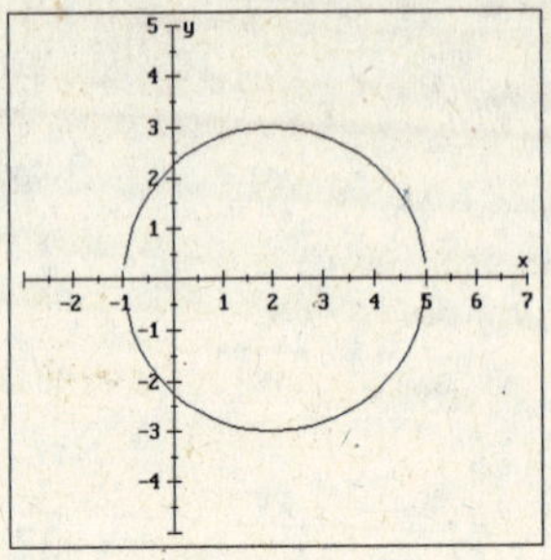

25. $(x - 2)^2 + (y - 4)^2 = 4$

$C\ (2, 4);\ r = \sqrt{4} = 2$

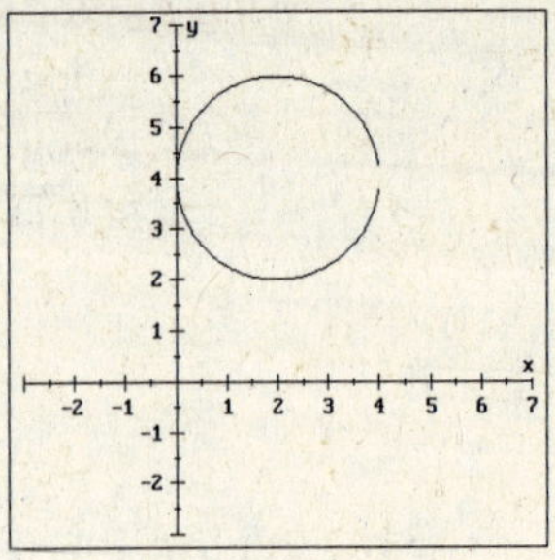

27. $(x + 3)^2 + (y - 1)^2 = 16$

$C\ (-3, 1);\ r = \sqrt{16} = 4$

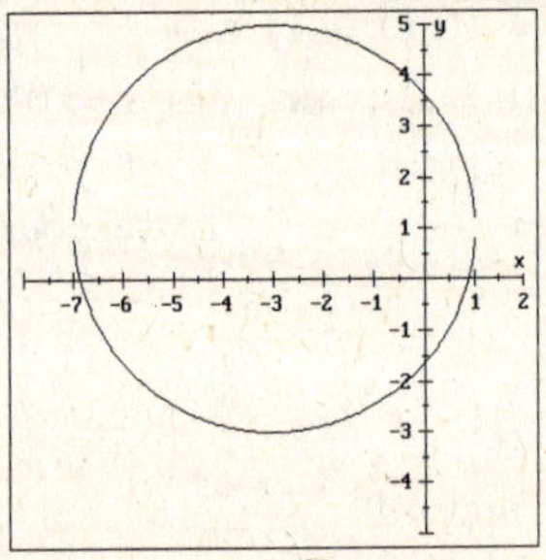

29.
$$x^2 + y^2 + 2x - 8 = 0$$
$$x^2 + 2x + y^2 = 8$$
$$x^2 + 2x + 1 + y^2 = 8 + 1$$
$$(x + 1)^2 + y^2 = 9$$

$C\ (-1, 0);\ r = \sqrt{9} = 3$

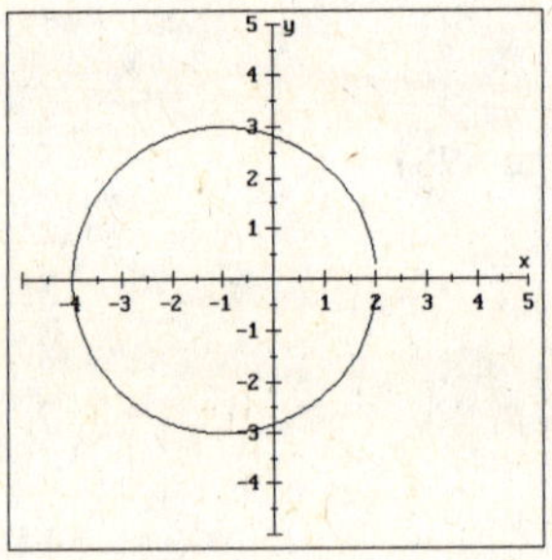

31.
$$9x^2 + 9y^2 - 12y = 5$$
$$x^2 + y^2 - \frac{4}{3}y = \frac{5}{9}$$
$$x^2 + y^2 - \frac{4}{3}y + \frac{4}{9} = \frac{5}{9} + \frac{4}{9}$$
$$x^2 + \left(y - \frac{2}{3}\right)^2 = 1$$

$C\ \left(0, \frac{2}{3}\right);\ r = \sqrt{1} = 1$

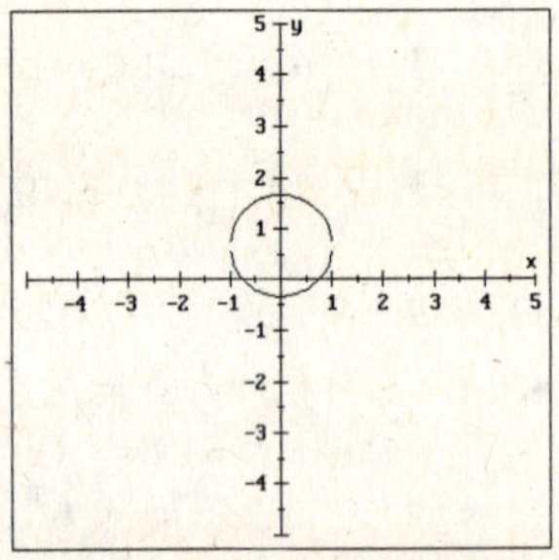

© 2015 Cengage Learning. All Rights Reserved. May not be scanned, copied or duplicated, or posted to a publicly accessible website, in whole or in part.

33.
$$x^2 + y^2 - 2x + 4y = -1$$
$$x^2 - 2x + y^2 + 4y = -1$$
$$x^2 - 2x + 1 + y^2 + 4y + 4 = -1 + 1 + 4$$
$$(x-1)^2 + (y+2)^2 = 4$$
C $(1, -2)$; $r = \sqrt{4} = 2$

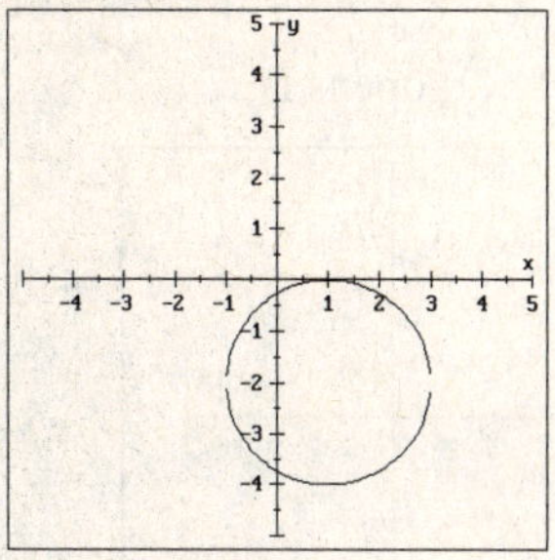

35.
$$x^2 + y^2 + 6x - 4y = -12$$
$$x^2 + 6x + y^2 - 4y = -12$$
$$x^2 + 6x + 9 + y^2 - 4y + 4 = -12 + 9 + 4$$
$$(x+3)^2 + (y-2)^2 = 1$$
C $(-3, 2)$; $r = \sqrt{1} = 1$

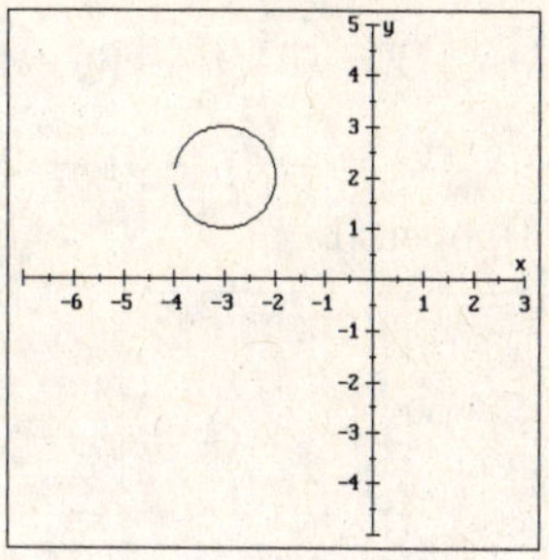

37. $(x-h)^2 + (y-k)^2 = r^2$
$(x-0)^2 + (y-0)^2 = 3^2$
$\boxed{x^2 + y^2 = 9}$
$\boxed{x^2 + y^2 - 9 = 0}$

39. $(x-h)^2 + (y-k)^2 = r^2$
$(x-2)^2 + (y-4)^2 = 7^2$
$\boxed{(x-2)^2 + (y-4)^2 = 49}$
$x^2 - 4x + 4 + y^2 - 8y + 16 = 49$
$x^2 + y^2 - 4x - 8y + 20 = 49$
$\boxed{x^2 + y^2 - 4x - 8y - 29 = 0}$

41. $(x-h)^2 + (y-k)^2 = r^2$
$(x-(-4))^2 + (y-7)^2 = 10^2$
$\boxed{(x+4)^2 + (y-7)^2 = 100}$
$x^2 + 8x + 16 + y^2 - 14y + 49 = 100$
$x^2 + y^2 + 8x - 14y + 65 = 100$
$\boxed{x^2 + y^2 + 8x - 14y - 35 = 0}$

43. $(x-h)^2 + (y-k)^2 = r^2$
$(x-0)^2 + (y-0)^2 = \left(4\sqrt{5}\right)^2$
$\boxed{x^2 + y^2 = 80}$
$\boxed{x^2 + y^2 - 80 = 0}$

45. $x = y^2$
$x = (y-0)^2 + 0$; V $(0, 0)$; opens R

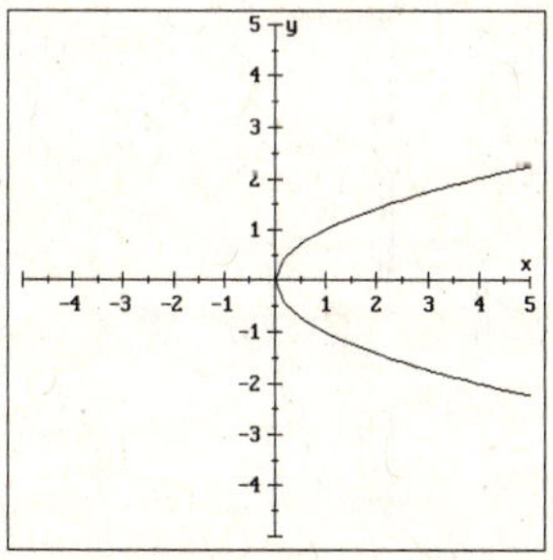

47. $x = -\frac{1}{4}y^2$
$x = -\frac{1}{4}(y-0)^2 + 0$; V $(0, 0)$; opens L

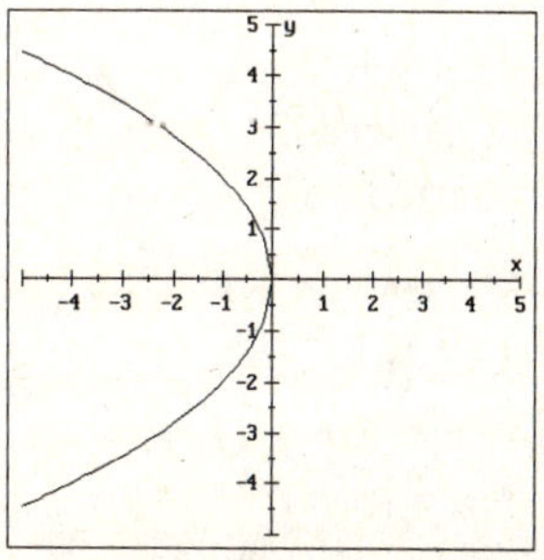

© 2015 Cengage Learning. All Rights Reserved. May not be scanned, copied or duplicated, or posted to a publicly accessible website, in whole or in part.

49. $y^2 + 4x - 6y = -1$

$$4x = -y^2 + 6y - 1$$
$$x = -\frac{1}{4}y^2 + \frac{3}{2}y - \frac{1}{4}$$
$$x = -\frac{1}{4}(y^2 - 6y) - \frac{1}{4}$$
$$x = -\frac{1}{4}(y^2 - 6y + 9) - \frac{1}{4} + \frac{9}{4}$$
$$x = -\tfrac{1}{4}(y-3)^2 + 2$$

V $(2, 3)$; opens L

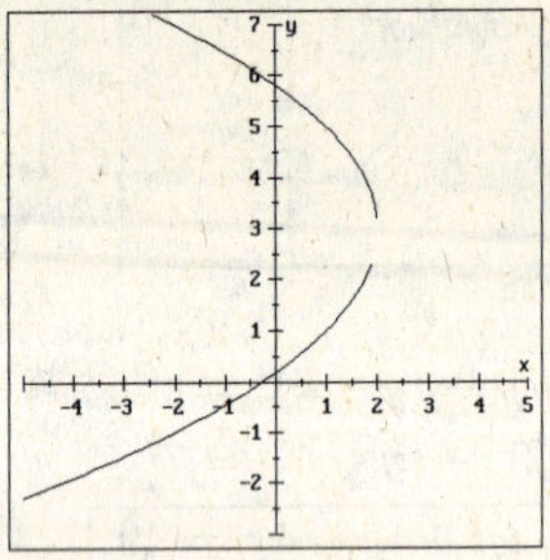

51. $y = -x^2 - x + 1$

$$y = -(x^2 + x) + 1$$
$$y = -\left(x^2 + x + \frac{1}{4}\right) + 1 + \frac{1}{4}$$
$$y = -\left(x + \frac{1}{2}\right)^2 + \frac{5}{4}$$

V $\left(-\frac{1}{2}, \frac{5}{4}\right)$; opens D

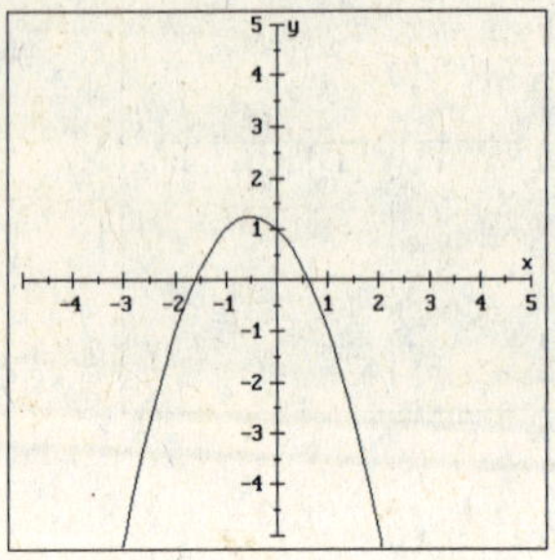

53. $x^2 + (y+3)^2 = 1$

C $(0, -3)$; $r = \sqrt{1} = 1$

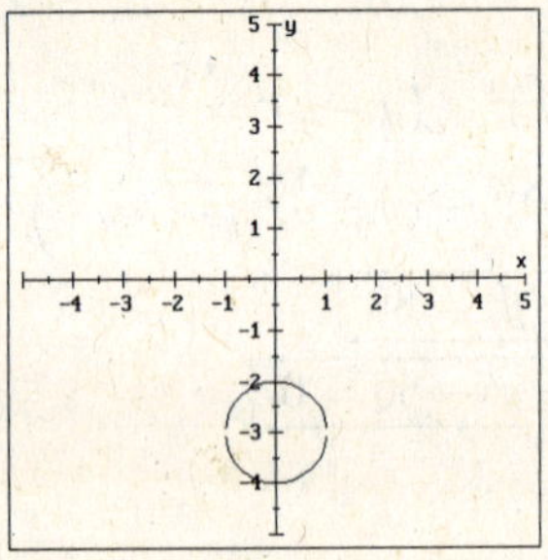

55. $y = 2(x-1)^2 + 3$

V $(1, 3)$; opens U

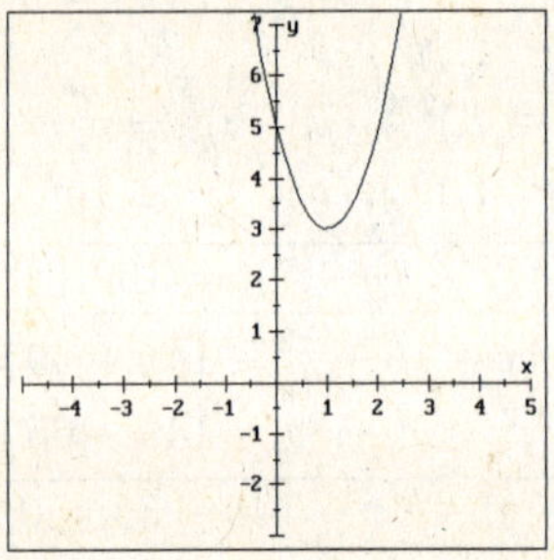

57. $y = x^2 + 4x + 5$

$$y = x^2 + 4x + 4 + 5 - 4$$
$$y = (x+2)^2 + 1$$

V $(-2, 1)$; opens U

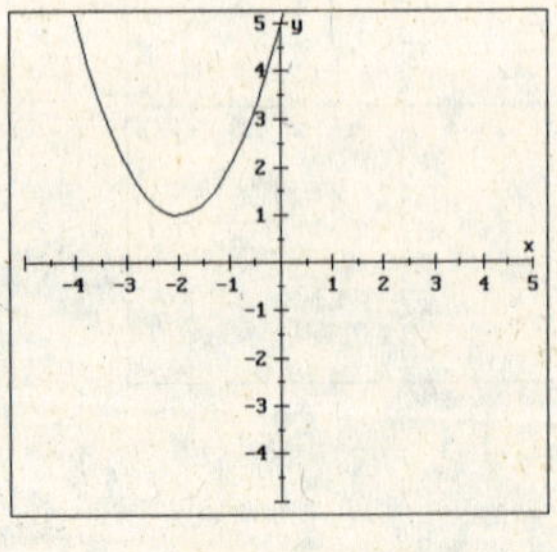

59. $y = 3(x+1)^2 - 2$

V $(-1, -2)$; opens U

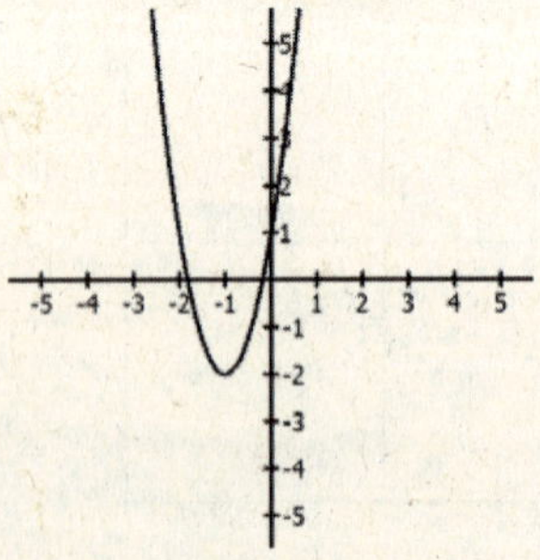

© 2015 Cengage Learning. All Rights Reserved. May not be scanned, copied or duplicated, or posted to a publicly accessible website, in whole or in part.

61. $3x^2 + 3y^2 = 16$

$$3y^2 = 16 - 3x^2$$

$$y^2 = \frac{16 - 3x^2}{3}$$

$$y = \pm\sqrt{\frac{16 - 3x^2}{3}}$$

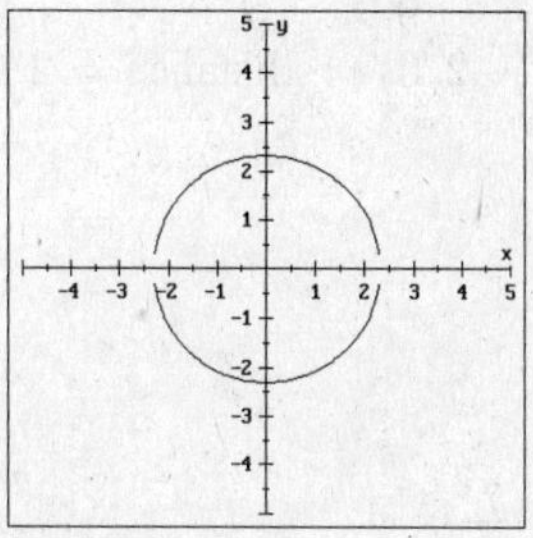

63. $(x+1)^2 + y^2 = 16$

$$y^2 = 16 - (x+1)^2$$

$$y = \pm\sqrt{16 - (x+1)^2}$$

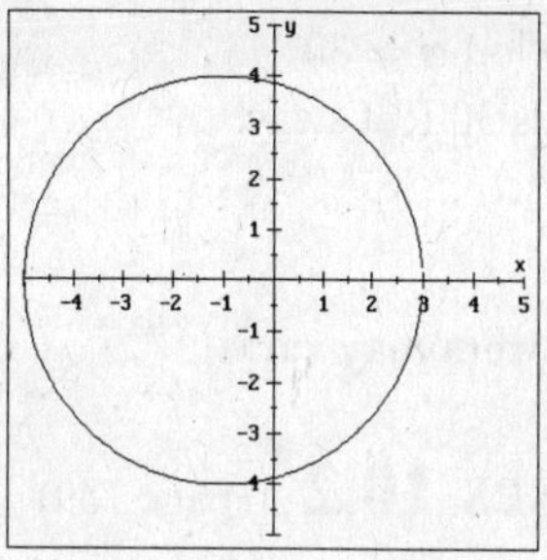

65. $x = 2y^2$

$$y^2 = \frac{x}{2}$$

$$y = \pm\sqrt{\tfrac{x}{2}}$$

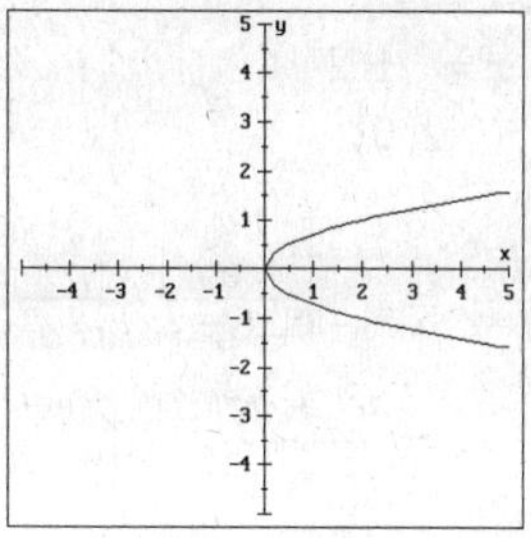

67. $x^2 - 2x + y = 6$

$$y = 6 - x^2 + 2x$$

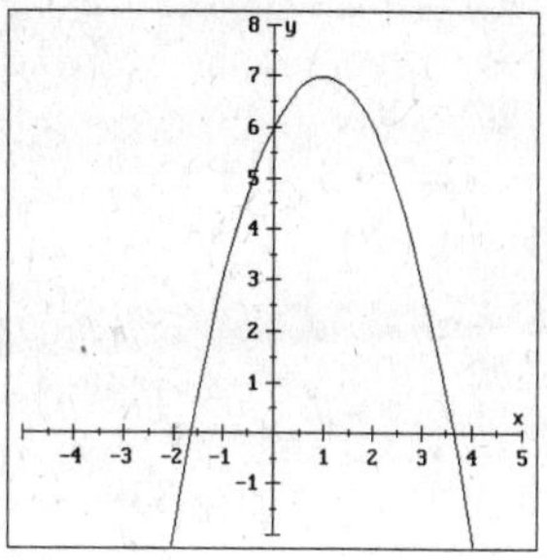

69. The radius of the larger gear is $\sqrt{16} = 4$. Centers: 7 units apart $\Rightarrow$ smaller gear $r = 3$.

$$(x-h)^2 + (y-k)^2 = r^2$$

$$(x-7)^2 + (y-0)^2 = 3^2 \Rightarrow (x-7)^2 + y^2 = 9$$

71. $x^2 + y^2 - 8x - 20y + 16 = 0$

$$x^2 - 8x + y^2 - 20y = -16$$

$$x^2 - 8x + 16 + y^2 - 20y + 100 = -16 + 16 + 100$$

$$(x-4)^2 + (y-10)^2 = 100$$

center: $(4, 10)$; radius $= 10$

$$x^2 + y^2 + 2x + 4y - 11 = 0$$

$$x^2 + 2x + 1 + y^2 + 4y + 4 = 11 + 1 + 4$$

$$(x+1)^2 + (y+2)^2 = 16$$

center: $(-1, -2)$; radius $= 4$

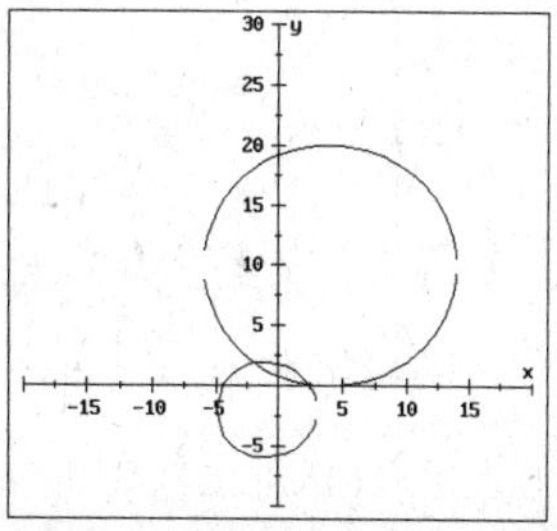

Since the ranges overlap (see graph), they can not be licensed for the same frequency.

© 2015 Cengage Learning. All Rights Reserved. May not be scanned, copied or duplicated, or posted to a publicly accessible website, in whole or in part.

73. Set $y = 0$:
$y = 30x - x^2$
$0 = 30x - x^2$
$0 = x(30 - x)$
$x = 0$ or $x = 30$
It lands 30 feet away.

75. Find the vertex:
$2y^2 - 9x = 18$
$-9x = -2y^2 + 18$
$x = \frac{2}{9}y^2 - 2$
$x = \frac{2}{9}(y - 0)^2 - 2$
vertex: $(-2, 0) \Rightarrow$ distance $= 2$ AU

77-79. Answers may vary.

Exercises 10.2 (page 709)

1.
$$\frac{x^2}{4} + \frac{y^2}{9} = 1$$
$$\frac{0^2}{4} + \frac{y^2}{9} = 1$$
$$\frac{y^2}{9} = 1$$
$$y^2 = 9$$
$$y = \pm 3$$
$(0, 3), (0, -3)$

3.
$$\frac{x^2}{4} + \frac{y^2}{9} = 1$$
$$\frac{x^2}{4} + \frac{0^2}{9} = 1$$
$$\frac{x^2}{4} = 1$$
$$x^2 = 4$$
$$x = \pm 2$$
$(2, 0), (-2, 0)$

5.
$$5x^{-3}y^4(7x^3 + 2y^{-4}) = 35x^0y^4 + 10x^{-3}y^0 = 35y^4 + \frac{10}{x^3}$$

7.
$$\frac{x^{-2} + y^{-2}}{x^{-2} - y^{-2}} = \frac{x^{-2} + y^{-2}}{x^{-2} - y^{-2}} \cdot \frac{x^2y^2}{x^2y^2} = \frac{y^2 + x^2}{y^2 - x^2}$$

9. ellipse; sum

11. center

13. $(0, 0)$; major axis; $2b$

15. $\frac{x^2}{4} + \frac{y^2}{9} = 1$
C $(0, 0)$; move 2 horiz. and 3 vert.

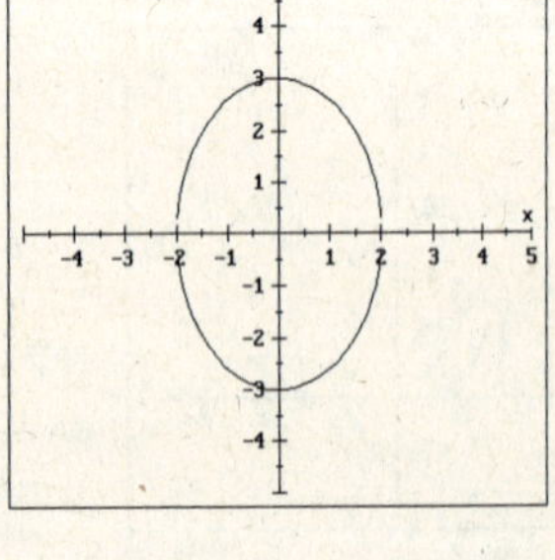

17. $\frac{(x-2)^2}{9} + \frac{(y-1)^2}{4} = 1$
C $(2, 1)$; move 3 horiz. and 2 vert.

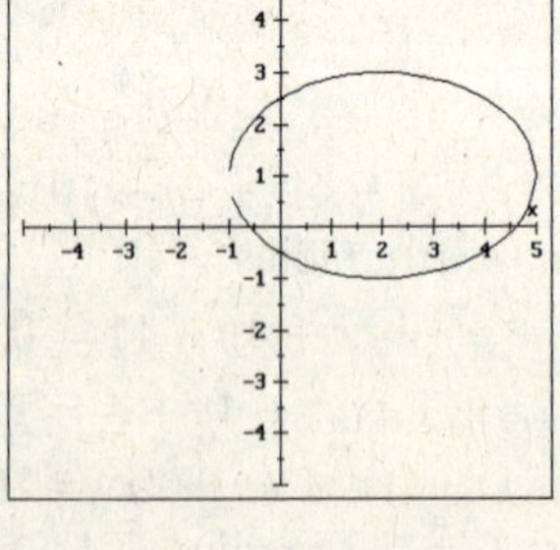

© 2015 Cengage Learning. All Rights Reserved. May not be scanned, copied or duplicated, or posted to a publicly accessible website, in whole or in part.

19. $x^2 + 9y^2 = 9$

$$\frac{x^2}{9} + \frac{9y^2}{9} = \frac{9}{9}$$

$$\frac{x^2}{9} + \frac{y^2}{1} = 1$$

C $(0, 0)$; move 3 horiz. and 1 vert.

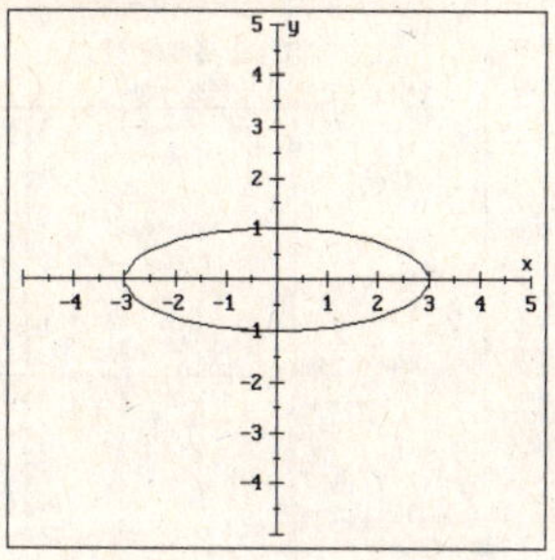

21. $(x+1)^2 + 4(y+2)^2 = 4$

$$\frac{(x+1)^2}{4} + \frac{4(y+2)^2}{4} = \frac{4}{4}$$

$$\frac{(x+1)^2}{4} + \frac{(y+2)^2}{1} = 1$$

C $(-1, -2)$; move 2 horiz. and 1 vert.

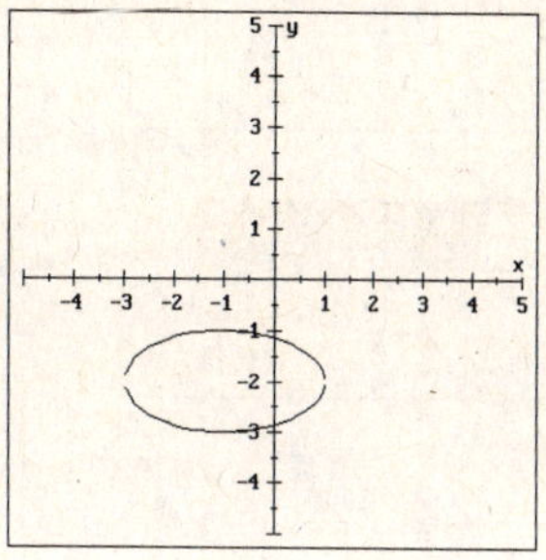

23.

$$x^2 + 4y^2 - 4x + 8y + 4 = 0$$
$$x^2 - 4x + 4(y^2 + 2y) = -4$$
$$x^2 - 4x + 4 + 4(y^2 + 2y + 1) = -4 + 4 + 4$$
$$(x-2)^2 + 4(y+1)^2 = 4$$
$$\frac{(x-2)^2}{4} + \frac{(y+1)^2}{1} = 1$$

C $(2, -1)$; move 2 horiz. and 1 vert.

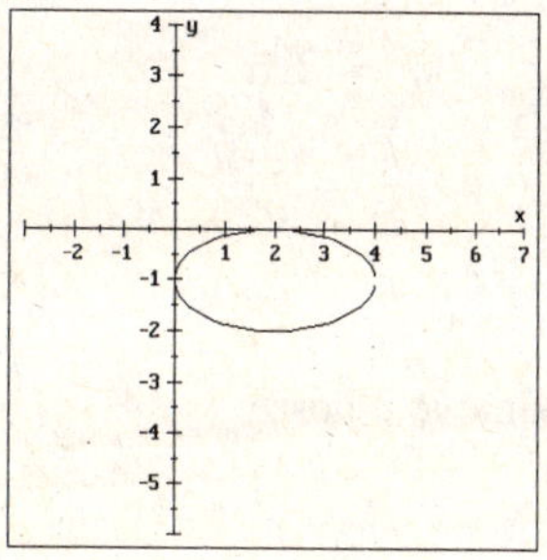

25.

$$9x^2 + 4y^2 - 18x + 16y = 11$$
$$9(x^2 - 2x) + 4(y^2 + 4y) = 11$$
$$9(x^2 - 2x + 1) + 4(y^2 + 4y + 4) = 11 + 9 + 16$$
$$9(x-1)^2 + 4(y+2)^2 = 36$$
$$\frac{(x-1)^2}{4} + \frac{(y+2)^2}{9} = 1$$

C $(1, -2)$; move 2 horiz. and 3 vert.

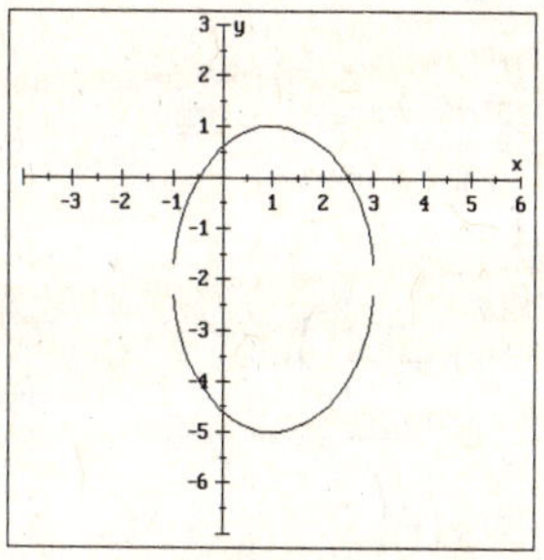

© 2015 Cengage Learning. All Rights Reserved. May not be scanned, copied or duplicated, or posted to a publicly accessible website, in whole or in part.

27. $\dfrac{x^2}{9}+\dfrac{y^2}{16}=1$

C $(0,0)$; move 3 horiz. and 4 vert.

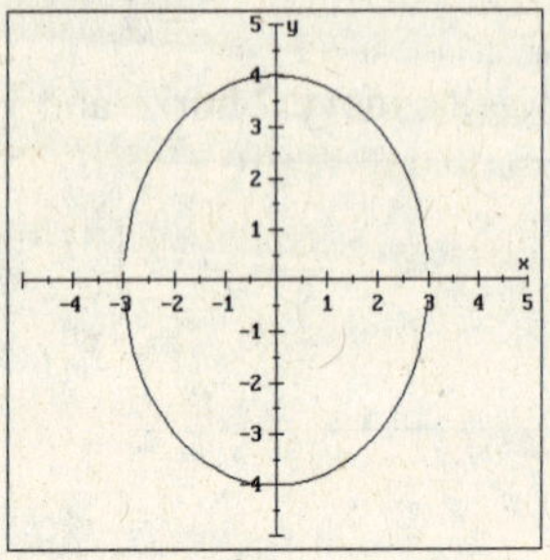

29. $\dfrac{(x-2)^2}{16}+\dfrac{y^2}{25}=1$

C $(2,0)$; move 4 horiz. and 5 vert.

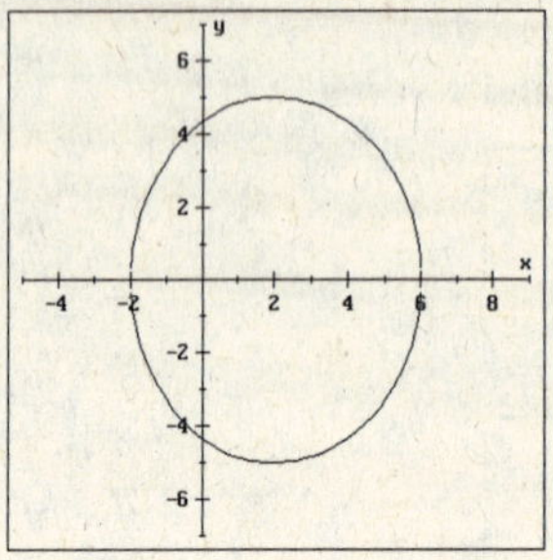

31. $16x^2+4y^2=64$

$$\frac{16x^2}{64}+\frac{4y^2}{64}=\frac{64}{64}$$

$$\frac{x^2}{4}+\frac{y^2}{16}=1$$

C $(0,0)$; move 2 horiz. and 4 vert.

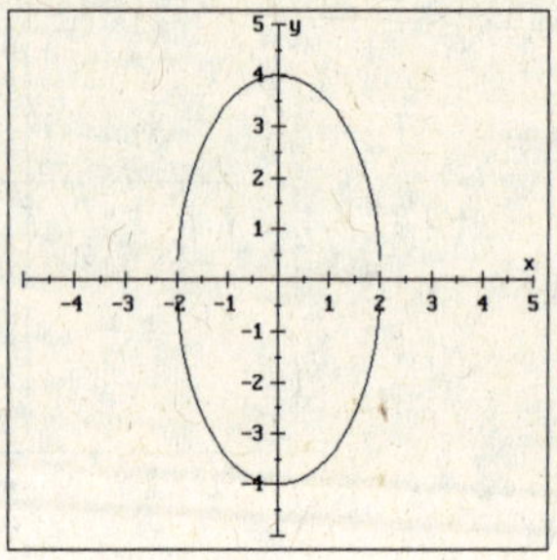

33. $25(x+1)^2+9y^2=225$

$$\frac{25(x+1)^2}{225}+\frac{9y^2}{225}=\frac{225}{225}$$

$$\frac{(x+1)^2}{9}+\frac{y^2}{25}=1$$

C $(-1,0)$; move 3 horiz. and 5 vert.

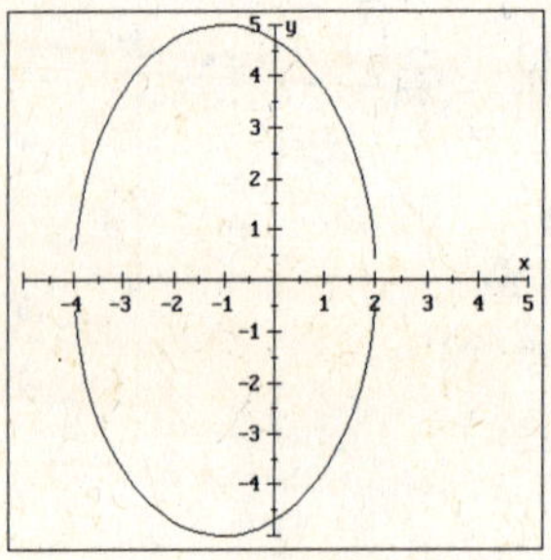

35. $\dfrac{x^2}{9}+\dfrac{y^2}{4}=1$

$$\frac{y^2}{4}=1-\frac{x^2}{9}$$

$$y^2=4\left(1-\frac{x^2}{9}\right)$$

$$y=\pm\sqrt{4\left(1-\frac{x^2}{9}\right)}$$

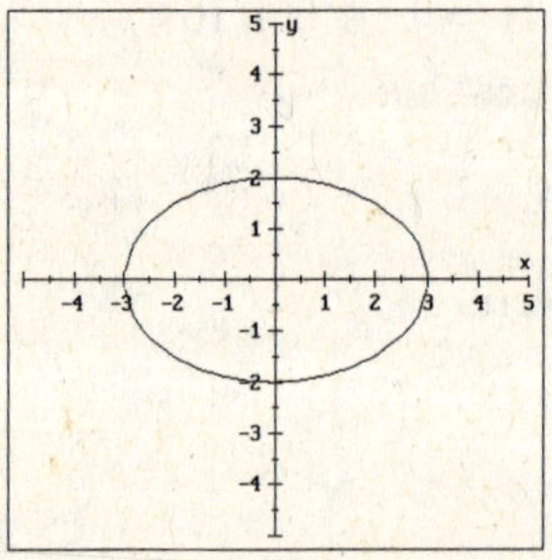

© 2015 Cengage Learning. All Rights Reserved. May not be scanned, copied or duplicated, or posted to a publicly accessible website, in whole or in part.

37. $\dfrac{x^2}{4}+\dfrac{(y-1)^2}{9}=1$

$\dfrac{(y-1)^2}{9}=1-\dfrac{x^2}{4}$

$(y-1)^2=9\left(1-\dfrac{x^2}{4}\right)$

$y-1=\pm\sqrt{9\left(1-\dfrac{x^2}{4}\right)}$

$y=1\pm\sqrt{9\left(1-\dfrac{x^2}{4}\right)}$

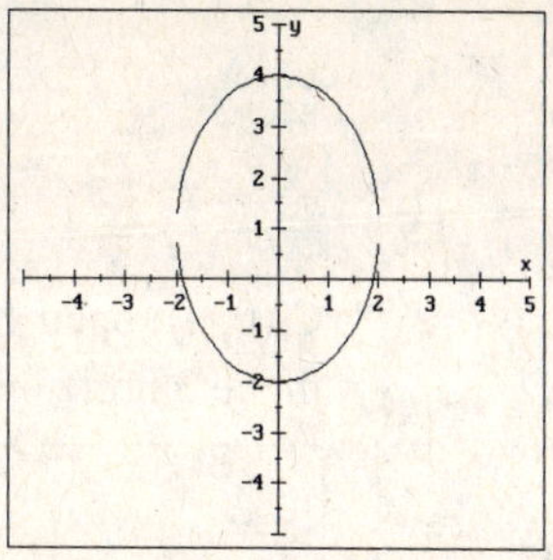

39. $a=24/2=12,\ b=10/2=5$

$\dfrac{x^2}{a^2}+\dfrac{y^2}{b^2}=1$

$\dfrac{x^2}{12^2}+\dfrac{y^2}{5^2}=1$

$\dfrac{x^2}{144}+\dfrac{y^2}{25}=1$

41. Note: $a=50/2=25,\ b=20$

$\dfrac{x^2}{a^2}+\dfrac{y^2}{b^2}=1$

$\dfrac{x^2}{625}+\dfrac{y^2}{400}=1$

$10{,}000\left(\dfrac{x^2}{625}+\dfrac{y^2}{400}\right)=10{,}000(1)$

$16x^2+25y^2=10{,}000$

$25y^2=10{,}000-16x^2$

$25y^2=16(625-x^2)$

$y^2=\frac{16}{25}(625-x^2)$

$y=\frac{4}{5}\sqrt{625-x^2}$

[just the top half of the ellipse, so no $\pm$]

43. $25x^2+4y^2=100$

$\dfrac{25x^2}{100}+\dfrac{4y^2}{100}=\dfrac{100}{100}$

$\dfrac{x^2}{4}+\dfrac{y^2}{25}=1\Rightarrow \text{area}=\pi ab=\pi(5)(2)=10\pi$ square units

45. **Answers may vary.**

47. It is a circle.

Exercises 10.3 (page 720)

1. $\dfrac{x^2}{4}-\dfrac{y^2}{9}=1$

$\dfrac{x^2}{4}-\dfrac{0^2}{9}=1$

$\dfrac{x^2}{4}=1$

$\dfrac{x^2}{4}=1$

$x^2=4$

$x=\pm 2$

$(2,0),(-2,0)$

© 2015 Cengage Learning. All Rights Reserved. May not be scanned, copied or duplicated, or posted to a publicly accessible website, in whole or in part.

SECTION 10.3

3. $-5x^4 + 10x^3 - 15x^2 = -5x^2(x^2 - 2x + 3)$

5. $14a^2 - 15ab - 9b^2 = (2a - 3b)(7a + 3b)$

7. hyperbola; difference

9. center

11. $(\pm a, 0)$; y-intercepts

13. $\dfrac{x^2}{9} - \dfrac{y^2}{4} = 1$
C $(0,0)$; open horiz.;
move 3 horiz. and 2 vert.

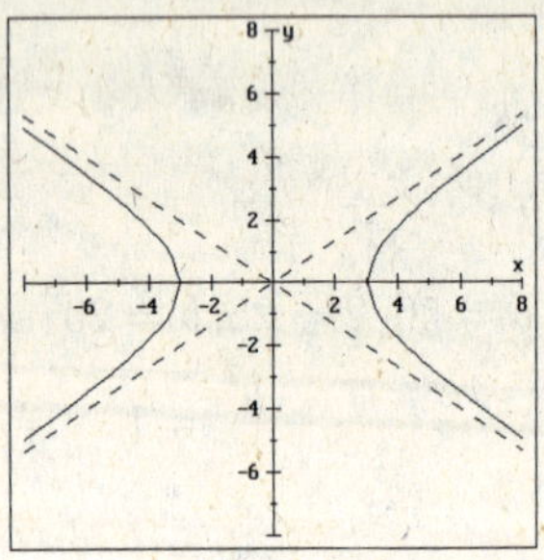

15. $\dfrac{y^2}{4} - \dfrac{x^2}{9} = 1$
C $(0,0)$; open vert.;
move 3 horiz. and 2 vert.

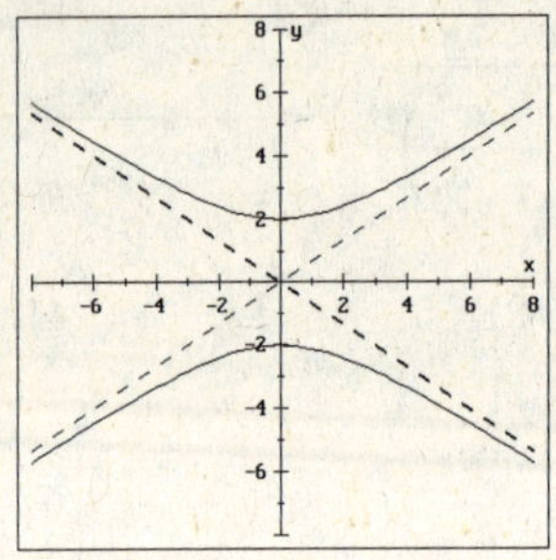

17. $\dfrac{(x-2)^2}{9} - \dfrac{y^2}{16} = 1$
C $(2,0)$; open horiz.;
move 3 horiz. and 4 vert.

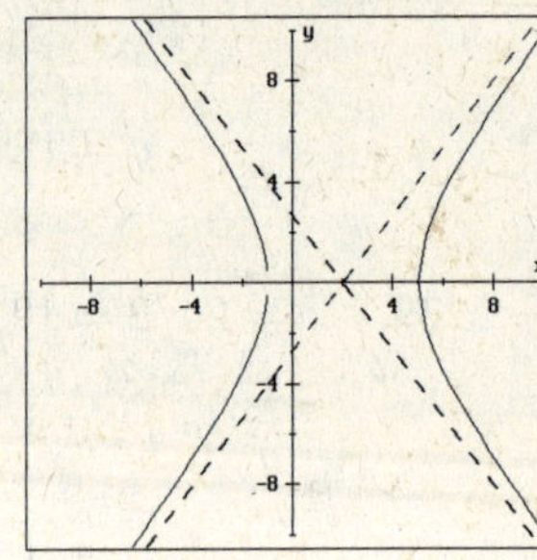

19. $\dfrac{(y+1)^2}{1} - \dfrac{(x-2)^2}{4} = 1$
C $(2,-1)$; open vert.; move 2 horiz. and 1 vert.

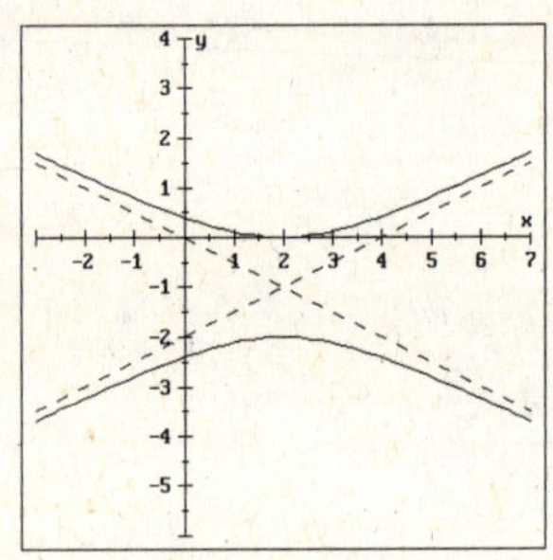

21.

$$4x^2 - y^2 + 8x - 4y = 4$$
$$4x^2 + 8x - y^2 - 4y = 4$$
$$4(x^2 + 2x) - (y^2 + 4y) = 4$$
$$4(x^2 + 2x + 1) - (y^2 + 4y + 4) = 4 + 4 - 4$$
$$4(x+1)^2 - (y+2)^2 = 4$$
$$\frac{(x+1)^2}{1} - \frac{(y+2)^2}{4} = 1$$

C $(-1,-2)$; opens horiz.; move 1 horiz. and 2 vert.

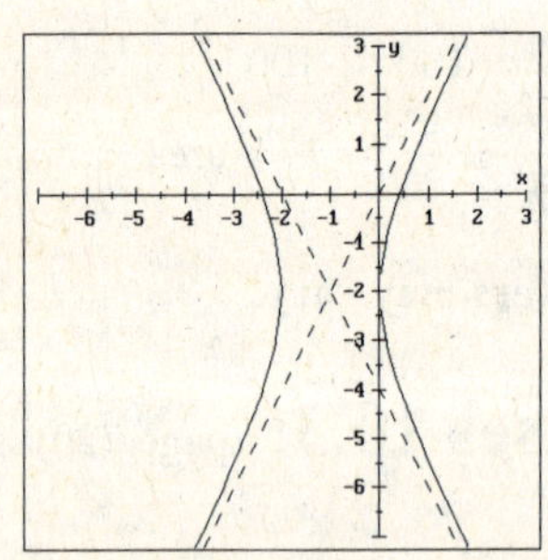

© 2015 Cengage Learning. All Rights Reserved. May not be scanned, copied or duplicated, or posted to a publicly accessible website, in whole or in part.

23.
$$
\begin{aligned}
4y^2 - x^2 + 8y + 4x &= 4 \\
4y^2 + 8y - x^2 + 4x &= 4 \\
4\left(y^2 + 2y\right) - \left(x^2 - 4x\right) &= 4 \\
4\left(y^2 + 2y + 1\right) - \left(x^2 - 4x + 4\right) &= 4 + 4 - 4 \\
4(y+1)^2 - (x-2)^2 &= 4 \\
\frac{(y+1)^2}{1} - \frac{(x-2)^2}{4} &= 1
\end{aligned}
$$
C $(2, -1)$; opens vert.; move 2 horiz. and 1 vert.

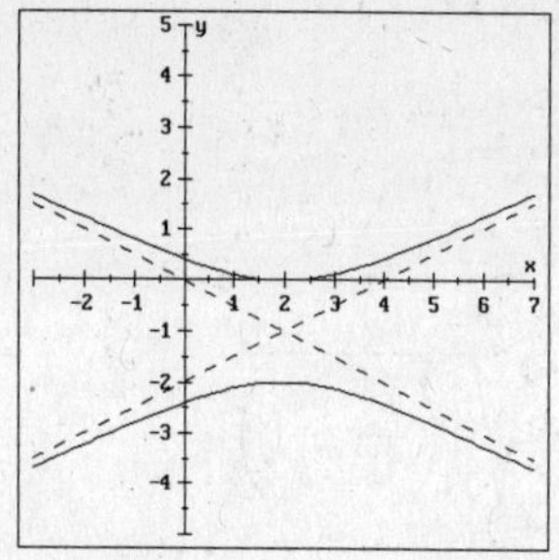

25. $xy = 10$

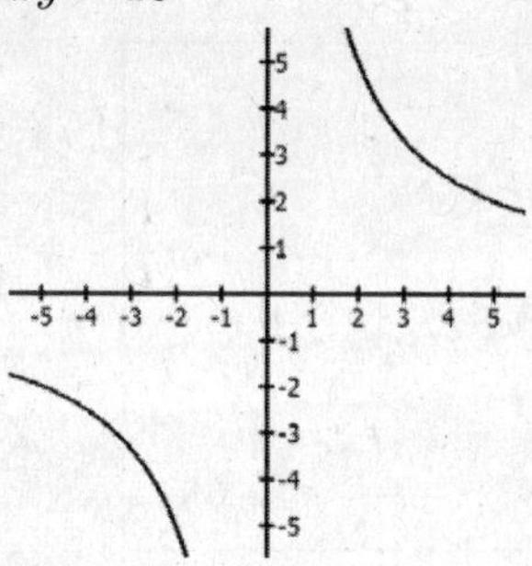

27. $xy = -12$

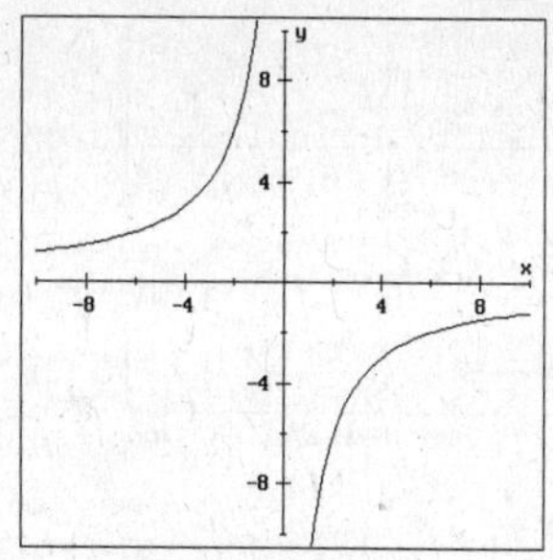

29. $25x^2 - y^2 = 25$
$$
\begin{aligned}
\frac{25x^2}{25} - \frac{y^2}{25} &= 1 \\
\frac{x^2}{1} - \frac{y^2}{25} &= 1
\end{aligned}
$$
C $(0, 0)$; open horiz.;
move 1 horiz. and 5 vert.

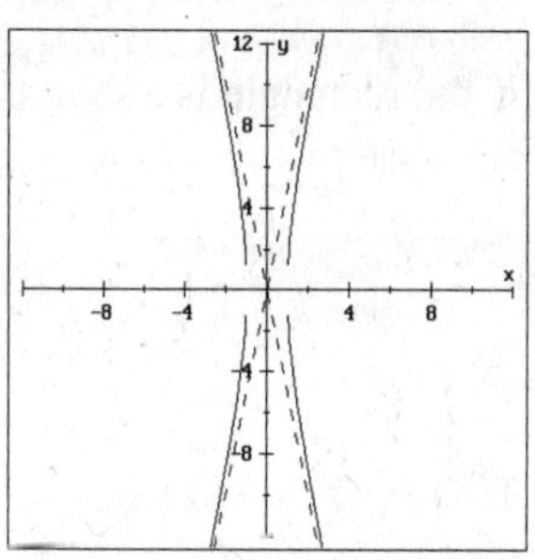

31. $4(x+3)^2 - (y-1)^2 = 4$
$$
\begin{aligned}
\frac{4(x+3)^2}{4} - \frac{(y-1)^2}{4} &= \frac{4}{4} \\
\frac{(x+3)^2}{1} - \frac{(y-1)^2}{4} &= 1
\end{aligned}
$$
C $(-3, 1)$; open horiz.;
move 1 horiz. and 2 vert.

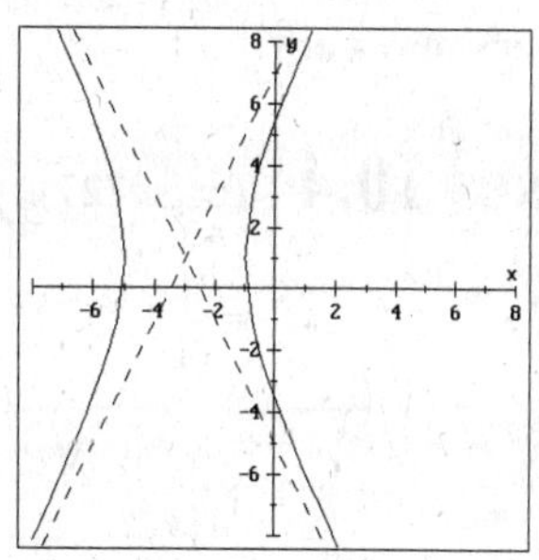

© 2015 Cengage Learning. All Rights Reserved. May not be scanned, copied or duplicated, or posted to a publicly accessible website, in whole or in part.

33. $\dfrac{x^2}{9} - \dfrac{y^2}{4} = 1$

$\dfrac{y^2}{4} = \dfrac{x^2}{9} - 1$

$y^2 = 4\left(\dfrac{x^2}{9} - 1\right)$

$y = \pm\sqrt{4\left(\dfrac{x^2}{9} - 1\right)}$

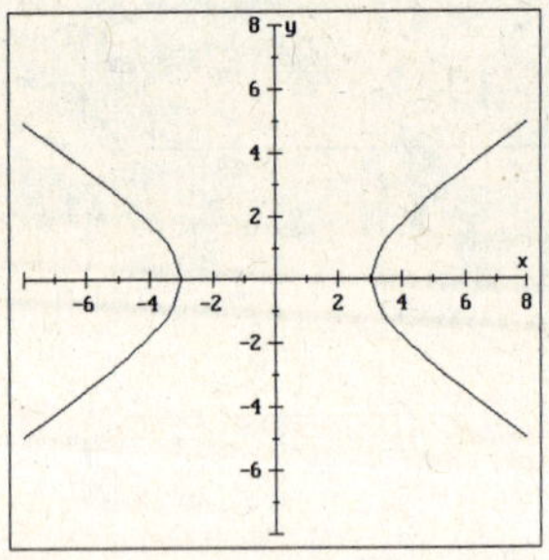

35. $\dfrac{x^2}{4} - \dfrac{(y-1)^2}{9} = 1$

$\dfrac{(y-1)^2}{9} = \dfrac{x^2}{4} - 1$

$(y-1)^2 = 9\left(\dfrac{x^2}{4} - 1\right)$

$y - 1 = \pm\sqrt{9\left(\dfrac{x^2}{4} - 1\right)}$

$y = 1 \pm \sqrt{9\left(\dfrac{x^2}{4} - 1\right)}$

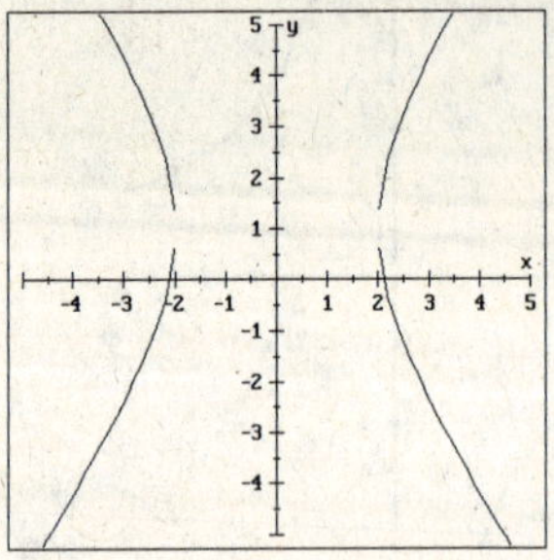

37. $9y^2 - x^2 = 81$

$\dfrac{9y^2}{81} - \dfrac{x^2}{81} = \dfrac{81}{81}$

$\dfrac{y^2}{9} - \dfrac{x^2}{81} = 1$

distance $= \sqrt{9} = 3$ units

39. $x^2 - 4y^2 = 4$

$\dfrac{x^2}{4} - \dfrac{4y^2}{4} = \dfrac{4}{4}$

$\dfrac{x^2}{4} - \dfrac{y^2}{1} = 1$

$a = 2 \Rightarrow 2a = 4$

The smallest distance is 4 units.

41. **Answers may vary.**

43. If $a = b$, the rectangle is a square.

Exercises 10.4 (page 727)

1. 0, 1, 2

3. 0, 1, 2, 3, 4

5. $\sqrt{45x^2} - 7\sqrt{80x^2} = \sqrt{9x^2}\sqrt{5} - 7\sqrt{16x^2}\sqrt{5} = 3x\sqrt{5} - 7(4x)\sqrt{5} = -25x\sqrt{5}$

7. $\dfrac{\sqrt{18x^3} - x\sqrt{8x}}{\sqrt{32x} - \sqrt{8x}} = \dfrac{\sqrt{9x^2}\sqrt{2x} - x\sqrt{4}\sqrt{2x}}{\sqrt{16}\sqrt{2x} - \sqrt{4}\sqrt{2x}} = \dfrac{3x\sqrt{2x} - 2x\sqrt{2x}}{4\sqrt{2x} - 2\sqrt{2x}} = \dfrac{x\sqrt{2x}}{2\sqrt{2x}} = \dfrac{x}{2}$

9. graphing; substitution

11. four

© 2015 Cengage Learning. All Rights Reserved. May not be scanned, copied or duplicated, or posted to a publicly accessible website, in whole or in part.

13. $\begin{cases} 8x^2 + 32y^2 = 256 \\ x = 2y \end{cases}$

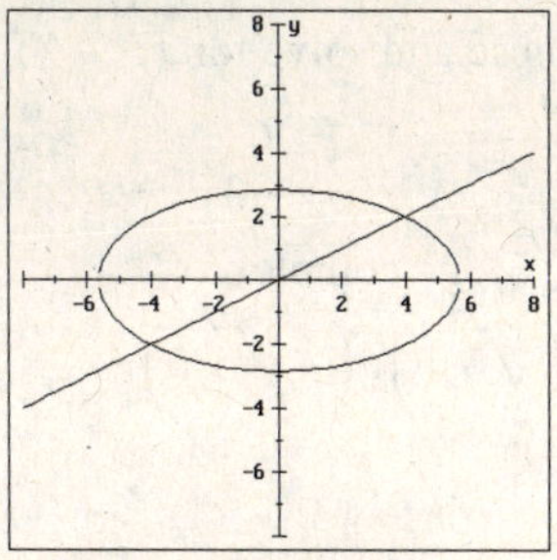

$(-4, -2), (4, 2)$

15. $\begin{cases} x^2 + y^2 = 10 \\ y = 3x^2 \end{cases}$

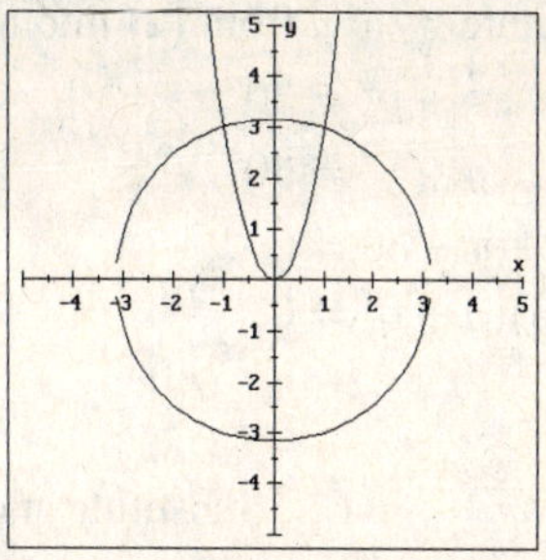

$(-1, 3), (1, 3)$

17. $\begin{cases} (1) \quad 25x^2 + 9y^2 = 225 \\ (2) \quad 5x + 3y = 15 \end{cases}$

Substitute $x = -\frac{3}{5}y + 3$ from (2) into (1):

$$\begin{aligned} 25x^2 + 9y^2 &= 225 \\ 25\left(-\tfrac{3}{5}y + 3\right)^2 + 9y^2 &= 225 \\ 25\left(\tfrac{9}{25}y^2 - \tfrac{18}{5}y + 9\right) + 9y^2 &= 225 \\ 9y^2 - 90y + 225 + 9y^2 &= 225 \\ 18y^2 - 90y &= 0 \\ 18y(y-5) &= 0 \end{aligned}$$

$18y = 0$ **or** $y - 5 = 0$

$y = 0 \qquad y = 5$

Substitute these and solve for x:

$$\begin{aligned} 5x + 3y &= 15 & 5x + 3y &= 15 \\ 5x + 3(0) &= 15 & 5x + 3(5) &= 15 \\ 5x &= 15 & 5x &= 0 \\ x &= 3 & x &= 0 \end{aligned}$$

Solutions: $(3, 0), (0, 5)$

19. $\begin{cases} (1) \quad x^2 + y^2 = 5 \\ (2) \quad x + y = 3 \end{cases}$

Substitute $x = 3 - y$ from (2) into (1):

$$\begin{aligned} x^2 + y^2 &= 5 \\ (3-y)^2 + y^2 &= 5 \\ 9 - 6y + y^2 + y^2 &= 5 \\ 2y^2 - 6y + 4 &= 0 \\ 2(y-2)(y-1) &= 0 \end{aligned}$$

$y - 2 = 0$ **or** $y - 1 = 0$

$y = 2 \qquad y = 1$

Substitute these and solve for x:

$$\begin{aligned} x &= 3 - y & x &= 3 - y \\ x &= 3 - 2 & x &= 3 - 1 \\ x &= 1 & x &= 2 \end{aligned}$$

Solutions: $(1, 2), (2, 1)$

21. $\begin{cases} (1) \quad x^2 + y^2 = 20 \\ (2) \quad y = x^2 \end{cases}$

Substitute $x^2 = y$ from (2) into (1):

$$\begin{aligned} x^2 + y^2 &= 20 \\ y + y^2 &= 20 \\ y^2 + y - 20 &= 0 \\ (y+5)(y-4) &= 0 \end{aligned}$$

$y + 5 = 0$ **or** $y - 4 = 0$

$y = -5 \qquad y = 4$

Substitute these and solve for x:

$$\begin{aligned} x^2 &= y & x^2 &= y \\ x^2 &= -5 & x^2 &= 4 \\ &\text{complex} & x &= \pm 2 \end{aligned}$$

Solutions: $(2, 4), (-2, 4)$

© 2015 Cengage Learning. All Rights Reserved. May not be scanned, copied or duplicated, or posted to a publicly accessible website, in whole or in part.

SECTION 10.4

23. $\begin{cases} (1) & x^2 + y^2 = 30 \\ (2) & y = x^2 \end{cases}$

Substitute $x^2 = y$ from (2) into (1):

$$x^2 + y^2 = 30$$
$$y + y^2 = 30$$
$$y^2 + y - 30 = 0$$
$$(y - 5)(y + 6) = 0$$

$y - 5 = 0$ **or** $y + 6 = 0$

$y = 5 \qquad y = -6$

Substitute these and solve for x:

$x^2 = y \qquad x^2 = y$

$x^2 = 5 \qquad x^2 = -6$

$x = \pm\sqrt{5} \qquad$ complex

Solutions: $\left(\sqrt{5}, 5\right), \left(-\sqrt{5}, 5\right)$

25.
$$\begin{array}{rcr} x^2 + y^2 & = & 34 \\ x^2 - y^2 & = & 16 \\ \hline 2x^2 & = & 50 \\ x^2 & = & 25 \\ x & = & \pm 5 \end{array}$$

Substitute and solve for y:

$x^2 + y^2 = 34 \qquad x^2 + y^2 = 34$

$5^2 + y^2 = 34 \qquad (-5)^2 + y^2 = 34$

$y^2 = 9 \qquad y^2 = 9$

$y = \pm 3 \qquad y = \pm 3$

Solutions: $(5, 3), (5, -3), (-5, 3), (-5, -3)$

27.
$$\begin{array}{rcr} x^2 + y^2 & = & 20 \\ x^2 - y^2 & = & -12 \\ \hline 2x^2 & = & 8 \\ x^2 & = & 4 \\ x & = & \pm 2 \end{array}$$

Substitute and solve for y:

$x^2 + y^2 = 20 \qquad x^2 + y^2 = 20$

$2^2 + y^2 = 20 \qquad (-2)^2 + y^2 = 20$

$y^2 = 16 \qquad y^2 = 16$

$y = \pm 4 \qquad y = \pm 4$

Solutions: $(2, 4), (2, -4), (-2, 4), (-2, -4)$

29. $\begin{cases} x^2 + y^2 = 25 \\ 12x^2 + 64y^2 = 768 \end{cases}$

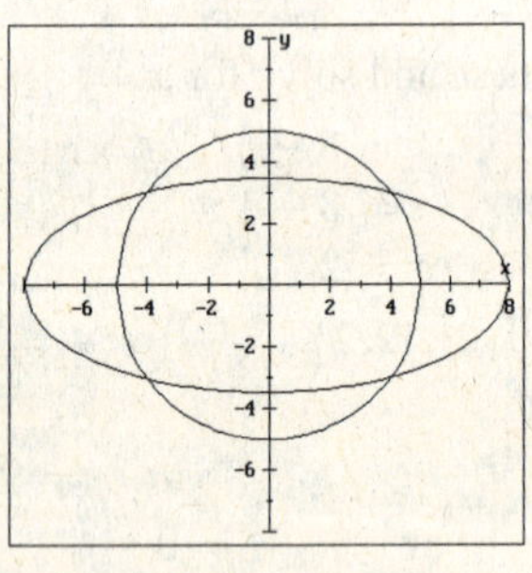

$(-4, 3), (4, 3), (4, -3)$
$(-4, -3)$

31. $\begin{cases} x^2 - 13 = -y^2 \\ y = 2x - 4 \end{cases}$

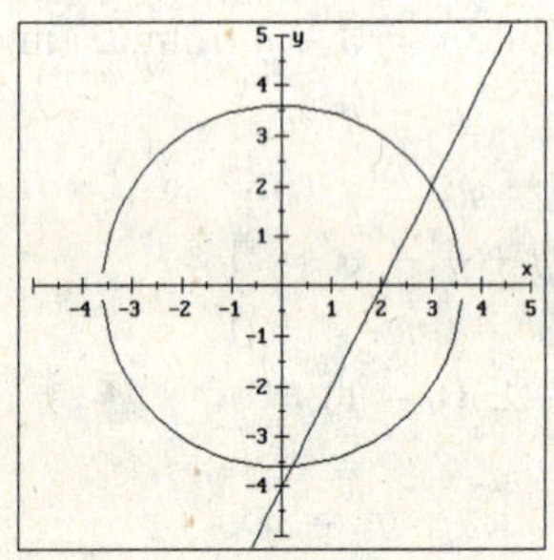

$\left(\frac{1}{5}, -\frac{18}{5}\right), (3, 2)$

© 2015 Cengage Learning. All Rights Reserved. May not be scanned, copied or duplicated, or posted to a publicly accessible website, in whole or in part.

33. $y = x^2 - 5 \Rightarrow \quad x^2 = y + 5$
$x^2 + y^2 = 25 \Rightarrow \quad -x^2 = y^2 - 25$

$$0 = y^2 + y - 20$$

$0 = y^2 + y - 20$
$0 = (y + 5)(y - 4)$
$y + 5 = 0$ **or** $y - 4 = 0$
$y = -5 \qquad y = 4$

Substitute these and solve for x:

$y = -5$	$y = 4$
$x^2 = y + 5$	$x^2 = y + 5$
$x^2 = -5 + 5$	$x^2 = 4 + 5$
$x^2 = 0$	$x^2 = 9$
$x = 0$	$x = \pm 3$

Solutions: $(0, -5), (3, 4), (-3, 4)$

35. $y^2 = 40 - x^2 \Rightarrow \quad y^2 = 40 - x^2$
$y = x^2 - 10 \Rightarrow \quad y = -10 + x^2$

$$y^2 + y = 30$$

$y^2 + y = 30$
$y^2 + y - 30 = 0$
$(y + 6)(y - 5) = 0$
$y + 6 = 0$ **or** $y - 5 = 0$
$y = -6 \qquad y = 5$

Substitute these and solve for x:

$y = -6$	$y = 5$
$x^2 = 40 - y^2$	$x^2 = 40 - y^2$
$x^2 = 40 - 36$	$x^2 = 40 - 25$
$x^2 = 4$	$x^2 = 15$
$x = \pm 2$	$x = \pm\sqrt{15}$

$(2, -6), (-2, -6), \left(\sqrt{15}, 5\right), \left(-\sqrt{15}, 5\right)$

37. $\begin{cases} (1) & x^2 + y^2 = 36 \\ (2) & 49x^2 + 36y^2 = 1764 \end{cases}$

Substitute $x^2 = 36 - y^2$ from (1) into (2):

$$49x^2 + 36y^2 = 1764$$
$$49(36 - y^2) + 36y^2 = 1764$$
$$1764 - 49y^2 + 36y^2 = 1764$$
$$-13y^2 = 0$$

$y = 0$

Substitute this and solve for x:

$x^2 = 36 - y^2$
$x^2 = 36 - 0$
$x^2 = 36$
$x = \pm 6$

Solutions: $(6, 0), (-6, 0)$

39. $x^2 - y^2 = -5 \Rightarrow (\times 2) \quad 2x^2 - 2y^2 = -10$
$3x^2 + 2y^2 = 30 \Rightarrow \quad 3x^2 + 2y^2 = 30$

$$5x^2 = 20$$
$$x^2 = 4$$
$$x = \pm 2$$

Substitute and solve for y:

$x^2 - y^2 = -5$	$x^2 - y^2 = -5$
$2^2 - y^2 = -5$	$(-2)^2 - y^2 = -5$
$-y^2 = -9$	$-y^2 = -9$
$y^2 = 9$	$y^2 = 9$
$y = \pm 3$	$y = \pm 3$

Solutions: $(2, 3), (2, -3), (-2, 3), (-2, -3)$

41. $\frac{3}{x} - \frac{1}{y} = \frac{1}{2} \Rightarrow (\times 2) \quad \frac{6}{x} - \frac{2}{y} = 1$
$\frac{2}{x} + \frac{2}{y} = 1 \Rightarrow \quad \frac{2}{x} + \frac{2}{y} = 1$

$$\frac{8}{x} = 2$$
$$8 = 2x$$
$$4 = x$$

Substitute and solve for y:

$\frac{2}{x} + \frac{2}{y} = 1$
$\frac{2}{4} + \frac{2}{y} = 1$
$\frac{2}{y} = \frac{2}{4}$
$8 = 2y$
$4 = y$

Solution: $(4, 4)$

© 2015 Cengage Learning. All Rights Reserved. May not be scanned, copied or duplicated, or posted to a publicly accessible website, in whole or in part.

SECTION 10.4

43. $\begin{cases} (1) & xy = -\frac{9}{2} \\ (2) & 3x + 2y = 6 \end{cases}$

Substitute $x = -\frac{9}{2y}$ from (1) into (2):

$$3\left(-\tfrac{9}{2y}\right) + 2y = 6$$
$$-\tfrac{27}{2y} + 2y = 6$$
$$-27 + 4y^2 = 12y$$
$$4y^2 - 12y - 27 = 0$$
$$(2y + 3)(2y - 9) = 0$$

$2y + 3 = 0$ **or** $2y - 9 = 0$

$y = -\frac{3}{2}$ $\qquad$ $y = \frac{9}{2}$

Substitute these and solve for x:

$x = -\dfrac{9}{2y}$ $\qquad$ $x = -\dfrac{9}{2y}$

$x = -\dfrac{9}{2(-\frac{3}{2})}$ $\qquad$ $x = -\dfrac{9}{2(\frac{9}{2})}$

$x = -\frac{9}{-3} = 3$ $\qquad$ $x = -\frac{9}{9} = -1$

Solutions: $(3, -\frac{3}{2}), (-1, \frac{9}{2})$

45. $\begin{cases} (1) & xy = \frac{1}{6} \\ (2) & y + x = 5xy \end{cases}$

Substitute $x = \frac{1}{6y}$ from (1) into (2):

$$y + \tfrac{1}{6y} = \tfrac{5y}{6y}$$
$$6y^2 + 1 = 5y$$
$$6y^2 - 5y + 1 = 0$$
$$(2y - 1)(3y - 1) = 0$$

$2y - 1 = 0$ **or** $3y - 1 = 0$

$y = \frac{1}{2}$ $\qquad$ $y = \frac{1}{3}$

Substitute these and solve for x:

$x = \frac{1}{6y}$ $\qquad$ $x = \frac{1}{6y}$

$x = \dfrac{1}{6(\frac{1}{2})}$ $\qquad$ $x = \dfrac{1}{6(\frac{1}{3})}$

$x = \frac{1}{3}$ $\qquad$ $x = \frac{1}{2}$

Solutions: $(\frac{1}{3}, \frac{1}{2}), (\frac{1}{2}, \frac{1}{3})$

47. $\begin{cases} x^2 - 6x - y = -5 \\ x^2 - 6x + y = -5 \end{cases}$

$\begin{cases} y = x^2 - 6x + 5 \\ y = -x^2 + 6x - 5 \end{cases}$

$(1, 0), (5, 0)$

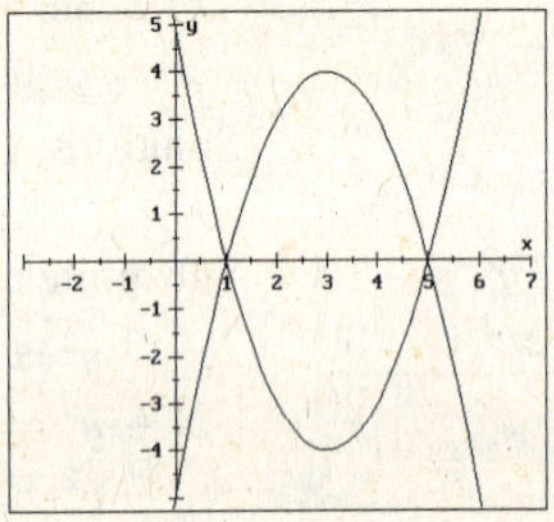

49. Let the integers be x and y. Then the equations are

$\begin{cases} (1) & xy = 32 \\ (2) & x + y = 12 \end{cases}$

Substitute $x = \frac{32}{y}$ from (1) into (2):

$$\tfrac{32}{y} + y = 12$$
$$32 + y^2 = 12y$$
$$y^2 - 12y + 32 = 0$$
$$(y - 4)(y - 8) = 0$$

$y - 4 = 0$ **or** $y - 8 = 0$

$y = 4$ $\qquad$ $y = 8$

Substitute these and solve for x:

$x = \frac{32}{y} = \frac{32}{4} = 8$ $\qquad$ $x = \frac{32}{y} = \frac{32}{8} = 4$

The integers are 8 and 4.

© 2015 Cengage Learning. All Rights Reserved. May not be scanned, copied or duplicated, or posted to a publicly accessible website, in whole or in part.

51. Let $l =$ the length of the rectangle, and $w =$ the width of the rectangle. Then the equations are:

$$\begin{cases} (1) & lw = 63 \\ (2) & 2l + 2w = 32 \end{cases}$$

Substitute $l = \frac{63}{w}$ from (1) into (2):

$$2\left(\tfrac{63}{w}\right) + 2w = 32$$
$$\tfrac{126}{w} + 2w = 32$$
$$126 + 2w^2 = 32w$$
$$2w^2 - 32w + 126 = 0$$
$$2(w-7)(w-9) = 0$$

$$w - 7 = 0 \quad \textbf{or} \quad w - 9 = 0$$
$$w = 7 \qquad\qquad w = 9$$

Substitute these and solve for l:

$$l = \tfrac{63}{w} = \tfrac{63}{7} = 9 \qquad l = \tfrac{63}{w} = \tfrac{63}{9} = 7$$

The dimensions are 7 cm by 9 cm.

53. Let $r =$ Rania's rate, and let $p =$ the amount Rania invested.
Then Jerome invested $p + 150$ at a rate of $r + 0.015$. The equations are

$$\begin{cases} (1)\ pr = 67.50 \Rightarrow p = \frac{67.5}{r} \\ (2)\ (p+150)(r+0.015) = 94.5 \end{cases}$$

Substitute $p = \frac{67.5}{r}$ from (1) into (2):

$$\left(\tfrac{67.5}{r} + 150\right)(r + 0.015) = 94.5$$
$$67.5 + \tfrac{1.0125}{r} + 150r + 2.25 = 94.5$$
$$67.5r + 1.0125 + 150r^2 + 2.25r = 94.5r$$
$$150r^2 - 24.75r - 1.0125 = 0$$
$$12{,}000r^2 - 1980r + 81 = 0$$
$$(100r - 9)(120r - 9) = 0$$

$$100r - 9 = 0 \quad \textbf{or} \quad 120r - 9 = 0$$
$$r = 0.09 \qquad\qquad r = 0.075$$

Substitute and solve for p:

$$p = \tfrac{67.5}{r} = \tfrac{67.5}{0.09} = 750 \text{ or } p = \tfrac{67.5}{r} = \tfrac{67.5}{0.075} = 900$$

Rania invested \$750 at 9% or \$900 at 7.5%.

55. Let $r =$ Jim's rate and $t =$ Jim's time. Then his brother's rate was $r - 17$ and his time was $t + 1.5$.

$$\begin{cases} (1)\ rt = 306 \Rightarrow t = \frac{306}{r} \\ (2)\ (r-17)(t+1.5) = 306 \end{cases}$$

Substitute $t = \frac{306}{r}$ from (1) into (2):

$$(r - 17)\left(\tfrac{306}{r} + 1.5\right) = 306$$
$$306 + 1.5r - \tfrac{5202}{r} - 25.5 = 306$$
$$306r + 1.5r^2 - 5202 - 25.5r = 306r$$
$$1.5r^2 - 25.5r - 5202 = 0$$
$$3r^2 - 51r - 10{,}404 = 0$$
$$(3r + 153)(r - 68) = 0$$

$$3r + 153 = 0 \quad \textbf{or} \quad r - 68 = 0$$
$$r = -153/3 \qquad\qquad r = 68$$

Substitute and solve for t:

$$t = \tfrac{306}{r} = \tfrac{306}{68} = 4.5$$

Jim drove for 4.5 hours at 68 miles per hour.

57. **Answers may vary.**

59. $0, 1, 2, 3, 4$

Exercises 10.5 (page 733)

1. $f(1) = 1^2 = 1$; $f(2) = 2^2 = 4$
$f(3) = 3^2 = 9$; increasing

3. $f(-3) = (-3)^2 = 9$; $f(-2) = (-2)^2 = 4$
$f(-1) = (-1)^2 = 1$; decreasing

© 2015 Cengage Learning. All Rights Reserved. May not be scanned, copied or duplicated, or posted to a publicly accessible website, in whole or in part.

SECTION 10.5

5. $$\begin{aligned}(6x-10)+(3x+10)&=180\\9x&=180\\x&=20\end{aligned}$$

7. domains

9. constant; $f(x)$

11. step

13. increasing on $(-\infty, 0)$, decreasing on $(0, \infty)$

15. decreasing on $(-\infty, 0)$, constant on $(0, 2)$ increasing on $(2, \infty)$

17. $f(x) = \begin{cases} -1 \text{ when } x \le 0 \\ x \text{ when } x > 0 \end{cases}$

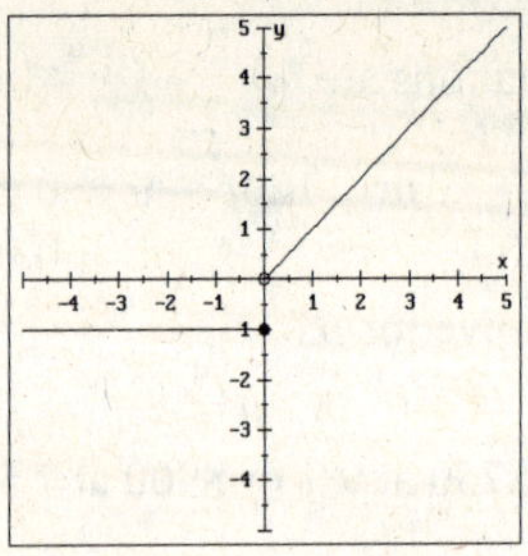

constant on $(-\infty, 0)$
increasing on $(0, \infty)$

19. $f(x) = \begin{cases} -x \text{ if } x \le 0 \\ x \text{ if } 0 < x < 2 \\ -x \text{ if } x \ge 2 \end{cases}$

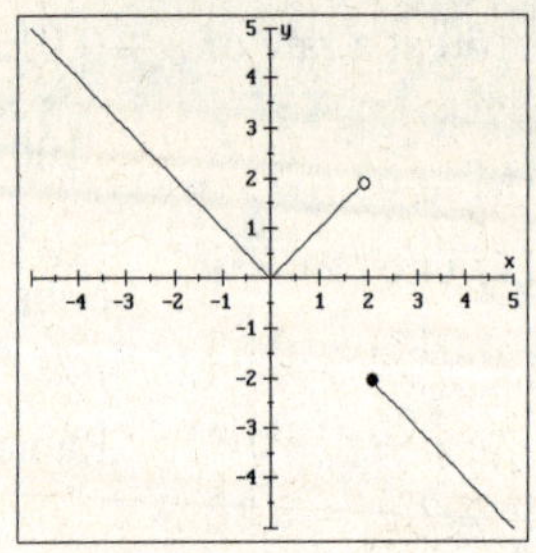

decreasing on $(-\infty, 0)$
increasing on $(0, 2)$
decreasing on $(2, \infty)$

21. $f(x) = -[[x]]$

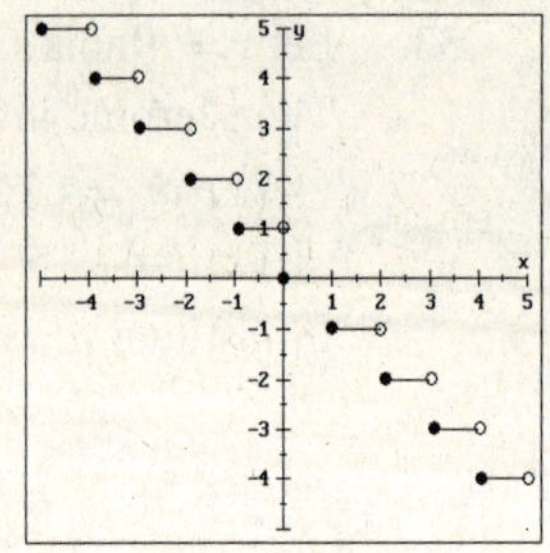

23. $f(x) = 2[[x]]$

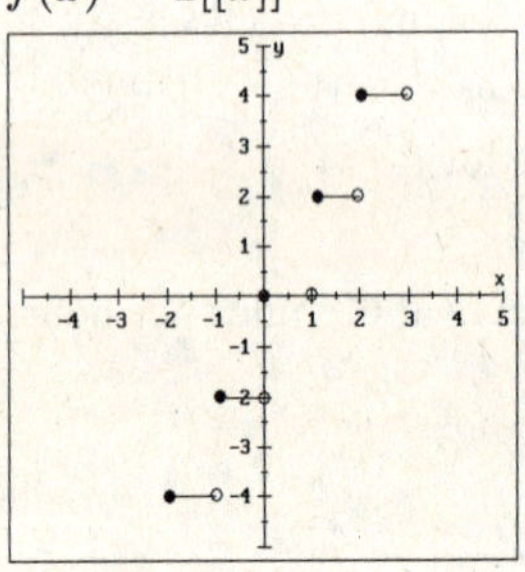

25. $f(x) = \begin{cases} -1 & \text{if } x < 0 \\ 0 & \text{if } x = 0 \\ 1 & \text{if } x > 0 \end{cases}$

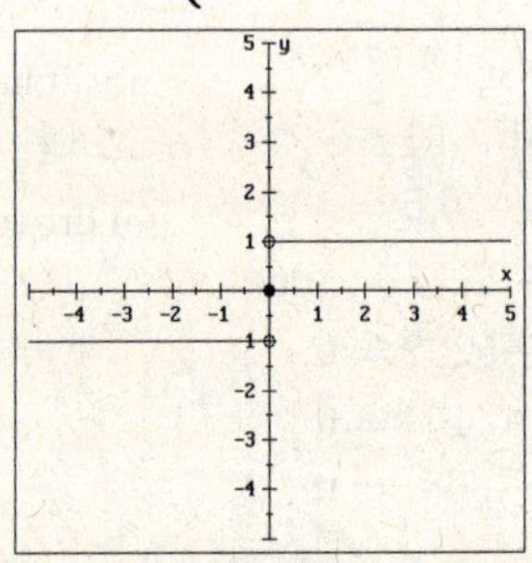

27. Find y when $x = 2.5$.
cost = \$30

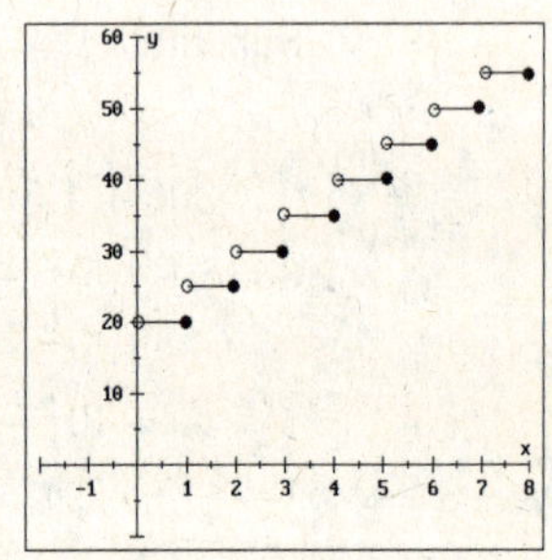

© 2015 Cengage Learning. All Rights Reserved. May not be scanned, copied or duplicated, or posted to a publicly accessible website, in whole or in part.

29. After 2 hours, B is cheaper.

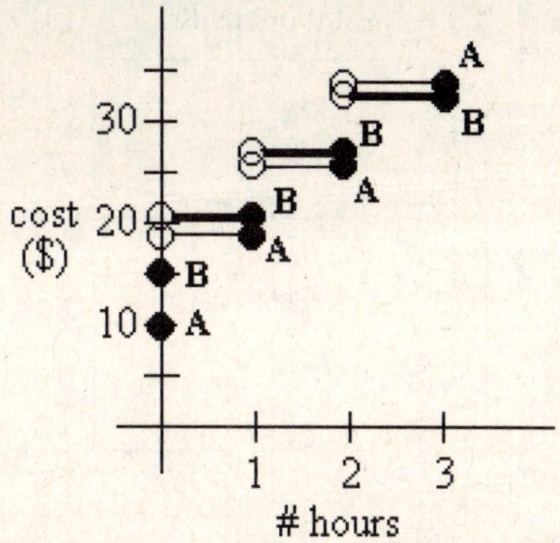

31. **Answers may vary.**

33. $f(x) = \begin{cases} x & \text{if } x < 6 \\ -x & \text{if } x > 6 \end{cases}$

Chapter 10 Review (page 739)

1. $(x-1)^2 + (y+2)^2 = 9$

C $(1, -2)$; $r = \sqrt{9} = 3$

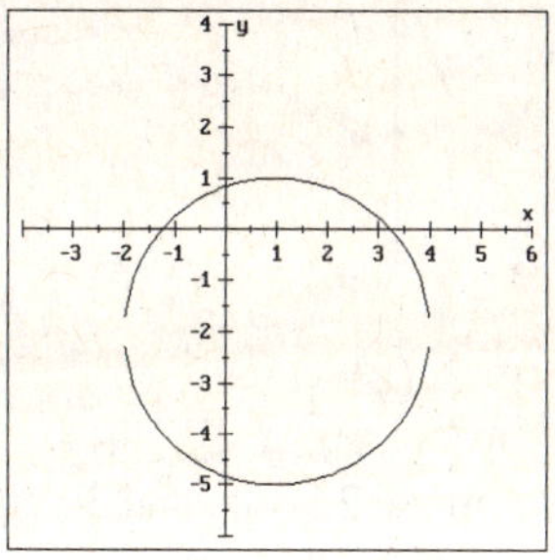

2. $x^2 + y^2 = 16$

C $(0, 0)$; $r = \sqrt{16} = 4$

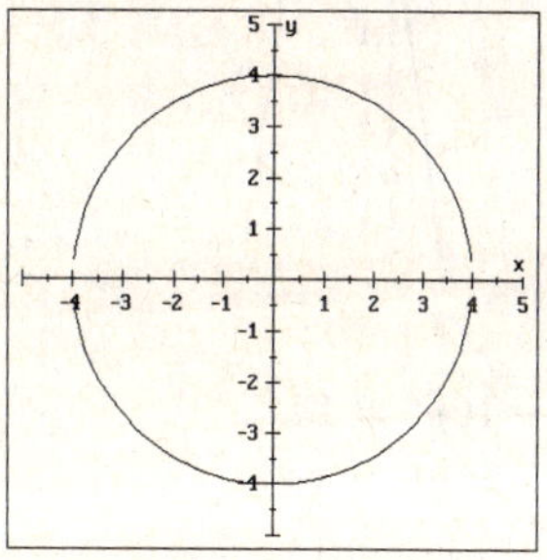

3.

$$x^2 + y^2 + 4x - 2y = 4$$
$$x^2 + 4x + y^2 - 2y = 4$$
$$x^2 + 4x + 4 + y^2 - 2y + 1 = 4 + 4 + 1$$
$$(x+2)^2 + (y-1)^2 = 9$$

C $(-2, 1)$; $r = \sqrt{9} = 3$

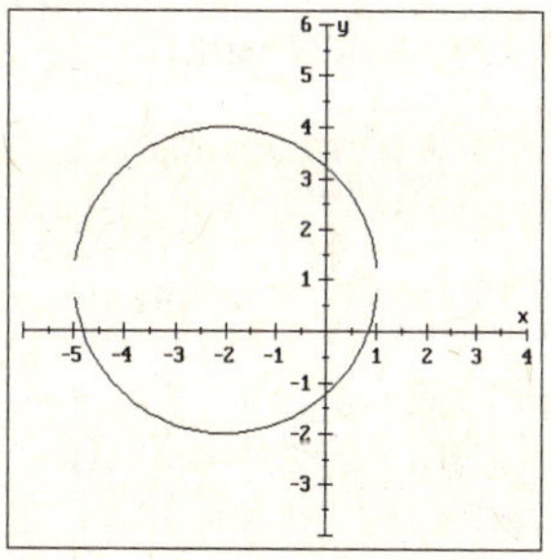

© 2015 Cengage Learning. All Rights Reserved. May not be scanned, copied or duplicated, or posted to a publicly accessible website, in whole or in part.

4. $x = -3(y-2)^2 + 5$
V $(5, 2)$; opens L

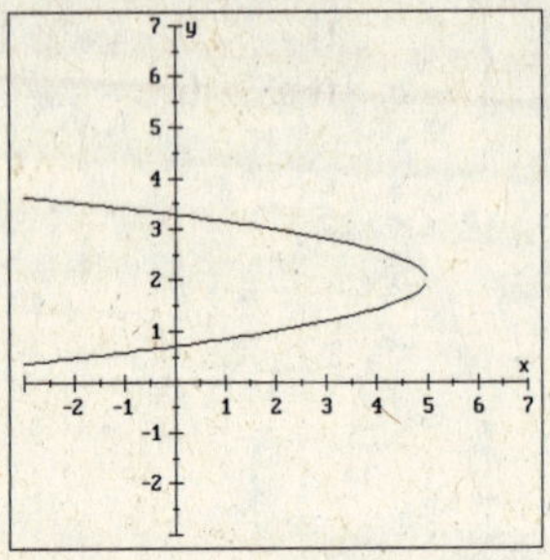

5. $x = 2(y+1)^2 - 2$
V $(-2, -1)$; opens R

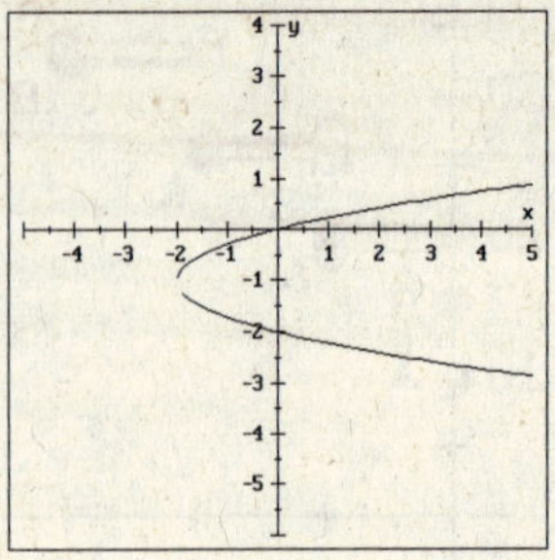

6. $y = -2(x-1)^2 + 3$
V $(1, 3)$; opens D

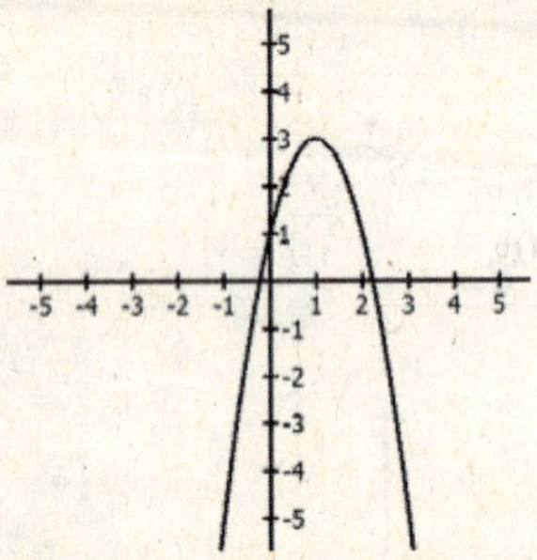

7. $y = (x+2)^2 - 3$
V $(-2, -3)$; opens U

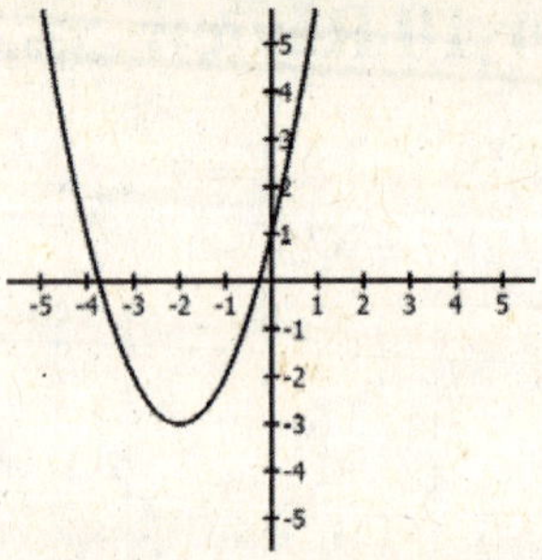

8.
$$9x^2 + 16y^2 = 144$$
$$\frac{9x^2}{144} + \frac{16y^2}{144} = \frac{144}{144}$$
$$\frac{x^2}{16} + \frac{y^2}{9} = 1$$
C $(0, 0)$; move 4 horiz. and 3 vert.

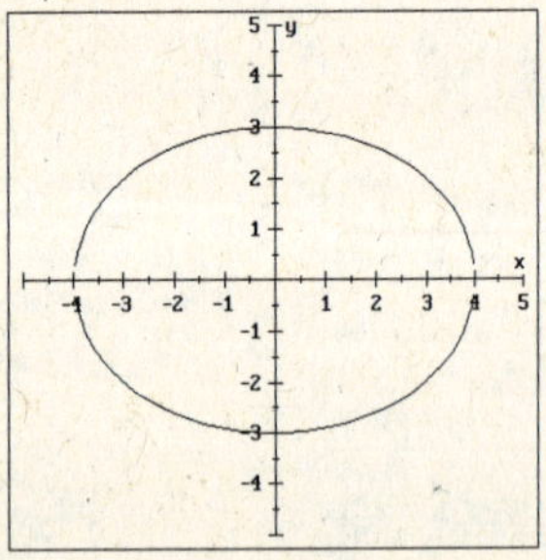

9. $\dfrac{(x-2)^2}{4} + \dfrac{(y-1)^2}{9} = 1$
C $(2, 1)$; move 2 horiz. and 3 vert.

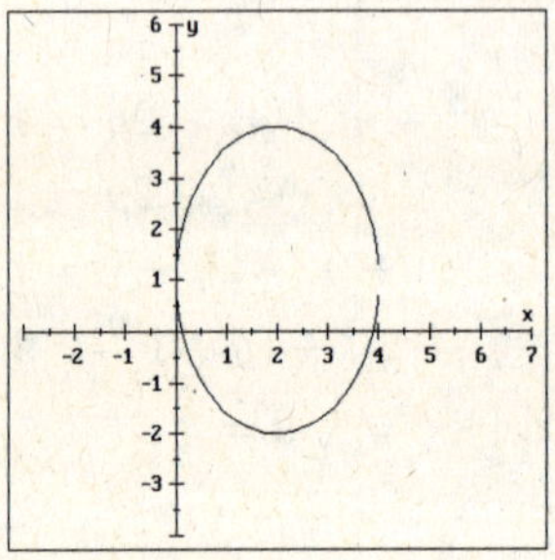

© 2015 Cengage Learning. All Rights Reserved. May not be scanned, copied or duplicated, or posted to a publicly accessible website, in whole or in part.

10.
$$\begin{aligned}4x^2+9y^2+8x-18y&=23\\4x^2+8x+9y^2-18y&=23\\4(x^2+2x)+9(y^2-2y)&=23\\4(x^2+2x+1)+9(y^2-2y+1)&=23+4+9\\4(x+1)^2+9(y-1)^2&=36\\\frac{(x+1)^2}{9}+\frac{(y-1)^2}{4}&=1\end{aligned}$$
C $(-1,1)$; move 3 horiz. and 2 vert.

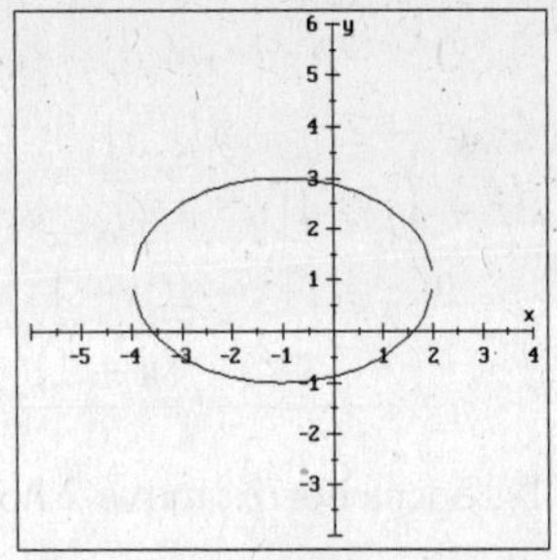

11.
$$\begin{aligned}9x^2-y^2&=-9\\\frac{9x^2}{-9}-\frac{y^2}{-9}&=\frac{-9}{-9}\\\frac{y^2}{9}-\frac{x^2}{1}&=1\end{aligned}$$
C $(0,0)$; opens vert.; move 1 horiz. and 3 vert.

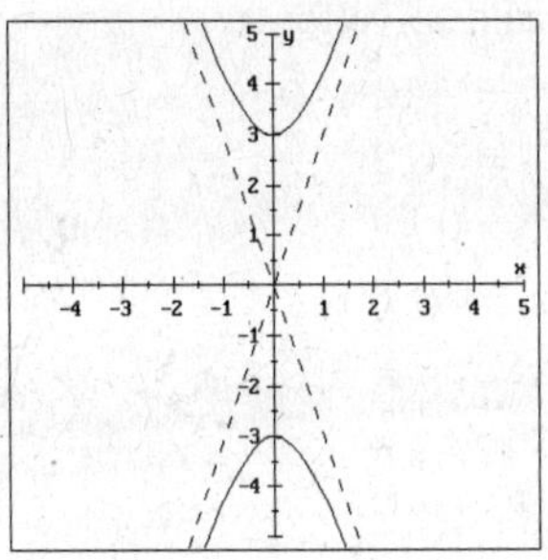

12. $xy=9$

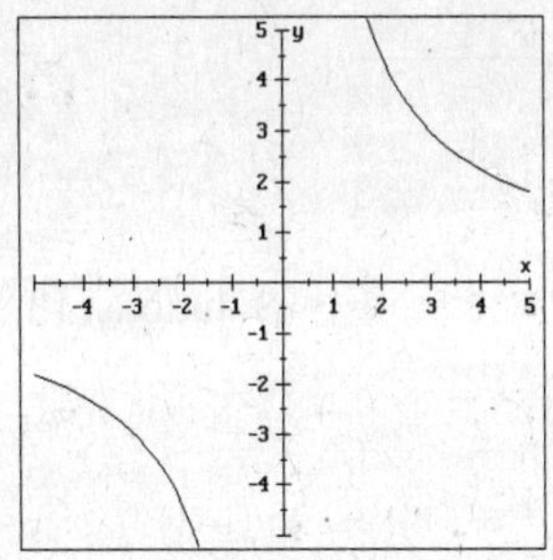

13.
$$\begin{aligned}2y^2-4x^2+8x-8y&=8\\y^2-4y-2x^2+4x&=4\\(y^2-4y+4)-2(x^2-2x)&=4+4\\(y^2-4y+4)-2(x^2-2x+1)&=8-2\\(y-2)^2-2(x-1)^2&=6\\\frac{(y-2)^2}{6}-\frac{2(x-1)^2}{6}&=1\\\frac{(y-2)^2}{6}-\frac{(x-1)^2}{3}&=1\Rightarrow \text{hyperbola}\end{aligned}$$

© 2015 Cengage Learning. All Rights Reserved. May not be scanned, copied or duplicated, or posted to a publicly accessible website, in whole or in part.

14.

$$\begin{aligned}
9x^2 - 4y^2 - 18x - 8y &= 31 \\
9x^2 - 18x - 4y^2 - 8y &= 31 \\
9(x^2 - 2x) - 4(y^2 + 2y) &= 31 \\
9(x^2 - 2x + 1) - 4(y^2 + 2y + 1) &= 31 + 9 - 4 \\
9(x-1)^2 - 4(y+1)^2 &= 36 \\
\frac{(x-1)^2}{4} - \frac{(y+1)^2}{9} &= 1
\end{aligned}$$

C $(1, -1)$; opens horiz.; move 2 horiz. and 3 vert.

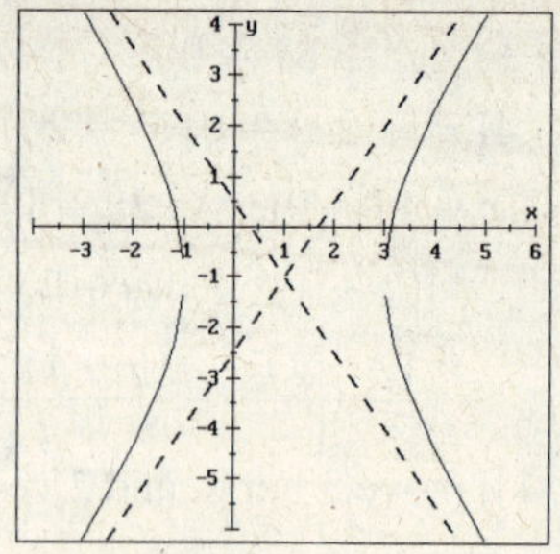

15.

$$\begin{array}{rcr}
x^2 + y^2 &=& 25 \\
4x^2 - y^2 &=& 20 \\
\hline
5x^2 &=& 45 \\
x^2 &=& 9 \\
x &=& \pm 3
\end{array}$$

Substitute and solve for y:

$$\begin{aligned}
x^2 + y^2 &= 25 & x^2 + y^2 &= 25 \\
3^2 + y^2 &= 25 & (-3)^2 + y^2 &= 25 \\
y^2 &= 16 & y^2 &= 16 \\
y &= \pm 4 & y &= \pm 4
\end{aligned}$$

Solutions: $(3, 4), (3, -4), (-3, 4), (-3, -4)$

16.

$$\begin{array}{lr}
\frac{x^2}{4} - \frac{y^2}{2} = -3 \Rightarrow \times 4 & x^2 - 2y^2 = -12 \\
\frac{x^2}{3} + y^2 = 11 \Rightarrow \times (-3) & -x^2 - 3y^2 = -33 \\
\hline
 & -5y^2 = -45 \\
 & y^2 = 9 \\
 & y = \pm 3
\end{array}$$

Substitute and solve for x:

$$\begin{aligned}
x^2 - 2y^2 &= -12 \\
x^2 - 2(9) &= -12 \\
x^2 &= 6 \\
x &= \pm\sqrt{6}
\end{aligned}$$

Solutions: $\left(\sqrt{6}, 3\right), \left(\sqrt{6}, -3\right), \left(-\sqrt{6}, 3\right), \left(-\sqrt{6}, -3\right)$

17. increasing on $(-\infty, -2)$; constant on $(-2, 1)$; decreasing on $(1, \infty)$

18. $f(x) = \begin{cases} x & \text{if } x \le 1 \\ -x^2 & \text{if } x > 1 \end{cases}$

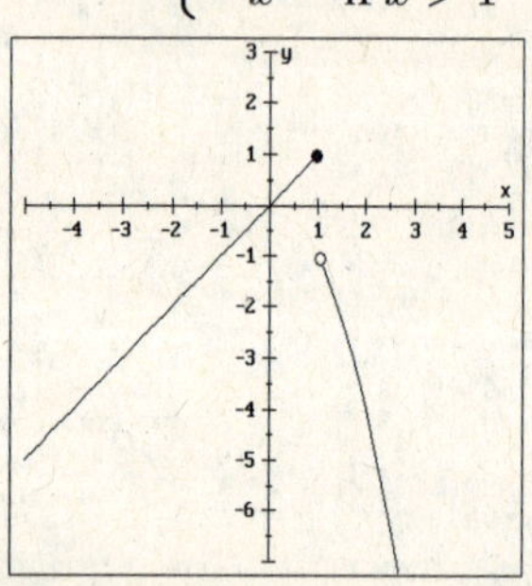

19. $f(x) = 3[[x]]$

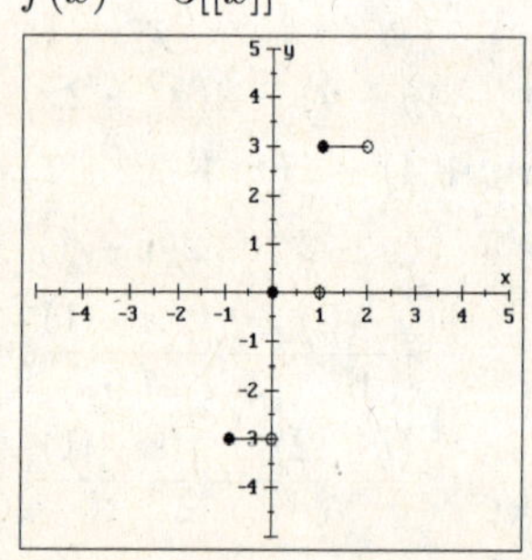

© 2015 Cengage Learning. All Rights Reserved. May not be scanned, copied or duplicated, or posted to a publicly accessible website, in whole or in part.

Chapter 10 Test (page 745)

1. $(x+5)^2+(y-2)^2=9$

Center: $(-5,2)$; radius $=\sqrt{9}=3$

2.
$$\begin{aligned} x^2+y^2+8x-4y &= 5 \\ x^2+8x+y^2-4y &= 5 \\ x^2+8x+16+y^2-4y+4 &= 5+16+4 \\ (x+4)^2+(y-2)^2 &= 25 \end{aligned}$$

Center: $(-4,2)$; radius $=\sqrt{25}=5$

3. $(x+1)^2+(y-2)^2=9$

C $(-1,2)$; $r=\sqrt{9}=3$

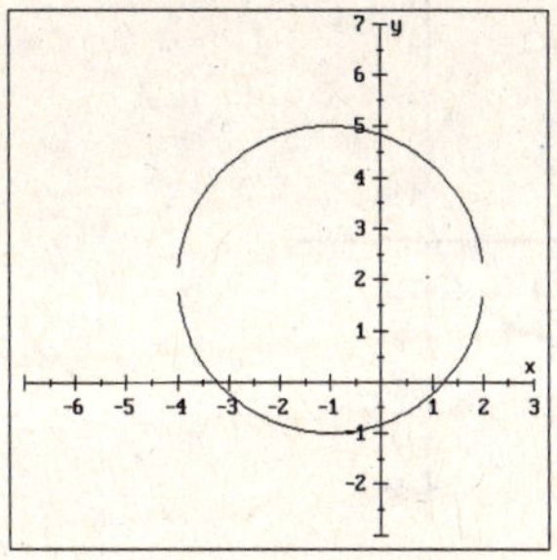

4. $x=(y-2)^2-1$

V $(-1,2)$; opens R

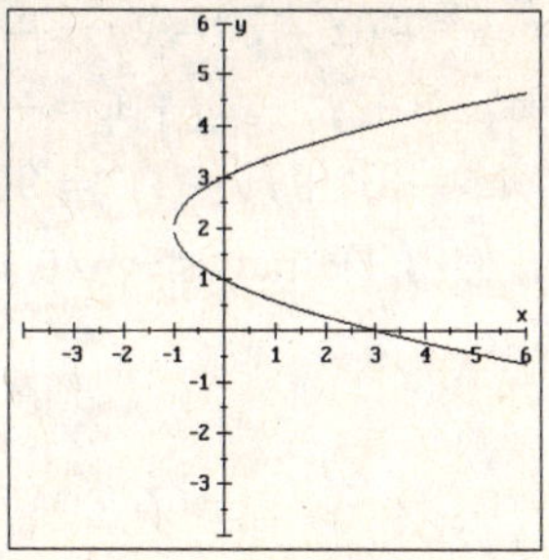

5.
$$\begin{aligned} 9x^2+4y^2 &= 36 \\ \frac{9x^2}{36}+\frac{4y^2}{36} &= \frac{36}{36} \\ \frac{x^2}{4}+\frac{y^2}{9} &= 1 \end{aligned}$$

C $(0,0)$; move 2 horiz. and 3 vert.

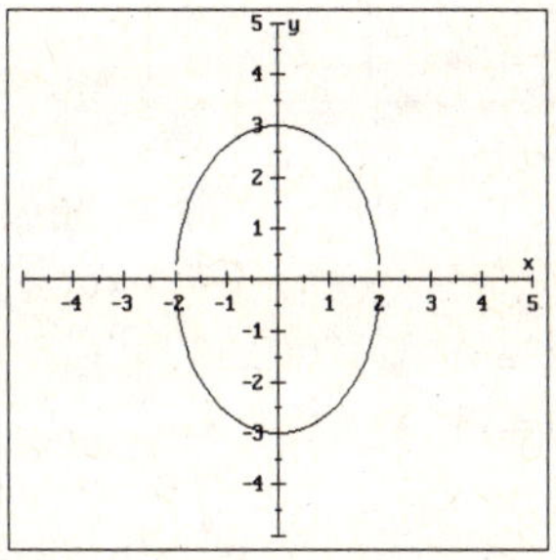

6. $\frac{(x-2)^2}{9}-y^2=1$

C $(2,0)$; opens horiz; move 3 horiz and 1 vert

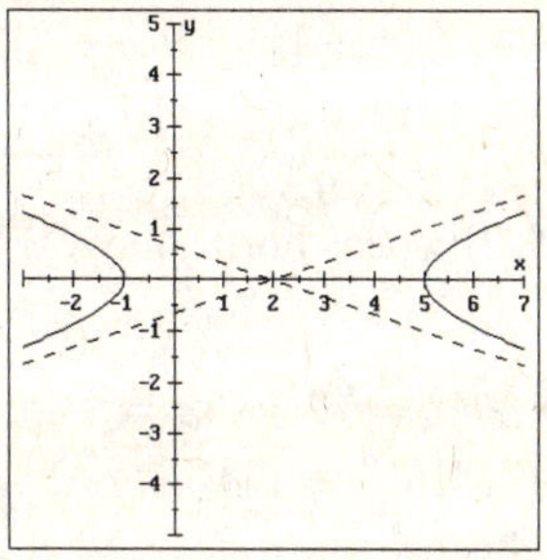

© 2015 Cengage Learning. All Rights Reserved. May not be scanned, copied or duplicated, or posted to a publicly accessible website, in whole or in part.

7.
$$\begin{aligned} 4x^2 + y^2 - 24x + 2y &= -33 \\ 4(x^2 - 6x) + (y^2 + 2y) &= -33 \\ 4(x^2 - 6x + 9) + (y^2 + 2y + 1) &= -33 + 36 + 1 \\ 4(x-3)^2 + (y+1)^2 &= 4 \\ \frac{(x-3)^2}{1} + \frac{(y+1)^2}{4} &= 1 \end{aligned}$$

C $(3, -1)$; move 1 horiz. and 2 vert.

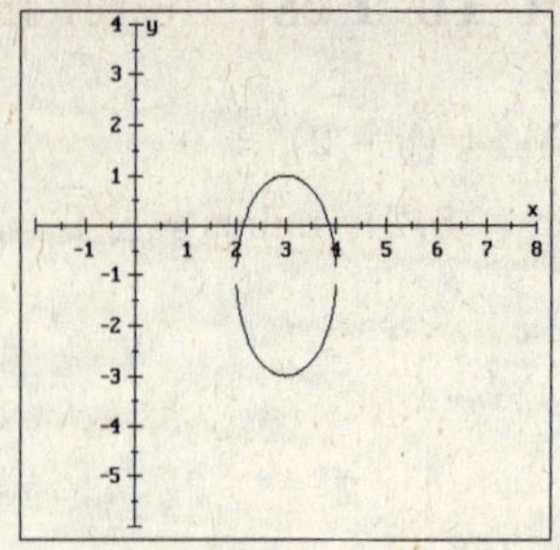

8.
$$\begin{aligned} x^2 + y^2 + 6x - 2y &= -1 \\ x^2 + 6x + y^2 - 2y &= -1 \\ x^2 + 6x + 9 + y^2 - 2y + 1 &= -1 + 9 + 1 \\ (x+3)^2 + (y-1)^2 &= 9 \end{aligned}$$

Center: $(-3, 1)$; radius $= \sqrt{9} = 3$

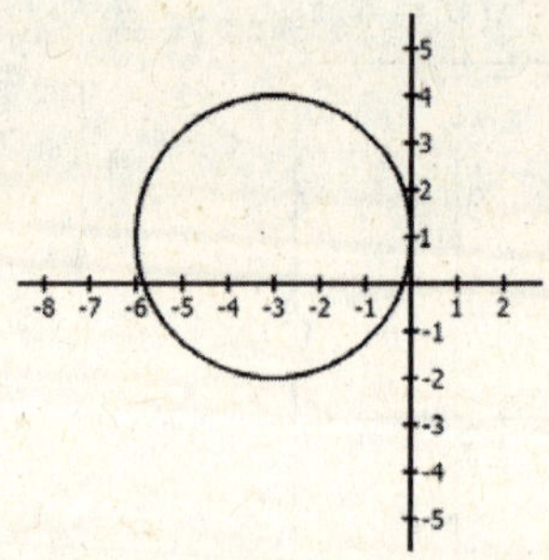

9.
$$\begin{aligned} x^2 - 9y^2 + 2x + 36y &= 44 \\ x^2 + 2x - 9y^2 + 36y &= 44 \\ (x^2 + 2x) - 9(y^2 - 4y) &= 44 \\ (x^2 + 2x + 1) - 9(y^2 - 4y + 4) &= 44 + 1 - 36 \\ (x+1)^2 - 9(y-2)^2 &= 9 \\ \frac{(x+1)^2}{9} - \frac{(y-2)^2}{1} &= 1 \end{aligned}$$

C $(-1, 2)$; opens horiz.; move 3 horiz. and 1 vert.

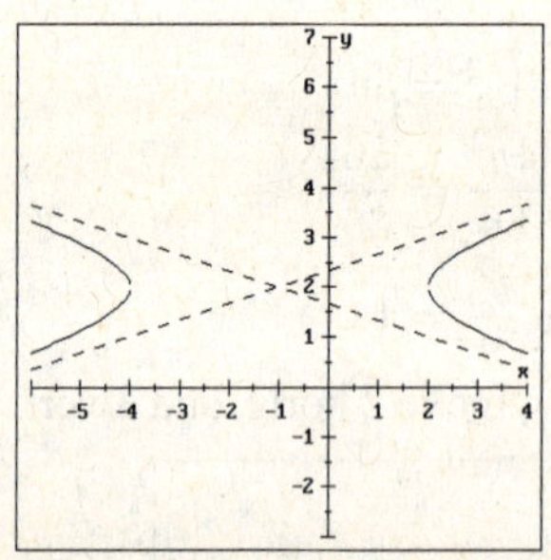

10. (1) $3x + 4y = 12 \Rightarrow 3x = -4y + 12 \Rightarrow (3x)^2 = (-4y + 12)^2 \Rightarrow 9x^2 = 16y^2 - 96y + 144$

(2) $9x^2 + 16y^2 = 144$

Substitute $9x^2 = 16y^2 - 96y + 144$:
$$\begin{aligned} 9x^2 + 16y^2 &= 144 \\ 16y^2 - 96y + 144 + 16y^2 &= 144 \\ 32y^2 - 96y &= 0 \\ 32y(y-3) &= 0 \end{aligned}$$

$32y = 0$ **or** $y - 3 = 0$

$y = 0$ $\qquad y = 3$

Substitute these and solve for y:

$$\begin{aligned} 3x + 4y &= 12 \\ 3x + 4(0) &= 12 \\ x &= 4 \end{aligned} \qquad \begin{aligned} 3x + 4y &= 12 \\ 3x + 4(3) &= 12 \\ x &= 0 \end{aligned}$$

Solutions: $(4, 0), (0, 3)$

© 2015 Cengage Learning. All Rights Reserved. May not be scanned, copied or duplicated, or posted to a publicly accessible website, in whole or in part.

11. $\begin{cases} (1) & x^2 + y^2 = 25 \\ (2) & 4x^2 - 9y = 0 \end{cases}$

Substitute $x^2 = 25 - y^2$ from (1) into (2):

$$4x^2 - 9y = 0$$
$$4(25 - y^2) - 9y = 0$$
$$100 - 4y^2 - 9y = 0$$
$$-4y^2 - 9y + 100 = 0$$
$$4y^2 + 9y - 100 = 0$$
$$(y - 4)(4y + 25) = 0$$

$y - 4 = 0$ **or** $4y + 25 = 0$

$y = 4$ $\qquad y = -\frac{25}{4}$

Substitute these and solve for x:

$x^2 = 25 - y^2$ $\qquad x^2 = 25 - y^2$

$x^2 = 25 - 4^2$ $\qquad x^2 = 25 - \left(-\frac{25}{4}\right)^2$

$x^2 = 25 - 16 = 9$ $\qquad x^2 = 25 - \frac{625}{16} = -\frac{225}{16}$

$x = \pm 3$ $\qquad x$ is nonreal.

Solutions: $(3, 4), (-3, 4)$

12. increasing: $(-3, 0)$; decreasing: $(0, 3)$

13. $f(x) = \begin{cases} -x^2 & \text{when } x < 0 \\ -x & \text{when } x \geq 0 \end{cases}$

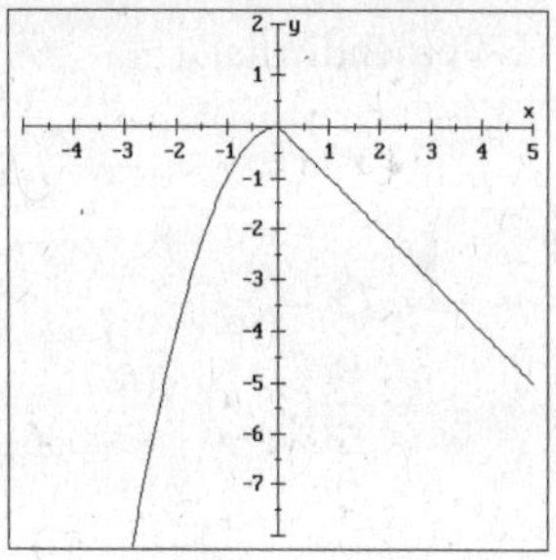

14. $f(x) = 2[[x]]$

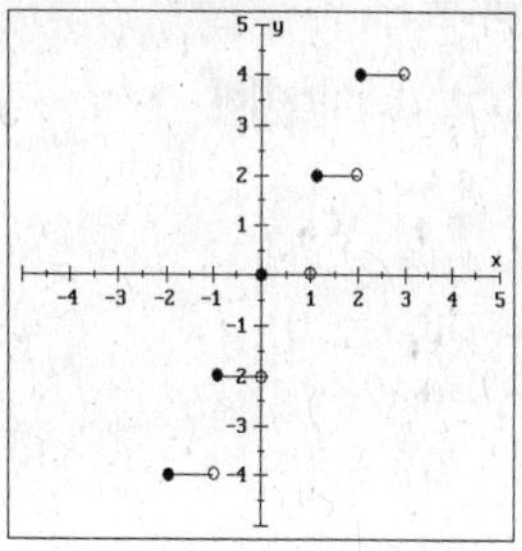

Cumulative Review Exercises (page 746)

1. $(2x - 5y)(4x + y) = 8x^2 + 2xy - 20xy - 5y^2 = 8x^2 - 18xy - 5y^2$

2. $(a^n + 1)(a^n - 3) = a^n a^n - 3a^n + a^n - 3 = a^{2n} - 2a^n - 3$

3. $\dfrac{7a - 14}{a^2 - 7a + 10} = \dfrac{7(a-2)}{(a-2)(a-5)} = \dfrac{7}{a-5}$

4. $\dfrac{a^4 - 5a^2 + 4}{a^2 + 3a + 2} = \dfrac{(a^2 - 4)(a^2 - 1)}{(a+2)(a+1)} = \dfrac{(a+2)(a-2)(a+1)(a-1)}{(a+2)(a+1)} = (a-2)(a-1)$

$= a^2 - 3a + 2$

5. $\dfrac{a^2 - a - 6}{a^2 - 4} \div \dfrac{a^2 - 9}{a^2 + a - 6} = \dfrac{a^2 - a - 6}{a^2 - 4} \cdot \dfrac{a^2 + a - 6}{a^2 - 9} = \dfrac{(a-3)(a+2)}{(a+2)(a-2)} \cdot \dfrac{(a+3)(a-2)}{(a+3)(a-3)} = 1$

© 2015 Cengage Learning. All Rights Reserved. May not be scanned, copied or duplicated, or posted to a publicly accessible website, in whole or in part.

CUMULATIVE REVIEW EXERCISES

6. $$\frac{2}{a-2}+\frac{3}{a+2}-\frac{a-1}{a^2-4}=\frac{2}{a-2}+\frac{3}{a+2}-\frac{a-1}{(a+2)(a-2)}$$
$$=\frac{2(a+2)}{(a-2)(a+2)}+\frac{3(a-2)}{(a+2)(a-2)}-\frac{a-1}{(a+2)(a-2)}$$
$$=\frac{2(a+2)+3(a-2)-(a-1)}{(a+2)(a-2)}$$
$$=\frac{2a+4+3a-6-a+1}{(a+2)(a-2)}=\frac{4a-1}{(a+2)(a-2)}$$

7. $$\begin{aligned}3x-4y&=12\\-4y&=-3x+12\\y&=\frac{3}{4}x-3\\m&=\frac{3}{4}\end{aligned}\qquad\begin{aligned}y&=\frac{3}{4}x-5\\m&=\frac{3}{4}\end{aligned}$$
Parallel

8. $$\begin{aligned}y&=3x+4\\m&=3\end{aligned}\qquad\begin{aligned}x&=-3y+4\\3y&=-x+4\\y&=-\frac{1}{3}x+\frac{4}{3}\\m&=-\frac{1}{3}\end{aligned}$$
Perpendicular

9. $$\begin{aligned}y-y_1&=m(x-x_1)\\y-3&=-6(x-0)\\y-3&=-6x\\y&=-6x+3\end{aligned}$$

10. $$m=\frac{y_2-y_1}{x_2-x_1}=\frac{4-(-5)}{-5-8}=\frac{9}{-13}=-\frac{9}{13}$$
$$\begin{aligned}y-y_1&=m(x-x_1)\\y-(-5)&=-\frac{9}{13}(x-8)\\y+5&=-\frac{9}{13}x+\frac{72}{13}\\y&=-\frac{9}{13}x+\frac{7}{13}\end{aligned}$$

11. $2x-3y<6$

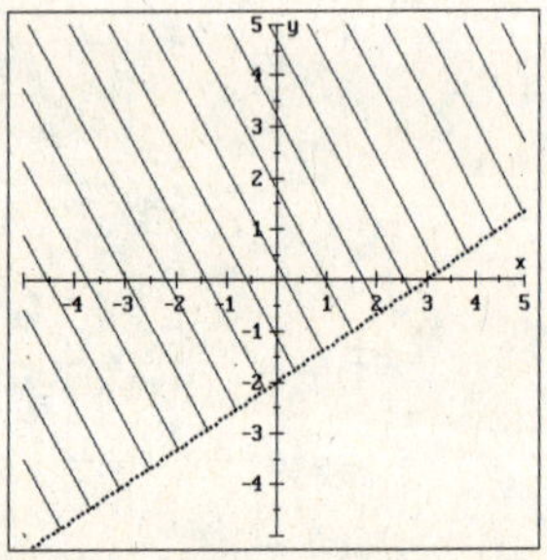

12. $y\geq x^2-4$

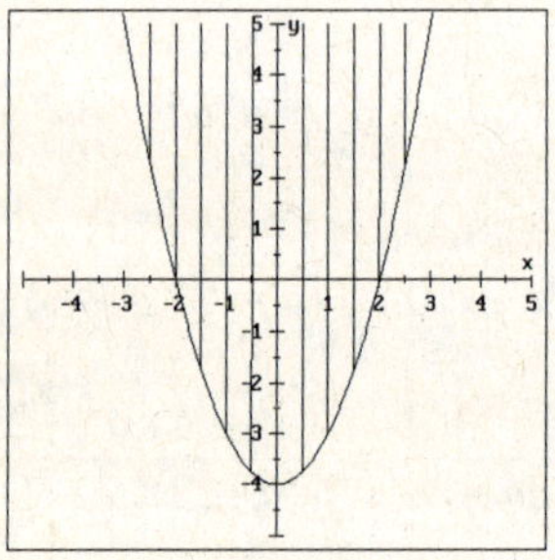

13. $\sqrt{98}+\sqrt{8}-\sqrt{32}=\sqrt{49}\sqrt{2}+\sqrt{4}\sqrt{2}-\sqrt{16}\sqrt{2}=7\sqrt{2}+2\sqrt{2}-4\sqrt{2}=5\sqrt{2}$

14. $$12\sqrt[3]{648x^4}+3\sqrt[3]{81x^4}=12\sqrt[3]{216x^3}\sqrt[3]{3x}+3\sqrt[3]{27x^3}\sqrt[3]{3x}=12(6x)\sqrt[3]{3x}+3(3x)\sqrt[3]{3x}$$
$$=72x\sqrt[3]{3x}+9x\sqrt[3]{3x}=81x\sqrt[3]{3x}$$

© 2015 Cengage Learning. All Rights Reserved. May not be scanned, copied or duplicated, or posted to a publicly accessible website, in whole or in part.

CUMULATIVE REVIEW EXERCISES

15.

$$\begin{aligned} \sqrt{3a+1} &= a-1 \\ \left(\sqrt{3a+1}\right)^2 &= (a-1)^2 \\ 3a+1 &= a^2-2a+1 \\ 0 &= a^2-5a \\ 0 &= a(a-5) \end{aligned}$$

$a = 0$ **or** $a - 5 = 0$

extraneous $\quad a = 5$

16.

$$\begin{aligned} x - \frac{2x}{x-5} &= 1 - \frac{10}{x-5} \\ (x-5)\left(x - \frac{2x}{x-5}\right) &= (x-5)\left(1 - \frac{10}{x-5}\right) \\ x(x-5) - 2x &= 1(x-5) - 10 \\ x^2 - 5x - 2x &= x - 5 - 10 \\ x^2 - 8x + 15 &= 0 \\ (x-3)(x-5) &= 0 \end{aligned}$$

$x - 3 = 0$ **or** $x - 5 = 0$

$x = 3 \quad x = 5$ (extraneous)

17.

$$\begin{aligned} 10a^2 + 21a - 10 &= 0 \\ (2a+5)(5a-2) &= 0 \end{aligned}$$

$2a + 5 = 0$ **or** $5a - 2 = 0$

$a = -\frac{5}{2} \quad a = \frac{2}{5}$

18.

$$2x^2 - 2x + 5 = 0$$

$$a = 2, b = -2, c = 5$$

$$\begin{aligned} x &= \frac{-b \pm \sqrt{b^2-4ac}}{2a} \\ &= \frac{-(-2) \pm \sqrt{(-2)^2 - 4(2)(5)}}{2(2)} \\ &= \frac{2 \pm \sqrt{4-40}}{4} \\ &= \frac{2 \pm \sqrt{-36}}{4} \\ &= \frac{2}{4} \pm \frac{6i}{4} = \frac{1}{2} \pm \frac{3}{2}i \end{aligned}$$

19. $(f \circ g)(x) = f(g(x)) = f(4x+1) = (4x+1)^2 - 3 = 16x^2 + 8x + 1 - 3 = 16x^2 + 8x - 2$

20.

$$\begin{aligned} y &= 2x^3 - 1 \\ x &= 2y^3 - 1 \\ x + 1 &= 2y^3 \\ \frac{x+1}{2} &= y^3 \\ \sqrt[3]{\frac{x+1}{2}} &= y \end{aligned}$$

$$y = f^{-1}(x) = \sqrt[3]{\frac{x+1}{2}} = \frac{\sqrt[3]{4x+4}}{2}$$

21. $y = \left(\frac{1}{2}\right)^x$

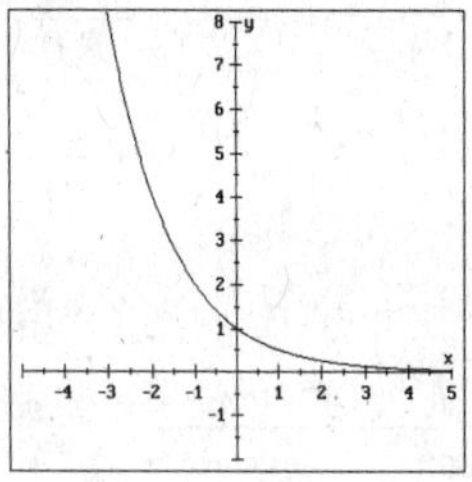

22. $y = \log_5 x \Rightarrow 5^y = x$

© 2015 Cengage Learning. All Rights Reserved. May not be scanned, copied or duplicated, or posted to a publicly accessible website, in whole or in part.

CUMULATIVE REVIEW EXERCISES

23.
$$
\begin{aligned}
2^{x+1} &= 8^{x^2-1} \\
2^{x+1} &= \left(2^3\right)^{x^2-1} \\
2^{x+1} &= 2^{3x^2-3} \\
x+1 &= 3x^2-3 \\
0 &= 3x^2-x-4 \\
0 &= (3x-4)(x+1) \\
x = \tfrac{4}{3} &\text{ or } x = -1
\end{aligned}
$$

24.
$$
\begin{aligned}
2\log 5 + \log x - \log 4 &= 2 \\
\log 5^2 + \log x - \log 4 &= 2 \\
\log \tfrac{25x}{4} &= 2 \\
10^2 &= \tfrac{25x}{4} \\
400 &= 25x \\
16 &= x
\end{aligned}
$$

25. $x^2 + (y+1)^2 = 9$

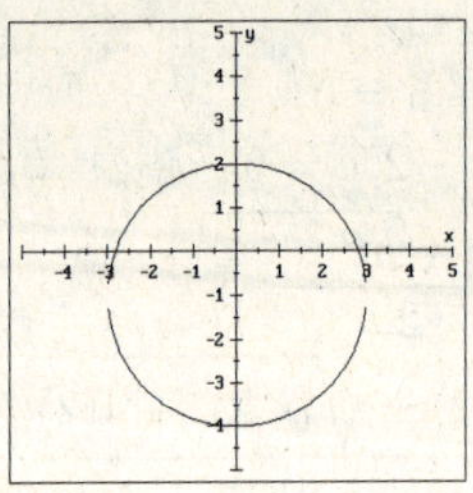

26. $x^2 - 9(y+1)^2 = 9$

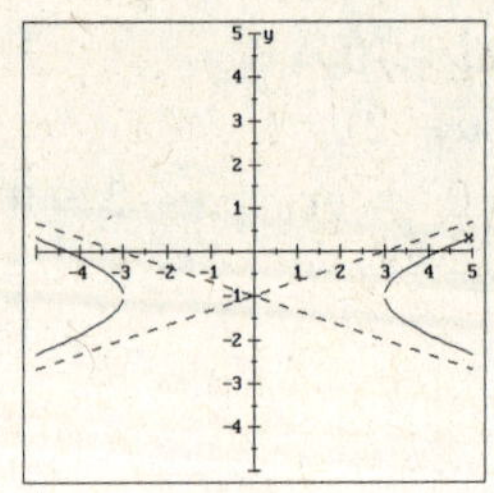

27.
$$
\begin{aligned}
&(1) & x + 2y + 3z &= 4 \\
&(2) & 5x + 6y + 7z &= 8 \\
&(3) & 9x + 10y + 11z &= 12
\end{aligned}
$$

$$
\begin{array}{rr}
-5\cdot(1) & -5x - 10y - 15z = -20 \\
(2) & 5x + 6y + 7z = 8 \\
\hline
(4) & -4y - 8z = -12
\end{array}
\qquad
\begin{array}{rr}
-9\cdot(1) & -9x - 18y - 27z = -36 \\
(3) & 9x + 10y + 11z = 12 \\
\hline
(5) & -8y - 16z = -24
\end{array}
$$

$$
\begin{array}{rr}
-2\cdot(4) & 8y + 16z = 24 \\
(5) & -8y - 16z = -24 \\
\hline
 & 0 = 0
\end{array}
$$

Dependent equations

If $z = z$, then

$$
\begin{aligned}
8y + 16z &= 24 \\
8y &= -16z + 24 \\
y &= -2z + 3
\end{aligned}
$$

$$
\begin{aligned}
x + 2y + 3z &= 4 \\
x + 2(-2z+3) + 3z &= 4 \\
x - 4z + 6 + 3z &= 4 \\
x &= z - 2
\end{aligned}
$$

Solution: $(z-2, -2z+3, z)$

28.
$$
\begin{array}{rcl}
x^2 + y^2 &=& 5 \\
x^2 - y^2 &=& 3 \\
\hline
2x^2 &=& 8 \\
x^2 &=& 4 \\
x &=& \pm 2
\end{array}
$$

Substitute and solve for y:

$$
\begin{aligned}
x^2 + y^2 &= 5 \\
2^2 + y^2 &= 5 \\
y^2 &= 1 \\
y &= \pm 1
\end{aligned}
\qquad
\begin{aligned}
x^2 + y^2 &= 5 \\
(-2)^2 + y^2 &= 5 \\
y^2 &= 1 \\
y &= \pm 1
\end{aligned}
$$

Solutions: $(2,1), (2,-1), (-2,1), (-2,-1)$

© 2015 Cengage Learning. All Rights Reserved. May not be scanned, copied or duplicated, or posted to a publicly accessible website, in whole or in part.

29. $\begin{cases} (1) & x^2 + y^2 = 25 \\ (2) & 4x^2 - 9y = 0 \Rightarrow x^2 = \frac{9}{4}y \end{cases}$

Substitute $x^2 = \frac{9}{4}y$ from (2) into (1):

$$x^2 + y^2 = 25$$
$$\frac{9}{4}y + y^2 = 25$$
$$9y + 4y^2 = 100$$
$$4y^2 + 9y - 100 = 0$$
$$(4y + 25)(y - 4) = 0$$

$4y + 25 = 0$ **or** $y - 4 = 0$

$y = -\frac{25}{4}$ $\qquad y = 4$

Substitute these and solve for x:

$x^2 = \frac{9}{4}y$ $\qquad x^2 = \frac{9}{4}y$

$x^2 = \frac{9}{4}(-\frac{25}{4})$ $\qquad x^2 = \frac{9}{4}(4)$

$x^2 = \pm\sqrt{-\frac{225}{16}}$ $\qquad x^2 = 9$

complex $\qquad x = \pm 3$

Solutions: $(3, 4), (-3, 4)$

Exercises 11.1 (page 757)

1. $6 \cdot 5 \cdot 4 \cdot 3 \cdot 2 \cdot 1 = 720$

3. $\dfrac{6 \cdot 5 \cdot 4 \cdot 3 \cdot 2 \cdot 1}{4 \cdot 3 \cdot 2 \cdot 1} = \dfrac{720}{24} = 30$

5. $(x + 3y)^2 = (x + 3y)(x + 3y)$
$= x^2 + 6xy + 9y^2$

7. $(x - 3y)^2 = (x - 3y)(x - 3y)$
$= x^2 - 6xy + 9y^2$

9. $\log_9 81 = x \Rightarrow 9^x = 81 \Rightarrow x = 2$

11. $\log_{25} x = \frac{1}{2} \Rightarrow 25^{1/2} = x \Rightarrow x = 5$

13. one

15. Pascal's

17. 8!

19. 1

21. $(a + b)^3 = 1a^3 + 3a^2b + 3ab^2 + 1b^3 = a^3 + 3a^2b + 3ab^2 + b^3$

23. $(a - b)^4 = 1a^4 + 4a^3(-b) + 6a^2(-b)^2 + 4a(-b)^3 + 1(-b)^4$
$= a^4 - 4a^3b + 6a^2b^2 - 4ab^3 + b^4$

25. $4! = 4 \cdot 3 \cdot 2 \cdot 1 = 24$

27. $-8! = -1 \cdot 8! = -8 \cdot 7 \cdot 6 \cdot 5 \cdot 4 \cdot 3 \cdot 2 \cdot 1$
$= -40{,}320$

29. $3! + 4! = 3 \cdot 2 \cdot 1 + 4 \cdot 3 \cdot 2 \cdot 1$
$= 6 + 24 = 30$

31. $3!(4!) = 3 \cdot 2 \cdot 1 \cdot 4 \cdot 3 \cdot 2 \cdot 1 = 144$

33. $\dfrac{9!}{11!} = \dfrac{9!}{11 \cdot 10 \cdot 9!} = \dfrac{1}{11 \cdot 10} = \dfrac{1}{110}$

35. $\dfrac{49!}{47!} = \dfrac{49 \cdot 48 \cdot 47!}{47!} = 49 \cdot 48 = 2352$

37. $\dfrac{9!}{7!\,0!} = \dfrac{9 \cdot 8 \cdot 7!}{7! \cdot 1} = 9 \cdot 8 = 72$

39. $\dfrac{5!}{3!(5-3)!} = \dfrac{5!}{3!2!} = \dfrac{5 \cdot 4 \cdot 3!}{3! \cdot 2 \cdot 1} = \dfrac{5 \cdot 4}{2 \cdot 1} = 10$

© 2015 Cengage Learning. All Rights Reserved. May not be scanned, copied or duplicated, or posted to a publicly accessible website, in whole or in part.

41. $$(x+y)^3 = x^3 + \frac{3!}{1!(3-1)!}x^2y + \frac{3!}{2!(3-2)!}xy^2 + y^3 = x^3 + \frac{3!}{1!2!}x^2y + \frac{3!}{2!1!}xy^2 + y^3$$
$$= x^3 + \frac{3\cdot 2!}{1!2!}x^2y + \frac{3\cdot 2!}{2!1!}xy^2 + y^3$$
$$= x^3 + \frac{3}{1}x^2y + \frac{3}{1}xy^2 + y^3$$
$$= x^3 + 3x^2y + 3xy^2 + y^3$$

43. $$(a+b)^6 = a^6 + \frac{6!}{1!(6-1)!}a^5b + \frac{6!}{2!(6-2)!}a^4b^2 + \frac{6!}{3!(6-3)!}a^3b^3 + \frac{6!}{4!(6-4)!}a^2b^4 + \cdots + \frac{6!}{5!(6-5)!}ab^5 + \frac{6!}{6!(6-6)!}b^6$$
$$= a^6 + \frac{6!}{1!5!}a^5b + \frac{6!}{2!4!}a^4b^2 + \frac{6!}{3!3!}a^3b^3 + \frac{6!}{4!2!}a^2b^4 + \frac{6!}{5!1!}ab^5 + \frac{6!}{6!0!}b^6$$
$$= a^6 + \frac{6\cdot 5!}{1!5!}a^5b + \frac{6\cdot 5\cdot 4!}{2!4!}a^4b^2 + \frac{6\cdot 5\cdot 4\cdot 3!}{3!3!}a^3b^3 + \frac{6\cdot 5\cdot 4!}{4!2!}a^2b^4 + \frac{6\cdot 5!}{5!1!}ab^5 + \frac{6!}{6!0!}b^6$$
$$= a^6 + \frac{6}{1}a^5b + \frac{30}{2}a^4b^2 + \frac{120}{6}a^3b^3 + \frac{30}{2}a^2b^4 + \frac{6}{1}ab^5 + b^6$$
$$= a^6 + 6a^5b + 15a^4b^2 + 20a^3b^3 + 15a^2b^4 + 6ab^5 + b^6$$

45. $$(a-b)^4 = a^4 + \frac{4!}{1!(4-1)!}a^3(-b) + \frac{4!}{2!(4-2)!}a^2(-b)^2 + \frac{4!}{3!(4-3)!}a(-b)^3 + (-b)^4$$
$$= a^4 + \frac{4!}{1!3!}(-a^3b) + \frac{4!}{2!2!}a^2b^2 + \frac{4!}{3!1!}(-ab^3) + b^4$$
$$= a^4 - \frac{4\cdot 3!}{1!3!}a^3b + \frac{4\cdot 3\cdot 2!}{2!\cdot 2\cdot 1}a^2b^2 - \frac{4\cdot 3!}{3!1!}ab^3 + b^4$$
$$= a^4 - \frac{4}{1}a^3b + \frac{12}{2}a^2b^2 - \frac{4}{1}ab^3 + b^4$$
$$= a^4 - 4a^3b + 6a^2b^2 - 4ab^3 + b^4$$

47. $$(x-2y)^3 = x^3 + \frac{3!}{1!(3-1)!}x^2(-2y) + \frac{3!}{2!(3-2)!}x(-2y)^2 + (-2y)^3$$
$$= x^3 + \frac{3!}{1!2!}\cdot(-2x^2y) + \frac{3!}{2!1!}\cdot 4xy^2 - 8y^3$$
$$= x^3 - \frac{3\cdot 2!}{1!2!}\cdot 2x^2y + \frac{3\cdot 2!}{2!1!}\cdot 4xy^2 - 8y^3$$
$$= x^3 - \frac{3}{1}\cdot 2x^2y + \frac{3}{1}\cdot 4xy^2 - 8y^3$$
$$= x^3 - 6x^2y + 12xy^2 - 8y^3$$

© 2015 Cengage Learning. All Rights Reserved. May not be scanned, copied or duplicated, or posted to a publicly accessible website, in whole or in part.

49. $$\begin{aligned}(2x+y)^3 &= (2x)^3 + \frac{3!}{1!(3-1)!}(2x)^2y + \frac{3!}{2!(3-2)!}2xy^2 + y^3 \\ &= 8x^3 + \frac{3!}{1!2!}\cdot 4x^2y + \frac{3!}{2!1!}\cdot 2xy^2 + y^3 \\ &= 8x^3 + \frac{3\cdot 2!}{1!2!}\cdot 4x^2y + \frac{3\cdot 2!}{2!1!}\cdot 2xy^2 + y^3 \\ &= 8x^3 + \frac{3}{1}\cdot 4x^2y + \frac{3}{1}\cdot 2xy^2 + y^3 \\ &= 8x^3 + 12x^2y + 6xy^2 + y^3\end{aligned}$$

51. $$\begin{aligned}(2x+3y)^3 &= (2x)^3 + \frac{3!}{1!(3-1)!}(2x)^2(3y) + \frac{3!}{2!(3-2)!}2x(3y)^2 + (3y)^3 \\ &= 8x^3 + \frac{3!}{1!2!}\cdot 4x^2(3y) + \frac{3!}{2!1!}\cdot 2x(9y^2) + 27y^3 \\ &= 8x^3 + \frac{3\cdot 2!}{1!2!}\cdot 12x^2y + \frac{3\cdot 2!}{2!1!}\cdot 18xy^2 + 27y^3 \\ &= 8x^3 + \frac{3}{1}\cdot 12x^2y + \frac{3}{1}\cdot 18xy^2 + 27y^3 \\ &= 8x^3 + 36x^2y + 54xy^2 + 27y^3\end{aligned}$$

53. In the 3rd term, the exponent on b is 2.
Variables: a^2b^2
Coef. $= \frac{n!}{r!(n-r)!} = \frac{4!}{2!2!} = 6$
Term $= 6a^2b^2$

55. In the 5th term, the exponent on y is 4.
Variables: x^2y^4
Coef. $= \frac{n!}{r!(n-r)!} = \frac{6!}{4!2!} = 15$
Term $= 15x^2y^4$

57. In the 4th term, the exponent on $-y$ is 3.
Variables: $x^1(-y)^3 = -xy^3$
Coef. $= \frac{n!}{r!(n-r)!} = \frac{4!}{3!1!} = 4$
Term $= 4(-xy^3) = -4xy^3$

59. In the 3rd term, the exponent on $-y$ is 2.
Variables: $x^6(-y)^2 = x^6y^2$
Coef. $= \frac{n!}{r!(n-r)!} = \frac{8!}{2!6!} = 28$
Term $= 28x^6y^2$

61. In the 3rd term, the exponent on y is 2.
Variables: $(4x)^3y^2 = 64x^3y^2$
Coef. $= \frac{n!}{r!(n-r)!} = \frac{5!}{2!3!} = 10$
Term $= 10(64x^3y^2) = 640x^3y^2$

63. In the 2nd term, the exponent on $-3y$ is 1.
Variables: $x^3(-3y)^1 = -3x^3y$
Coef. $= \frac{n!}{r!(n-r)!} = \frac{4!}{1!3!} = 4$
Term $= 4(-3x^3y) = -12x^3y$

65. In the 4th term, the exponent on -5 is 3. Variables: $(2x)^4(-5)^3 = (16x^4)(-125) = -2000x^4$
Coef. $= \frac{n!}{r!(n-r)!} = \frac{7!}{3!4!} = 35$; Term $= 35(-2000x^4) = -70{,}000x^4$

© 2015 Cengage Learning. All Rights Reserved. May not be scanned, copied or duplicated, or posted to a publicly accessible website, in whole or in part.

67. In the 5th term, the exponent on $-3y$ is 4. Variables: $(2x)^1(-3y)^4 = 2x(81y^4) = 162xy^4$

$$\text{Coef.} = \frac{n!}{r!(n-r)!} = \frac{5!}{4!1!} = 5;\ \text{Term} = 5(162xy^4) = 810xy^4$$

69. $8(7!) = 8 \cdot 7 \cdot 6 \cdot 5 \cdot 4 \cdot 3 \cdot 2 \cdot 1 = 40{,}320$

71. $\dfrac{7!}{5!(7-5)!} = \dfrac{7!}{5!2!} = \dfrac{7 \cdot 6 \cdot 5!}{5! \cdot 2 \cdot 1} = \dfrac{7 \cdot 6}{2 \cdot 1} = 21$

73. $\dfrac{5!(8-5)!}{4!7!} = \dfrac{5!3!}{4!7!} = \dfrac{5!}{7!} \cdot \dfrac{3!}{4!} = \dfrac{5!}{7 \cdot 6 \cdot 5!} \cdot \dfrac{3!}{4 \cdot 3!} = \dfrac{1}{7 \cdot 6} \cdot \dfrac{1}{4} = \dfrac{1}{168}$

75. $11! = 39{,}916{,}800$

77. $20! = 2.432902008 \times 10^{18}$

79.
$$\begin{aligned}(3+2y)^4 &= 3^4 + \frac{4!}{1!(4-1)!}3^3(2y) + \frac{4!}{2!(4-2)!}3^2(2y)^2 + \frac{4!}{3!(4-3)!}3(2y)^3 + (2y)^4 \\ &= 81 + \frac{4!}{1!3!} \cdot 27(2y) + \frac{4!}{2!2!} \cdot 9(4y^2) + \frac{4!}{3!1!} \cdot 3(8y^3) + 16y^4 \\ &= 81 + \frac{4 \cdot 3!}{1!3!} \cdot 54y + \frac{4 \cdot 3 \cdot 2!}{2! \cdot 2 \cdot 1} \cdot 36y^2 + \frac{4 \cdot 3!}{3!1!} \cdot 24y^3 + 16y^4 \\ &= 81 + \frac{4}{1} \cdot 54y + \frac{12}{2} \cdot 36y^2 + \frac{4}{1} \cdot 24y^3 + 16y^4 \\ &= 81 + 216y + 216y^2 + 96y^3 + 16y^4\end{aligned}$$

81.
$$\begin{aligned}\left(\frac{x}{2} - \frac{y}{3}\right)^3 &= \left(\frac{x}{2}\right)^3 + \frac{3!}{1!(3-1)!}\left(\frac{x}{2}\right)^2\left(-\frac{y}{3}\right) + \frac{3!}{2!(3-2)!}\left(\frac{x}{2}\right)\left(-\frac{y}{3}\right)^2 + \left(-\frac{y}{3}\right)^3 \\ &= \frac{x^3}{8} - \frac{3!}{1!2!} \cdot \frac{x^2}{4} \cdot \frac{y}{3} + \frac{3!}{2!1!} \cdot \frac{x}{2} \cdot \frac{y^2}{9} - \frac{y^3}{27} \\ &= \frac{x^3}{8} - \frac{3 \cdot 2!}{1!2!} \cdot \frac{x^2y}{12} + \frac{3 \cdot 2!}{2!1!} \cdot \frac{xy^2}{18} - \frac{y^3}{27} \\ &= \frac{x^3}{8} - \frac{3}{1} \cdot \frac{x^2y}{12} + \frac{3}{1} \cdot \frac{xy^2}{18} - \frac{y^3}{27} \\ &= \frac{x^3}{8} - \frac{x^2y}{4} + \frac{xy^2}{6} - \frac{y^3}{27}\end{aligned}$$

83.
$$\begin{aligned}(x-y)^5 &= x^5 + \frac{5!}{1!(5-1)!}x^4(-y) + \frac{5!}{2!(5-2)!}x^3(-y)^2 + \frac{5!}{3!(5-3)!}x^2(-y)^3 \\ &\qquad + \frac{5!}{4!(5-4)!}x(-y)^4 + \frac{5!}{5!(5-5)!}(-y)^5 \\ &= x^5 - \frac{5!}{1!4!}x^4y + \frac{5!}{2!3!}x^3y^2 - \frac{5!}{3!2!}x^2y^3 + \frac{5!}{4!1!}xy^4 - \frac{5!}{5!0!}y^5 \\ &= x^5 - \frac{5 \cdot 4!}{1!4!}x^4y + \frac{5 \cdot 4 \cdot 3!}{2!3!}x^3y^2 - \frac{5 \cdot 4 \cdot 3!}{3!2!}x^2y^3 + \frac{5 \cdot 4!}{4!1!}x^2y^4 - \frac{5!}{5!0!}y^5 \\ &= x^5 - \frac{5}{1}x^4y + \frac{20}{2}x^3y^2 - \frac{20}{2}x^2y^3 + \frac{5}{1}xy^4 - y^5 \\ &= x^5 - 5x^4y + 10x^3y^2 - 10x^2y^3 + 5xy^4 - y^5\end{aligned}$$

© 2015 Cengage Learning. All Rights Reserved. May not be scanned, copied or duplicated, or posted to a publicly accessible website, in whole or in part.

85. $\left(\frac{x}{2}+\frac{y}{3}\right)^4$

$$= \left(\frac{x}{2}\right)^4 + \frac{4!}{1!(4-1)!}\left(\frac{x}{2}\right)^3\left(\frac{y}{3}\right) + \frac{4!}{2!(4-2)!}\left(\frac{x}{2}\right)^2\left(\frac{y}{3}\right)^2 + \frac{4!}{3!(4-3)!}\left(\frac{x}{2}\right)\left(\frac{y}{3}\right)^3 + \left(\frac{y}{3}\right)^4$$

$$= \frac{x^4}{16} + \frac{4!}{1!3!}\cdot\frac{x^3}{8}\cdot\frac{y}{3} + \frac{4!}{2!2!}\cdot\frac{x^2}{4}\cdot\frac{y^2}{9} + \frac{4!}{3!1!}\cdot\frac{x}{2}\cdot\frac{y^3}{27} + \frac{y^4}{81}$$

$$= \frac{x^4}{16} + \frac{4\cdot 3!}{1!3!}\cdot\frac{x^3y}{24} + \frac{4\cdot 3\cdot 2!}{2!2!}\cdot\frac{x^2y^2}{36} + \frac{4\cdot 3!}{3!1!}\cdot\frac{xy^3}{54} + \frac{y^4}{81}$$

$$= \frac{x^4}{16} + \frac{4}{1}\cdot\frac{x^3y}{24} + \frac{12}{2}\cdot\frac{x^2y^2}{36} + \frac{4}{1}\cdot\frac{xy^3}{54} + \frac{y^4}{81}$$

$$= \frac{x^4}{16} + \frac{x^3y}{6} + \frac{x^2y^2}{6} + \frac{2xy^3}{27} + \frac{y^4}{81}$$

87.

1
1 1
1 2 1
1 3 3 1
1 4 6 4 1
1 5 10 10 5 1
1 6 15 20 15 6 1
1 7 21 35 35 21 7 1
1 8 28 56 70 56 28 8 1
1 9 36 84 126 126 84 36 9 1

89. In the 3rd term, the exponent on $\sqrt{3}y$ is 2. Variables: $(\sqrt{2}x)^4(\sqrt{3}y)^2 = (4x^4)(3y^2) = 12x^4y^2$

Coef. $= \frac{n!}{r!(n-r)!} = \frac{6!}{2!4!} = 15$; Term $= 15(12x^4y^2) = 180x^4y^2$

91. In the 2nd term, the exponent on $-\frac{y}{3}$ is 1. Variables: $\left(\frac{x}{2}\right)^3\left(-\frac{y}{3}\right)^1 = \left(\frac{x^3}{8}\right)\left(-\frac{y}{3}\right) = -\frac{x^3y}{24}$

Coef. $= \frac{n!}{r!(n-r)!} = \frac{4!}{1!3!} = 4$; Term $= 4\left(-\frac{x^3y}{24}\right) = -\frac{x^3y}{6} = -\frac{1}{6}x^3y$

93. In the 3rd term, the exponent on 5 is 2.
Variables: $x^4(5)^2 = 25x^4$
Coef. $= \frac{n!}{r!(n-r)!} = \frac{6!}{2!4!} = 15$
Term $= 15(25x^4) = 375x^4$

95. In the 4th term, the exponent on b is 3.
Variables: $a^{n-3}b^3$
Coef. $= \frac{n!}{r!(n-r)!} = \frac{n!}{3!(n-3)!}$
Term $= \frac{n!}{3!(n-3)!}a^{n-3}b^3$

97. In the 5th term, the exponent on $-b$ is 4. Variables: $a^{n-4}(-b)^4 = a^{n-4}b^4$

Coef. $= \frac{n!}{r!(n-r)!} = \frac{n!}{4!(n-4)!}$; Term $= \frac{n!}{4!(n-4)!}a^{n-4}b^4$

© 2015 Cengage Learning. All Rights Reserved. May not be scanned, copied or duplicated, or posted to a publicly accessible website, in whole or in part.

99. In the rth term, the coefficient on b is $r-1$. Variables: $a^{n-(r-1)}b^{r-1} = a^{n-r+1}b^{r-1}$

$$\text{Coef.} = \frac{n!}{r!(n-r)!} = \frac{n!}{(r-1)![n-(r-1)]!} = \frac{n!}{(r-1)!(n-r+1)!}$$

$$\text{Term} = \frac{n!}{(r-1)!(n-r+1)!}a^{n-r+1}b^{r-1}$$

101-103. Answers may vary.

105. $\frac{n!}{0!(n-0)!} = \frac{n!}{0!n!} = \frac{n!}{1\cdot n!} = \frac{n!}{n!} = 1$

107. $1, 1, 2, 3, 5, 8, 13, \ldots$; Beginning with 2, each number is the sum of the previous two numbers.

109. $\left(a - \frac{1}{a}\right)^9 = \left(a - a^{-1}\right)^9$. The desired term occurs when the exponent is 5.

The $(r+1)$th term of $(a-a^{-1})^9$ is $\frac{9!}{r!(9-r)!}a^{9-r}\left(-a^{-1}\right)^r = \frac{9!}{r!(9-r)!}a^{9-r}(-a)^{-r}$.

But $\frac{9!}{r!(9-r)!}a^{9-r}(-a)^{-r} = (-1)^{-r}\frac{9!}{r!(9-r)!}a^{9-r}a^{-r} = (-1)^{-r}\frac{9!}{r!(9-r)!}a^{9-2r}$.

If $9-2r=5$, then $r=2$.

The term is $(-1)^{-2}\frac{9!}{2!(9-2)!}a^{9-2}a^{-2} = \frac{9!}{2!7!}a^5 = \frac{9\cdot 8\cdot 7!}{2\cdot 1\cdot 7!} = 36a^5$. Coefficient $= 36$

Exercises 11.2 (page 766)

1. $2(1) = 2, 2(2) = 4, 2(3) = 6, 2(4) = 8$

3. $1+4=5, 2+4=6, 3+4=7, 4+4=8$

5. $5+(1-1)2 = 5+0(2) = 5, 5+(2-1)2 = 5+1(2) = 7, 5+(3-1)2 = 5+2(2) = 9,$
$5+(4-1)2 = 5+3(2) = 11$

7. $5(3x^2-6x+4)+3(4x^2+6x-2) = 15x^2-30x+20+12x^2+18x-6 = 27x^2-12x+14$

9.
$$\begin{aligned}\frac{3a+4}{a-2}+\frac{3a-4}{a+2} &= \frac{(3a+4)(a+2)}{(a-2)(a+2)}+\frac{(3a-4)(a-2)}{(a+2)(a-2)}\\ &= \frac{3a^2+10a+8+3a^2-10a+8}{(a+2)(a-2)} = \frac{6a^2+16}{(a+2)(a-2)}\end{aligned}$$

11. sequence; finite; infinite

13. arithmetic; difference

15. arithmetic mean

17. series

19. $1+2+3+4+5+6+7$

21. $a_2 = 3(2)-2 = 4$

23. $a_{30} = 3(30)-2 = 88$

25. $3, 5, 7, 9, 11$

27. $-5, -8, -11, -14, -17$

29. $4, 7, 10, 13, 16$
$a_n = a_1 + (n-1)d$
$a_{15} = 4 + (15-1)3$
$a_{15} = 4 + (14)3 = 46$

31. $-7, -9, -11, -13, -15$
$a_n = a_1 + (n-1)d$
$a_{30} = -7 + (30-1)(-2)$
$a_{30} = -7 + (29)(-2) = -65$

© 2015 Cengage Learning. All Rights Reserved. May not be scanned, copied or duplicated, or posted to a publicly accessible website, in whole or in part.

SECTION 11.2

33. $a_1 = 2, d = 5$
$a_n = a_1 + (n-1)d$
$a_{50} = 2 + (50-1)(5)$
$a_{50} = 2 + 49(5) = 247$

35. $a_1 = 5, d = -6$
$a_n = a_1 + (n-1)d$
$a_{49} = 5 + (49-1)(-6)$
$a_{49} = 5 + 48(-6) = -283$

37. $a_n = a_1 + (n-1)d$
$a_5 = a_1 + (5-1)d$
$29 = 5 + 4d$
$24 = 4d$
$6 = d$
$5, 11, 17, 23, 29$

39. $a_n = a_1 + (n-1)d$
$a_6 = a_1 + (6-1)d$
$-39 = -4 + 5d$
$-35 = 5d$
$-7 = d$
$-4, -11, -18, -25, -32$

41. Form an arithmetic sequence with a 1st term of 2 and a 5th term of 11:
$a_n = a_1 + (n-1)d$
$a_5 = a_1 + (5-1)d$
$11 = 2 + 4d$
$9 = 4d$
$\frac{9}{4} = d$
$2, \boxed{\frac{17}{4}, \frac{13}{2}, \frac{35}{4}}, 11$

43. Form an arithmetic sequence with a 1st term of 10 and a 6th term of 20:
$a_n = a_1 + (n-1)d$
$a_6 = a_1 + (6-1)d$
$20 = 10 + 5d$
$10 = 5d$
$2 = d$
$10, \boxed{12, 14, 16, 18}, 20$

45. $a_1 = 1, d = 3, n = 30$
$a_n = a_1 + (n-1)d = 1 + 29(3) = 88$
$S_n = \frac{n(a_1 + a_n)}{2} = \frac{30(1+88)}{2} = 1335$

47. $a_1 = -5, d = 4, n = 17$
$a_n = a_1 + (n-1)d = -5 + 16(4) = 59$
$S_n = \frac{n(a_1 + a_n)}{2} = \frac{17(-5+59)}{2} = 459$

49. $\sum_{k=1}^{4}(3k) = 3(1) + 3(2) + 3(3) + 3(4) = 3 + 6 + 9 + 12$

51. $\sum_{k=4}^{6}k^2 = 4^2 + 5^2 + 6^2 = 16 + 25 + 36$

53. $\sum_{k=1}^{4}6k = 6(1) + 6(2) + 6(3) + 6(4) = 6 + 12 + 18 + 24 = 60$

55. $\sum_{k=3}^{4}(k^2 + 3) = (3^2 + 3) + (4^2 + 3) = 9 + 3 + 16 + 3 = 31$

© 2015 Cengage Learning. All Rights Reserved. May not be scanned, copied or duplicated, or posted to a publicly accessible website, in whole or in part.

57.
$$\begin{aligned} a_n &= a_1 + (n-1)d \\ a_6 &= a_1 + (6-1)d \\ -83 &= a_1 + 5(7) \\ -83 &= a_1 + 35 \\ -118 &= a_1 \end{aligned}$$
$-118, -111, -104, -97, -90$

59.
$$\begin{aligned} a_n &= a_1 + (n-1)d \\ a_7 &= a_1 + (7-1)d \\ 16 &= a_1 + 6(-3) \\ 16 &= a_1 - 18 \\ 34 &= a_1 \end{aligned}$$
$34, 31, 28, 25, 22$

61.
$$\begin{aligned} a_n &= a_1 + (n-1)d \\ a_{19} &= a_1 + (19-1)d \\ 131 &= a_1 + 18d \\ a_n &= a_1 + (n-1)d \\ a_{20} &= a_1 + (20-1)d \\ 138 &= a_1 + 19d \end{aligned}$$

$a_1 + 18d = 131 \Rightarrow \times(-1)$ $\quad -a_1 - 18d = -131$
$a_1 + 19d = 138 \Rightarrow$ $\quad a_1 + 19d = 138$
$d = 7$

Substitute and solve for a_1:
$$\begin{aligned} a_1 + 18d &= 131 \\ a_1 + 18(7) &= 131 \\ a_1 + 126 &= 131 \\ a_1 &= 5 \Rightarrow 5, 12, 19, 26, 33 \end{aligned}$$

63.
$$a_n = a_1 + (n-1)d$$
$$a_{30} = a_1 + (30-1)d = 7 + 29(12) = 7 + 348 = 355$$

65.
$$\begin{aligned} a_n &= a_1 + (n-1)d \\ a_2 &= a_1 + (2-1)d \\ -4 &= a_1 + d \\ a_n &= a_1 + (n-1)d \\ a_3 &= a_1 + (3-1)d \\ -9 &= a_1 + 2d \end{aligned}$$

$a_1 + d = -4 \Rightarrow \times(-1)$ $\quad -a_1 - d = 4$
$a_1 + 2d = -9 \Rightarrow$ $\quad a_1 + 2d = -9$
$d = -5$

Substitute and solve for a_1:
$$\begin{aligned} a_1 + d &= -4 \\ a_1 + (-5) &= -4 \\ a_1 &= 1 \end{aligned}$$

Find the desired term:
$$\begin{aligned} a_n &= a_1 + (n-1)d \\ a_{37} &= a_1 + (37-1)d \\ &= 1 + 36(-5) \\ &= 1 - 180 = \boxed{-179} \end{aligned}$$

67.
$$\begin{aligned} a_n &= a_1 + (n-1)d \\ a_{27} &= a_1 + (27-1)d \\ 263 &= a_1 + 26(11) \\ 263 &= a_1 + 286 \\ -23 &= a_1 \end{aligned}$$

69.
$$\begin{aligned} a_n &= a_1 + (n-1)d \\ a_{44} &= a_1 + (44-1)d \\ 556 &= 40 + 43d \\ 516 &= 43d \\ 12 &= d \end{aligned}$$

71. Form an arithmetic sequence with a 1st term of 10 and a 3rd term of 19:
$$\begin{aligned} a_n &= a_1 + (n-1)d \\ a_3 &= a_1 + (3-1)d \\ 19 &= 10 + 2d \\ 9 &= 2d \\ \frac{9}{2} &= d \end{aligned}$$
$10, \boxed{\tfrac{29}{2}}, 19$

73. Form an arithmetic sequence with a 1st term of -4.5 and a 3rd term of 7:
$$\begin{aligned} a_n &= a_1 + (n-1)d \\ a_3 &= a_1 + (3-1)d \\ 7 &= -4.5 + 2d \\ 11.5 &= 2d \\ 5.75 &= d \end{aligned}$$
$-4.5, \boxed{1.25}, 7$

© 2015 Cengage Learning. All Rights Reserved. May not be scanned, copied or duplicated, or posted to a publicly accessible website, in whole or in part.

75. $a_n = a_1 + (n-1)d$
$a_2 = a_1 + (2-1)d$
$7 = a_1 + d$
$a_n = a_1 + (n-1)d$
$a_3 = a_1 + (3-1)d$
$12 = a_1 + 2d$

$$\begin{array}{ll} a_1 + d = 7 \Rightarrow \times(-1) & -a_1 - d = -7 \\ \underline{a_1 + 2d = 12} \Rightarrow & \underline{a_1 + 2d = 12} \\ & d = 5 \end{array}$$

Substitute and solve for a_1:
$a_1 + d = 7$
$a_1 + 5 = 7$
$a_1 = 2, d = 5, n = 12$
$a_n = a_1 + (n-1)d = 2 + 11(5) = 57$
$$S_n = \frac{n(a_1 + a_n)}{2} = \frac{12(2+57)}{2} = 354$$

77. $f(n) = 2n + 1 \Rightarrow f(1) = 3$
$f(n) = 2n + 1 = 31$
$2n = 30$
$n = 15$
$$S_n = \frac{n(a_1 + a_n)}{2} = \frac{15(3+31)}{2} = 255$$

79. $a_1 = 1, d = 1, n = 50$
$a_n = a_1 + (n-1)d = 1 + 49(1) = 50$
$$S_n = \frac{n(a_1 + a_n)}{2} = \frac{50(1+50)}{2} = 1275$$

81. $a_1 = 1, d = 2, n = 50$
$a_n = a_1 + (n-1)d = 1 + 49(2) = 99$
$$S_n = \frac{n(a_1 + a_n)}{2} = \frac{50(1+99)}{2} = 2500$$

83. $\sum_{k=4}^{4} (2k + 4) = 2(4) + 4 = 8 + 4 = 12$

85. $a_1 = 60, d = 50 \Rightarrow 60, 110, 160, 210, 260, 310; n = 121$
$a_n = a_1 + (n-1)d = 60 + (121-1)(50) = 60 + 120(50) = \$6060$

87. $a_1 = 1, d = 1, n = 150, a_n = 150 \Rightarrow S_n = \frac{n(a_1+a_n)}{2} = \frac{150(1+150)}{2} = 11{,}325$ bricks

89. After 1 sec.: $s = 16(1)^2 = 16$; After 2 sec.: $s = 16(2)^2 = 64$; After 3 sec.: $s = 16(3)^2 = 144$
During 2nd second $\Rightarrow$ falls $64 - 16 = 48$ ft; During 3rd second $\Rightarrow$ falls $144 - 64 = 80$ ft
The sequence of the amounts fallen during each second is $16, 48, 80 \Rightarrow a_1 = 16, d = 32$
$a_n = a_1 + (n-1)d = 16 + (12-1)(32) = 16 + 11(32) = 368$ ft

91. **Answers may vary.**

93. $\sum_{n=1}^{6} \left(\frac{1}{2}n + 1\right)$: $\frac{3}{2}, 2, \frac{5}{2}, 3, \frac{7}{2}, 4$

95. Form an arithmetic sequence with a 1st term of a and a 3rd term of b:
$a_n = a_1 + (n-1)d$
$b = a_1 + (3-1)d$
$b = a + 2d$
$b - a = 2d$
$$\frac{b-a}{2} = d \Rightarrow \text{mean} = a_1 + \frac{b-a}{2} = a + \frac{b-a}{2} = \frac{2a}{2} + \frac{b-a}{2} = \frac{a+b}{2}$$

© 2015 Cengage Learning. All Rights Reserved. May not be scanned, copied or duplicated, or posted to a publicly accessible website, in whole or in part.

97. $\sum_{k=1}^{5} 5k = 5(1) + 5(2) + 5(3) + 5(4) + 5(5) = 5(1 + 2 + 3 + 4 + 5) = 5 \sum_{k=1}^{5} k.$

99. $\sum_{k=1}^{n} 3 = \sum_{k=1}^{n} 3k^0 = 3(1)^0 + 3(2)^0 + \cdots + 3(n)^0 = 3 + 3 + \cdots + 3 = 3n$

Exercises 11.3 (page 774)

1. $2(4^2) = 2(16) = 32$

3. $4(3)^{5-1} = 4(3)^4 = 4(81) = 324$

5. $7\left(\frac{1}{3}\right)^{4-1} = 7\left(\frac{1}{3}\right)^3 = 7\left(\frac{1}{27}\right) = \frac{7}{27}$

7.
$$x^2 - 4x - 5 \le 0$$
$$(x - 5)(x + 1) \le 0$$
$x - 5$ – – – – – – – – – 0++++
$x + 1$ – – – 0++++++++++
$\longleftarrow [\longrightarrow] \longrightarrow$ at -1 and 5
solution set: $[-1, 5]$

9.
$$\frac{x - 4}{x + 3} \ge 0$$
$x - 4$ – – – – – – – – 0++++
$x + 3$ – – – 0++++++++++
$\longleftarrow) \longrightarrow [\longrightarrow$ at -3 and 4
solution set: $(-\infty, -3) \cup [4, \infty)$

11. geometric

13. common ratio

15. $S_n = \dfrac{a_1 - a_1 r^n}{1 - r}$

17. $9 \cdot 3 = 27$

19. $\dfrac{0.5}{0.2} = \dfrac{5}{2}$, or 2.5

21. $\dfrac{54}{18} = 3, 2 \cdot 3 = 6$

23. 4, 12, 36, 108, 324
$$a_n = a_1 r^{n-1}$$
$$a_8 = 4(3)^{8-1} = 4(3^7) = 8748$$

25. $-5, -1, -\frac{1}{5}, -\frac{1}{25}, -\frac{1}{125}$
$$a_n = a_1 r^{n-1}$$
$$a_8 = -5\left(\frac{1}{5}\right)^{8-1} = -5\left(\frac{1}{5}\right)^7 = -\frac{1}{15{,}625}$$

27. $a_1 = 1, r = 3$
$$a_n = a_1 r^{n-1}$$
$$a_6 = 1(3)^{6-1} = 1(3)^5 = 243$$

29. $a_1 = 2, r = \frac{1}{2}$
$$a_n = a_1 r^{n-1}$$
$$a_7 = 2\left(\frac{1}{2}\right)^{7-1} = 2\left(\frac{1}{2}\right)^6 = \frac{1}{32}$$

31.
$$a_n = a_1 r^{n-1}$$
$$32 = 2r^{3-1}$$
$$32 = 2r^2$$
$$16 = r^2$$
$$\pm 4 = r, \text{ so } r = 4\ (r > 0)$$
$2, 8, 32, 128, 512$

33.
$$a_n = a_1 r^{n-1}$$
$$50 = 2r^{3-1}$$
$$50 = 2r^2$$
$$25 = r^2$$
$$\pm 5 = r, \text{ so } r = -5\ (r < 0)$$
$2, -10, 50, -250, 1250$

© 2015 Cengage Learning. All Rights Reserved. May not be scanned, copied or duplicated, or posted to a publicly accessible website, in whole or in part.

SECTION 11.3

35. $a_1 = 2, a_5 = 162$

$$a_n = a_1 r^{n-1}$$
$$a_5 = a_1 r^{5-1}$$
$$162 = 2r^4$$
$$81 = r^4$$
$$\pm 3 = r \Rightarrow \text{choose } r = 3$$
$$2, \boxed{6, 18, 54}, 162$$

37. $a_1 = -4, a_6 = -12500$

$$a_n = a_1 r^{n-1}$$
$$a_6 = a_1 r^{6-1}$$
$$-12500 = -4r^5$$
$$3125 = r^5$$
$$5 = r$$
$$-4, \boxed{-20, -100, -500, -2500}, -12500$$

39. $a_1 = 2, a_3 = 128$

$$a_n = a_1 r^{n-1}$$
$$a_3 = a_1 r^{3-1}$$
$$128 = 2r^2$$
$$64 = r^2$$
$$\pm 8 = r \Rightarrow \text{choose } r = -8$$
$$2, \boxed{-16}, 128$$

41. $a_1 = 10, a_3 = 20$

$$a_n = a_1 r^{n-1}$$
$$a_3 = a_1 r^{3-1}$$
$$20 = 10r^2$$
$$2 = r^2$$
$$\pm\sqrt{2} = r \Rightarrow \text{choose } r = \sqrt{2}$$
$$10, \boxed{10\sqrt{2}}, 20$$

43. $a_1 = 120, r = \frac{1}{2}, n = 4;\ S_n = \dfrac{a_1 - a_1 r^n}{1 - r} = \dfrac{120 - 120\left(\frac{1}{2}\right)^4}{1 - \frac{1}{2}} = \dfrac{120 - 120\left(\frac{1}{16}\right)}{\frac{1}{2}} = \dfrac{112.5}{\frac{1}{2}} = 225$

45. $a_1 = 2, r = -3, n = 5;\ S_n = \dfrac{a_1 - a_1 r^n}{1 - r} = \dfrac{2 - 2(-3)^5}{1 - (-3)} = \dfrac{2 - 2(-243)}{4} = \dfrac{488}{4} = 122$

47. If the 3rd term is 50 and the 2nd term is 10, then the common ratio $r = 50 \div 10 = 5$.

$a_1 = a_2 \div r = 10 \div 5 = 2$

$2, 10, 50, 250, 1250$

49.

$$a_n = a_1 r^{n-1}$$
$$-4 = -64r^{5-1}$$
$$-4 = -64r^4$$
$$\tfrac{1}{16} = r^4$$
$$\pm\tfrac{1}{2} = r, \text{ so } r = -\tfrac{1}{2}\ (r < 0)$$
$$-64, 32, -16, 8, -4$$

51. $a_n = a_1 r^{n-1} = 4 \cdot 3^{9-1} = 4 \cdot 3^8 = 4 \cdot 6561 = 26{,}244$

53.

$$a_n = a_1 r^{n-1}$$
$$-81 = a_1(-3)^{8-1}$$
$$-81 = a_1(-3)^7$$
$$-81 = -2187a_1$$
$$\tfrac{1}{27} = a_1$$

55.

$$a_n = a_1 r^{n-1}$$
$$-1944 = -8r^{6-1}$$
$$-1944 = -8r^5$$
$$243 = r^5$$
$$3 = r$$

© 2015 Cengage Learning. All Rights Reserved. May not be scanned, copied or duplicated, or posted to a publicly accessible website, in whole or in part.

SECTION 11.3

57. $a_1 = -50, a_3 = 10 \qquad a_n = a_1 r^{n-1}$

$$a_3 = a_1 r^{3-1}$$
$$10 = -50r^2$$
$$-\tfrac{1}{5} = r^2 \text{ No such mean exists.}$$

59. $a_1 = 3, r = -2, n = 8;\ S_n = \dfrac{a_1 - a_1 r^n}{1-r} = \dfrac{3 - 3(-2)^8}{1-(-2)} = \dfrac{3 - 3(256)}{3} = \dfrac{-765}{3} = -255$

61. $a_1 = 3, r = 2, n = 7;\ S_n = \dfrac{a_1 - a_1 r^n}{1-r} = \dfrac{3 - 3(2)^7}{1-2} = \dfrac{3 - 3(128)}{-1} = \dfrac{-381}{-1} = 381$

63. If the 3rd term is $\frac{1}{5}$ and the 2nd term is 1, then the common ratio $r = \frac{1}{5} \div 1 = \frac{1}{5}$.

$a_1 = 1 \div \frac{1}{5} = 5, r = \frac{1}{5}, n = 4;\ S_n = \dfrac{a_1 - a_1 r^n}{1-r} = \dfrac{5 - 5\left(\frac{1}{5}\right)^4}{1 - \frac{1}{5}} = \dfrac{5 - \frac{1}{125}}{\frac{4}{5}} = \dfrac{\frac{624}{125}}{\frac{4}{5}} = \dfrac{156}{25}$

65. If the 4th term is 1 and the 3rd term is -2, then the common ratio $r = 1 \div (-2) = -\frac{1}{2}$.

$a_2 = -2 \div \left(-\frac{1}{2}\right) = 4;\ a_1 = 4 \div \left(-\frac{1}{2}\right) = -8;\ a_1 = -8, r = -\frac{1}{2}, n = 6$

$$S_n = \frac{a_1 - a_1 r^n}{1-r} = \frac{-8 - (-8)\left(-\frac{1}{2}\right)^6}{1 - \left(-\frac{1}{2}\right)} = \frac{-8 + \frac{1}{8}}{\frac{3}{2}} = \frac{-\frac{63}{8}}{\frac{3}{2}} = -\frac{21}{4}$$

67. Sequence of population: $500, 500(1.06), 500(1.06)^2, \ldots$

$a_1 = 500, r = 1.06, n = 6 \Rightarrow a_n = a_1 r^{n-1} = 500(1.06)^5 \approx 669$

69. Sequence of amounts: $10000, 10000(0.88), 10000(0.88)^2, \ldots$

$a_1 = 10000, r = 0.88, n = 16 \Rightarrow a_n = a_1 r^{n-1} = 10000(0.88)^{15} \approx \$1{,}469.74$

71. Sequence of values: $70000, 70000(1.06), 70000(1.06)^2, \ldots$

$a_1 = 70000, r = 1.06, n = 13 \Rightarrow a_n = a_1 r^{n-1} = 70000(1.06)^{12} \approx \$140{,}853.75$

73. Sequence of areas: $1, \frac{1}{2}, \frac{1}{4}, \ldots$

$a_1 = 1, r = \frac{1}{2}, n = 12 \Rightarrow a_n = a_1 r^{n-1} = 1\left(\frac{1}{2}\right)^{11} = \left(\frac{1}{2}\right)^{11} \approx 0.0005$

75. Sequence of amounts: $1000(1.03), 1000(1.03)^2, 1000(1.03)^3, \ldots$

$a_1 = 1030, r = 1.03, n = 4 \Rightarrow S_n = \dfrac{a_1 - a_1 r^n}{1-r} = \dfrac{1030 - 1030(1.03)^4}{1 - 1.03} = \dfrac{-129.2740743}{-0.03}$

$\approx \$4{,}309.14$

77. **Answers may vary.**

79. **Answers may vary.**

81. arithmetic mean

83. **Answers may vary.**

© 2015 Cengage Learning. All Rights Reserved. May not be scanned, copied or duplicated, or posted to a publicly accessible website, in whole or in part.

Exercises 11.4 (page 780)

1. $r = \frac{6}{2} = 3$

3. $r = \frac{10}{50} = \frac{1}{5}$

5. $r = \frac{\frac{1}{5}}{1} = \frac{1}{5}$

7. $y = -2x^3 + 5$
function

9. $3x = y^2 + 4$
not a function

11. infinite

13 $S_\infty = \frac{a_1}{1-r}$

15. $a_1 = 48, r = \frac{1}{4}$
$$S_\infty = \frac{a_1}{1-r} = \frac{48}{1-\frac{1}{4}} = \frac{48}{\frac{3}{4}} = 64$$

17. $a_1 = 200, r = \frac{1}{5}$
$$S_\infty = \frac{a_1}{1-r} = \frac{200}{1-\frac{1}{5}} = \frac{200}{\frac{4}{5}} = 250$$

19. $a_1 = 12, r = -\frac{1}{2}$
$$S_\infty = \frac{a_1}{1-r} = \frac{12}{1-\left(-\frac{1}{2}\right)} = \frac{12}{\frac{3}{2}} = 8$$

21. $a_1 = -81, r = -\frac{1}{3}$
$$S_\infty = \frac{a_1}{1-r} = \frac{-81}{1-\left(-\frac{1}{3}\right)} = \frac{-81}{\frac{4}{3}} = -\frac{243}{4}$$

23. $a_1 = -3, r = 2$
No sum since $|r| = 2 > 1$.

25. $a_1 = \frac{3}{4}, r = 2$
No sum since $|r| = 2 > 1$.

27. $0.\overline{1} = \frac{1}{10} + \frac{1}{100} + \frac{1}{1000} + \cdots \Rightarrow a_1 = \frac{1}{10}, r = \frac{1}{10} \Rightarrow S_\infty = \frac{a_1}{1-r} = \frac{\frac{1}{10}}{1-\frac{1}{10}} = \frac{\frac{1}{10}}{\frac{9}{10}} = \frac{1}{9}$

29. $-0.\overline{3} = -\frac{3}{10} - \frac{3}{100} - \frac{3}{1000} + \cdots \Rightarrow a_1 = -\frac{3}{10}, r = \frac{1}{10} \Rightarrow S_\infty = \frac{a_1}{1-r} = \frac{-\frac{3}{10}}{1-\frac{1}{10}} = \frac{-\frac{3}{10}}{\frac{9}{10}} = -\frac{1}{3}$

31. $0.\overline{12} = \frac{12}{100} + \frac{12}{10{,}000} + \frac{12}{1{,}000{,}000} + \cdots \Rightarrow a_1 = \frac{12}{100}, r = \frac{1}{100}$
$$S_\infty = \frac{a_1}{1-r} = \frac{\frac{12}{100}}{1-\frac{1}{100}} = \frac{\frac{12}{100}}{\frac{99}{100}} = \frac{12}{99} = \frac{4}{33}$$

33. $0.\overline{75} = \frac{75}{100} + \frac{75}{10{,}000} + \frac{75}{1{,}000{,}000} + \cdots \Rightarrow a_1 = \frac{75}{100}, r = \frac{1}{100}$
$$S_\infty = \frac{a_1}{1-r} = \frac{\frac{75}{100}}{1-\frac{1}{100}} = \frac{\frac{75}{100}}{\frac{99}{100}} = \frac{75}{99} = \frac{25}{33}$$

35. $a_1 = -54, r = -\frac{1}{3}$
$$S_\infty = \frac{a_1}{1-r} = \frac{-54}{1-\left(-\frac{1}{3}\right)} = \frac{-54}{\frac{4}{3}} = -\frac{81}{2}$$

37. $a_1 = -\frac{27}{2}, r = \frac{2}{3}$
$$S_\infty = \frac{a_1}{1-r} = \frac{-\frac{27}{2}}{1-\frac{2}{3}} = \frac{-\frac{27}{2}}{\frac{1}{3}} = -\frac{81}{2}$$

39. $S_\infty = \frac{a_1}{1-r} = \frac{1000}{1-0.8} = \frac{1000}{0.2} = 5{,}000$ moths

© 2015 Cengage Learning. All Rights Reserved. May not be scanned, copied or duplicated, or posted to a publicly accessible website, in whole or in part.

SECTION 11.4

41. Distance ball travels down $= 10 + 5 + 2.5 + \cdots = \dfrac{a_1}{1-r} = \dfrac{10}{1-\frac{1}{2}} = \dfrac{10}{\frac{1}{2}} = 20$

Distance ball travels up $= 5 + 2.5 + 1.25 + \cdots = \dfrac{a_1}{1-r} = \dfrac{5}{1-\frac{1}{2}} = \dfrac{5}{\frac{1}{2}} = 10$

Total distance $= 20 + 10 = 30$ m

43. **Answers may vary.**

45.
$$\begin{aligned} S_\infty &= \frac{a_1}{1-r} \\ 5 &= \frac{1}{1-r} \\ 5(1-r) &= 1 \\ 5 - 5r &= 1 \\ 4 &= 5r \\ \frac{4}{5} &= r \end{aligned}$$

47. $0.\overline{9} = \frac{9}{10} + \frac{9}{100} + \frac{9}{1000} + \cdots \Rightarrow a = \frac{9}{10}, r = \frac{1}{10} \Rightarrow S_\infty = \dfrac{a_1}{1-r} = \dfrac{\frac{9}{10}}{1-\frac{1}{10}} = \dfrac{\frac{9}{10}}{\frac{9}{10}} = \dfrac{9}{9} = 1$

49. No. $0.999999 = \dfrac{999{,}999}{1{,}000{,}000} < 1$

Exercises 11.5 (page 789)

1. $\dfrac{7!}{3!} = \dfrac{7 \cdot 6 \cdot 5 \cdot 4 \cdot 3!}{3!} = 840$

3. $\dfrac{5!}{1!} = \dfrac{5 \cdot 4 \cdot 3 \cdot 2 \cdot 1!}{1!} = 120$

5. $\dfrac{8!}{3!5!} = \dfrac{8 \cdot 7 \cdot 6 \cdot 5!}{3!5!} = \dfrac{336}{6} = 56$

7.
$$|2x - 7| = 17$$
$$\begin{aligned} 2x - 7 &= 17 \quad \textbf{or} \quad & 2x - 7 &= -17 \\ 2x &= 24 & 2x &= -10 \\ x &= 12 & x &= -5 \end{aligned}$$

9.
$$\begin{aligned} \frac{3}{x-5} &= \frac{8}{x} \\ 3x &= 8(x-5) \\ 3x &= 8x - 40 \\ -5x &= -40 \\ x &= 8 \end{aligned}$$

11. $p \cdot q$

13. permutation

15. $P(n,r) = \dfrac{n!}{(n-r)!}$

17. combination

19. $C(n,r) = \dfrac{n!}{r!(n-r)!}$

21. $P(5,5) = \dfrac{5!}{(5-5)!} = \dfrac{5!}{0!} = \dfrac{120}{1} = 120$

23. $P(6,2) = \dfrac{6!}{(6-2)!} = \dfrac{6!}{4!} = \dfrac{720}{24} = 30$

© 2015 Cengage Learning. All Rights Reserved. May not be scanned, copied or duplicated, or posted to a publicly accessible website, in whole or in part.

25. $P(4,2) \cdot P(5,2) = \dfrac{4!}{(4-2)!} \cdot \dfrac{5!}{(5-2)!} = \dfrac{4!}{2!} \cdot \dfrac{5!}{3!} = \dfrac{24}{2} \cdot \dfrac{120}{6} = 240$

27. $\dfrac{P(n,n)}{P(n,0)} = \dfrac{\frac{n!}{(n-n)!}}{\frac{n!}{(n-0)!}} = \dfrac{\frac{n!}{0!}}{\frac{n!}{n!}} = \dfrac{\frac{n!}{1}}{\frac{1}{1}} = n!$

29. $C(4,3) = \dfrac{4!}{3!(4-3)!} = \dfrac{4!}{3!1!} = \dfrac{24}{6 \cdot 1}$
$= \dfrac{24}{6} = 4$

31. $\dbinom{8}{5} = \dfrac{8!}{5!(8-5)!} = \dfrac{8!}{5!3!} = \dfrac{40{,}320}{120 \cdot 6} = \dfrac{40{,}320}{720} = 56$

33. $\dbinom{7}{4}\dbinom{7}{5} = \dfrac{7!}{4!(7-4)!} \cdot \dfrac{7!}{5!(7-5)!} = \dfrac{7!}{4!3!} \cdot \dfrac{7!}{5!2!} = \dfrac{7 \cdot 6 \cdot 5 \cdot 4!}{4!3!} \cdot \dfrac{7 \cdot 6 \cdot 5!}{5!2!} = \dfrac{210}{6} \cdot \dfrac{42}{2}$
$= 35 \cdot 21 = 735$

35. $\dfrac{C(38,37)}{C(19,18)} = \dfrac{\frac{38!}{37!(38-37)!}}{\frac{19!}{18!(19-18)!}} = \dfrac{\frac{38 \cdot 37!}{37!1!}}{\frac{19 \cdot 18!}{18!1!}} = \dfrac{\frac{38}{1}}{\frac{19}{1}} = \dfrac{38}{19} = 2$

37. $(x+y)^4 = \dbinom{4}{0}x^4y^0 + \dbinom{4}{1}x^3y^1 + \dbinom{4}{2}x^2y^2 + \dbinom{4}{3}x^1y^3 + \dbinom{4}{4}x^0y^4$
$= \dfrac{4!}{0!4!}x^4 + \dfrac{4!}{1!3!}x^3y + \dfrac{4!}{2!2!}x^2y^2 + \dfrac{4!}{3!1!}xy^3 + \dfrac{4!}{4!0!}y^4 = x^4 + 4x^3y + 6x^2y^2 + 4xy^3 + y^4$

39. $(a+b)^5 = \dbinom{5}{0}a^5b^0 + \dbinom{5}{1}a^4b^1 + \dbinom{5}{2}a^3b^2 + \dbinom{5}{3}a^2b^3 + \dbinom{5}{4}a^1b^4 + \dbinom{5}{5}a^0b^5$
$= \dfrac{5!}{0!5!}a^5 + \dfrac{5!}{1!4!}a^4b + \dfrac{5!}{2!3!}a^3b^2 + \dfrac{5!}{3!2!}a^2b^3 + \dfrac{5!}{4!1!}a^1b^4 + \dfrac{5!}{5!0!}a^0b^5$
$= a^5 + 5a^4b + 10a^3b^2 + 10a^2b^3 + 5ab^4 + b^5$

41. $(2x+y)^3 = \dbinom{3}{0}(2x)^3y^0 + \dbinom{3}{1}(2x)^2y^1 + \dbinom{3}{2}(2x)^1y^2 + \dbinom{3}{3}(2x)^0y^3$
$= \dfrac{3!}{0!3!} \cdot 8x^3 + \dfrac{3!}{1!2!} \cdot 4x^2y + \dfrac{3!}{2!1!} \cdot 2xy^2 + \dfrac{3!}{3!0!}y^3$
$= 8x^3 + 12x^2y + 6xy^2 + y^3$

43. $(4x-y)^3 = \dbinom{3}{0}(4x)^3(-y)^0 + \dbinom{3}{1}(4x)^2(-y)^1 + \dbinom{3}{2}(4x)^1(-y)^2 + \dbinom{3}{3}(4x)^0(-y)^3$
$= \dfrac{3!}{0!3!} \cdot 64x^3 - \dfrac{3!}{1!2!} \cdot 16x^2y + \dfrac{3!}{2!1!} \cdot 4xy^2 - \dfrac{3!}{3!0!}y^3$
$= 64x^3 - 48x^2y + 12xy^2 - y^3$

45. $C(12,0)C(12,12) = \dfrac{12!}{0!(12-0)!} \cdot \dfrac{12!}{12!(12-12)!} = \dfrac{12!}{0!12!} \cdot \dfrac{12!}{12!0!} = \dfrac{12!}{12!} \cdot \dfrac{12!}{12!} = 1 \cdot 1 = 1$

© 2015 Cengage Learning. All Rights Reserved. May not be scanned, copied or duplicated, or posted to a publicly accessible website, in whole or in part.

SECTION 11.5

47. $\dfrac{P(6,2)\cdot P(7,3)}{P(5,1)} = \dfrac{\frac{6!}{(6-2)!}\cdot\frac{7!}{(7-3)!}}{\frac{5!}{(5-1)!}} = \dfrac{\frac{6!}{4!}\cdot\frac{7!}{4!}}{\frac{5!}{4!}} = \dfrac{\frac{720}{24}\cdot\frac{5040}{24}}{\frac{120}{24}} = \dfrac{30\cdot 210}{5} = 1{,}260$

49. $C(n,2) = \dfrac{n!}{2!(n-2)!}$

51. $(3x-2)^4$

$$= \binom{4}{0}(3x)^4(-2)^0 + \binom{4}{1}(3x)^3(-2)^1 + \binom{4}{2}(3x)^2(-2)^2 + \binom{4}{3}(3x)^1(-2)^3 + \binom{4}{4}(3x)^0(-2)^4$$

$$= \frac{4!}{0!4!}\cdot 81x^4(1) + \frac{4!}{1!3!}\cdot 27x^3(-2) + \frac{4!}{2!2!}\cdot 9x^2(4) + \frac{4!}{3!1!}\cdot 3x(-8) + \frac{4!}{4!0!}\cdot 1(16)$$

$$= 81x^4 - 216x^3 + 216x^2 - 96x + 16$$

53. $\binom{5}{3}x^2(-5y)^3 = \dfrac{5!}{3!2!}x^2\left(-125y^3\right) = 10x^2\left(-125y^3\right) = -1{,}250x^2y^3$

55. $\binom{4}{1}\left(x^2\right)^3\left(-y^3\right)^1 = \dfrac{4!}{1!3!}x^6\left(-y^3\right) = -4x^6y^3$

57. $7\cdot 5 = 35$

59. $7! = 5{,}040$

61. $10\cdot 10\cdot 10\cdot 10\cdot 10\cdot 10 = 1{,}000{,}000$

63. $9\cdot 9\cdot 8\cdot 7\cdot 6\cdot 5 = 136{,}080$

65. $8\cdot 10\cdot 10\cdot 10\cdot 10\cdot 10\cdot 10 = 8{,}000{,}000$

67. $4!\cdot 5! = 24\cdot 120 = 2{,}880$

69. $25\cdot 24\cdot 23 = 13{,}800$

71. $P(10,3) = \dfrac{10!}{(10-3)!} = \dfrac{10!}{7!} = 720$

73. $9\cdot 10\cdot 10\cdot 1\cdot 1 = 900$

75. $C(14,3) = \dfrac{14!}{3!(14-3)!} = \dfrac{14!}{3!11!} = 364$

77. $C(5,3) = 10 \Rightarrow 5$ persons

79. $C(100,6) = \dfrac{100!}{6!(100-6)!} = \dfrac{100!}{6!94!} = \dfrac{100\cdot 99\cdot 98\cdot 97\cdot 96\cdot 95\cdot 94!}{6!94!} = 1{,}192{,}052{,}400$

80. $C(15,10)\cdot 5 = \dfrac{15!}{10!(15-10)!}\cdot 5 = \dfrac{15!}{10!5!}\cdot 5 = \dfrac{15\cdot 14\cdot 13\cdot 12\cdot 11\cdot 10!}{10!5!}\cdot 5 = 15{,}015$

81. $C(3,2)\cdot C(4,2) = \dfrac{3!}{2!1!}\cdot\dfrac{4!}{2!2!} = 3\cdot 6 = 18$

83. $C(12,2)\cdot C(10,3) = \dfrac{12!}{2!10!}\cdot\dfrac{10!}{3!7!} = 66\cdot 120 = 7{,}920$

© 2015 Cengage Learning. All Rights Reserved. May not be scanned, copied or duplicated, or posted to a publicly accessible website, in whole or in part.

85. Answers may vary.

87. Consider the two people who insist on standing together as one person. Then there are a total of 4 "persons" to be arranged. This can be done in $4! = 24$ ways. However, the two people who are standing together can be arranged in 2 different ways, so there are $24 \cdot 2 = 48$ arrangements.

Exercises 11.6 (page 795)

1. $C(52, 26)$

3.
$$\begin{aligned} 4^{3x} &= \frac{1}{64} \\ 4^{3x} &= 4^{-3} \\ 3x &= -3 \\ x &= -1 \end{aligned}$$

5.
$$\begin{aligned} 5^{x^2-2x} &= 125 \\ 5^{x^2-2x} &= 5^3 \\ x^2 - 2x &= 3 \\ x^2 - 2x - 3 &= 0 \\ (x+1)(x-3) &= 0 \end{aligned}$$
$x + 1 = 0$ **or** $x - 3 = 0$
$x = -1$ $\qquad$ $x = 3$

7.
$$\begin{aligned} 3^{x^2+4x} &= \frac{1}{81} \\ 3^{x^2+4x} &= 3^{-4} \\ x^2 + 4x &= -4 \\ x^2 + 4x + 4 &= 0 \\ (x+2)(x+2) &= 0 \end{aligned}$$
$x + 2 = 0$ **or** $x + 2 = 0$
$x = -2$ $\qquad$ $x = -2$

9. experiment

11. $\dfrac{s}{n}$

13. 0

15. **a.** 6
b. 52
c. $\dfrac{6}{52}; \dfrac{3}{26}$

17. $\{(1, H), (2, H), (3, H), (4, H), (5, H), (6, H), (1, T), (2, T), (3, T), (4, T), (5, T), (6, T)\}$

19. $\{A, B, C, D, E, F, G, H, I, J, K, L, M, N, O, P, Q, R, S, T, U, V, W, X, Y, Z\}$

21. rolls of 5: $\{(1, 4), (2, 3), (3, 2), (4, 1)\}$; Probability $= \dfrac{4}{36} = \dfrac{1}{9}$

23. rolls < 6: $\{(1, 1), (1, 2), (1, 3), (1, 4), (2, 1), (2, 2), (2, 3), (3, 1), (3, 2), (4, 1)\}$
Probability $= \dfrac{10}{36} = \dfrac{5}{18}$

25. $\dfrac{\text{\# diamonds}}{\text{\# cards}} = \dfrac{13}{52} = \dfrac{1}{4}$

27. $\dfrac{\text{\# red face cards}}{\text{\# cards in deck}} = \dfrac{6}{52} = \dfrac{3}{26}$

© 2015 Cengage Learning. All Rights Reserved. May not be scanned, copied or duplicated, or posted to a publicly accessible website, in whole or in part.

29. $\dfrac{\text{\# ways to get 6 diamonds}}{\text{\# ways to get 6 cards from the deck of 52}} = \dfrac{\binom{13}{6}}{\binom{52}{6}} = \dfrac{1716}{20{,}358{,}520} = \dfrac{33}{391{,}510}$

31. $\dfrac{\text{\# ways to get 5 clubs}}{\text{\# ways to get 5 cards from the 26 black cards}} = \dfrac{\binom{13}{5}}{\binom{26}{5}} = \dfrac{1287}{65780} = \dfrac{9}{460}$

33. $\dfrac{1}{6}$

35. $\dfrac{4}{6} = \dfrac{2}{3}$

37. $\dfrac{19}{42}$

39. $\dfrac{13}{42}$

41. $\dfrac{3}{8}$

43. $\dfrac{0}{8} = 0$

45. $\dfrac{\text{\# face} \cdot \text{\# ten}}{\text{\# ways to get 2 cards}} = \dfrac{12 \cdot 4}{52 \cdot 52} = \dfrac{48}{2704} = \dfrac{3}{169}$

47. rolls of 7: $\{(1,6), (2,5), (3,4), (4,3), (5,2), (6,1)\}$; Probability $= \dfrac{6}{36} = \dfrac{1}{6}$

49. $\dfrac{\text{\# yellow}}{\text{\# eggs}} = \dfrac{7}{12}$

51. $FFFF \Rightarrow \dfrac{1}{16}$

53. $PPFF, PFPF, PFFP, FPPF, FPFP, FFPP \Rightarrow \dfrac{6}{16} = \dfrac{3}{8}$

55. $SSSS \Rightarrow \dfrac{1}{16}$

57. $\dfrac{\binom{8}{4}}{\binom{10}{4}} = \dfrac{70}{210} = \dfrac{1}{3}$

59. $\dfrac{176}{282} = \dfrac{88}{141}$

61. $\dfrac{15}{71}$

63. **Answers may vary.**

65. $P(\text{lux and } 2^{\text{nd}} \text{ car}) = P(\text{lux}) \cdot P(2^{\text{nd}} | \text{lux}) = 0.2(0.7) = 0.14$

Chapter 11 Review (page 800)

1. $(5!)(2!) = 5 \cdot 4 \cdot 3 \cdot 2 \cdot 1 \cdot 2 \cdot 1 = 240$

2. $\dfrac{5!}{3!} = \dfrac{5 \cdot 4 \cdot 3!}{3!} = 5 \cdot 4 = 20$

3. $\dfrac{6!}{2!(6-2)!} = \dfrac{6!}{2!4!} = \dfrac{6 \cdot 5 \cdot 4!}{2 \cdot 1 \cdot 4!} = \dfrac{30}{2} = 15$

4. $\dfrac{12!}{3!(12-3)!} = \dfrac{12!}{3!9!} = \dfrac{12 \cdot 11 \cdot 10 \cdot 9!}{3 \cdot 2 \cdot 1 \cdot 9!} = \dfrac{1320}{6} = 220$

5. $(n-n)! = 0! = 1$

6. $\dfrac{10!}{6!} = \dfrac{10 \cdot 9 \cdot 8 \cdot 7 \cdot 6!}{6!} = 5040$

© 2015 Cengage Learning. All Rights Reserved. May not be scanned, copied or duplicated, or posted to a publicly accessible website, in whole or in part.

CHAPTER 11 REVIEW

7. $(x+y)^5 = x^5 + \frac{5!}{1!(5-1)!}x^4y + \frac{5!}{2!(5-2)!}x^3y^2 + \frac{5!}{3!(5-3)!}x^2y^3 + \frac{5!}{4!(5-4)!}xy^4 + y^5$

$= x^5 + \frac{5!}{1!4!}x^4y + \frac{5!}{2!3!}x^3y^2 + \frac{5!}{3!2!}x^2y^3 + \frac{5!}{4!1!}xy^4 + y^5$

$= x^5 + \frac{5\cdot 4!}{1!4!}x^4y + \frac{5\cdot 4\cdot 3!}{2\cdot 1\cdot 3!}x^3y^2 + \frac{5\cdot 4\cdot 3!}{3!\cdot 2\cdot 1}x^2y^3 + \frac{5\cdot 4!}{4!\cdot 1}xy^4 + y^5$

$= x^5 + \frac{5}{1}x^4y + \frac{20}{2}x^3y^2 + \frac{20}{2}x^2y^3 + \frac{5}{1}xy^4 + y^5$

$= x^5 + 5x^4y + 10x^3y^2 + 10x^2y^3 + 5xy^4 + y^5$

8. $(x-y)^4 = x^4 + \frac{4!}{1!(4-1)!}x^3(-y) + \frac{4!}{2!(4-2)!}x^2(-y)^2 + \frac{4!}{3!(4-3)!}x(-y)^3 + (-y)^4$

$= x^4 + \frac{4!}{1!3!}(-x^3y) + \frac{4!}{2!2!}x^2y^2 + \frac{4!}{3!1!}(-xy^3) + y^4$

$= x^4 - \frac{4\cdot 3!}{1!3!}x^3y + \frac{4\cdot 3\cdot 2!}{2!\cdot 2\cdot 1}x^2y^2 - \frac{4\cdot 3!}{3!1!}xy^3 + y^4$

$= x^4 - \frac{4}{1}x^3y + \frac{12}{2}x^2y^2 - \frac{4}{1}xy^3 + y^4 = x^4 - 4x^3y + 6x^2y^2 - 4xy^3 + y^4$

9. $(4x-y)^3 = (4x)^3 + \frac{3!}{1!(3-1)!}(4x)^2(-y) + \frac{3!}{2!(3-2)!}(4x)(-y)^2 + (-y)^3$

$= 64x^3 + \frac{3!}{1!2!}\cdot(-16x^2y) + \frac{3!}{2!1!}\cdot 4xy^2 - y^3$

$= 64x^3 - \frac{3\cdot 2!}{1!2!}\cdot 16x^2y + \frac{3\cdot 2!}{2!1!}\cdot 4xy^2 - y^3$

$= 64x^3 - \frac{3}{1}\cdot 16x^2y + \frac{3}{1}\cdot 4xy^2 - y^3 = 64x^3 - 48x^2y + 12xy^2 - y^3$

10. $(x+4y)^3 = x^3 + \frac{3!}{1!(3-1)!}x^2(4y) + \frac{3!}{2!(3-2)!}x(4y)^2 + (4y)^3$

$= x^3 + \frac{3!}{1!2!}\cdot 4x^2y + \frac{3!}{2!1!}\cdot 16xy^2 + 64y^3$

$= x^3 + \frac{3\cdot 2!}{1!2!}\cdot 4x^2y + \frac{3\cdot 2!}{2!1!}\cdot 16xy^2 + 64y^3$

$= x^3 + \frac{3}{1}\cdot 4x^2y + \frac{3}{1}\cdot 16xy^2 + 64y^3 = x^3 + 12x^2y + 48xy^2 + 64y^3$

11. In the 5th term, the exponent on y is 4.
Variables: x^1y^4
Coef. $= \frac{n!}{r!(n-r)!} = \frac{5!}{4!1!} = 5$
Term $= 5xy^4$

12. In the 4th term, the exponent on $-y$ is 3.
Variables: $x^2(-y)^3 = -x^2y^3$
Coef. $= \frac{n!}{r!(n-r)!} = \frac{5!}{3!2!} = 10$
Term $= 10(-x^2y^3) = -10x^2y^3$

13. 2nd term: The exponent on $-4y$ is 1. Variables: $(3x)^2(-4y)^1 = (9x^2)(-4y) = -36x^2y$
Coef. $= \frac{n!}{r!(n-r)!} = \frac{3!}{2!1!} = 3$; Term $= 3(-36x^2y) = -108x^2y$

© 2015 Cengage Learning. All Rights Reserved. May not be scanned, copied or duplicated, or posted to a publicly accessible website, in whole or in part.

14. 3rd term: The exponent on $3y$ is 2. Variables: $(4x)^2(3y)^2 = (16x^2)(9y^2) = 144x^2y^2$

Coef. $= \dfrac{n!}{r!(n-r)!} = \dfrac{4!}{2!2!} = 6$; Term $= 6(144x^2y^2) = 864x^2y^2$

15. $a_n = a_1 + (n-1)d = 7 + (8-1)5 = 7 + 7 \cdot 5 = 42$

16.

$$\begin{aligned} a_n &= a_1 + (n-1)d \\ 242 &= a_1 + (9-1)d \\ 242 &= a_1 + 8d \\ a_n &= a_1 + (n-1)d \\ 212 &= a_1 + (7-1)d \\ 212 &= a_1 + 6d \end{aligned}$$

$$\begin{aligned} a_1 + 8d = 242 &\Rightarrow \times(-1) & -a_1 - 8d &= -242 \\ \underline{a_1 + 6d = 212} &\Rightarrow & \underline{a_1 + 6d} &= \underline{212} \\ & & -2d &= -30 \\ & & d &= 15 \end{aligned}$$

Substitute and solve for a_1:

$$\begin{aligned} a_1 + 6d &= 212 \\ a_1 + 6(15) &= 212 \\ a_1 + 90 &= 212 \\ a_1 &= 122 \Rightarrow 122, 137, 152, 167, 182 \end{aligned}$$

17. Form an arithmetic sequence with a 1st term of 8 and a 4th term of 25:

$$\begin{aligned} a_n &= a_1 + (n-1)d \\ 25 &= 8 + (4-1)d \\ 17 &= 3d \\ \tfrac{17}{3} &= d \Rightarrow 8, \boxed{\tfrac{41}{3}, \tfrac{58}{3}}, 25 \end{aligned}$$

18. $a_1 = 11, d = 7, n = 20$

$a_n = a_1 + (n-1)d = 11 + 19(7) = 144$

$S_n = \dfrac{n(a_1 + a_n)}{2} = \dfrac{20(11 + 144)}{2} = 1{,}550$

19. $a_1 = 9, d = -\frac{5}{2}, n = 10$

$a_n = a_1 + (n-1)d = 9 + 9(-\frac{5}{2}) = -\frac{27}{2}$

$S_n = \dfrac{n(a_1 + a_n)}{2} = \dfrac{10(9 - \frac{27}{2})}{2} = -\dfrac{45}{2}$

20. $\sum_{k=1}^{5} 4k = 4(1) + 4(2) + 4(3) + 4(4) + 4(5) = 4 + 8 + 12 + 16 + 20 = 60$

21. $\sum_{k=1}^{4} (k^2 + 1) = (1^2 + 1) + (2^2 + 1) + (3^2 + 1) + (4^2 + 1) = 2 + 5 + 10 + 17 = 34$

22. $\sum_{k=1}^{4} (3k - 4) = (3(1) - 4) + (3(2) - 4) + (3(3) - 4) + (3(4) - 4) = -1 + 2 + 5 + 8 = 14$

23. $\sum_{k=10}^{10} 36k = 36(10) = 360$

© 2015 Cengage Learning. All Rights Reserved. May not be scanned, copied or duplicated, or posted to a publicly accessible website, in whole or in part.

CHAPTER 11 REVIEW

24. If the 5th term is $\frac{3}{2}$ and the 4th term is 3, then the common ratio $r = \frac{3}{2} \div 3 = \frac{1}{2}$.
$a_3 = a_4 \div r = 3 \div \frac{1}{2} = 6$
$a_2 = a_3 \div r = 6 \div \frac{1}{2} = 12$
$a_1 = a_2 \div r = 12 \div \frac{1}{2} = 24$
$24, 12, 6, 3, \frac{3}{2}$

25. $$a_n = a_1 r^{n-1} = \frac{1}{16}(2)^{5-1} = \frac{1}{16}(16) = 1$$

26. $a_1 = -6, a_4 = 384$
$$a_n = a_1 r^{n-1}$$
$$384 = -6r^{4-1}$$
$$-64 = r^3$$
$$-4 = r \Rightarrow -6, \boxed{24, -96}, 384$$

27. $a_1 = 240, r = \frac{1}{2}, n = 6;\ S_n = \dfrac{a_1 - a_1 r^n}{1-r} = \dfrac{240 - 240\left(\frac{1}{2}\right)^6}{1-\frac{1}{2}} = \dfrac{240 - 240\left(\frac{1}{64}\right)}{\frac{1}{2}} = \dfrac{\frac{945}{4}}{\frac{1}{2}} = \dfrac{945}{2}$

28. $a_1 = \frac{1}{8}, r = -2, n = 8;\ S_n = \dfrac{a_1 - a_1 r^n}{1-r} = \dfrac{\frac{1}{8} - \frac{1}{8}(-2)^8}{1-(-2)} = \dfrac{\frac{1}{8} - \frac{1}{8}(256)}{3} = \dfrac{-\frac{255}{8}}{3} = -\dfrac{85}{8}$

29. Sequence of amounts: $5000, 5000(0.8), 5000(0.8)^2, \ldots$
$a_1 = 5000, r = 0.8, n = 6 \Rightarrow a_n = a_1 r^{n-1} = 5000(0.8)^5 \approx \$1{,}638.40$

30. Sequence of amounts: $25700, 25700(1.18), 25700(1.18)^2, \ldots$
$a_1 = 25700, r = 1.18, n = 11 \Rightarrow a_n = a_1 r^{n-1} = 25700(1.18)^{10} \approx \$134{,}509.57$

31. $a_1 = 300, d = 75$
$$a_n = a_1 + (n-1)d$$
$$1200 = 300 + (n-1)75$$
$$1200 = 300 + 75n - 75$$
$$975 = 75n$$
$13 = n$; In the 13th year, or after 12 years of increases.

32. $16, 48, 80, \ldots;\ a_1 = 16,\ d = 32$
$$a_n = a_1 + (n-1)d = 16 + (10-1)(32) = 16 + 9(32) = 304 \text{ ft}$$
$$S_n = \frac{n(a_1 + a_n)}{2} = \frac{10(16+304)}{2} = 1600 \text{ ft}$$

33. $a_1 = 100, r = \frac{1}{5};\ S_\infty = \dfrac{a_1}{1-r} = \dfrac{100}{1-\frac{1}{5}} = \dfrac{100}{\frac{4}{5}} = 125$

34. $0.05 = \dfrac{5}{100} + \dfrac{5}{10{,}000} + \dfrac{5}{1{,}000{,}000} + \cdots \Rightarrow a_1 = \dfrac{5}{100}, r = \dfrac{1}{100}$
$$S_\infty = \frac{a_1}{1-r} = \frac{\frac{5}{100}}{1-\frac{1}{100}} = \frac{\frac{5}{100}}{\frac{99}{100}} = \frac{5}{99}$$

35. $17 \cdot 8 = 136$

36. $P(7,7) = \dfrac{7!}{(7-7)!} = \dfrac{7!}{0!} = 7! = 5{,}040$

© 2015 Cengage Learning. All Rights Reserved. May not be scanned, copied or duplicated, or posted to a publicly accessible website, in whole or in part.

37. $P(5,0) = \dfrac{5!}{(5-0)!} = \dfrac{5!}{5!} = 1$

38. $P(8,6) = \dfrac{8!}{(8-6)!} = \dfrac{8!}{2!} = \dfrac{40{,}320}{2} = 20{,}160$

39. $\dfrac{P(9,6)}{P(10,7)} = \dfrac{\frac{9!}{(9-6)!}}{\frac{10!}{(10-7)!}} = \dfrac{\frac{9!}{3!}}{\frac{10!}{3!}} = \dfrac{9!}{3!} \cdot \dfrac{3!}{10!} = \dfrac{9!}{10!} = \dfrac{9!}{10 \cdot 9!} = \dfrac{1}{10}$

40. $C(8,8) = \dfrac{8!}{8!(8-8)!} = \dfrac{8!}{8!0!} = \dfrac{8!}{8!} = 1$

41. $C(7,0) = \dfrac{7!}{0!(7-0)!} = \dfrac{7!}{0!7!} = \dfrac{7!}{7!} = 1$

42. $\dbinom{8}{6} = \dfrac{8!}{6!(8-6)!} = \dfrac{8 \cdot 7 \cdot 6!}{6!2!} = \dfrac{56}{2} = 28$

43. $\dbinom{10}{4} = \dfrac{10!}{4!(10-4)!} = \dfrac{10 \cdot 9 \cdot 8 \cdot 7 \cdot 6!}{4!6!} = \dfrac{5040}{24} = 210$

44. $C(6,3) \cdot C(7,3) = \dfrac{6!}{3!(6-3)!} \cdot \dfrac{7!}{3!(7-3)!} = \dfrac{6!}{3!3!} \cdot \dfrac{7!}{3!4!} = 20 \cdot 35 = 700$

45.
$$(x+2)^3 = \binom{3}{0}x^3 2^0 + \binom{3}{1}x^2 2^1 + \binom{3}{2}x^1 2^2 + \binom{3}{3}x^0 2^3$$
$$= \frac{3!}{0!3!} \cdot x^3 + \frac{3!}{1!2!} \cdot 2x^2 + \frac{3!}{2!1!} \cdot 4x + \frac{3!}{3!0!} \cdot 8 = x^3 + 6x^2 + 12x + 8$$

46. $5! = 120$

47. $5! \cdot 3! = 720$

48. $\dbinom{10}{3} = \dfrac{10!}{3!7!} = 120$

49. $\dbinom{5}{2}\dbinom{6}{2} = 10 \cdot 15$
$= 150$

50. $\dfrac{6}{16} = \dfrac{3}{8}$

51. $\dfrac{8}{16} = \dfrac{1}{2}$

52. $\dfrac{14}{16} = \dfrac{7}{8}$

53. roll of 5 $\Rightarrow \{(1,4), (2,3), (3,2), (4,1)\}$
$\dfrac{\text{\# roll of 5}}{\text{total \# rolls}} = \dfrac{4}{36} = \dfrac{1}{9}$

54. impossible $\Rightarrow 0$

55. $\dfrac{\text{\# tens}}{\text{total \# cards}} = \dfrac{4}{52} = \dfrac{1}{13}$

56. $\dfrac{\text{\# ways to get 3 aces} \cdot \text{\# ways to get 2 other cards}}{\text{\# ways to draw 5 cards}} = \dfrac{\binom{4}{3}\binom{48}{2}}{\binom{52}{5}} = \dfrac{4 \cdot 1128}{2{,}598{,}960} = \dfrac{4512}{2{,}598{,}960} = \dfrac{94}{54145}$

57. $\dfrac{\text{\# ways to get 5 diamonds}}{\text{\# ways to get 5 cards}} = \dfrac{\binom{13}{5}}{\binom{52}{5}} = \dfrac{1287}{2{,}598{,}960} = \dfrac{33}{66{,}640}$

© 2015 Cengage Learning. All Rights Reserved. May not be scanned, copied or duplicated, or posted to a publicly accessible website, in whole or in part.

Chapter 11 Test (page 805)

1. $\frac{9!}{7!} = \frac{9 \cdot 8 \cdot 7!}{7!} = 9 \cdot 8 = 72$

2. $0! = 1$

3. In the 2nd term, the exponent on $-y$ is 1.
Variables: $x^4(-y)^1 = -x^4y$
Coef. $= \frac{n!}{r!(n-r)!} = \frac{5!}{1!4!} = 5$
Term $= 5(-x^4y) = -5x^4y$

4. In the 3rd term, the exponent on $2y$ is 2.
Variables: $(x)^2(2y)^2 = 4x^2y^2$
Coef. $= \frac{n!}{r!(n-r)!} = \frac{4!}{2!2!} = 6$
Term $= 6(4x^2y^2) = 24x^2y^2$

5. $a_1 = 3, d = 7, n = 10; a_n = a_1 + (n-1)d = 3 + (10-1)(7) = 3 + 9(7) = 66$

6. $a_1 = -2, d = 5, n = 12; a_n = a_1 + (n-1)d = -2 + (12-1)(5) = -2 + 11(5) = 53$
$S_n = \frac{n(a_1 + a_n)}{2} = \frac{12(-2+53)}{2} = \frac{12(51)}{2} = 306$

7. Form an arithmetic sequence with a 1st term of 2 and a 4th term of 98:
$a_n = a_1 + (n-1)d$
$98 = 2 + (4-1)d$
$96 = 3d$
$32 = d \Rightarrow 2, \boxed{34, 66}, 98$

8. $\sum_{k=1}^{4}(3k-4) = (3(1)-4) + (3(2)-4) + (3(3)-4) + (3(4)-4) = -1 + 2 + 5 + 8 = 14$

9. $a_1 = -\frac{1}{9}, r = 3, n = 7; a_n = a_1r^{n-1} = -\frac{1}{9}(3)^{7-1} = -\frac{1}{9}(3)^6 = -\frac{1}{9}(729) = -81$

10. $a_1 = \frac{1}{27}, r = 3, n = 6; S_n = \frac{a_1 - a_1r^n}{1-r} = \frac{\frac{1}{27} - \frac{1}{27}(3)^6}{1-3} = \frac{\frac{1}{27} - \frac{1}{27}(729)}{-2} = \frac{-\frac{728}{27}}{-2} = \frac{364}{27}$

11. $a_1 = 3, a_4 = 648$
$a_n = a_1r^{n-1}$
$648 = 3r^{4-1}$
$216 = r^3$
$6 = r \Rightarrow 3, \boxed{18, 108}, 648$

12. $a_1 = 9, r = \frac{1}{3}$
$S_\infty = \frac{a_1}{1-r} = \frac{9}{1-\frac{1}{3}} = \frac{9}{\frac{2}{3}} = \frac{27}{2}$

13. $r = \frac{10}{5} = 2$; Since $|r| > 1$, the sum does not exist.

14. $P(7,3) = \frac{7!}{(7-3)!} = \frac{7!}{4!} = \frac{7 \cdot 6 \cdot 5 \cdot 4!}{4!} = 210$

15. $P(8,8) = \frac{8!}{(8-8)!} = \frac{8!}{0!} = 8! = 40{,}320$

© 2015 Cengage Learning. All Rights Reserved. May not be scanned, copied or duplicated, or posted to a publicly accessible website, in whole or in part.

16. $C(5,3) = \frac{5!}{3!2!} = \frac{5 \cdot 4 \cdot 3!}{3!2!} = \frac{20}{2} = 10$

17. $C(8,3) = \frac{8!}{3!5!} = \frac{8 \cdot 7 \cdot 6 \cdot 5!}{3 \cdot 2 \cdot 1 \cdot 5!} = \frac{56 \cdot 6}{6} = 56$

18. $C(6,0) \cdot P(6,5) = \frac{6!}{0!6!} \cdot \frac{6!}{1!} = 1 \cdot 6!$
$= 720$

19. $\frac{P(6,4)}{C(6,4)} = \frac{\frac{6!}{2!}}{\frac{6!}{4!2!}} = \frac{6!}{2!} \cdot \frac{4!2!}{6!} = 4! = 24$

20. $(x-3)^4 = \binom{4}{0}x^4(-3)^0 + \binom{4}{1}x^3(-3)^1 + \binom{4}{2}x^2(-3)^2 + \binom{4}{3}x^1(-3)^3 + \binom{4}{4}x^0(-3)^4$
$= \frac{4!}{0!4!} \cdot x^4(1) + \frac{4!}{1!3!} \cdot x^3(-3) + \frac{4!}{2!2!} \cdot x^2(9) + \frac{4!}{3!1!} \cdot x(-27) + \frac{4!}{4!0!} \cdot (81)$
$= x^4 - 12x^3 + 54x^2 - 108x + 81$

21. $\binom{8}{5} = \frac{8!}{5!3!} = 56$

22. $\text{\# ways to get 2 women} \cdot \text{\# ways to get 1 man} = \binom{4}{2}\binom{5}{1}$
$= 6 \cdot 5 = 30$

23. $\frac{1}{6}$

24. $\frac{8}{52} = \frac{2}{13}$

25. $\frac{\text{\# ways to get 5 hearts}}{\text{\# ways to get 5 cards}} = \frac{\binom{13}{5}}{\binom{52}{5}} = \frac{1287}{2{,}598{,}960} = \frac{33}{66{,}640}$

26. $\frac{\text{\# ways to get 2 heads}}{\text{\# ways to toss 5 times}} = \frac{\binom{5}{2}}{2^5} = \frac{10}{32} = \frac{5}{16}$

Cumulative Review Exercises (page 805)

1. $\begin{cases} 2x + y = 5 \\ x - 2y = 0 \end{cases}$

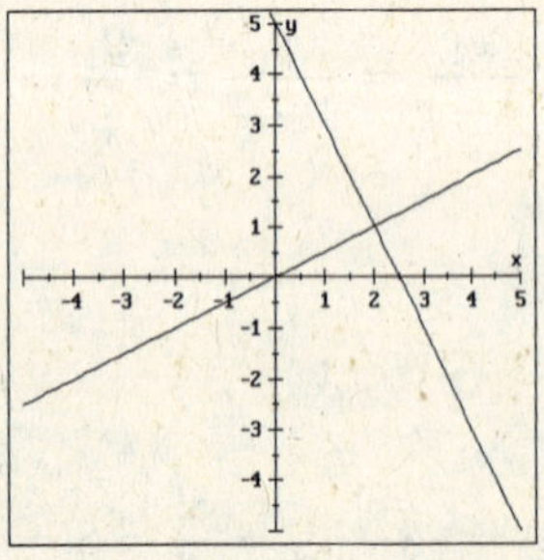

Solution: $(2, 1)$

2. $\begin{cases} (1) \quad 3x + y = 4 \\ (2) \quad 2x - 3y = -1 \end{cases}$

Substitute $y = 4 - 3x$ from (1) into (2):

$$2x - 3y = -1$$
$$2x - 3(4 - 3x) = -1$$
$$2x - 12 + 9x = -1$$
$$11x = 11$$
$$x = 1$$

Substitute this and solve for y:

$y = 4 - 3x = 4 - 3(1) = 1$

Solution: $(1, 1)$

© 2015 Cengage Learning. All Rights Reserved. May not be scanned, copied or duplicated, or posted to a publicly accessible website, in whole or in part.

CUMULATIVE REVIEW EXERCISES

3.
$$\begin{array}{l} x+2y=-2 \\ \underline{2x-\ \ y=\ \ 6} \Rightarrow \times (2) \end{array} \quad \begin{array}{l} x+2y=-2 \\ \underline{4x-2y=\ 12} \\ 5x\qquad = 10 \\ x \qquad = \ \ 2 \end{array} \quad \begin{array}{r} x+2y=-2 \\ 2+2y=-2 \\ 2y=-4 \\ y=-2 \end{array}$$

Solution: $\boxed{(2,-2)}$

4.
$$\begin{array}{l} \dfrac{x}{10}+\dfrac{y}{5}=\dfrac{1}{2} \Rightarrow \times 10 \quad x+2y=\ 5 \\ \underline{\dfrac{x}{2}-\dfrac{y}{5}=\dfrac{13}{10} \Rightarrow \times 10 \quad 5x-2y=13} \\ \qquad\qquad\qquad\qquad 6x \qquad = 18 \\ \qquad\qquad\qquad\qquad x \qquad = \ 3 \end{array} \quad \begin{array}{r} x+2y=5 \\ 3+2y=5 \\ 2y=2 \\ y=1 \end{array}$$

$y=1 \Rightarrow$ Solution: $\boxed{(3,1)}$

5.
$$\begin{vmatrix} 4 & -3 \\ 2 & -2 \end{vmatrix} = 4(-2)-(-3)(2) = -8+6=-2$$

6.
$$y = \frac{\begin{vmatrix} 4 & -1 \\ 3 & -7 \end{vmatrix}}{\begin{vmatrix} 4 & -3 \\ 3 & 4 \end{vmatrix}} = \frac{4(-7)-(-1)(3)}{4(4)-(-3)(3)} = \frac{-25}{25}=-1$$

7.
$$\begin{array}{ll} (1) & x+y+z=1 \\ (2) & 2x-y-z=-4 \\ (3) & x-2y+z=4 \end{array} \quad \begin{array}{ll} (1) & x+y+z=\ \ 1 \\ (2) & \underline{2x-y-z=-4} \\ (4) & 3x \qquad\quad =-3 \\ & x \qquad\quad\ =-1 \end{array} \quad \begin{array}{ll} (2) & 2x-\ y-z=-4 \\ (3) & \underline{x-2y+z=\ \ 4} \\ (5) & 3x-3y \qquad = \ \ 0 \end{array}$$

$$\begin{array}{r} 3x-3y=0 \\ 3(-1)-3y=0 \\ -3-3y=0 \\ -3y=3 \\ y=-1 \end{array} \qquad \begin{array}{r} x+y+z=1 \\ -1+(-1)+z=1 \\ -2+z=1 \\ z=3 \end{array}$$

$\boxed{\text{The solution is } (-1,-1,3).}$

8.
$$z = \frac{\begin{vmatrix} 1 & 2 & 6 \\ 3 & 2 & 6 \\ 2 & 3 & 6 \end{vmatrix}}{\begin{vmatrix} 1 & 2 & 3 \\ 3 & 2 & 1 \\ 2 & 3 & 1 \end{vmatrix}} = \frac{1\begin{vmatrix} 2 & 6 \\ 3 & 6 \end{vmatrix} - 2\begin{vmatrix} 3 & 6 \\ 2 & 6 \end{vmatrix} + 6\begin{vmatrix} 3 & 2 \\ 2 & 3 \end{vmatrix}}{1\begin{vmatrix} 2 & 1 \\ 3 & 1 \end{vmatrix} - 2\begin{vmatrix} 3 & 1 \\ 2 & 1 \end{vmatrix} + 3\begin{vmatrix} 3 & 2 \\ 2 & 3 \end{vmatrix}} = \frac{1(-6)-2(6)+6(5)}{1(-1)-2(1)+3(5)} = \frac{12}{12} = 1$$

© 2015 Cengage Learning. All Rights Reserved. May not be scanned, copied or duplicated, or posted to a publicly accessible website, in whole or in part.

CUMULATIVE REVIEW EXERCISES

9. $\begin{cases} 3x - 2y < 6 \\ y < -x + 2 \end{cases}$

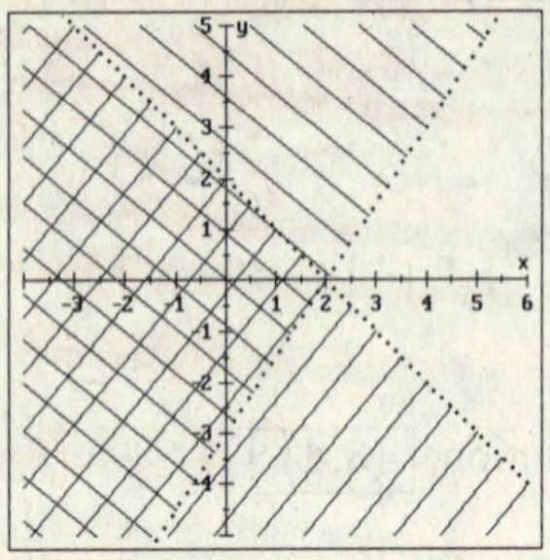

10. $\begin{cases} y < x + 2 \\ 3x + y \le 6 \end{cases}$

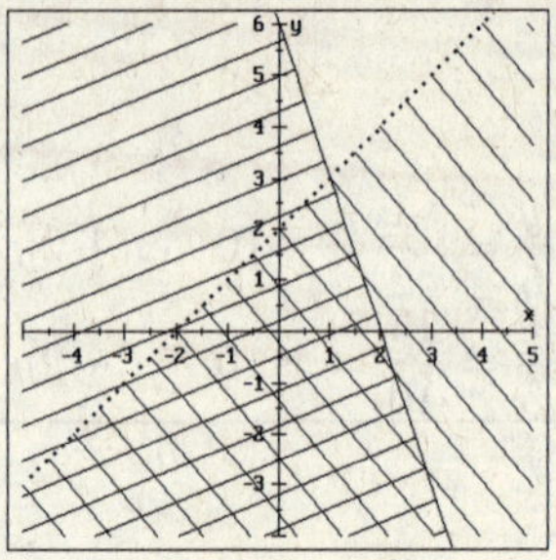

11. $y = \left(\frac{1}{2}\right)^x$

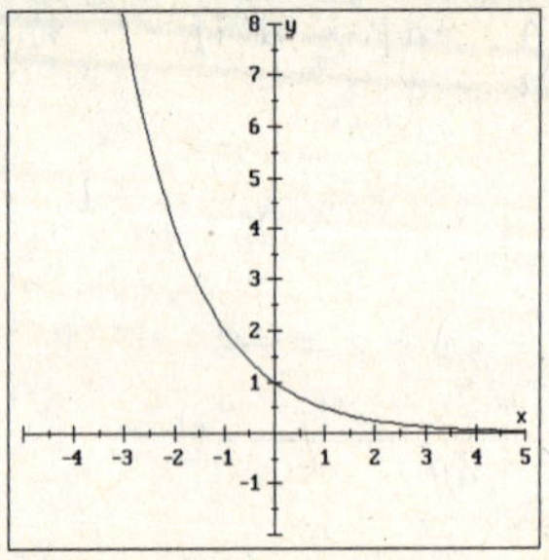

12. $y = \log_2 x \Rightarrow 2^y = x$

13. $\log_x 27 = 3 \Rightarrow x^3 = 27 \Rightarrow x = 3$

14. $\log_5 125 = x \Rightarrow 5^x = 125 \Rightarrow x = 3$

15. $\log_2 x = -4 \Rightarrow 2^{-4} = x \Rightarrow x = \frac{1}{16}$

16. $\log_5 x = 0 \Rightarrow 5^0 = x \Rightarrow x = 1$

17. $y = 2^x$

18. $y = \log_{10} 10^x = x \log_{10} 10 = x \cdot 1 = x$

19. $\log 98 = \log(14 \cdot 7) = \log 14 + \log 7 = 1.1461 + 0.8451 = 1.9912$

20. $\log 2 = \log \frac{14}{7} = \log 14 - \log 7 = 1.1461 - 0.8451 = 0.3010$

21. $\log 28 = \log(2 \cdot 14) = \log\left(\frac{14}{7} \cdot 14\right) = \log 14 + \log 14 - \log 7 = 2(1.1461) - 0.8451 = 1.4471$

22. $\log \frac{7}{5} = \log \frac{14}{10} = \log 14 - \log 10 = 1.1461 - 1 = 0.1461$

© 2015 Cengage Learning. All Rights Reserved. May not be scanned, copied or duplicated, or posted to a publicly accessible website, in whole or in part.

CUMULATIVE REVIEW EXERCISES

23.
$$\begin{aligned} 2^{x+5} &= 3^x \\ \log 2^{x+5} &= \log 3^x \\ (x+5)\log 2 &= x \log 3 \\ x\log 2 + 5\log 2 &= x\log 3 \\ 5\log 2 &= x\log 3 - x\log 2 \\ 5\log 2 &= x(\log 3 - \log 2) \\ \frac{5\log 2}{\log 3 - \log 2} &= x \end{aligned}$$

24.
$$\begin{aligned} \log 5 + \log x - \log 4 &= 1 \\ \log \frac{5x}{4} &= 1 \\ 10^1 &= \frac{5x}{4} \\ 40 &= 5x \\ 8 &= x \end{aligned}$$

25. $A = P\left(1+\frac{r}{k}\right)^{kt} = 9000\left(1+\frac{-0.12}{1}\right)^{1(9)} \approx \$2{,}848.31$

26. $\log_7 9 = \frac{\log 9}{\log 7} \approx 1.1292$

27. $\frac{5!7!}{6!} = \frac{5!\cdot 7\cdot 6!}{6!} = 5!\cdot 7 = 840$

28. $(3a-b)^4$
$$\begin{aligned} &= (3a)^4 + \frac{4!}{1!(4-1)!}(3a)^3(-b) + \frac{4!}{2!(4-2)!}(3a)^2(-b)^2 + \frac{4!}{3!(4-3)!}(3a)(-b)^3 + (-b)^4 \\ &= 81a^4 + \frac{4!}{1!3!}(-27a^3b) + \frac{4!}{2!2!}(9a^2b^2) + \frac{4!}{3!1!}(-3ab^3) + b^4 \\ &= 81a^4 - \frac{4\cdot 3!}{1!3!}(27a^3b) + \frac{4\cdot 3\cdot 2!}{2!\cdot 2\cdot 1}(9a^2b^2) - \frac{4\cdot 3!}{3!1!}(3ab^3) + b^4 \\ &= 81a^4 - \frac{4}{1}(27a^3b) + \frac{12}{2}(9a^2b^2) - \frac{4}{1}(3ab^3) + b^4 \\ &= 81a^4 - 108a^3b + 54a^2b^2 - 12ab^3 + b^4 \end{aligned}$$

29. In the 7th term, the exponent on $-y$ is 6.
Variables: $(2x)^2(-y)^6 = 4x^2y^6$
Coef. $= \frac{n!}{r!(n-r)!} = \frac{8!}{6!2!} = 28$
Term $= 28(4x^2y^6) = 112x^2y^6$

30.
$$\begin{aligned} a_1 = -11, d &= 6, n = 20 \\ a_n &= a_1 + (n-1)d \\ &= -11 + (19)(6) \\ &= -11 + 114 = 103 \end{aligned}$$

31. $a_1 = 6, d = 3, n = 20;\ a_n = a_1 + (n-1)d = 6 + (20-1)(3) = 6 + 19(3) = 63$
$S_n = \frac{n(a_1+a_n)}{2} = \frac{20(6+63)}{2} = \frac{20(69)}{2} = 690$

32. $a_1 = -3;\ a_4 = 30$:
$a_n = a_1 + (n-1)d$
$30 = -3 + (4-1)d$
$33 = 3d$
$11 = d \Rightarrow -3, \boxed{8, 19}, 30$

33.
$$\begin{aligned} \sum_{k=1}^{4} 2k^2 &= 2(1)^2 + 2(2)^2 + 2(3)^2 + 2(4)^2 \\ &= 2 + 8 + 18 + 32 = 60 \end{aligned}$$

© 2015 Cengage Learning. All Rights Reserved. May not be scanned, copied or duplicated, or posted to a publicly accessible website, in whole or in part.

34. $\sum_{k=3}^{6}(4k-5)=(4(3)-5)+(4(4)-5)+(4(5)-5)+(4(6)-5)=7+11+15+19=52$

35. $a_1=\frac{1}{27}, r=3, n=7; a_n=a_1r^{n-1}=\frac{1}{27}(3)^{7-1}=\frac{1}{27}(3)^6=\frac{1}{27}(729)=27$

36. $a_1=\frac{1}{64}, r=2, n=10; S_n=\frac{a_1-a_1r^n}{1-r}=\frac{\frac{1}{64}-\frac{1}{64}(2)^{10}}{1-2}=\frac{\frac{1}{64}-\frac{1}{64}(1024)}{-1}=\frac{-\frac{1023}{64}}{-1}=\frac{1023}{64}$

37. $a_1=-3, a_4=192$

$a_n=a_1r^{n-1}$

$192=-3r^{4-1}$

$-64=r^3$

$-4=r\Rightarrow -3, \boxed{12, -48}, 192$

38. $a_1=9, r=\frac{1}{3}$

$S_\infty=\frac{a_1}{1-r}=\frac{9}{1-\frac{1}{3}}=\frac{9}{\frac{2}{3}}=\frac{27}{2}$

39. $P(9,3)=\frac{9!}{(9-3)!}=\frac{9!}{6!}=\frac{9\cdot8\cdot7\cdot6!}{6!}=9\cdot8\cdot7=504$

40. $C(10,7)=\frac{10!}{7!(10-7)!}=\frac{10!}{7!3!}=\frac{10\cdot9\cdot8\cdot7!}{7!\cdot3\cdot2\cdot1}=\frac{720}{6}=120$

41. $\frac{C(8,4)C(8,0)}{P(6,2)}=\frac{\frac{8!}{4!4!}\cdot\frac{8!}{0!8!}}{\frac{6!}{4!}}=\frac{70\cdot1}{30}=\frac{7}{3}$

42. $C(n,n)=1$ is smaller than $P(n,n)=n!$.

43. $7!=5{,}040$

44. $\binom{9}{3}=\frac{9!}{3!6!}=84$

45. $\frac{6}{52}=\frac{3}{26}$

Appendix 1 (page A-4)

1. $y=x^2-1$

x-axis	y-axis	origin
$-y=x^2-1$	$y=(-x)^2-1$	$-y=(-x)^2-1$
not equivalent: no symmetry	$y=x^2-1$	$-y=x^2-1$
	equivalent: $\boxed{\text{symmetry}}$	not equivalent: no symmetry

3. $y=x^5$

x-axis	y-axis	origin
$-y=x^5$	$y=(-x)^3$	$-y=(-x)^3$
not equivalent: no symmetry	$y=-x^5$	$-y=-x^5$
	not equivalent: no symmetry	$y=x^5$
		equivalent: $\boxed{\text{symmetry}}$

© 2015 Cengage Learning. All Rights Reserved. May not be scanned, copied or duplicated, or posted to a publicly accessible website, in whole or in part.

5. $y = -x^2 + 2$

x-axis	y-axis	origin
$-y = -x^2 + 2$	$y = -(-x)^2 + 2$	$-y = (-x)^2 + 2$
not equivalent: no symmetry	$y = -x^2 + 2$	$-y = x^2 + 2$
	equivalent: symmetry	not equivalent: no symmetry

7. $y = x^2 - x$

x-axis	y-axis	origin
$-y = x^2 - x$	$y = (-x)^2 - (-x)$	$-y = (-x)^2 - (-x)$
not equivalent: no symmetry	$y = x^2 + x$	$-y = x^2 + x$
	not equivalent: no symmetry	not equivalent: no symmetry

9. $y = -\|x + 2\|$

x-axis	y-axis	origin
$-y = -\|x + 2\|$	$y = -\|-x + 2\|$	$-y = -\|-x + 2\|$
not equivalent: no symmetry	not equivalent: no symmetry	not equivalent: no symmetry

11. $\|y\| = x$

x-axis	y-axis	origin
$\|-y\| = x$	$\|y\| = -x$	$\|-y\| = -x$
$\|-1\|\|y\| = x$	not equivalent: no symmetry	$\|-1\|\|y\| = -x$
$\|y\| = x$		$\|y\| = -x$
equivalent: symmetry		not equivalent: no symmetry

13. $y = x^4 - 4$

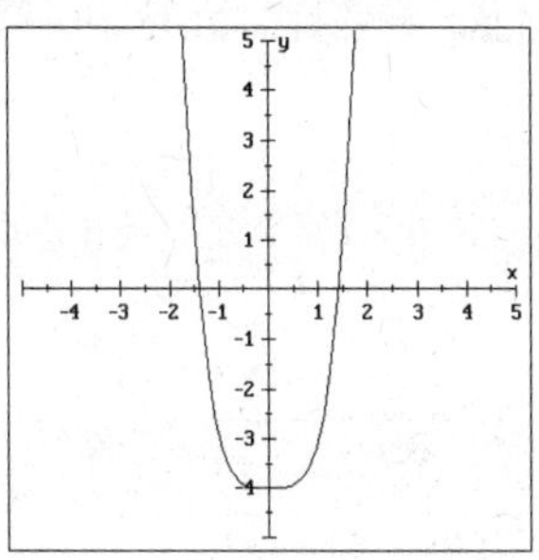

D $(-\infty, \infty)$; R $[-4, \infty)$

15. $y = -x^3$

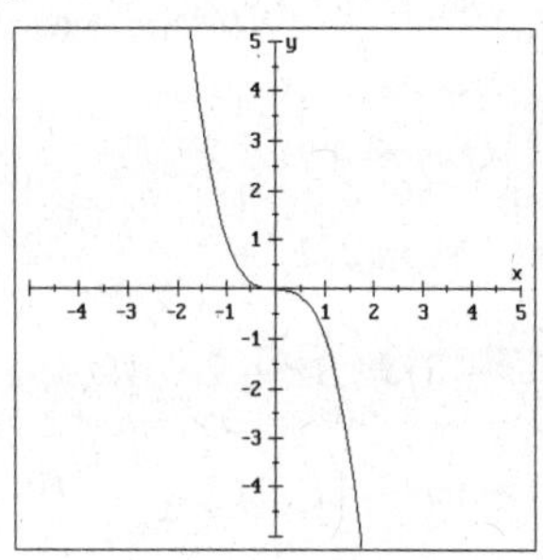

D $(-\infty, \infty)$; R $(-\infty, \infty)$

17. $y = x^4 + x^2$

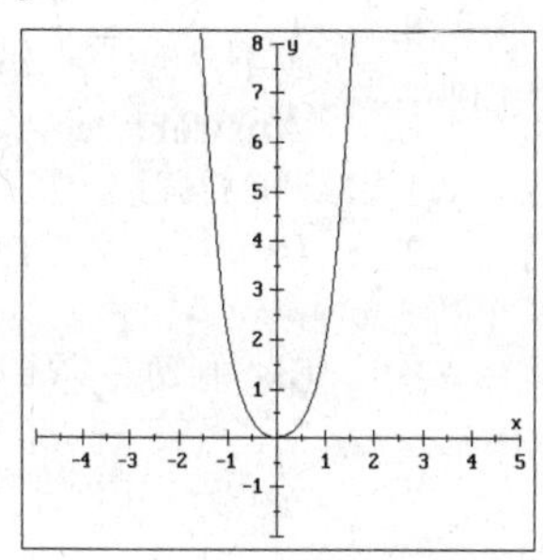

D $(-\infty, \infty)$; R $[0, \infty)$

© 2015 Cengage Learning. All Rights Reserved. May not be scanned, copied or duplicated, or posted to a publicly accessible website, in whole or in part.

19. $y = x^3 - x$

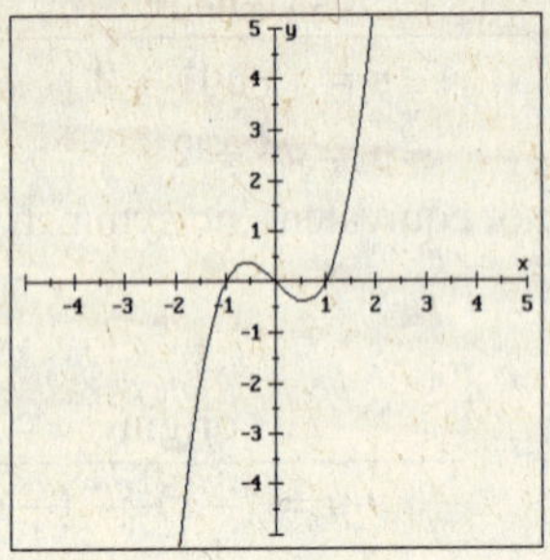

D $(-\infty, \infty)$; R $(-\infty, \infty)$

21. $y = \frac{1}{2}|x| - 1$

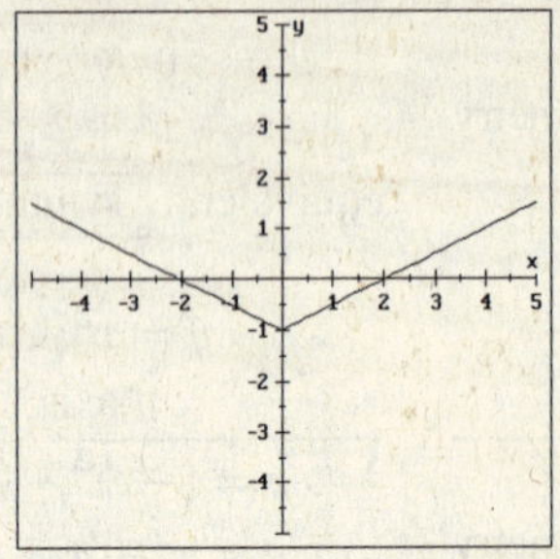

D $(-\infty, \infty)$; R $[-1, \infty)$

23. $y = -|x + 2|$

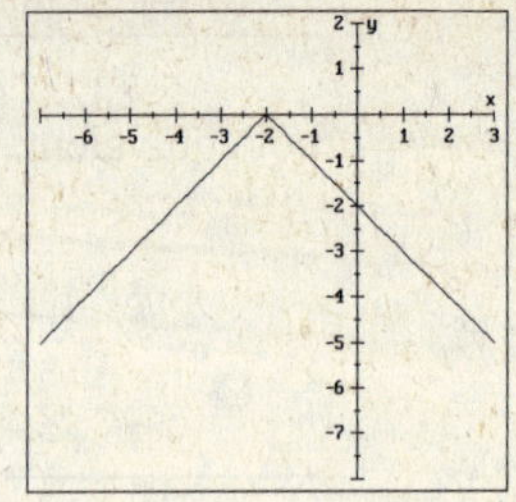

D $(-\infty, \infty)$; R $(-\infty, 0]$

Sample Final Examination (page A-7)

1. Prime numbers from 40 and 50: 41, 43, 47
Answer: a

2. Commutative property of $\times$: $ab = ba$
Answer: a

3. $\dfrac{c - ab}{bc} = \dfrac{6 - 3(-2)}{-2(6)} = \dfrac{12}{-12} = -1$
Answer: d

4.
$$\begin{aligned} \frac{1}{2} + \frac{3}{4} \div \frac{5}{6} &= \frac{1}{2} + \frac{3}{4} \cdot \frac{6}{5} \\ &= \frac{1}{2} + \frac{9}{10} \\ &= \frac{5}{10} + \frac{9}{10} = \frac{14}{10} = \frac{7}{5} \end{aligned}$$
Answer: e

5. $\left(\dfrac{a^2}{a^5}\right)^{-5} = \left(\dfrac{a^5}{a^2}\right)^{5} = \left(a^3\right)^5 = a^{15}$
Answer: c

6. $0.0000234 = 2.34 \times 10^{-5}$
Answer: a

7. $P(-1) = 2(-1)^2 - (-1) - 1 = 2(1) + 1 - 1 = 2 \Rightarrow$ **Answer: c**

8. $(3x + 2) - (2x - 1) + (x - 3) = 3x + 2 - 2x + 1 + x - 3 = 2x \Rightarrow$ **Answer: b**

9.
$$\begin{aligned} (3x - 2)(2x + 3) &= 6x^2 + 9x - 4x - 6 \\ &= 6x^2 + 5x - 6 \end{aligned}$$
Answer: d

10.
$$\begin{array}{r} x - 2 \\ 2x + 1 \overline{\big)\, 2x^2 - 3x - 2} \\ \underline{2x^2 + x \phantom{{}- 2}} \\ -4x - 2 \\ \underline{-4x - 2} \\ 0 \end{array}$$
Answer: c

© 2015 Cengage Learning. All Rights Reserved. May not be scanned, copied or duplicated, or posted to a publicly accessible website, in whole or in part.

SAMPLE FINAL EXAMINATION

11. $5x - 3 = -2x + 10$

$7x = 13$

$x = \frac{13}{7}$

Answer: e

12. Let x and $x + 2$ represent the integers.

$x + x + 2 = 44$ The integers are 21

$2x = 42$ and 23.

$x = 21$ Product $= 483$

Answer: d

13.

$$\begin{aligned} 2ax - a &= b + x \\ 2ax - x &= a + b \\ x(2a - 1) &= a + b \\ x &= \frac{a + b}{2a - 1} \end{aligned}$$

Answer: b

14.

$$|2x + 5| = 13$$

$2x + 5 = 13$ **or** $2x + 5 = -13$

$2x = 8$ $\quad 2x = -18$

$x = 4$ $\quad x = -9$

Sum of solutions $= 4 + (-9) = -5$

Answer: e

15.

$$\begin{aligned} -2x + 5 &> 9 \\ -2x &> 4 \\ \frac{-2x}{-2} &< \frac{4}{-2} \\ x &< -2 \end{aligned}$$

Answer: c

16. $|2x - 5| \le 9$

$$\begin{aligned} -9 \le 2x - 5 &\le 9 \\ -9 + 5 \le 2x - 5 + 5 &\le 9 + 5 \\ -4 \le 2x &\le 14 \\ -2 \le x &\le 7 \end{aligned}$$

Answer: d

17. $3ax^2 + 6a^2x = 3ax(x + 2a)$

Answer: e

18.

$$\begin{aligned} x^4 - 16 &= \left(x^2 + 4\right)\left(x^2 - 4\right) \\ &= \left(x^2 + 4\right)(x + 2)(x - 2) \end{aligned}$$

$x^2 + 4 + x + 2 + x - 2 = x^2 + 2x + 4$

Answer: d

19. $8x^2 - 2x - 3 = (4x - 3)(2x + 1)$

$4x - 3 + 2x + 1 = 6x - 2$

Answer: d

20.

$$\begin{aligned} 27a^3 + 8 &= (3a)^3 + 2^3 \\ &= (3a + 2)\left(9a^2 - 6a + 4\right) \end{aligned}$$

Answer: e

21.

$$6x^2 - 5x - 6 = 0$$

$$(2x - 3)(3x + 2) = 0$$

$2x - 3 = 0$ **or** $3x + 2 = 0$

$x = \frac{3}{2}$ $\quad x = -\frac{2}{3}$

Answer: d

22.

$$\begin{aligned} \frac{x^2 + 5x + 6}{x^2 - 9} &= \frac{(x + 2)(x + 3)}{(x + 3)(x - 3)} \\ &= \frac{x + 2}{x - 3} \end{aligned}$$

Answer: b

23.

$$\begin{aligned} \frac{3x + 6}{x + 3} - \frac{x^2 - 4}{x^2 + x - 6} &= \frac{3x + 6}{x + 3} - \frac{(x + 2)(x - 2)}{(x + 3)(x - 2)} = \frac{3x + 6}{x + 3} - \frac{x + 2}{x + 3} \\ &= \frac{3x + 6 - x - 2}{x + 3} = \frac{2x + 4}{x + 3} \end{aligned}$$

Answer: e

© 2015 Cengage Learning. All Rights Reserved. May not be scanned, copied or duplicated, or posted to a publicly accessible website, in whole or in part.

24. $\dfrac{y}{x+y}+\dfrac{x}{x-y}=\dfrac{y(x-y)}{(x+y)(x-y)}+\dfrac{x(x+y)}{(x-y)(x+y)}=\dfrac{xy-y^2+x^2+xy}{(x+y)(x-y)}=\dfrac{x^2+2xy-y^2}{(x+y)(x-y)}$

Answer: a

25. $\dfrac{\frac{1}{x}+\frac{1}{y}}{\frac{1}{y}}=\dfrac{\left(\frac{1}{x}+\frac{1}{y}\right)xy}{\frac{1}{y}\cdot xy}=\dfrac{y+x}{x}$

Answer: c

26.
$$\begin{aligned}\frac{2}{y+1}&=\frac{1}{y+1}-\frac{1}{3}\\ 3(y+1)\cdot\frac{2}{y+1}&=3(y+1)\left(\frac{1}{y+1}-\frac{1}{3}\right)\\ 6&=3-(y+1)\\ 6&=3-y-1\\ y&=-4\end{aligned}$$

Answer: c

27.
$$\begin{aligned}x&=0 & y&=0\\ 2x+3y&=6 & 2x+3y&=6\\ 2(0)+3y&=6 & 2x+3(0)&=6\\ 3y&=6 & 2x&=6\\ y&=2 & x&=3\end{aligned}$$

Sum $=2+3=5$

Answer: c

28. $m=\dfrac{y_2-y_1}{x_2-x_1}=\dfrac{-1-(-2)}{5-3}=\dfrac{1}{2}$

Answer: d

29.
$$\begin{aligned}2x-3y&=4 & 3x+2y&=1\\ -3y&=-2x+4 & 2y&=-3x+1\\ y&=\frac{2}{3}x-\frac{4}{3} & y&=-\frac{3}{2}x+\frac{1}{2}\\ m&=\frac{2}{3} & m&=-\frac{3}{2}\end{aligned}$$

perpendicular

Answer: b

30. $m=\dfrac{y_2-y_1}{x_2-x_1}=\dfrac{7-5}{6-(-2)}=\dfrac{2}{8}=\dfrac{1}{4}$

$$\begin{aligned}y-y_1&=m(x-x_1)\\ y-7&=\frac{1}{4}(x-6)\\ y-7&=\frac{1}{4}x-\frac{3}{2}\\ y&=\frac{1}{4}x+\frac{11}{2}\end{aligned}$$

Answer: b

31.
$$\begin{aligned}g(t+1)&=(t+1)^2-3\\ &=t^2+2t+1-3\\ &=t^2+2t-2\end{aligned}$$

Answer: d

32.
$$\begin{aligned}d&=kt\\ 12&=k(3)\\ 4&=k\end{aligned}$$

Answer: b

33. $x^{a/2}x^{a/5}=x^{a/2+a/5}=x^{5a/10+2a/10}=x^{7a/10}\Rightarrow$ **Answer: b**

34.
$$\begin{aligned}\left(x^{1/2}+2\right)\left(x^{-1/2}-2\right)&=x^{1/2}x^{-1/2}-2x^{1/2}+2x^{-1/2}-4\\ &=x^0-2x^{1/2}+2x^{-1/2}-4\\ &=1-2x^{1/2}+2x^{-1/2}-4=-3-2x^{1/2}+2x^{-1/2}\end{aligned}$$

Answer: c

© 2015 Cengage Learning. All Rights Reserved. May not be scanned, copied or duplicated, or posted to a publicly accessible website, in whole or in part.

35. $\sqrt{112a^3} = \sqrt{16a^2}\sqrt{7a} = 4a\sqrt{7a} \Rightarrow$ **Answer: e**

36. $\sqrt{50} - \sqrt{98} + \sqrt{128} = \sqrt{25}\sqrt{2} - \sqrt{49}\sqrt{2} + \sqrt{64}\sqrt{2} = 5\sqrt{2} - 7\sqrt{2} + 8\sqrt{2} = 6\sqrt{2}$
Answer: b

37. $$\frac{3}{2-\sqrt{3}} = \frac{3\left(2+\sqrt{3}\right)}{\left(2-\sqrt{3}\right)\left(2+\sqrt{3}\right)} = \frac{3\left(2+\sqrt{3}\right)}{4+2\sqrt{3}-2\sqrt{3}-3} = \frac{3\left(2+\sqrt{3}\right)}{1} = 6+3\sqrt{3}$$
Answer: b

38. $$\begin{aligned} d = \sqrt{(x_2-x_1)^2+(y_2-y_1)^2} = \sqrt{(6-(-2))^2+(-8-3)^2} &= \sqrt{8^2+(-11)^2} \\ &= \sqrt{64+121} = \sqrt{185} \end{aligned}$$
Answer: a

39. $$\begin{aligned} \sqrt{x+7} - 2x &= -1 \\ \sqrt{x+7} &= 2x-1 \\ \left(\sqrt{x+7}\right)^2 &= (2x-1)^2 \\ x+7 &= 4x^2-4x+1 \end{aligned}$$
$$\begin{aligned} 0 &= 4x^2-5x-6 \\ 0 &= (x-2)(4x+3) \end{aligned}$$
$x-2=0$ **or** $4x+3=0$
$x=2$ $\quad x=-\frac{3}{4}$
extraneous
Answer: a

40. $y > 3x+2$

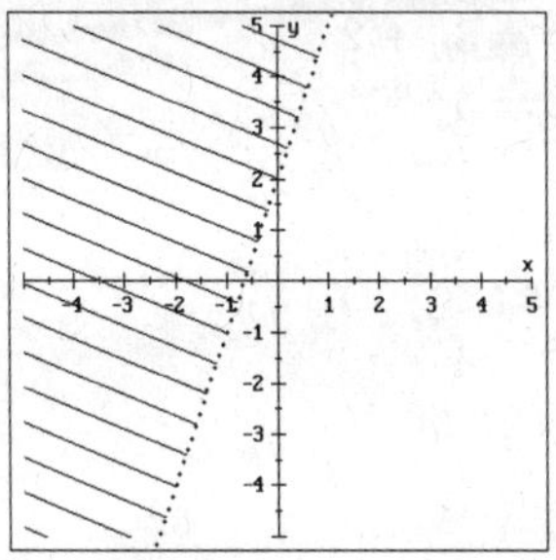

Answer: d

41. $x = \dfrac{-b \pm \sqrt{b^2-4ac}}{2a}$
Answer: d

© 2015 Cengage Learning. All Rights Reserved. May not be scanned, copied or duplicated, or posted to a publicly accessible website, in whole or in part.

42. $$\begin{aligned}(2+3i)^2 &= (2+3i)(2+3i)\\ &= 4+6i+6i+9i^2\\ &= 4+12i+9(-1)\\ &= 4+12i-9\\ &= -5+12i\end{aligned}$$

Answer: a

43. $$\begin{aligned}\frac{i}{3+i} &= \frac{i(3-i)}{(3+i)(3-i)}\\ &= \frac{3i-i^2}{9-3i+3i-i^2}\\ &= \frac{3i-(-1)}{9-(-1)}\\ &= \frac{1+3i}{10} = \frac{1}{10}+\frac{3}{10}i\end{aligned}$$

Answer: d

44. $$\begin{aligned}y &= 2x^2+4x-3\\ y+3 &= 2\left(x^2+2x\right)\\ y+3+2 &= 2\left(x^2+2x+1\right)\\ y &= 2(x+1)^2-5\end{aligned}$$

Vertex: $(-1,-5)$

Answer: c

45. $$\begin{aligned}\frac{2}{x} &< 3\\ \frac{2}{x}-3 &< 0\\ \frac{2}{x}-\frac{3x}{x} &< 0\\ \frac{2-3x}{x} &< 0\end{aligned}$$

$2-3x$ ++++++++++0 − − −

x − − − 0+++++++++++

solution set: $(-\infty,0)\cup\left(\frac{2}{3},\infty\right)$

Answer: c

46. $f(3)=2(3)^2+1=2(9)+1=19$

Answer: b

47. $$\begin{aligned}y &= 3x+2\\ x &= 3y+2\\ x-2 &= 3y\\ \frac{x-2}{3} &= y\end{aligned}$$

Answer: c

48. $(x+2)^2+(y-4)^2=16$

Answer: a

© 2015 Cengage Learning. All Rights Reserved. May not be scanned, copied or duplicated, or posted to a publicly accessible website, in whole or in part.

SAMPLE FINAL EXAMINATION

49. $\frac{4}{x}+\frac{2}{y}=\ 2 \Rightarrow 4m+2n=\ 2 \Rightarrow \times 3\quad 12m+6n=\ 6$

$\frac{2}{x}-\frac{3}{y}=-1 \Rightarrow 2m-3n=-1 \Rightarrow \times 2\quad 4m-6n=-2$

$$\begin{aligned} 16m &= 4 \\ m &= \tfrac{1}{4} \end{aligned}$$

Solve for n:

$$\begin{aligned} 4m+2n &= 2 \\ 4\left(\tfrac{1}{4}\right)+2n &= 2 \\ 1+2n &= 2 \\ 2n &= 1 \\ n &= \tfrac{1}{2} \end{aligned}$$

Solve for x:

$$\begin{aligned} m &= \frac{1}{x} \\ \frac{1}{4} &= \frac{1}{x} \\ 4 &= x \end{aligned}$$

Solve for y:

$$\begin{aligned} n &= \frac{1}{y} \\ \frac{1}{2} &= \frac{1}{y} \\ 2 &= y \end{aligned}$$

Solution: $\boxed{(4,2)}$ Sum $=4+2=6$

Answer: a

50. $y=\dfrac{\begin{vmatrix} 4 & 5 \\ 8 & 3 \end{vmatrix}}{\begin{vmatrix} 4 & 6 \\ 8 & -9 \end{vmatrix}}=\dfrac{12-40}{-36-48}=\dfrac{-28}{-84}=\dfrac{1}{3}$

Answer: b

51. $\begin{vmatrix} 2 & -3 \\ 4 & 4 \end{vmatrix}=8-(-12)=20$

Answer: b

52. $z=\dfrac{\begin{vmatrix} 1 & 1 & 4 \\ 2 & 1 & 6 \\ 3 & 1 & 8 \end{vmatrix}}{\begin{vmatrix} 1 & 1 & 1 \\ 2 & 1 & 1 \\ 3 & 1 & 2 \end{vmatrix}}=\dfrac{1\begin{vmatrix} 1 & 6 \\ 1 & 8 \end{vmatrix}-1\begin{vmatrix} 2 & 6 \\ 3 & 8 \end{vmatrix}+4\begin{vmatrix} 2 & 1 \\ 3 & 1 \end{vmatrix}}{1\begin{vmatrix} 1 & 1 \\ 1 & 2 \end{vmatrix}-1\begin{vmatrix} 2 & 1 \\ 3 & 2 \end{vmatrix}+1\begin{vmatrix} 2 & 1 \\ 3 & 1 \end{vmatrix}}=\dfrac{1(2)-1(-2)+4(-1)}{1(1)-1(1)+1(-1)}=\dfrac{0}{-1}=0$

Answer: a

53. $\log_a N=x \Rightarrow a^x=N$

Answer: a

54. $\log_2\frac{1}{32}=x \Rightarrow 2^x=\frac{1}{32} \Rightarrow x=-5$

Answer: c

55. $\log 7+\log 5=\log(7\cdot 5)=\log 35$

Answer: d

56. $b^{\log_b x}=x$

Answer: b

© 2015 Cengage Learning. All Rights Reserved. May not be scanned, copied or duplicated, or posted to a publicly accessible website, in whole or in part.

57. $\log y + \log(y+3) = 1$

$$\begin{aligned} \log y(y+3) &= 1 \\ 10^1 &= y^2 + 3y \\ 0 &= y^2 + 3y - 10 \\ 0 &= (y+5)(y-2) \end{aligned}$$

$y + 5 = 0$ **or** $y - 2 = 0$

$y = -5$ $\quad y = 2$

extraneous

Answer: b

58. In the 3rd term, the exponent on b is 2.

Variables: a^4b^2

Coef. $= \dfrac{n!}{r!(n-r)!} = \dfrac{6!}{2!4!} = 15$

Term $= 15a^4b^2$

Answer: c

59. $P(7,3) = \dfrac{7!}{(7-3)!} = \dfrac{7!}{4!} = \dfrac{7 \cdot 6 \cdot 5 \cdot 4!}{4!}$

$= 7 \cdot 6 \cdot 5 = 210$

Answer: b

60. $C(7,3) = \dfrac{7!}{3!(7-3)!}$

$= \dfrac{7!}{3!4!}$

$= \dfrac{7 \cdot 6 \cdot 5 \cdot 4!}{3 \cdot 2 \cdot 1 \cdot 4!} = \dfrac{7 \cdot 6 \cdot 5}{6} = 35$

Answer: a

61. $a_1 = 2, d = 3, n = 100$

$$\begin{aligned} a_n &= a_1 + (n-1)d \\ a_{100} &= 2 + (100-1)(3) \\ &= 2 + 99(3) \\ &= 2 + 297 = 299 \end{aligned}$$

Answer: d

62. $a_1 = 1, r = \dfrac{1}{3}$

$S_\infty = \dfrac{a_1}{1-r} = \dfrac{1}{1-\frac{1}{3}} = \dfrac{1}{\frac{2}{3}} = \dfrac{3}{2}$

Answer: c

63. $\dfrac{2}{6} = \dfrac{1}{3}$

Answer: b

© 2015 Cengage Learning. All Rights Reserved. May not be scanned, copied or duplicated, or posted to a publicly accessible website, in whole or in part.